[redacted]



| Years ago | Epoch | Period | Era |
| --- | --- | --- | --- |
| 0,000 YEARS AGO | HOLOCENE | QUATERNA | CENOZOI |
| | PLEISTOCENE | | |
| 2 MILLION YEARS | PLIOCENE | TERTIARY | |
| 7 MY | MIOCENE | | |
| 25 MY | OLIGOCENE | | |
| 37 MY | EOCENE | | |
| 54 MY | PALEOCENE | | |
| 65 MILLION YEARS | | CRETACEOUS | MESOZOIC |
| 135 MILLION YEARS | | JURASSIC | |
| 190 MILLION YEARS | | TRIASSIC | |
| 230 MILLION YEARS | | PERMIAN | PALAEOZOIC |
| 280 MILLION YEARS | | CARBONIFEROUS | |
| 350 MILLION YEARS | | DEVONIAN | |
| 390 MILLION YEARS | | SILURIAN | |
| 440 MILLION YEARS | | ORDOVICIAN | |
| 500 MILLION YEARS | | CAMBRIAN | |
| 570 MILLION YEARS | | PRECAMBRIAN | |

# NEW ZEALAND ADRIFT

By the same author:

*Paekakariki — Wellington. New Zealand Geological Survey Tour Guide*
*Rugged Landscape*
*A Tramper's Geology of the Tararuas*
*Wellington's Restless Coast*
*The Anatomy of a Marlborough Fault Line*

# NEW ZEALAND ADRIFT

The theory of continental drift
in a New Zealand setting

Graeme R. Stevens

**Illustrations by Diane L. Stevens**



REED

*First published 1980*
*Reprinted 1981*

**A. H. & A. W. REED LTD**
68-74 Kingsford-Smith Street, Wellington 3
53 Myoora Road, Terrey Hills, Sydney 2084
also
16-18 Beresford Street, Auckland
85 Thackeray Street, Christchurch 2



ISBN 0 589 01184 7

Design and typography by David J. Kingston
Type set by Filmset Type Ltd, Auckland
Printed by Dai Nippon Printing Company (Hong Kong) Ltd

TO

*My wife Diane and our children*
*Peter, Robert and Katrina,*
*and to my mother and father*

# CONTENTS

# LIST OF ILLUSTRATIONS

Figures not individually acknowledged are the work of the author.
References to sources are included in Suggestions For Further Reading; pp. 413–424.

# PREFACE

THIS WORK is not a scientific treatise on the topic of continental drift. It has been written for enjoyment and to meet the needs of the general reader who is interested in how our earth has evolved, but who at the same time is not inclined to wade through the chapter and verse of the extensive, and often daunting, geological, geophysical and oceanographic literature. To this end I have attempted to avoid technicalities and jargon wherever possible. I have, however, used a few basic technical terms, as these provide convenient "labels" for ideas. Nonetheless, these have been kept to a minimum and they have been defined both in the text at their first appearance and in the glossary (pp. 425–433).

In a work such as this it is impossible, and in my opinion not desirable, to attempt to dot the i's and cross the t's scientifically in everything I wish to say. Thus in a number of instances I have had to sacrifice strict accuracy for simplicity and, to avoid confusing the reader with a multiplicity of theories, I have often chosen to present one point of view, although wherever possible I have indicated that other alternative explanations have been advanced. Also, in some of the more controversial areas of the text I have omitted, for readability's sake, a great many let-out words such as "probably", "possibly" and "maybe".

In my writing I have tried to convey something of the excitement and stimulation to be found in investigating the processes and structure of our planet. Continental drift, and its modern version "plate tectonics", have provided a unifying concept for the whole of the earth sciences. For many years the earth sciences were fragmented and pigeon-holed. For example, there were compartments for the study of strata, volcanic rocks and igneous bodies (that is, once-molten rocks) such as granite. Other compartments dealt with the structure and shape of the land. Palaeontology, the study of fossils, seemed to be endless lists of unpronouncable and mysterious names. Earthquakes and volcanoes were exciting, but were off in a corner by themselves. How all these phenomena were related was cloaked in obscurity. Continental drift and plate tectonics have, however, provided a central theme, linking the whole spectrum of geological observations.

The book is based on lecture courses given under the auspices of the Departments of University Extension of Victoria and Massey Universities, and I wish to thank the students of these courses for providing me with a "sounding board" for many of my ideas of presentation.

Assembly of this work has been a weekend and evening activity and as a result has stretched over some four years. During this time there has been a gradual evolution in geological thought on pre-drift reconstructions of the South-west Pacific region. For example, there has been a progression from simple fits such as Fig. 15.6 to computer-assisted models such as Figs 13.14, 18.6, 18.22. Reconstructions postulating a progressively expanding earth have also been put forward (e.g. Figs 15.1-15.5). As at the time of writing no one particular reconstruction has gained general acceptance, a selection of the many that have been put forward has been used at various places in the text, to illustrate specific points. I hope, however, that this does not confuse the reader.

The entire book is based on already published information and although I have decided not to reference every statement, as in my opinion this would detract from the overall readability of the text, a reference list for each chapter is given on pp. 413–424, with distinction being made between technical and non-technical publications.

As this book is appearing in the period during which metric measurements are being introduced, both imperial and metric units have been used, but where appropriate the conversions have been "rounded off" or taken to the nearest whole number.

The illustrations have been chosen to complement the text and to highlight salient ideas. Although most have been taken from technical publications, all have been modified in various ways to (I hope) make their message clearer. Expanded captions have been provided so that the illustrations can be studied in their own right, independent of the main text. Sources for the illustrations are individually acknowledged on pp. xii–xvii.

Preparation of the line drawings has been very much a joint effort with my wife, Diane. Her artistic skill has turned my sketches into imaginative illustrations and, as the text and drawings were done simultaneously, her questions and comments often brought the text into far more comprehensible form. My debt to her is immense. Diane's creativity is indeed appreciated, as is also her support and encouragement over the time it has taken to write this book.

Our children, Peter, Robert and Katrina, have also been most helpful and co-operative as their parents struggled with "the book".

I wish to express my thanks to all those who have placed photographic material at my disposal—particularly to the photographers of the New Zealand Geological Survey (Lloyd Homer, Bruce Scott and Diane Russell).

My sincere thanks are also due to the New Zealand Geological Survey and the Department of Scientific and Industrial Research for many years of support of my research activities. Indeed, the earth sciences in general in New Zealand owe much to the liberal attitudes of these organisations.

Many of my Geological Survey colleagues have assisted me in various ways—by patiently attending to my requests for information and by reading parts of the text. Their kind assistance is indeed appreciated.

Personnel of the Cartographic Section, Science Information Division, Department of Scientific and Industrial Research, have provided much-appreciated advice on the finer points of drawing for publication.

The staff of A. H. and A. W. Reed have been most helpful in all phases of the work. In particular I would like to thank Elizabeth Caffin, David Elworthy (formerly Editorial Department), Lindsay Cuthbertson and David Kingston (Production Department) for their kindly advice, patience and encouragement.

Maureen Weir, Normandale, Lower Hutt, kindly typed the manuscript. Elizabeth Currie of Wadestown, Wellington and Pam Gates of Normandale read the entire text to weed out any inconsistencies and obscurities. The generous assistance of these three friends is very much appreciated.

Finally I wish to thank my wife, Diane, and my parents for their help and encouragement over the years.

GRAEME R. STEVENS
NORMANDALE
LOWER HUTT
NEW ZEALAND

# INTRODUCTION

BECAUSE OF the comparative shortness of our individual lives, we human beings tend to take for granted the inherent stability of land and sea. We just cannot visualise land and sea ever changing.

It took our ancestors a long time to accept that earthquakes, landslides and volcanoes can alter the land and sea in various ways—creating or destroying land. These effects, however, usually occur rather rapidly, within part of a human lifespan, and so are reasonably capable of human comprehension.

Other movements on a global scale and occurring infinitesimally slowly, over millions of years, are beyond human ken; the idea that continents can move and oceans open and close is as strange to most of us as ideas of Martians and flying saucers.

Although, as we shall see, speculations on moving continents ("continental drift") have a long history, as recently as 20 years ago few scientists were prepared to seriously consider the idea that the continents were moving about on the surface of our planet.

Like all ideas that appear to challenge the prevailing "commonsense" view of the world about us, the subject of continental drift became a highly charged emotional issue and the resulting furore was exceedingly bitter. Scorn and ridicule were heaped on the proponents of continental drift. As arguments for stability of the earth prevailed, the concept of continental drift took on a rather comic aspect and became the source of much geological humour.

However, almost at the eleventh hour, continental drift was rescued from impending obscurity by a post-Second World War resurgence of research into the earth sciences, using a galaxy of new tools, often developed from wartime apparatus, that provided entirely new insights into the relations and make-up of continents and oceans.

This initiated a swing of the pendulum of scientific opinion that led to the now widespread acceptance of the theory of continental drift.

This turnabout marks one of the most profound revolutions in scientific thought in modern times—one which has led to a new understanding of the earth and the forces that shape it.

This change of thought compares with the scientific upheavals that occurred when Copernicus showed that our planet was not at the centre of the Universe; Isaac Newton's formulation of the laws of gravity; Charles Darwin postulating the slow evolution of living things; Niels Bohr describing the atom, and Albert Einstein enunciating the concepts of relativity. However, most of these revolutionary theories involve concepts and speculations so far beyond our daily experience that in many instances they remain remote from the ordinary citizen. On the other hand, the new understanding of the earth does not lean so heavily on difficult mathematical formulations and the like, and it very much concerns phenomena that surround our daily lives. It answers questions that have remained unanswered over the centuries.

From countless scientific meetings and a veritable avalanche of research papers has come recognition at last of a notion propounded in detail over 60 years ago by a long-unheeded German meteorologist and theorist named Alfred Wegener.

The new view of the earth is of continents drifting majestically from place to place, of mountains and island chains forming like rumples in rugs pushed together and of oceans opening and closing. The earth's seemingly rigid crust is now thought to consist of a patchwork quilt of great rafts or "plates" that are much like huge ice floes jostling about on a frozen sea. Sliding over a hot, semi-plastic layer below, the rigid plates slowly move, carrying the continents and oceans with them. The plates grind and crash together, causing earthquakes and volcanic eruptions.

They crack and pull apart, usually in the ocean basins where they are thinnest, and in the cracks, molten rock wells up and solidifies. Along other edges plates over-ride each other and the plate that is pushed downward is consumed within the earth's interior.

No doubt all this sounds like utter fantasy—as it did to most scientists a mere decade ago. However, in this book I hope to show what makes the continental drift theory so fascinating and appealing and how it can explain many of the features of New Zealand geology. In the process it is necessary to introduce a certain amount of technical information, to set the scene and provide background. Nonetheless, as many excellent text books are available (see Suggestions For Further Reading; pp. 413–424), only the bare bones are presented—with, I hope, an absolute minimum of the jargon that unfortunately is all too common in science and technology.

CHAPTER ONE

# CONTINENTS AND OCEANS

READERS have probably wondered: why is there "land" and "sea"? Why does land float up, as it were, above the sea? How do landmasses maintain their "freeboard" above the sea? The answers to these questions lie in the fundamental structure of our earth.

If we look at many of the rocks that are found on land we see that often they are light-coloured or sometimes a light rusty colour. These light

*Fig. 1.1*

**Fig. 1.1** A typical sialic rock is granite, composed essentially of quartz and feldspar (the light-coloured minerals in the specimen) with subordinate mica (dark-coloured minerals in the specimen). This specimen, 95 mm (4 in) wide, is from Fiordland. Some areas of New Zealand, including some that are geologically the oldest, are very rich in granites. These areas are mainly in the South Island and include Stewart Island, Fiordland, the West Coast, and North-west Nelson. Many other areas in New Zealand are composed of rocks, originally laid down as sediments on the sea floor, which were derived from the erosion of granite landscapes. The bulk composition of these sediments therefore tends to approach that of granite.

**Fig. 1.2.** "Granite Country", southern Stewart Island. The Gog Magog hills in the foreground are composed of large masses of granite originally formed some 140 million years ago.

*Fig. 1.2*

colourings are because the rocks are rich in light-coloured minerals such as quartz (silicon dioxide, or "silica") and various feldspars (compounds of aluminium, sodium, calcium, water and quartz). Iron minerals may also be present: mixtures of red iron oxide (ferric oxide) or green iron oxide (ferrous oxide) with water and other compounds. These iron minerals, even if present in only very small traces, can give reddish or greenish colours to the rocks.

**Fig. 1.3.** A typical simatic rock is basalt, a volcanic rock composed of minerals rich in manganese, magnesium and iron. Many oceanic islands have been built up by successive eruptions of basalt lava, produced by localised melting of the sima making up the sea floor.

Basalt lavas have been erupted onto the land, or into former seas, in a number of areas in New Zealand—notably the Bay of Islands area and Banks Peninsula (formerly a twin-volcano island off the Canterbury Coast).

The specimen illustrated is basalt lava from the Lyttelton volcano, that was last active 12-10 million years ago. The specimen, obtained from the Lyttelton rail tunnel (1867), is 84 mm (3¼ in) in length.

**Sial and sima**

The rocks of the land, rich in silica and aluminium, are often called the "sialic" rocks and the layer of such rocks making up the lands of the world, the "sial" (SI = silica, AL = aluminium). Sometimes, as the rocks are often granites themselves, or approach granites in overall composition, they may be collectively called the "granitic" rocks (Figs 1.1, 1.2).

Rocks lining the deep, abyssal floors of the oceans and forming many of the oceanic islands (e.g. the Hawaiian Islands), on the other hand, are completely different. The rocks are dark-coloured and usually very heavy. They are composed of minerals rich in manganese, magnesium and iron. Although many of these metallic elements are combined with silica, there is much less silica than in the sialic rocks. The oceanic rocks are called "simatic" rocks and the layer lining the ocean "sima" (SI = silica; MA = manganese and magnesium). As many of the simatic rocks are basalt, or close to basalt in overall composition, the name "basaltic rock" is also used as a general term just like "granitic rock" (Figs 1.3, 1.4). Sometimes sialic and simatic rocks are melted up together and the resulting mixture produces "intermediate" rocks, of which the volcanic rock andesite, named after the Andes, is a typical representative (Figs 1.5, 1.6).

When we compare the weights (or more correctly the "densities") of sialic and simatic rocks we find that compared with water, which is assumed to have a density of 1.0, the sial has a density of 2.7 and sima 3.3. There is thus a fundamental difference in the density of land rocks (sial) and sea-floor rocks (sima) as well as rock composition (Fig. 1.7).

**Floating continents**

As geologists gradually became aware of this fundamental difference between land and sea, it became very obvious that the lands "floated" above the seas because they were composed of lighter rock (sial). The floors of the oceans, on the other hand, remained at low levels in relation to the lands because they were composed of heavy, dense rock (sima).

It was also realised that it was physically impossible to simply interchange land and ocean permanently. A landmass could not suddenly become ocean permanently nor could an ocean become land. It was recognised that exceptions to this general rule did occur, however. A piece of sial may have vast thicknesses of basalt (i. e. sima) piled up on it and, weighed down by the heavy rock, the sial may be bowed down below sea level. If a cork is floating in a glass of water, the only way of keeping the cork submerged is to keep a finger on it. Once the finger is removed the cork pops up to the surface again. The loading of basalt etc., on to land may act as a "giant's finger" and hold the land down for a while, but as the weight is taken away (e.g., removed by erosion) the land rises up again.

The concept of light continental rocks "floating" on dense oceanic rocks became embodied in the principle of "isostasy", which envisages the earth's surface as being in a state of balance (or

**Fig. 1.4.** Between 12 and 8 million years ago twin basalt volcanoes existed in the sea off the Canterbury coast (cf. Figs 16.35, 16.36), on the site of the modern Banks Peninsula. The Canterbury Plains, formed of material carried by rivers from Ice Age glaciers that existed in the Southern Alps between 2 million and 10,000 years ago, have been built out to link the Banks Peninsula volcanoes to the mainland.

This satellite view, extending from the Rakaia River in the south to the Hurunui River in the north, shows the twin volcanoes of Banks Peninsula, with their extinct craters carved out by erosion to form Lyttelton and Akaroa harbours.

equilibrium)—with mountains, for example, because of their greater thickness of sial, floating higher (i.e. being more buoyant) than lowland areas, with thinner sial.

## Ups and downs

On a world scale we can recognise two major groups of topographic features (Fig. 1.8):

(i) Continental platforms
(ii) Ocean basins

The continental platforms comprise the continents themselves, rising above the sea, but also included are features that often surround the emergent continents, but yet are covered to shallow depths by the sea and provide a link between the continents and the ocean basins. "Continental margin" is the name given to this intermediate zone, which includes the continental shelf and continental slope. The continental shelf, as its name implies, is a bench-like area around the continents that has been cut by the sea at various times in the past (pp. 31–34). The width of the continental shelf ranges from zero to 1500 km (932 miles), with an average of 78 km (48 miles), and the ocean depth at the shelf edge ranges from 20 to 550 m (65-1800 ft), with an average of 133 m (440 ft). The continental slopes drop off abruptly from the edges of the continental shelves at angles of 2-3.5° and continue down to the ocean basin floors.

Fig. 1.5

The continental slope is the true edge of the continents—*not* the present day coastline, which is only related to very recent geological events (particularly in the last two million years; see p. 34).

**Fig. 1.5.** Andesite is a typical "intermediate" rock, formed by the melting together of sima and sial. Andesite rocks abound in the Coromandel Peninsula and the central North Island. Recently active andesite volcanoes are Mounts Egmont, Ruapehu, Tongariro and Ngauruhoe. The specimen illustrated is a lava blob from the 1974 eruption of Ngauruhoe (width of specimen: 165 mm, 6½ in).

**Fig. 1.6.** Mount Egmont, viewed from the south-east. Egmont has had a long history of andesite eruptions, starting about 70,000 years ago and ending in A.D. 1755.

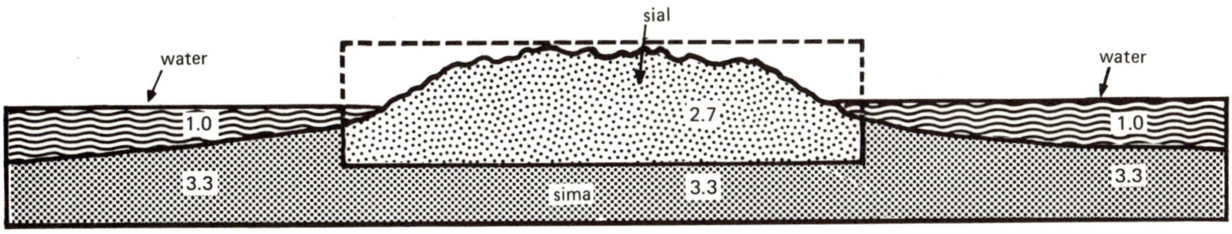


**Fig. 1.7.** Sketch to show the differences in density between the sial of the continents and the sima of the oceans.

The flooded continental shelf and slope constitute 10.9 percent of the area of the whole earth. This is about one-quarter of the total area of the earth that is structurally continental.

The ocean basins descend to abyssal depths of some 3 to 6 km (2-3.75 miles) and contain vast areas of monotonously flat sea floor. However, oceanic islands, of volcanic origin, pepper the sea floor in some areas. The ocean basins are traversed by long, narrow, sinuous oceanic ridges and are often margined by deep submarine trenches, both of which we will read a lot about later (Chapter 7). In general terms, the continental platforms consist of sial, and so stand above the abyssal ocean basins, which are floored with sima.

The continental margins, however, lie at intermediate depths between the continental lands and the ocean basins. These are areas where land has been worn down to low levels by long-continued erosion, and then the sea has finished off the giant planing process. Thus in these areas the sial has been "thinned" and so has only part of the buoyancy of the adjacent high and dry land.

A piece of ocean floor may become involved in enormous squeezing and warping movements of the earth's crust, as though it were caught between the jaws of a great vice. It is often squeezed and buckled upwards to form land. This has happened in certain regions of New Zealand, New Caledonia and Cyprus (pp. 135–143, 297).

Temporary floodings of large areas of land have occurred at times throughout geological history, usually when land has been eroded down to a very low level (i.e. the sial has been thinned) and for various reasons the sea has risen above its previous levels. Such marine transgressions are, however, usually short-lived and in time the submerged land rises again, bringing up with it the sea-floor muds, sands and other sediments that have meanwhile accumulated on its surface.

Large parts of New Zealand have had this sort of history in comparatively recent times (i.e. over the last few million years, which is "recent" in terms of geological time measured in hundreds and thousands of millions of years). The Wanganui-Taranaki and Hawkes Bay regions, for example, have risen up out of the sea over the last one or two million years, and for this reason many shell deposits, looking just like those found on any beach today, can be seen in road and other cuttings there (Fig. 1.9).

On a broader scale, New Zealand is part of a large piece of continental crust half the size of Australia, only one-tenth of which is above the sea at present. This situation is, however, rather unusual and is related to the past geological history of the New Zealand region. As we will learn later (p. 270), what we are seeing now are the eroded stumps of an ancient landmass—a large ancestral New Zealand—that was worn down to sea level. The sial was thus substantially thinned by erosion and, lacking the necessary buoyancy, was later inundated by the sea, except for the pieces that are now New Zealand, New Caledonia, Lord Howe Island, Norfolk Island and other outlying islands. The material is still nonetheless continental, and in the future is capable of becoming dry land under the influence of regional earth movements.

**Permanence *v.* drifting continents**

Because of the differences in density it is not possible to sink a piece of sial permanently below the denser material of the deep ocean floor. Atlantis-style submergences of land in deep ocean basins are not physically possible.

As these geological "facts of life" became known they added weight to theories of the earth that envisaged the continents and oceans as having been completely fixed in position throughout geological time. Such theories of "permanence" pictured the oceans as being immensely old features. The Pacific, for example, was thought to be the deep and extensive scar left on the earth by the tearing away of material to form the moon (cf. Fig. 6.7). (Dating of lunar material has shown that the moon and earth were formed almost simultaneously, thus effectively disposing of the Pacific "moon-scar" theory.) Until the 1950s belief in the permanence of continents and oceans was so deeply ingrained in geological and geophysical thought that the idea that the continents have been continually changing their positions, that oceans had opened and closed in the geological past, was thought to be almost in the realm of flying saucers and travelling to the moon, and equally laughable.

Recent discoveries on both land and sea have, however, completely reversed the old ideas. Now it is quite clear that the continents are very old and the oceans, on the other hand, very young. The implications of this have led to a great revolution in geological and geophysical thought and in our ideas of evolution of the earth. This is what the theory of continental drift is all about. Continents are envisaged as moving about the globe, oceans opening and closing in their wake. As originally conceived, the theory pictured the continents as being great rafts of sial sliding around on underlying layers of sima. The rafts of sial—the continents—were thought of as the "king-pins" and the title of the theory ("continental drift") conveys this impression. The Atlantic, for example, was thought to have been left in the wake of the westward drift of South America, and the Andes were envisaged as having been formed by the crumpling of the leading edge of the continent, as it was pushed up against the floor of the Pacific Ocean.

The modern version of the theory—and it is still very much a theory—places less emphasis on the continents. Continents are not thought of as individual units, but as parts of large composite rigid "plates" of crustal material, containing pieces of both sial and sima. The continents merely ride as passengers on the plates. The plates move over a deeper ductile layer which is capable of yielding under pressure (i.e., it is "plastic").

The name "continental drift" has therefore

**Fig. 1.8.** The earth's surface may be subdivided into four categories, shown on this map:
(i) Continents (heavy dot pattern)
(ii) Continental margins (continental platforms and shelves) (light dot pattern)
(iii) Oceanic ridges and rises (squared pattern)
(iv) Abyssal ocean floor (no pattern)

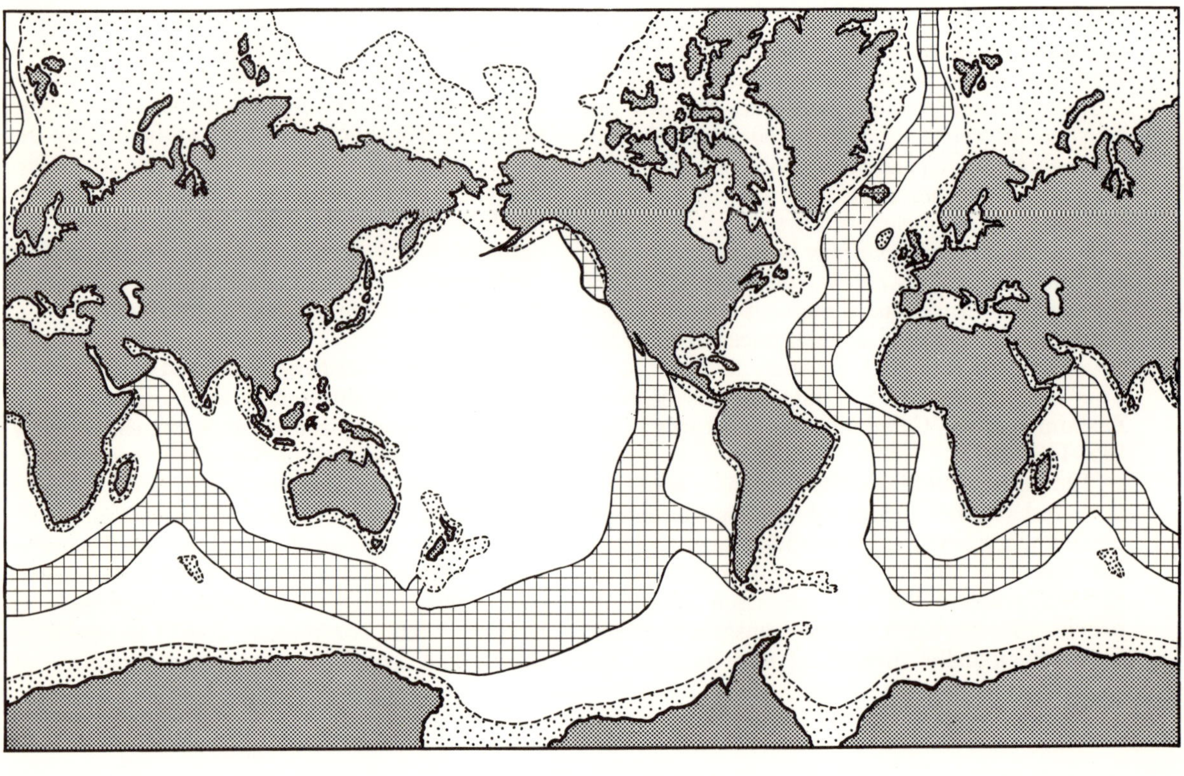

**Fig. 1.9.** The terrain of modern New Zealand was originally part of an ancestral land mass created some 140-100 million years ago. Erosion reduced this land to a low level and then some 60-30 million years ago it was flooded by the sea, and large areas remained more or less submerged until about one million years ago. In the meantime, the old rocks of the ancestral land were covered by great thicknesses of muds and sands. The Wanganui district is one such region with a long history of submergence. Here earth movements that lifted up the region about a million years ago raised the old sea floor to form land, and subsequently, down-cutting rivers and streams have exposed many layers of the sea floor sediments. This photo, inland from Wanganui, looks across the many flat plateaux and hills all rising to the same height ("summit accordance") that chart the course of the old sea floor.

been discarded as the general term for the theory and the more accurate "plate tectonics" (i.e. plate movements) is used instead. However, as this latter term is rather a jaw-breaker, "continental drift" will be used in this book to include all ideas of continental mobility, as opposed to ideas of continental stability. It must be remembered, however, that a lot more than the continents themselves have been involved in past movements of the earth's crust.

The new ideas of continental drift have spawned a host of new terms. Many (but not all!) will be introduced as we go along. The terms sima and sial are still used to some extent to recognise the fundamental difference in composition between the continents and oceans. Today, however, there is greater emphasis on the strength of the earth's materials and the way they respond to stresses and strains, and the terms "lithosphere" (for the outer rigid layer of the earth, coined from *lithos*, the Greek for "stone") and "asthenosphere" (for the inner plastic layer, derived from *asthenes*, the Greek for "weak") have been introduced.

Acceptance of the theory of continental drift has provided earth scientists with a novel and exciting way of looking at our globe. In the succeeding chapters we will be tracing the history of the development of ideas of moving continents, and showing how the present theory can be applied to interpreting the geological history of the Southwest Pacific, and New Zealand in particular.

CHAPTER TWO

# THE EARLY DAYS

AS EXPLORATION proceeded and people became aware of the fact that certain countries, now separated by deep and wide oceans, had the same, or nearly the same, plants and animals, there was speculation as to how this came about. Ancient landmasses, now submerged without a trace beneath the oceans, were proposed, as well as strips of land—"land bridges" or "isthmian links"—and these provided the most fashionable explanations, probably because most people were rather overwhelmed by the prospect of the continents actually moving.

**Earth catastrophes**

Most of the early writers were firm believers in the idea that the earth's features had been shaped by catastrophes of enormous proportions. Awesome underground fires were postulated as a cause of mountain upheavals, and the biblical Deluge, flooding over pre-existing land, was seen as the reason for the separation of the modern landmasses.

Francis Bacon, who lived in the time of James I some 300 years ago, pointed out in *Novum Organum,* published in 1620, the general conformity of the coastlines of South America and Africa, both of which tapered southwards. There was no hint, however, that he thought of them as once having been close together.

Other ideas concerning the origin of the earth's geography were expressed by the French moralist Francois Placet, who in 1666 published a booklet, *La Corruption du grand et petit Monde*. In this work Placet proposed that before the biblical flood the earth was one landmass and that the submergence of "Atlantis" had separated off the New World from the Old World.

The German naturalist and explorer Alexander von Humboldt, writing in 1801, remarked on the parallelism of the coasts of eastern South America and West Africa and their complementary geology, but he saw this only as the two sides of a huge ancient valley scooped out by the sea at some time in the past. The thought did not occur to him that the two landmasses themselves may have actually moved apart.

**Snider and Taylor**

We had to wait until 1858 for the first unequivocal suggestion that the continents have been moved into their present positions by tremendous subterranean forces. In this year an Italian, Antonio Snider-Pellegrini, published *La Creation et ses Mysteres Devoiles*, in which it was proposed that initially the earth had cooled and crystallised from the molten state to produce one major landmass, and that this situation had led to an imbalance of the globe and eventually to extensive cracking and the pulling apart of the New World from the Old World, and the flooding of the intervening area by the biblical flood (Fig. 2.1). The "jig-saw" fit of the coastline of South America and Africa was cited as evidence to support such a contention.

Snider's ideas, as they involved a series of catastrophic events, were soon put to one side, however, as people began to realise that natural processes were generally gradual in effect, not catastrophic, and further steps in the evolution of ideas of a moving earth were not taken until 1910. In this year an American, F. B. Taylor, published a work in which he analysed the patterns of the world's mountain belts and sought to explain them in terms of "mighty creeping movements" of the earth's crust, rucking up lines of folds (the mountain chains) rather like the buckles that appear in a table cloth that has been pushed across the surface of a table. Taylor did not, however, provide very much evidence to back up his ideas, and probably for this reason alone his paper had very little impact.

**Gondwana and Laurasia**

Another development that occurred about this

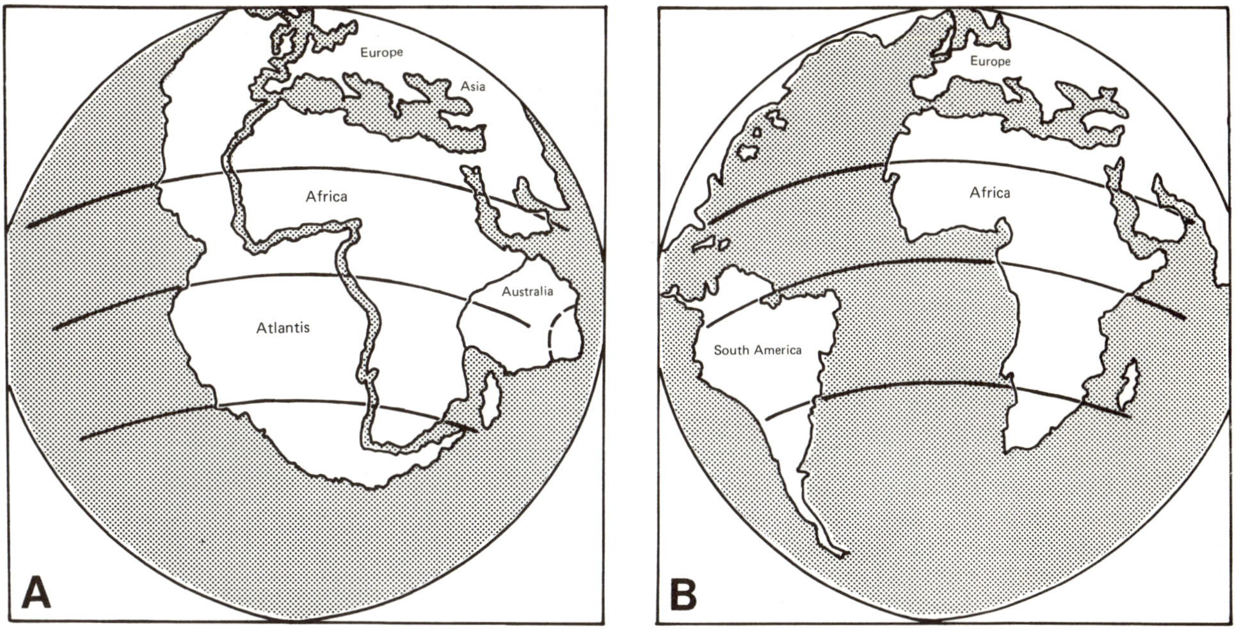


**Fig. 2.1.** Diagrams published by Antonio Snider in 1858 to illustrate his concept of the evolution of the world.

A. As the world appeared "Before the Separation", in the interval from Adam to Noah.

B. "After the Separation", which took place during the Deluge.

time was the publication by the eminent Austrian geologist, Eduard Suess, of a four-volume work entitled *The Face of the Earth* (originally published in German, 1885-1909). This extensive and thorough treatment of global geology had a great influence on early twentieth century geological thinking (including Taylor's) and, amongst other things, Suess drew attention to the fact that in the geological past, South America, Central and Southern Africa, Madagascar, Peninsular India and Australia shared identical or near identical floras and faunas and were in his opinion remnants of a former super-continent that he termed "Gondwana", after an ancient tribe (the Gonds) which had lived in central Peninsular India.

Suess thought, however, that South America, Africa, India and Australia had in the past been linked by pieces of land ("land bridges"), that have now been submerged to form ocean floor. Today we know that Africa and the rest have actually *moved* apart. Suess had nevertheless planted in geologists' minds the idea of the former existence of super-continents, such as Gondwana (often called Gondwanaland).

Subsequently, because of the discovery of other geological similarities, the boundaries of Gondwanaland came to include all of the now widely separated continents of the Southern Hemisphere, as well as India.

Eduard Suess also recognised the former existence of two ancient landmasses in the Northern Hemisphere. "Laurentia" consisted of the heartland of North America from Texas to the Arctic and eastward to Greenland. "Angora Land" included eastern Siberia and northern China.

These names survived until 1924, when the German geologist Rudolph Staub modified Suess's names and coined "Laurasia" *(Laur*entia + Eur*asia*) for a single ancient Northern Hemisphere landmass (rather than two).

CHAPTER THREE

# ALFRED WEGENER THE VISIONARY

IT WAS left to Alfred Wegener, a German meteorologist (Fig. 3.1), to spell out the first coherent theory of continental movements or "drifting" in the geological past.

As we have seen, the striking parallelism of the coastlines on opposite sides of the Atlantic Ocean began to excite interest as soon as the first maps were compiled of the newly discovered western continents. Alfred Wegener had noticed the match whilst a student in 1903, but at that time, and later, he was busily writing a doctoral thesis on astronomy, revising tables of planetary motions.

Shortly thereafter Wegener accompanied an expedition to Greenland as meteorologist. On his return to Germany he was appointed a university lecturer in meteorology, practical astronomy and cosmic physics. His first book, published in 1911, was on the thermodynamics of the atmosphere. At this point Fate took a hand.

**A Christmas gift**

A new atlas had just been published, and was given to Wegener as a present for Christmas, 1910. The new atlas had maps showing bathymetric contours, and sufficient detail and accuracy to enable him to examine the "fit" of the Atlantic continents at the 500-fathom (3000 ft, 900 m) bathymetric contour, that he considered to be the true edge of the continents. Then in 1911 he came quite by accident upon a booklet describing the geological similarities between South Africa and Brazil. The booklet presented arguments for a former land bridge across the Atlantic. Wegener, however, saw the problem differently; to him this evidence, added to that of the "fit" he had observed in the atlas, suggested the possibility of continental drift.

His imagination was fired and within four months he had formulated his hypothesis and was looking for a publisher. His book was published in mid-1912.

**Fig. 3.1.** Alfred Wegener (1880-1930), at the age of twenty-seven.

The First World War found Wegener as an artillery officer. He was wounded twice and in 1915, while he was on convalescent leave, he wrote a book, *Die Enstehung der Kontinente und Ozeane* (*The Origin of Continents and Oceans*). The first edition, a slim volume of 94 pages with no index, never had a wide circulation and is now quite a collector's piece.

In 1919 Wegener became the Director of the Meteorological Research Department of the Marine Observatory in Hamburg. In the next

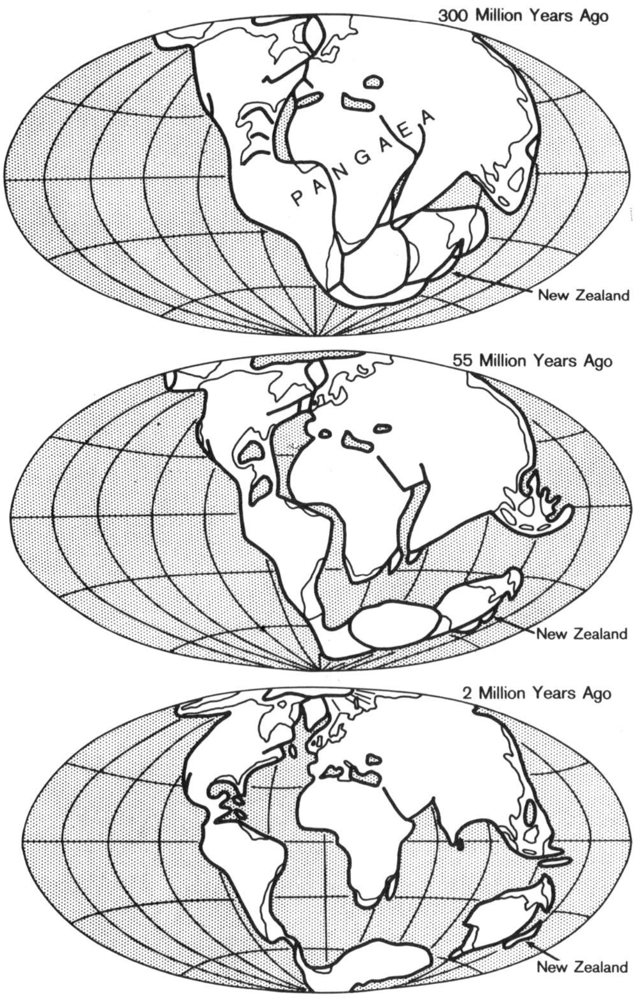


**Fig. 3.2.** Diagrams published by Alfred Wegener in 1929 to illustrate his concept of "Continental Drift", spanning the late Carboniferous (300 million years ago), Eocene (55 million years ago) and Quaternary (2 million years ago).

year he published the second edition of his book, a thoroughly reorganised version incorporating much new material.

This second edition excited interest and controversy in Europe, but there was little reaction from the English-speaking world until 1922, when reviews were published in English and American literature. Then in 1924 the third edition of the book was translated into English, French, Spanish, Swedish and Russian, and as a consequence was widely read for the first time.

**Pangaea**

Wegener proposed that all the landmasses of the earth were once parts of a single super-sized continent which he called "Pangaea", coined from the Greek meaning "all the earth". Pangaea occupied half the earth's surface, the other half being covered by the primaeval Pacific Ocean. He then envisaged that the super-continent had gradually split up, opening gaps that became the modern oceans—a process of "Continental Drift". According to Wegener's timetable, South America and Africa first began to separate in the Jurassic period (about 150 million years ago), followed a bit later by North America and Europe. The North Atlantic Ocean remained, however, fairly narrow right up to the Quaternary period, some two million years ago (Fig. 3.2).

Wegener pictured the New World landmass (North and South America), for example, drifting westwards as an enormous raft-like slab, leaving the strings of Caribbean Islands and the Scotia Arc (the islands between South America and Antarctica) as fragmentary bits and pieces lagging behind in the wake, as it were. The American Cordilleran chain (Rockies, Andes etc.), on the other hand, was thought of as being pushed up, almost as a bow wave, as the continental slab drifted towards the Pacific.

Other mountain chains were thought of by Wegener in a similar way. The Himalayas, for example, were thought of as representing a collision zone between India and Asia. India was pictured as having drifted northwards from a position close to Madagascar and South Africa, to collide eventually with the Asiatic mainland.

Whereas Suess had thought of Gondwanaland and Laurasia as former continents, much of which had now sunk, Atlantis-like, beneath the waves, Wegener thought of them as lands that had fractured and drifted apart. Wegener, using the newly established principle of isostasy (p. 2), argued that it was physically impossible for such great land areas to sink out of sight into the sea.

**"Flight from the poles"**

To explain how continents could move, Wegener put forward three possible mechanisms.

First, he envisaged a *"Polflucht"* (or "pole-flight"), a force to move the continents towards the equator. This force is a by-product of the earth's own rotation. As the earth spins on its axis the motion tends to make its outer layers fly off. This same centrifugal force is experienced by passengers in a car or on a train when they slide in their seats as it rounds a curve. The earth's rate of spin, however, is not fast enough to cause pieces of the crust to fly off into space. But Wegener believed that it could push material—including the continents—away from the poles and towards the Equator. As proof of this pushing (*"Polfluchtkraft"*, or "pole-fleeing force") Wegener pointed out that the earth has a prominent equatorial bulge. In other words, the earth's shape is not that of a circle, as it is commonly drawn, but rather that of a sphere that has been flattened at the poles (it is an "oblate spheroid"). Therefore Wegener thought there had been a movement of material towards the equator.

The second mechanism Wegener proposed to account for the westerly drift of North and South America, was simply that of the gravitational pull of the moon and sun—the same pulling that causes the ebb and flow of the ocean tides.

The tidal forces of the sun and moon act as a brake on the rotating earth. Wegener concluded that on the eastward-rotating earth this drag would have most effect on the protruding light sialic continental blocks and would tend to make them lag behind the more rapidly rotating sima. This lagging behind in a race to the east would give the continents an apparent westward motion. Wegener pictured the continental fragments as lagging behind at different rates and so appearing to drift westward for different distances.

A third proposal for the motive force was simple gravity. Other workers had pointed out that gravity measurements had indicated that some parts of the earth bulged locally more than others. Wegener thought that the existence of these local bulges may produce enough of a slope so that continents could sometimes slide downhill to new positions.

Geophysicists admitted that Wegener's *Polflucht* and tidal and gravitational forces were real enough, but their calculations showed that he had vastly exaggerated their strength. Taken individually or together, these forces exert no more than a fraction of the power that Wegener attributed to them. Forces with some million times more power would have been required to nudge the continents a few miles, to say nothing of moving them across the globe over the course of geological time.

Wegener himself realised that the Achilles heel of his theory was its attempt to explain what force could be powerful enough to push continents through the ocean floors. However, in the 1929 edition of his book (the last) he wrote:

"The Newton of the drift theory has not yet appeared. His absence need cause no anxiety; the theory is still young..."

Wegener noted that the behaviour of falling bodies and orbiting planets was observed and recorded well before Newton formulated the laws of gravity that explained their behaviour. In the same way, he was confident that sooner or later continental drift would be satisfactorily explained.

The closest that Wegener came to present-day thinking on the mechanism for continental drift was in a brief discussion of possible convection currents within the earth, included as an addition to the 1929 edition of his book.

Convection currents arise in any fluid heated on a stove. Hot material rises over the flame, spreads and cools at the top, and sinks down the sides to be reheated and rise again. Wegener cited recently published proposals that hot rock within the earth could flow in this manner—rising under the continents, spreading and cooling to sink beneath the oceans. Such a flow, he believed, could have broken up the original supercontinents.

Theories of convection currents were taken up and greatly amplified and refined by later workers, as we shall see later (Chapter 10).

**Rejection**

Wegener read widely in the fields of climatology, geology, geophysics, geodesy and biology, and presented a mass of supporting evidence drawn from past climates, rock structures, past (i.e. fossil) and present plant and animal distributions, and the various physical properties of the earth. As a complete outsider (his only previous publications had been in astronomy and meteorology), Wegener probably had the great advantage of being able to take a broad view of the subject he was tackling, whilst the experts couldn't (or wouldn't) see the wood for the trees. However, his very lack of background may also have provided a convenient excuse for people to dismiss his views as those of a crank and heretic. As has been remarked, he was "... a member of the wrong trade union".

The scientific establishment heaped scorn and ridicule on Wegener and his theory. "A pipe dream...—a beautiful fairy story..." they chanted. Geophysicists almost to a man completely opposed the theory, as they could not possibly see any likely mechanism for sliding around blocks of land the size of continents. Geologists, too, although conceding that perhaps Wegener had a point about the close geological relationships between some of the landmasses, were, like the geophysicists, rather daunted by the mechanical problem and thought that the relationships could equally (and perhaps better) be explained by means of land bridges. Also, geologists thought that *if* Wegener's ideas were worth consideration at all, they *might* possibly have some application back in the depths of the geological past, some hundreds of millions of years ago. But a glance at Wegener's maps will show (Fig. 3.2) that he had large amounts of continental drift (e.g. the opening up of virtually the entire North Atlantic Ocean) occurring in the Quaternary, the most recent period of geological time, that spans the last two million years. This was, for most geologists, completely unacceptable, and reflected badly on the theory as a whole as it was quite clear, and had been even at the turn of the present century, that geological processes were very slow and gradual and required vast amounts of time. The opening of most of the North Atlantic within the space of two million years would have necessitated extremely rapid movements, for which there was no evidence at all.

**Death on the ice cap**

In 1929, when Wegener revised his book for the fourth time, he included some material in answer to his many critics. But unfortunately there was little time left for him to fully defend himself. While the debate over his theory swirled around him he went ahead and organised another expedition to Greenland. One of his objectives was to measure the thickness of the ice cap.

This expedition, consisting of 21 scientists and technicians, departed early in 1930 to spend 18 months making geophysical and meteorological observations. Wegener set up three year-round observing stations in an unexplored part of Greenland. One station, named Eismitte ("Mid-ice") was located in the centre of the ice cap at an elevation of 3000 metres (10,000 ft). Here, instruments were deployed and underground chambers hollowed into the ice where men could live and for the first time take measurements throughout the winter in central Greenland.

Late in October 1930 Wegener visited Eismitte. There, on 1 November, the party celebrated his 50th birthday—with a precious apple for each man—before Wegener left for the west coast in the company of a Greenlander, Rasmus Villumsen. The two men departed with two sledges and 17 dogs, but never arrived at the coast.

The following spring a search was made. At a site 195 km (121 miles) from Eismitte, Wegener's skis were found standing upright in the snow. Digging into the snow at this point the search party found his body, carefully sewn between two sleeping bag covers. He was fully dressed, lying upon his sleeping bag and a reindeer skin, and covered by his fur clothing and another reindeer skin. His clothing was in perfect order and he showed no ill effects from hunger or cold. He clearly had not died on the march, but in camp and, in all likelihood, from a heart attack. The search party placed his body back in the snow just as they had found it, and they erected over it a large cairn of ice blocks topped with a sledge. Of Rasmus Villumsen, who had so carefully laid Wegener's body to rest, the search party could find no trace except for some fragmentary tracks.

With the death of its major spokesman, continental drift was soon all but forgotten, as was Wegener himself. If it was mentioned at all it was usually with contempt. Geology professors occasionally held it up to their students as a classic example of scientific blundering. Ridiculous "matches" provided an endless source of merriment. One professor said that he fully expected to one day find one half of a fossil in South Africa and the matching other half in Brazil!

A few geologists used Wegener's ideas to explain away some of their pet geological problems, but very much tongue-in-cheek.

Nonetheless, in the late 1920s and 1930s two men laboured to keep the theory alive: Arthur Holmes in the United Kingdom and Alexander du Toit in South Africa.

**Fig. 3.3.** Arthur Holmes, photographed in 1950.

**Arthur Holmes**

Arthur Holmes (Fig. 3.3) was a distinguished British geologist who held the Chair of Geology at Edinburgh University for thirteen years.

He was a lifelong supporter of the hypothesis of continental drift. He was also a total scientist, whose grasp of essential principles and store of factual knowledge were all-embracing. His influential textbook, *Principles of Physical Geology* (published in 1944), was a masterful compendium that went through many editions. Through the medium of its pages, and in scientific papers, Holmes saw to it that the theory of continental drift was kept in the public eye.

Holmes decided that on balance the geological evidence around the globe favoured continental drift. Nonetheless, like most other geologists, he was worried about how it could occur. He had an advantage over most other geologists, however, in that he was among the earliest and greatest pioneers in developing methods using radioactivity as a basis for determining ages of rocks and of the earth itself. Knowing what he knew about radioactive heat sources within the earth, Holmes was unimpressed by the arguments for a solid rigid earth. He took up the idea first mooted in 1881 (and also considered by Wegener in the last edition of his book, but not pursued to any great extent; see p. 12) that the mantle—the layer of the earth underlying the crust—was stirred by slow-moving circulating currents ("convection currents"), fuelled by radioactive heat and capable of providing a driving force for continental drift.

Where currents rose to the earth's surface under a slab of sial, strong pulling-apart movements of the underlying mantle would split the sial, creating a new ocean. Convergence of two currents at the continental borders would, according to Holmes, promote a localised sinking of the sea floor and formation of deep oceanic trenches. The combined effects of outward and downward motion would cause mountain-building along the leading edges of the continental fragments. These ideas were published by Holmes in 1931 but, as we shall see in Chapter 10, they still have much relevance today.

**Alexander du Toit**

Alexander du Toit (Fig. 3.4) was one of southern Africa's leading geologists. In the course of almost 25 years of often arduous field work traversing the length and breadth of South Africa, du Toit acquired an intimate and encyclopaedic knowledge of that country's various rock formations and fossils. Intrigued by Wegener's ideas, which at that time had just appeared in English literature, du Toit spent some five months of 1926 in Brazil, Uruguay and Argentina looking for evidence to support the supposed "fit" between Africa and South America. Du Toit's trained eye found evidence aplenty. Often he could match layer for layer, fossil for fossil. His observations were published in 1927 in a paper entitled "A Geological Comparison of South America with South Africa". Du Toit found so many matching geological features that Wegener in the 1929 edition of his book commented, "Du Toit's paper made an extraordinary impression on me, since up till then I had hardly dared to expect so close a geological correspondence between the two continents."

As well as those between Africa and South America, du Toit found geological continuities between other southern continents and reconstructed an ancient seaway, the Samfrau seaway

(S-AMerica-aFRica-AUstralia) that in Palaeozoic times, some 250 million years ago, had stretched from Argentina to Australia across the Cape Province of South Africa (Fig. 3.5).

Du Toit's research culminated in 1937 in publication of a book, *Our Wandering Continents*. Like Holmes, du Toit was a resolute supporter of the idea that convection currents stirring the earth's mantle provided the main motive power for continental drift. However, in du Toit's view there were originally two super-continents, not one (Pangaea) as envisaged by Wegener. The two super-continents, furthermore, had their own built-in convection current systems, and so were capable of independent development. The Southern Hemisphere continents, with the addition of India, formed a southern landmass (Gondwana) and the Northern Hemisphere landmasses a northern Laurasia. According to du

**Fig. 3.4.** Alexander du Toit, photographed in 1944.

**Fig. 3.5.** A reconstruction of Gondwanaland published by Alexander du Toit in 1937. Du Toit recognised the existence of an ancient seaway that traversed the Gondwana continents in the Palaeozoic era some 400-230 million years ago. This feature was called the Samfrau Geosyncline (coined from South America, South Africa, Australia).

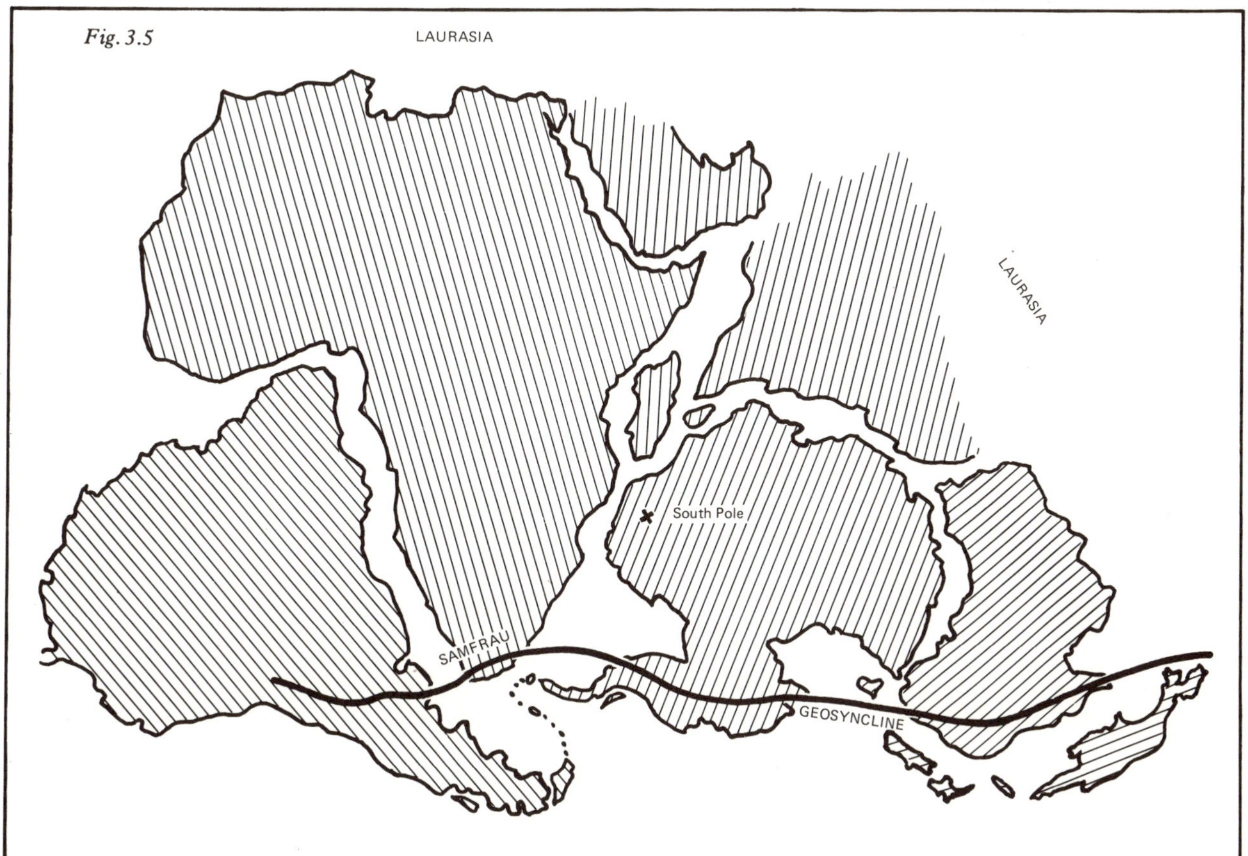

Fig. 3.6. Lester King, photographed in 1965.

Fig. 3.6

Toit, both landmasses had different and largely independent histories and from at least 250 million years ago were separated by an extensive sea, called the Tethys, which was not broken up until 40 or 50 million years ago as Africa and India drifted northwards toward Eurasia (Fig. 3.5).

Throughout the years, while Wegener's theory was "beyond the pale", so to speak, southern Africa provided a fertile breeding ground for "continental drifters". In the 1930s, 1940s and 1950s a number of New Zealand geologists went to South Africa and adjoining countries (Kenya, Rhodesia etc.) and after rubbing shoulders with African geology became convinced "drifters". This is probably because in southern Africa the geologist is brought face to face with the fact that the geology runs out to the coast and is there abruptly cut off to reappear, as du Toit found, on the other side of the Atlantic. As a glance at Figs 3.5, 5.7 will show, the southern half of Africa has been carved out of Gondwanaland on all sides and so the coastline tears abruptly across the underlying geological "grain" of the country.

### Lester King

An expatriate New Zealander in South Africa, Lester King (Fig. 3.6), continued where du Toit left off and in effect "carried the flag" for continental drift throughout the 1940s and 1950s at a time when it was at its lowest ebb.

King, a graduate of Victoria University of Wellington, had studied under Professor (later Sir Charles) Cotton and, later, as a lecturer in the 1930s, had carried out detailed studies of the New Zealand landscape. In South Africa, where he has recently retired as Professor of Geology at the University of Natal, he achieved international fame for similar studies of the African landscape. But throughout his career he was an indefatigable supporter of the continental drift theory and at every opportunity he interpreted his landscape studies in the light of past continental movements. His work culminated in a massive 700-page treatise entitled *Morphology of the Earth*, published in 1962.

Fig. 3.7

Fig. 3.7. S. Warren Carey, photographed in 1970.

**Warren Carey**

A contemporary of Lester King, and another ardent continental drifter, Warren Carey of the University of Tasmania (Fig. 3.7) recalls in *The Expanding Earth*, published in 1977, that in 1937 he had included many continental drift concepts in the draft of his doctoral thesis, but these had to be omitted at the eleventh hour because he realised that they were too radical for acceptance then, and would have cost him his degree. Even as late as 1955 he was having trouble getting his views published in the scientific literature. To circumvent this "censorship", Carey arranged in 1956 a Continental Drift Symposium in Hobart. The volume containing the symposium papers, published in 1958, provided a forum for Carey's (and King's) views at a time when winds of change were stirring in the corridors of power of the geological establishment.

It is interesting to note, therefore, that the ideas of continental drift were largely kept alive by the activities of people like Lester King and Warren Carey and other geologists from Africa, South America, India and Australia—all fragments of old Gondwanaland, where the evidence in favour of continental drift is most compelling.

CHAPTER FOUR

# CAN THE CONTINENTS MOVE?

Fig. 4.1

ALTHOUGH geology could provide good evidence for continental drift—by the matching of rock structures, strata and fossils—the over-riding handicap that dogged the continental drift theory right from the start was the question of a mechanism. How could the continents move? What was the driving force?

Until about the mid 1950s, although the theory had a few geological and biological adherents, it is probably fair to say that it had none in geophysics. This was because at that stage it was thought that the entire earth, except for the core, was composed of hard, rigid rocks, similar to those we now see on the surface of the land. The geophysicists said that if the continents were to be pushed around, how could they slide if they and the underlying rocks were hard and rigid? If movement was to occur the continents would have to slide over some sort of slippery layer somewhere underneath. Until the 1950s and 1960s there was no suggestion that such a layer existed.

*Fig. 4.2*

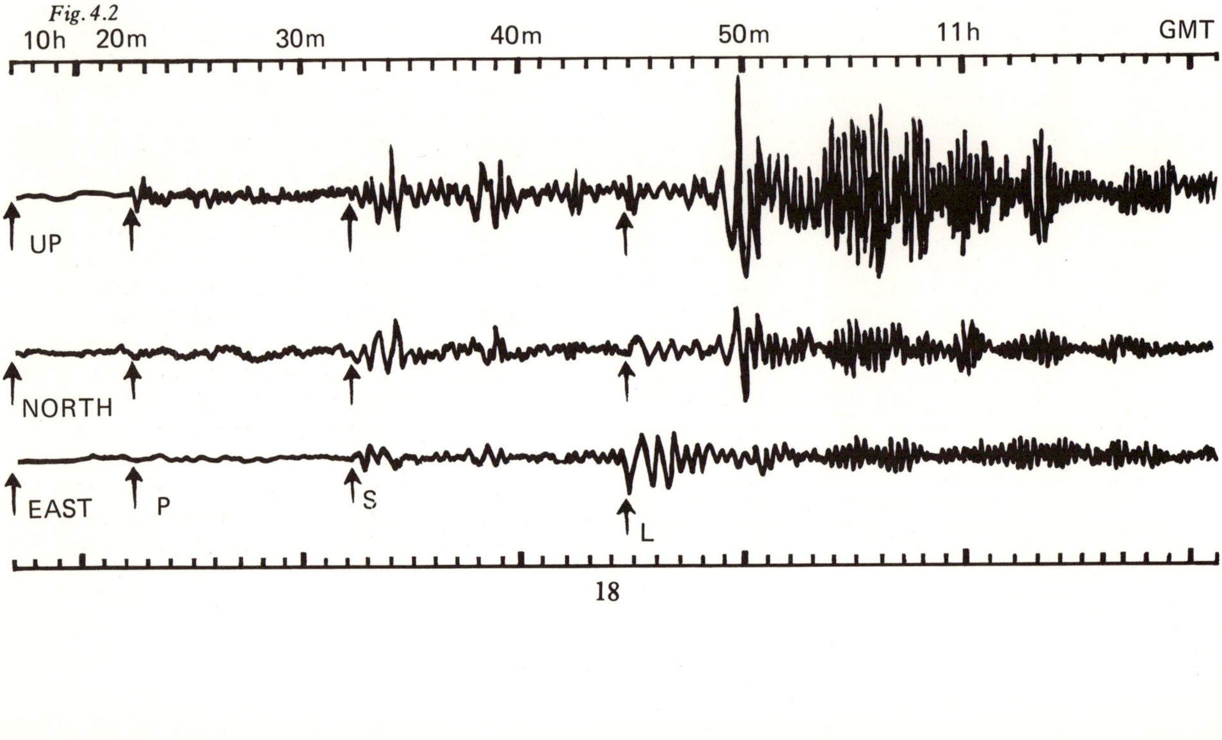

**Earthquakes to the rescue**

Large earthquakes, although terribly destructive, do nonetheless provide us with information about deeper structures within the earth. When an earthquake occurs, rocks break and shatter deep within the earth and vibrations pass out from the earthquake's centre—like ripples passing out from a stone thrown into a pond. The vibrations pass through the various layers of the earth and can be recorded by sensitive instruments called seismographs (Greek, *seio*, to shake) (Fig. 4.1).

To most people earthquakes are made up of a lot of shaking movements, just as though ocean waves were passing across the earth's surface. This is not, however, strictly true, as study of seismograph records shows that earthquakes are composed of three distinct components, each travelling at a different speed (Fig. 4.2).

First are the pressure waves, "Primary Waves" or "P waves". In these the shock waves that pass through the earth are vibrating in a longitudinal direction, or along the direction in which the waves are travelling. They are compressional or "push-pull" in effect: if an earthquake is coming from the south and its P waves pass under a statue, the statue wobbles in a north-south direction. P waves are like the vibrations that travel down a soft spring when it is tapped at one end. The P waves are the fastest of the earthquake waves.

The second type of vibration wave is called the "Secondary Wave", "Shear Wave" or "S wave". The vibrations of the S waves travel in a transverse direction, at right angles to the direction the earthquake is travelling. They are shaking waves and, as the S waves come after the P waves, the statue that at first wobbled in a north-south direction with the passage of the P waves, would wobble in an east-west direction as the S waves passed under it. S waves are like the vibrations of a taut string when it is plucked.

The third type of earthquake wave is the "Long Wave", "Love Wave" or "L wave". These waves, like the P waves, are longitudinal waves—that is, they vibrate in the same direction as the direction in which the 'quake is moving. The L waves are the slowest of the earthquake waves and have a different "amplitude" or distance between the individual vibrations. The P and S waves have a short distance between vibrations and if we could separate out the vibrations individually they would feel like a juddering movement, almost as though we were in a car going at speed over a badly corrugated road. The L waves, on the other hand, have a large amplitude, and continue for very much longer than the P and S waves. As L waves pass they buckle the earth's surface into a series of ridges and troughs, just like waves at sea, so that a

**Fig. 4.1.** A Willmore seismometer with its cover removed. This is the type of instrument used for recording local earthquakes at many of New Zealand's seismograph stations. The heavy black cylindrical magnet is supported by flat leaf springs, and tends to stay in the same position when vertical ground movements displace the frame of the instrument. A small pick-up coil converts the relative movement of the frame and the magnet into an electric current that can be recorded in a variety of ways. The springs can be disconnected and the seismometer laid on its side to measure horizontal movements. A fully-equipped station has three seismometers with their axes at right angles.

**Fig. 4.2.** When an earthquake occurs three major types of vibrational waves are produced. Two of these travel rapidly through the body of the earth, whereas the third travels along its surface.

One of the two body waves is a compressional or "push and pull" wave. It travels fastest, being the first impulse recorded by distant seismographs, and is called the primary or P wave.

The other body wave is a shear or "shake" wave. This travels more slowly than the P wave, and is called the secondary or S wave.

P and S waves start from the origin of an earthquake at the same time and travel on identical paths. The P wave travels at some 25,750 km (16,000 miles) per hour, but the S wave travels at about 14,484 km (9000 miles) per hour.

The time interval between the arrival of the P and S waves enables the distance of the origin to be calculated. Measurement of the distance from three or more seismograph stations pin-points the location of the quake.

Surface waves travel along the earth's surface and are known as L waves because they have long vibration periods.

The seismogram illustrated is redrawn from one recorded by the Seismological Observatory, Geophysics Division, D.S.I.R., Wellington, on 30 January 1963. The earthquake occurred in the South Atlantic Ocean, near the South Sandwich Islands, at 10.10 a.m. G.M.T. The P wave took 12 minutes to reach Wellington, arriving at 10.22, the slower S wave arrived at 10.32 and the first L wave at 10.45.

From the 10-minute interval between the arrival of the P and S waves seismologists calculated that the earthquake originated 9012 km (5600 miles) from Wellington.

The seismogram includes three traces: one records up and down motion and others horizontal motions in north-south and east-west directions.

*Fig. 4.3*

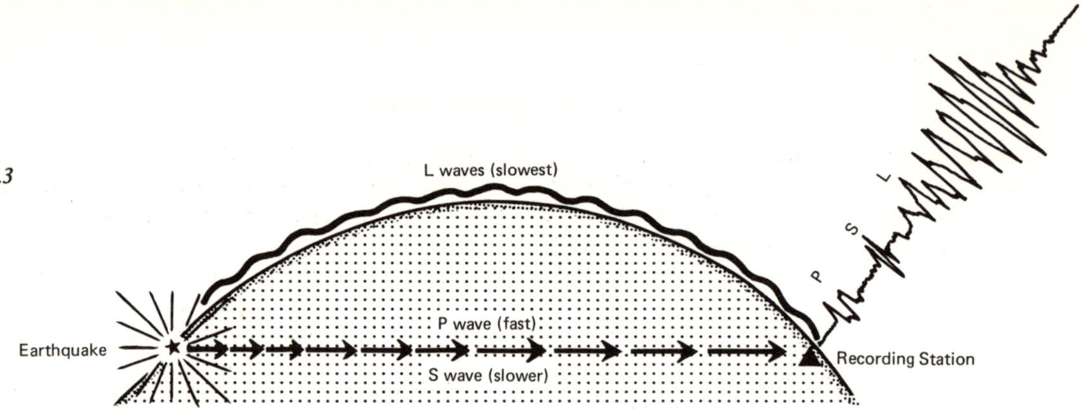


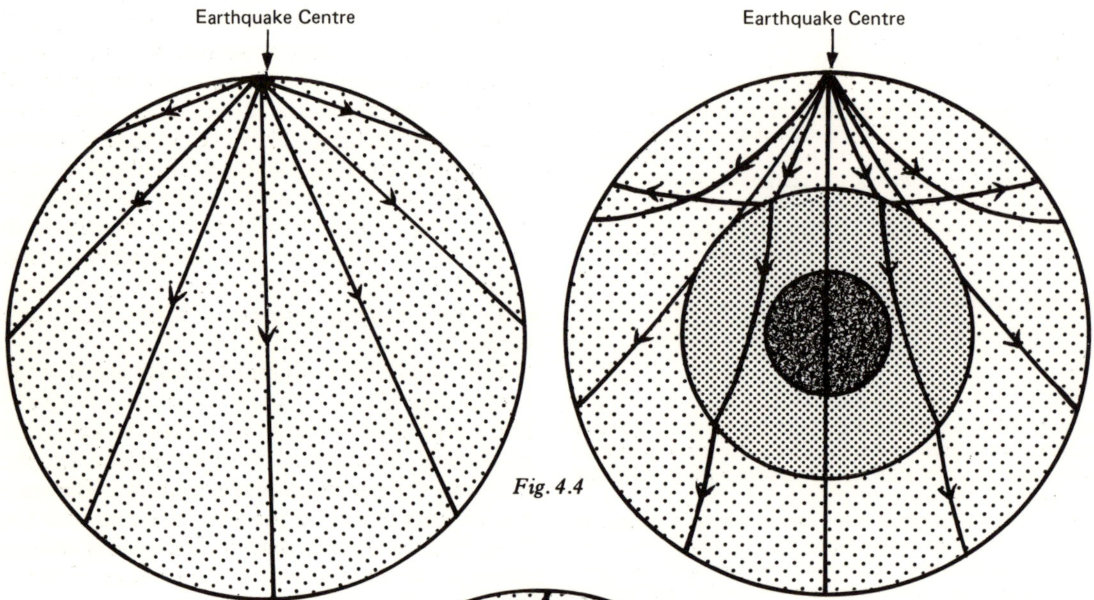


*Fig. 4.4*

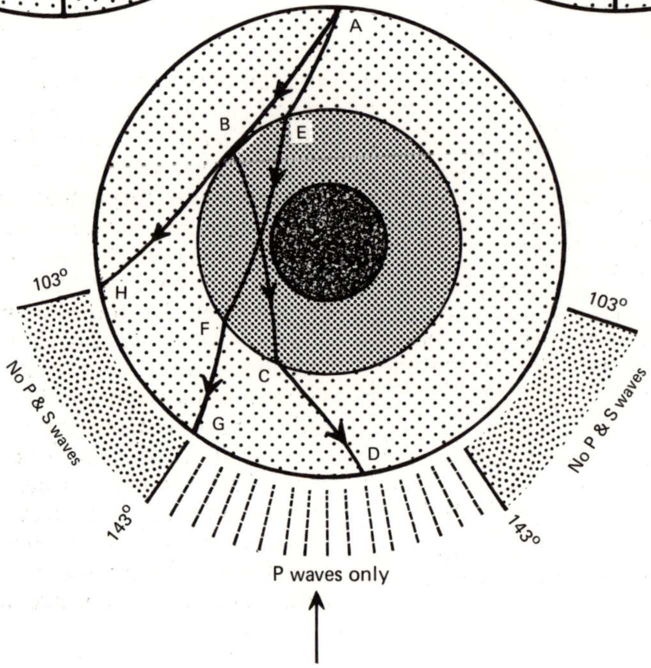


*Fig. 4.5*

swaying motion is produced. These are the waves that do the damage. Observers out in the prairie country of Western North America have seen L waves of an earthquake travelling towards them and the passing vibrations have momentarily rippled the surfaces of fields, like the rippling of a pond or the sea.

The importance of the various earthquake waves to the theory of continental drift is that they provide the means of studying the internal structure of the earth (i.e. they give scientists an X-ray picture of the earth's interior). The P and S waves travel through the interior of the earth, while the L waves travel along the surface, taking, as it were, the long way round the earth (Fig. 4.3). Thus at any seismograph station the P and S waves are recorded first (and the distance they are apart is an indication of how far they have travelled—as the P waves, being faster, steadily outdistance the S waves), followed by the L waves. As the P and S waves travel through the deeper parts of the earth they encounter rocks of varying composition and density. The waves respond to these variations in various ways and by speeding up or slowing down. By studying the characteristics of the waves seismologists have been able to build up over the years a picture of the interior of the earth, showing that it is made up of a series of roughly concentric layers of different rock types (Fig. 4.4). For example, a change in the velocity of earthquake waves, interpreted as being related to a change in rock composition, occurs at the base of the earth's crust (Fig. 4.6). This seismic boundary or "discontinuity" has been recognized for many years as the "Mohorovicic Discontinuity" (named after a Jugoslav seismologist), or "Moho" for short (pp. 89, 135, 174).

For some time it was known that if an earthquake occurred on the opposite side of the globe the P and S waves would travel through the earth to be received on the other side. But then it was found that some regions received weakened P waves and no S waves. In other words a "shadow zone" existed, in which S waves were somehow blacked out (Fig. 4.5).

It was known from experimental evidence that S waves were not capable of being transmitted through liquids. P waves, on the other hand, could get through—but only in a weakened form. The existence of the "shadow zone" was therefore taken to show that the earth had a liquid or semi-liquid core, and it was decided that the core probably consisted of molten nickel-iron. Still later work has established the presence of an inner and outer core—the outer one of molten nickel-iron and the inner one of solid material (Fig. 4.6).

To come back to our story: continental drift, if it was ever to be accepted, had to have a mechanism. It was all very well for geologists to be able to do jig-saw puzzles and to show that South American geology, for example, fitted that of South Africa. But the geophysicists could always come up with the comment, "This is all very well, but just *how* are you going to *move* the continents around?" For many years there was no

**Fig. 4.3.** Section through part of the earth showing the paths followed by the P, S and L waves of an earthquake originating at the focus and recorded at a seismograph station.

**Fig. 4.4.** The paths of earthquake waves through the earth's interior provide information about its structure and the physical properties of its concentric layers.

Energy from the source of an earthquake fans out through the earth in all directions. If the earth were composed of the same material throughout, the paths of the earthquake waves would follow straight lines, as in the left-hand diagram, and they would all travel at the same speed. Measurements show however that the wave speeds, and therefore the physical properties of the material through which they are travelling, change abruptly at certain levels (right-hand diagram), thus revealing the presence of concentric layers.

P waves can pass through both solids and liquids. S waves, which cannot pass through liquids, do not penetrate the earth's core and a "shadow zone" is created (Fig. 4.5), showing that at least the outer core is in the liquid state.

**Fig. 4.5.** The earth's "shadow zone".

Because there is a sudden increase in density at the boundary between core and mantle, P waves are strongly refracted (i.e. "bent") as they pass. Thus a P wave, originating at A and just impinging on the core at B, will be strongly refracted towards the earth's centre, because the angle of incidence is large and the wave is entering a medium where it travels more slowly. Then on leaving the core at C it is once more refracted on entering more rigid material. Finally, it will reach the earth's surface at D.

A P wave such as AE will hit the core at a smaller angle of incidence and will thus be refracted less on entering and leaving the core and come back to the surface at G. The smallest angle such a wave can have is 143°, whereas a wave that just brushes the core (ABH) reaches the surface 103° from the origin A.

S waves cannot pass through the core, and so over the whole of the area between the two 103° and 143° marks, no direct waves of any kind are recorded and between the two 143° marks only direct P waves emerge.

answer to this criticism and the absence of a viable mechanism condemned continental drift in most people's eyes.

The geophysicists needed only to point to the earthquake information then available—for as far as was then known, the P and S waves could travel without interruption from the surface of the earth right down to the core. If a slippery and presumably almost-molten layer existed somewhere under the earth's crust, on which the continents could slide, its near liquid state would surely affect the passage of the P and S waves.

Prior to the 1950s all available evidence—mainly from P and S waves—suggested the earth was composed of hard, solid rocks from the surface right down to the core. During the 1940s and 1950s, however, there was a steady improvement in the global coverage of seismograph stations, and sensitivity of seismographs. Part of this improvement was the scientific spin-off from the programme to monitor the testing of nuclear weapons, as it became necessary to develop new seismographs capable of distinguishing between naturally generated earthquake waves and those from nuclear explosions.

**Benio Gutenberg**

In 1952 Benio Gutenberg, a world-famous seismologist from the California Institute of Technology, was studying the seismograph records obtained from a very large earthquake that had occurred in Kamchatka, eastern U.S.S.R. Very large 'quakes set the earth ringing like a gigantic bell and vibrations bounce backwards and forwards from side to side within the earth. Very good sets of records can therefore be obtained and interpretations checked and cross-checked, as they all relate to the same event.

After examination of the Kamchatka records Gutenberg thought that he had picked up some irregularities in the P and S waves, but confirmation had to wait until the 1960s as, fortunately for humanity, world-shaking 'quakes occur fairly infrequently. In 1960 and 1964 there occurred the very large Chilean and Alaskan earthquakes and study of these records confirmed Gutenberg's hunch that something *did* happen to the P and S waves beneath the crust. It was worked out that both waves momentarily faltered in a narrow zone just below the crust. This zone

**Fig. 4.6.** Study of earthquake waves has shown that the earth is composed of a series of concentric layers of differing physical and chemical properties. The rocks of the thin crust are cool and rigid. Mantle rock, which is hot, is capable of very slow movement. Mantle and crust are separated by a "low-velocity layer" that has properties approaching that of a liquid (see pp. 173–174; Fig. 4.7). Evidence from earthquake waves indicates that the outer core is molten nickel iron, while the inner core is of the same material but solid.

This diagram pictures the earth's layers as simple concentric shells and does not attempt to show the active processes that go on in the interior (cf. Figs 10.2, 10.16).

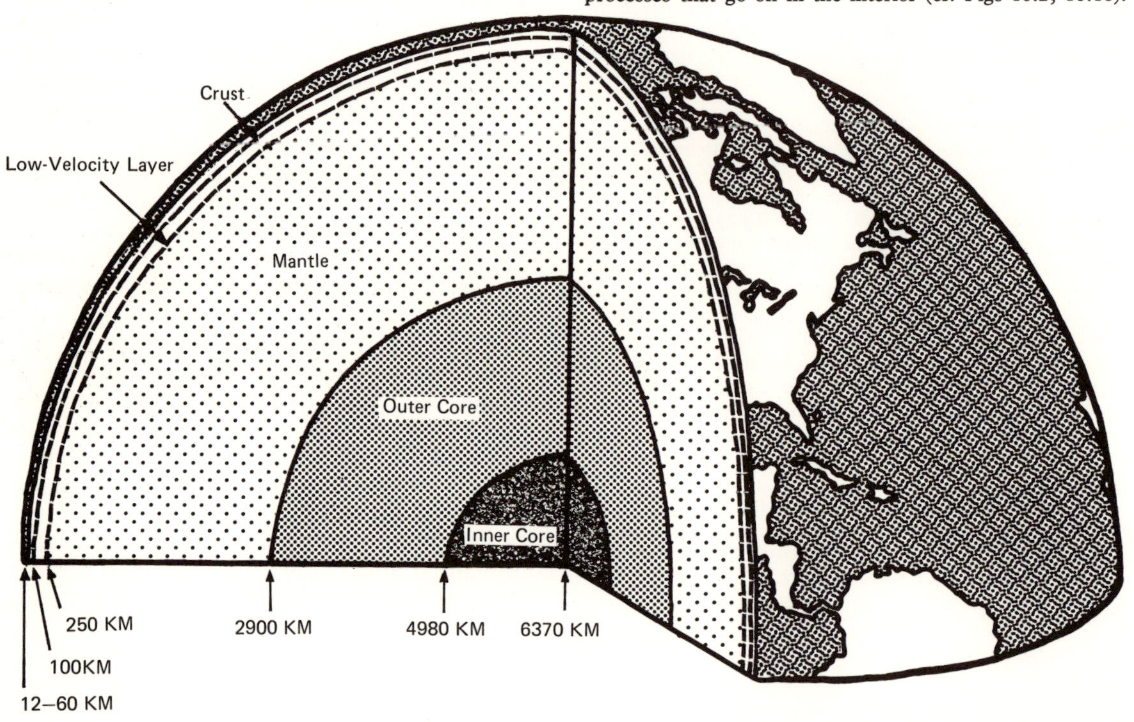

was defined by further work, helped enormously by the records obtained from the nuclear bomb monitoring network, and it was found to be between 60 and 150km (37-94 miles) below the earth's surface. Subsequent experimental work indicated that the zone is probably a layer of rock material that is in a "plastic" state. It is so hot that it is like hot toffee (or, perhaps more correctly, a slush of crystals interspersed with melted material), but the zone as a whole is prevented

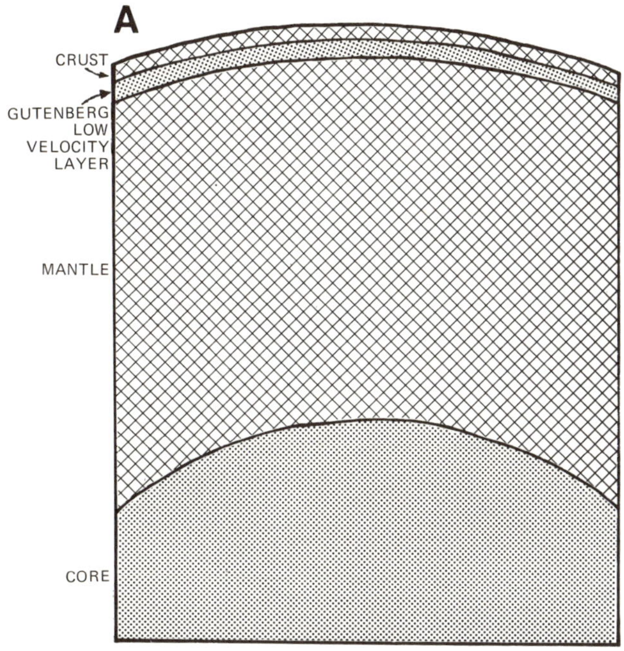


**Fig. 4.7.** A. Diagram showing the position of the Gutenberg Low-Velocity Layer in the upper part of the earth's mantle layer. The layer consists of a zone of lower rigidity and greater plasticity than the average, which allows sliding movements of the crust to occur.
B. The plastic zone, or Gutenberg Layer, of the earth's mantle occupies an ill-defined region some 60-150 km (37-94 miles) below the surface. In this zone the temperature approaches the melting point of the rock. The lithosphere is brittle rock; whereas the asthenosphere can flow and relieve stress. In this diagrammatic cross-section of the upper portion of the earth the ocean is to the left. Dots indicate earthquake centres. Broken arrows mark possible movement of continents over ocean basin. Solid arrows mark possible slippage of lithosphere as a whole over the asthenosphere.
Melting occurs in the plastic zone in areas where stress is relieved; such molten material passes upwards as magma (solid black).

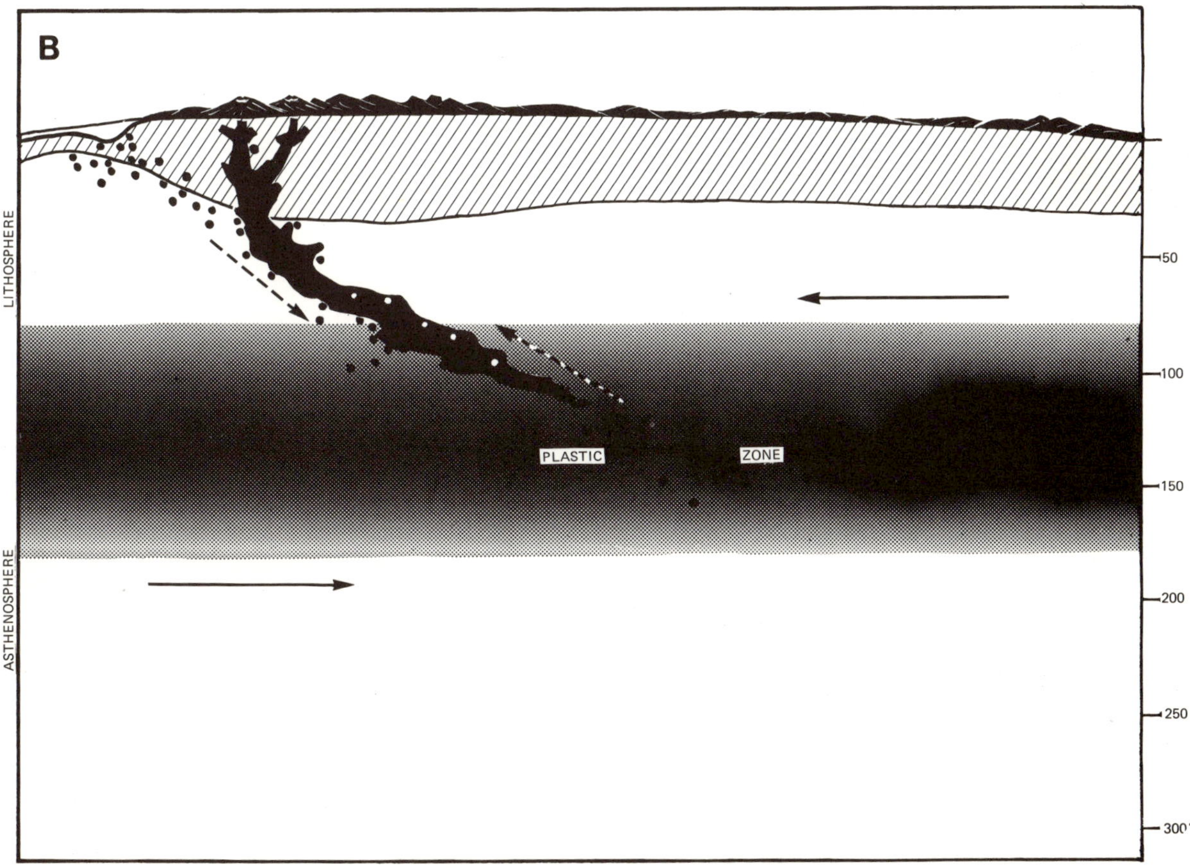

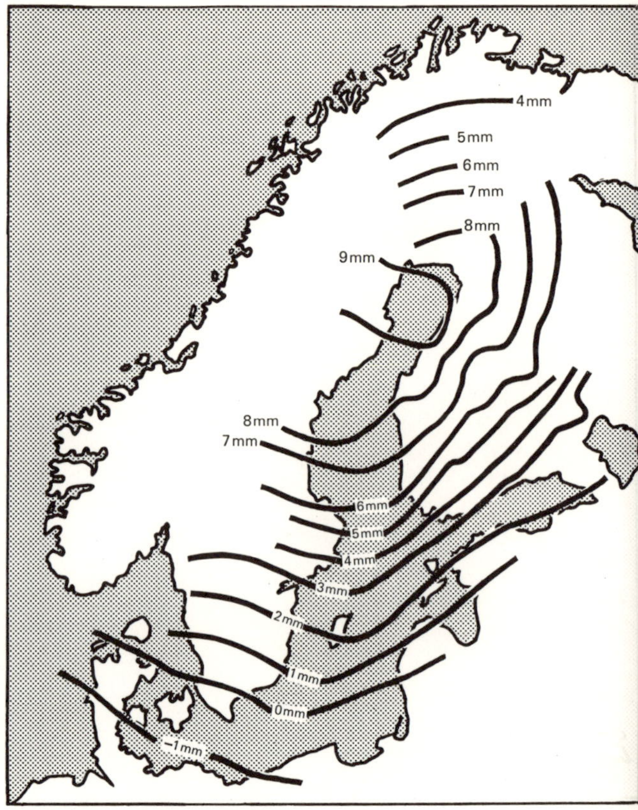


**Fig. 4.8.** Diagram showing the rapid uplift Scandinavia has been undergoing since the removal of the vast weight of the Ice Age glaciers. The contours show the uplift of areas expressed as mm per year. Note that areas south of the Baltic are sinking.

from melting by the immense pressure under which it is held.

The plastic zone was called the Gutenberg Low-Velocity Layer, as both the P and S waves momentarily slow down as they pass through it (Fig. 4.7). The plastic nature of the layer means that it will respond to pressures both above and below it by flowing very slowly, just like a glacier. Thus the rocks overlying the Gutenberg layer could, over long periods of time, slide over the "slush"; and so a practical mechanism for continental drift was demonstrated.

The plastic layer also responds to shifting loads on the earth's crust. This phenomenon can be illustrated by a modern example. During the various Ice Ages that occurred between two million years and 10,000 years ago, vast ice sheets built up on the North American and European continents. A particularly thick part of the ice sheet (2500-3000 m thick) covered Scandinavia, drowning the landscape in ice, and the weight of ice was such that the Scandinavian crust was pushed down some 200 m (660 ft). This amount of depression of the crust was accommodated by the squeezing out of material from the underlying plastic zone which then flowed to either side. The areas around Scandinavia thus rose. However, the warming of the climate that occurred at the end of the Ice Ages melted the ice sheets, and consequently removed the load on Scandinavia. This allowed the material that had been squeezed to either side to flow back again so that Scandinavia has been rising for the last 10,000 years, and is still going up (Fig. 4.8). On the other hand the areas beneath which the excess material squeezed out from under Scandinavia had lain, and which had risen during the Ice Ages, now began to sink as material flowed out from under them. Thus as Scandinavia has been rising since the Ice Ages, so the "Low Countries" of Europe

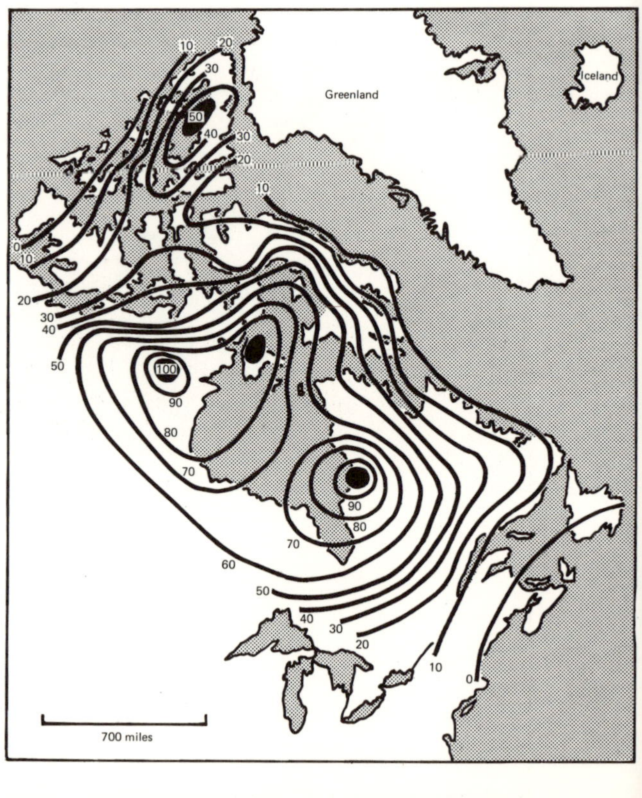


**Fig. 4.9.** Diagram showing the total amount (in metres) North America has risen in the last 6000 years.

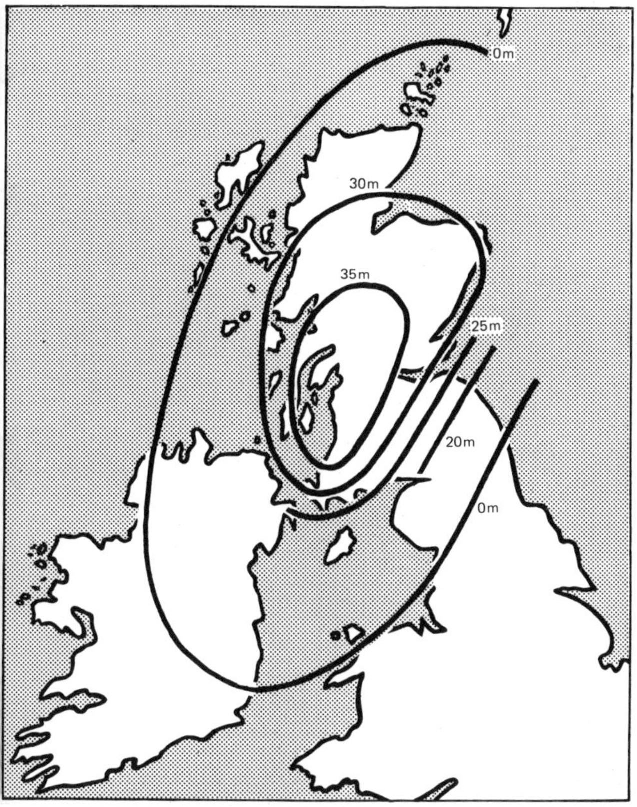


**Fig. 4.10.** Diagram showing the total amount (in metres) the British Isles have been uplifted in the 10,000 years or so that have elapsed since the Ice Age glaciers melted and disappeared.

(Holland and Belgium) have been sinking. The Dutch have therefore had to battle against a subsiding crust as well as the vagaries of the sea. The British Isles and northern North America, like Scandinavia, are also rising, but as they were covered with thinner ice sheets (750-1000 m), the rate of recovery of the land is less dramatic (Figs 4.9, 4.10).

Thus the continents are like great barges. When a barge is loaded, it sinks and water beneath it flows aside to make room for its greater submerged volume. When unloaded, the water presses back in to lift the barge back to a height commensurate with its reduced weight and greater buoyancy. The low-velocity layer of the earth flows like the water under the laden barge, but very, very slowly—as in the case of Scandinavia, which is still recovering 10,000 years after losing its load of ice.

A similar phenomenon, but on a very much smaller scale, occurs when very large dams like the Kariba Dam are filled, the earth's crust being bent down by the imposed weight of water.

**Transcurrent faults**

The other major point of criticism put forward by the geophysicists, apart from that of absence of a mechanism, was that if continental movements had occurred in the geological past, some hundreds of millions of years ago, why weren't they happening today? When this criticism was posed not a shred of evidence was available to show that any movements were occurring today or even had occurred in geologically recent times, a few millions of years ago.

At the time Alfred Wegener had tried to answer this criticism by saying that methods such as astronomical position-finding, or the transmission of radio time signals across lines of longitude, carried out over a number of years, might provide some indication that the widening of the Atlantic was still occurring. By 1929 there seemed to be some evidence from these sources that Greenland was moving westwards with respect to Europe (satellite measurements have made the job easier today—see pp. 152, 165). But in Wegener's time there was still considerable scepticism. As with the vexed question of a mechanism, it was not until the 1950s that an answer was in sight.

In the 1950s the existence of large-scale horizontal or "transcurrent" faulting began to be

*Fig. 4.11*

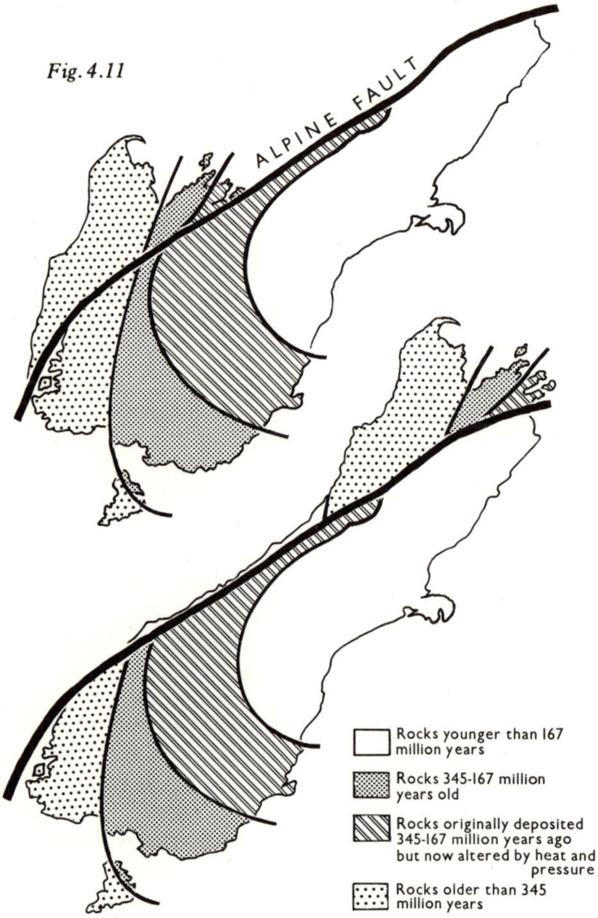


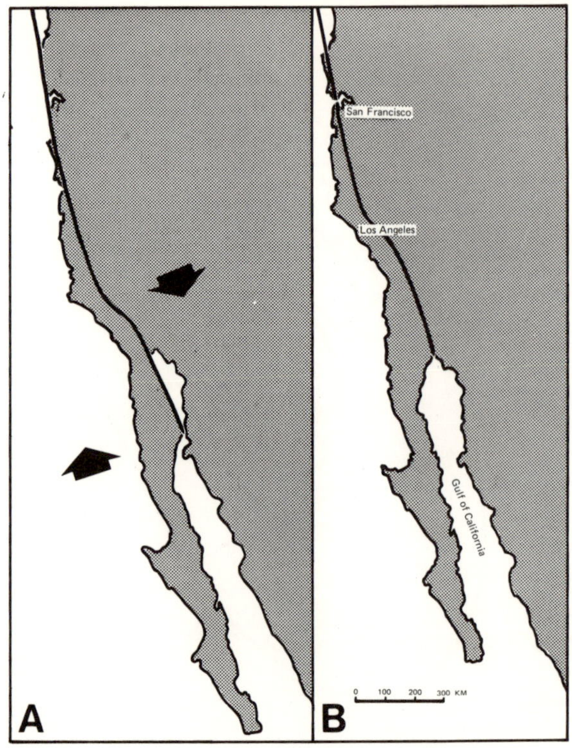


**Fig. 4.12.** The long narrow California peninsula, the southern part of which extends into Mexico (Baja California), once occupied a position 315 km (200 miles) north of its present position. Horizontal movements along large faults, of which the San Andreas Fault is the largest, have over the last 25 million years progressively slid the peninsula southwards.

**Fig. 4.11.** Gradual movement along the Alpine Fault, spread over many millions of years, has torn New Zealand's South Island in two and moved for a distance of some 450 km (280 miles) rocks that were once adjacent. The upper diagram depicts the situation before movement took place (some people say about 140 million years ago, others 9 million years ago), and the lower diagram the situation as it is today.

realised by geologists. In this type of earth movement large blocks of country slide horizontally, often for great distances, along lines of weakness in the crust, called faults. One such large horizontal movement was recognised at about this time in New Zealand. It was found that the belts of rocks and fossils that occur in Southland and Otago continue in Nelson and Marlborough, but that the two areas are separated by a very large fault, the Alpine Fault, and that progressive movements along its line, spread over millions of years, have slid the two areas apart (Fig. 4.11). Examples were also found in other countries, such as California (Fig. 4.12), Scotland (Fig. 4.13), and Japan, and it was realised that here was the process of continental drift occurring right under our very noses. Slabs of crust—albeit not the size of continents, but large enough to prove the point—were being slid around, and it could be proved, by precise

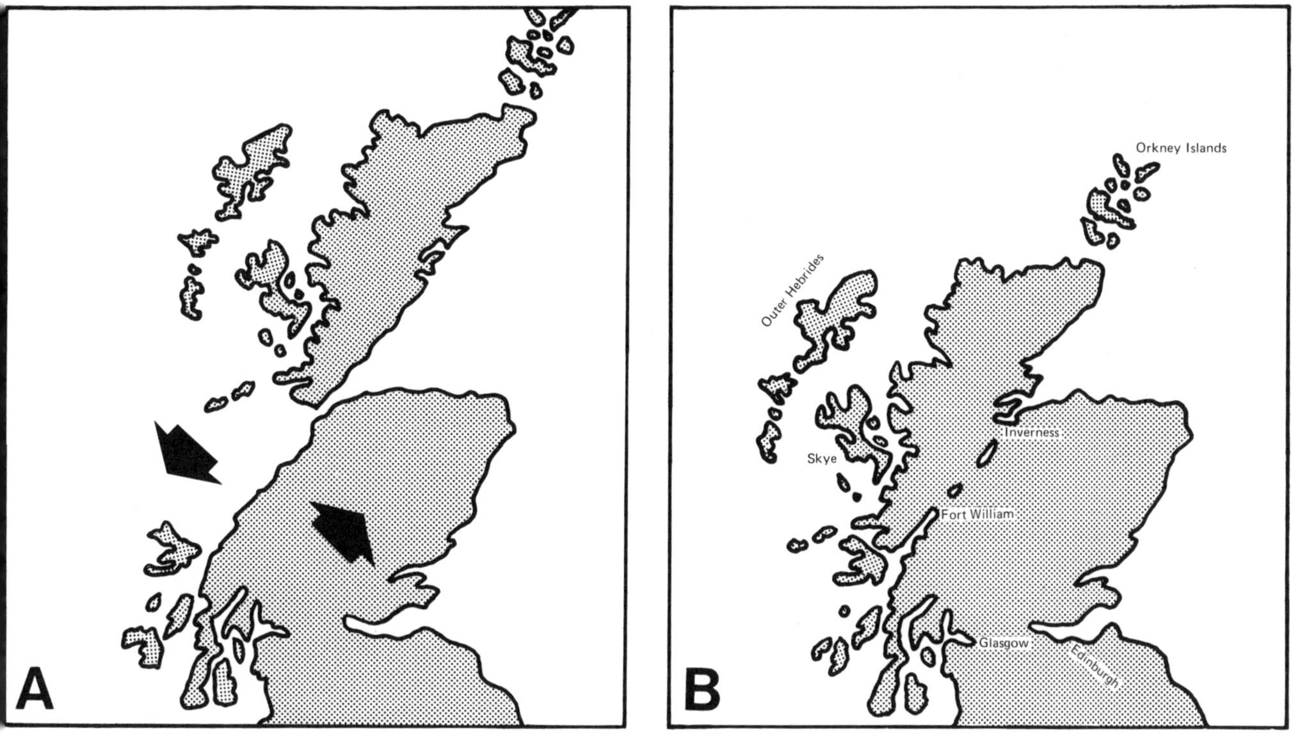


**Fig. 4.13.** The northern part of Scotland has been moved some 105 km (65 miles) by sliding along a large fault, the Great Glen Fault, the line of which is marked by linear valleys and lochs (including Loch Ness) and is traversed by the Caledonian Canal.
A: Before fault movement
B: After fault movement
Most of the movements along the Great Glen Fault occurred some 350 million years ago.

surveying, that the large California fault, the San Andreas Fault, is still moving today (and the Alpine Fault also, to judge from work in progress in New Zealand).

Thus in the 1950s the two major geophysical obstructions to continental drift were cleared away and, as luck would have it, discoveries at about the same time in the field of fossil magnetism (see Chapter 7) really clinched matters.

CHAPTER FIVE

# THE STORY IN THE ROCKS

HAVING reviewed some of the major objections that were raised against Wegener's theory of continental drift, we can now turn to some of the geological and topographical information that had originally impressed Wegener, du Toit, Holmes, King, Carey and other pioneers of the theory.

**Continental jig-saw puzzles**

Alfred Wegener was most impressed by the "jig-saw" fit of both sides of the Atlantic and modern research, coupled with improved dating methods, has amply confirmed and refined the reality of the fit.

The fit of South America against Africa is extraordinarily good. In Wegener's own words: "It is just as if we were to refit the torn pieces of a newspaper by matching their edges and then check whether the lines of print run smoothly across." The rocks and fossils of South Africa's Cape Province, laid down in seas some 250 million years ago, trend in an east-west direction and are sharply cut off by the South-west African coast. They re-appear, however, on the other side of the Atlantic, in Argentina's Buenos Aires province and Brazil's Parana and Sao Paulo provinces. Both southern South Africa, and the area of northern Argentina and southern Brazil contain widespread lake and river deposits with abundant fossils of large land and river animals. One of these animals, called *Mesosaurus* (Fig. 5.1), was rather like the present-day alligator (i.e. it inhabited fresh water, not the sea). It is known, however, only from South Africa and southern South America. By all accounts it could not cross extensive areas of sea, but yet the two regions are now separated by the entire width of the South Atlantic Ocean.

Still older rocks can be matched between Brazil and West Africa (Fig. 5.2). In Wegener's day the matching was mainly based on a visual comparison of the rocks (which were admittedly of some

**Fig. 5.1.** *Mesosaurus*, a small reptile that lived in the Permian, some 250 million years ago, is found only in fossil beds in South Africa and Brazil.

*Mesosaurus* had a long, slender head with delicate needle-like teeth that suited it for scavenging minute creatures from brackish coastal shallows. Palaeontologists are agreed that *Mesosaurus* was not a sea-going creature capable of swimming the present Atlantic, yet well-preserved skeletons lie in mudstones on both sides of that ocean. The lined pattern denotes areas of occurrences.

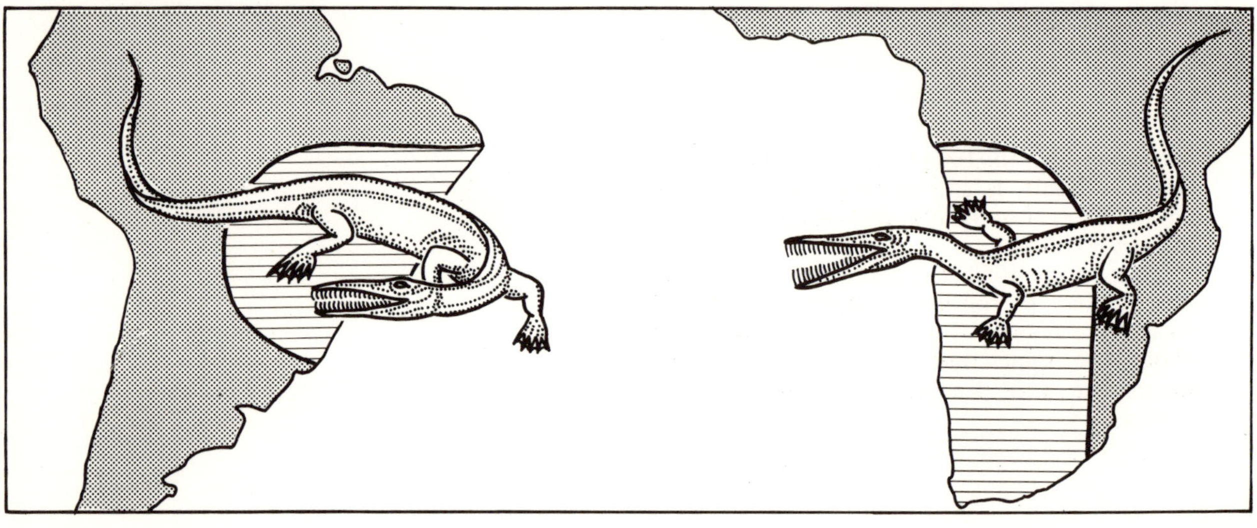

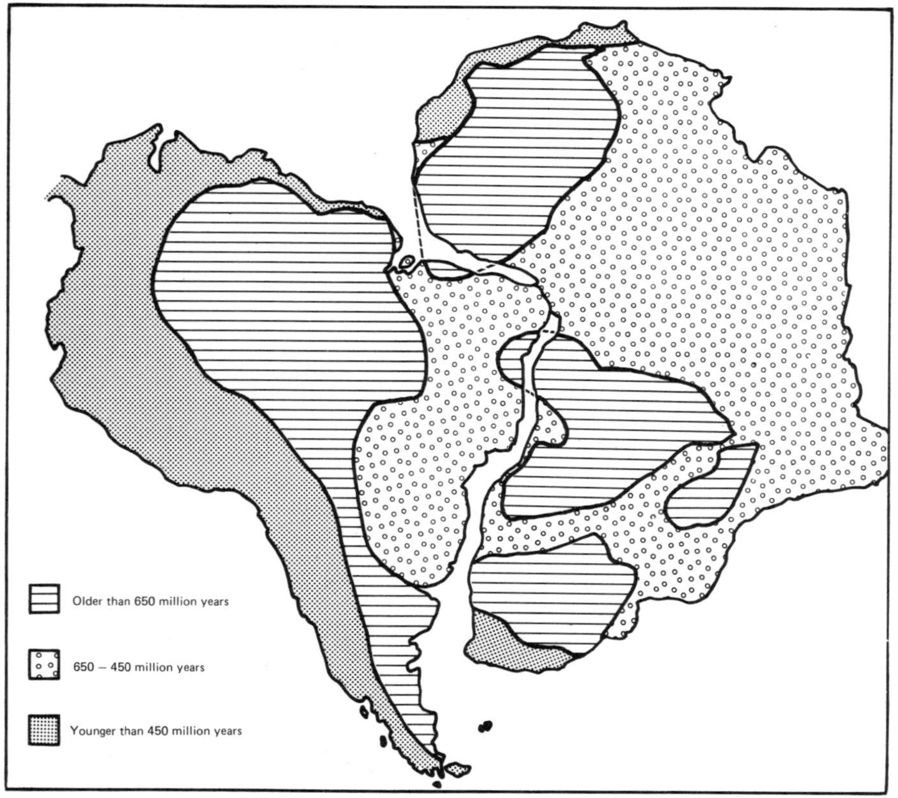


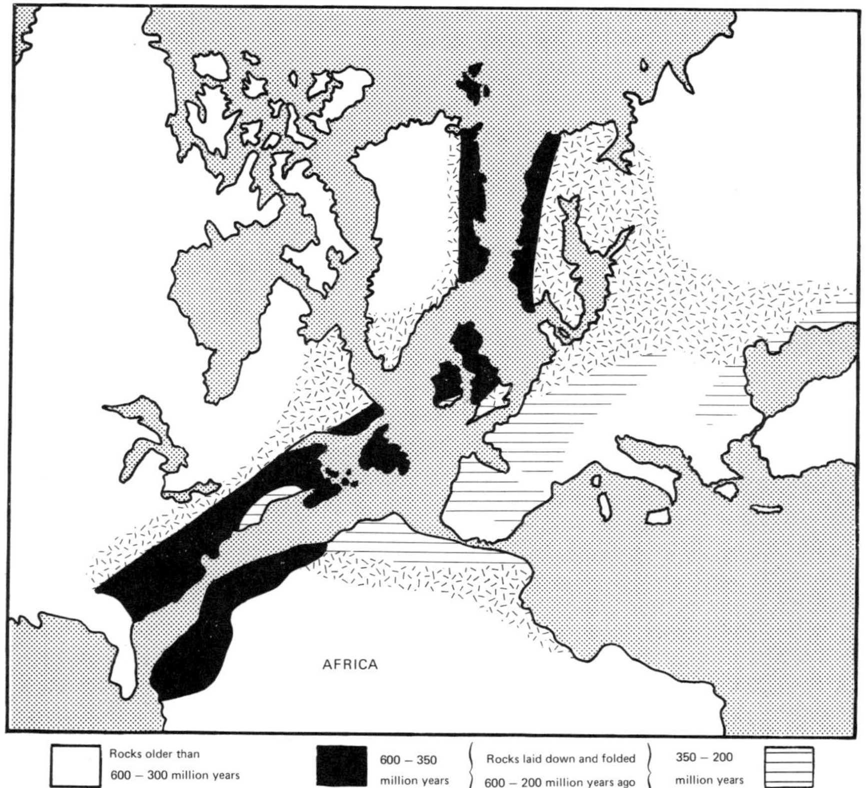


**Figs 5.2, 5.3.** The dating of rocks, both by means of fossils and by radioactivity, has enabled a clear picture to be built up of correlations across the South Atlantic (Fig. 5.2) and North Atlantic (Fig. 5.3).

Fig. 5.2 shows how South America and Africa fitted together up to some 200-150 million years ago. The lined areas are ancient continental blocks, called cratons, that are at least 2000 million years old. The other patterned areas are younger zones of geological activity, mostly troughs filled with sediments and volcanic rocks that were folded, compressed and intruded by hot materials forming granites. Note that the opening of the South Atlantic has left two parts of African cratons isolated in Brazil.

Fig. 5.3. This reconstruction of the North Atlantic region matches belts that had similar geological activity in ancient times. The speckled pattern denotes very old craton rocks (cf. Fig. 5.2), older than about 600 million years. The solid black and lined patterns show sediment-filled troughs and folded mountains that formed in the Palaeozoic, 600-230 million years ago. The North Atlantic began to open about 150 million years ago.

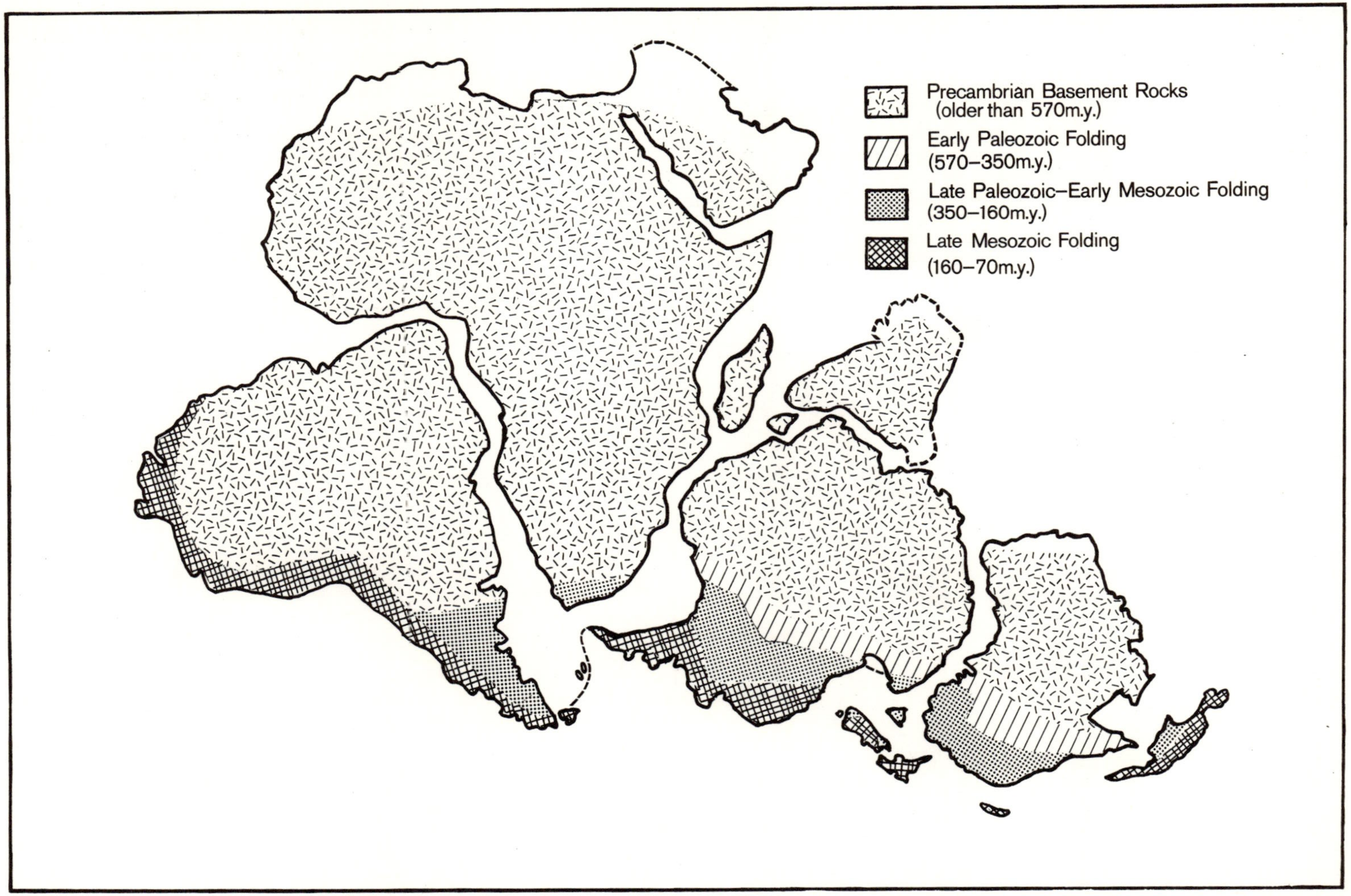


**Fig. 5.4.** Study of rocks and fossils and determination of radioactive ages has enabled correlations to be made between the geological zones of Gondwanaland.

very distinctive types). However, with the development of highly refined and precise methods of dating rocks by radioactive means (using various radioactive minerals), it became possible to work out patterns of rocks of various ages and to trace these across the gap of the Atlantic. Thus it was found that in Sao Luis, a small town on the Brazilian coast, one part of the town was sitting on rocks between 650 and 450 million years old and the remainder on very ancient rocks, a lot older than 650 million years, the continuation of which could be picked up in Ghana, the Ivory Coast, Liberia and Sierra Leone (Fig. 5.2).

The geological match across the North Atlantic, another thing that impressed Wegener, is also very good (Fig. 5.3). During the period between 650 and 350 million years ago (Cambrian-Devonian) the North Atlantic Ocean did not exist and shallow seas lapped onto the edges of the continents that are now separated by many hundreds of kilometres of ocean. Very distinctive sediments were laid down in the shallow seas of this time and marine creatures of various characteristic types lived and died in the same environment, to become preserved as fossils. Thus we find similar rocks and fossils dating back to this period of time in Spitzbergen, Norway and the British Isles, and on the other side of the Atlantic, in Eastern Greenland, Labrador, Newfoundland, Nova Scotia and the Appalachian region of the U.S.A. Earth movements that disrupted these rocks also match, and the Great Glen Fault that bisects the Scottish Highlands (Fig. 4.13) lines up with the Cabot Fault that cuts Newfoundland, separates Nova Scotia from New Brunswick and then follows the coasts of Maine and Massachusetts as far as Rhode Island. Both faults were active in late Devonian and early Carboniferous times, about 350 million years ago, when they were responsible for large movements rather like those of the more modern Alpine and San Andreas faults (Figs 4.11, 4.12).

Between 350 and 170 million years ago (Carboniferous-Triassic) many of the North Atlantic lands were above the sea and it was not until the Jurassic, some 150 million years ago, when the North Atlantic Ocean began to open, that sea reappeared.

Geological matches can also be made right around what was originally the southern edge of Gondwanaland (Fig. 5.4). This was one of the jig-saw fits that Alex du Toit had noted in 1937 and incorporated in his bold concept of a once continuous seaway, the "Samfrau Geosyncline", linking South America, South Africa, Antarctica and Australia (Fig. 3.5). Since du Toit's time, however, increased knowledge of the geology of Antarctica, New Zealand, New Caledonia and New Guinea have amply confirmed the reality of such a picture and, as we shall see, in Jurassic times, some 150 million years ago, shallow-water marine creatures were able to migrate from

**Fig. 5.5.** Landmasses are surrounded by shallow or "continental shelf" waters and there is often a reasonably steep descent ("continental slope") down into deep abyssal water. Continental shelves around New Zealand vary greatly in character. They range from narrow fringes less than 8 km (5 miles) wide to extensive submarine plains, many tens of kilometres wide; from comparatively smooth platforms to areas of irregular relief. The example illustrated in this diagram is typical of the sea floor off the south Taranaki coast.

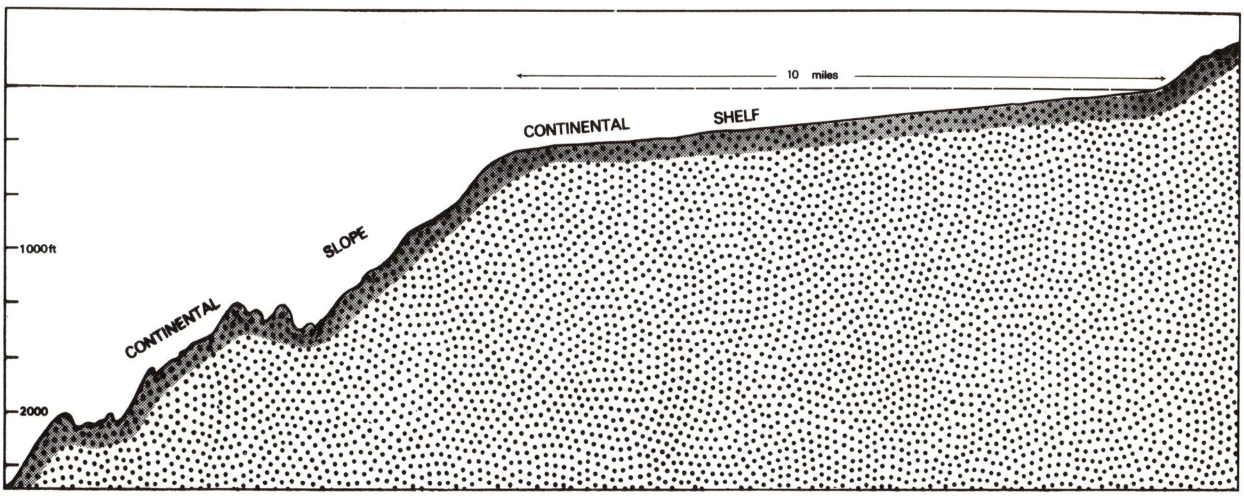

**Fig. 5.6.** Computer fit of the Atlantic continents. The fit was made at the edge of the continental shelf, using the 500-fathom (915 m) contour line of the continental slope. The edge of the continental shelf is dotted and the black areas mark gaps or overlaps. Although the pieces of the jig-saw puzzle have been apart for some 150 million years the overall fit is very good and the average mismatch is no more than one degree. Many of the gaps may have been original inlets or lakes, between continental highlands, and some of the overlaps may result from a build-up of sediment in local areas during the 150 million years that have elapsed since separation. A notable example of an overlap with this latter origin is the Niger delta, responsible for a considerable overlap onto the north-eastern corner of Brazil.

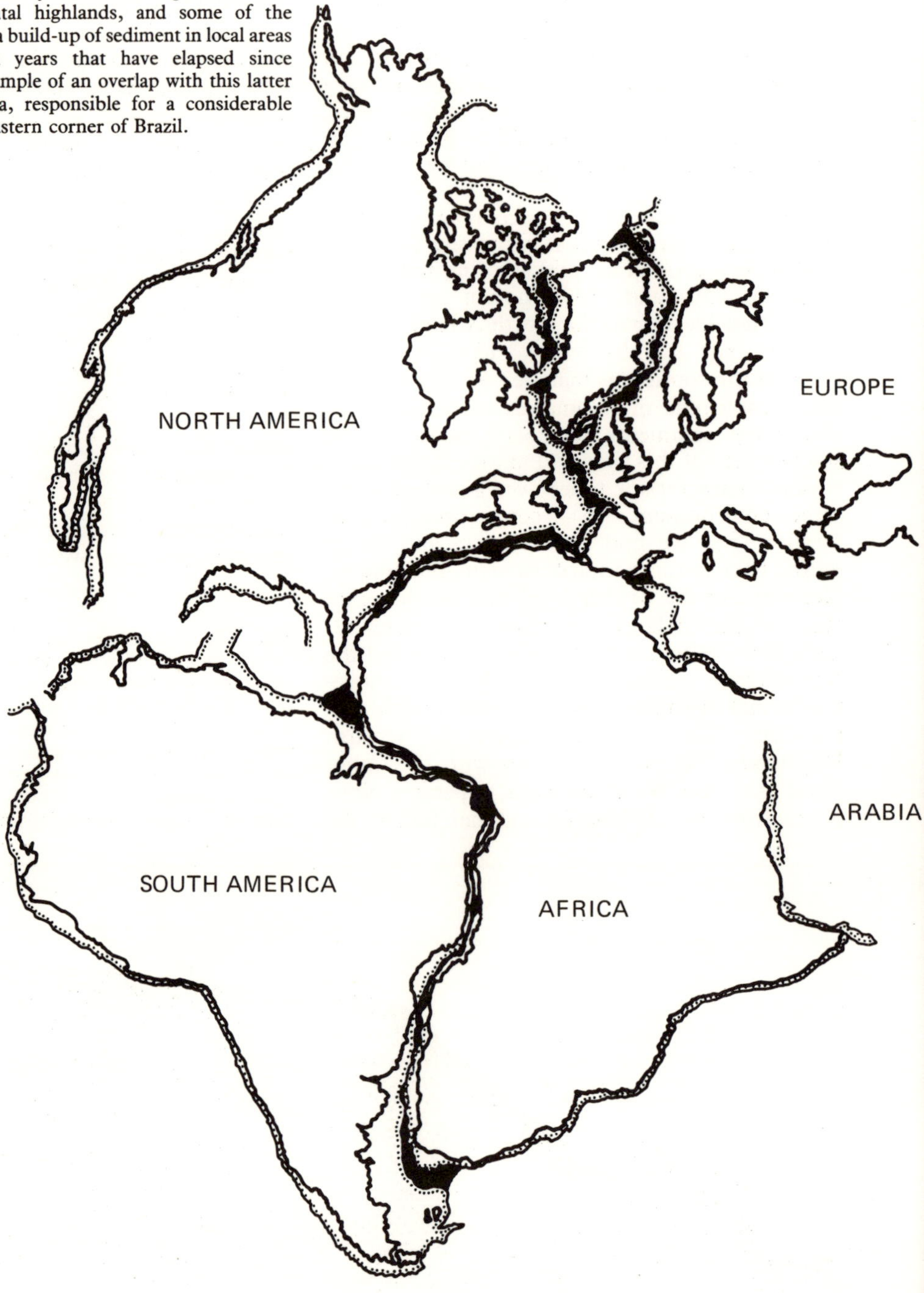

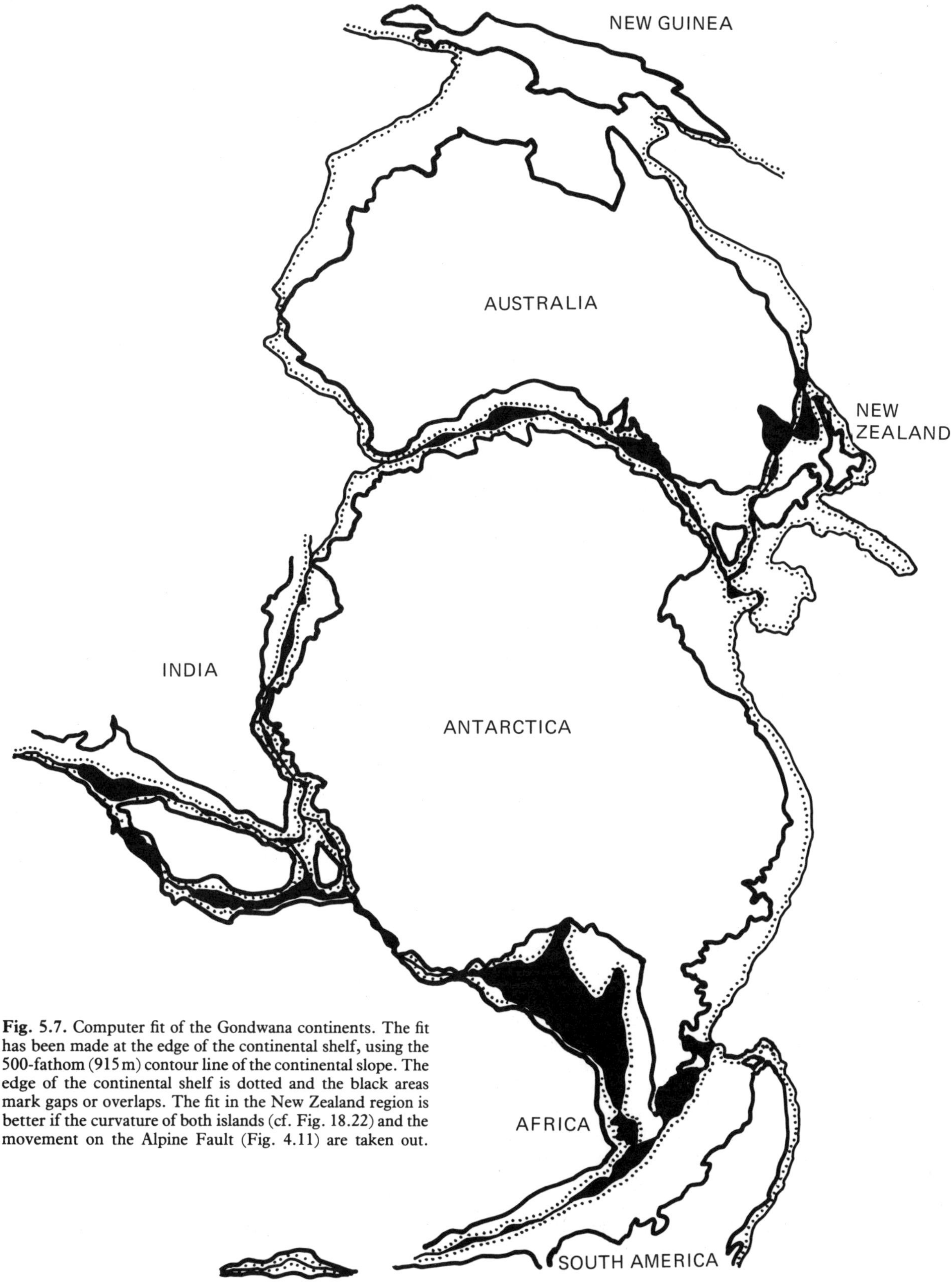


**Fig. 5.7.** Computer fit of the Gondwana continents. The fit has been made at the edge of the continental shelf, using the 500-fathom (915 m) contour line of the continental slope. The edge of the continental shelf is dotted and the black areas mark gaps or overlaps. The fit in the New Zealand region is better if the curvature of both islands (cf. Fig. 18.22) and the movement on the Alpine Fault (Fig. 4.11) are taken out.

Arabia via the Himalayas, Western Australia and Indonesia into New Caledonia and New Zealand and thence to West Antarctica and southern South America. Such a migration route, whilst feasible with the continents assembled as in Fig. 5.4 (as it merely involves migration around the coastal edge of Gondwanaland), would have been impossible, or improbable, in terms of modern geography.

**Computers**
It is also perhaps appropriate that in this computer age the precision of the "fit" across the Atlantic was subjected to scrutiny by computer. In this exercise it was necessary to feed into the computer geographic details of the exact edge of the various continents and, although it might seem strange, this is *not* the seaward edge of the present landmasses. Rather, it is the edge of the "continental shelf", the area of shallow water, sometimes wide, sometimes narrow, at the margin of the earth's continents (pp.3–6).

The continental shelves are the edges of the continents that have been flooded by the sea. We know this as over the years marine geologists have been able to trace the rocks of the land out under the sea to the very edge of the continental shelf, and often they outcrop as jagged bluffs on the continental slope (Fig. 5.5).

During the Ice Age or Pleistocene epoch, which began two million years ago and ended about 10,000 years ago, world-wide sea levels steadily dropped in the various periods during which ice accumulated on the land. This was because moisture was falling on the land as snow and ice and being held there as ice sheets and glaciers, rather than draining back to the sea. Thus the sea, deprived of significant amounts of moisture—the North American ice sheet, which extended as far south as southern Illinois, reached thicknesses of 750-1000 m (2500-3500 ft) and the Scandinavian sheet three times as much—steadily shrank until at various times, depending on the extent of the build-up of ice on the land, waves were eating away at the edges of the continents some 120-180 m (400-600 ft) below modern sea level. This erosion at low glacial levels created the sharp edges to the continental shelves we can see today, where the continents give way to the oceanic depths.

At the end of the Ice Age, some 10,000 years ago (and at other warmer spells between cold intervals), as the ice sheets and glaciers progressively melted, the water that had been "locked up" on the land gradually returned to the sea, sea level rose and the waves flooded in over the edges of the continents, to carve out the present "continental shelves" that loom so large in many oil prospecting programmes (as they provide favourable conditions for the accumulation of oil; see p. 380).

In assembling the continental jig-saw the matching of the continental outlines should therefore be made at the very edge of the continental shelf. The various geographic plots of the edge of the shelf were fed into the computer, which was then programmed to find the best possible fit. The result is shown in Fig. 5.6, and as can be seen the "fit" is amazingly good, especially when it is considered that the pieces of the jig-saw have been apart for some 150 million years. The few gaps might well have been original lakes or inland seas. The overlaps, when they are looked into closely, are seen to be areas in which muds, sands and other sediments have accumulated over the last 150 million years and have therefore padded out the continental shelves at these points.

A similar computer fit for the Southern Hemisphere landmasses, grouped together as Gondwana, has also yielded a very good result (Fig. 5.7) and, interestingly enough, has confirmed the reality of du Toit's original Gondwana reconstruction (Fig. 3.5). Du Toit had based his reconstruction on a matching of the rocks, but at that time knowledge of Antarctica was scanty and the matching-up of the various rocks and fossils of the other countries was also very much in its infancy. The overall growth of our geological knowledge, and explorations in Antarctica over the last decade, have added to the precision of the matching and as a consequence the disposition of the Gondwana lands is now fairly well known (Fig. 5.4).

At various times in the geological past Laurasia and Gondwana had largely independent histories, but some 350-150 million years ago, before the final break-up, they fitted together in a way similar to that illustrated in Fig. 5.8. Laurasia and Gondwana were separated by a vast east-west seaway, the western part of which was the ancestor of the modern Mediterranean. This ancient sea was called the Tethys by Eduard Suess, the name being derived from Greek

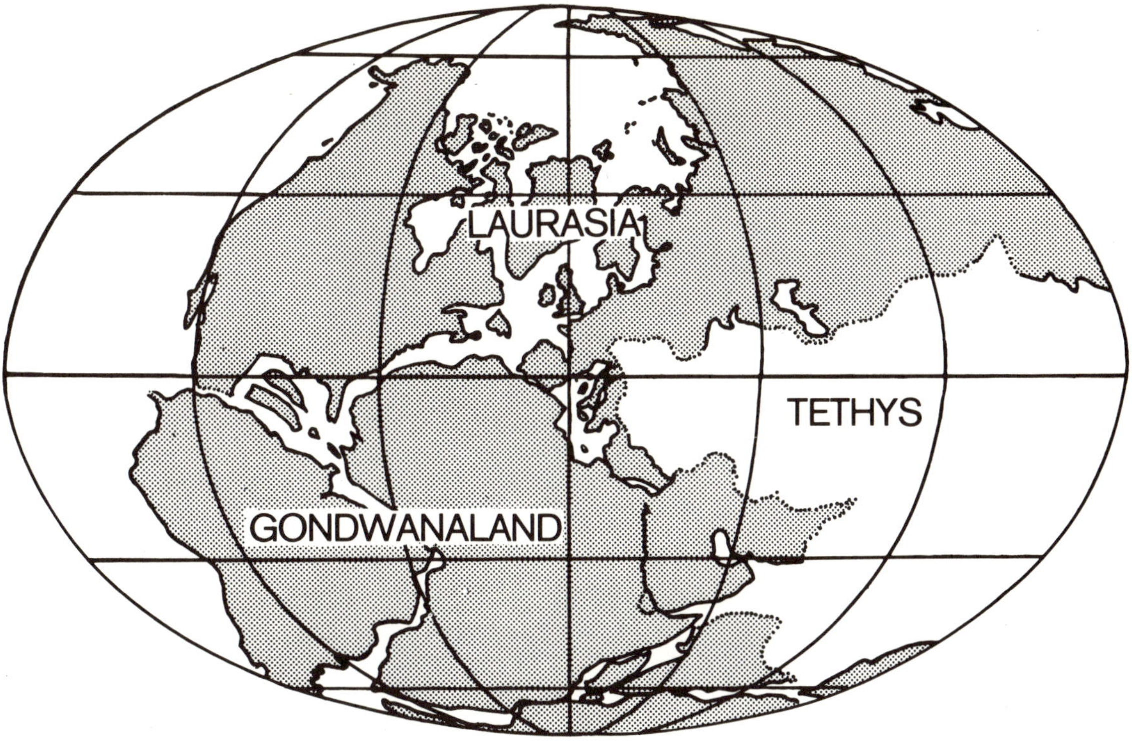


**Fig. 5.8.** Before starting to finally break up some 150 million years ago, Laurasia and Gondwana may have fitted together in this way and the world may have looked something like this 350-150 million years ago.

mythology: Tethys was the wife of Oceanus and mother of the seas. Although there is a difference of opinion over how extensive the Tethys was (cf. Fig. 8.39), the assembly in Fig. 5.8 shows one version of how Laurasia and Gondwana fitted together.

## Ancient Ice Ages

As geological studies continued in the latter part of the nineteenth and the early part of the twentieth centuries, it was found that lake and river deposits in South Africa and southern South America laid down some 350-250 million years ago (Carboniferous-Permian) contained layers of material that had unmistakably been deposited by glaciers and ice sheets. Similar glacial deposits of the same age were found in India and Australia. Antarctica remained an enigma, however, with its geological secrets shrouded by snow and ice. This gap was filled in 1961 when one of the American "Deep Freeze" expeditions found Carboniferous-Permian glacial deposits in the Horlick Mountains, near the South Pole.

In all the countries having these ancient glacial deposits it was found, by studying the way the material had been laid down and the way the underlying rocks had been scored and scratched by the movement of large rocks encased in ice sheets, that the ice had apparently moved in from the sea. This observation did not make any sense. This was because ice caps and glaciers require an adjacent area of high country to act as a collecting ground for snow. Here the snow is compacted into hard glacier ice which then flows outwards as glaciers or ice sheets. It thus follows that ice sheets and glaciers usually push out from the land onto the sea, not *vice versa*. Also, in the case of India the ice was apparently moving northwards, coming *from* the direction of the modern equator. In Africa there was a similar paradox, as here the ice was apparently spreading out from a centre in the vicinity of Southern Rhodesia and one tongue of ice had pushed northwards, towards the modern equator, and crossed it to penetrate into the Congo (Fig. 5.9).

As geologists became aware of this 250 million-year-old glaciation, and plotted the

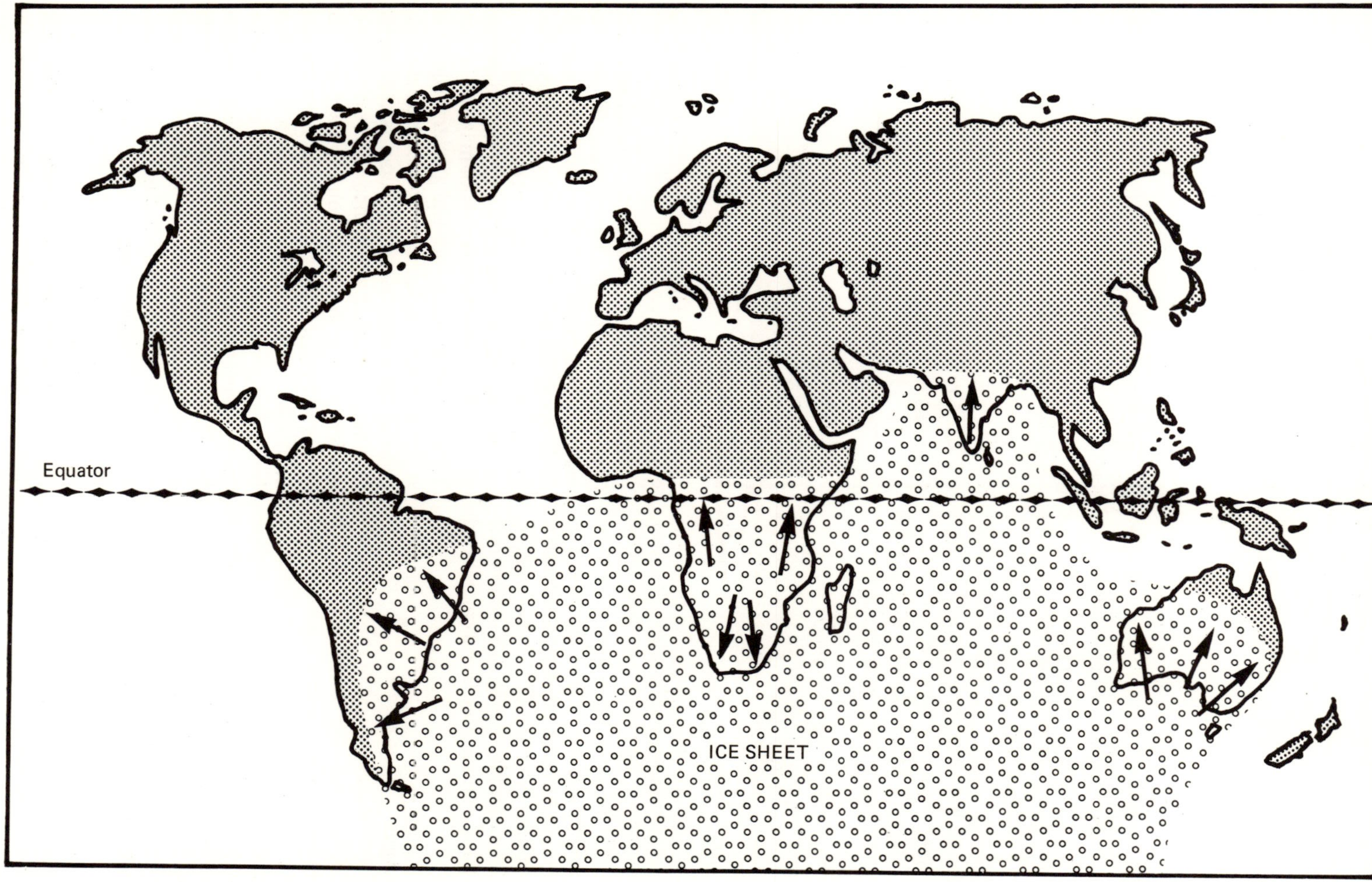


*Fig. 5.9*

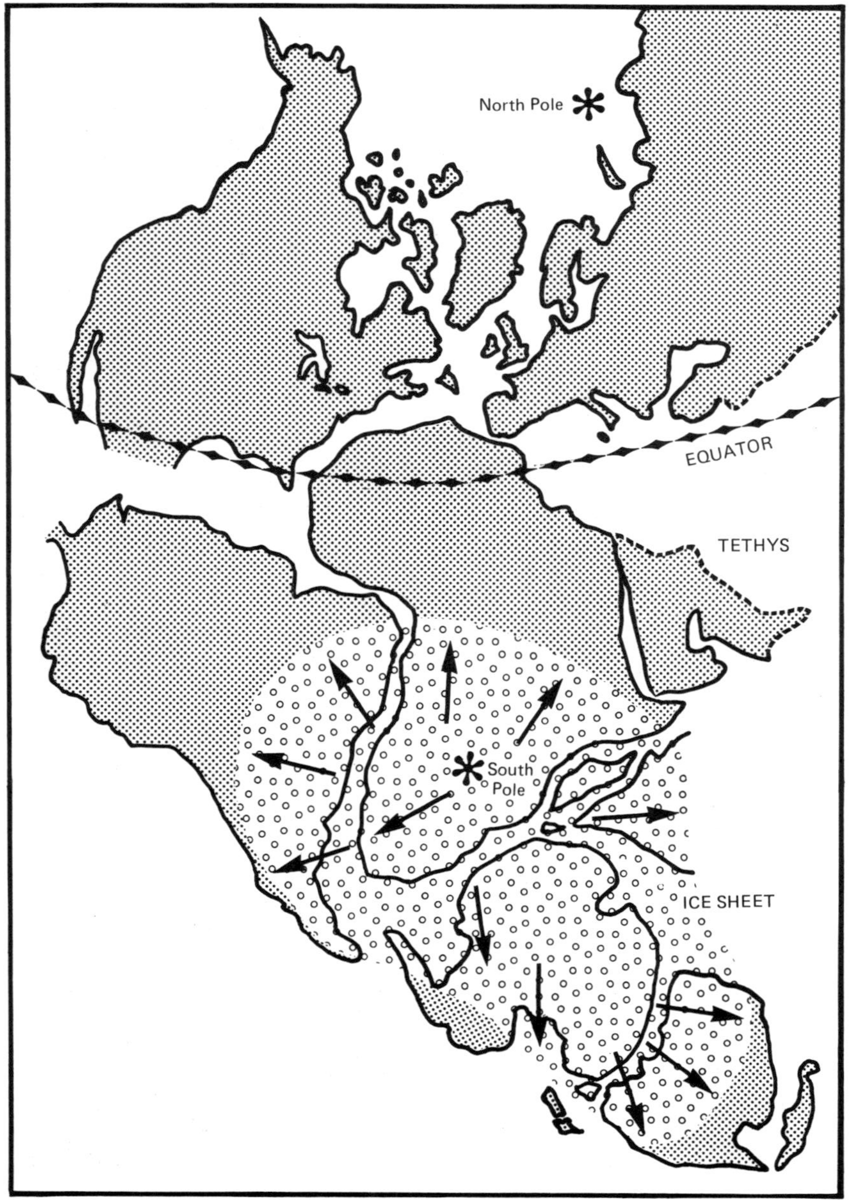


**Figs 5.9, 5.10.** The distribution of remnants of the Carboniferous-Permian glaciation (350-230 million years ago) is shown in relation to present-day geography in Fig. 5.9. The directions of ice movement recorded on the various continents are shown by the arrows. The hypothetical ice sheet that would be required to glaciate the landmasses in the way indicated is depicted by the circle pattern. As can be seen, such an ice sheet, if it was indeed present, would have to span most of the South Atlantic and Indian oceans. Fig. 5.10 depicts the situation if the continents are restored to the positions it is thought they originally occupied in Carboniferous-Permian times. Note that the continents that have remnants of the Carboniferous-Permian glaciation preserved on them were at that time close to the South Pole. Note also that the directions of ice movement are consistent with a spread of ice outwards from the South Pole.

New Zealand was on the edge of the immense ice sheet (cf. Fig. 5.11), but at that time, however, little if any land was present in the New Zealand region and ice therefore did not accumulate.

Countries like England, France and the Netherlands were at that time very much closer to the equator than they are today and experienced hot, arid conditions.

*Fig. 5.10*

localities on a map of the modern world, it became very clear that it was a physical impossibility to have a huge ice sheet pushing across vast areas of sea, some tropical and subtropical, to impinge onto the widely separated landmasses. If, however, the landmasses were pushed together, as they once were when they formed parts of Gondwana, the glaciation immediately made sense. Gondwana at that time had straddled the South Pole, and a large ice cap similar in size to the modern one now covering Antarctica had built up, and so its deposits left tell-tale evidence on the now separated individual lands (Fig. 5.10).

Supplementary evidence also came to light. If large parts of Gondwana were close to the South Pole some 250 million years ago, it also followed that other parts of Gondwana and the adjacent Laurasia probably had climates at that time vastly different from today. For example, places like Northern England and Northern Germany would at that time have been almost equatorial in position (Fig. 5.10). Therefore these countries, which now have anything but a warm climate, some 250 million years ago had a hot climate sufficient to create large salt lakes that were probably like the modern Dead Sea (Israel) and the Great Salt Lake (U.S.A.) as well as huge areas of drifting sand and steamy swamps, choked with vegetation.

As geologists were able to trace the glacial deposits and to use them as evidence for past climates, so they were able to trace the equatorial deposits—preserved as beds of red desert sands, layers of salt and other minerals (gypsum, potash, magnesia etc.) laid down by the evaporation of great quantities of water from large lakes and extensive shallow arms of the sea. Lush tropical vegetation, living and dying in extensive swamp lands, produced the bulk of the coal beds of Europe, North America and Asia.

The "evaporite" mineral deposits of Europe and North America, quite apart from their value as sources of salt, gypsum, potash and other industrial minerals, are also associated with oil and gas accumulations—notably the North Sea fields that are now very much in the news. The shallow tropical seas covering much of Europe, North America and Asia at that time probably supported enormous swarms of plankton, including many waxy algae and single-celled microscopic animals, the remains of which, when distilled by natural processes (pressure and heat deep beneath the earth's surface), produce oil and gas. Thus the position of the equatorial region at this time had a decisive influence on the formation of many of the world's foremost energy resources.

By studying the remains of animal and plant life preserved in the rocks, geologists were able to prove that the Northern Hemisphere deposits of desert sands, salts, oil and coal were being laid down at the same time that ice sheets were scouring their way across the Gondwana lands.

The discovery of the glacial beds and the corresponding belts of equatorial deposits brought the realisation that their geographic distribution was at variance with modern geography and was really only compatible with a past geography completely different from that of today. This one bit of evidence, so impressive and decisive in its implications for past geography, was in many respects the trump card of the Continental Drift Theory and remained as a major piece of evidence throughout the "dark days" of the theory. Today, even with all the new information from studies of earth's magnetism and the ocean, it still remains a most impressive geological fact.

### Glossopteris

Although Antarctica figured prominently in the reconstructions of Wegener and du Toit, for many years its exact position in the Gondwana jig-saw remained rather enigmatic. Modern Antarctic research, however, has greatly clarified its geological features, and equivalents of rocks and fossils known from South Africa, Australia and South America have now been found by various expeditions; as may be seen from Fig. 5.4, belts of the various Gondwana rocks can now be traced throughout the length and breadth of the super-continent.

New fossil discoveries added considerably to the overall picture. It was found that layers of leaves and sometimes flowers and seeds of a distinctive plant, called *Glossopteris*, often occurred associated with the 250 million-year-old glacial beds in the Gondwana countries (Fig. 5.11). These plants (Figs 5.12, 5.13) had thick, fleshy, fibrous, tongue-shaped leaves and owing to their tough construction were preserved in great quantities, forming in some instances thick layered beds. *Glossopteris* was first found in India and Australia, and discoveries followed in South Africa, South America and Tasmania. Then

Scott's expedition found specimens in cliffs flanking the Beardmore Glacier, Eastern Antarctica.

These specimens from Antarctica were probably the hardest-won geological finds in history. They had been collected by Dr Edward Wilson, scientist and artist, from the cliffs of a massive flat-lying sandstone formation, the Beacon Sandstone, flanking the Beardmore Glacier, a great ice river 160 km (100 miles) long

**Fig. 5.11.** A hardy plant called *Glossopteris* (cf. Figs 5.12, 5.13) lived on ice-free areas in and around the ice sheet that covered extensive areas of Gondwanaland in the Carboniferous-Permian, 350-230 million years ago (cf. Figs 5.9, 5.10). This map shows the localities where *Glossopteris* has been recorded. Isolated leaves are known from Southland, New Zealand (Fig. 5.13). *Glossopteris* was restricted to Gondwanaland.

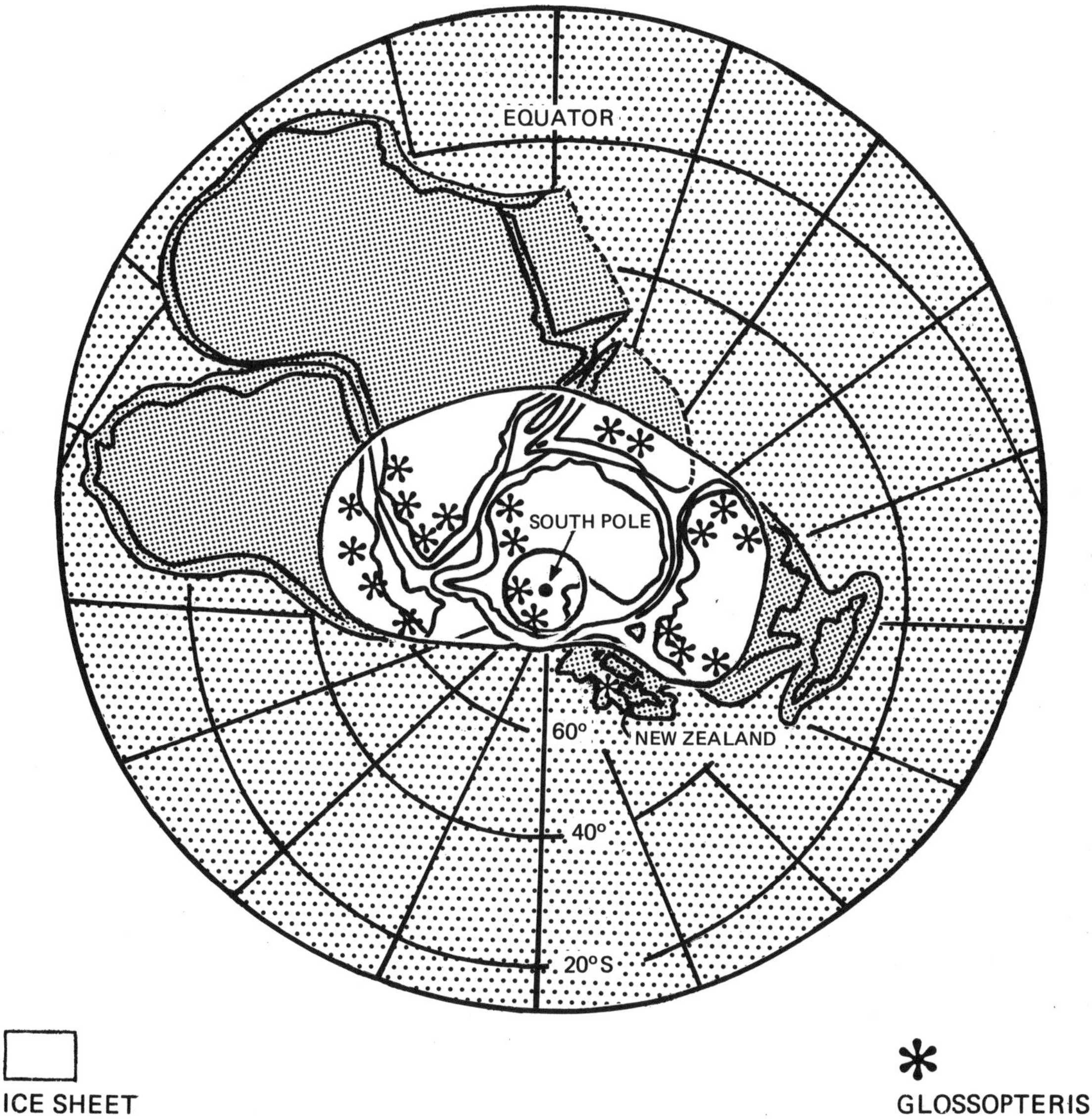

Fig. 5.12

**Fig. 5.12.** These *Glossopteris* specimens were collected by a New Zealand Antarctic Expedition from the Beacon Sandstone, close to an area sampled by Wilson, Scott and companions on their ill-fated trip to the South Pole, 1911-1912. The slab is 400 mm in length.

Fig. 5.13

**Fig. 5.13.** A solitary *Glossopteris* leaf, found in marine rocks some 235 million years old, exposed in the Takitimu Hills, Western Southland. The leaf is 65 mm at its greatest width.

and many kilometres wide, that flows down through the Transantarctic Mountains from the polar ice sheet. The specimens, along with others, totalling some 16 kilos (35 lb) in weight, were obtained on the ill-fated return journey from the South Pole. They were carried on the expedition's sledge right to the end, until Scott and his companions could walk no more. The next summer the bodies were found in their tent, the rock collection on the sledge nearby.

As the years progressed it was realised that there was a close link between *Glossopteris* and the glacial deposits, and that it was a plant which, although obviously very plentiful, only grew adjacent to the Gondwana ice sheets, as it was not known from other areas. Thus it was probably very much like the hardy sedges, rushes, mountain daisies and snow grass that grow in profusion close to our modern mountain glaciers and ice fields. *Glossopteris* probably grew in similar situations, in little ice-free hollows and on bare slopes, amongst and adjacent to the tongues of Gondwana ice.

For many years the relationship of New Zealand to Gondwana, if it had any relationship at all, was not very clear and some geologists had indeed put forward the idea that New Zealand had once been an appendage of the Asiatic landmass of Laurasia and therefore had not in fact been part of Gondwana. Recent geological work, however, both in New Zealand and in the Ross Sea sector of Antarctica, has provided key links (cf. Fig. 5.4). Then in 1970 a single leaf of *Glossopteris* was discovered in Southland (Fig. 5.13), proving that New Zealand had been part of the Gondwana super-continent 250 million years ago. New Zealand, however, unlike countries such as Australia, Tasmania, India, Africa and South America, which even at that time were large landmasses, was then merely a patch of sea floor, perhaps with a few strings of volcanic islands, lying between the Antarctic and Australian coasts (Fig. 13.1).

New Zealand was thus the repository for much of the material being eroded off the Australian and Antarctic landmasses (Fig. 13.14), but as we had very little land poking above the sea we did not have ice caps building up and leaving thick deposits such as those seen in Australia and elsewhere. We do, however, have preserved in the rocks laid down on the sea bed at that time a very good record of marine life. Study of the various shellfish preserved as fossils shows that the climate of the time was often cold, but that periods of normal warmer climate (interglacials) separated the colder (glacial) periods. Similar climatic fluctuations also occurred in the most recent Ice Age (or Pleistocene epoch) that began two million years ago and ended 10,000 years ago (p. 320).

Although New Zealand did not have enough

land to support ice sheets, the waters of the ocean covering it were probably at times full of icebergs, breaking off and floating away from the edges of the ice sheets on Australia, Tasmania and Antarctica. Some of the rocks that were laid down on the sea bed that covered Southland at this time contain large blocks of apparently "foreign" rocks, which appear to have been "dropped into" the muds and sands mantling the sea bed. It has been suggested that these rocks have been dropped from melting icebergs and it is quite possible that the single *Glossopteris* leaf known from Southland (Fig. 5.13) was carried from the adjacent lands in the same way.

### Lystrosaurus

Another milestone in the Gondwana story was the discovery of the reptile *Lystrosaurus* in Antarctica. In 1967 a New Zealand geologist, Peter J. Barrett, then working as a graduate student at the Institute of Polar Studies, Ohio State University (and now at Victoria University, Wellington, New Zealand), discovered reptilian bone fragments that provided a starting point for the investigations that were to lead to the finding of *Lystrosaurus*. On the strength of Barrett's discovery a large fossil-hunting expedition was mounted in 1969 and found *Lystrosaurus* and a number of other reptiles.

*Lystrosaurus* lived in Triassic times, about 200 million years ago, and was a heavily-built reptile with short, squat limbs, broad feet and large, heavy, beaked skull (Fig. 5.14). *Lystrosaurus* was a denizen of the land. Although it shows certain features such as the high position of the nostrils, indicating that it may have been aquatic to the extent of living in and around rivers and lakes, it was like the modern hippopotamus and tapir and quite incapable of venturing across even relatively narrow stretches of deep water.

Before being found in Antarctica *Lystrosaurus* was already known from South Africa, India and Western China (Fig. 5.15). In terms of modern world geography it was therefore possible for *Lystrosaurus* to walk from South Africa via India into Western China without getting its feet wet. However, its discovery in Antarctica confronted scientists with the problem of getting a heavy reptile, which would sink like the proverbial stone in any deep water, across to a continent now completely surrounded by oceans. It was not as though *Lystrosaurus* was alone in Antarctica—rather, it was accompanied by a number of other reptiles and amphibia, some of which, such as the weasel-like reptile *Thrinaxodon*, were undoubted land creatures. Furthermore these creatures are identical to those found with *Lystrosaurus* in South Africa and India. Thus an entire reptile and amphibian fauna migrated to Antarctica and without a doubt travelled over continuous dry land. Furthermore, the close affinities of the faunas point to a direct, not a round-about

**Fig. 5.14.** *Lystrosaurus* was a land reptile that lived in the Triassic period some 200 million years ago. It was about as large as a medium-sized dog, and had a heavy body, short limbs and broad feet. Its skull was massive and beaked and was without teeth except for a single large tusk-like tooth projecting downwards on either side of the upper jaw.

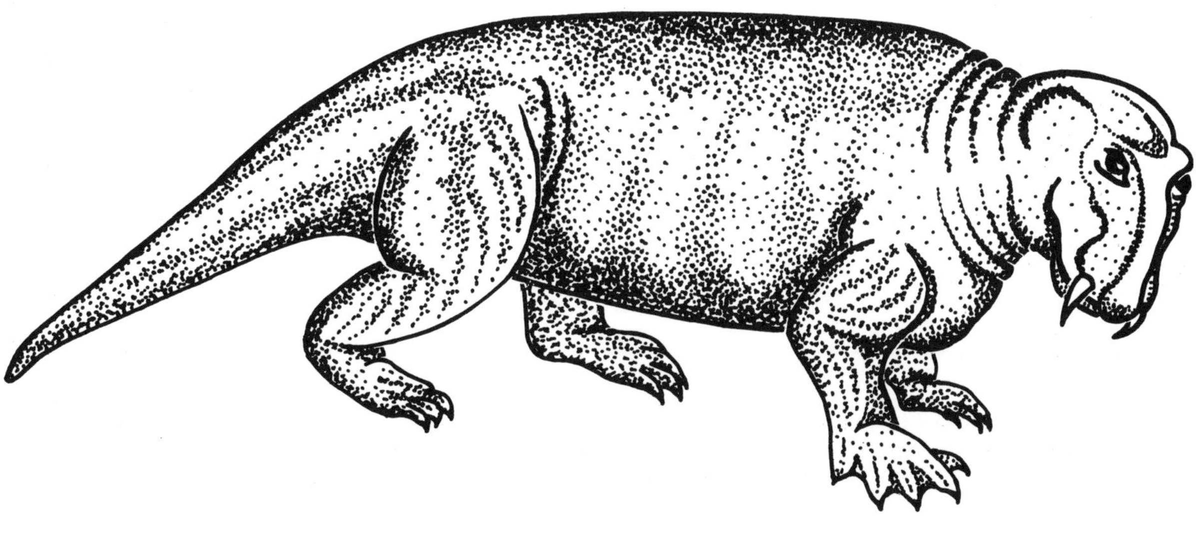

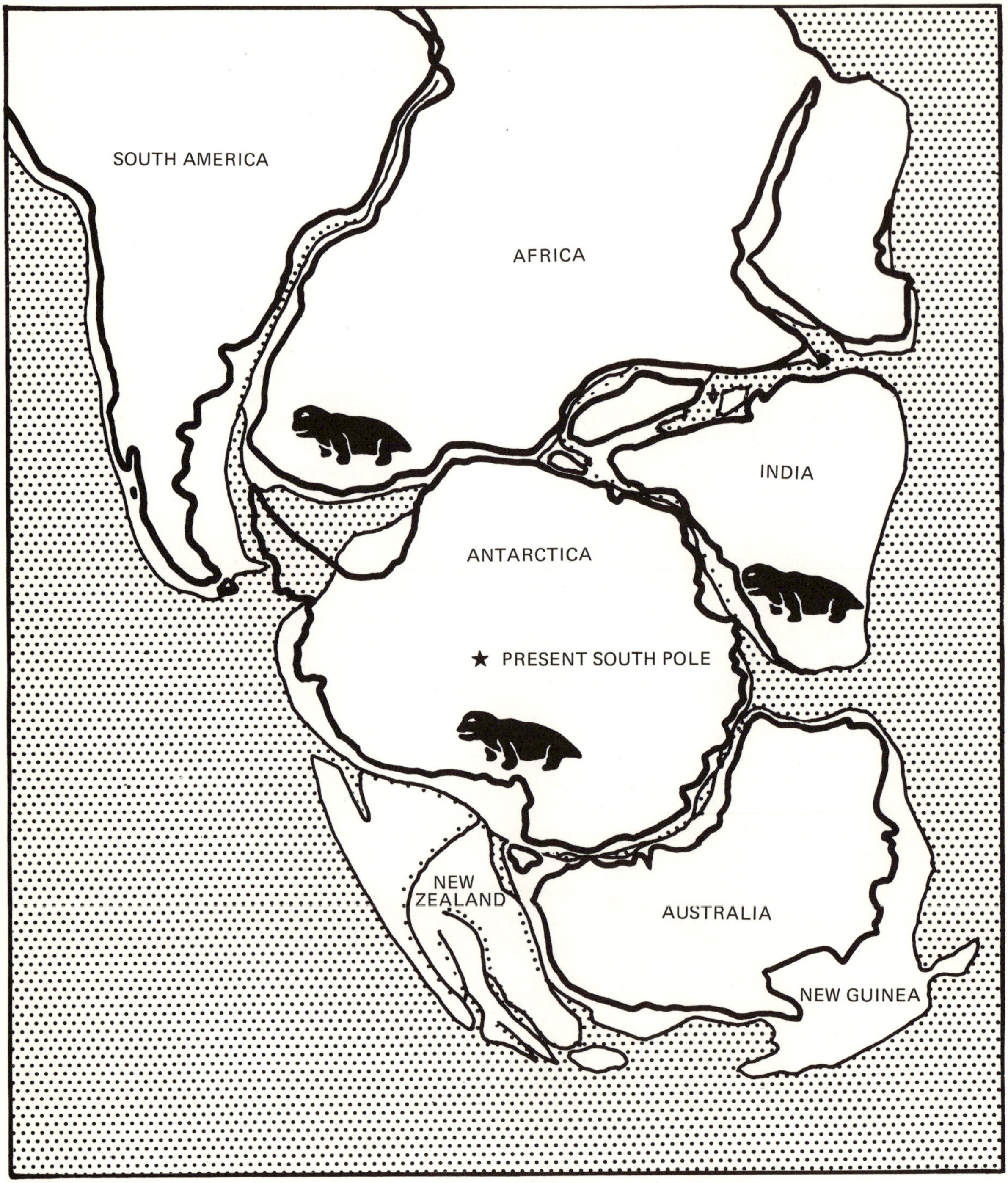


**Fig. 5.15.** *Lystrosaurus* is known from Antarctica, South Africa, India and Western China. While it could have walked across dry land to get to Africa, India and China, it was not possible for it to walk to Antarctica, unless some form of continental drift had occurred, as depicted in this diagram. New Zealand, New Caledonia, and associated submarine rises, ridges and plateaux are shown fitted between eastern Australia and Antarctica (cf. Fig. 13.14).

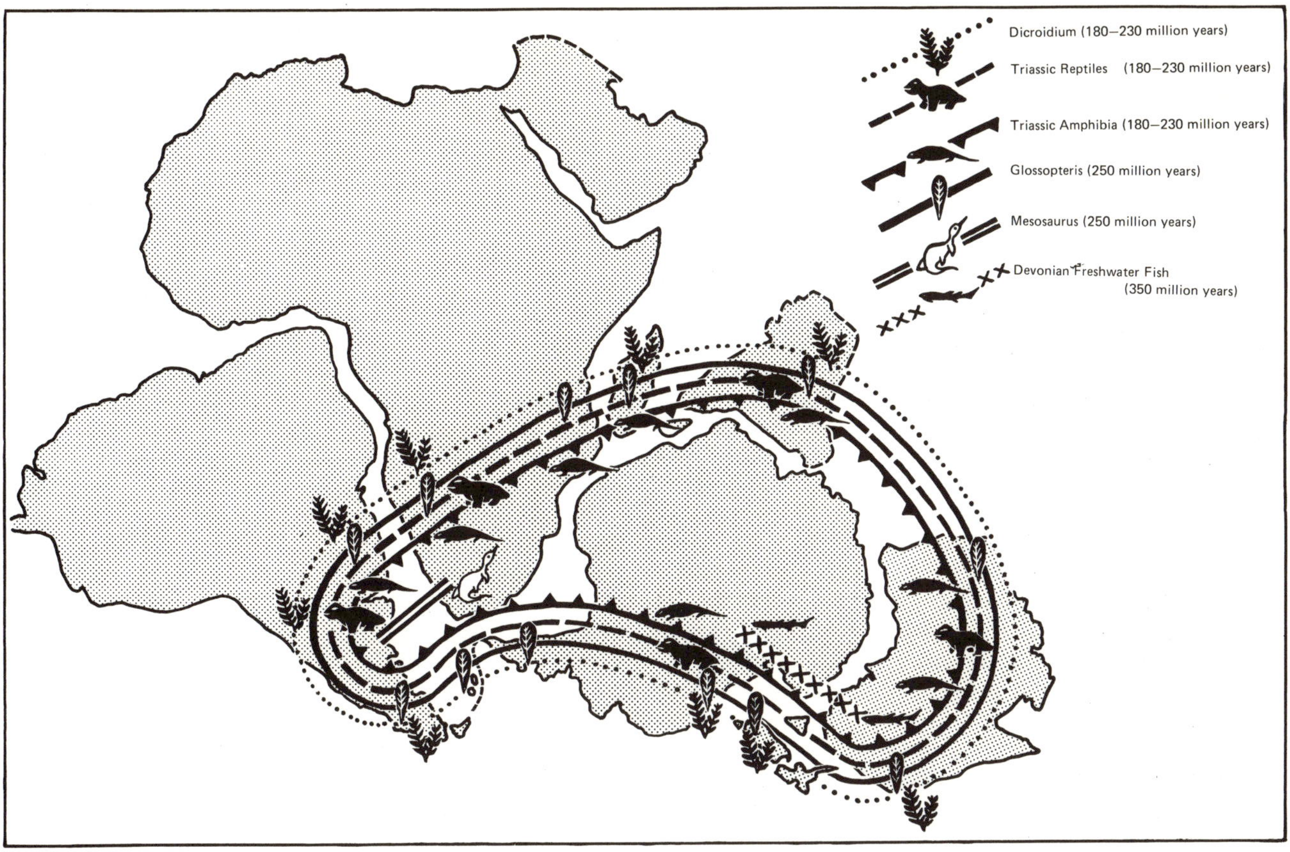


**Fig. 5.16.** In the text *Glossopteris, Mesosaurus* and *Lystrosaurus* have been mentioned as examples of some of the fossil links that bind the fragments of Gondwanaland together. However, other fossil links are also apparent and some of these are summarised in this diagram. All of the plants and animals illustrated were inhabitants of either the land or fresh water, and could not have spread to the various continents if they had been separated by oceans as they are today.

Fossil fresh-water fish of a very distinctive type occur in river and lake deposits laid down some 350 million years ago in Eastern Antarctica (near McMurdo Sound) and South-east Australia. *Mesosaurus*, an alligator-like reptile, occurs only in river and lake deposits laid down some 250 million years ago in South Africa and South America (Fig. 5.1), while the plant *Glossopteris* occurs in beds of the same age in all of the Gondwana countries (Fig. 5.11).

Early Triassic reptiles like *Lystrosaurus* and its allies (Figs 5.14, 5.15), as well as amphibia, occur in most of the Gondwana countries, although not apparently in New Zealand, which was mainly under the sea at that time. However, a Triassic plant *Dicroidium* achieved a distribution similar to that of *Glossopteris* in the Permian and is found on all Gondwana countries, including New Zealand.

connection. This rules out, for example, any long migration route such as that extending from Africa, India and China over the Bering Strait land bridge (which on occasions in the past provided a link between Asia and North America) then down through the Isthmus of Panama and across a land bridge following the curving island chains between South America and Antarctica.

It could mean only one thing—that dry land must have existed in Triassic times between South Africa, India and Antarctica; an area now occupied by the Indian Ocean. As there is no evidence of a sunken Atlantis-like land in the Indian Ocean, the simplest way out of the dilemma is to slide Antarctica close to South Africa and India, as in du Toit's Gondwana reconstruction (cf. Figs 3.5, 5.4, 5.15). *Lystrosaurus*, and its reptilian and amphibian companions, are the latest discoveries of a whole host of fossil finds which, when coupled with our knowledge of how the rocks link up, bind Gondwanaland together (Fig. 5.16).

CHAPTER SIX

# LOST CONTINENTS AND LAND BRIDGES

WEGENER'S ideas received an enthusiastic reception amongst many zoologists and botanists. As exploration proceeded apace in the late nineteenth and early twentieth centuries, and as detailed zoological, botanical and geological evidence accumulated, various specialists in these fields were for the first time able to draw conclusions about the relationships of the world's animals and plants, both modern and extinct (the latter preserved as fossils in the rocks). It therefore became evident that many continents, now widely separated by deep oceans, had very closely related animals and plants, both living and fossil. The over-riding problem was how to get these land-bound creatures from one continent to the next without "getting their feet wet".

Before continental drift came on the scene it was fashionable to get animals and plants around the world by means of "sunken continents" and "land bridges". Thus there was thought to have been a "lost continent" (called "Tasmantis") in the Tasman Sea, between New Zealand and Australia (Fig. 6.1), which had sunk without a trace beneath the waves. "Lemuria" was another "lost continent", linking Madagascar with India, as these two lands have so many animals and plants in common, including the only living monkey-like lemurs (Fig. 6.2).

In the late 1800s quite a number of these "lost continents" were proposed, to provide migration routes for animals and plants (Figs 6.2-6.5). As the early 1900s progressed, however, and geologists became aware of the differences between continental (sial) and oceanic (sima) rocks, particularly the density differences, the sizes of some of these mythical lands—some of which were postulated to have filled entire oceans—were progressively reduced (Figs 6.5, 6.6). This was because it had become obvious that continents generally floated high, and were therefore land because their rocks, together constituting the sial, are lighter than the rocks of the oceans (sima). Thus it is a physical impossibility to push a continental-sized landmass down under the sea, and keep it there, unless it is loaded down with a lot of heavy "ballast" in the form of great thicknesses of volcanic lava (and even then it is doubtful whether this would have any major effect).

If it *is* possible to load a landmass with so much heavy volcanic material that it becomes submerged and is thus prevented from popping up again, it still means that a situation would arise in which a mass of lighter sial material lay beneath heavier sima material. Thus there would be an easily measured difference in the pull of gravity in regions with lighter sialic rocks underlying heavier simatic rocks compared with regions composed solely of simatic rocks (Fig. 1.7), making them very easy to identify. Even if a sialic mass was broken up, individual fragments would still float and be capable of being recognised. But as exploration proceeded it was in fact found that the few continental island-fragments (e.g. the Seychelles) that were identified were so few and far between that it would not be possible to construct any sort of ocean-bridging landmass out of them.

Although land bridges were still in vogue well into the middle of the present century, some geologists nonetheless realised that there were huge areas of ocean completely free of sial. The great German geologist Hans Stille, writing in the early 1900s, called these areas "*Ur-Ozeane*", or ancient oceans (Fig. 6.7). They were thought to be extremely primitive and to have always been oceans throughout the earth's history (the Pacific ancient-ocean was thought to have originated when the moon pulled away from the earth). Thus land bridges, if they were to be built, would have to be built around such ancient oceans.

As all these facts became known, so the people who wished to cling to the idea of the permanency of the modern oceans and landmasses, but yet still

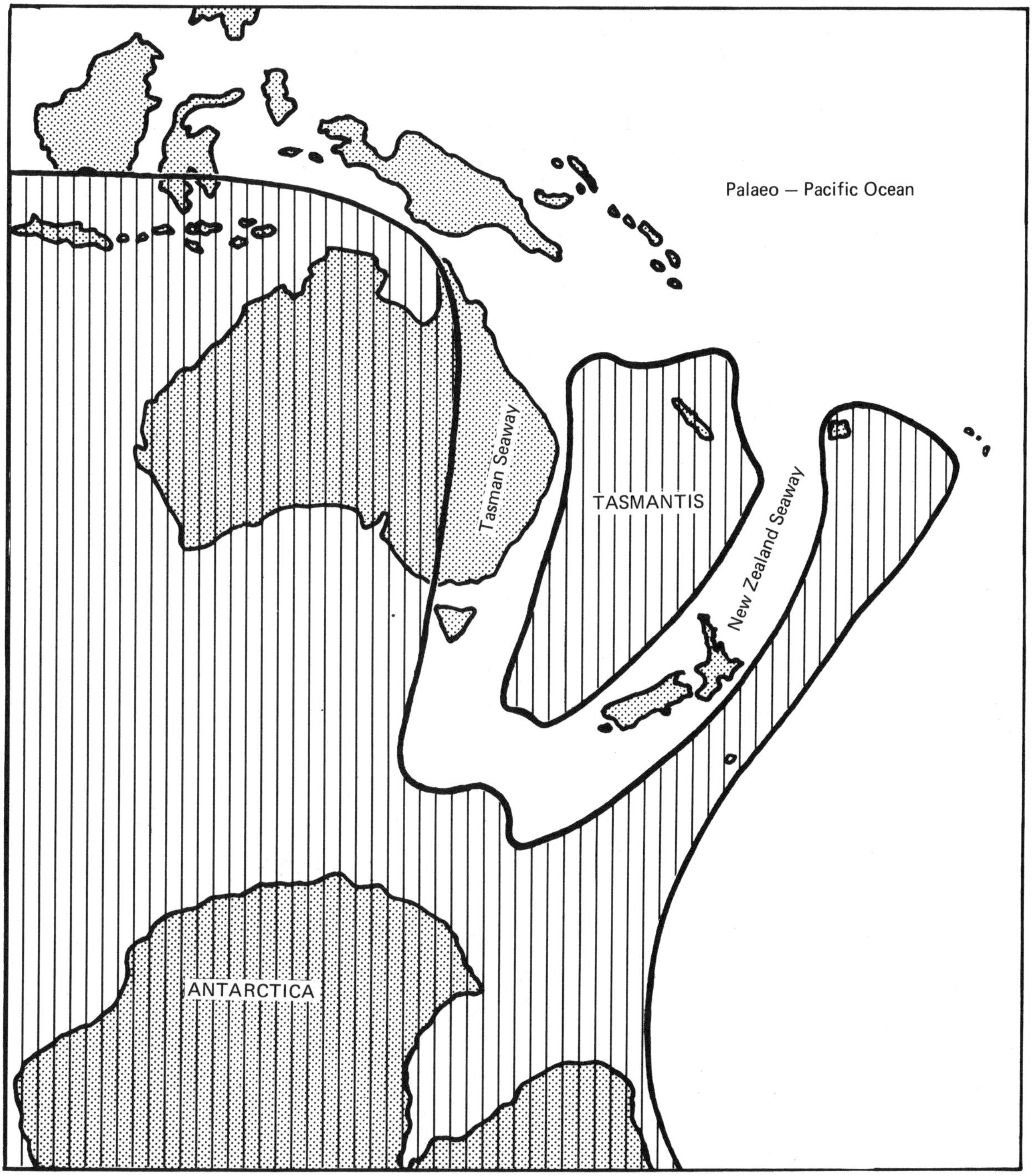


**Fig. 6.1.** This palaeogeographic map was published in 1952 to show how coastal marine animals could migrate to New Zealand from eastern Australia around the shores of "Tasmantis", a continent now submerged beneath the Tasman Sea. The map represented the geography of the South-west Pacific in the early Devonian, 375 million years ago. The lined areas are land, the clear areas sea.

In terms of modern ideas, it is now easier to place New Zealand close to East Australia, so that both countries can share their coastal animals, and then to gradually move New Zealand across to its modern position, in the process opening up the Tasman Sea (cf. Fig. 13.14).

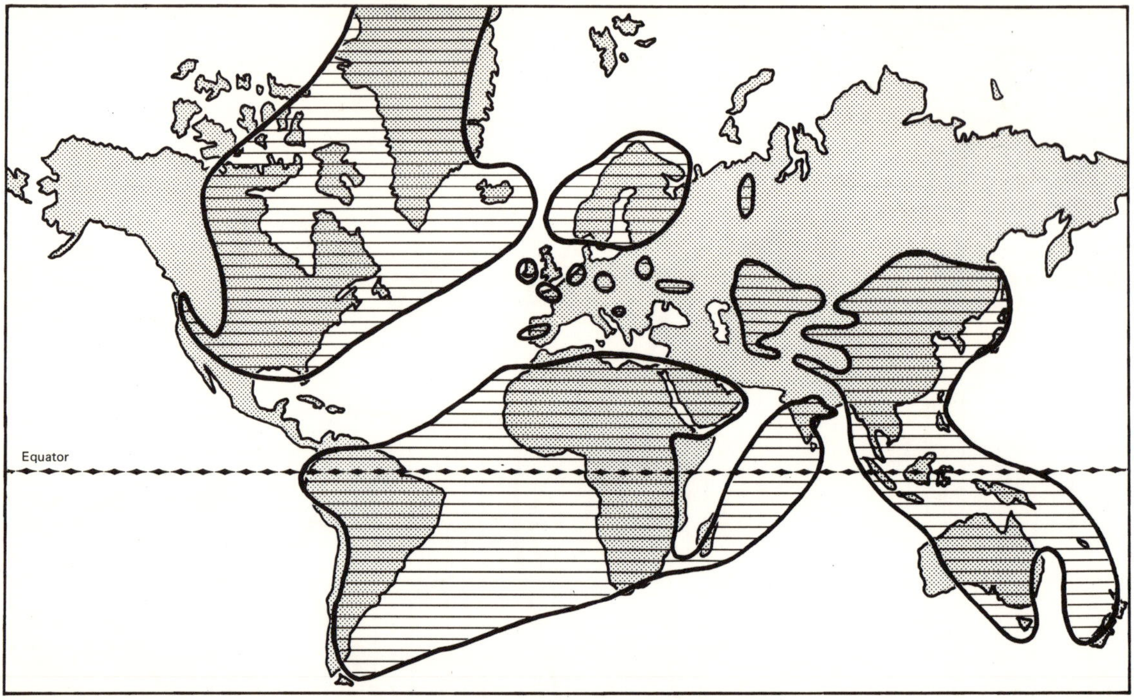


*Fig. 6.2*

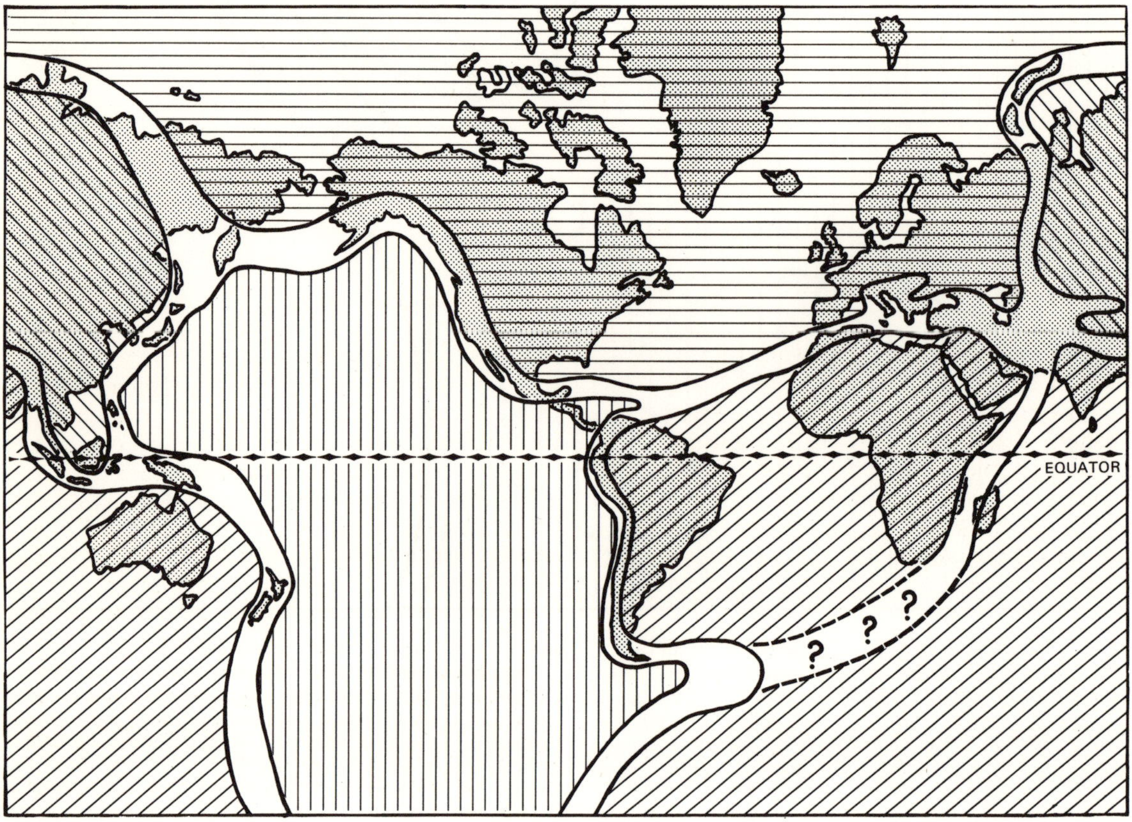


*Fig. 6.3*

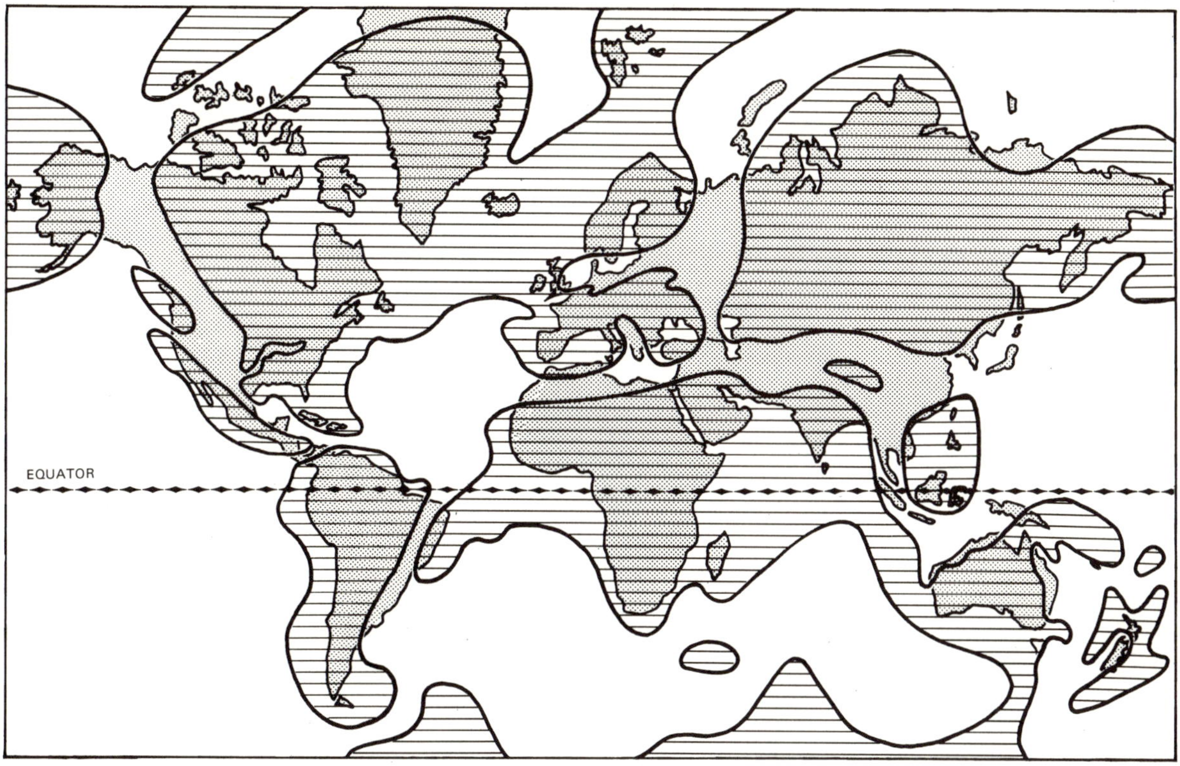


*Fig. 6.4*

**Fig. 6.2.** In 1887 Melchior Neumayr published the first palaeogeographic map. The map, of which this is a simplified version, showed the world in the Jurassic period, some 150 million years ago. Land is shown as patterned areas, sea as clear areas. The equator is indicated.

The southern Atlantic was spanned by an immense Brazilian-Ethiopian continent and a peninsula (named "Lemuria" as it was thought to provide a migratory route for the monkey-like lemurs) links Madagascar and India.

**Fig. 6.4.** In 1923 Charles Schuchert published this map for the early Permian, 280-260 million years ago, in which he envisaged a large Gondwana landmass spanning the South Atlantic and South Indian Ocean and continuing to Australia and the Tasman. The lined areas are land and the clear areas sea.

had to account for animal and plant migration, had to drastically pare down the "lost continents" and "land bridges" which had been proposed. They thought that by reducing them to the absolute minimum they *might* be able to avoid the consequences of the physical constraints imposed by the fundamental density differences between sial and sima. However, even with the barest bones of bridges—they were called "isthmian links", as they were thought to be similar to the modern Isthmus of Panama—the sheer length of the links required (e.g., stretching from Madagascar to India) still involved a mass of sial that could not by any stretch of the imagination be simply and conveniently lost in the ocean depths. Thus by the time Alfred Wegener came on the scene, the old stand-by of "land bridges" was already becoming suspect and the time was ripe for many biologists to adopt continental drift as a means of getting animals and plants around the world.

**Fig. 6.3.** The Mesozoic world, some 200-100 million years ago, as envisaged by Emile Haug in 1907. Haug thought that there were five vast continents (patterned areas), separated by seaways (geosynclines), shown as clear areas in the map. New Zealand was linked by a seaway to New Guinea, Antarctica and South America.

Although the mechanism for continental drifting was at that time unknown, many biologists were content to leave the unravelling of this problem to the geologists and geophysicists. As far as the biologists were concerned, they knew from the way certain countries shared the same or nearly the same animals and plants that some type of land migration route had been present in the past, even though deep oceans now separate the countries. As land bridges were suspect, continental drift became a viable alternative.

**Figs 6.5.** (Upper), **6.6.** (Lower).
At the turn of the century geologists envisaged large ocean-spanning continental masses (Fig. 6.5; cf. Fig. 6.2), but by the 1900s these had been considerably thinned-down to "isthmian" links (Fig. 6.6). The hypothetical land links are denoted by the lined pattern.

In the 1920s, 1930s and 1940s many biological distributions were therefore explained in terms of continental movements in geological past ages. However in many respects over-zealous use of drifting by some biologists brought the theory into disrepute. For example, to explain the distribution of a weevil found in two widely separated areas, a continent was blithely slid almost half-way around the world—in spite of the fact that there was little or no geological evidence for such a movement. As a result biological (and palaeontological) evidence supposedly supporting continental drift became rather suspect.

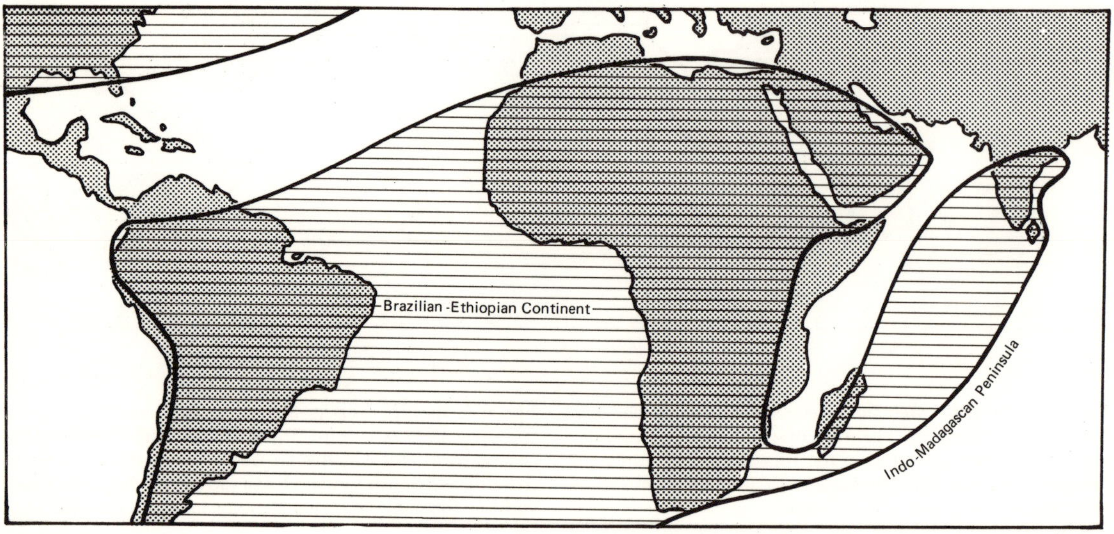


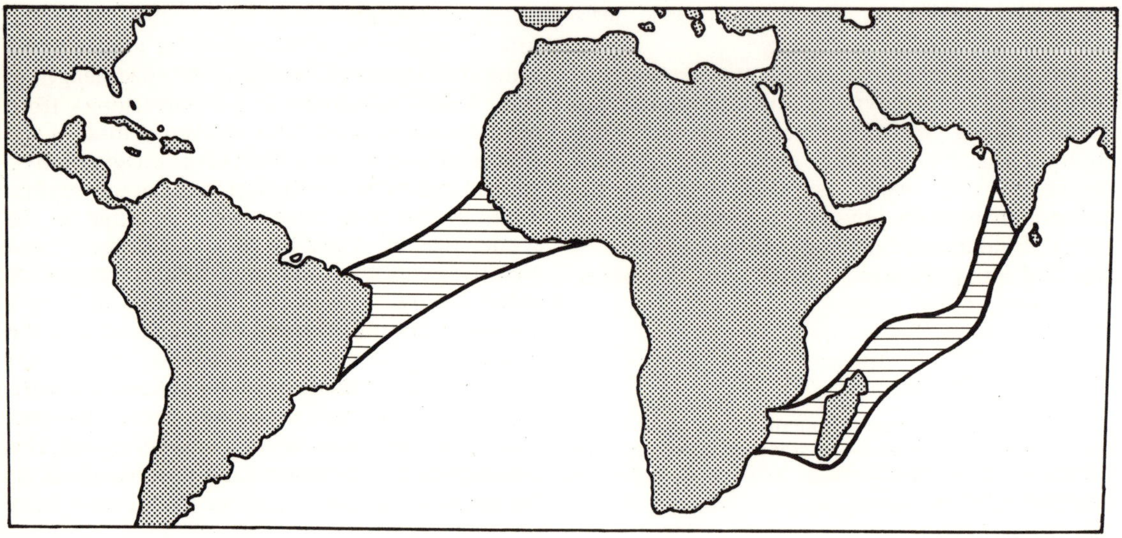

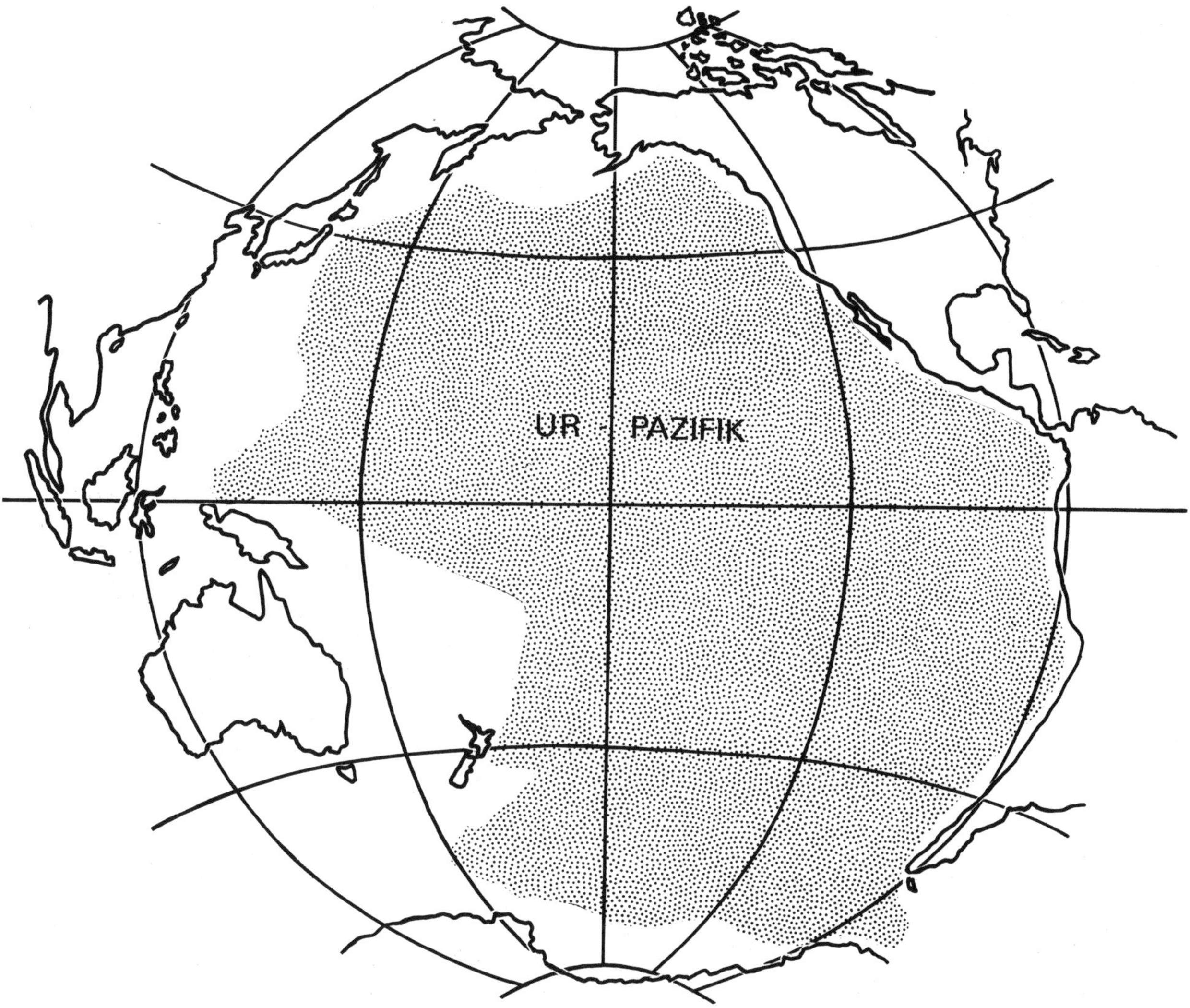


**Fig. 6.7.** The Pacific Ocean was one of Hans Stille's "Ur-Ozeane", or ancient oceans, and the patterned area shows the part he classified in this way.

A little later on, however, it was realised that in studying the distribution of plants and animals the biologists (and palaeontologists) had overlooked an entire field of inquiry: that of the possibility of chance dispersal—i.e., the effect of chance happenings like storms and currents on the distribution of animals and plants.

It was found, for example, that some echinoderms, like the sea eggs, starfishes and brittle stars so common along many New Zealand coasts, also occurred on the Chatham Islands, various islands across the South Pacific between New Zealand and South America, and sometimes all the way around the world in the vicinity of 40° and 50°S Latitude (Fig. 6.11). This pattern of distribution was not the result of migration of the echinoderms around the shores of a "lost continent" or "land bridge", but rather the result of natural transport by currents in the sea.

**Swimmers and floaters**

Most people are no doubt familiar with the fact that certain creatures, such as oysters, are to be found only in shallow coastal waters—the waters covering the continental shelves (p. 3). Other completely different creatures inhabit the floor of the deep sea.

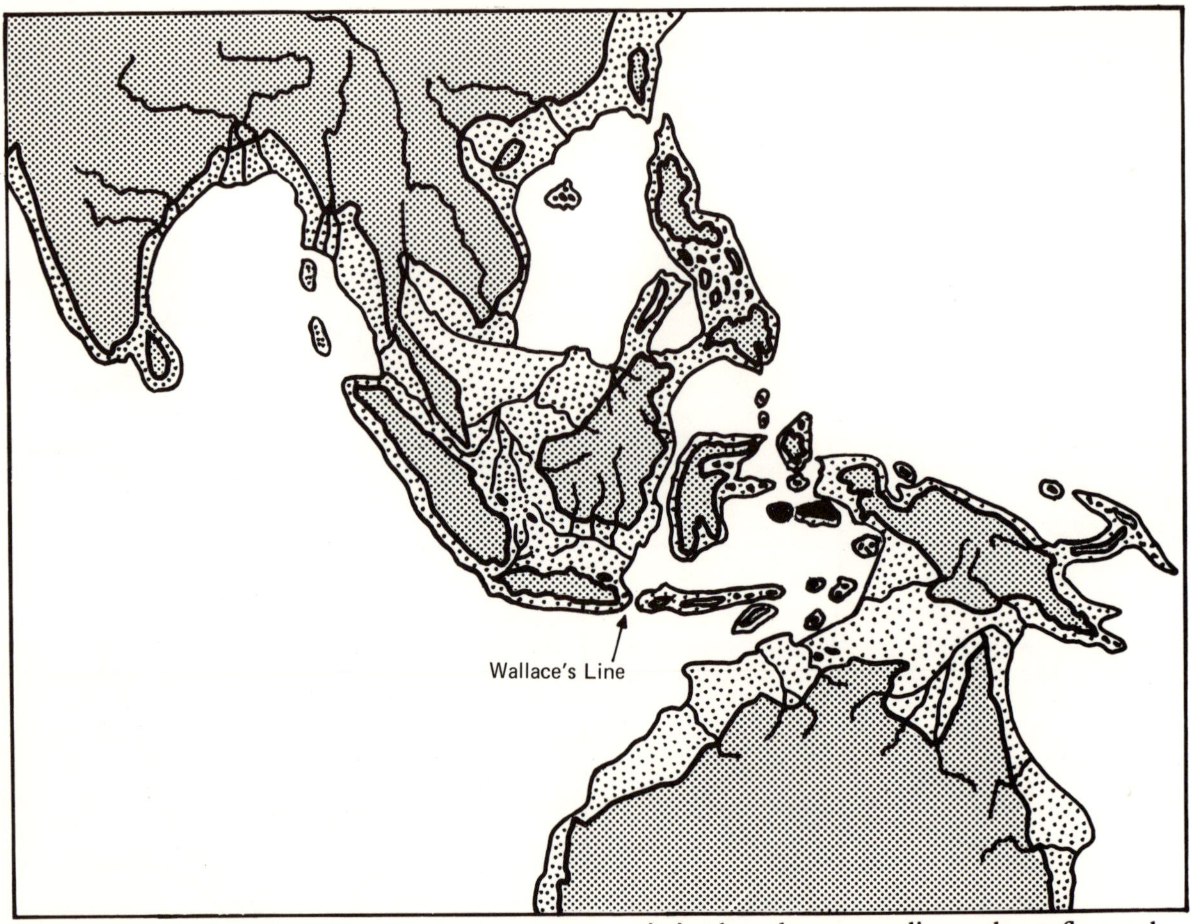


**Fig. 6.8.** During the various glacial episodes of the "Ice Age", or Pleistocene, world-wide sea level fell to often very low levels, sometimes 183-238 m (600-775 ft) below modern sea level. These global falls in sea level drastically altered the shapes of many of the present landmasses and dry land linked some countries now separated by sea. The myriad of islands between South-east Asia, Sumatra, Java and Borneo were joined by land, and animals and plants could migrate from the Asian mainland to populate areas now isolated by sea. Thus zoologists recognise an Oriental (or Indian) fauna that extends southwards into Indonesia. However, even during the times of greatest sea level drop an impassable sea channel existed between the islands of Bali and Lombok, and the Oriental faunas could not progress beyond this point.

Other Indonesian islands eastwards from the 38 km (24 miles) Lombok strait are populated by faunas with marked Australian affinities. The faunal (and floral) boundary extending through the Lombok Strait is named Wallace's Line, after the biologist Alfred Russel Wallace, a contemporary of Charles Darwin. In this map the present landmasses are shown by a fine dot pattern and the continental shelves, that were dry land during the Ice Age, are shown by a coarse dot pattern. Note how many of the rivers had to flow very much further to reach the sea.

Animals such as oysters live on the sea floor and are called "benthic" creatures. Other animals are very light and can float (they are called "planktonic"), whilst other animals have to swim to stay afloat (they are called "nektoplanktonic").

The planktonic and nektoplanktonic creatures live in the top few metres of the sea that are warm and sunlit, and, as they are either floating or swimming, the depth of the underlying water is immaterial—they can live over either shallow or deep water. Planktonic and nektoplanktonic sea creatures can therefore be blown about all over the world by winds, waves and currents, and so the modern or past distribution of such creatures can be explained very readily. Sea-floor (or benthic) creatures populating the coastal waters, on the other hand, have very definite limits to their distribution.

The creatures—both animals and plants—that swim or float in the sea can drift almost anywhere at the whim of currents, storms, winds etc., the main factor limiting their distribution often being temperature. Many bethnic or sea-floor-dwelling creatures, on the other hand, are restricted to the shallow coastal "continental shelf" seas.

This is because most of the shelf inhabitants are dependent on light—either to warm the water or to allow growth of food plants etc. The depths of the sea beyond the continental shelves are dark and often cold and generally support a different assemblage of plants and animals, adapted to such conditions. Should a shelf creature be swept down into the deep oceanic waters, perhaps as a result of a storm, it will probably perish very quickly.

Oysters live in coastal waters, attached to rocks and other firm objects (so that extensive areas of soft muds are inhospitable to them), feeding by straining out from the water microscopic plants and animals (plankton). When oysters reproduce they release swarms of minute larvae ("spat"). These larvae are minute blobs of jelly-like protoplasm composed of actively dividing cellular material, and fringed by hair-like processes ("cilia") that beat in sequence so that undulating waves of movement pass around the edge of the larva, rather like minute rowing oars, enabling it to swim through the water. Once released from the parent, oyster larvae swim actively, but as their very limited food supply is used up (it is in the form of globules of food material incorporated in the protoplasm), they become less active and begin to sink towards the sea floor. The sinking of the larvae is also a result of an increase in weight, as all the time during their active swimming phase the beginning of a shell is starting to grow and this minute shell, adding to the weight of the larvae, also tends to make them sink.

The oyster larvae start to sink towards the sea floor after swimming in the upper layers of the sea for some days. The larval life of an oyster may vary between a few hours and one or two weeks. *Ostrea lutraria*, living on the New Zealand coast, has a very short larval life—a few hours, or 1-2 days. This short larval life is interpreted as an adaptation to life on exposed coasts, where longer periods would allow the larvae to be swept off shore.

As the larvae land on the sea floor they may sink into soft mud and die. To survive they need to settle onto something firm (rocks, other shellfish, wood). Once settled, the larval oysters cement their tiny shells onto the firm object and start their growth towards adulthood.

A single oyster may release several million larvae over one season, and although it is impossible to be precise, only a small percentage of these find suitable attachment places, quite apart from the toll that various predators take when the larvae are at their vulnerable swimming stage.

Up to now we have assumed that the larvae have managed to stay in shallow continental shelf waters. But if they are swept by currents or storms out over deep water, when the time comes for them to sink to the sea floor they find that they go down into the dark cold depths, where they perish. Thus areas of deep sea are barriers to both adult and young coastal creatures.

We can therefore divide the fauna and flora of the sea into various groups on the basis of their spreading ability.

The planktonic and nektoplanktonic groups can spread widely, and areas of deep water are no problem to them. However, as we have seen, most sea-floor creatures that inhabit the continental shelves can migrate only along continuous shorelines or areas of shallow water. Deep water presents an insurmountable barrier. During the Ice Ages, when sea level was up to 180 m (600 ft) below its present level, continental shelves were extensively exposed around the lands of South-east Asia and there were stretches of continuous shallow water extending from Thailand and Malaysia and linking Singapore, Sumatra, Java and Bali (Fig. 6.8). Thus at this time many plants and animals were able to migrate along these shorelines—both on the land and in the sea—so that all these places have many creatures in common. But between Bali and Lombok, two islands in the Indonesian chain extending eastwards from Java (Fig. 6.8), there is a deep strait, the Lombok Strait, often traversed by ships on Far Eastern cruises. The depth of the strait is such that even during the low sea levels of the Ice Ages it was still sufficiently wide and deep to form a barrier to migration. The strait therefore forms a sharp dividing line between two different groups of animals and plants. This dividing line was recognised by Alfred Russel Wallace, a contemporary of Charles Darwin, who incidentally arrived at the idea of evolution by means of natural selection at about the same time as Darwin. In his honour the dividing line at the Lombok Strait is called "Wallace's Line" (Figs 6.8, 16.28).

By studying the distribution, both today and in

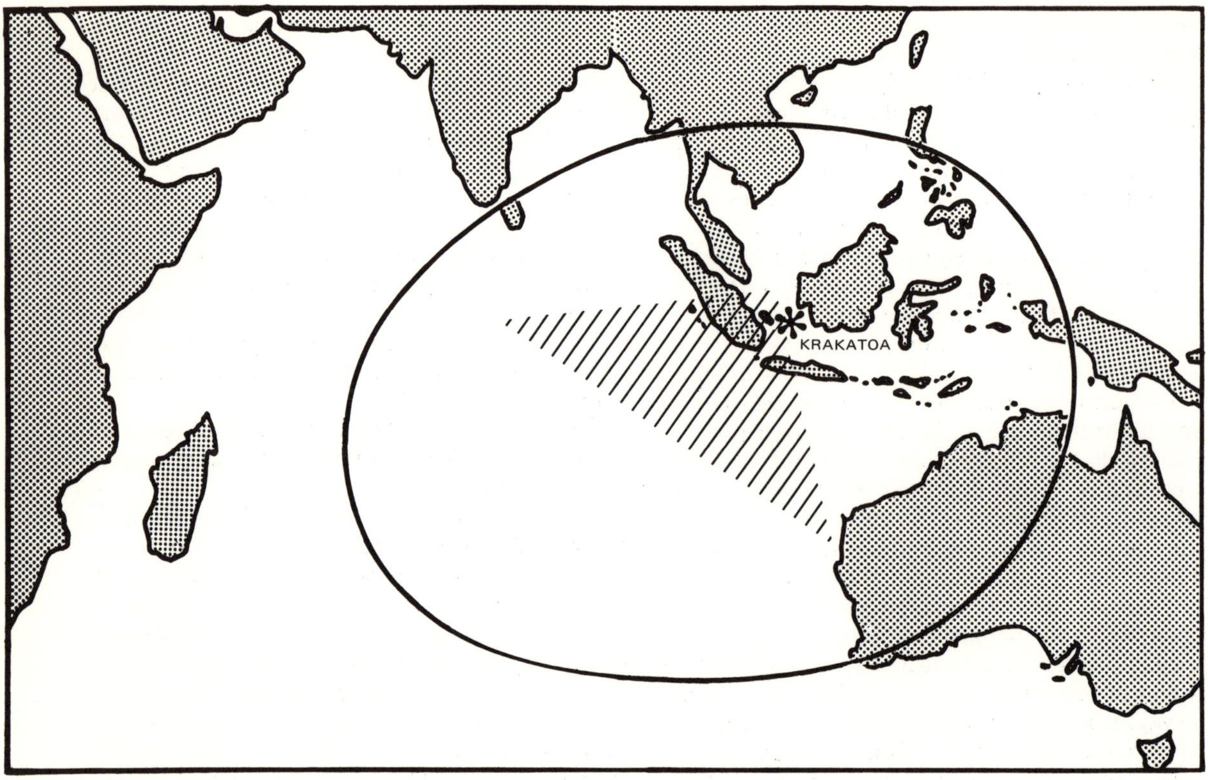


*Fig. 6.9*

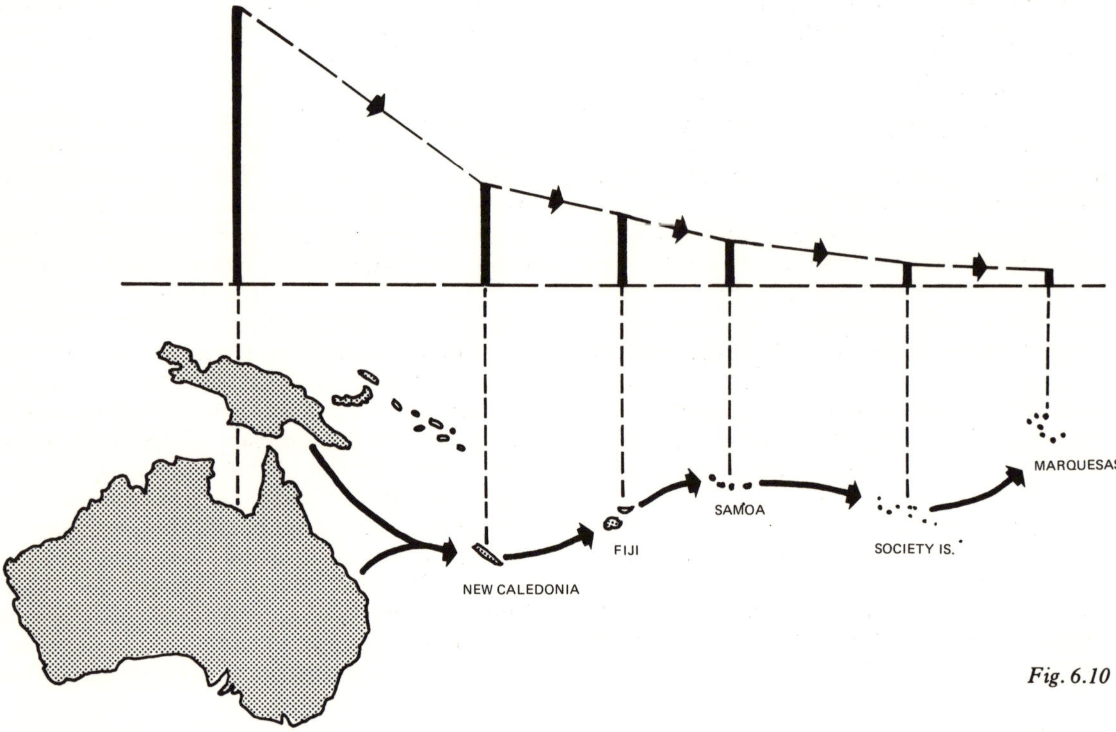


*Fig. 6.10*

**Fig. 6.9.** The cataclysmic eruption of Krakatoa in 1883 affected an enormous area of South-east Asia and the Indian Ocean. This diagram shows the region covered by volcanic ash (lined area) and the elliptical line encloses the area over which the great explosions that accompanied the eruption were heard. Vibrations were felt as far away as Tasmania. The explosions demolished a volcanic island that was originally 445 m (1460 ft) high, leaving a chasm in the sea floor some 182 m (600 ft) deep. Giant sea waves (tsunamis) that accompanied the eruption reached heights of 35 m (115 ft) and drowned 36,000 people along the coasts of Java and Sumatra.

the past, of shellfish and other creatures that are normally restricted to shallow sea waters and/or land, we can chart the extent of links between landmasses at various times. Through such a study we can say, for example, that 550-400 million years ago New Zealand had close links with Victoria (Australia), and was therefore presumably fairly close to it, whereas between 400 and 350 million years ago we lost our links with Victoria (and presumably moved away from it) and instead gained links with Tasmania and New South Wales.

The idea of continental drift was mooted at a time (the early 1920s) when enough was known about the relationships of plants and animals, both modern and fossil, for people to realise that if plants and animals of Brazil and Ghana, for example, really do relate to one another, then some transportation method must be found. For years "land bridges" and "lost continents" had been a convenient way out, in that they did not involve upsetting the modern geography of the world. Even in 1925 one biologist working in Polynesia had postulated the existence of a whole network of land bridges between individual islands in Polynesia, along which animals and plants could migrate, and raised and lowered sea level by 3000-5500 m (10,000-18,000 ft) in quick succession! However, as geologists and geophysicists learned more about the earth's detailed structure, and the fundamental differences between oceans and continents, both "land bridges" and "sunken continents" were relegated further and further into the limbo of lost scientific causes.

**Fig. 6.10.** This diagram shows how a group of forest-dwelling weevils have managed to island-hop from west to east, crossing wide areas of sea by means of chance dispersal (as the islands named have always been separated by sea). The vertical bars in the upper half of the diagram are in proportion to the number of kinds (or genera) of weevils at each place.

The decreasing lengths of successive bars indicate that a sifting-out has occurred at each island-hop, fewer and fewer types of weevil managing to make each successive hop. Even so, these insects seem to be particularly good at getting themselves around. This is an example of "sweepstakes" dispersal.

Thus when the idea of continental drift came on the scene it was welcomed with open arms by many zoologists, botanists and palaeontologists. It provided a very neat solution to their problems of explaining how their particular pet creatures got around the world, apparently leap-frogging from country to country. Even at that time, however, enough was known about the underlying fundamental geological structures (the patterns or "geological grain" of a country, rather like the grain of a lump of wood) to realise that certain continents, if they were to fit at all, slotted into other continents at certain specific points and that there was little or no room for manoeuvring. The geological structure—strata, folds and similar features—acted like the lines of print running across pieces of an originally whole page of newspaper, providing clues to the fitting together of the pieces.

Nonetheless this did not deter some hardy souls from rushing in where angels feared to tread and moving continents about with gay abandon, with little or no regard to the geological patterns, merely to get an animal or plant from one place to another.

These activities tended to reflect badly on the theory of continental drift, and as time went on it was found that it was all rather unnecessary anyway, as other means of migration were available to many animals and plants.

The zoologists, botanists and palaeontologists who had made use of continental drift in such an extravagant way had overlooked the really major influence of chance on the distribution of animals and plants.

## Playing dice with Nature

The decisive influence of chance, loading the dice one way or another in determining whether animals and plants make it across a stretch of unfriendly ocean, is illustrated by the example of Krakatoa, the volcanic island in Indonesia lying between Java and Sumatra, which blew up in 1883 (Fig. 6.9).

Prior to the explosion Krakatoa had been dormant for more than 200 years. But as it resumed activity, molten volcanic rock rose from deep within the earth and a large subterranean chamber full of molten rock was created. It is thought that at some stage the chamber roof partially collapsed, allowing millions of tonnes of sea water to pour in. Contact of the cold water with the molten rock produced vast quantities of steam and there were four enormous explosions, the greatest of which was heard 4800 km (3000 miles) away in Australia (Fig. 6.9). A huge glowing cloud of hot pumice and ash rose 80 km (50 miles) into the air and as a result the world experienced brilliant sunsets and sunrises for many years afterwards as the sun's rays reflected off the fine dust in the atmosphere. Before the explosion Krakatoa was clothed in lush tropical forest and no doubt had the great variety of animals and plants that are found in neighbouring Java and Sumatra. However, the explosion sterilised the island, so that it had to be completely colonised all over again by animals and plants migrating across the open sea. The nearest island not destroyed by the eruption is over 18 km (11 miles) distant and yet in only three years there were 11 types of fern and 15 types of flowering plants growing on Krakatoa. Animals soon followed and within 25 years there were 263 types of animal resident on the island. By 1928, 45 years after the eruption, virtually a full range of animals and plants had re-established themselves on the island's blasted and scorched surface. This had been accomplished purely and simply by natural means, with chance dispersal calling the shots. The following sequence of events is a sample of what probably happened.

Soon after the island had cooled down sufficiently for it to support life, a log, that had perhaps been cast adrift from Java or Sumatra, may have floated in from the sea. The log may have had a great range of tiny animals and plants inhabiting its surface and living in cracks and wrinkles. Also clinging to the log may have been seeds and spores of larger plants and eggs of larger animals. Soon after the stranding of the log the "hitch-hikers" may have emerged, explored their new-found land and eventually, with passing time, multiplied and formed colonies. Thus we have a means of transporting animals and plants across large areas of sea that would normally form impenetrable barriers. We need only one such log, filled with its complement of passengers, to ground every few months or years and then perhaps over a period of tens of years we may end up with a reasonably representative fauna and flora. As an illustration of the potential of floating logs in this regard, logs from South America regularly drift to Macquarie Island and Tasmania (probably via the West Wind Drift, see Figs 6.11, 6.12).

Many animals and plants die almost instantaneously in salt water and even their eggs or seeds often have very limited tolerance, if any. For example, seeds of the New Zealand beech, *Nothofagus*, die after only a brief immersion in sea water. Yet zoologists and botanists find the same varieties on adjacent islands separated by deep sea and, furthermore, with no possibility of the presence of some form of linking land in the past.

Chance dispersal provides the answer. I have used floating logs as an example, but there are many other avenues open to would-be migrants. Bird transport is another example: the carrying of seeds, spores, and eggs in the stomachs and intestines of birds, to be eventually dropped on a distant land. For example, the plants *Astelia* (with succulent berries which are attractive to birds), *Acaena* and *Uncinia* (with hooked seeds, which enable rides to be hitched on passing animals) have been widely distributed throughout the southern lands and to many volcanic islands which have always been isolated, such as Hawaii.

The population of isolated islands by plants and animals is very much a gamble, with nature loading the dice. Fig. 6.10 shows the distribution of weevils; animals which would not last a minute in salt water. Yet we see that over the years, presumably by means of chance dispersal, they have been able to distribute themselves from New Guinea and Australia across a fair stretch of the South-west Pacific. The migration of these creatures has been a true process of "island-hopping" as the various islands on the map, with the exception of New Caledonia, are volcanic in origin, built up from the sea floor, and there is therefore no possibility of land links having been present in the past to act as migration routes. Thus dispersal has been across the sea gaps, but if we look closely at the weevils present on each successive island we find that the overall number of varieties present steadily falls as we go further and further away from the source lands (Australia and New Guinea). This is a classic example of a "sweepstakes migration route" — so named because at each island hop only a proportion of the weevil population manages to get across the barrier. At each throw of Nature's dice only the winners get through.

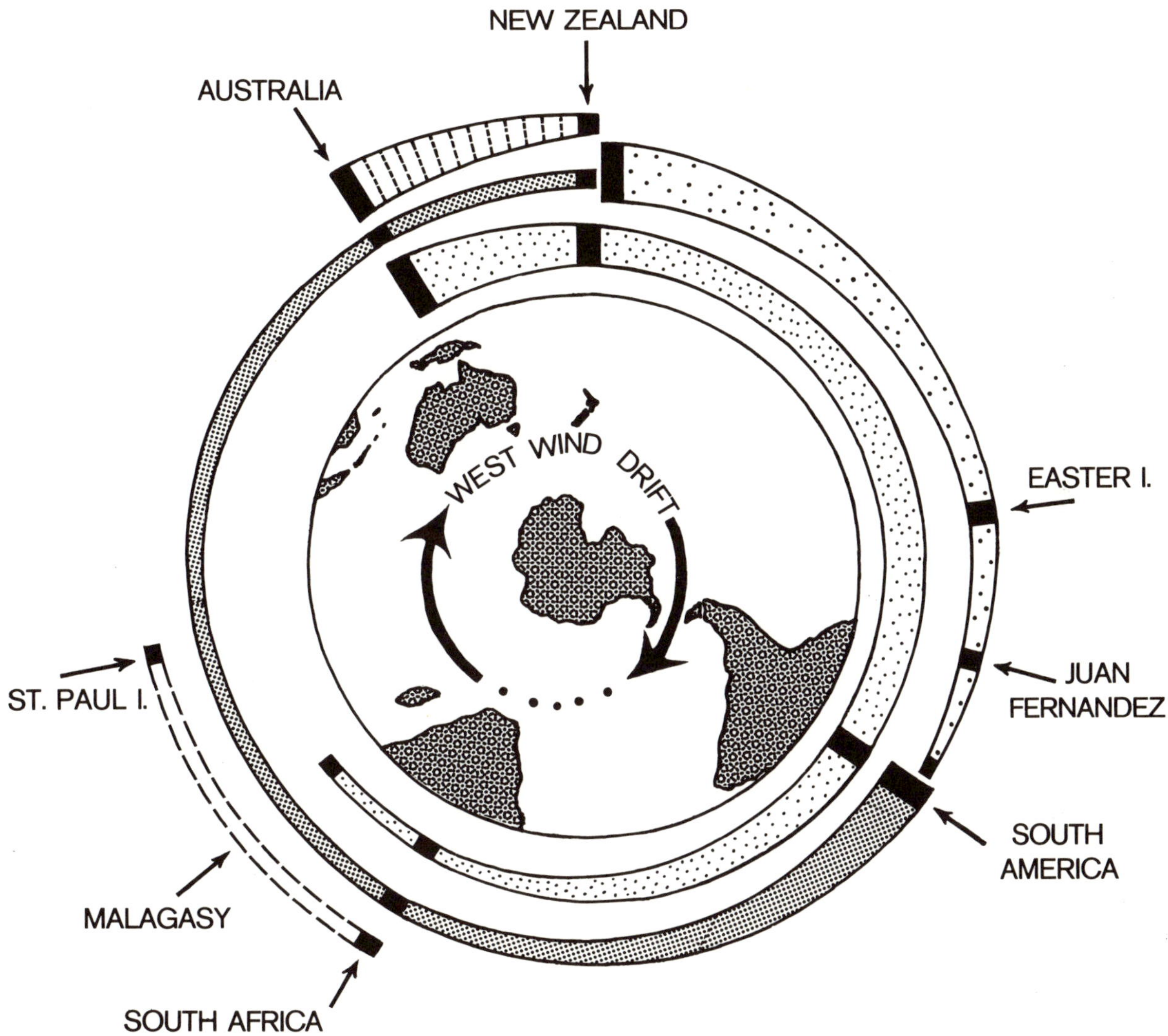


**Fig. 6.11.** Some of the commonest and most conspicuous of the echinoderms (sea stars, sea eggs etc.) seen on New Zealand coasts are also shared by many of the other southern continents and islands. This is not a result of land bridges linking the various lands in the geological past, but rather is the product of natural migration across the open oceans via the West Wind Drift (Fig. 6.12). This is facilitated either by long-lived floating larvae or by hitch-hiking of the adults on floating seaweed. Some examples of distribution by this means are illustrated in this diagram. In each example the height of the vertical bar is proportional to the number of varieties (species) of a particular kind (genus) of echinoderm present in the indicated land areas. An example of each type of distribution is included i.e. Australia-New Zealand; New Zealand-South America; Australia-New Zealand-South America-South Africa; South America-South Africa-Australia-New Zealand; South Africa-St Paul Island. As in Fig. 6.10 this is an example of "sweepstakes" dispersal.

**Fig. 6.12.** A research programme is under way to trace global air circulation patterns using strong plastic balloons, each with an electronics package, that drift with the winds at constant altitudes. One experiment involved release of a balloon at Christchurch, New Zealand, holding it at a constant altitude of 12 km (40,000 ft) and tracking it by satellite for 102 days. This provided a convincing demonstration of the strength and efficiency of the prevailing westerlies of the Southern Hemisphere that play an enormously important role in the dispersal of plants and animals between the southern lands no longer in direct contact. After release, the balloon reached South America in just over five days, and in the 102 days it was tracked it circled the world eight times and repeatedly crossed all of the southern continents and most islands. The stars indicate days elapsed from time of launch.

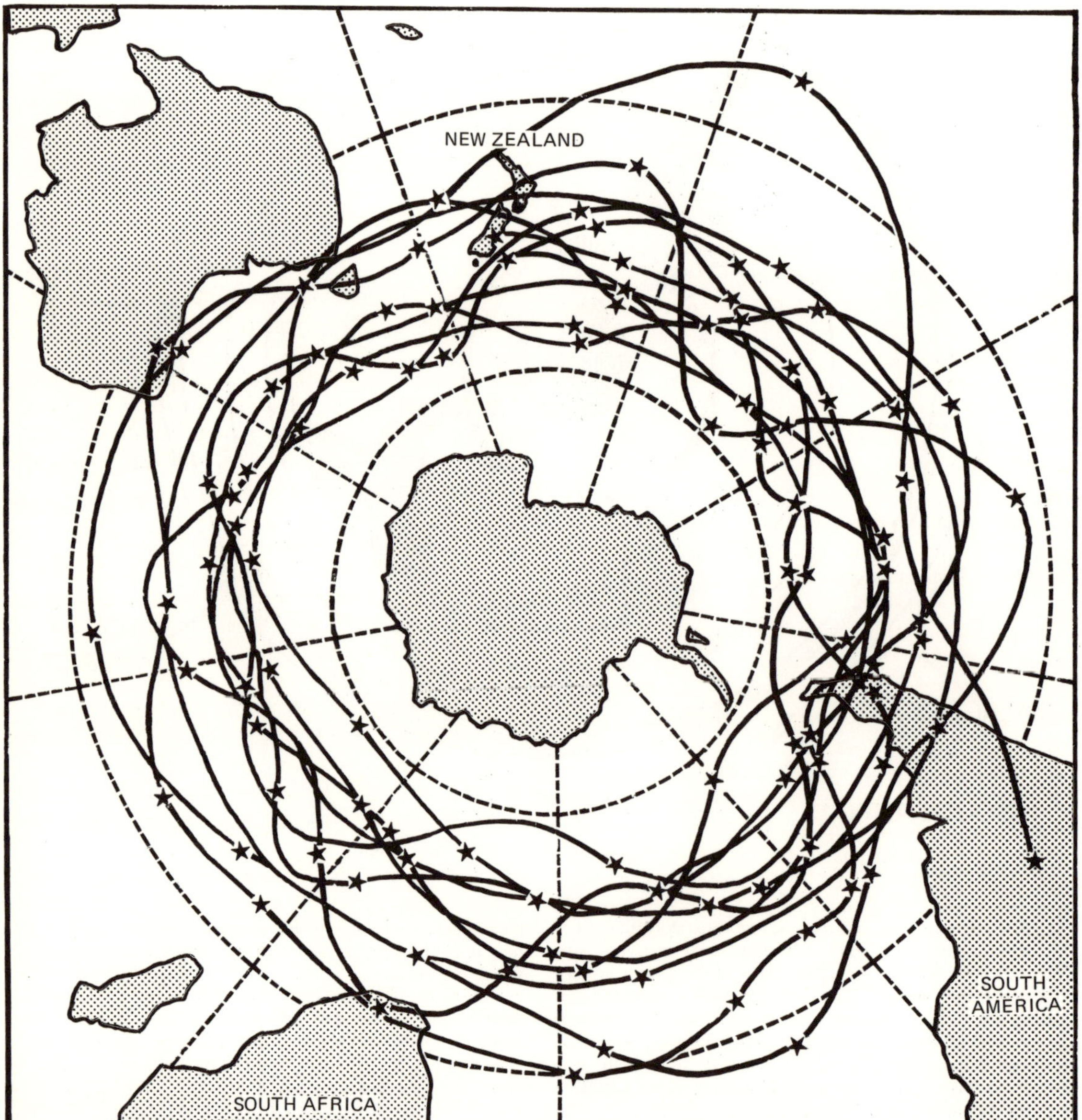

**The Roaring Forties**

Chance dispersal has had a notable effect on the flora and fauna in the New Zealand area. Taking just one example: in the echinoderms (starfish, brittle stars, sea eggs etc.) it is found that while some varieties are purely local in distribution and encompass the mainland of New Zealand and its outlying islands (Chatham Islands, Antipodes Island, Auckland Island, Campbell Island and The Snares) others are virtually circumpolar in distribution. Thus in some instances we find groups of similar echinoderms living in New Zealand, the Kermadec Islands, Easter Island, Juan Fernandez, southern South America (Patagonia and Tierra del Fuego), Falkland Islands, southern South Africa, St Paul Island, Kerguelen Islands and Australia (Fig. 6.11).

One explanation previously offered to explain a similar distribution in plants postulated a series of land bridges linking the landmasses, but as we have seen it soon became apparent that this was unacceptable geologically. By the same token, however, the distribution pattern has had nothing to do with continental drift.

We in New Zealand straddle the "Roaring Forties", a belt in the vicinity of 40°S Latitude and 50°S Latitude characterised by strong westerly winds, which often blow for weeks on end. The force of the winds can drive the surface water of the sea along at considerable speeds, so that quite apart from the wind, there is also a rapid movement of surface water in an easterly direction. The influence of the "West Wind Drift", as it is called, is dramatically illustrated by tracks of weather balloons (Fig. 6.12), and equally dramatic is its effectiveness as transportation agent, as shown by the way animals and plants have been distributed to the various countries lying in its path.

To return to our example of the echinoderms. The eggs of a number of echinoderms hatch out to produce a larva that is free-swimming and contains substantial quantities of yolk. Such a vitellaria or "yolk-larva" is shown in Fig. 6.13. Vitellariae have a simple cylindrical or barrel-shaped body, without arms or other projections, and they are opaque owing to the presence of globules of yolk material in the tissues. The body is encircled by rings of hair-like threads (cilia). The cilia beat rhythmically and so propel the vitellaria through the water.

The yolk supply in the vitellaria's body enables it to swim for a considerable time before exhausting its food supply and, although the time varies, some vitellariae can stay swimming in the upper layers of the sea for as long as eight weeks.

Although this does not sound very long, it nonetheless may be long enough for the larva to travel many thousands of miles, pushed along by the West Wind Drift. This is because the driving force of the wind (as demonstrated by the balloon in Fig. 6.12) is sufficient to move the uppermost layers of the sea (in which the larvae live) independently of the deeper layers. Corresponding oceanic current systems are also set up in the deeper layers of the sea, but these are generally slower than the movements of the surface layers. Oceanic currents often move at a rate of 5-10 km (4-9 miles) per day, but some may attain 50 km (30 miles) per day. Thus, while currents may account for some short island-hopping movements, it is likely that the boost the West Wind Drift gives to the surface layers is the key to long distance transport.

Apart from long-lived larvae, another way the echinoderms get around is for the adults to "hitch a ride" on masses of bull kelp and similar large seaweeds (this mode of life is called "epiplanktonic"). Kelp and other brown seaweeds have grape-shaped, air-filled flotation structures attached to their stems and their tissues, too, are spongy and air-filled. Thus, if they are ripped free by storms they are capable of drifting for considerable distances—e.g. they can drift between Kerguelen Island and Australia, a distance of some 8000 km (5000 miles). It is possible therefore that both mechanisms operate, and perhaps they might combine—with adults going part of the way on a journey clinging to seaweed and then releasing larvae that swim in the surface layers and continue on to the final destination. The vitellaria larvae, once their yolk has been used up, cease swimming and settle to the bottom. There, providing they have settled into waters of a suitable depth, they develop into the adult echinoderm.

To judge from fossil occurrences, at least some of our New Zealand echinoderms may have travelled from the Egyptian-Indian area in Oligocene times, 30 million years ago, spread throughout Australasia and then encircled the Southern Hemisphere at least once, riding the West Wind Drift, evolving as they went, and eventually arriving back in New Zealand where they started, but from the other direction (i.e. from the west) and as different forms.

I have used echinoderms as an example, but other groups of shallow-water marine animals have similar means of migration—either by means of free-swimming (nektoplanktonic) larvae or hitch-hiking (epiplanktonic) adults.

Land creatures may also get blown across quite large areas of sea, and there are a number of examples of moths, butterflies and birds being blown across from Australia and surviving and breeding here in New Zealand during this century (e.g. Painted Lady Butterfly, Monarch Butterfly, Bogong Moth, White-faced Heron, Welcome Swallow, White Eye). The important thing to emphasise is that these migrations are accomplished without the benefit of land bridges or continental drift.

As I hope I have shown, chance dispersal is a force to be reckoned with in natural environments. Therefore the significance of distribution patterns of plants and animals, both today and in the geological past, has to be weighed up very carefully and each example considered in ecological terms. Although fossil plants and animals can provide us with many clues to past geography, the information thus obtained must always be linked with the available geological evidence, as it is probably fair to say that the fundamental geological features—the matching of strata and rock structures—remain the primary basis for correlation. For example, although there are good fossil links binding Gondwanaland together (cf. Fig. 5.16), there are equally good rock ties that provide back-up for the fossils (Fig. 5.4).

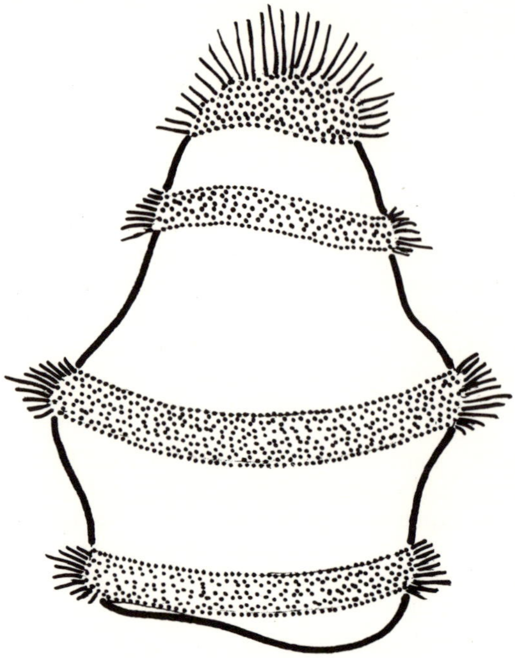

**Fig. 6.13.** A vitellaria or "yolk-larva" of an echinoderm (in this instance a brittle star). The actual size of the larva is about one millimetre in length. It has a simple body, cylindrical or barrel-shaped, and is opaque owing to the presence of yolk material in the tissues.

Rings of fine hair-like cilia encircle the larva and these enable it to swim freely through the sea water.

## CHAPTER SEVEN

# FOSSIL MAGNETISM

THE 1950s saw a great change in the fortunes of the Continental Drift Theory. During this decade the first glimmer of a mechanism for drifting began to appear and the existence was proved of large-scale horizontal fault movements, involving the sliding of huge crustal blocks. In this way two of the major geophysical objections to the theory were removed.

Almost as if to atone for their previous total opposition to the Continental Drift Theory, geophysicists at about this time developed a powerful new tool for gauging the actual amounts and speeds of continental movements in the past. Soon the wheel was to go full circle and geophysicists were to become the greatest proponents of the drift theory.

The discovery that really clinched the issue as far as many people were concerned was the use of palaeomagnetism in mapping past geographies of the world. Palaeomagnetism is the study of the earth's magnetic field throughout geological time. The fact that some rocks have magnetic properties has been known for a long time, and it could be said that palaeomagnetism is the oldest branch of geophysics. The Chinese knew about the magnetic properties of lodestone, a rock rich in the iron mineral magnetite that responds to the earth's magnetic field (hence the coining of the name "magnetite"), and in the second century B.C. were using lodestone as a magnetic compass. This crude compass was simply a spoon-shaped piece of lodestone rotating on a smooth board.

Travellers of the fourteenth century brought news of the discovery back to Europe, where it was adopted as a navigational aid, making possible the

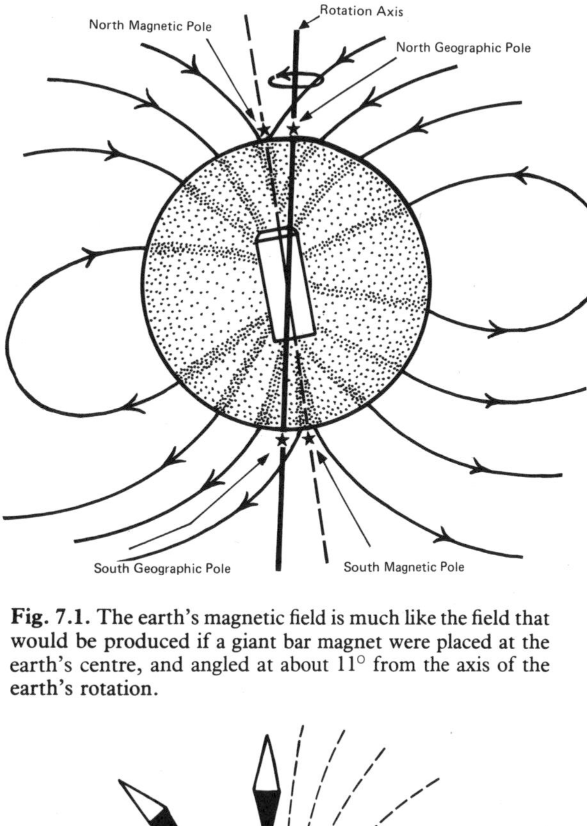


**Fig. 7.1.** The earth's magnetic field is much like the field that would be produced if a giant bar magnet were placed at the earth's centre, and angled at about 11° from the axis of the earth's rotation.

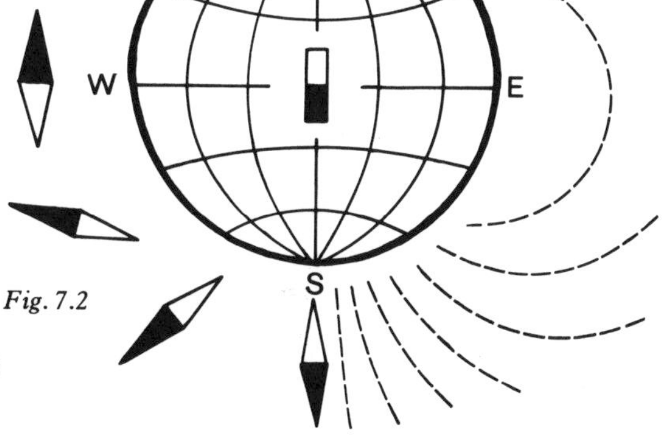


*Fig. 7.2*

**Fig. 7.2.** A suspended magnetised needle, free to pivot vertically, will align itself in the direction of the earth's magnetic field. The inclination (or dip) of the needle varies from vertical over the pole to horizontal at the equator. Measurements of the angle of inclination of the magnetisation in rocks can therefore be used to determine the latitude in which the rocks gained their magnetisation.

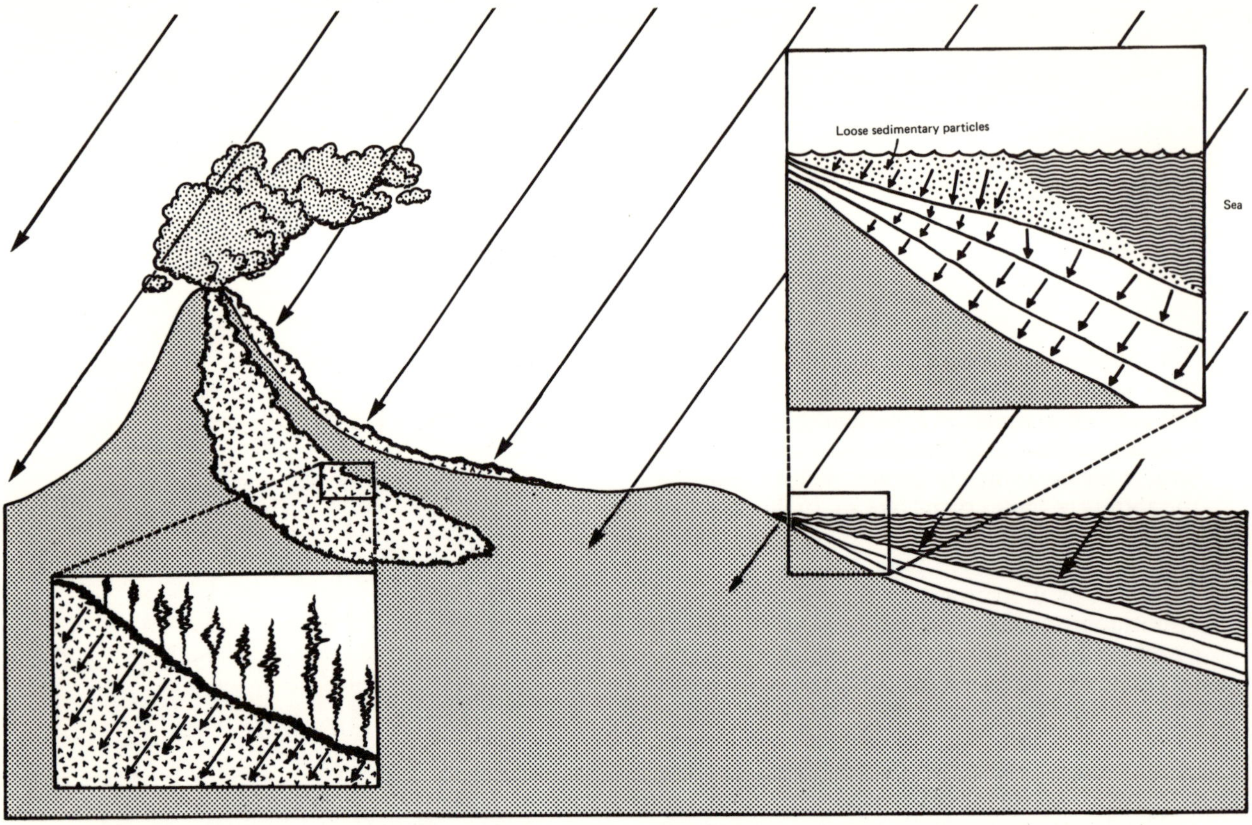


great feats of sailors such as Columbus and Magellan during the Age of Exploration.

It remained for William Gilbert, physician to Queen Elizabeth I, to explain in his book *De Magnete*, published in 1600, how the magnetic compass works. He offered the proposition that "the whole Earth is a big magnet", whose field acts on the small magnet of the compass needle to align it in the north-south direction. This explanation is still appropriate, and the earth's magnetic field is very much like the field that would be produced if a giant bar magnet were placed at the earth's centre and slightly inclined (11°) from the axis of rotation (Fig. 7.1).

However, we now know that most magnetic minerals lose their magnetism if they are heated above about 500°C. As this temperature is probably exceeded below depths of about 20-30 km (13-19 miles) in the earth, it is fairly obvious that there is *not* anything like a giant bar magnet at the centre of the earth, even though this notion gives a fairly accurate representation of the situation.

Another way to make a magnetic field is with electricity. Dynamos in power plants make electricity by using an electrical conductor, propelled by steam or falling water, rotating in a magnetic field. Of course there is no such thing happening inside the earth, but as the earth spins there is likely to be interaction between the solid mantle, the liquid outer core and the solid inner core (Fig. 4.6) sufficient to act as a giant dynamo.

The "lines of force" of the magnetic field shown in Fig. 7.1 imply the presence of a magnetic force

**Fig. 7.3.** On the left side of the diagram, lava freshly erupted from a volcano is cooling, and as it passes through the Curie Point the various iron mineral particles it contains become magnetised in the direction of the earth's then prevailing magnetic field. The direction of magnetism thus "frozen" into the lava is very stable and is not affected by subsequent changes in the earth's magnetic field.

On the right side of the diagram, volcanic and other material is being eroded and transported as loose sedimentary particles to the sea. Iron mineral particles in the sedimentary material, although they may already be magnetised, are tumbled about, losing their previous alignments, and as they settle to the sea floor and form sedimentary layers, gradually realign themselves in the new direction of the earth's magnetic field. As further material is laid down on top the sediments are compacted and hardened and the sedimentary magnetic particles are fixed into positions parallel to the earth's magnetic field.

so that a magnetic needle that is free to swing rotates into a position parallel to the local lines of magnetic force.

The magnetic needle of a compass does not point to true geographic north (the axis around which the earth spins), but rather to the magnetic North Pole.

The angle (or declination) between magnetic and geographic north would be zero everywhere, making a navigator's life simpler, if the magnetic field, shown in Fig. 7.1, lined up perfectly with the earth's axis of rotation. However, as it does not, the angle of declination varies depending on where you are in the world.

A magnetised needle free to swing vertically measures the inclination, or dip, of the magnetic field—its angle of deviation from the horizontal. At the magnetic poles the needle is vertical (Fig. 7.2), which is how some early explorers located these two points on the earth's surface. Near the equator, however, where the inclination is zero, the dip needle is horizontal.

Thus an ordinary magnetic compass works perfectly well away from the poles as the dip of the magnetic field is such that the compass needle lies almost flat and so swings freely. Near the poles, however, because the dip of magnetism is almost vertical (Fig. 7.2) the compass needle is pointing almost straight down and jams against the compass card.

**Rock magnetism and fossil compasses**

In the late nineteenth and early twentieth centuries studies were made of the natural magnetic properties of a wide range of rocks, particularly recent lava flows. It was not until the early 1950s, however, that highly sensitive instruments became available for precisely measuring the magnetisation of ancient rocks.

Palaeomagnetism uses rocks as fossil compasses. When molten rock is poured out over the earth's surface as volcanic lava it contains a great variety of minerals, all mixed up together. Depending on the type of lava, various iron minerals will be present. Basalt lava, the type that flowed out from volcanic vents around the Bay of Islands between 2.3 million years and 10,000 years ago, and from the various Auckland volcanoes between 40,000 and 225 years ago, is particularly rich in iron minerals, as indicated by the red colours that predominate in the basalt outcrops in both areas.

Iron oxide minerals (combinations of iron and oxygen) have the property of being able to record the directions of the earth's magnetism. The minerals magnetite (chemical formula: $Fe_3O_4$) and titanomagnetite (chemical formula: $Fe_2TiO_4$; i.e. magnetite in combination with titanium) are the most common magnetic recorders in volcanic rocks.

Materials that are normally magnetic lose their magnetic properties above a certain temperature. This can be demonstrated by heating an ordinary magnet. The magnet can be made to pick up some nails, for example, and then it can be taken and held, with the nails attached, in the hottest part of a gas welding torch. As the magnet gets hotter and hotter its magnetism steadily weakens and the nails fall away one by one until eventually when it is white hot it has no magnetism at all. The temperature above which a material has no magnetic properties is called the Curie Point, and it varies with the material's composition. Magnetite has a Curie Point of 575°C and the various titanomagnetite minerals have Curie Points of between 400 and 600°C. Without going into too much detail the phenomenon of the Curie Point can be explained by saying that once it is exceeded the atomic particles within the iron oxide grains, loosely arranged in grids or "lattices", are vibrating so much under the influence of the additional heat that the iron atoms are unable to form stable magnetic patterns with neighbouring atoms. In other words, the randomly orientated individual magnetic fields cancel out one another.

When they first burst forth from volcanic vents, lava flows are usually very hot; temperatures between 1000 and 1200°C are typical of many basalt lavas. As soon as the lava encounters the atmosphere, however, it starts to lose heat. Heat passes out into the surrounding air and ground and very soon the lava starts to congeal and form a solid skin. The interior of the lava flow may nonetheless stay very hot for a considerable length of time. In time, however, the various parts of the flow cool, and their temperatures fall towards the Curie Points for the diverse iron minerals they contain. Above the Curie Point the iron minerals (magnetite, titanomagnetite, etc.) are randomly orientated throughout the hot lava, but as the Curie Point is approached they begin to respond to the direction of the pull of the earth's magnetic field. Then, as they continue to cool and pass their particular Curie Point, they become aligned parallel to the earth's magnetism, the direction of which is in effect "frozen in" (Fig. 7.3).

As we have seen (p. 59, Fig. 7.2), the dip or angle ("inclination") of magnetism varies with latitude, being parallel to the earth's surface at the equator and perpendicular at the poles. Therefore lavas that poured out onto the earth's surface in North Auckland, for example, have a completely different angle of magnetism frozen into them, compared with lavas in Antarctica (e.g. those from Mt. Erebus, near Scott Base).

Rocks that were poured out as lavas tens and hundreds of millions of years ago can be sampled from a land that is now at the equator. Yet when the direction of magnetism frozen into the lava is analysed, it corresponds not to that of the present latitude of the continent, but rather to one that may be vastly different—e.g. corresponding to a latitude very close to the pole. This means that either the pattern of the earth's magnetism has shifted in the past, causing the positions of the poles to alter, or the particular landmass has been moving.

**Polar wandering**

At first, in the early 1950s when the first detailed palaeomagnetic results were obtained, geophysicists were rather timid about referring to the possibility of their indicating past continental movements—and probably this timidity was fully justified in the climate of opinion at that time, as continental drift was still a rather heretical idea, particularly in geophysical circles. Thus the movements were merely described as "polar

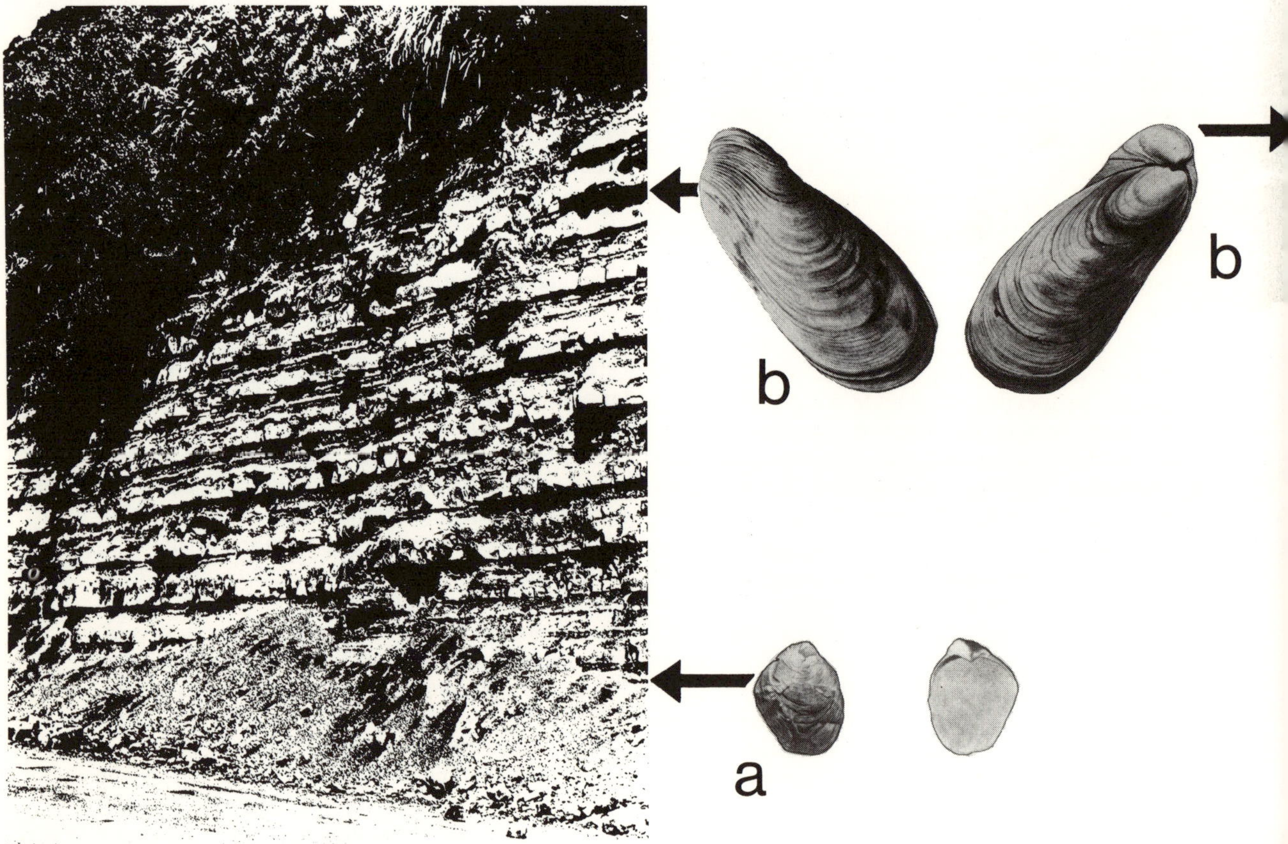

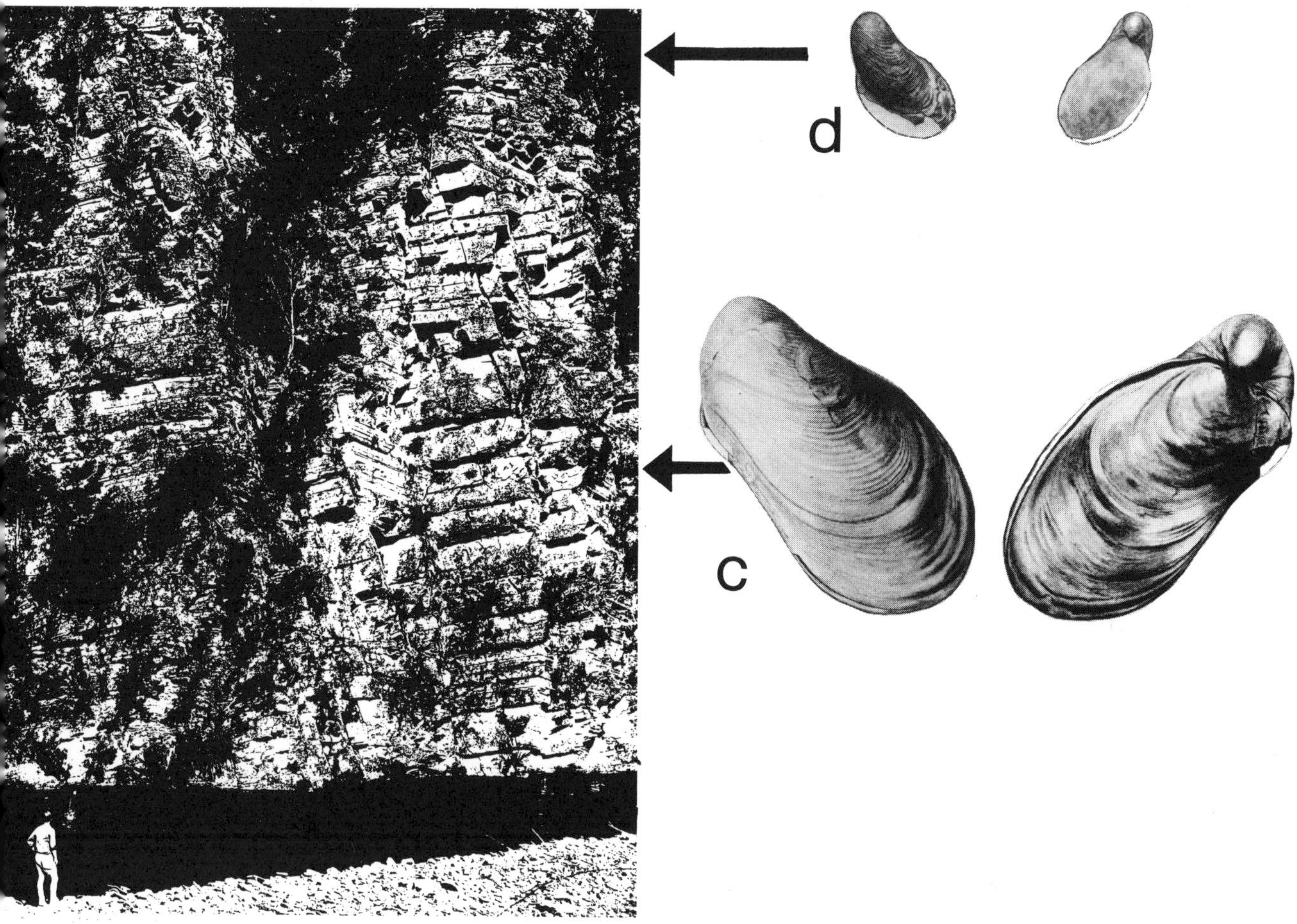


Fig. 7.4. In one area we can show that fossil "a" is overlain by fossil "b". We can trace fossil "b" to another area and there find it overlain by rocks with fossils "c" and "d". Therefore we can say that fossil "a" is the oldest and fossil "d" the youngest. If fossil "c" is accompanied by some radioactive minerals these can be dated and the resulting date, say 140 million years, applied to fossil "c". Thus we can say that fossil "c" represents an age of about 140 million years and fossil "d" a younger age and fossils "a" and "b" older ages.

wandering". As time went on, however, and as more and more results came to light, it became evident that if the poles *had* been wandering they had been "wandering" for vast distances all over the globe! In the meantime the climate of opinion was swinging towards acceptance of continental drift, and so the alternative explanation of moving continents became more acceptable. Also, more information was coming to hand about the true nature of polar wandering. As the earth is not a perfect sphere, but rather is pear-shaped and not evenly balanced (most of the land being in the Northern Hemisphere) it does not spin smoothly, but instead wobbles somewhat, like a spinning top. The wobbles cause the poles to "move"—that is, the geographic poles (the axis on which the globe is rotating) and magnetic poles tend to apparently move about and diverge so that today

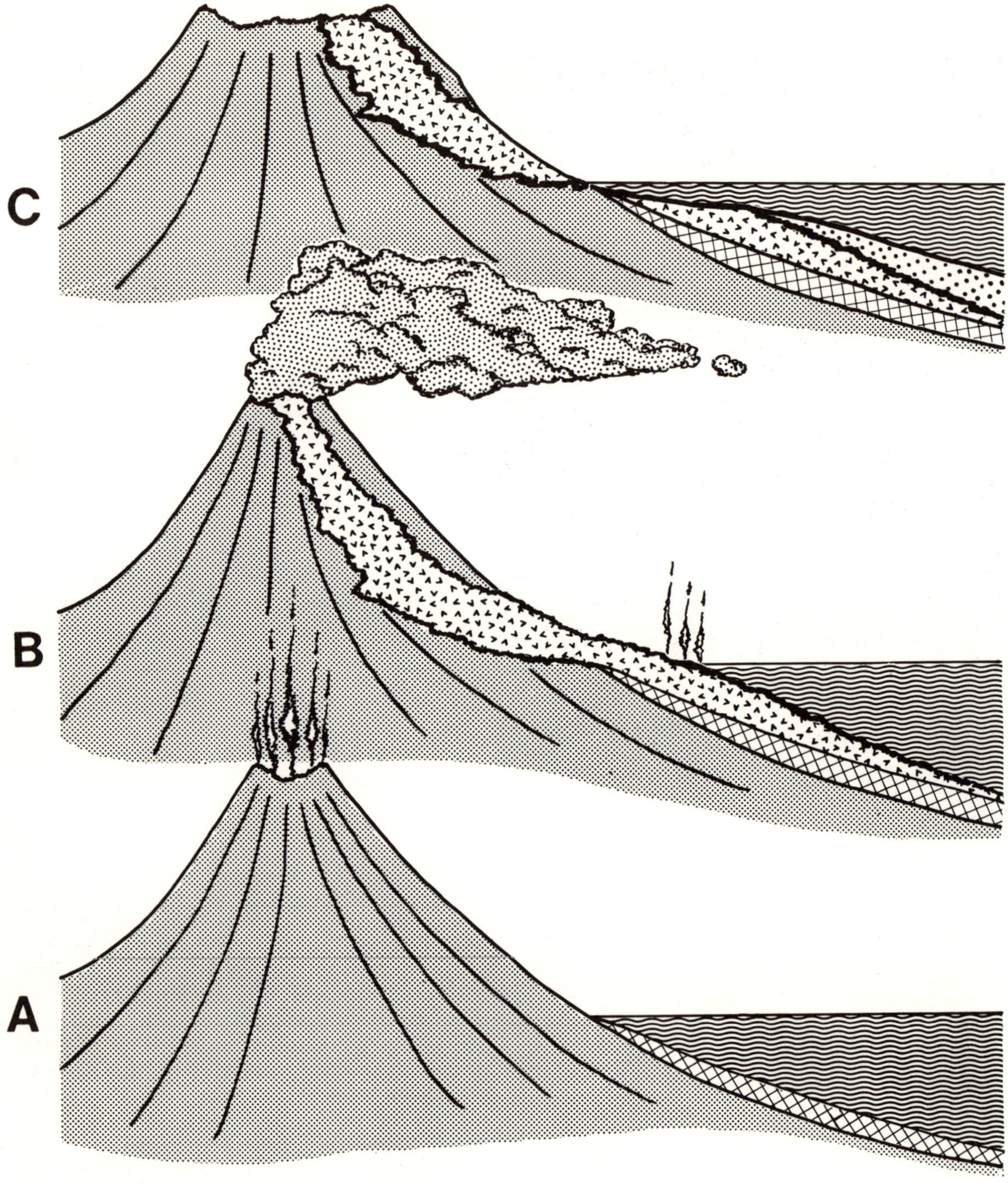


**Fig. 7.5.** Stage A: Sediments are laid down in the sea off-shore from a quiescent volcano. The sediments contain the remains of marine organisms living at that particular time.

Stage B: The volcano bursts into activity and lava flows down its flanks into the sea, where it forms a layer overlying the sediments deposited during Stage A.

Stage C: Volcanic activity dies down again and sedimentation is renewed, new fossil-bearing sediments covering up the lava erupted during Stage B.

The ages of the fossils deposited in the sediments during Stages A & C will bracket the age of the lava. A radioactive date obtained from the lava will provide an absolute age for the lava as well as a minimum age for the Stage A fossils and a maximum age for Stage C fossils.

in New Zealand there is a difference of some 11½° between them. It was soon established that the earth's wobble was not of sufficient magnitude to explain the extensive "polar wanderings" revealed by the palaeomagnetic work. The conclusion therefore was inescapable: the earth's magnetic pattern had remained essentially unchanged throughout geological time (except for the comparatively small "wobbles"), but the landmasses themselves had been moving. At last science had a tool that could be used to chart accurately the movements of landmasses in the past.

But first the tool had to be refined, so that it could give reliable results. Sophisticated laboratory techniques had to be worked out, and it is interesting to note that a number of the researchers who worked on this in the late 1940s and early 1950s had obtained their expertise whilst working on aspects of magnetic mines for the Royal Navy during the Second World War.

The palaeomagnetism tool also had to be calibrated, so that the magnetic changes of the past could be related to the geological time scale.

Two major methods of dating are available. The first employs fossils and the second uses natural radioactivity as a type of clock.

**Ages from fossils**

As we have seen, fossils are the remains of extinct plants and animals preserved in rocks that have been laid down in the sea or in lakes, rivers and swamps.

In the geological past animals and plants have changed, or evolved, so that as we go back in time and turn back the clock millions of years at a time, we pass from creatures that we can passably recognise as being related to modern forms, to creatures that become progressively more and more bizarre. As geological time passed and generations of these extinct creatures lived and died in the seas, lakes, rivers and swamps, some of them were preserved in the sediments being laid down at that time. To give some idea of what happens, we may assume that at a particular time groups of a certain creature lived on the sea floor. This creature may have been a plant or an animal, but the main thing is that it was sufficiently distinctive for people to be able to recognise it easily. We need not worry about its proper scientific name—we can just call it "a". As individuals of "a" die, their remains are soon buried by the sands and muds that accumulate on the sea bed. In time, therefore, layers of sands and muds build up that contain remains of the creature "a". Sooner or later, however, the group of creatures we have called "a" may die out, or perhaps change progressively into another form.

If the creature dies out, another will move in to occupy its "niche" in the environment. On the other hand, if "a" evolves to produce another form, the resulting creature will probably be distinctive enough to be readily separated from "a".

Either way the result is the same: sands and muds containing the remains of "a" give way to sediments with the new creatures—we can call them "b". Thus layers containing "a" are overlain by those containing "c" and if time goes on, and other changes occur, we might get a sequence: a—b—c—d. With time the old sea-floor muds and sands are buried deeper and deeper and as this happens, their individual particles are squeezed together so that hard compact rock layers (mudstone, sandstone etc.) are created. However, the organic remains—the skeletons and shells of animals, the outlines of leaves and the pieces of wood—all stay in the rocks as fossils. In time the old muds and sands become involved in earth movements, during which they are squeezed and folded and pushed up to form land. The original layers, now tilted up at various angles and sculpted by erosion, will now reveal their internal structure and fossil content in hillsides and riverbeds. By observing which layer overlies which we can say that "d" is the youngest fossil and "a" the oldest (Fig. 7.4). This "Law of Superposition", of which I have given a very simple example (skipping over all the "ifs and buts" associated with its application in practice), can be used to establish a succession of fossils. By using this fossil succession a series of fossil "ages", marking stages in the evolution of life on the earth, have been established. Thus we can recognise an age of ancient life or "Palaeozoic", an age of middle life or "Mesozoic" and an age of recent life or "Cenozoic".

So by studying the succession of life we can put various fossils in order of their appearance on the earth, just as we put fossils a-b-c-d in order, with fossil "d" the youngest and "a" the oldest. In doing this we have merely put the fossils in relative chronological order. It is as though we knew that Kupe had arrived in New Zealand, followed by Tasman and then by Cook in that *relative* order but had no idea of the *exact* dates. The assignment of actual dates to the succession of fossils had to await the discovery of radioactivity. This discovery gave geologists the chance to calibrate the fossil time scale.

**Ages from radioactivity**

Many quite ordinary rocks contain small grains and flakes of minerals possessing radioactive elements. Radioactive elements have the property of gradually changing from one state to the next. This change occurs at a constant rate and can be measured precisely. Some radioactive elements have very short lives, measured in millionths of a second, as for example some of the elements created momentarily in nuclear reactors. Other radioactive elements, however, have very long lives, measured in tens of thousands and millions of years. These have provided the geologist with a clock, for as soon as they are formed in a rock they gradually change at a constant rate. The radioactive elements change or "decay" to yield "daughter" products and by measuring the ratio of "parent" material to "daughter" material and relating that to the known decay rate, it is possible to work out an age for the formation of the radioactive substance and hence an age for the enclosing rock.

The decay rates of radioactive elements are worked out in terms of what is known as the "half-life", i.e. the amount of time required for the radioactivity of a given amount of radioactive substance to decay to one half of its original value.

An example of the use of radioactive dating is that using a radioactive variety (or "isotope") of the element potassium. Potassium is a common and widespread element in many rocks and minerals, and so any dating method based on its use is likely to have wide application. Three varieties or "isotopes" of potassium are known and these are labelled as $^{39}K$ (K is the chemical symbol for potassium, derived from the Latin *Kalium*), $^{40}K$, and $^{41}K$. Of these only potassium 40 is radioactive and therefore useful for dating purposes. Potassium 40 decays by two entirely different processes, each with its own characteristic half-life. It can decay to give either calcium 40, or argon 40. The decay route to argon 40 is the one that is used for dating purposes and this has a half-life of 11,900 million years, more than ample for all geological dating, as the oldest rocks known on the surface of the earth are 3800 million years old, and the oldest meteorites 4600 million years old.

Young rocks—those between 30,000 years and about a million years old—do present difficulties, however, as not enough decay has occurred to produce sufficient argon 40 to be measured precisely.

The method, generally known as "K-Ar" dating, is nonetheless a very popular one and can be used to date lava flows, ash beds, igneous rocks that have crystallised deep in the earth, and sediments that have been "cooked up" (or metamorphosed) by contact with such rocks.

Limestones and sandstones that have originated on the sea floor sometimes contain the potassium-bearing green-coloured mineral glauconite which is formed at the time of sedimentation and can also be dated by the K-Ar method.

Other methods use the decay of rubidium 87 to strontium 87, uranium 238 to lead 206 and thorium 232 to lead 208. A variety of carbon, carbon 14, is used to date carbon-bearing materials (wood, charcoal, shells, etc.) but it can only extend back to about 38,000 years ago.

To return, however, to the assignment of precise dates to the various stages in the succession of fossil life. By dating minerals obtained from the rock enveloping the fossils we automatically obtain a date for the fossil itself. Sometimes we can do just this, but often it is not so simple. Our fossil-bearing rock may not itself contain radioactive substances, but instead may overlie a lava flow and be in turn covered by another lava. By dating both lava flows we can "bracket" the age of the fossil bed.

Once a particular fossil or group of fossils has been dated in this way, this age can be transferred to other areas in the world containing the same fossils, or near relatives, so that in the field we can pick up a fossil and say, "This is 250 million years old." You will probably realise that the fossil by itself cannot give that answer, but it does mean that at some time, somewhere in the world, somebody has been able to relate that particular fossil, or a near relative, to dates obtained by radioactive dating.

Radioactive dating methods are often very time-consuming, need expensive sophisticated instruments and are a laboratory rather than a field tool. Fossils on the other hand provide us with a ready-to-hand almost "instant" age in the field without fuss and expense.

**Ages of lava flows**

Getting back to lava flows and fossil compasses, most lavas, as they pass out from volcanic vents, flow over previously deposited sediments that have been laid down in rivers, lakes or the sea. Such sediments probably provided homes for various animals or plants and therefore the chances are very good that fossil remains have been left behind in the sediment. The lava thus flows over material that has been "labelled" by its contained fossils. Once it has cooled the lava flow may in time be covered by muds and sands and again these may contain fossil remains. The lava flow is therefore both underlain and overlain by fossil-bearing sediments (Fig. 7.5).

Study of the fossils and assignment of ages to them can therefore provide two age limits—one younger than the lava flow (obtained from the overlying fossils), and the other older than the lava flow (from the underlying fossils). For example, the ages obtained from fossils under (i.e. older than) the lava may all lie in the Lower Jurassic, some 172 million years ago; whereas fossils over the lava (i.e. younger than the lava) are all of an Upper Jurassic age, some 162 million years ago. On this basis the lava flow can best be assigned to the Middle Jurassic, between 172 and 162 million years ago. Alternatively the radioactive minerals occurring within the lava flow may be separated out and analysed and then, providing nothing has affected the lava since its deposition (e.g. later heating-up), a date for the actual lava may be obtained—and the fossil dates can then serve as an independent check.

Some sedimentary rocks can also take on magnetism. If there has been volcanic activity nearby, or if there is another source of magnetic mineral grains, the sedimentary particles settling through the ocean onto the sea floor will contain some magnetic grains (often chips of the mineral magnetite). As these minute grains settle through the sea water they become aligned in the direction of the earth's magnetic field. This orientation is thus preserved in the sediment, which in this way "remembers" the magnetic field that existed at the time of deposition (Fig. 7.3).

**Magnetic tracking of continents**

In the geological past many lands have had very lively histories of volcanic action, with massive lava flows erupting over the landscape at frequent intervals. Lands like England, France and India, although they now lack active volcanoes, have had long periods of vulcanism in the past. By criss-crossing a continent, sampling a lava of a certain age here and another lava of another age there, earth scientists have been able to build up a series of compass "fixes" for many of the world's continents, providing us with a sequence of glimpses into the chapters of the geological history of particular areas.

India, for example, was once a close neighbour of the Southern Hemisphere continents, forming part of Gondwanaland (see Fig. 5.4), and over a long period of time has moved from latitudes south of the equator to its present position between 10° and 30°N Latitude (Fig. 7.6). Some 250 million years ago India was at about 70°S Latitude. As it moved from this position and travelled slowly northwards to the equator and beyond, episodes of volcanic activity occurred, separated by periods during which shallow seas flooded in over the land, depositing fossil-bearing sediments. Thus the movements of India over the last 250 million years (and beyond) have been documented by both fossil beds and lava flows (Fig. 7.6). Studies of the fossils have provided dates and much information on past climates, while sampling of the lavas and measurement of the palaeomagnetic directions has provided a series of compass "fixes" for India at various times in its long geological history. Admittedly the story is fragmentary, as the amount of volcanic activity varied considerably—it occurred in spasmodic spurts rather than continuously—but a decipherable pattern can still be obtained. Therefore by studying the sequences of fossil life and tying them into the fossil magnetism, a reasonably complete picture of India's wandering movements has been built up (Fig. 7.6).

Some 250 million years ago India was close to 70°S Latitude and experienced widespread glaciation (Fig. 5.11). Then a gradual northward drift occurred, the rate of movement varying, depending on what was happening in the adjacent oceans and landmasses. Over the last 50 million years, however, palaeomagnetic measurements have shown that India moved from 30°S Latitude to 20°N, a distance of some 5500 km (3400 miles), giving a rate of drift of 102 mm (4 in) per year.

Similar latitude determinations were made for other continents at various times in the geological

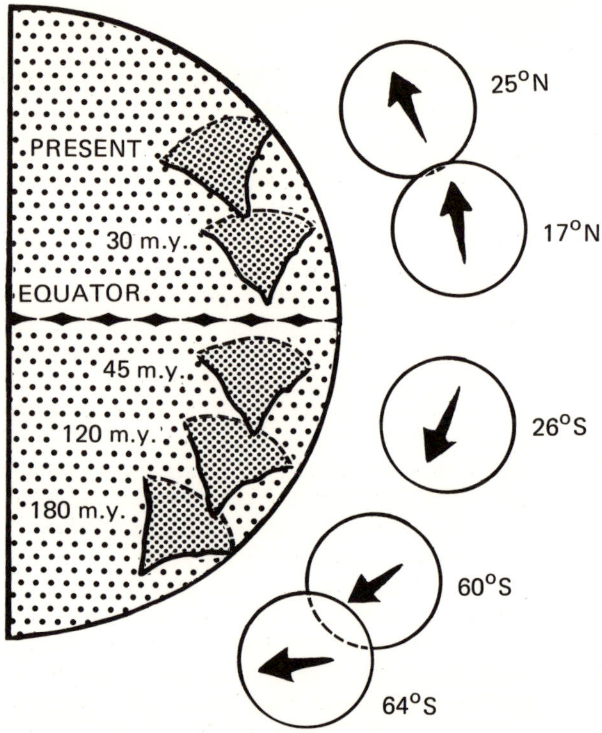


**Fig. 7.6.** Palaeomagnetism has helped track the movements of India over the past 180 million years.

past and the resulting "palaeolatitudes" plotted up as in Fig. 7.7. However, obviously the latitude lines had once been continuous and a method had to be developed to match up the various landmasses by means of their palaeomagnetism. This method was arrived at as a by-product of a research project carried out in England in the middle 1950s. As part of the project, samples had been obtained from British and European lavas of widely different ages, and palaeomagnetic measurements of them showed that the apparent position of the North Pole had steadily changed throughout geological time. In Precambrian time, some 900 million years ago, the North Pole had a position close to Hawaii. After this it had slowly moved northwards until about 250 million years ago it was between Japan and Kamchatka. Then, in the period between 250 million years ago and the present day, the North Pole had moved via eastern Siberia to its present position.

Thus the North Pole, as viewed from Europe, had apparently "wandered" in a curved path (Fig. 7.8), and, as the amount of difference between the modern pole position and various fossil positions steadily increases as the geological age increases, movement of either the poles or the continents was clearly suggested.

The polar movements thus obtained were compared with the available geological information on past climates. For example, support for the palaeomagnetists' estimate of latitudes for Europe of between 20°N and 20°S during times past was provided by the presence in European rocks some 350-300 million years old of thick coal beds that accumulated in tropical swamps, and extensive limestones bearing the skeletons of tropical corals. Similar evidence is present in 250-million-year rocks in the form of layers of various salts that were left behind by the evaporation of great saline lakes, and by intercalations of desert sands.

An even greater surprise was to come. The European work was followed up by systematic study of North American lavas. At times in the geological past vast outpourings of volcanic lava had occurred in various places in the U.S.A., particularly in the west, in the vicinity of the Colorado and Columbia rivers. Here huge volcanic plateaux had been built up, with layer upon layer of lava, often separated by sediments. Then in comparatively recent times the large rivers of the region have cut massive gorges through the plateaux, so that the sequences of lavas are now clearly exposed, like the layers of a cake or sandwich. Particularly fine examples are seen in the walls of the Grand Canyon (cut by the

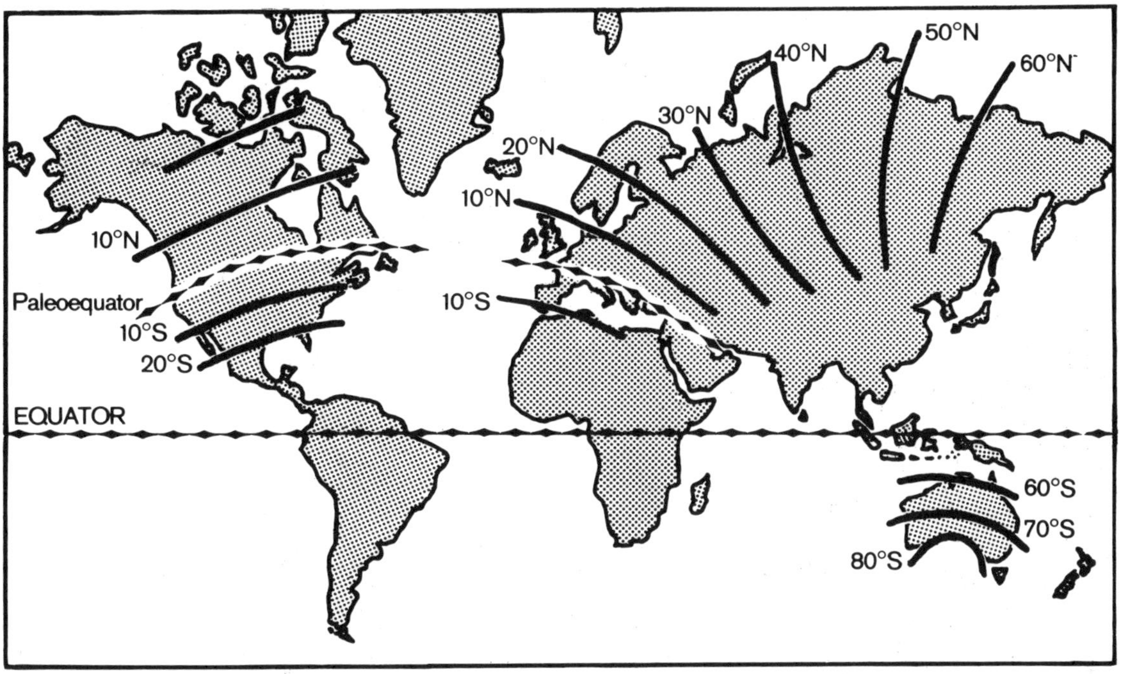


**Fig. 7.7.** Palaeomagnetic studies have allowed palaeolatitudes to be drawn for various countries (i.e. those with suitable rocks) at times in the geological past. This map shows some results for the Permian period, about 240 million years ago. Note that the equator at that time ("Palaeo-equator" in the diagram) was some distance north of the modern equator and that Australia was virtually at the South Pole. Similar maps can be drawn for other periods of geological time.

Colorado) and the canyons cut by the Snake and Columbia rivers in the states of Washington, Oregon and Idaho.

Once again a series of palaeomagnetic measurements was obtained that spanned numerous geological ages, and the apparent "wandering" path of the North Pole, as viewed from North America, was also found to be on a curve (Fig. 7.8). The curve was very similar to the European one, although that part of it for the period between about 900 million and 200 million years ago was slightly off-set to the west. But the curve for the period from 200 million years ago (Triassic period) to the present merged with the European curve.

Although nearly all geophysicists had at that time been brought up with the belief that continental drift was unlikely, if not impossible, the palaeomagnetists involved in this work soon realised that the anomaly between the two curves would disappear if North America and Europe had once been close together, as Alfred Wegener had proposed on the basis of matching geology (Fig. 3.2). The absence of any anomaly in the curves from the Triassic period onward implied that the continents had moved apart to assume their present positions at some point during that length of time (i.e. within the last 200 million years). Thus the curves could be used to pinpoint the exact times landmasses moved apart.

Before long, other similar results had been obtained from many of the world's continents, and by envisaging movement of the landmasses so that their polar wandering paths merged with those of other lands, it was possible to reconstruct their past positions and to produce diagrams like Fig. 7.9, that are based on palaeomagnetic determinations alone. As a glance at this figure will show, the assembly of the continents thus obtained is very similar to that proposed by Alexander du Toit in 1937, based on the geological information then available (Fig. 3.5). Current practice is to blend the palaeomagnetic information with all other available data so that continental assemblies are arrived at by consensus.

Australia has a rich heritage of lavas suitable for palaeomagnetic sampling and its movements throughout geological time have been fairly

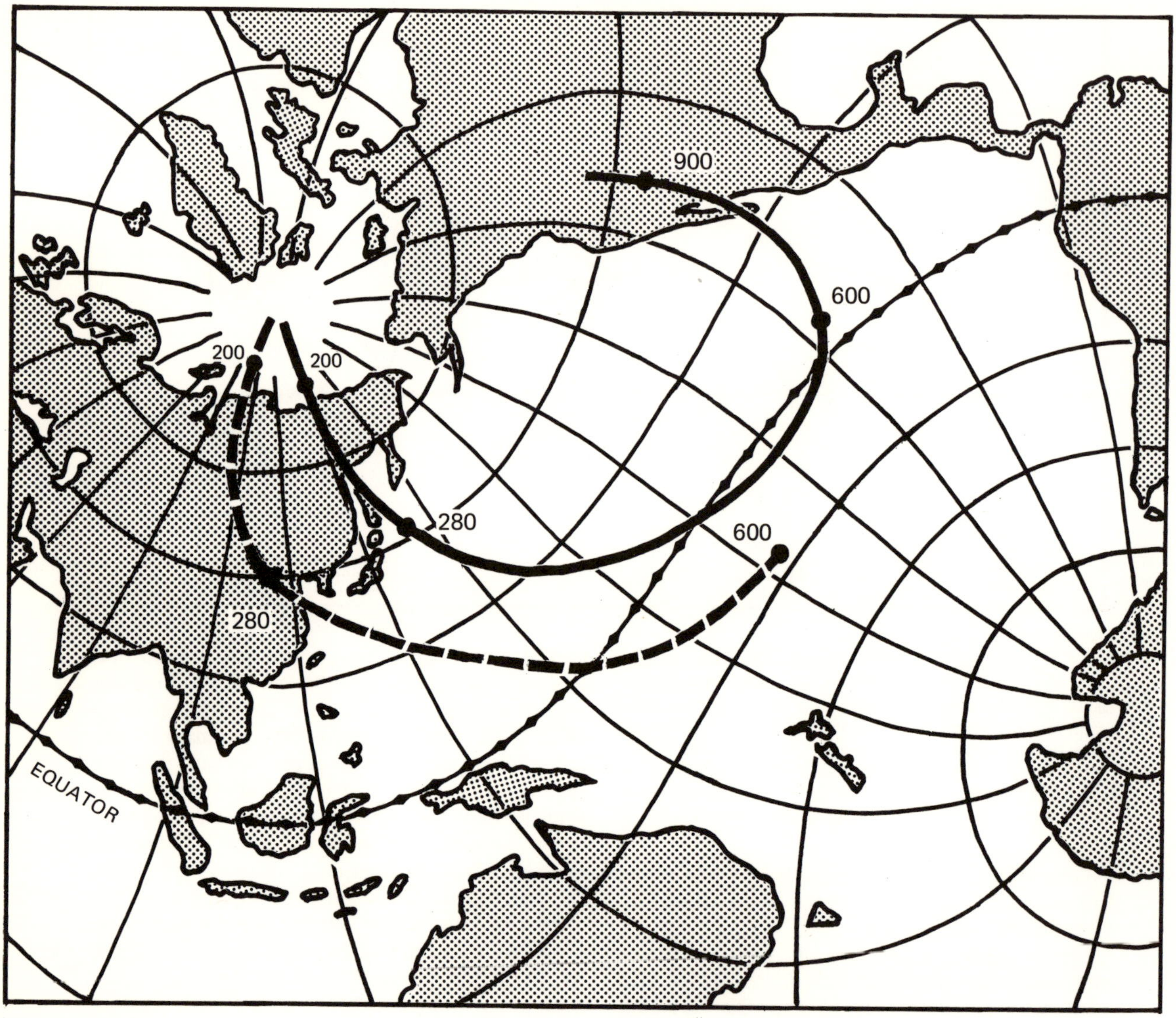


**Fig. 7.8.** This diagram shows a simplified plot of the apparent wanderings of the North Magnetic Pole over about the last 900 million years, based on measurements made in Europe (continuous line) and North America (dashed line). Note that until about 200 million years ago the curves are roughly parallel, but displaced about 30° of longitude. The figures on the curves are in millions of years. Most geophysicists believe that the pole didn't actually wander, as the figure suggests. Rather it remained fixed, close to the earth's axis of rotation, while the "map"—that is the outer layers of the earth—drifted over the interior.

**Fig. 7.9.** An assembly of the world's continents based on palaeomagnetic determinations alone. The assembly is for the Jurassic period, some 150 million years ago.

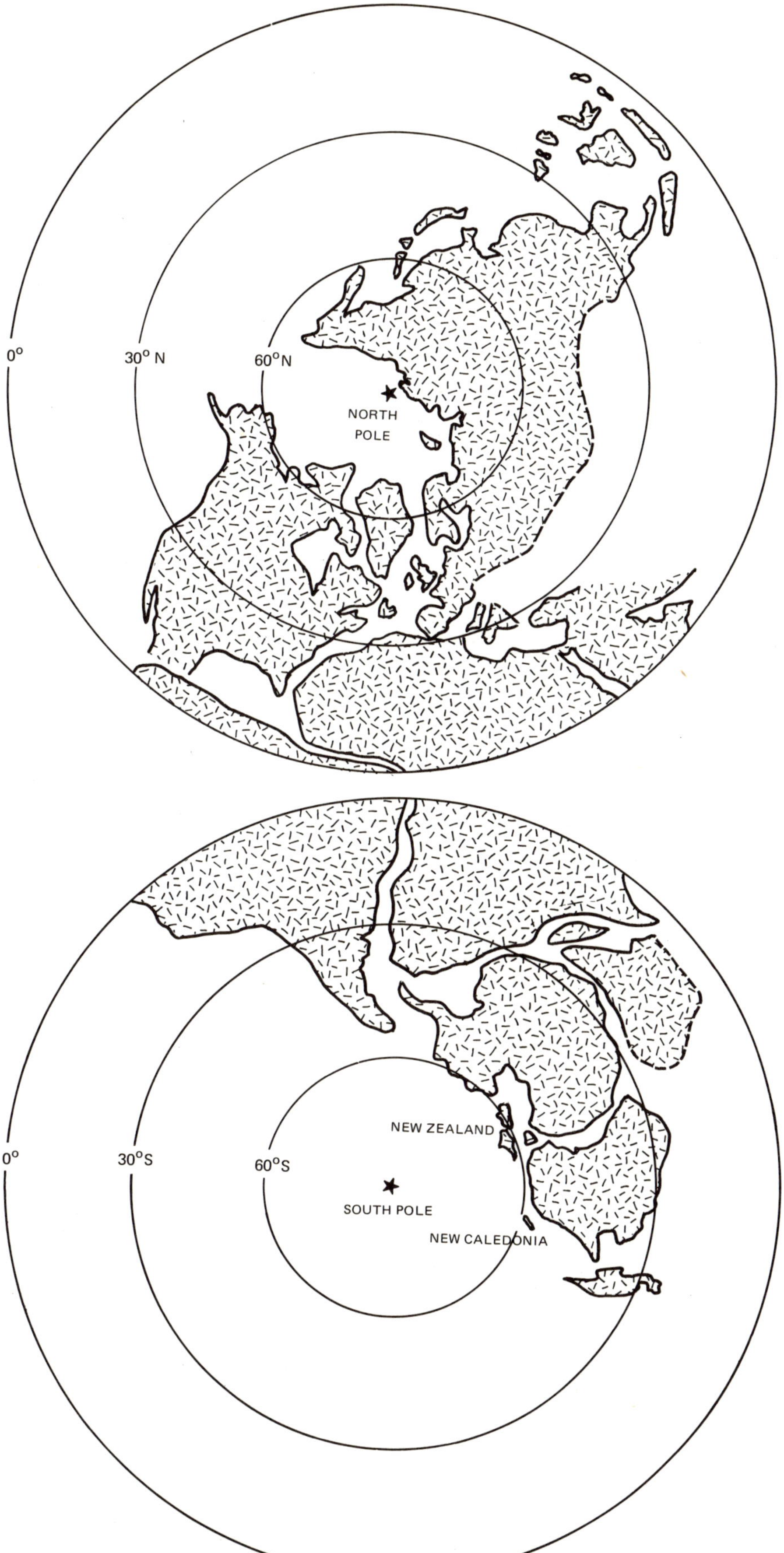
0°
30° N
60° N
NORTH POLE
0°
30°S
60°S
NEW ZEALAND
SOUTH POLE
NEW CALEDONIA

*Fig. 7.9*

thoroughly explored (Fig. 7.10). New Zealand's palaeomagnetism, on the other hand, has been little studied. As all the available evidence suggests, however, that New Zealand was close to the east coast of Australia for most, if not all, of the time extending from Cambrian to Jurassic time, spanning 570-140 million years ago, it is probably reasonable to assume that New Zealand tagged along as Australia gyrated around the Southern Hemisphere (Fig. 7.10). The record of ancient life preserved in New Zealand's rocks also supports this assumption.

Thus the New Zealand fossils (pp. 185–187), and the Australian palaeomagnetic data, indicate that in the Cambrian, Silurian and Devonian New Zealand probably lay at about 10–20°S Latitude, but after this it drifted gradually southwards until in the Permian, 280–230 million years ago, it was close to 70°S Latitude, and southern cool water shellfish lived in the seas then covering New Zealand (pp. 191–192). After this New Zealand gradually drifted northwards again until the middle and late Jurassic, 160–140 million years ago, it was in mid-latitudes (i.e. 40–50°S Lat.), close to its present-day position. At this time warm temperate shellfish lived in New Zealand seas (pp. 214–215).

However in the latest Jurassic and early Cretaceous rifting and rotating movements accompanying the start of the break-up of Gondwanaland moved New Zealand southwards again. Consequently throughout most of the Cretaceous New Zealand lay close to the South Pole—and southern cool water shellfish again lived in its seas (pp. 220, 278). Then in the latest Cretaceous and early Tertiary movements accompanying the opening of the Tasman Sea, and the pulling away of Australia from Antarctica, gradually moved New Zealand northwards towards its modern position.

The few palaeomagnetic results from New Zealand document the southward drift of New Zealand in the Cretaceous and its subsequent return to middle latitudes. Results from Glenham, Southland, show that in the middle Jurassic, some 160 million years ago, this part of New Zealand was at its present latitude, 46°S. By contrast, results from Mount Somers (inland Canterbury) and the Chatham Islands, both now at about 44°S Lat., show that in the middle Cretaceous, 95 million years ago these parts of New Zealand were at about 80°S Lat. Other palaeomagnetic results from the same areas show that New Zealand then moved gradually northwards, coming to lie at 62°S Lat., about 74 million years ago (i.e. almost at the end of the Cretaceous). Northward movement continued throughout Tertiary time.

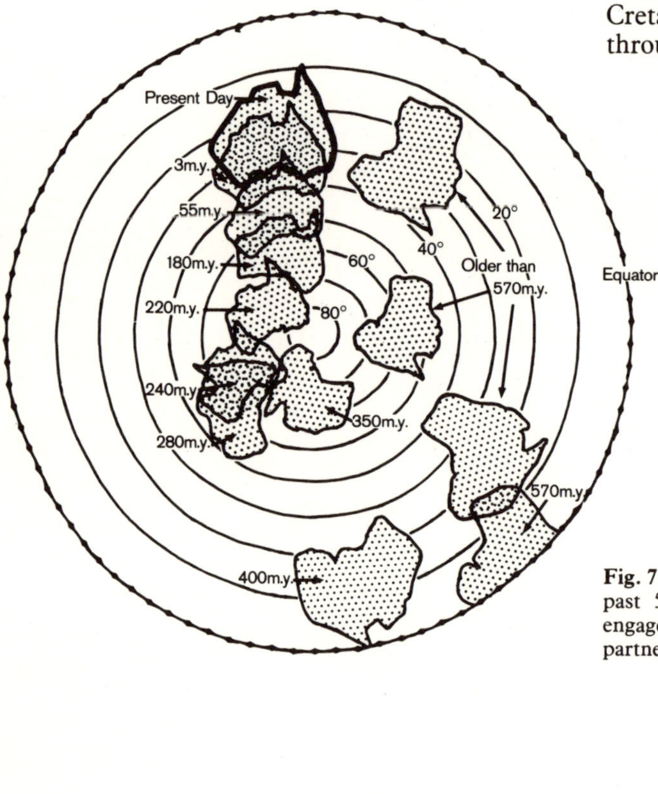


**Fig. 7.10.** Palaeomagnetic studies have revealed that over the past 570 million years and beyond, Australia has been engaged in a global-sized waltz, with the South Pole as a partner, as summed up in this diagram.

CHAPTER EIGHT

# SECRETS FROM DAVY JONES'S LOCKER

WHEN Alfred Wegener was formulating his ideas on continental drift in the early 1900s, he believed that the ocean floors were made up of seemingly endless underwater plains, and this idea of oceanic "landscape" was probably a fair sample of the thoughts then prevailing. Evidence was accumulating, however, to indicate that this view was too simplified.

H.M.S. *Challenger*, a 2000-ton, three-masted corvette with auxiliary steam power, embarked on the first detailed oceanographic cruise in 1872. During the following three and a half years the *Challenger* investigated every ocean except the Arctic. At numerous points her scientists took depth soundings (by the laborious method of sending down a weighted line), and sampled the bottom sediments. They measured the temperature of the water, sampled its flora and fauna at regular intervals from top to bottom, and also recorded winds and currents. The vast accumulation of information collected by *Challenger* eventually filled fifty large royal quarto volumes, published between 1885 and 1895.

The Second World War gave a great boost to oceanographic exploration, as, apart from the need to provide accurate maps of the sea floor on which to base naval activities, the wartime developments of precise automatic depth recording systems, using a type of underwater radar to echo-sound the sea bed, provided the

**Fig. 8.1.** Oceanographic research has revealed a globe-encircling system of oceanic mountain chains (oceanic "ridges" and "rises").

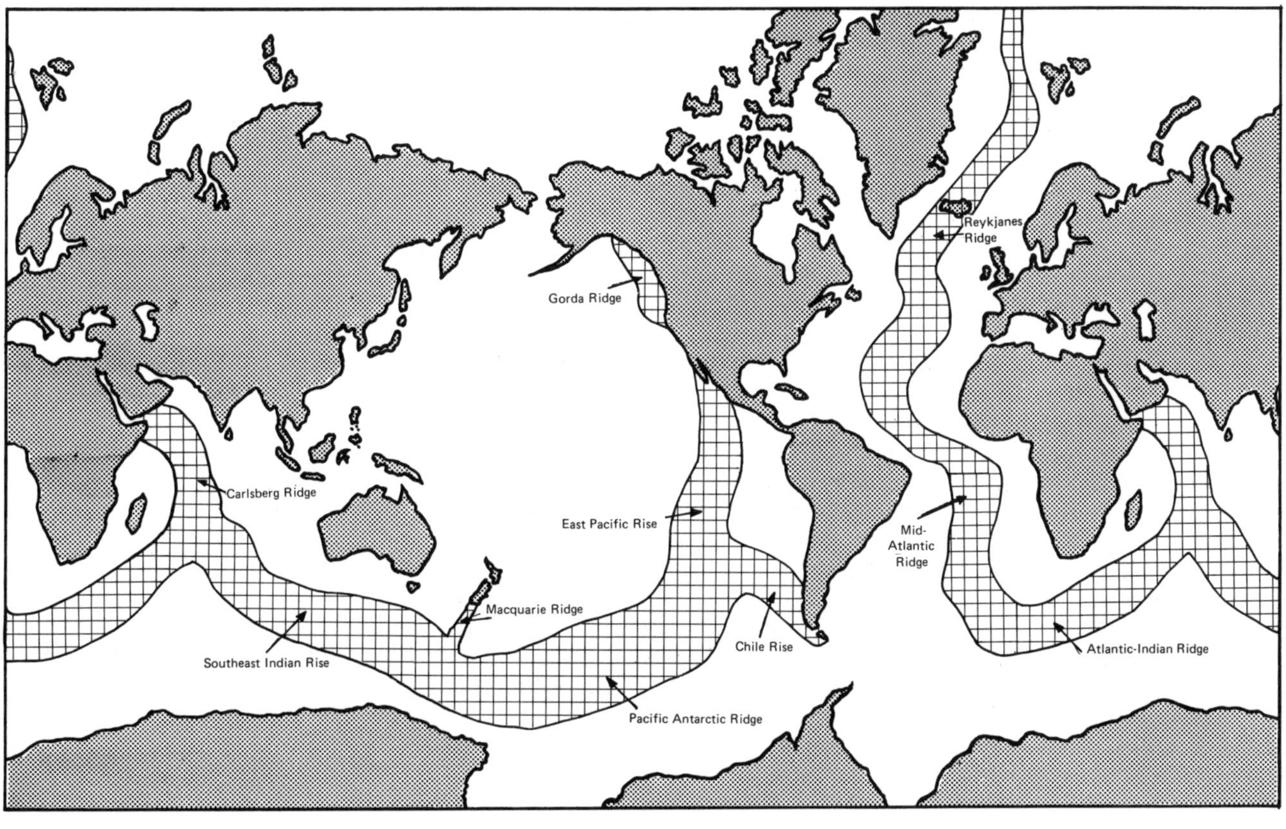

scientists with sophisticated mapping tools. In time therefore it was found that smooth abyssal plains, lying between 4000 and 5000 m (13,000-17,000 ft) below sea level are indeed very widespread, as Wegener and his contemporaries thought, and make up some 50 percent of the ocean floor's surface. On the other hand, the remaining 50 percent was found to contain enormous world-encircling systems of rugged submarine mountain chains, as well as deep trenches and fracture zones and tens of thousands of volcanic islands.

The world's oceans in fact conceal a vast range of topographic relief. The deepest spot in the ocean, in the Marianas Trench south of Guam, extends to 11,516 m (almost 38,000 ft) below sea level, while the highest oceanic mountain, capped by the volcano Mauna Kea in the Hawaiian Islands, rises to a height of 4205 m (almost 14,000 ft) above sea level. The overall range of topographic relief on the ocean floor is thus a spectacular 15,721 m (51,565 ft), which is about double the range of relief on the continents, namely 8848 m (29,015 ft) from sea level to the summit of Mount Everest. Mauna Kea is in fact the world's greatest mountain, with 10,203 m (33,475 ft) of relief from base to summit.

**Mountains under the sea**

As oceanographic ships roamed across the oceans of the world compiling maps of the "landscape" of the sea floor, it soon became evident that entire mountain ranges existed beneath the sea. These gigantic features caught the scientists' imagination, and soon they were tracing their course throughout the world's oceans (Figs 1.8, 8.1). Although the general term "oceanic ridge" is often used to cover all these sea-floor features, in fact a distinction is made between the high oceanic mountain chains (oceanic ridges) and broader lower features that arch upwards from the sea floor (oceanic rises).

It was found that the world was encircled by a system of oceanic ridges and rises. The system and its branches were found to be some 60,000 km (about 37,300 miles) in length, about 1000 km (630 miles) in average width, and ranging up to 4000 m (about 13,000 ft) in height above the sea floor (Fig. 8.2). A very massive ridge, the Mid-Atlantic Ridge, runs as a giant seam-like structure for the entire length of the Atlantic Ocean, and in the 1950s this particular ridge was subjected to fairly detailed exploration. During this time very precise echo-sounding surveys found that a deep trough or axial valley ran along the crestline of the ridge.

The axial trough is quite often a major feature in its own right, even although sometimes it is dwarfed by adjacent ridge peaks. Along the Mid-Atlantic Ridge, in many ways a "type example" of an oceanic ridge, the axial trough is flanked by walls 600 to 2000 m (2000-7000 ft) high and the trough floor may be 12 to 50 km wide (7½-32 miles). Similar features occur on land, notably in the "Rift Valley" system of East Africa, which has a comparable scale of relief, as a glance at Fig. 8.2 will show. The two features—the oceanic axial trough and the rift valley of the land—have both been formed in the same way, as we will see on pp. 121–131.

As is shown in Fig. 8.1, some of the ridges are fairly well centred in oceans like the Atlantic, that have fractured "jig-saw" pattern coastlines. The East Pacific Rise however lies to the eastern side of the Pacific Ocean. Other ridge systems make for the coastlines and strike across the land as rift valleys or zones of disturbed fractured rock. The oceanic ridges, unlike the mountain ranges of the land, are neither straight nor smoothly curved but are sliced into short segments by transverse fracture zones, criss-crossing and offsetting segments.

The East Pacific Rise, for example, is offset by several large east-west fracture zones that run across the Pacific floor from the American coast. These fracture zones in the Pacific are very straight and thousands of kilometres in length (e.g. Fig. 9.23). The Pacific-Antarctic Ridge, south of New Zealand, is similarly offset by huge fracture zones that can be traced onto the New Zealand continental shelf (Fig. 18.5).

The oceanic ridges are the scene of continual earthquake activity. Some 5 percent of the total earthquake energy released throughout the world is concentrated along the oceanic ridges. In fact, some parts of the ridge systems were first pin-pointed by their earthquake activity, long before they were mapped by sounding techniques (Fig. 9.1).

Another feature of the oceanic ridges is that they are the site of the release of vast amounts of subterranean heat. In the 1950s devices were developed for measuring the amounts of heat coming out of the sea floor, flowing up from deep

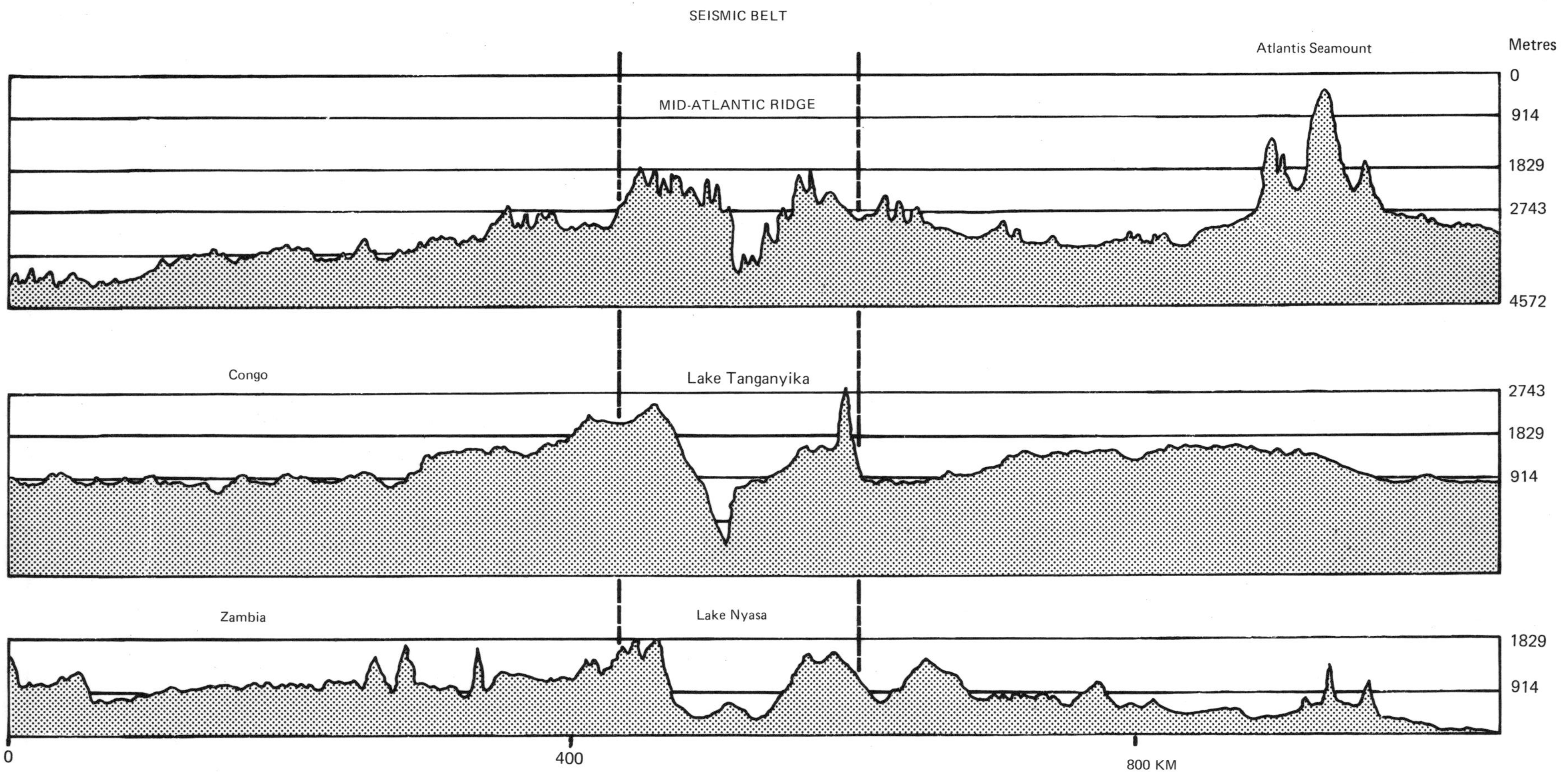


**Fig. 8.2.** Cross-sections across the Mid-Atlantic Ridge (upper) and the East African Rift Valley (middle, in the vicinity of Lake Tanganyika; lower, in the vicinity of Lake Nyasa) to show the great similarity between the features. Note the central rift valley, developed along the axis of the ridge system.

Aleutian Trench
Kurile Trench
Japan Trench
Izu Trench
Ryukyu Trench
Marianas Trench
Philippine & Mindanao Trenches
Mindoro & Palawan Trenches
Yap Trench
Palau Trench
Celebes Tr.
0°
0°
Vityaz Tr.
Banda Tr.
Solomon Tr.
Java Trench
Timor Tr.
San Cristobal Tr.
New Hebrides Trench
Tonga - Kermadec Trench

*Fig. 8.3*

within the earth's crust. When surveys were made with these devices, it was found that the ridges in general, and the rift valleys in particular, were marked by a very high rate of heat flow—far higher than the surrounding ocean floor. Also associated with the oceanic ridges are strings of volcanic islands, and these are composed of a volcanic rock called basalt, that flows out, usually in a fairly tranquil manner, except for spectacular "fire fountains" of burning volcanic gases.

### Trenches and island arcs

Submarine trenches—huge gashes in the sea floor—stretch the land-dweller's imagination to the limit. A typical trench is V-shaped; some are 3200 km (2000 miles) long and 10 km (6½ miles) deep, narrowing to a width of only 5 km (3 miles) at the bottom (Figs 8.3-8.6). By comparison, the Grand Canyon, scooped out by the Colorado River, in the western United States, is only 320 km (200 miles) long and 1.6 km (1 mile) deep.

The world's oceans are furrowed by some twenty trenches, all of which lie close to land (Fig. 9.2). Trenches are often curving, although some are straight for very long distances. Chains of volcanoes are closely associated with the trenches and are usually situated up to 200 km (130 miles) from the inside curve of the trenches, where they form "island arcs" (see Fig. 8.3). Unlike the frequently more tranquil basalt volcanoes along the oceanic ridges, those of the

**Fig. 8.3.** Many of the deep-sea trenches occur in the Western Pacific, as shown on this map. Other trenches flank the Pacific coasts of Central and Southern America, and border Macquarie Island (south of New Zealand), the South Sandwich Islands (south Atlantic) and the West Indies (see Fig. 9.2). Chains of volcanic islands flank the trenches, forming island arcs.

**Fig. 8.4.** The Tonga Trench, north of New Zealand, reaches a depth of almost 10,000 m. Most cross-sections, on land and sea, are drawn with a vertical exaggeration, to accentuate their major features. The lower profile has been drawn with a vertical exaggeration of about 20 (i.e. the vertical scale is about 20 times that of the horizontal) and for comparison a natural scale profile is shown at the top.

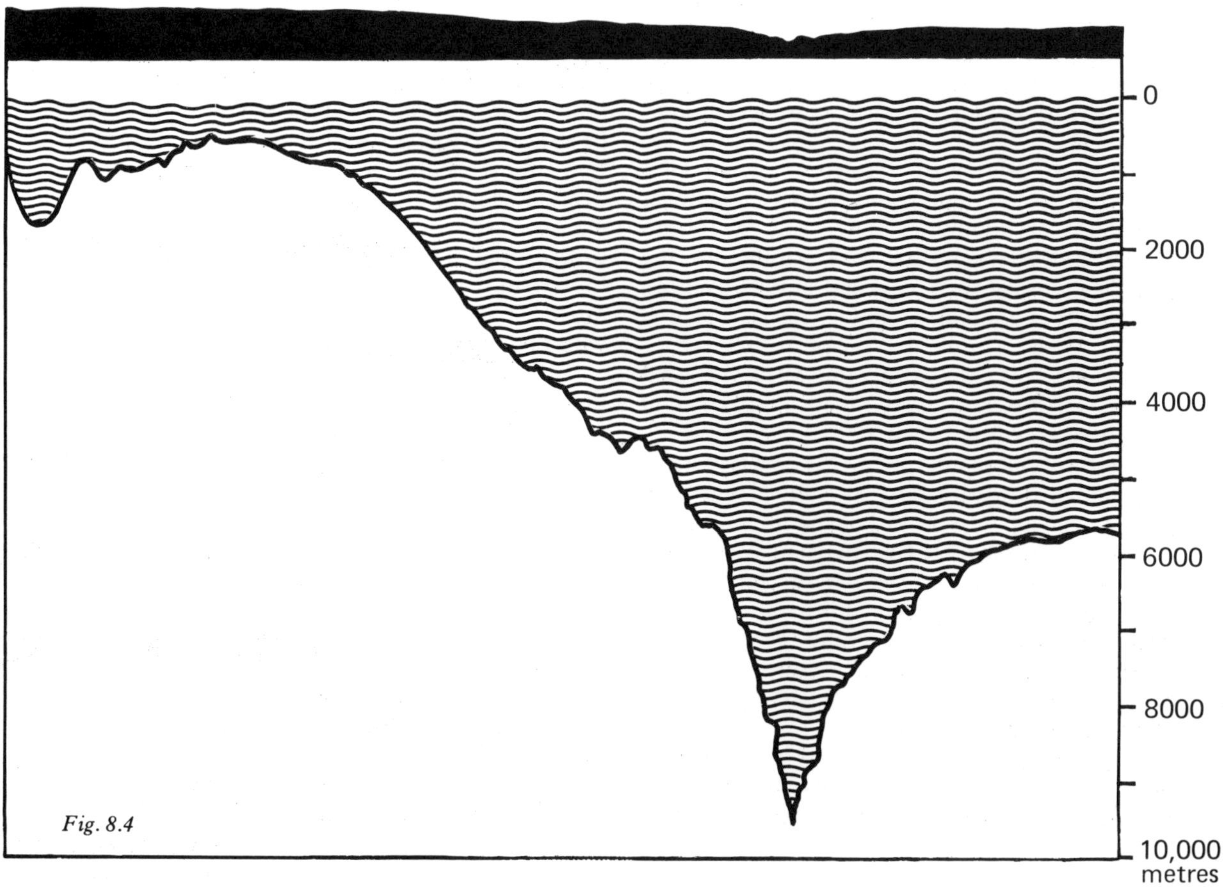


*Fig. 8.4*

island arcs are composed of a volcanic rock called andesite, that often erupts with great violence. An extensive trench and volcanic arc system extends from Tonga through the Kermadec Islands and into the North Island of New Zealand, where it is marked by andesite volcanoes such as Ruapehu, Ngauruhoe and Egmont (Fig. 8.6).

The trenches and island arcs mark zones of great crustal instability. They are associated with the greatest oceanic depths, the largest anomalies of the earth's gravity field, the most extensive vulcanism, and the most intense shallow earthquake activity. Furthermore, almost all of the known deep earthquakes occur in island arcs and trenches.

A Dutch geophysicist, Felix A. Vening Meinesz, had the idea that submarines, cruising in the tranquil depths of the sea, well beyond the reach of the turbulent waves, would provide ideal stable platforms on which to use pendulum devices to measure the pull of the earth's gravity at sea. He was able to talk the Dutch navy into allowing him to travel on its submarines and in 1923 he tested his device during a submarine voyage from Holland to the East Indies. Numerous other voyages followed and within 20 years he had measured gravity in oceans around the world.

His most celebrated discovery was of long curved strips of anomalous gravity readings, thousands of kilometres long, and 50-100km (32-64 miles) wide, that run parallel to the trenches, inside their concave curves. Within these belts of anomalous gravity measurements, Vening Meinesz found that the pull of the earth's natural gravity field was very much less than that of the surrounding areas—they were belts of "negative gravity anomalies".

Vening Meinesz interpreted the negative strips as marking down-buckled furrows of light sial projecting deep into the heavier sima, a situation potentially so unstable that he thought it could only be maintained by the use of considerable force, capable of dragging down, and holding down, the light sial. This force may be in the form of subterranean currents, or "convection currents" acting within the mantle, about which we shall hear on pp. 154–164.

The Pacific Ocean is encircled by a "ring of fire", or in more prosaic terms, the "Circum-Pacific Mobile Belt", which is characterised by intense earthquake and volcanic activity (Figs 9.1, 9.3). The Pacific margin accounts for some 80 percent of the seismic energy released in earthquakes. The oceanic ridges account for another 5 percent, the mountain belts extending from the Mediterranean to the Himalayas for about 14 percent, and shocks elsewhere for 1 percent or less (Fig. 9.1).

A number of the Pacific earthquakes, particularly if they occur close to or in ocean basins, trigger tsunamis or giant ocean shock waves, which radiate outward from the earthquake centre at speeds of 500 to 800km (300-500 miles) per hour and build up to heights of up to 30 metres (100ft) if they sweep across shallow bays or along constricted channels.

**Benioff Zones**

Earthquake shocks originating at depths of some tens of kilometres have been well known for many years. In the 1920s and 1930s, however, as the science of earthquake location improved, it was found that earthquakes could be traced right down to depths of 700km (435 miles). This discovery led to speculation about the origin of such deep shocks, as it was thought that at such depths the rocks would not be capable of tearing apart along faults. Rather, it was thought that the rocks would be so hot and under such pressure that they would flow, like warm toffee. At that time therefore the deep shocks were presumed to come from sudden, almost explosive, changes in the form of deep-seated materials, such as sudden crystallisation or the collapse of originally relatively open crystal structures to form denser, more compact, varieties.

As time went on, however, it was shown that the deep earthquake shocks are indeed caused by fracturing and shearing of solid rocks and, further, that along the Pacific margins the deeper the earthquake's source the further inland was its point of origin, projected onto the earth's surface. From this it was concluded that giant fault zones dip from the ocean floor and extend under the land at angles of 30° to 60°. These dipping earthquake-generating zones are called "Benioff Zones", named after a famous seismologist, Hugo Benioff, and are thought to separate continental crust from oceanic crust (Figs 8.6, 9.14, 19.15).

Where the Benioff Zones intersect the Gutenberg Low Velocity Channel at depths of 100 or

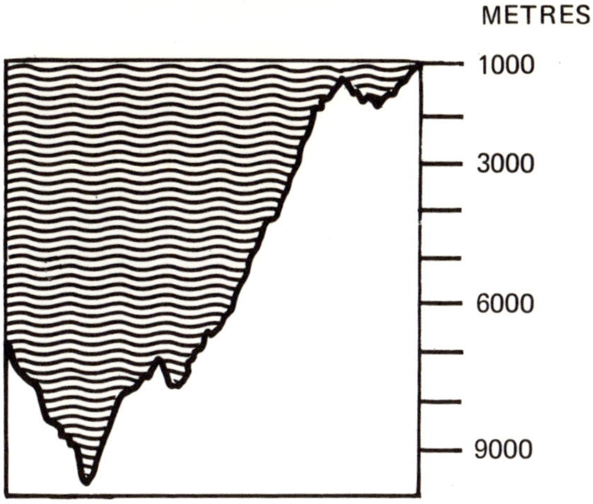


**Fig. 8.5.** The honour of discovering the deepest spot in the oceans was the subject of some rivalry between oceanographic expeditions both before and after World War II. Eventually the search narrowed to the Philippines Trench and adjacent Mindinao Trench, but in 1950 H.M.S. *Challenger*, the modern successor of the pioneering 19th century oceanographic vessel, sounded 10,863 m (35,640 ft) in the Marianas Trench, south of Guam and in 1959 the Russian survey ship *Vitiaz* recorded 11,055 m (36,204 ft) in the same vicinity. In 1960 the United States bathyscape *Trieste* descended to a depth of 10,912 m (35,802 ft) in this general area. This diagram shows two echo-sounding profiles in the Marianas Trench.

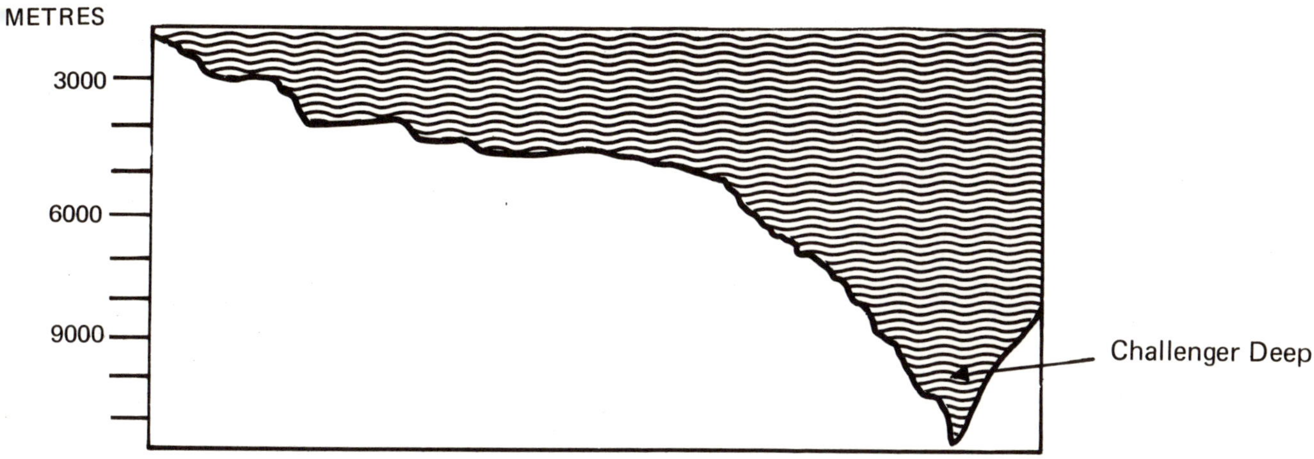


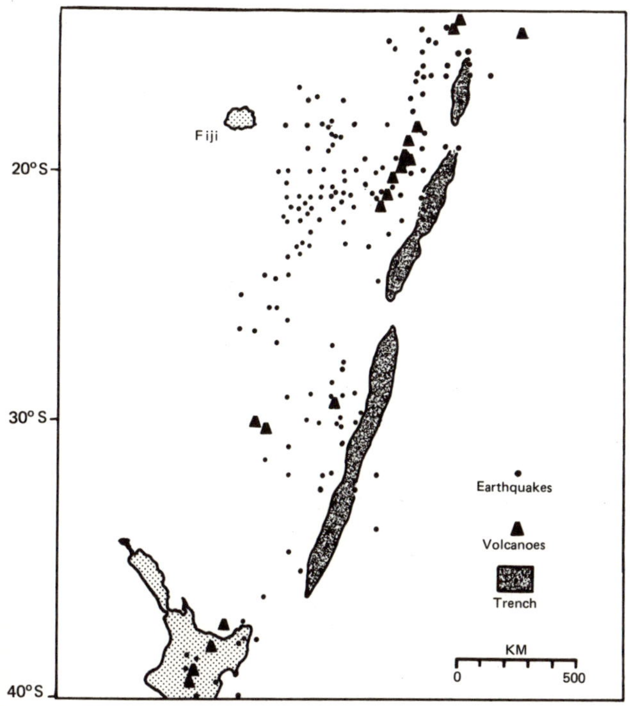


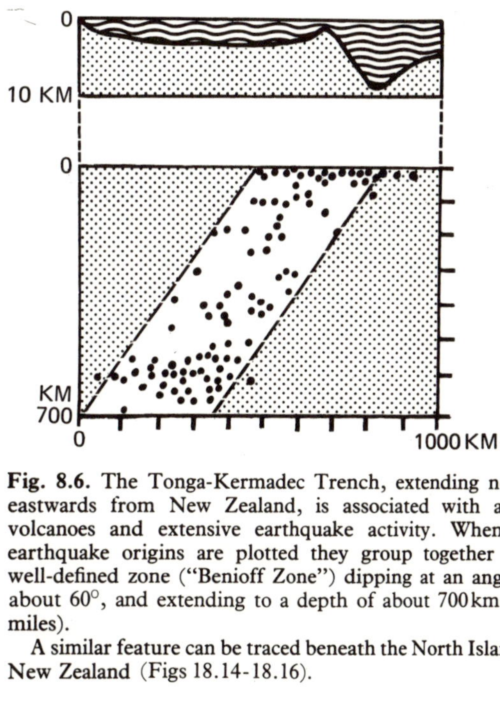


**Fig. 8.6.** The Tonga-Kermadec Trench, extending north-eastwards from New Zealand, is associated with active volcanoes and extensive earthquake activity. When the earthquake origins are plotted they group together in a well-defined zone ("Benioff Zone") dipping at an angle of about 60°, and extending to a depth of about 700 km (435 miles).

A similar feature can be traced beneath the North Island of New Zealand (Figs 18.14-18.16).

200 km (60-130 miles) is the site of the melting-up of crustal materials to form volcanic magmas. The magmas force their way upwards to the earth's surface, where they erupt, to produce the volcanic chains of the island arcs and the Pacific margins (Figs 9.3, 9.14).

The North Island of New Zealand straddles a Benioff Zone, and the volcanic activity of the Taupo-Rotorua region, and many of the earthquakes felt in that area, result from the presence of such a deep crustal structure (Figs 18.14-18.16). The situation is complicated in New Zealand, however, by the presence of other continental structures. The picture is simpler and hence clearer further to the north-east, where the New Zealand situation passes into the Tonga-Kermadec island chain which is almost a "type example" of a Benioff Zone (Figs 8.6, 18.9).

**Magnetic footprints on the sea floor**

During the Second World War the problem of submarine detection from the air led to the development of an extremely sensitive device called the Airborne Magnetic Detector. It was housed in a teardrop capsule which could be lowered from a blimp or scouting plane to trail below and behind the craft, well clear of its own magnetic field. The device was effective for submarines at shallow depth, and after the war geologists looking for oil-bearing formations found it useful for aerial magnetic surveys, since it could delineate, in rough terms, regional distributions of igneous and sedimentary rocks. Then it was thought that it might be possible to learn something about the ocean floor by simply towing a magnetometer behind a ship. If the towline was about 100 m long, the pod-shaped magnetometer being towed would be clear of the magnetic field caused by the ship's metal.

The first magnetometers were tried out in the early 1950s and over the years oceanographers have been trailing them behind their ships to record the variations in the magnetism of the ocean floor. The layers of soft muds and sands mantling the sea floor have little if any magnetism and the magnetic readings picked up by the sea-going magnetometers come from the underlying hard rocks, the so-called "basement" rocks of the ocean floor. As the oceanographers plotted their magnetic observations, they found that there were various fluctuations—sometimes the magnetic response was strong, at other times it was weak. Further study showed that the strengths and weaknesses were related to the fact that at times the direction of the magnetism was "normal" (i.e. the magnetic minerals that are tiny fossil compasses "frozen" into the rocks were all aligned so that they pointed northwards), and was therefore strong, whereas at other times the direction of magnetism was "reversed", i.e. the magnetic minerals were pointing southwards, and the resulting signals thus lacked the strength of the signal coming from the "normal" magnetised rocks.

In Chapter 7 we dealt with a different sort of fossil magnetism—the *angle*, dip, or inclination at which the magnetic minerals in the rocks were locked into place during their cooling processes, this being related to latitude (Fig. 7.2). The phenomenon being measured by the oceanographers' magnetometers is another property of the earth's magnetic field—that of "polarity", i.e. whether the magnetic field is related to the North Pole, which is the "normal" magnetic field (and hence all compass needles point towards the north), or whether it is "reverse" and related to the South Pole.

As we have seen on pp. 59–61 the earth's magnetic field is generated by the spinning motions of the earth itself. In general terms the earth's body acts as a self-sustaining dynamo. As the earth spins, electricity is generated within the fluid core and is conducted to the surface, where the centrifugal or spinning force (the Coriolis Force) distributes the magnetic field as a grid or network of electrically charged particles aligned around the earth's axis of rotation (Fig. 7.1). Changes in polarity of a magnetic field is a phenomenon that is poorly understood. Polarity changes have however been produced in laboratory experiments through the interactions of several dynamo systems, and it is thought that there might be some such interactions between the separate dynamo systems that probably exist between solid core and liquid core on the one hand, and solid mantle and crust on the other (see Fig. 10.4). Such interactions may periodically cause the earth's magnetic field to change or flip over from being normal (or north-orientated) to reverse (or south-orientated).

As modern man has not lived through a polarity reversal it is difficult to get an insight into what exactly happens when a flip occurs. However, very precise dating methods (but still with some margin

**Fig. 8.7.** Sometimes a sequence of lava flows captures a record of a polarity change. In this diagram the sea has cut a cliff face across the flank of a volcano, exposing a cross-section through the lava flow sequence. Radioactive dating of the successive flows determines that the lavas range in age from 180,000 years to 1.29 million years.

Palaeomagnetic determinations show that a reversal occurred between 580,000 and 900,000 years ago.

A glance at Fig. 8.8 will show that, while this record has picked up the magnetic flip known as the Jaramillo Polarity Event (890,000-950,000 years ago), it has however missed the flip marking the Matuyama-Brunhes Epoch boundary (690,000 years ago). However, other lava sequences elsewhere, spanning different age ranges, may have captured this unrecorded "flip", and by sampling numerous sequences in various countries, the complete record may be put together.

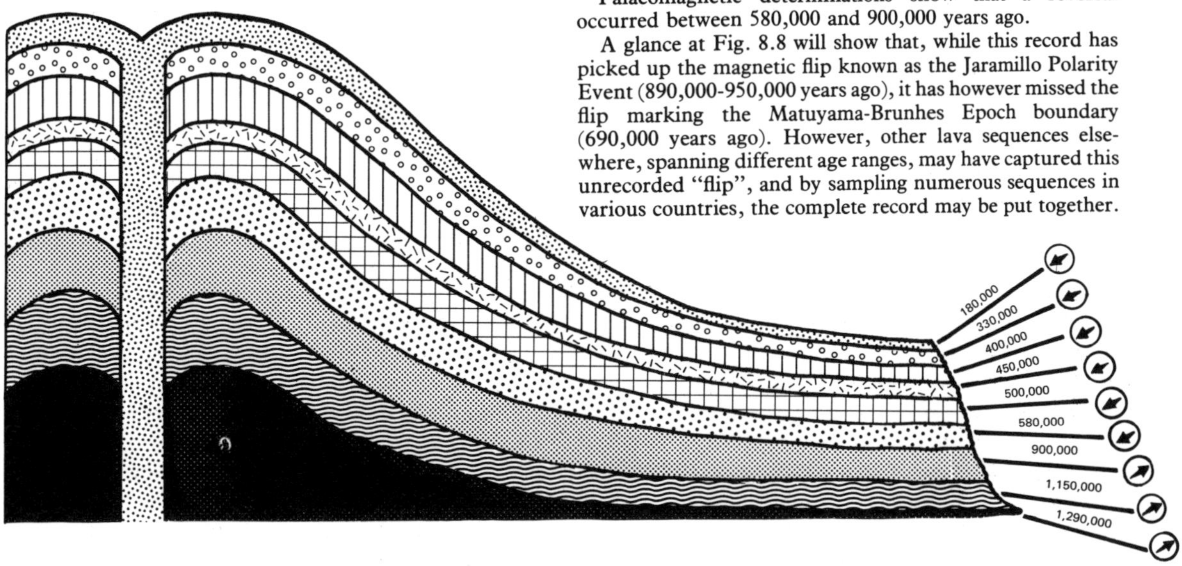


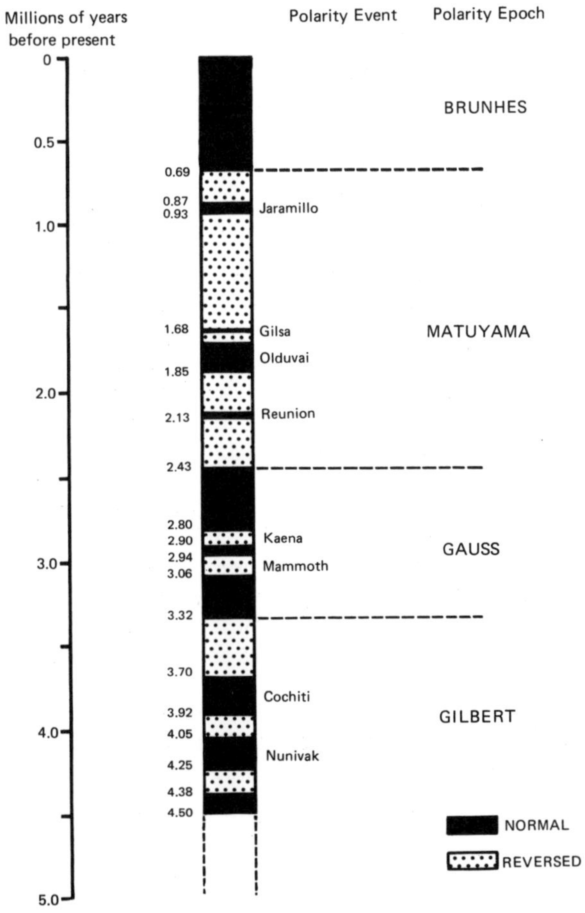


of error) have enabled calculations to be made of what probably happens. It appears as though a "flip" does not occur overnight, as it were, so that one day the compass needle is pointing northwards and the next day it is pointing southwards (imagine the consternation amongst navigators if this happened!). Rather, it is thought that some 2000 years before a reversal, the strength of the magnetic field drops by some 80 percent. Then it takes the polarity another 3000 years in which to change its direction.

Thus the flips take place over a period of some

**Fig. 8.8.** The time sequence of flips of the earth's magnetic field (i.e. polarity changes) has been worked out by combining magnetic studies of the type illustrated in Fig. 8.7 with radioactive dating.

Within the time scale broad "epochs" can be distinguished, when the field was predominantly of one polarity, while "events" are very short periods of opposite polarity within epochs. This scale may not yet be complete and a number of short events may still remain undiscovered. For example, a very small period of reversed polarity appears to have occurred between 30,000 and 20,000 years ago.

Beyond 4.5 million years ago the dating methods become too imprecise to define the events properly. But nevertheless, broad divisions can be distinguished.

5000 years, which is but an instant on the scale of geological time. Polarity transitions therefore provide very sharp time markers indeed and can be used to divide off chunks of geological time (Fig. 8.8).

During periods of reversed magnetism the strength of the magnetic field remains at about one-fifth of its usual value. Compared with periods of normal polarity, reversals are fairly brief in duration, 10,000-100,000 years being thought to be the usual length (see Fig. 8.8).

The diminution of the strength of the earth's magnetic field that accompanies polarity reversals may have an effect on the animals and plants living on our planet during such time.

The earth's magnetic field holds in place a zone of charged particles—the Van Allen Belt—that acts as a shield against cosmic rays. The particles in the belt largely absorb the harmful cosmic radiation and reduce it to a very low level. If, however, during a reversal period the strength of the magnetic field is markedly reduced, it has been suggested that many of the particles in the Van Allen Belt will gradually spin out into space and be dissipated. The amount of radiation from cosmic rays falling on the earth would then increase dramatically. Such radiation would induce genetic changes ("mutations") in plants and animals. Existing forms may die out, to give way to new forms with fundamentally different genetic compositions. Suggestions that such a thing may have been happening in the past come from studies of cores of sediments from the sea floor that often, according to some people at least, show major changes in microscopic plant and animal life in the sea that seem to coincide with the flips in the magnetic field. Further evidence is needed, however, before it will be possible to judge the full extent to which reversals may have affected life on the earth.

As for possible effects on human beings of a magnetic reversal in the future, we can do a bit of crystal-ball gazing. We know from the geological record that the most recent reversal ended some 800,000 years ago. Judging from the periodicity of reversals in the past, we probably have a few tens of thousands of years to go to the next one. However, even if a reversal is closer than this, it takes several thousand years to accomplish, and so by then we may have developed the technology to cope with increased cosmic radiation, if it does in fact eventuate.

**Magnetic flips**

The existence of reversed magnetism in some volcanic rocks has been known since the French physicist Bernard Brunhes detected its presence in 1906. Brunhes, however, and other workers in the 1920s and 1930s, thought that the phenomenon was purely local and of no global significance. It was left to a group of British, Australian and American workers to make the breakthrough. In the early 1960s these workers systematically sampled many of the world's young lava flows and determined the exact nature and timing of the magnetic flips in the geological past. To do this they started with very young lava flows. These were chosen because it is certain they have not been disturbed by later earth movements.

The group began its research in the Hawaiian Islands, where the superb lava flows exposed on the flanks of volcanoes provide magnetic records commencing with flows that occurred in the years 1955, 1935 and 1907 and then extending right back through the last five million years. Samples were taken from each flow, and then the direction of magnetism was measured and the age determined by the potassium-argon technique (p. 66).

The initial investigations into magnetic reversals were greatly aided by advances in the precision dating of rocks that had been taking place at about the same time. Refinements in analytical techniques meant that radioactive potassium-argon dating, the most suitable method for volcanic rocks, could be applied to very young rocks, in the range of thousands rather than millions of years. Also, the potassium-argon clock starts to run at exactly the same time as the magnetic record is frozen into volcanic rock.

Thus by diligent collecting researchers were able to build up a range of dated volcanic rocks, from a variety of countries, and a sequence of magnetic flips, related to geological time, was established.

Starting with lavas that had been erupted in modern times, and had "normal" polarity directions, the researchers sampled older and older lavas, until they encountered one with "reversed" polarity. This meant that a polarity flip had been "captured" at this point in time (Fig. 8.7).

Four major normal and reversed sequences are defined by the magnetic and radioactive-clock data for the past 4.5 million years (Fig. 8.8). These major groupings have been called geomagnetic polarity epochs and have been named after people who have made significant contributions to our knowledge of the earth's magnetic field. Thus the normal epoch that began some 800,000 years ago and continues today has been named after Bernard Brunhes. Superimposed on the polarity epochs are short flips in polarity that are called polarity events. Events are named after the localities from where they were first recognised. The Olduvai event is a flip of normal polarity that occurred about 1.9 million years ago, during a period of reversed polarity (the Matuyama epoch). It was first picked up during work associated with the dating of the various remains of fossil man discovered in Olduvai Gorge, Tanzania.

New Zealand, with its abundance of volcanoes over the past 4.5 million years, is well equipped to take advantage of the polarity time scale and the epochs and some of the events (Fig. 8.8) have been identified in various parts of the country. Fossil as well as radioactive dating has been used to check the polarity sequences in areas as widely separated as Northland and the Wairarapa together with a relatively new method, using crystals of volcanic glass as age indicators. This method, called fission-track dating, uses as an indication of age the amount of damage caused to tiny crystals of volcanic glass by the passage of radioactive particles.

When crystals of volcanic glass form inside a lava, they often pick up as they crystallise various minute particles of radioactive materials that are also floating around in the molten lava. These minute foreign bodies remain within the crystal of volcanic glass as "inclusions". However, like all radioactive substances they decay steadily with time, giving off charged radiation particles which pass out from the inclusion through the glass

**Fig. 8.9.** Readings of the magnetism of the volcanic rocks forming the ocean floor show that huge areas are magnetised in a stripe-like pattern. Combination and comparison of these patterns reveal that the magnetic patterns were laid out symmetrically on either side of the crests of ridges, as shown here diagrammatically. An actual example is illustrated in Fig. 8.14.

The symmetry is very striking and is difficult to explain in terms of familiar volcanic processes. According to the theory of sea-floor spreading, convection currents in the earth's mantle (the layer below the crust) bring up molten volcanic material to form a mid-ocean ridge. The material then cools, becomes magnetised and is then carried by the currents outward, away from the ridge. Thus a natural magnetic tape recorder is created, successive lots of lava recording changes in the earth's magnetic field.

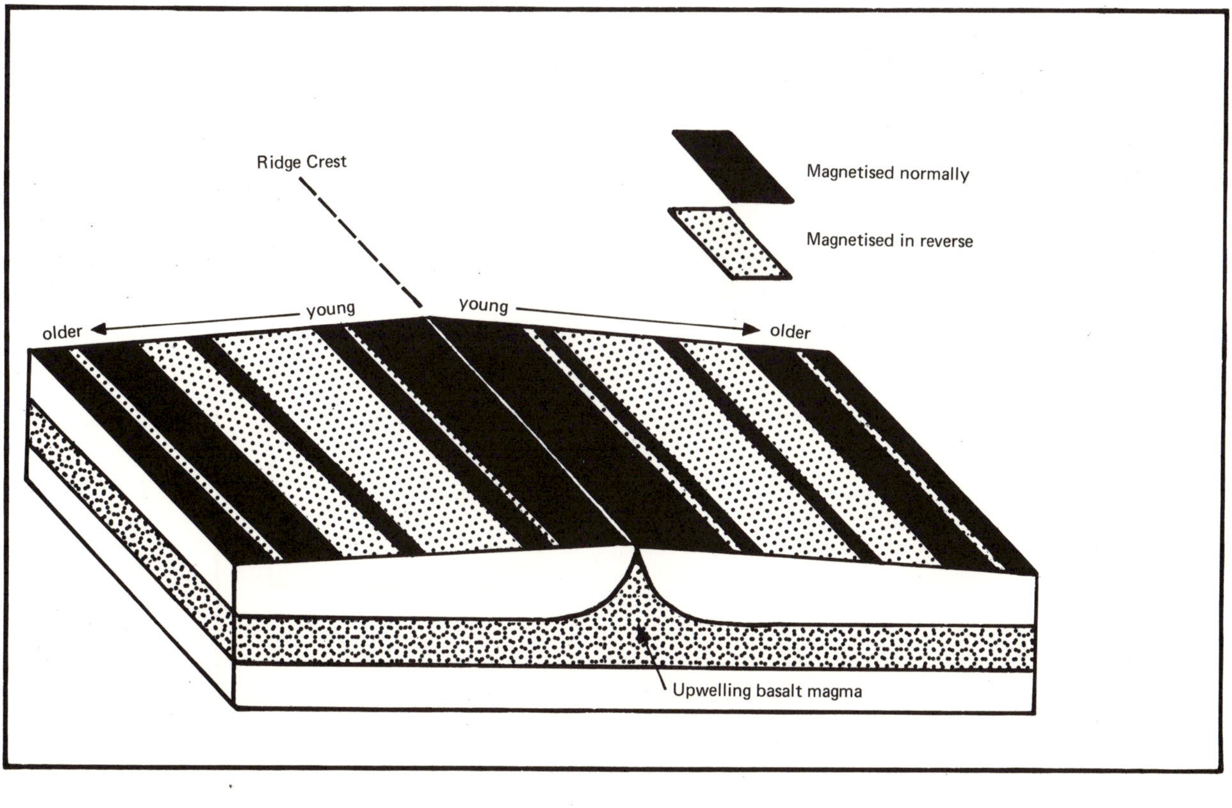

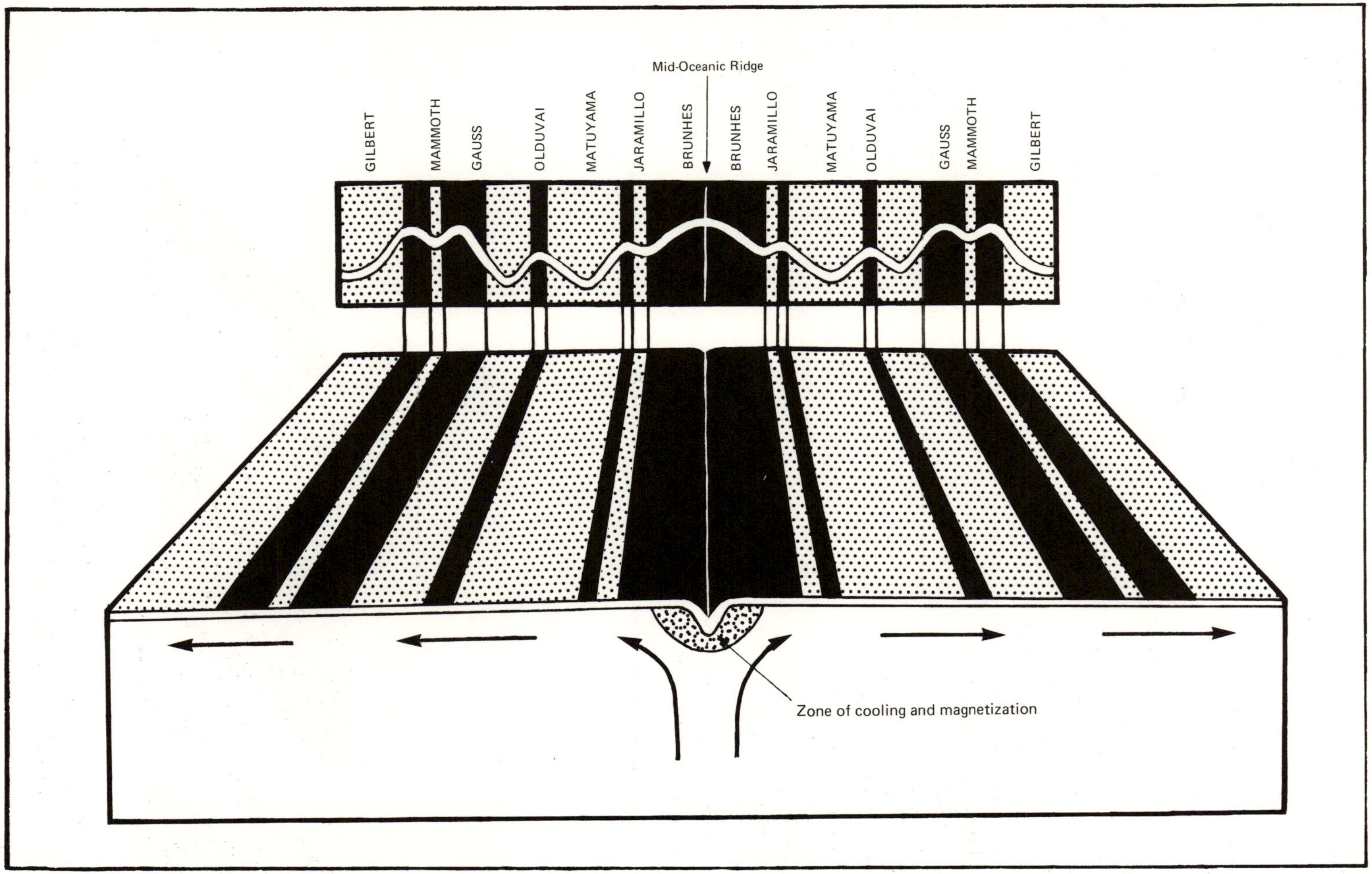


**Fig. 8.10.** Study of the succession of magnetic reversals, combined with radioactive dating, has provided a magnetic time scale that can be applied to the magnetic stripes on the sea floor, recording flips in the earth's magnetic field. Thus the spacing of the polarity bands to either side of the crest of an oceanic ridge provides a measure of the amount of sea-floor spreading. To either side of the Mid-Atlantic Ridge, for example, the sea-floor has been created at a rate of about one centimetre per year.

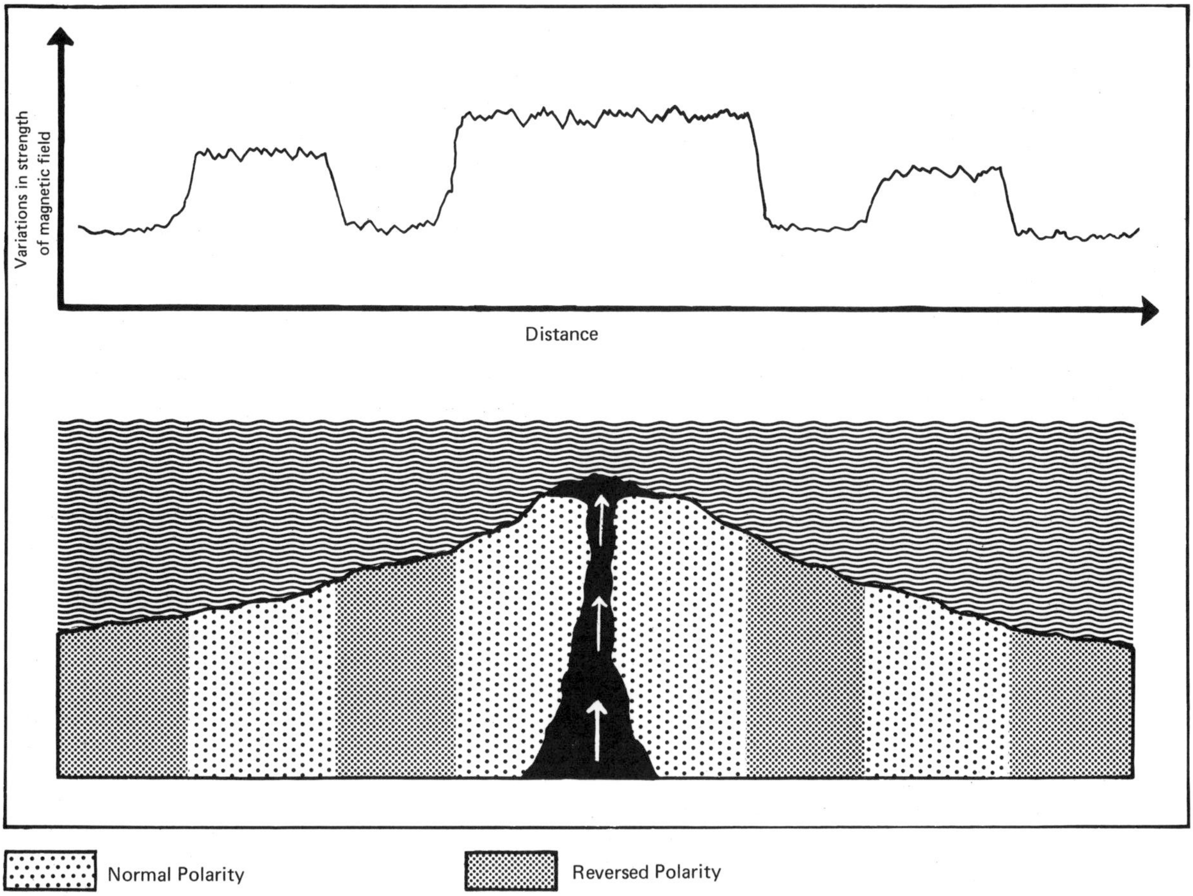


**Fig. 8.11.** Detail of part of the sea floor to either side of the crest of an oceanic ridge showing how the ridge splits along its axis, leading to the welling up of molten volcanic rock, that as it cools preserves a record of the earth's magnetic field. As these rocks are pulled away from the ridge axis, they form bands of frozen magnetism parallel to the ridges and these variations can be picked up by a magnetometer (upper diagram).

crystal, leaving a trail of damage that, after suitable treatment, can be seen as a track. In essence the method uses the number of tracks as a measure of the amount of time elapsed—a recently formed crystal, for example, having relatively few tracks, compared with one much older. This dating method is potentially most promising in New Zealand, as many of our younger geological deposits contain an abundance of volcanic glass, resulting from eruptions that have showered the countryside with ash.

Polarity changes that occurred more than 4.5 million years ago are not so well documented. We know that reversals were still frequent in the period of time back to at least 150 million years, but they are more difficult to date as the error in radioactive dating methods increases with the age of the rocks.

However, by using the magnetic information obtained from the oceans, the time scale has been extended backwards into time, but not with quite the same degree of precision and detail as that shown in Fig. 8.8, although it is being continually refined by new researches.

## Sea-floor spreading

We can now return to the striped magnetic patterns that the oceanographers picked up on the ocean floors. The magnetic patterns were found to be parallel bands or zones extending for hundreds and thousands of kilometres, in which the intensity of the earth's magnetic field is higher or lower than the average for the region (Fig. 8.9).

For many years such features were left virtually unexplained, but with the discovery of magnetic reversals it was realised that the presence of normal and reversed magnetised rock formations in the crust of the earth, beneath the sediments of the sea floor, which would add to or subtract from the earth's overall magnetic field, could account for such findings.

Analysis of the sea-floor magnetic patterns revealed that many of them showed a remarkable symmetry on either side of the crests of certain mid-oceanic ridges (Fig. 8.9). This symmetrical "striping" effect was very difficult to explain in terms of the familiar volcanic processes. Successive eruptions on the sea floor would produce heaps of lava materials, arranged in a haphazard fashion, not in symmetrical bands.

A special type of volcanic process had to be envisaged and this introduced the idea of "sea-floor spreading". The core of this idea is that deep convection currents (p. 154) are operating within the earth's mantle, the layer beneath the crust. These currents are thought of as moving molten volcanic rock upwards to form a mid-oceanic ridge. The molten material would solidify and the tiny magnetic particles within it would take up the direction of magnetism then prevailing, normal or reversed (Figs 8.10, 8.11).

Meanwhile, the underlying convection currents, relentlessly moving in an outward direction on either side of the mid-oceanic ridge, would split the newly solidified material down the middle, to form a rift valley, and move the two parts to either side. Fresh molten volcanic rock would then well up into the rift valley and solidify, and if in the meantime the polarity of the earth's magnetic field had flipped over, the congealed lavas would have frozen within them a polarity opposite to that of the previous lavas. This process can be repeated over and over again throughout millions of years.

Thus the lavas that originally congealed along the line of the primordial mid-oceanic ridge have been separated into two parts, and progressively moved further and further away from the ridge crest, new sea floor forming in the area between them (Fig. 8.12). As the new sea floor has been formed during various polarity epochs, it is therefore striped with patterns of normally and reversely magnetised rock, arranged essentially symmetrically on either side of the crest line of the mid-oceanic ridges.

A magnetic time scale can thus be fitted to the sea-floor stripes and an estimate obtained of the rate at which sea-floor spreading has been occurring, and the ages of the oceans (Figs 8.13-8.15).

South, south-east and east of New Zealand the floor of the South-west Pacific Ocean is striped with magnetic patterns all the way across to the crest of the Pacific-Antarctic Ridge and East Pacific Rise (Fig. 8.29). Dating of the magnetic patterns closest to New Zealand shows they were formed some 80 million years ago; at that time they were at the crest of the Pacific-Antarctic Ridge or East Pacific Rise and the sea floor now present in between has been formed over that period of time (Fig. 18.5).

**Dating sea-floor sediments**

An important application of the polarity time scale is in determining the ages of deep-sea sediments. As minute particles of sedimentary material sink slowly downward through the quiet depths of the sea, they may become magnetised in the direction of the earth's magnetic field as they sink. Thus when such particles become incorporated in the layers of sea floor sediment, they provide a record, often of unparalleled completeness, of the various flips of the earth's magnetic field (Fig. 7.3).

Oceanographers have piston-driven devices that can drive long hollow tubes down through the various sedimentary layers mantling the sea bed, so that a "core" sample of the layers can be obtained. This core, encased in the sampling tube, can be raised to the surface, where it can be pushed out, like toothpaste from a tube, to provide a continuous, relatively undisturbed, sample of the sea-bed layers (Fig. 8.16).

The sea-floor layers thus sampled can be dated by means of the microscopic plant and animal remains they contain and at the same time the magnetism can be studied (Figs 8.17-8.21). Study and dating of deep sea cores has complemented that of magnetised volcanic rocks on land, so that we now have a very complete and precise picture of when the various flips have occurred in the earth's magnetic field.

**Mohole and its aftermath**

Modern piston-assisted coring devices can penetrate up to 30 m (100 ft) into the sea-floor sediments. However, in many areas of the ocean this is only scratching the surface, as echo-sounding of the bottom layers shows the existence of many hundreds of metres of sediment laid down over the volcanic rock floor (the so-called

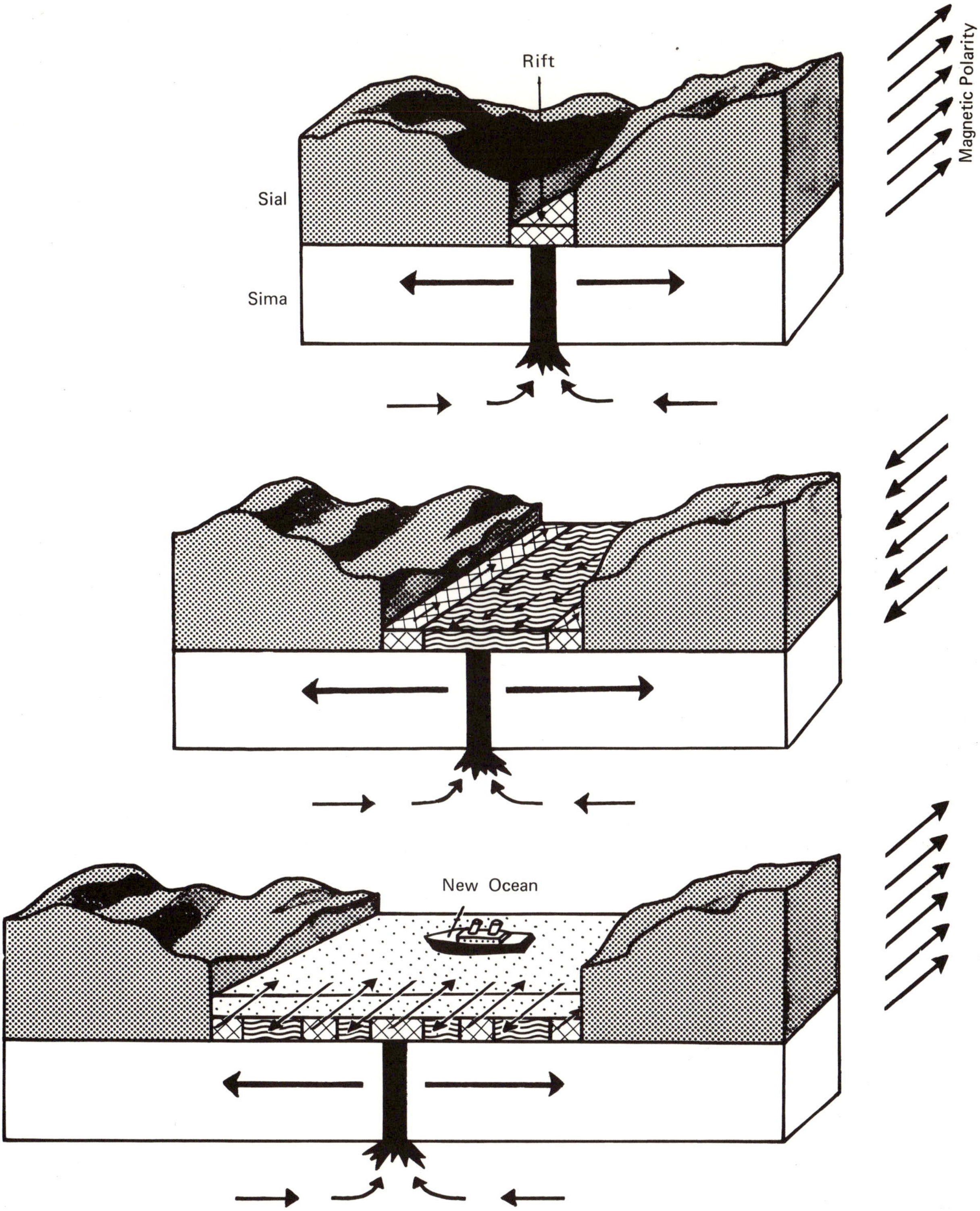


**Fig. 8.12.** If splitting occurs between two continents, a new ocean can open up, whose floor will be paved with a striped pattern of magnetic variations.

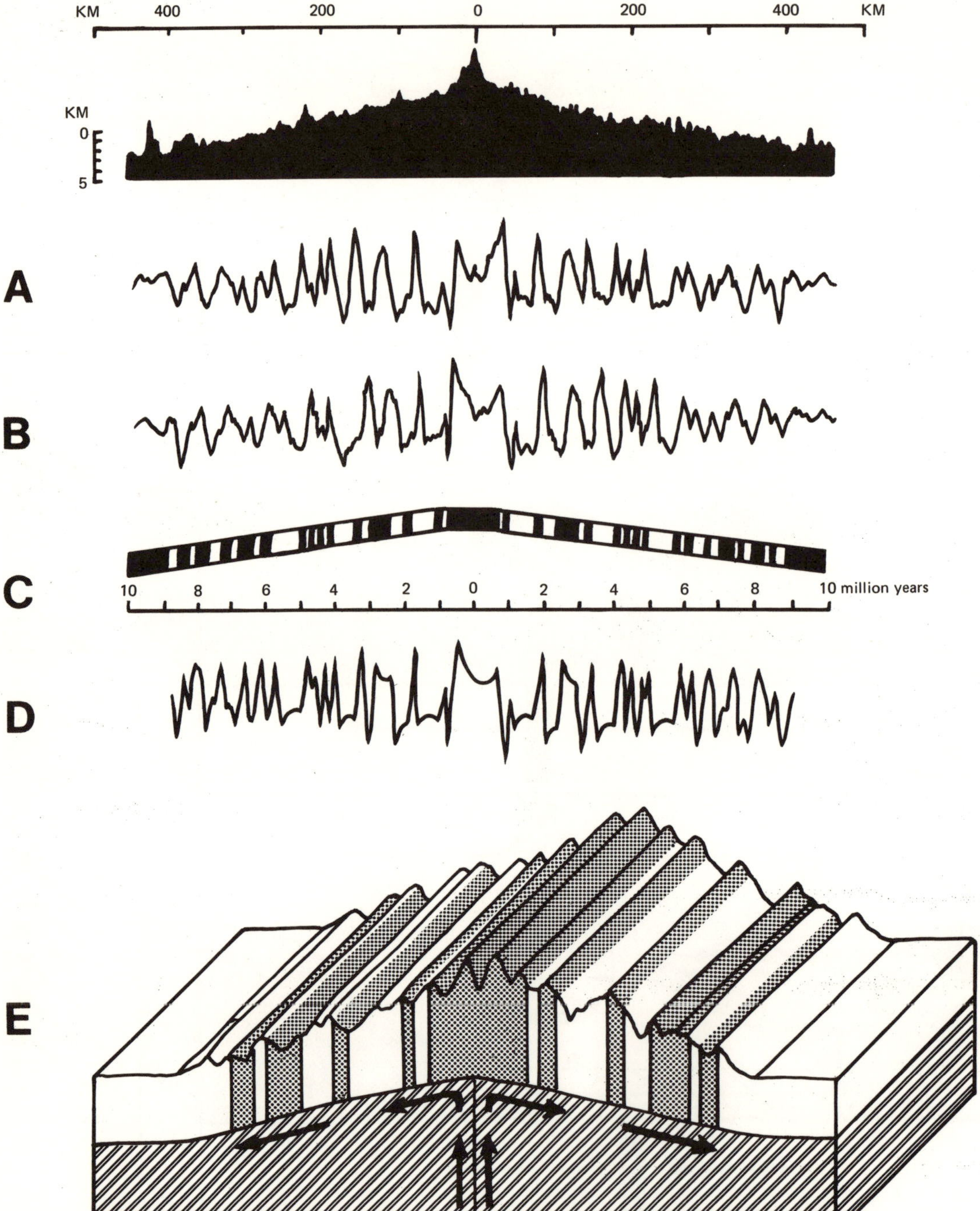


**Fig. 8.13.** This diagram shows a cross-section of the Pacific-Antarctic Ridge, at a point about half-way between New Zealand and South America, the topography of which is profiled in black at the top. A magnetic profile was run across the ridge, using a magnetometer towed behind a ship. The resulting profile is A. As a test for overall symmetry the profile was then drawn as a mirror image, as in B. Then a magnetic time scale was fitted, as in C, and assumed rates of sea-floor spreading worked out, which were then fed into a computer, to give a theoretical profile as in D. Profile E shows the overall interpretation of the data.

"basement"). In order to sample these deeper layers, a ship was needed which was capable of staying put over a spot in the ocean, drilling a hole and taking samples, right down to the hard "basement" rocks. The development of such a ship arose out of a still-born project to drill into the earth's mantle.

This project was put forward in the late 1950s with the aim of drilling through the crust of the earth, and penetrating the Moho (Mohorovicic Discontinuity), thought to mark the boundary between crust and mantle.

Since the Moho is known to be about 35 km (22 miles) below the continents, drilling down to it on land was considered impossible (the deepest drill hole on land is 9.59 km, almost 6 miles, in Texas). But as the Moho is only some 5 km (3 miles) under many ocean floors it was thought feasible to drill through the sea floor.

At about the time the project was mooted, a group of American oil companies had converted a large ex-U.S. Navy barge for drilling into sea-floor sediments through about 500 m (1600 ft) of water. The barge was held over the drill hole by mooring lines attached to large anchored buoys. This floating barge (named *Cuss I* after the four oil companies: Continental, Union, Shell, and Superior) was very successful, and techniques were developed for drilling holes 3 km (almost 2 miles) into the sea floor. This was still a far cry, however, from drilling in 3-4 km of water and then a comparable distance into the sea floor. Nonetheless it was a start, and it encouraged the United States National Science Foundation to charter the barge for an experimental deep-sea drilling programme.

The buoy mooring system would of course not be very practical in the deep ocean beyond the continental shelf. But large diesel outboard motors, developed in the United States during the Second World War to propel U.S. Navy barges, were available and for deep-water drilling these were installed on the four corners of the barge. It was thought that by aiming the motors in suitable directions it would be possible to hold the barge in position over a deep-sea drill-hole. Buoys moored to the bottom and held 30 m (100 ft) below the surface, where they would be free from wave action, served as echo-sounding reference points for the man controlling the motors.

The barge *Cuss I* was accordingly modified in this way and in 1961 drilled some test holes in almost 3600 m (12,000 ft) of water off Guadalupe Island, Mexico, penetrating 200 m (660 ft) into the sea-floor sediments and recovering some cores up to 3 m (10 ft) in length from the underlying basalt.

Although this was a big step forward, drilling some 5 km (3 miles) through the ocean floor was still another thing altogether. However, it was the time of "big science" with massive funds being made available for space and atomic studies, and, as one earth scientist remarked: "The ocean's bottom is at least as important as the moon's behind."

Thus "Project Mohole" was mounted and construction started on an enormous two-deck platform, 71 m × 76 m (230 × 250 ft), supported on two submarine hulls each 113 m (375 ft) long. Serious financial problems soon arose, however, as estimates of the cost rose from $5 million to $68 million and then to $125 million. Eventually, in 1966, the project ran out of funds and Mohole died. Yet from the ashes of Mohole was to rise, phoenix-like, the highly successful Deep Sea Drilling Project.

By 1962 it had become evident that little if any sediment sampling would eventuate from the Mohole project and that a separate scheme for drilling and sampling ocean sediments would be necessary. Proposals were then put forward to the United States National Science Foundation to use some of the technology developed for the Mohole in a somewhat less ambitious project using one of a new class of large drill ships that were then becoming commercially available. In 1967 with Mohole dead, funds were made available to adapt a drill ship for deep ocean work. The ship was named *Glomar Challenger*, "Glomar" being coined from Global Marine Inc., the operator on behalf of the U.S. National Science Foundation, and "Challenger" from H.M.S. *Challenger*, the British corvette that in the 1870s had carried out the world's first major oceanographic expedition.

### The New *Challenger*

The new *Challenger* has little in common with its canvas-rigged predecessor. Rather, it looks like a cross between an oil tanker and oil derrick (Fig. 8.18). The 122 m (400 ft)-long hull has bridge and living quarters at the stern. Amidships, rising above an open hole (6 × 6.7 m; 20 × 22 ft) that passes right through the ship, is the derrick, its top rising 61 m (200 ft) above the sea. Aft, more than

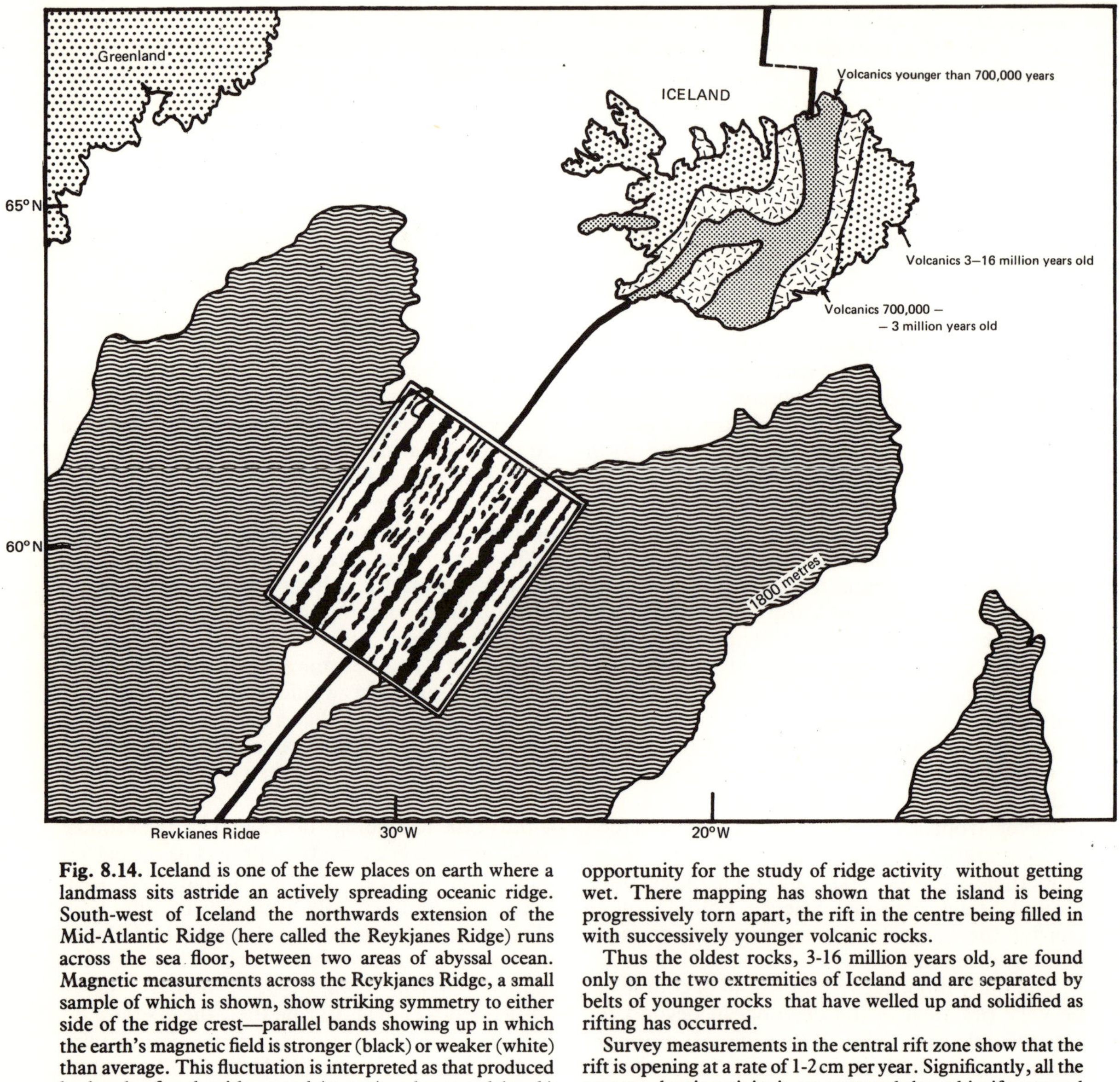


**Fig. 8.14.** Iceland is one of the few places on earth where a landmass sits astride an actively spreading oceanic ridge. South-west of Iceland the northwards extension of the Mid-Atlantic Ridge (here called the Reykjanes Ridge) runs across the sea floor, between two areas of abyssal ocean. Magnetic measurements across the Reykjanes Ridge, a small sample of which is shown, show striking symmetry to either side of the ridge crest—parallel bands showing up in which the earth's magnetic field is stronger (black) or weaker (white) than average. This fluctuation is interpreted as that produced by bands of rock with normal (strong) and reversed (weak) magnetism, as in Figs 8.9-8.11.

The ridge, presumably spreading in the way depicted in Fig. 8.11, continues through Iceland, providing a unique opportunity for the study of ridge activity without getting wet. There mapping has shown that the island is being progressively torn apart, the rift in the centre being filled in with successively younger volcanic rocks.

Thus the oldest rocks, 3-16 million years old, are found only on the two extremities of Iceland and are separated by belts of younger rocks that have welled up and solidified as rifting has occurred.

Survey measurements in the central rift zone show that the rift is opening at a rate of 1-2 cm per year. Significantly, all the recent volcanic activity is concentrated along this rift zone and the most recent eruptions—Surtsey (1963) and Heimaey (1973)—hit world headlines (Fig. 8.15).

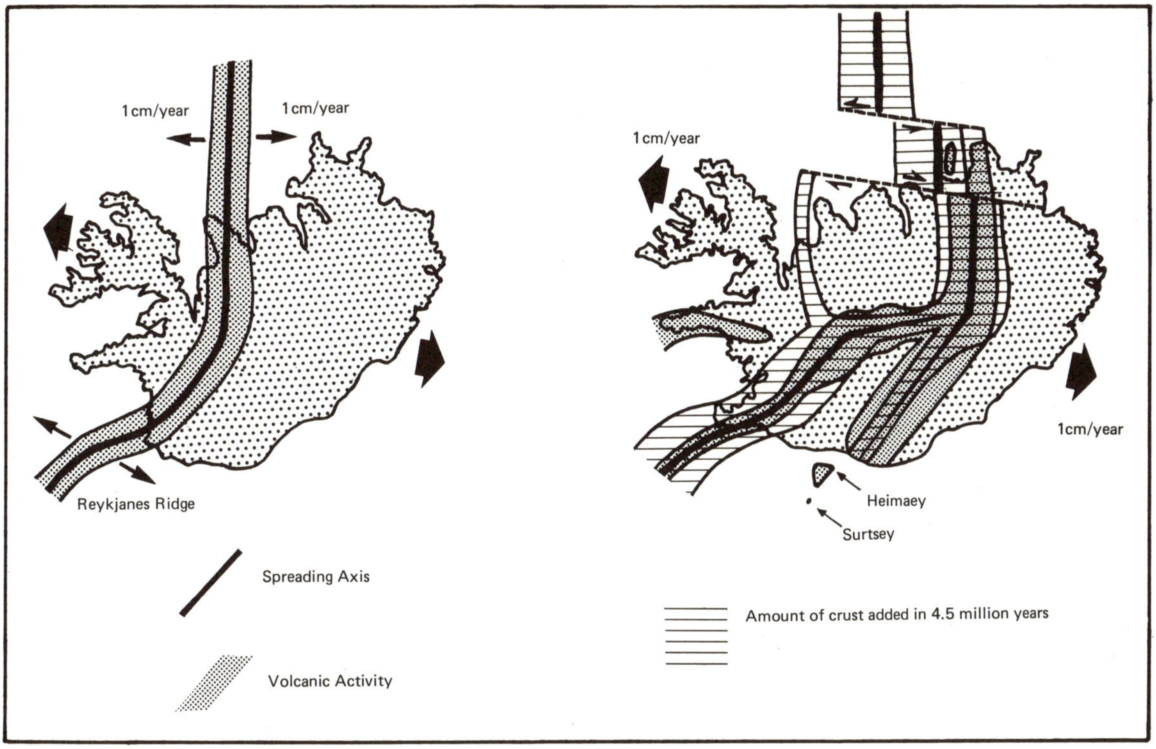


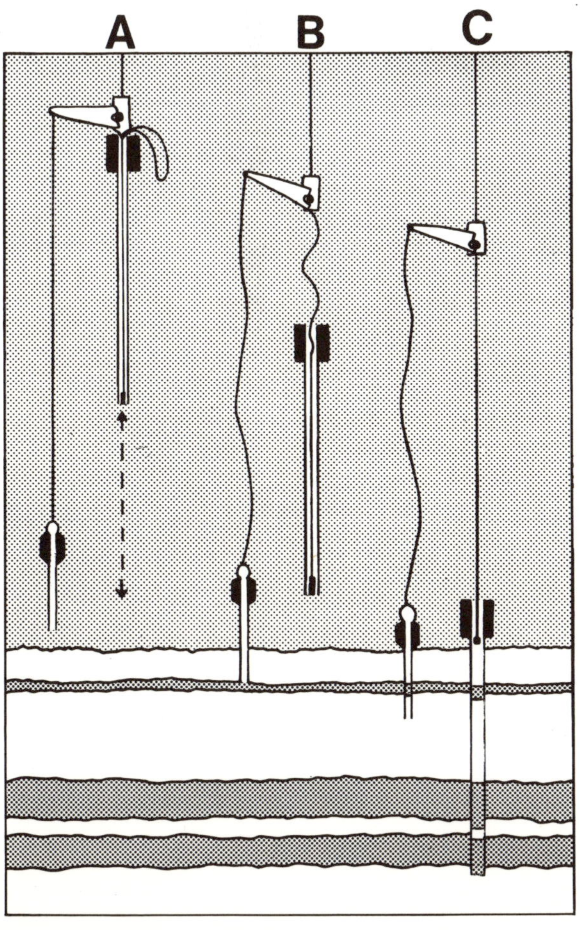


**Fig. 8.15.** The left-hand diagram shows Iceland 4.5 million years ago, with spreading occurring at a rate of about one centimetre per year to either side of the Reykjanes Ridge. The right-hand diagram shows Iceland today split by rift valleys floored with new crust. Comparable splitting movements can be traced back to at least 16 million years ago, when presumably the island was first formed.

**Fig. 8.16.** Oceanographers can obtain relatively undisturbed cores of the sediment layers on the sea floor by employing coring devices similar to the one illustrated in this diagram.

The apparatus is lowered from the ship in such a way that a trigger release dangles below the core barrel. The weighted trigger release strikes the sea floor first (A). This releases the heavy core barrel (B) that, falling freely under its own weight, drives itself deep into the sedimentary layers (C).

The ship then winches up the apparatus, with the enclosed core sample.

Many modern core samplers are piston-assisted, to enable greater penetration to be achieved. Cores 30m (100ft) in length have been obtained in this way.

7 km (4 miles) of drilling pipes in 29 m (95 ft) lengths are stacked on the deck, with more below in the holds (Fig. 8.19). It has accommodation for 70 scientists and crew members and has six separate laboratories for studying various aspects of deep-sea cores.

Like the experimental Mohole barge *Cuss I*, *Glomar Challenger* uses supplementary propellors to maintain its position above the drill hole. Besides its twin screws at the stern, *Challenger* has two pairs of smaller propellors, each pair pointing in opposite directions. Through her bow, below the waterline, are two tunnels, each fitted with a propellor that can be used to push the bow to one side or the other. There are two similar tunnels through the stern. In this way the ship can be pushed broadside, or can be spun without going fore or aft, by thrusting the bow sideways in one direction and the stern in the opposite direction. As the ship must be able to remain stationary over one spot in the deep ocean for several weeks or more, while a well is drilled to a depth of perhaps 3000 m (10,000 ft) below the ocean bed, a computer-based technology called "dynamic-positioning" has been developed.

As the ship approaches the target, her precise position is checked by signals received from navigation satellites in orbit around the earth. A sonic beacon is dropped to the sea bed. Where the water is thousands of metres deep it may take an hour for the beacon to sink to the bottom. Once it comes to rest, sonar signals are transmitted from the ship and bounced back from the beacon, to be received by four hydrophones suspended beneath the ship. By comparing these arrival times the computer can work out the direction the ship is drifting and issue appropriate commands to the ship's main engines as well as to the four propellors in the bow and stern tunnels. The precise amount of thrust is then provided from the various propellors to keep the ship exactly in position.

Also linked to the computer is a wind-reading anemometer. As a gust of wind develops, the appropriate propellors react by pushing the ship against it with just the right amount of counter-thrust. With the power of winds and currents effectively neutralised, the ship is held in precise position over a well until drilling is completed. Dynamic positioning is so effective that the *Glomar Challenger* can maintain its position, and keep on drilling, in the midst of battering waves 5 m (15 ft) high, with 30-40 knot winds. Thus it has been able to operate in some of the world's stormiest seas—for example, those between New Zealand and Antarctica.

The *Challenger* is also equipped with gyroscopically controlled balance tanks that sharply reduce pitching and rolling—and also, no doubt, serve to limit seasickness amongst the scientists and crew.

So that drillers can effectively and rapidly control on a rolling deck the enormous lengths of heavy drill pipe needed to probe the ocean floor, the *Challenger* has an automatic pipe-stacking device that lies forward of the derrick. The ship receives daily satellite photographs of cloud patterns which are used to give advance warning of unsettled weather conditions.

A major difference from drilling on land is the need to provide for the heaving of the ship. This motion would lift the drill bit from the rock into which it was chewing, then drop it back with shuddering force, leading to possible damage. However, shock-absorbing devices (called "bumper subs") in the form of telescopic sections of pipe, have been installed a short distance above the drill bit, that allow for a few metres of upward and downward motion, but yet through interlocking teeth continue to transmit rotation to the bit.

Once the sonic beacons are in place on the sea floor, and the ship's computer locked in, the first section of drill pipe is ready to be lowered. This section carries on its end the drilling bit and bumper subs. The bit, 23.5 cm (9.25 inches) in diameter, is spiked with revolving tungsten carbide teeth, or studded with up to 800 carats of small diamonds, depending on the type of sediment to be penetrated. The bit bores a hole larger than the 12.7 cm (5 in) drill pipe, to allow for sea water to be pumped down through the pipe and back up the hole, outside the pipe, to flush out chips of rock as drilling proceeds.

The first length of drill pipe is lowered down into the hole passing through the centre of the ship. A clamp at deck level then holds the top of the pipe while it dangles freely from the ship. Meanwhile, another length of pipe is hoisted up from the pipe rack on the forward part of the ship and pulled up inside the derrick. It is then screwed on to the top of the clamped pipe. Once this connection has been made, the clamp is released and the newly attached length of pipe lowered to

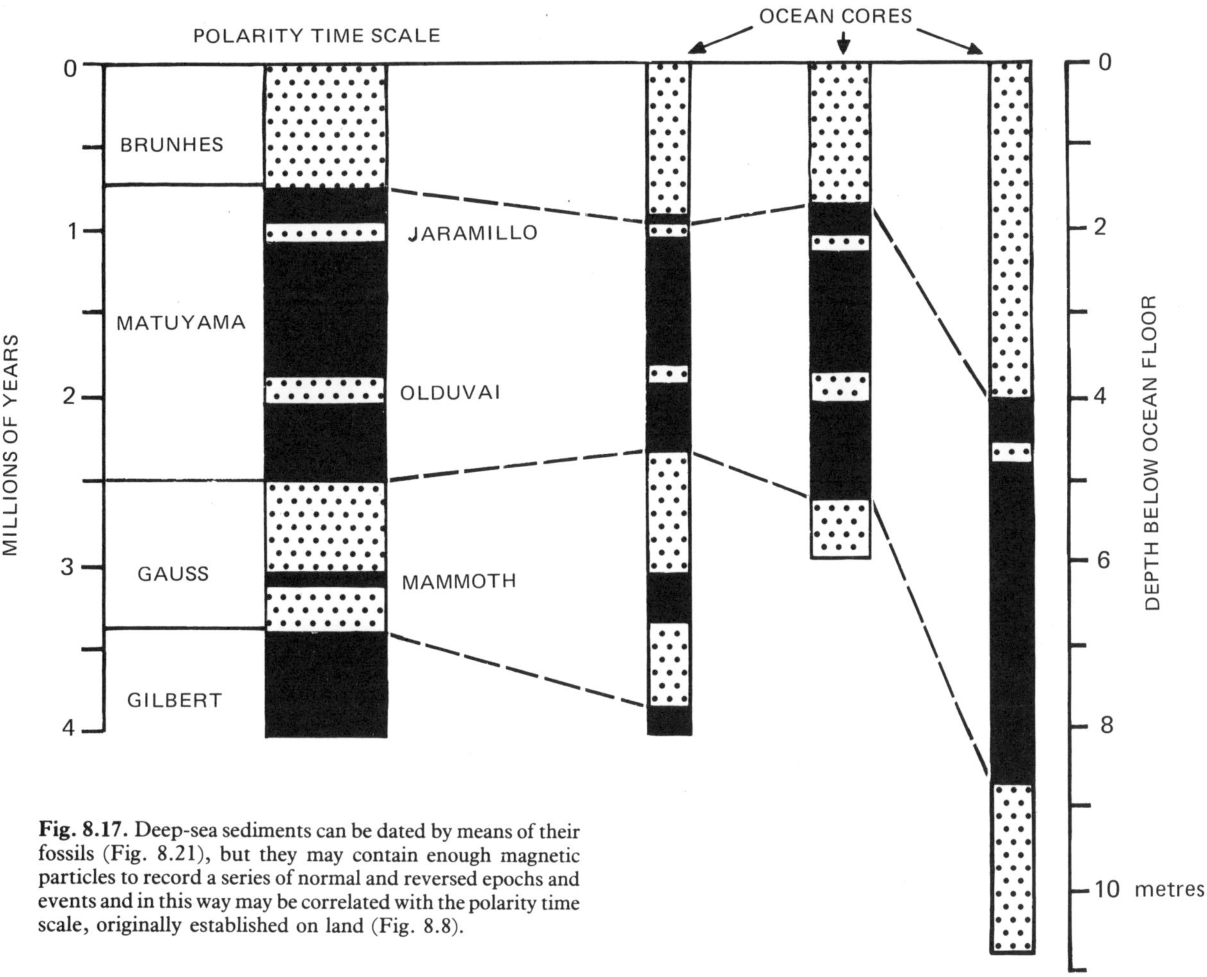


**Fig. 8.17.** Deep-sea sediments can be dated by means of their fossils (Fig. 8.21), but they may contain enough magnetic particles to record a series of normal and reversed epochs and events and in this way may be correlated with the polarity time scale, originally established on land (Fig. 8.8).

deck level. This process goes on, clamping and screwing numerous 29 m (95 ft) sections of pipe (Fig. 8.20), until the bit reaches the sea floor. In 3350 m (11,000 ft) of water it takes about four hours of clamping and screwing pipes to get the bit onto the sea floor. At this stage large motors are attached to the upper end of the "string" of pipe. The motors turn the entire length of pipe and the drill bit at its lower end that starts to chew into the rocks below.

As the drill begins turning, an enormous block and tackle inside the derrick slowly lowers the pipe as it bores down, but at the same time supports enough of the weight to keep the pipe from buckling. New pipe sections are added as required.

As the drilling bit penetrates the sediment on the sea floor, cores are taken. Drilling is temporarily stopped while a thin wire running through the drill pipes pulls up a removable centre piece in the drill bit. A 63 mm (2½ in)-wide core tube, 10 metres in length, is then sent down in its place. The core tube rides on roller bearings and is therefore not forced to turn as the rotating bit bores on. Thus a central core of sediment or rock rises up into the core tube.

The filled tube is hauled up again and the 10 m (35 ft) of core, enclosed in a plastic inner tube, is carefully pushed out of the core tube.

The core is then cut into 1.5 m (5 ft) lengths, which are split lengthwise. Half the split sample is sealed and placed in refrigerated containers

**Fig. 8.19.** View of the *Glomar Challenger*, looking forward from one of the scientific laboratories. The legs of the drilling derrick frame the photo. On the derrick floor the drilling crew are running lengths of pipe into the hole. Beyond them is the automated drill pipe stacker.

**Fig. 8.20.** The *Glomar Challenger*'s drill crew have run a length of drill pipe down below the derrick floor. Its top is now being clamped and on the left the next length of pipe has been brought into position by the automatic pipe stacker, ready to be hoisted into position and screwed into place.

**Fig. 8.18.** The deep-sea drilling ship *Glomar Challenger*, 122 m (400 ft) long and 10,500 tonnes displacement. The drilling derrick amidships towers 61 m (200 ft) above the waterline. The entire forward half of the ship is occupied by an automatic pipe stacker, that stores more than 7 km (4 miles) of drill pipe. Living quarters and six scientific laboratories occupy the upper decks aft of the drilling derrick.

**Fig. 8.21.** A selection of microscopic fossils washed out from a sediment sample taken from one of the *Glomar Challenger*'s deep-sea cores. This particular core is from the Indian Ocean and is of Middle Eocene age (45 million years ago).

The rod-like structures are pieces of "spicules" that form a loose supporting mesh inside a sponge. The delicate helmet-shaped shells with various protuberances are the siliceous skeletons of minute plants, called Radiolaria. The other more globular shells are the limey shells secreted by single-celled animals called Foraminifera, marine relatives of the *Amoeba*, a favourite classroom biology subject.

The largest of the Foraminifera in this photo are a bit smaller than a pin-head.

The Foraminifera and Radiolaria are members of the plankton, floating in the sea, and when they die their empty shells sink to the sea floor, to be incorporated in the sedimentary material.

**Fig. 8.22.** Core material being examined in one of the scientific laboratories on board *Glomar Challenger*.

**Fig. 8.23.** A re-entry funnel, with three sonar reflectors attached, being lowered into the sea beside *Glomar Challenger*.

below decks. At the end of the cruise the containers are shipped to special core laboratories, where they are placed in permanent storage, to provide study material for future generations of scientists.

Each core receives immediate study in the six scientific laboratories on board ship. One of the first tasks is to determine its age. Pinches of material are taken from the two ends of the core for microscopic examination. If there are tiny fossils in these top and bottom samples they will tell what span of time is represented by the core. A single core may span a million or more years and contain billions of tiny extinct plankton creatures (Fig. 8.21).

The cores are then studied to determine their composition, grain size and general mineral content (Fig. 8.22). The layering structure is described and colour changes documented (by colour photographs, as colours will change in storage). Such studies continue on shore, after the cruise is over, and include X-ray analysis, palaeomagnetic determination and various sophisticated chemical and physical tests.

At the conclusion of drilling, the hole is plugged with cement and the crew begins the laborious business of retrieving the drill pipe. Reversing the original operation, one section of pipe is hauled back after another, unscrewed and stacked on the pipe rack. As it is stowed away, each pipe is inspected and treated for corrosion, since a flaw in

**Fig. 8.24.** The *Glomar Challenger* uses "dynamic positioning" to hold itself on station above a drill hole. Sonar beacons placed on the sea floor are locked into shipboard computers that control the vessel's motors so that it holds its position without anchors.

When a drill bit is worn out it is possible to pull up the drill string, change the bit, and return to the same bore hole, through a re-entry funnel placed on the sea floor.

High resolution scanning sonar is used to locate the funnel and to guide the drill bit to it.

This artist's concept shows one of the sonar beacons used for "dynamic positioning" of the ship and a sonar scanner at the end of the drill string searching for the three sonar reflectors on the re-entry funnel. The relative positions of bit and funnel are shown at the surface on a Positive Position Indicator Scope. The drill bit can be guided towards the re-entry funnel by means of water jets.

# DYNAMIC POSITIONING AND RE-ENTRY

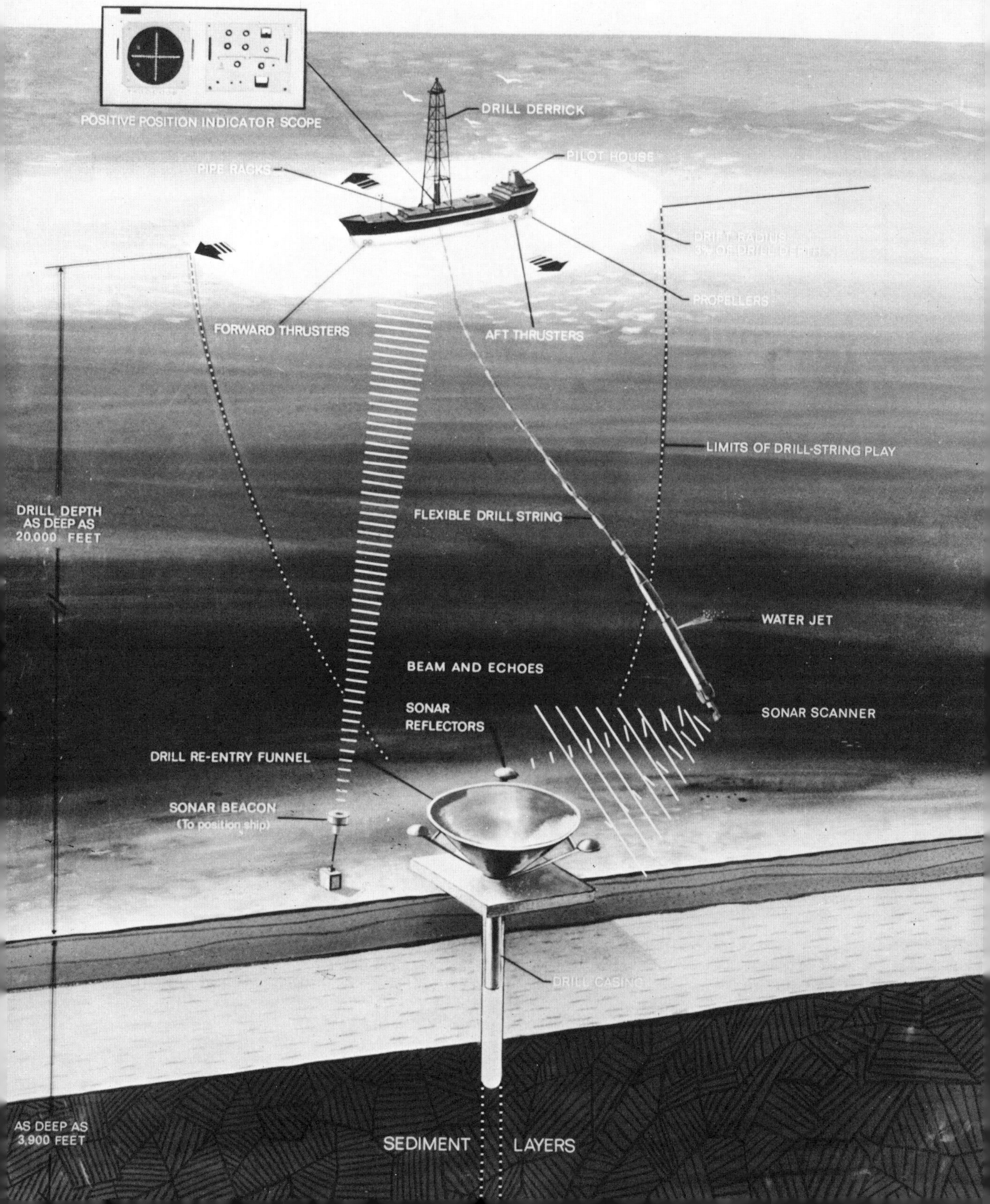

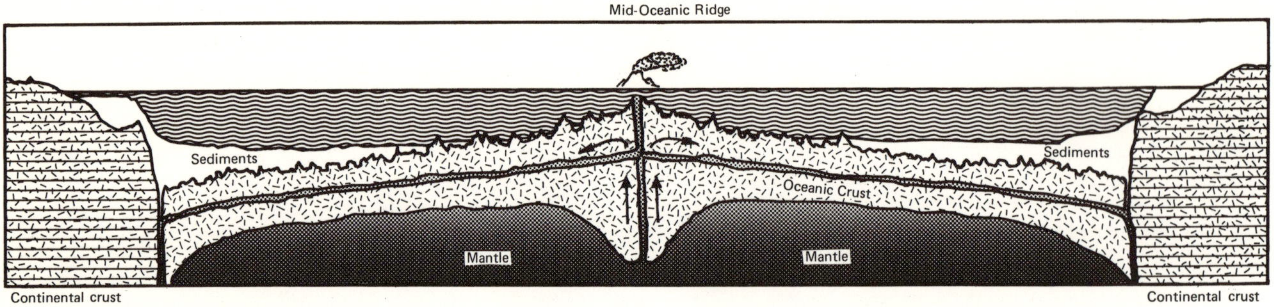


*Fig. 8.25*

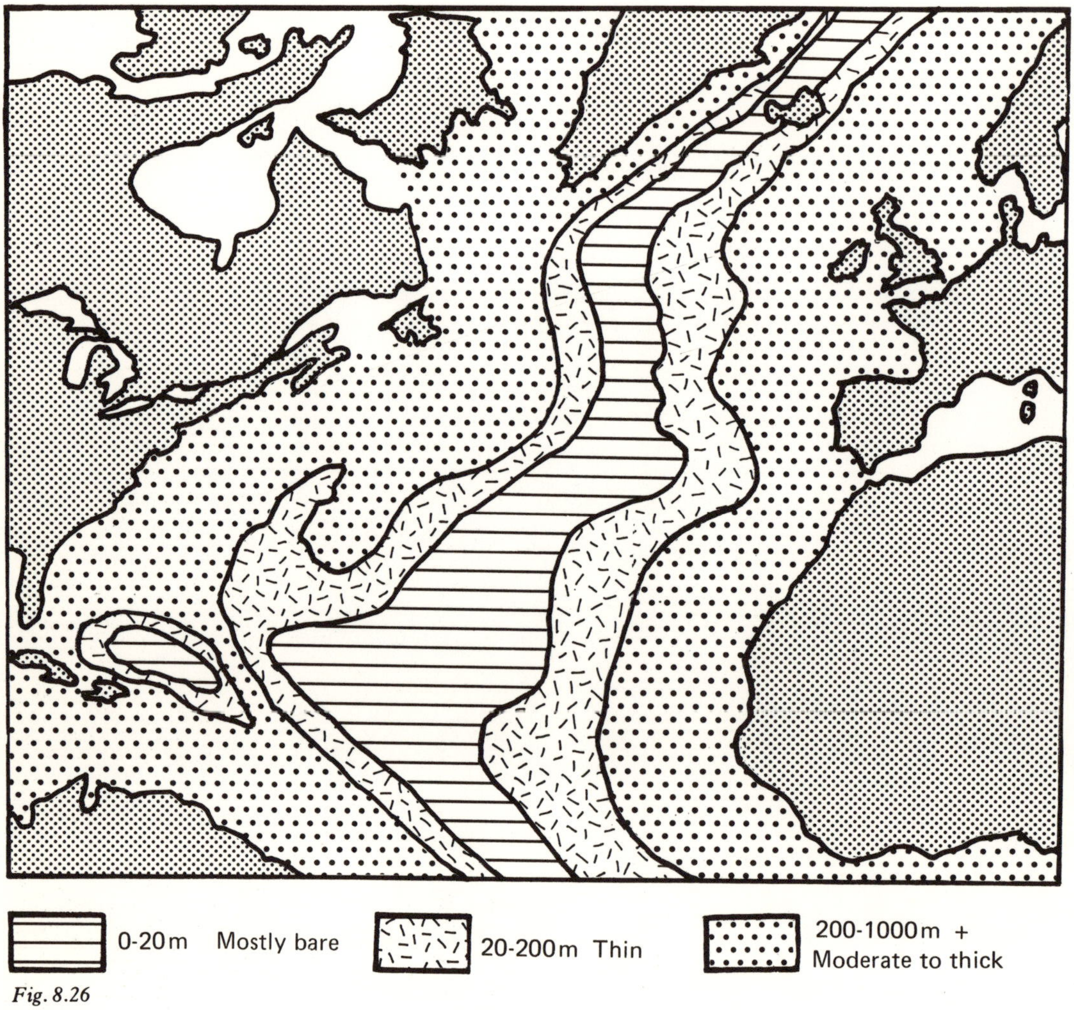


*Fig. 8.26*

**Fig. 8.26.** Sediment thicknesses in the North Atlantic Ocean as revealed by echo-sounding and deep-sea drilling. This simplified map shows how bands of increasing sediment thickness parallel the Mid-Atlantic Ridge, in proportion to the age of the sea floor. However, the age of the sea floor is not the only influence on the thickness of sediment. Other factors, such as proximity to land, depth of water, number of organisms living in the sea, latitude, currents etc. are also important.

**Fig. 8.25.** Study of deep-sea cores obtained by the *Glomar Challenger* has shown that close to the mid-oceanic ridges the "basement" rocks underlying the sea floor are only thinly veneered with sediment, and furthermore that this sediment is very young. However, as the distance increases away from the mid-oceanic ridge, so both the thickness of sediment and the age span represented by the various layers increase. This has been interpreted to mean that the sea floor close to the mid-oceanic ridges has been formed comparatively recently, and so has not had a chance to accumulate very much in the way of a sediment covering. On the other hand, the sea floor some distance away from the ridges has been formed for a much longer period of time and is therefore mantled by many layers of sediment. The age of the oldest layer of sediment (i.e. the bottom-most layer, immediately overlying the basement rocks) increases as the distance away from the mid-oceanic ridge increases. This cross-section is typical of most parts of the Mid-Atlantic Ridge, although the general principle applies to all ridges. Molten rock is believed to be continually moving up along the central portions of the ridge, cooling and being pushed slowly aside, resulting in progressively older ocean floor away from the ridge (and a thicker sedimentary cover—see Fig. 8.26).

one section might cause loss of a whole drill string in future drilling. Only the inexpensive sonar beacons are left on the sea floor.

Until 1970 *Glomar Challenger* could drill only until such a time as the drill bit wore out. There was no way of re-entering the drill-hole. Now however a large funnel (re-entry cone) is left over the hole once the drill pipe is removed. Equipped with sonar beacons, the funnel gives off signals that serve as guides for the drill's return to the hole (Figs 8.23, 8.24). The drill barrel itself is equipped with a remote-controlled jet or thruster that lets the ship's crew manoeuvre the drill into the funnel and into the hole. This system has vastly increased the ability to drill and recover materials from greater depths below the sea floor and to penetrate very hard rock.

Before the re-entry system was developed the *Challenger* could get only a few tens of metres into

**Fig. 8.27.** A sketch showing the approximate age of the floor of the Pacific Ocean, based on drilling results from the *Glomar Challenger*. The ages have been obtained from fossil-dating of the bottom-most sediment layer, immediately overlying the hard volcanic "basement" rocks.

The numerals indicate approximate ages in million of years. The zone of very young sea floor off the American coast (7 million years and younger) corresponds to the crest of the East Pacific Rise. The contact between young and old rocks east of the Philippines marks the zone of present-day destruction of crust along the Japan and Marianas trenches (see Figs 8.3, 9.2).

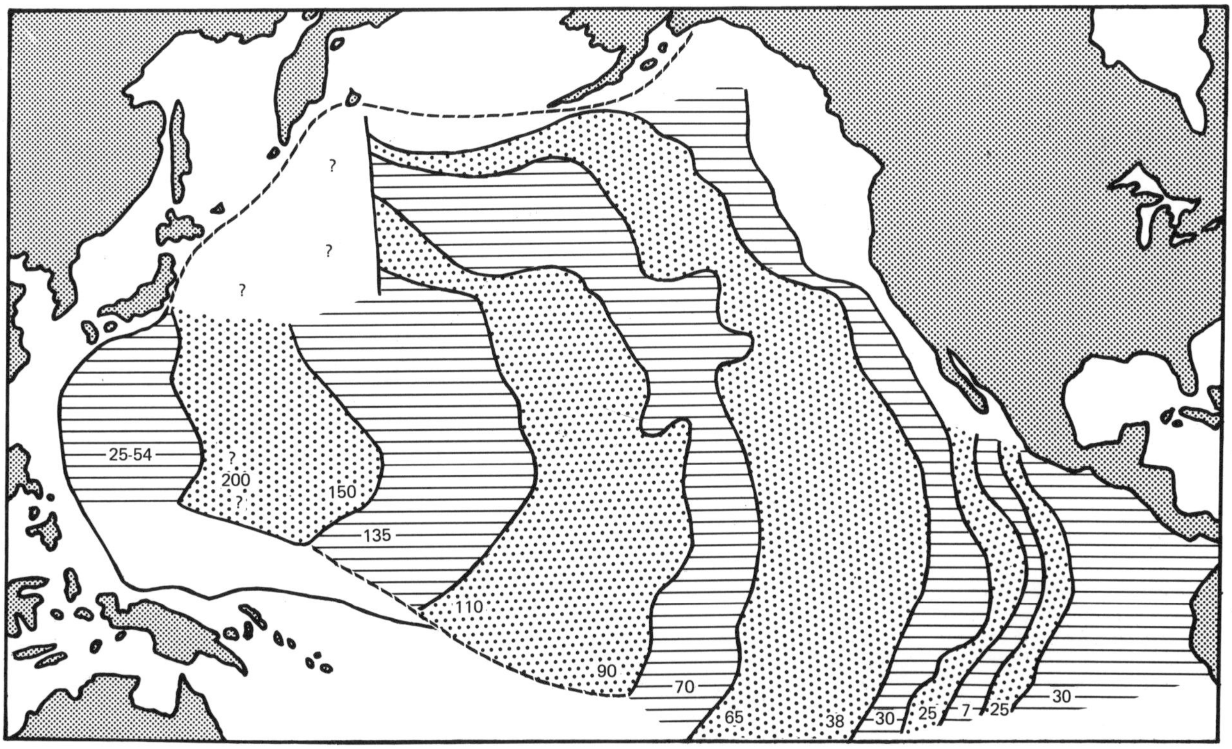

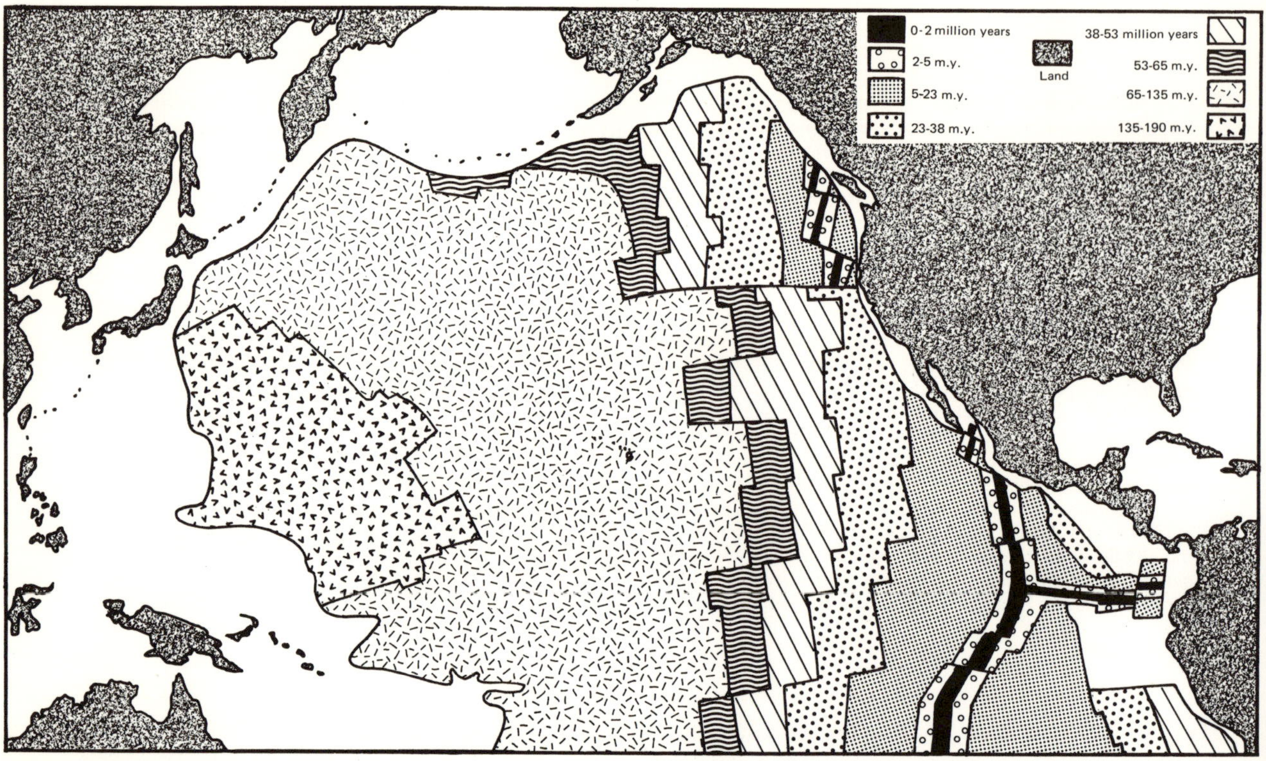

0-2 million years
2-5 m.y.
5-23 m.y.
23-38 m.y.
Land
38-53 million years
53-65 m.y.
65-135 m.y.
135-190 m.y.


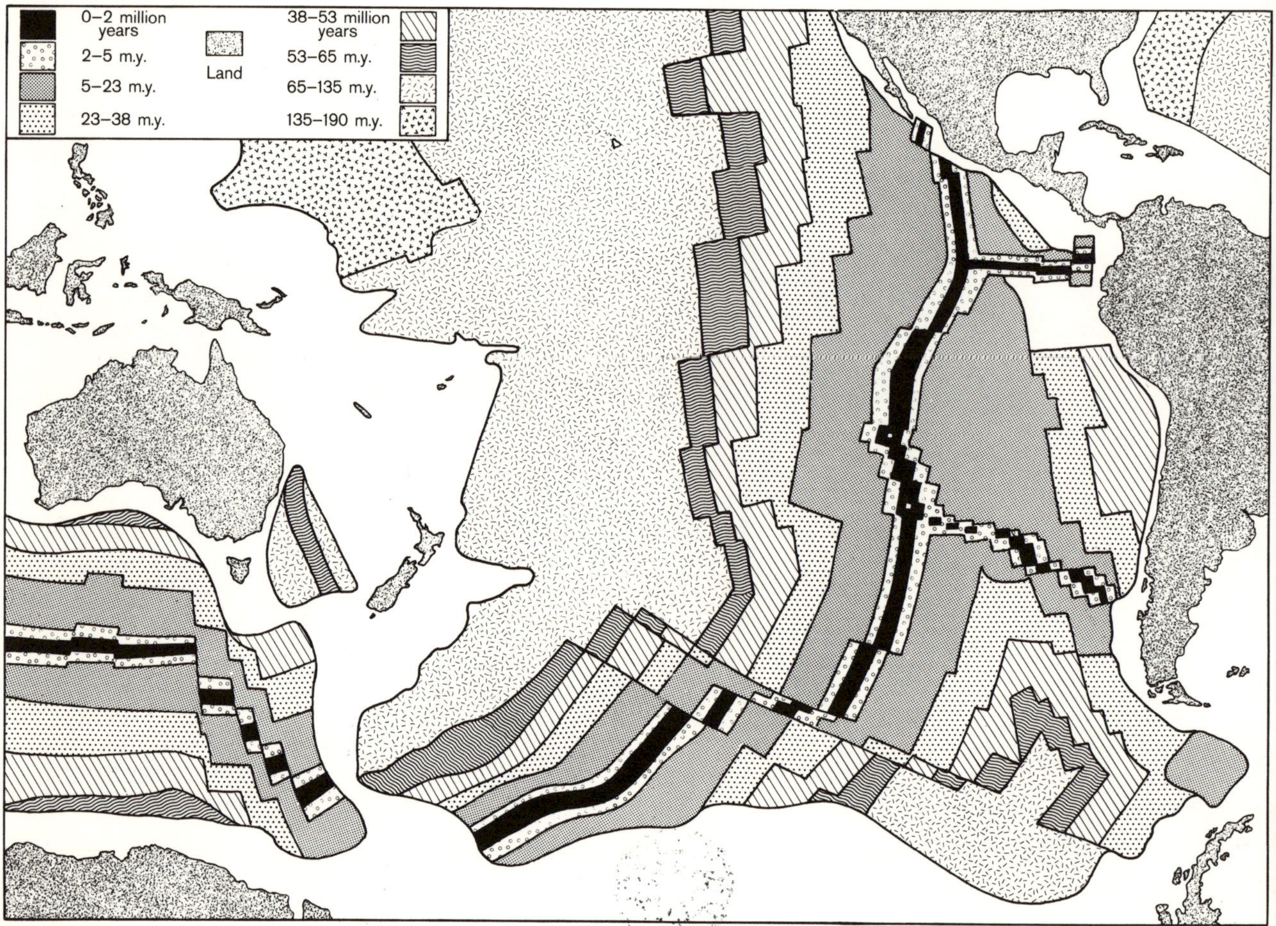

0–2 million years
2–5 m.y.
5–23 m.y.
23–38 m.y.
Land
38–53 million years
53–65 m.y.
65–135 m.y.
135–190 m.y.

**Fig. 8.28.** Individual magnetometer readings can be combined to give an overall picture of the magnetic striping of the sea floor and then when this is related to a magnetic time scale, appropriate ages can be assigned to various sectors of the stripe pattern. The ages thus assigned can then be "spot-checked" by the *Glomar Challenger*'s drill. Comparison with Fig. 8.27 shows that the agreement between the magnetic and sediment ages is very good.

the hard volcanic "basement" rocks before the drill bit became blunt and would not cut. Thus the whole emphasis was on obtaining sediment cores. Now however a new phase has begun, with the aim of drilling deeply into the volcanic basement, to find out more about the sea floor spreading process.

Over the period 1968-1975 the *Glomar Challenger* cruised the oceans of the world and drilled 573 holes in the deep sea floor. The total length of cores obtained in this period is more than 65 km (40 miles). A recent hole in the South Atlantic set a drilling record: a hole 1315 m (4315 ft) deep was drilled in a water depth of 4558 m (14,954 ft), giving a total length of drill string of 5873 m (19,269 ft or 3.65 miles). Currently preparations are underway to drill even deeper—to penetrate 2000-3000 m (7000-10,000 ft) into the sea floor.

Up to 1975 the U.S. Government had provided U.S.$68 million in funding for the project. However by no means has it been an "All-American Show". It has been a consistent policy to involve the world's scientists in all cruises and New Zealanders have participated in many, particularly those in the South-west Pacific and Southern Ocean. The results obtained so far fill some 40 fat quarto-size volumes, many approaching 1000 pages in length. The project has also given rise to countless individual scientific papers.

With the advent of the re-entry system, the Deep Sea Drilling Project has been re-structured to make it even more international in scope and financial contributions have been made by the United Kingdom, West Germany, Japan and France. To date the record is seven re-entries and three bit changes in the same hole and a continuous core of 600 m (2000 ft) of basalt obtained.

Perhaps the most important single scientific accomplishment of the Deep Sea Drilling Project has been virtual verification of the hypothesis of sea-floor spreading.

In terms of the hypothesis, the sediments overlying the lava "basement" should be thin at the mid-oceanic ridge and thicken progressively away from the ridge. While echo-sounding of the oceans had suggested that this was probably true, confirmation had to await the *Glomar Challenger*'s drill. Not only should the sediments thicken away from the ridge, but those immediately above the basement should be progressively older with distance from the ridge.

Thus many of the *Challenger*'s cruises were specifically designed to test this hypothesis. Very early in the project it was found in the Atlantic that the sedimentary cover near the Mid-Atlantic Ridge is indeed thin and comparatively young. It thickens and the bottom parts get progressively older towards the continents on either side of the Atlantic (Figs 8.25, 8.26). The most ancient sediments are found close to the edge of the continental slope, and are 150 million years old (i.e. of Jurassic age). Thus the Atlantic started to open at about this time.

In the Pacific the pattern was found to be more complex as the oceanic ridge is not centred, as in the Atlantic, but rather is in the eastern part of the Pacific. But here again it was found that the sediments are thinnest and youngest near the East Pacific Rise and get older and thicker to the west (Fig. 8.27). The oldest sediments found were slightly older than those of the Atlantic (about 200 million years old).

As ship-borne and air-borne magnetometers criss-crossed the oceans of the world, global patterns of sea-floor magnetic stripes were built up and with the fitting of time scales to these patterns, it was possible to assign ages to sectors of the sea floor. Again the *Glomar Challenger* with her drill was in a unique position to test the validity of such patterns. Fig. 8.28 is a map compiled from numerous magnetometer tracks and interpreted in terms of age, while Fig. 8.27 is a map based on the ages of the oldest sediments overlying the volcanic basement, as sampled by the *Challenger*. Although of course many, many hundreds of holes would be needed to test the magnetic ages

**Fig. 8.29.** A map of sea-floor ages based on magnetometer readings and comparable to that of Fig. 8.28, can also be compiled for the South Pacific. However detailed testing of the ages by drilling has not been carried out to any great extent.

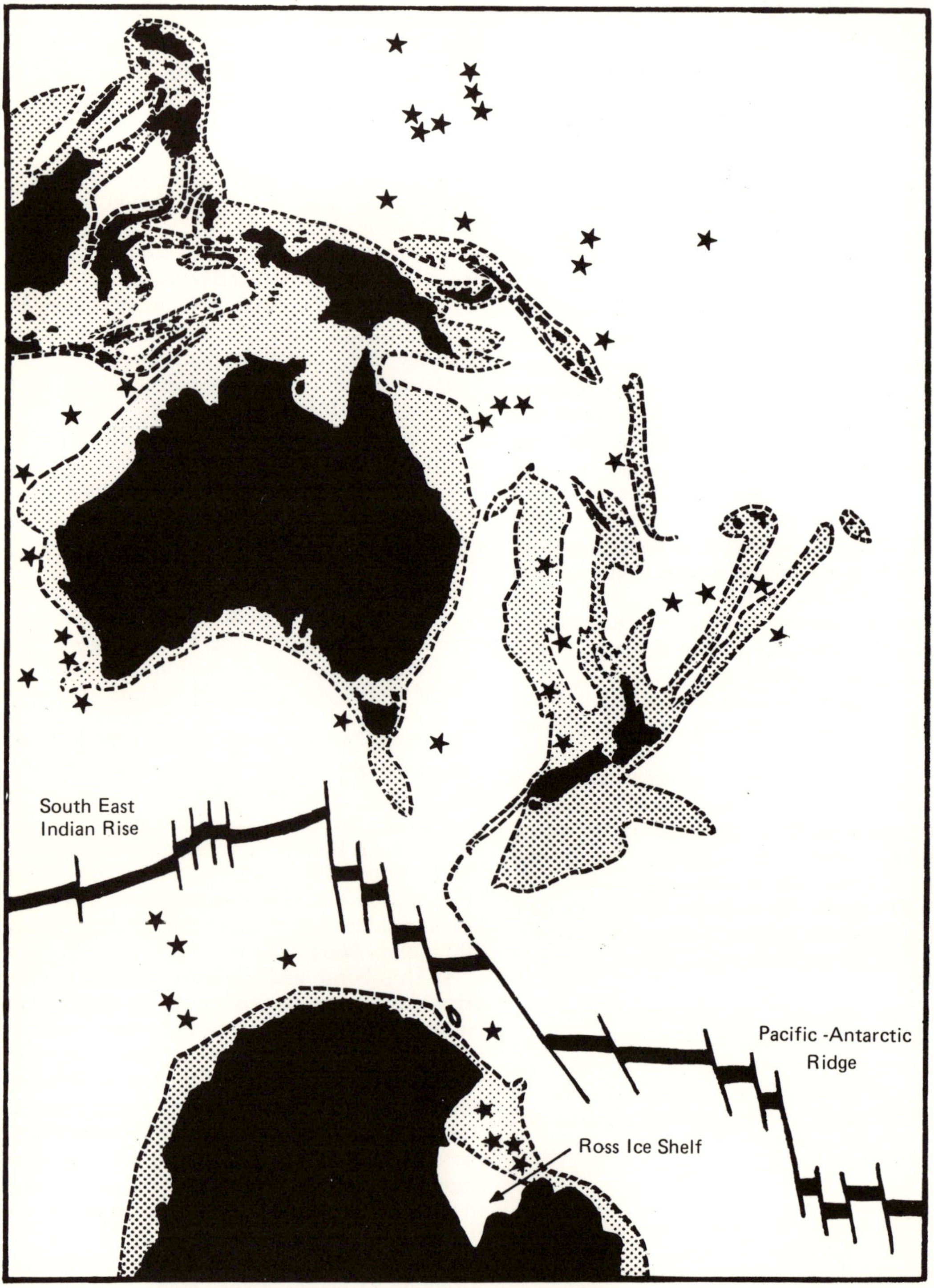
South East
Indian Rise
Pacific -Antarctic
Ridge
Ross Ice Shelf

**Fig. 8.30.** Holes drilled by the *Glomar Challenger* in the New Zealand-Australian-Antarctic region. The drill hole sites are indicated by the stars. Sea floor down to depths of 2000 m (6560 ft) is shown by the dotted pattern.

thoroughly, you can see that the two independent maps do agree pretty well.

The *Glomar Challenger* has also been very busy in New Zealand waters. During 1971-73 some 35 holes were drilled in the South-west Pacific around New Zealand. Then in 1972-73 the *Challenger* ventured into Antarctic waters and drilled five holes in the Ross Sea region south of New Zealand and another five holes south of Australia. These holes confirmed the interpretations gained from study of the sea-floor magnetic patterns that the Tasman Sea had opened up between Australia and New Zealand during the period 80-60 million years ago. On the other hand the ocean between Australia and Antarctica did not start to open until 55 million years ago (Fig. 8.29).

The *Glomar Challenger*'s programme in the South-west Pacific has been aided by work carried out over the past 15 years by New Zealand's Oceanographic Institute (D.S.I.R.) (Fig. 8.31).

Related to the story shown by the deep-sea drill holes is the discovery that the ages of oceanic islands tend to increase in direct proportion to their distance away from a mid-oceanic ridge system. This is beautifully illustrated by the various volcanic islands in the north and south Atlantic Ocean (Fig. 8.32). The age relationships can be explained simply by the idea that at various times volcanic islands were formed along the crests of mid-oceanic ridges (as Iceland, Tristan da Cunha and Bouvet Island are today) and as new sea floor is created, the piece of sea floor with the particular island on it is progressively moved further and further away from its original position on the mid-oceanic ridge.

## Project FAMOUS

Just as the *Glomar Challenger* has been used to test the hypothesis of sea-floor spreading, so a project was conceived to get a closer look at the processes operating under the sea on a mid-oceanic ridge. A lot has been learned from Iceland, sitting astride the northern continuation of the Mid-Atlantic Ridge (Figs 8.14, 8.15), but according to the sea-floor spreading hypothesis, things similar to those seen in Iceland should also be occurring along the undersea crest of the ridge.

In the North Atlantic the crest-line of the Mid-Atlantic Ridge is cleft by a great central rift valley (Fig. 8.2). Part of this rift, south-west of the Azores, was selected for detailed examination and for some three years twenty ships criss-crossed the area mapping, sampling and photographing. Even the *Glomar Challenger* made a contribution by drilling into the nearby sea-floor. Then an ambitious project, dubbed FAMOUS (French-American Mid-Ocean Undersea Study; although

**Fig. 8.31.** The New Zealand Oceanographic Institute's research ship *Tangaroa*.

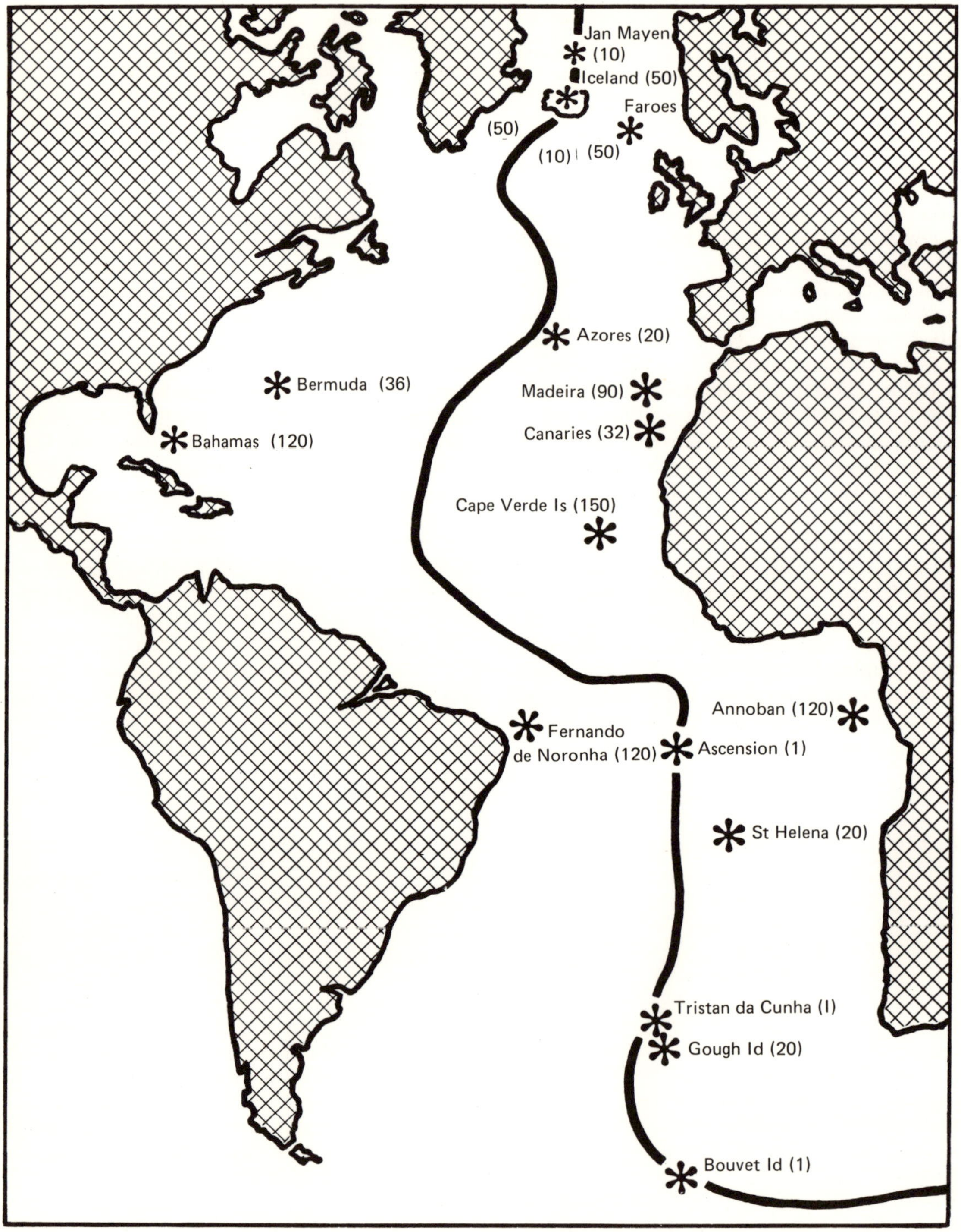


**Fig. 8.32.** The ages of islands in the Atlantic Ocean increase progressively away from the Mid-Atlantic Ridge (solid black line). The islands are indicated by stars and their ages in terms of millions of years are indicated in brackets.

scientific support also came from Canada and the United Kingdom), got under way. The aim of the project was to explore the rift valley with the help of the French bathyscape *Archimède* and diving saucer *Cyana* and the American submersible *Alvin* (Fig. 8.33). Altogether, in 1973 and 1974 a total of 51 dives was made into the rift valley, at water depths of more than 2750 m (9000 ft).

The rift valley was found to be some 30 km (20 miles) in width from edge to edge and 1500 m (5000 ft) or more in depth, forming a valley on the scale of the Grand Canyon. The main feature narrows down to an innermost valley which is only 1-3 km across (1000 yards to two miles). This inner valley is margined by steep walls on both sides, higher than 300 m (1000 ft) in places.

The floor of the inner valley was found to be strewn with haystack-like features 5 m (15 ft) high of fresh rubbly lava, with many individual lava flows resembling toothpaste squeezed out from a tube (Fig. 8.34).

Two prominent hills 400 and 210 m (1300 and 700 ft) in height, also composed of fresh lava and therefore obviously young volcanoes, also rose from the floor.

Rocks brought up from the hills and "haystacks" were found to be very young, with fresh mirror-like glassy surfaces resulting from quick chilling of the molten lava (at 1204°C; 2200°F) by cold sea water (4°C,: 39°F) under a pressure of about one tonne per square centimetre (about two tons per square inch).

Samples taken progressively further away from the central axis of the rift valley became more and more sea-worn, altered by the chemicals of the sea water, and coated with minerals. This alone indicated that the volcanic rocks were getting older and older to either side of the rift axis, and this was confirmed by dating samples at the base of the marginal cliffs as 100,000 years old. Such dating verifies that new sea floor is being created along the ridge axis, as envisaged by the sea-floor spreading hypothesis, and that both sides are moving apart by 2-3 cm (1 in) per year, as predicted by studies of the magnetic stripe patterns.

The underwater explorers also found numerous fissures and fractures, some 6-10 m (20-30 ft) wide, mute testimony to the tremendous forces pulling apart the Mid-Atlantic Ridge (Fig. 8.35).

Since FAMOUS, other dives have been made to 3660 m (12,000 ft) in the Cayman Trough, part of the submarine trench system in the Caribbean, to sample basalt layering exposed in the near-vertical cliffs of the trough's walls.

**Conveyor belts in the sea**

A vast range of sophisticated techniques, involving magnetic studies, deep-sea drilling and underwater exploration have wrested a great number of secrets from Davy Jones's Locker and it is likely that another rich harvest will accrue from the new drilling and exploration programmes of the *Glomar Challenger* and the American and French submersibles.

In summary, all these studies have shown that the sea floor to either side of the mid-oceanic ridges is very young and even at their margins the world's oceans are much younger than many of the continental rocks. Thus earth scientists have had to do an "about-face" and accept that the oceans are young features, with fairly short geological histories, whereas the continents are very ancient, with in most instances very long geological histories. To explain the "new oceanography", the theory of sea-floor spreading made its debut, envisaging new sea floor being continuously created at the mid-oceanic ridges. According to this theory fresh lava is continually welling up along huge cracks in the rift valleys at the crests of the mid-oceanic ridges. This lava congeals, but is soon split asunder by renewed movement. New lava again wells up, and the already congealed strips are pushed to one side. Over periods of millions of years, new sea floor is created and the ocean in which this is occurring expands, and the landmasses to either side of the ocean get further and further apart.

It is perhaps obvious however that expansion of the oceans cannot go on indefinitely—our earth would get bigger and bigger and bigger... (although some geologists do nonetheless believe that a certain amount of expansion has occurred).

Deep-sea drilling has shown that the oldest piece of ocean is no older than the Jurassic, some 150 million years ago, or slightly older (Fig. 8.27). Thus the ocean that was present before this time has had to be disposed of somehow. It is thought that the globe-encircling systems of deep submarine trenches are the sites of "sinks" or "subduction zones", where the sea floor is being pushed down below the earth's surface and into the mantle (Fig. 8.36). As the sea floor is pushed down and is over-ridden by the adjacent crustal

*Fig. 8.33*

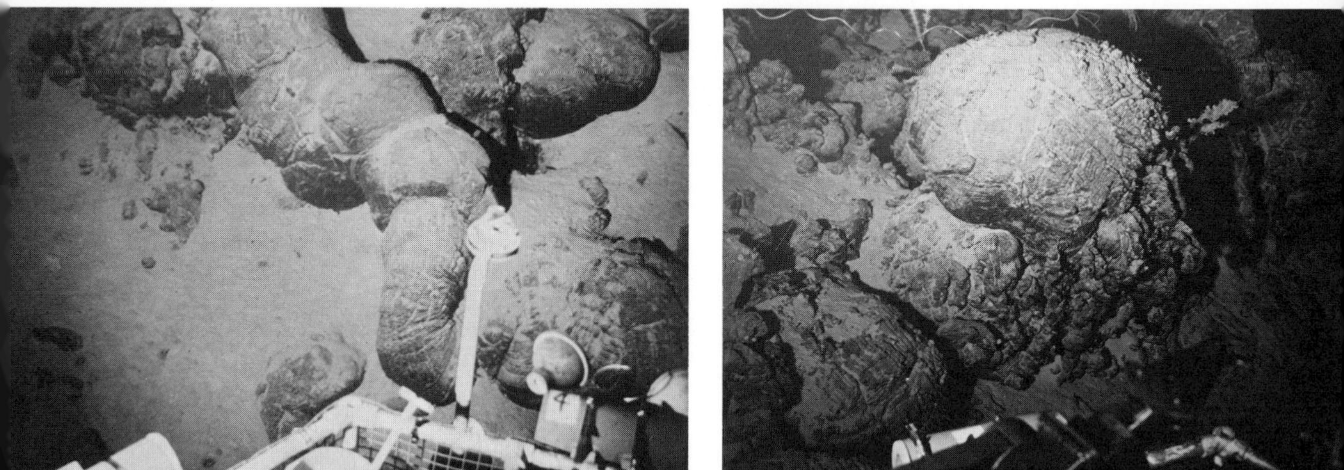

*Fig. 8.34*

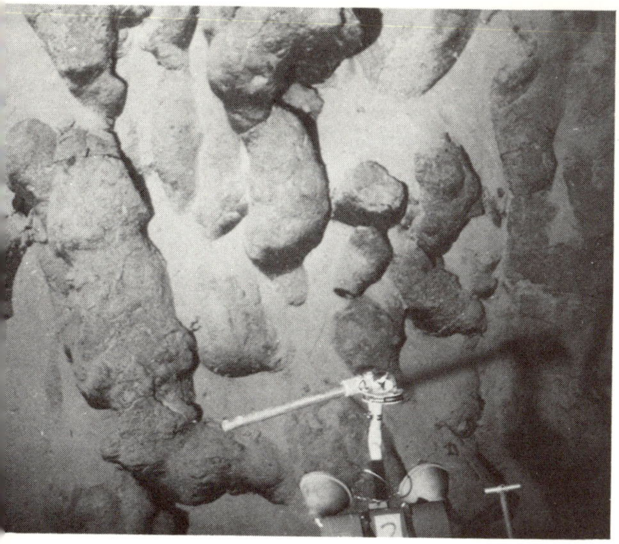

*Fig. 8.35*

**Fig. 8.33.** The deep-diving submarine *Alvin* of the Woods Hole Oceanographic Institution being prepared for a dive.

*Alvin*, along with two French submersibles, the bathyscape *Archimede* and the diving saucer *Cyana*, developed by Jacques Cousteau, took part in the boldest attempt yet to study sea-floor spreading—the French-American Mid-Ocean Undersea Study (FAMOUS).

The venture, reminiscent of Jules Verne, was to explore the central rift valley of the Mid-Atlantic Ridge, and observe at first hand the evidence of the activity that is tearing the earth's surface apart at the seams.

In 1974 the submersibles of the FAMOUS expedition explored the central rift valley of the Mid-Atlantic Ridge, in a region south-west of the Azores.

**Fig. 8.34.** Diving into the central part of the rift valley furrowing the crest of the Mid-Atlantic Ridge, where the earth is literally turning itself inside out, the submersibles of the FAMOUS expedition found formations of "toothpaste lava" that had welled up out of fresh cracks in the sea floor.

These long snake-like features, thicker than a person's thigh, took shape as molten lava at 1204°C (2200°F) surged up, only to be quickly congealed by cold sea water (4°C, 39°F) under intense pressure. The lava then pushed forth again and again, a metre or so at a time, rather like toothpaste being forced from a tube, so that growth rings, striations and pauses mark the surface of the lava tubes.

The left photo shows a toothpaste-like lava tube 0.3 m (1 ft) wide flowing down an undersea slope, as viewed from *Alvin* (lights, a compass and collecting basket are in the foreground).

The right photo shows pillow-like blobs of congealed lava, about 0.6 m (2 ft) high, lying on the surface of a haystack-like volcanic mound, formed when molten lava flowed from a fissure.

Both photos were taken in water depths of about 2600 m (8500 ft).

**Fig. 8.35.** The submersibles of the FAMOUS expedition found the rift valley of the Mid-Atlantic Ridge to be some 30 km (20 miles) across from ridge top to ridge top and some 1500 m (5000 ft) deep—forming a gash on the scale of the Grand Canyon. The rift narrows at its base to form a trough 1-3 km across (0.6-2.0 miles).

The floor of this innermost valley consists of a network of jagged fissures, running parallel to the axis of the rift, and increasing in width away from the centre.

The fissures are associated with numerous haystack-like volcanic mounds, formed where hot lava has welled up and congealed, and rough fields of blocky lava debris. Some areas of the sea floor adjacent to fissures were found to be heavily encrusted with manganese and iron ore—showing the manner in which mineral concentrations may form in rift zones.

The left photo shows part of a pile of lava blobs, produced by the breaking up of hot lava on coming into contact with cold sea water, lining a massive fissure in the sea floor.

The right photo shows part of another fissure; some of the fissures are 6-10 m (20-30 ft) wide and 100 m (320 ft) or more long.

Both photos were taken in water depths of about 2600 m (8500 ft).

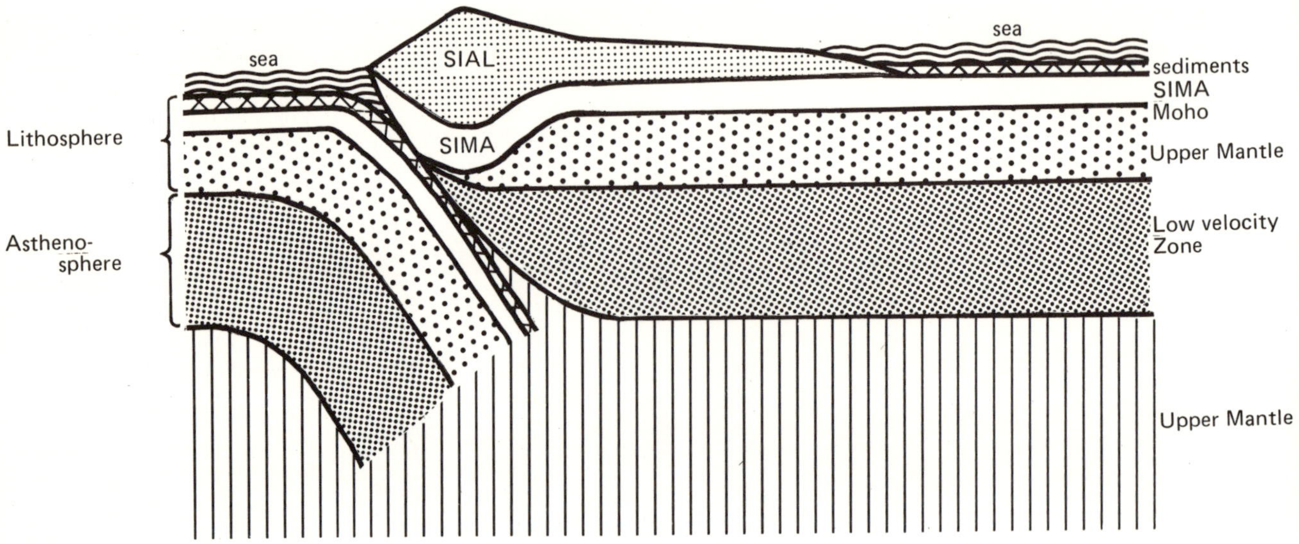


*Fig. 8.36*

*Fig. 8.37*

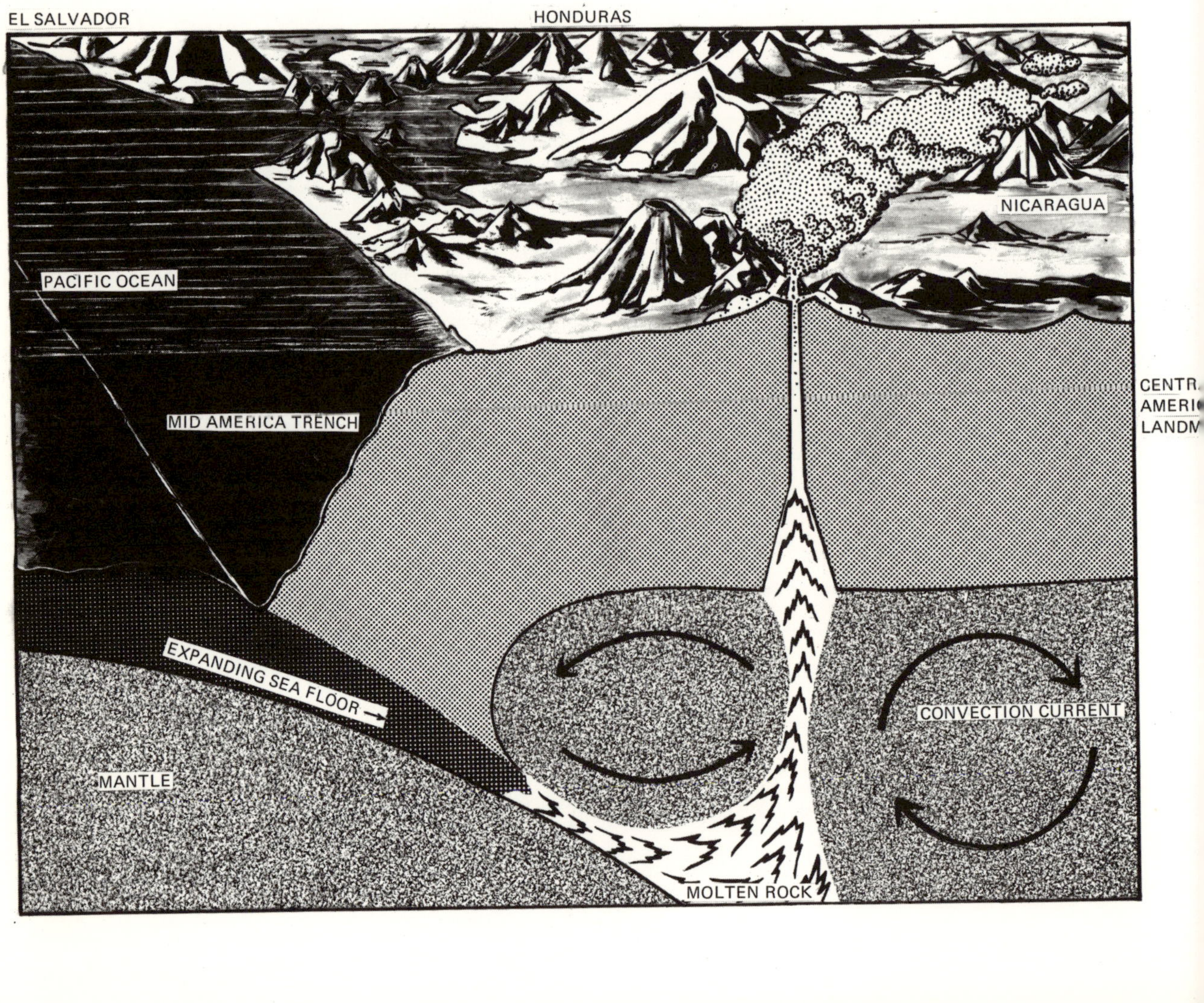

**Fig. 8.36.** A diagrammatic cross-section through the earth's crust and upper mantle in the vicinity of a subduction zone situated at the edge of a continent (e.g. Central and South America).

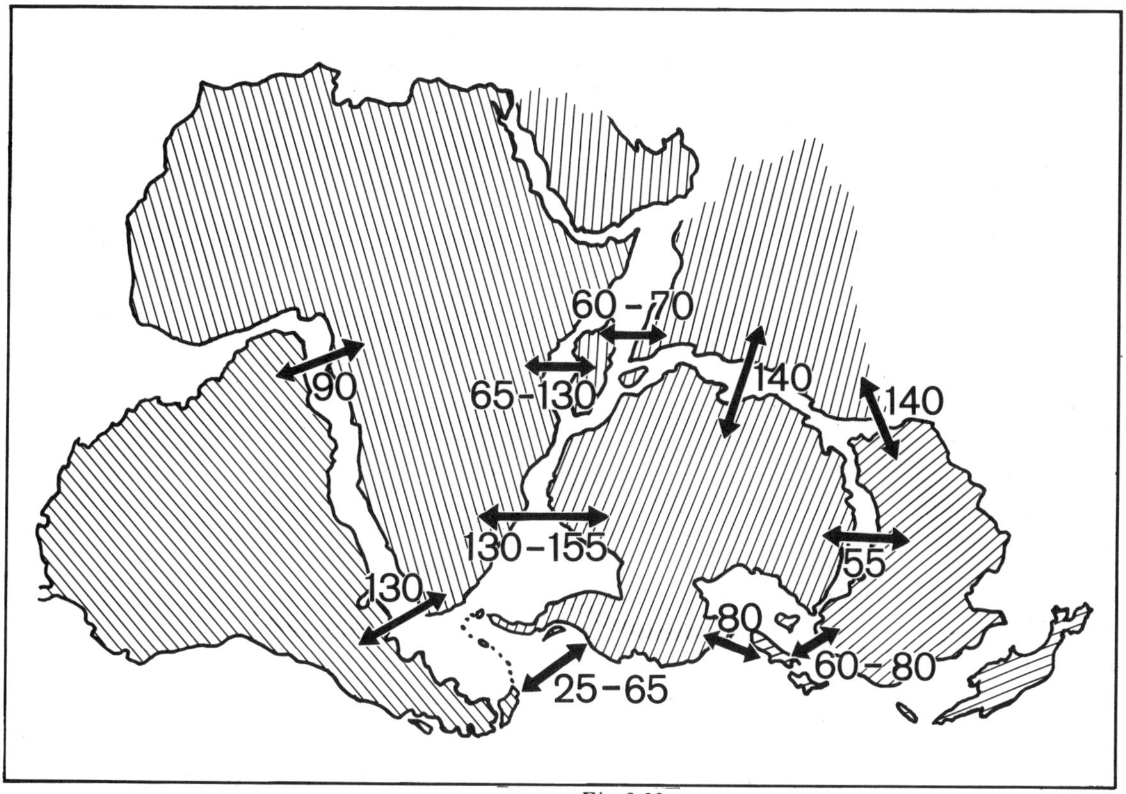


*Fig. 8.38*

**Fig. 8.38.** Sea-floor spreading information has allowed geologists to draw up a timetable for the separation of the world's continents. This diagram shows the times, in millions of years from the present day, at which the various continents separated from Gondwanaland.

slab, the tremendous friction thus developed results in many earthquake shocks. These earthquakes can be traced downwards into the earth as a sloping zone—the Benioff Zone—extending to depths of some 700km. As the descending slab of sea floor is pushed further and further down it enters into progressively hotter regions, until in time it begins to melt. Also dragged down with the sea-floor slab are the covering layers of sediment, much of it derived from adjacent continents and essentially sialic in composition. Thus as melting starts to take place, a molten rock (magma) is produced that contains both simatic (from the old sea floor) and sialic components, giving rise to mixed or "intermediate" rock types. These magmas find their way to the surface, where they give vent to volcanic activity of a particularly violent nature. Destructive eruptions occur at frequent intervals and massive cones are built up, that are in turn frequently torn asunder by awesome explosions. These volcanoes and the distinctive volcanic rocks they produce are called "andesite", a name coined after the Andes of South America, a mountain chain with many active, dormant and extinct volcanoes developed along the margins of an extensive submarine trench system (Fig. 8.37).

**Fig. 8.37.** An artist's impression of a real-life subduction zone. This is an attempt to show visually the processes that go on in a subduction zone, in this instance the one bordering Central America.

The general picture therefore is that new sea floor is being created along the mid-oceanic ridges and is being swallowed up along the oceanic

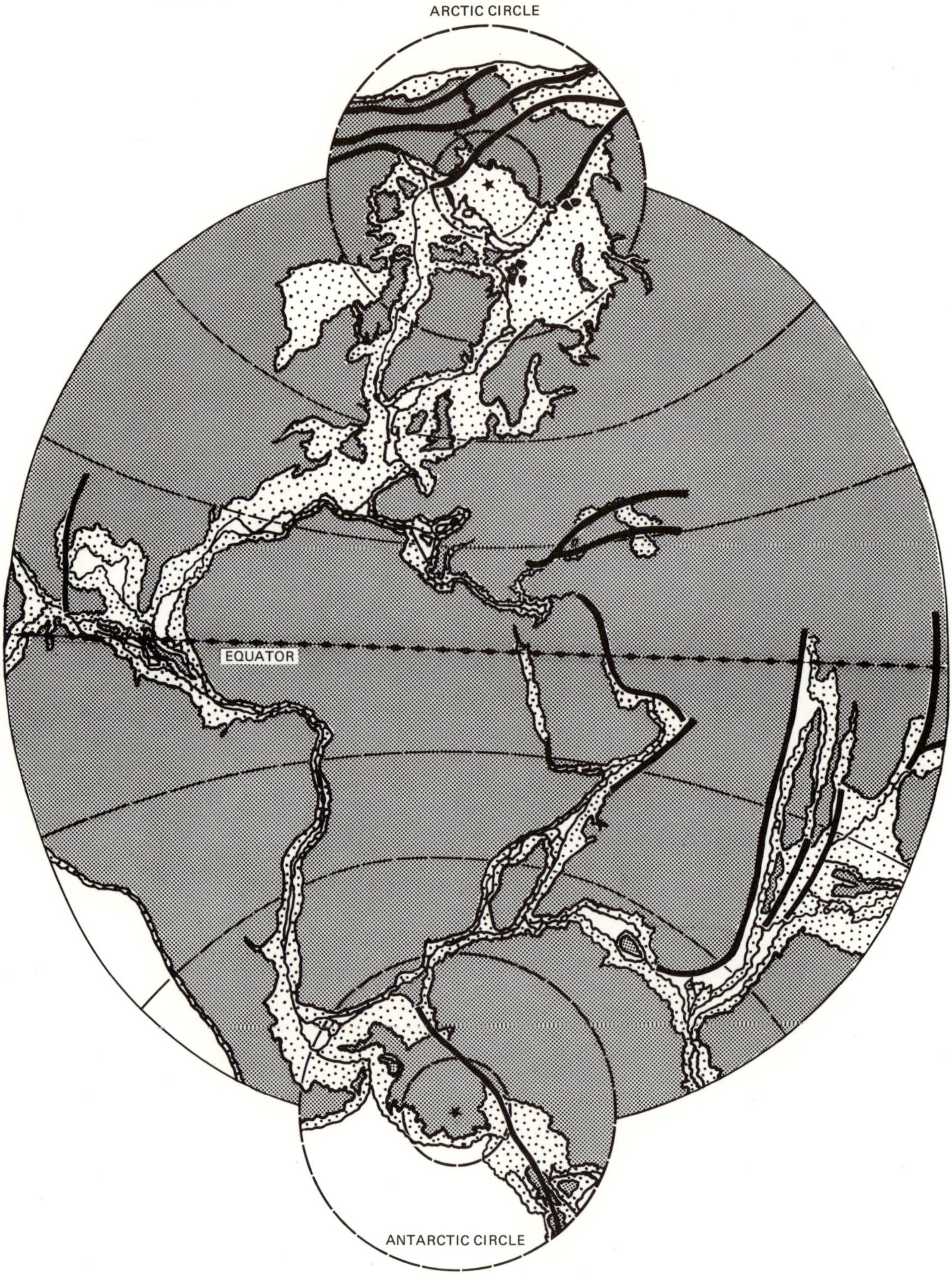


**Figs 8.39, 8.40.** A reconstruction of what the world may have looked like about 180 million years ago (Jurassic).

This reconstruction has assumed that at this time the earth's globe was only 80 percent of its present size. For ease of recognition present-day continents are shown by the fine dot pattern and continental shelves by the coarse dot. The distribution of land and sea was, however, very different from that of today and a large shallow sea (the Tethys) covered much of the Mediterranean countries and continued as seaways across Iran, Arabia, Northern India, South-east Asia, Australia, New Zealand, West Antarctica and South America.

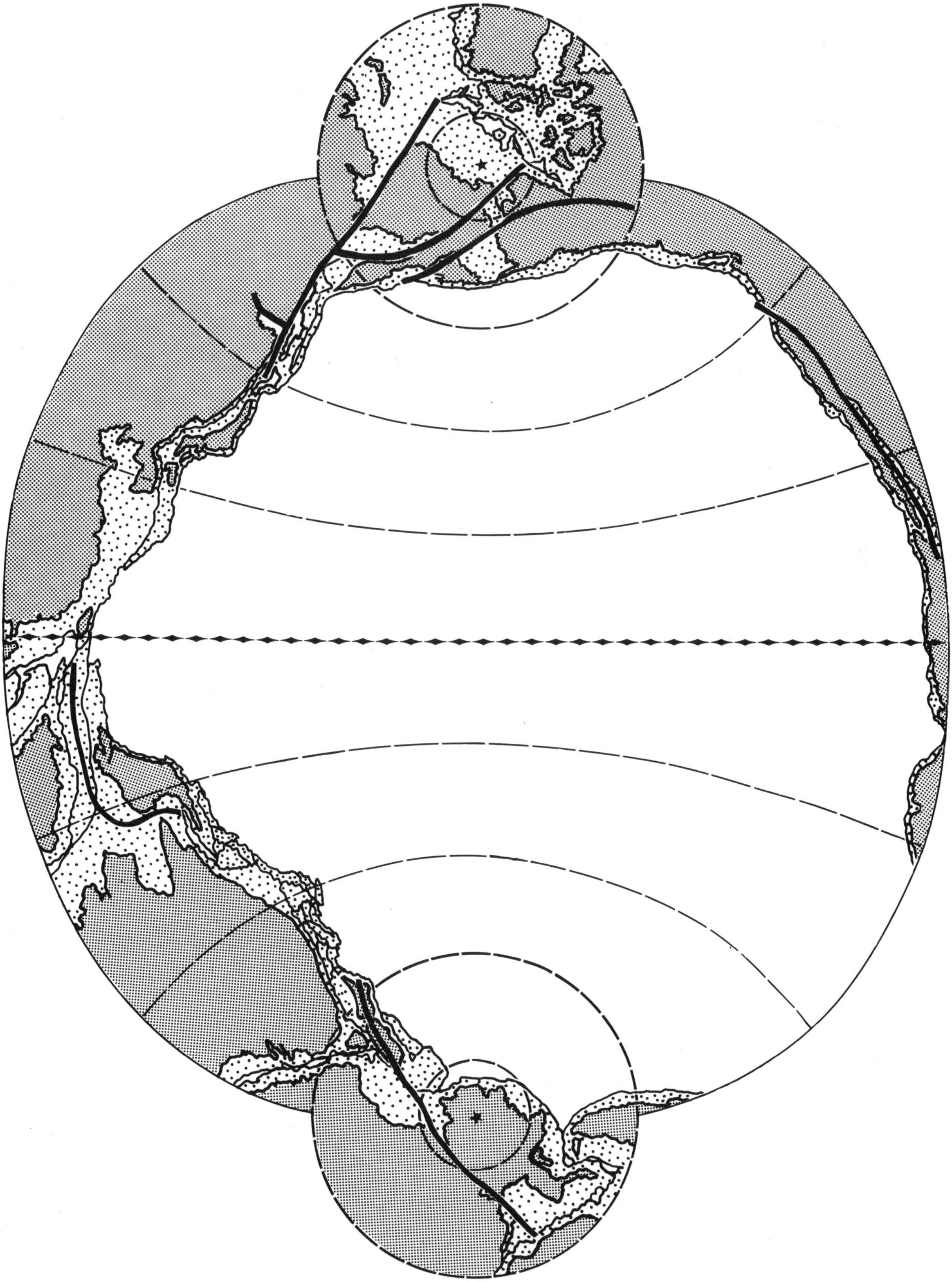

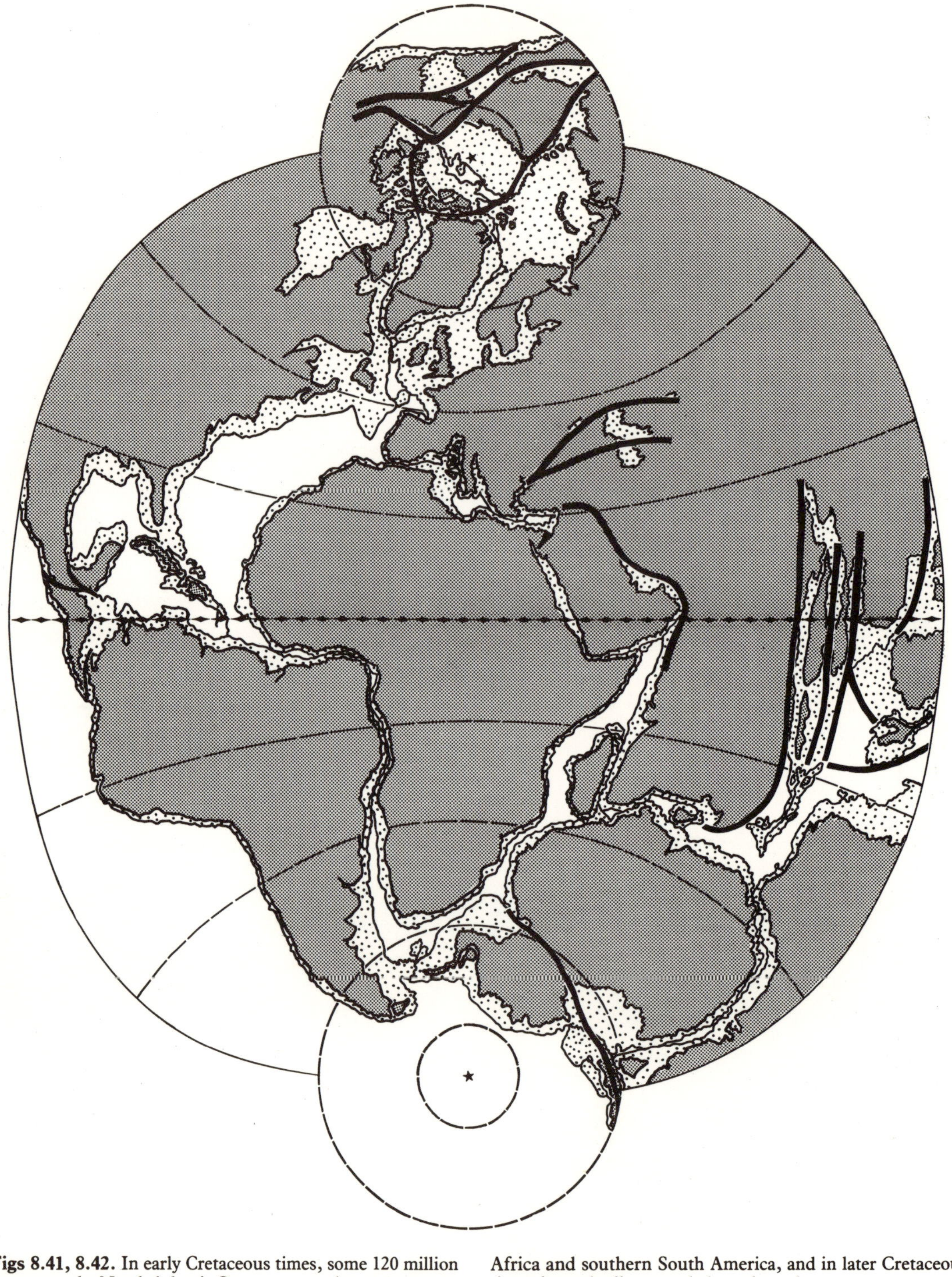

**Figs 8.41, 8.42.** In early Cretaceous times, some 120 million years ago, the North Atlantic Ocean was starting to open up, a split first occurring between Africa and North America, but gradually extending northwards towards Canada and Northern Europe.

A rather narrower split also appeared between southern Africa and southern South America, and in later Cretaceous times it gradually extended northwards.

An open seaway extended down the eastern side of Africa, heralding the opening of the Indian Ocean.

This reconstruction assumes that at this time the earth's globe was about 87 percent of its present size.

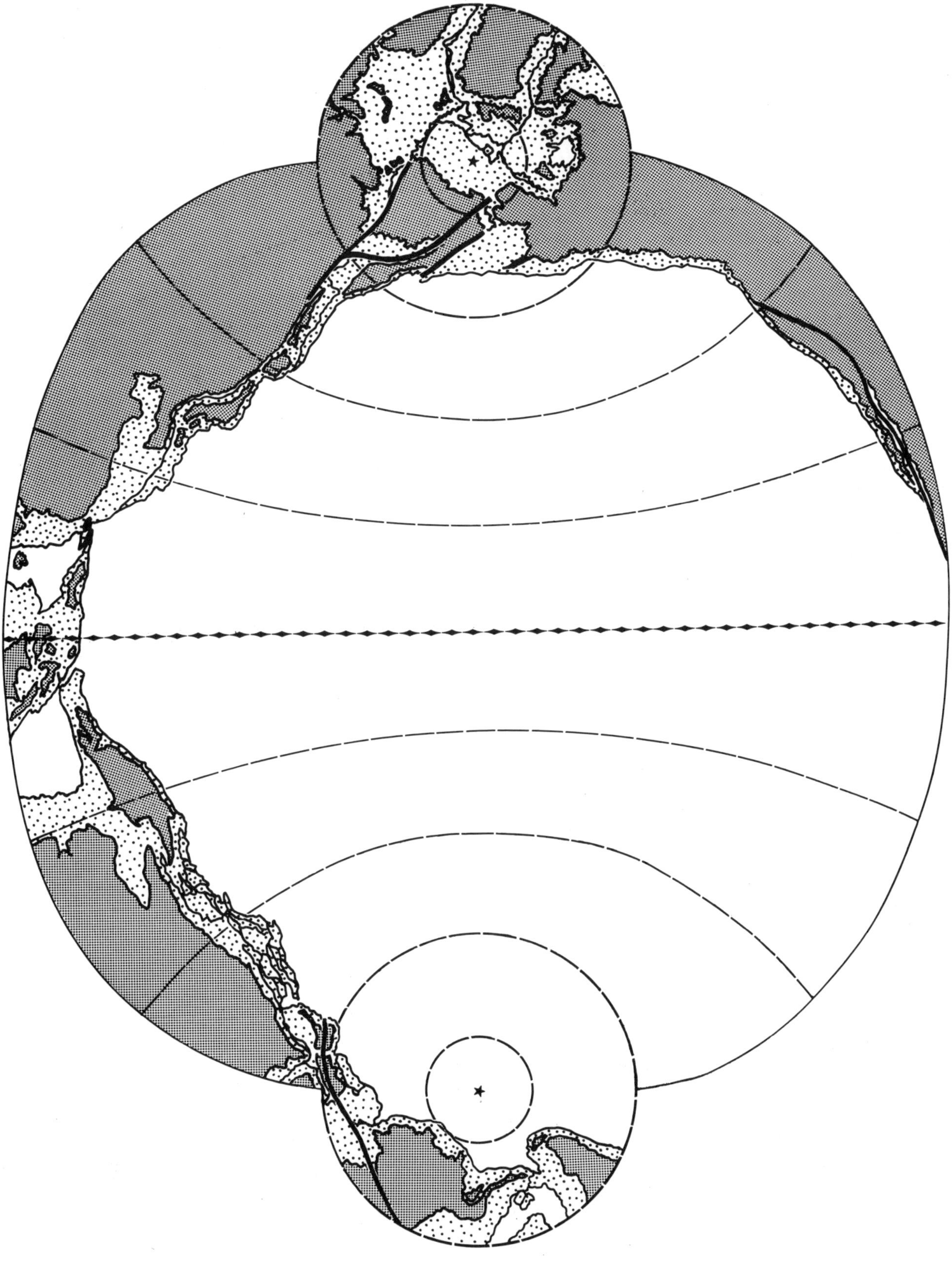

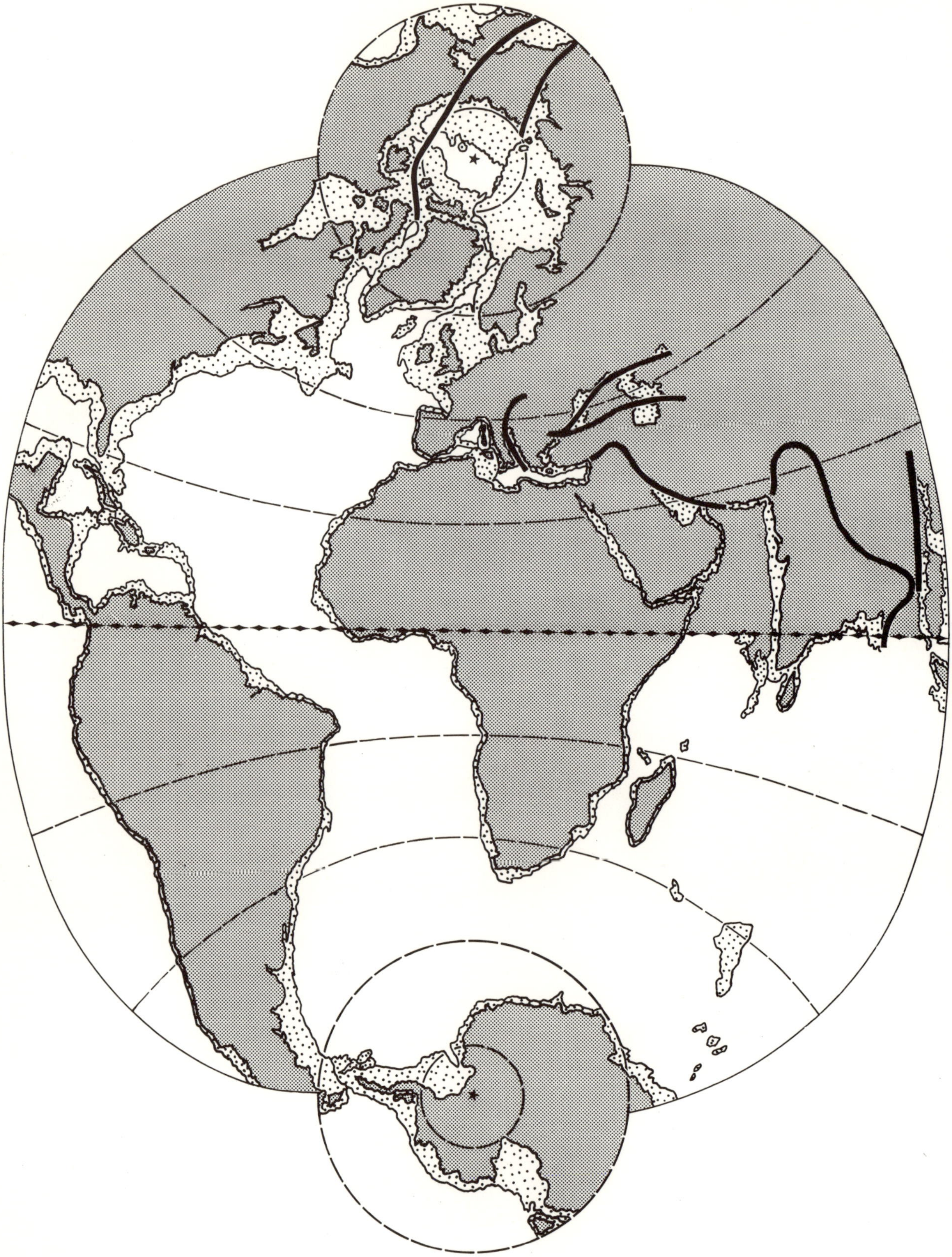

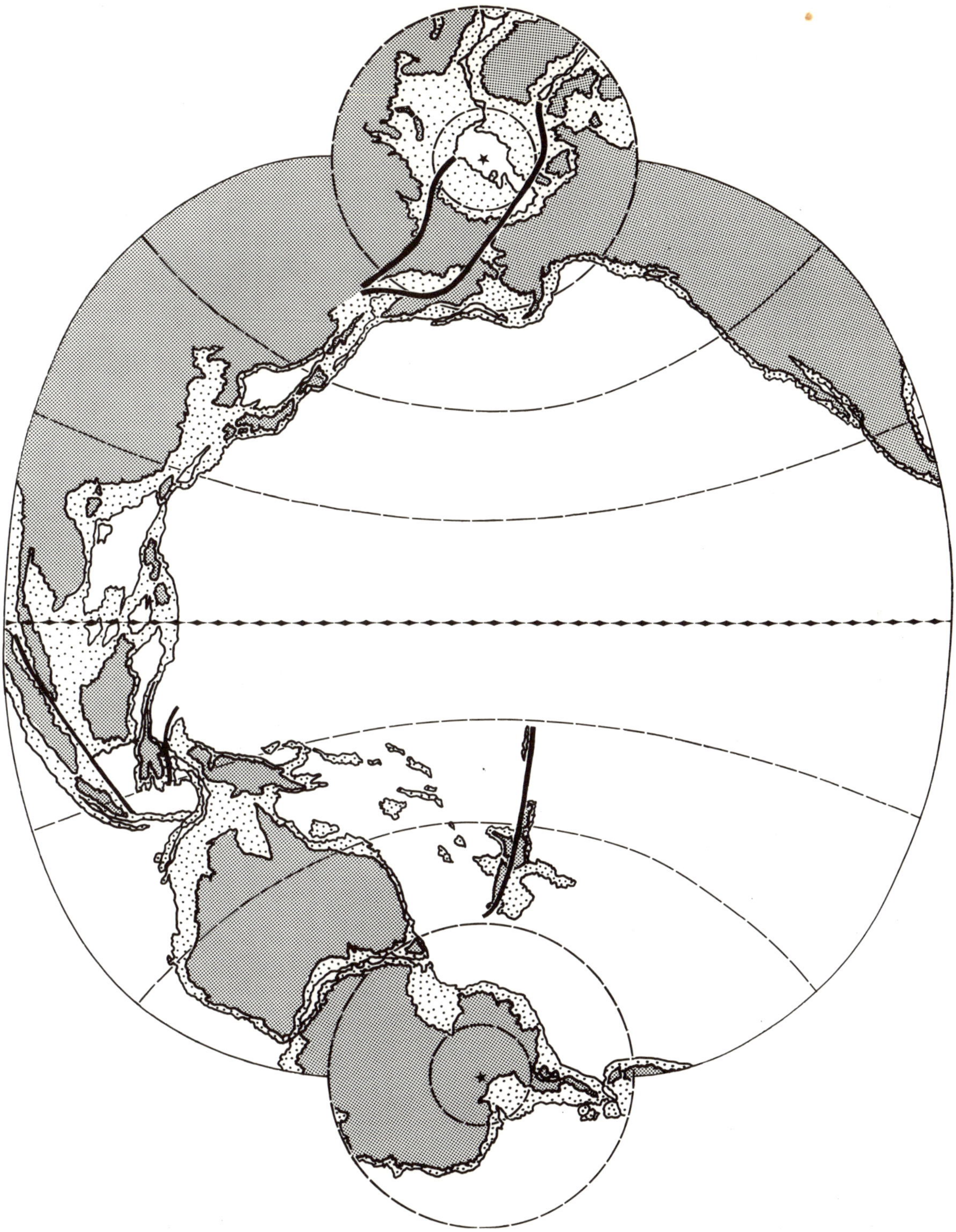

**Figs 8.43, 8.44.** By Paleocene times, about 60 million years ago, the North and South Atlantic oceans and the Indian Ocean were steadily opening, but the Arctic Circle countries were still largely in contact. The Arctic Ocean was starting to open. India has drifted northwards and was astride the equator. The Tasman Sea had started to open 80 million years ago and by the time represented by this diagram had opened up to its full extent.

Australia and Antarctica were still in contact, but began to separate 55 million years ago. The earth's globe was about 93 percent of its present size.

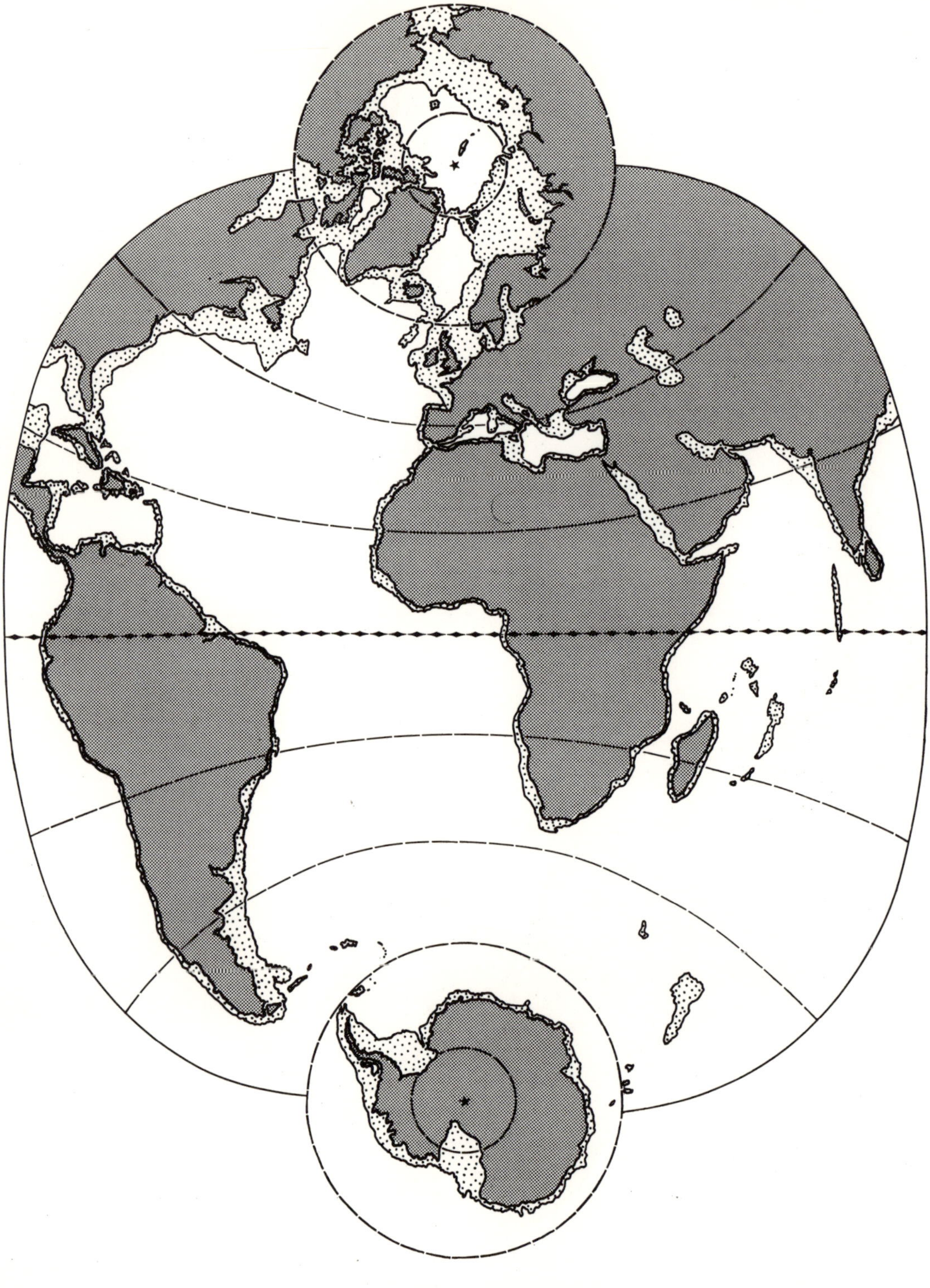

**Figs 8.45, 8.46.** Between Paleocene times and the present day, illustrated in these diagrams, all the oceans of the world opened out to their modern sizes. Africa, Arabia and India drifted northwards and wedged themselves against Europe and Eurasia, pushing up mountain chains like the Alps, Carpathians, Caucasus, Zagros and Himalayas. Australia and New Zealand have drifted northwards and Antarctica southwards.

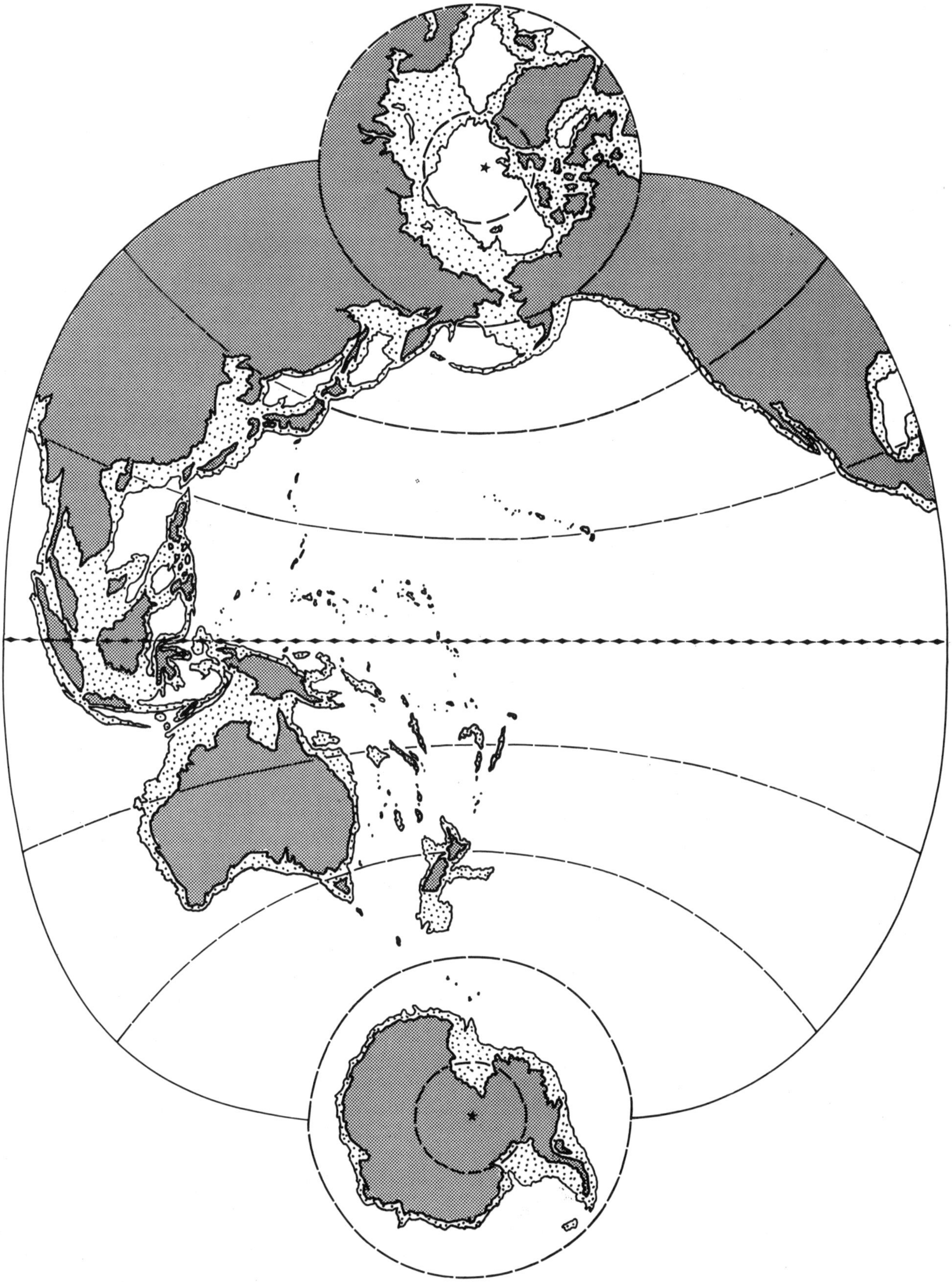

trenches. In between, the sea floor slides along rather like a conveyor belt, carrying islands and continents alike as "passengers". The moving sea floor also acts as a natural "tape recorder", recording the flips of magnetic polarity throughout geological time, and thus providing the means to date precisely and to identify various segments of the sea floor. It has been found, for example, by matching up the magnetic signatures of various slabs of the ocean floor, that huge areas of the Pacific Ocean have moved laterally thousands of kilometres, along very large fault lines (Figs 8.28, 8.29, 18.5). The total displacement along the Mendocino fracture zone, for example, is 1160 km (721 miles).

The magnetic signatures, plus deep-sea drilling, have also enabled large areas of the ocean to be zoned according to age, and rates of spreading over various periods of geological time have been worked out (Figs 8.28, 8.29). In some areas spreading appears to have been continuous throughout at least the last 70 million years (i.e. the Tertiary part of geological time), in other areas spreading has been episodic and in still others, largely dormant for long periods.

The highest rates of sea-floor spreading seem to have occurred to either side of the East Pacific Rise. To judge from the magnetic patterns, new sea floor has been added to the Pacific at a rate of about 10 cm per year (about 4 in). The Atlantic on the other hand has been opening up at a rate of about 2 cm per year (about 1 in) in the north and 3 cm per year (about 1½ in) in the south. The Indian Ocean has also been opening at a rate of about 3 cm per year.

All this led to the construction of timetables for the separation of the various continents (Fig. 8.38). In many instances the times can be checked by referring to events on land (e.g. a change from a long history of terrestrial deposits to one of marine sediments, may herald the onset of a split).

Thus the Tasman Sea between Australia and New Zealand is thought to have opened up between 80 and 60 million years ago, with sea-floor spreading occurring at a rate of 6-9 cm per year (2½-3½ in). The South-west Pacific Ocean between the Campbell Plateau and the East Pacific Rise, on the other hand, started to open at the same time, but at a slower rate, 6 cm per year, and is presumably still opening today. The Southern Ocean between Australia and Antarctica has opened up over the past 55 million years at a rate of 4 cm per year (1½ in) (Figs 16.14-16.21).

Reconstructions of what the world may have looked like at various times in the geological past have also been attempted, again using sea-floor spreading information, but combining it with palaeomagnetic determinations (Figs 8.39-8.46). However, there is no agreement on whether the earth has kept its diameter constant, or has expanded with time; if such expansion has occurred, this will alter the positioning of many landmasses. Also, although there is a lot of information available from the Atlantic and some other areas in the Northern Hemisphere, many regions in the Southern Hemisphere, particularly those around Australia and New Zealand, are not so well endowed and as a consequence there is still some room for argument regarding continental positions.

The concept of sea-floor spreading dealt a crushing blow to the proponents of a stable fixed earth and carried the implication of crustal movements on a scale never before seriously contemplated, even by Alfred Wegener, who had limited movements to the continents, that were pictured as "drifting" like gigantic icebergs.

As time went on however, the straightforward conveyor-belt model of sea-floor spreading ran into a number of problems. One such dilemma was that posed by the way in which some continents, such as Africa and Antarctica, are encircled by oceanic ridges (Fig. 8.1). If new sea floor, generated at the oceanic ridges, is moving slowly to either side, carrying or pushing the continental blocks as "passengers", then where for example is Africa drifting? Is it drifting eastwards from the Atlantic ridge, or westwards from the Indian Ocean ridge (the Carlsberg Ridge)? Or is it being squeezed between two vice-like descending slabs of ocean floor—one derived from the Atlantic Ocean, the other from the Indian Ocean? Yet Africa is not rimmed by deep trenches, folded mountain ranges or other evidence of squeezing-together forces having been at work. Antarctica, like Africa, is also surrounded by oceanic ridges, but again it has no signs of crustal compression along its edges.

These and other problems prompted major revision of the concept of sea-floor spreading and in 1967-68 it was incorporated into a far more fundamental and all-embracing theory of the earth called "Plate Tectonics" or more grandly "The New Global Tectonics".

CHAPTER NINE

# MOVING PLATES

As oceanographers busied themselves around the globe and as more and more marine magnetic observations were made, it began to become very clear in the early 1960s that huge areas of most, if not all, of the world's oceans were striped by magnetic patterns. It was then gradually realised that the sea floor must be acting as a rigid slab or crustal "plate". This was because the sea floors were ornamented by magnetic pattern "birth marks" and furthermore the patterns seemed to survive intact and undistorted from the moment of their creation at the oceanic ridge, through to their being swallowed up by a submarine trench, at the other end of the sea-floor "conveyor belt". In other words, the plates are being created along one edge and are being crumpled or swallowed up along another edge. Thus all of the "action" as it were, is confined to the edges of the rigid plates. It is just as though the crustal plates are like huge ice floes, jostling one another in a polar sea with some edges of floes being added to by freshly frozen sea ice and other edges being crumpled or pushed under by contact with other floes.

New developments in earthquake research (seismology) helped bring the idea of crustal plates into finer focus. Over the years, as earthquake disasters have come and gone around the world, purse strings have been loosened and international co-operation strengthened by the desire to find out more about earthquakes. Therefore, for some time now, increasing amounts of money have been made available for the establishment of world-wide chains of recording stations, to pinpoint earthquake epicentres, from which vibrations radiate outwards through the earth. In recent years also, it has become important to be able to distinguish between natural shocks and man-made shocks coming from nuclear weapons. Thus government money has supported the construction of gigantic instruments called "seismic arrays", capable of eavesdropping on nuclear tests held in distant countries, but at the same time providing science with extremely valuable earthquake information.

The largest seismic array (in Montana, U.S.A.), built at a cost of 10 million dollars, uses 525 seismographs arranged in 21 star-shaped clusters and buried 60 m (200 ft) underground. All the seismographs are linked by underground cable to a computer complex. The arms of each cluster are 3.5 km (2¼ miles) long and the complete array extends over a circular area about 200 km (125 miles) in diameter, thus dwarfing even the most impressive radio telescope.

The seismic arrays have been sited in the middle of stable areas of very ancient continental rocks, such as Scandinavia, central Scotland, Alaska, central and northern Canada, central and northern U.S.A., central India, central Australia and central Brazil. These areas are known as "seismically quiet" areas, because no earth movements have occurred in these regions for many hundreds of millions of years. Thus the instruments, with their very high degree of sensitivity, can be "tuned" to record and pin-point vibrations coming from a great distance, without having to overcome "background noise" from nearby earth movements. The various vibrations that are received by the individual seismographs of the array are channelled into the computer complex. There the records from the individual instruments are compared and cross-checked with the records obtained by all the other seismographs of the array.

This "filtering" of the information removes much of the extraneous "background noise" and puts the event that is being recorded into finer focus—whether it is a natural earthquake or vibrations from a nuclear bomb test.

As well as the arrays, a world-wide seismological network has been gradually extended over the years and this too has facilitated the building-up of a very precise picture of world seismicity. The picture that emerged was full of

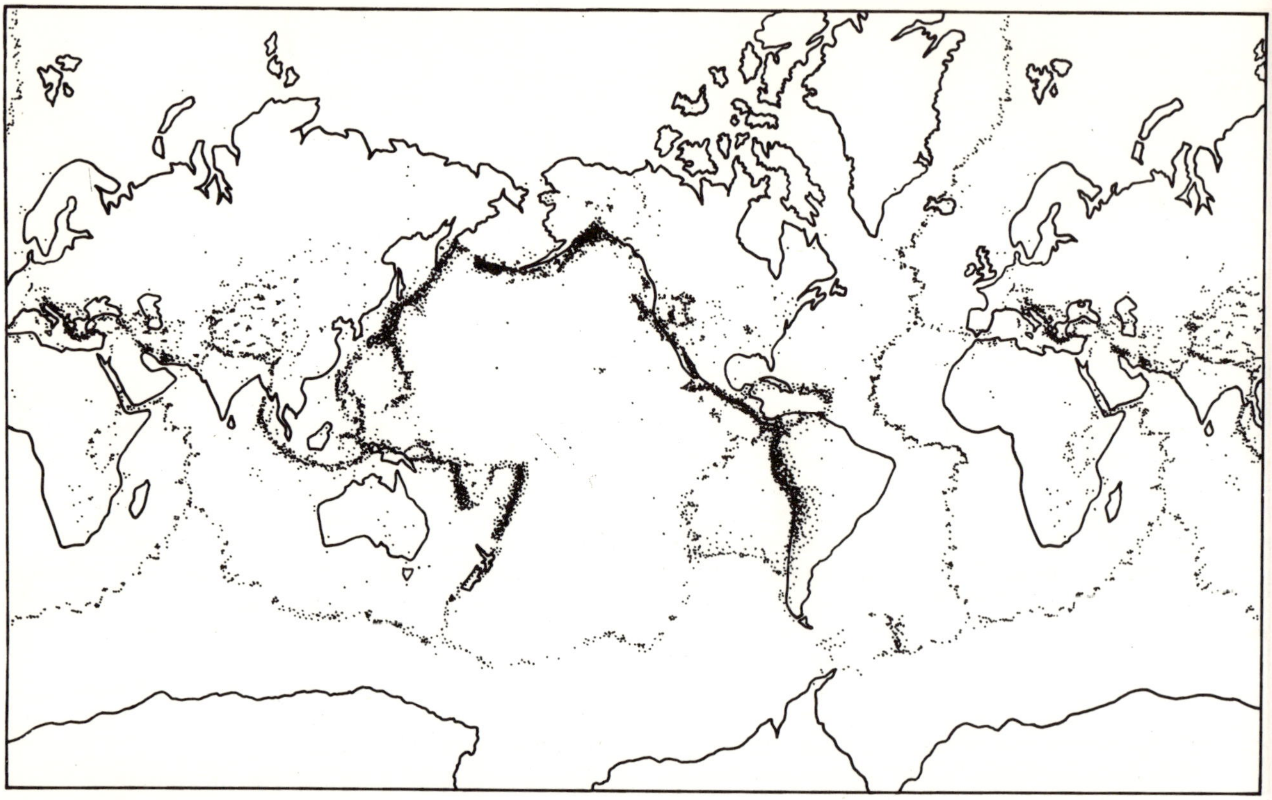

significance for global geology, as it showed with great clarity that the earthquake (seismic) belts of the world are extremely narrow (Fig. 9.1). The seismic belts run along the tops of the mid-oceanic ridges, around the margins of the Pacific (where they are often associated with submarine trenches; Fig. 9.2) and through the mountain chains (Himalayas, Caucasus, Carpathians, Alps, Pyrenees etc.) that have been formed in the last few million years on the site once occupied by the ancient Tethyan seaway (Fig. 9.17). These narrow seismic belts surround large areas that are free of all but a few sporadic and often small earthquakes.

It was realised that as earthquakes are generated by deep-seated crustal processes—involving mountain building, fault movements, volcanic activity etc.—all the global geological "action" was concentrated along such belts (Fig. 9.3). The large areas delineated by the seismic belts are termed "plates", and as they are essentially free of earthquakes, they are thought to be moving as rigid crustal slabs, like ice floes or paving stones. It is only along the edges of such plates that any earthquake-generating activity occurs, as the plates slowly slide and rotate, shearing, jostling and over-riding one another.

The moving plates are not just the continents, as Alfred Wegener postulated, nor individual ocean floors. Most include continental and oceanic crust coupled together to form pieces in a global jig-saw puzzle.

**Fig. 9.1.** World-wide distribution of earthquakes 1961-1967. Many of the earthquakes are distributed in quite narrow zones that outline the plate boundaries (cf. Fig. 9.4). The areas where plates are converging are marked by broad belts of concentrated earthquake activity that also coincide with the distribution of submarine trenches and volcanoes (Figs 9.2, 9.3).

The areas where plates are diverging (i.e. where sea floor spreading is taking place) are marked by a very narrow, sometimes step-like pattern of belts of only moderate earthquake activity, related to rifting and fracturing along the crests of oceanic ridges. Most areas away from plate boundaries have only a diffuse pattern of moderate earthquake activity.

The earth is thought to be made up of seven major and at least eight minor plates (Fig. 9.4) like great rafts jostling one another on an ocean. A new name has been coined for this global activity: "Plate Tectonics". "Tectonics", meaning the building of geological features, comes from the same Greek root as "architect"; "plate tectonics" is the creation of global features by the interaction of crustal plates.

So far we have traced the history of theories relating to our mobile earth from the days of Alfred

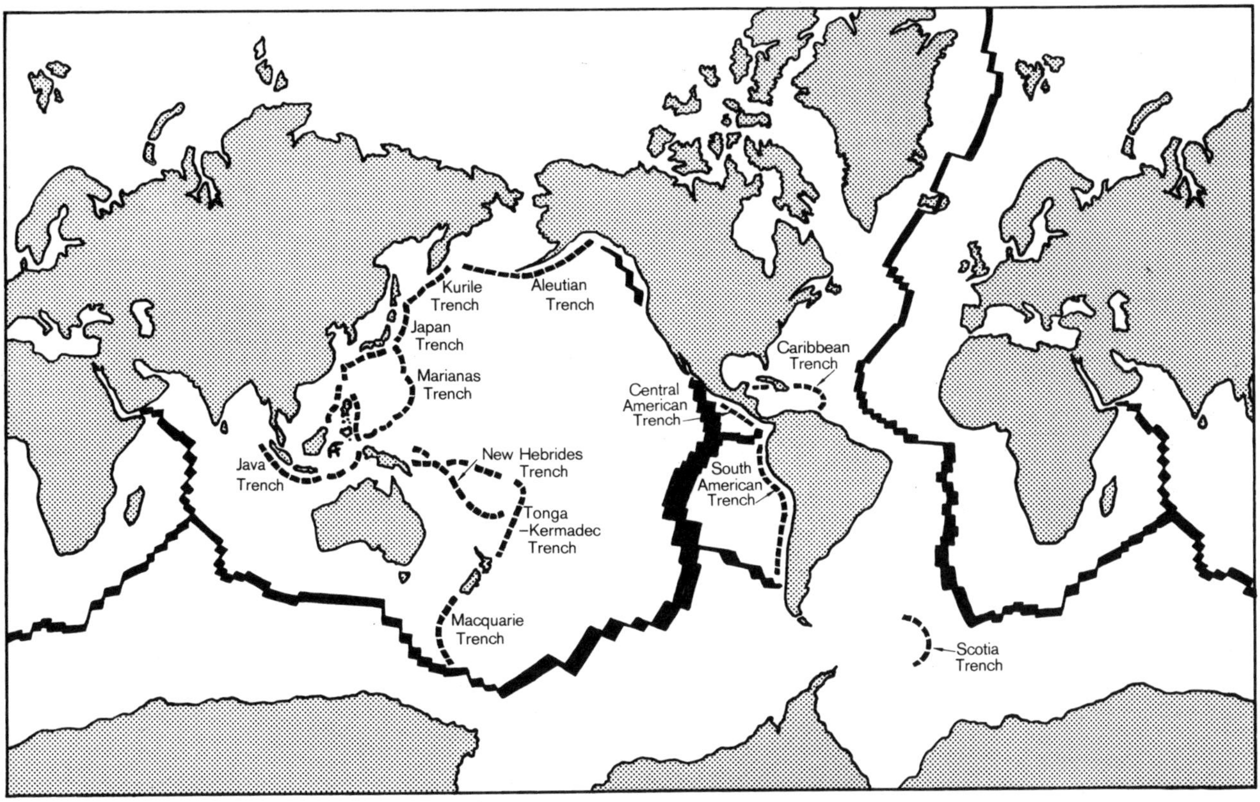


**Fig. 9.2.** Submarine trenches mark the regions of the world where oceanic crust is being pushed or drawn down into the mantle ("subduction zones"). This consumption of crust takes place where two plates are converging on one another (Fig. 9.4), the resulting under-thrusting dragging the ocean floor down to form a steep-sided trench (Figs 8.4, 8.5). A network of trenches edges the Western Pacific (cf. Fig. 8.3 for details), but trenches also flank the Pacific coasts of Central and South America and margin the Caribbean islands and islands of the Scotia Arc (South Georgia and South Sandwich Islands).

This map also shows the oceanic ridges as broad belts, disrupted at intervals by transverse fractures (cf. Fig. 9.15).

Wegener, when iceberg-like drifting of the continents was envisaged, through the early 1960s, when the importance of sea-floor spreading was recognised, to the present day when "plate tectonics" is all the rage. The world, as pictured in terms of the "new global tectonics", can be divided up into zones of divergence, where neighbouring plates are pulling apart, and zones of convergence, where plates are jostling and over-riding one another. We can also recognise zones where plates are gliding past one another along large faults (transform faults), and fossil subduction zones (or "collision zones") where continents, riding as passengers on plates, have been brought into collision.

## Zones of divergence

The oceanic ridges are zones of tension, or pulling-apart, where new lithosphere material (oceanic sima) is added to the plate margins. It has been estimated that oceanic ridges generate fresh basalt at a rate of about 50 billion tonnes per year. To give an idea of what this involves, we can mention the embryonic oceanic ridge which appeared some 80 million years ago between Australia and New Zealand, then almost nudging one another. Fresh sea floor was generated on either side of the ridge, and in the period of time between 80 and 60 million years ago the entire Tasman Sea was created and sea-floor spreading rafted New Zealand away from Australia.

As the crustal plates move apart, fissures repeatedly open up along the axes of the oceanic ridges and hot lava wells up and congeals in the fissures, gaining as it does so a magnetic "signature", recording the polarity of the earth's magnetic field at that time.

Each new fissure will presumably open along the hottest, weakest part (i.e. the centre) of the materials filling earlier fissures. This "halving" of the lava material filling the fissures leads to a roughly symmetrical pattern in the magnetic strips on the ridge flanks, although because of the great variety of possible plate movements and

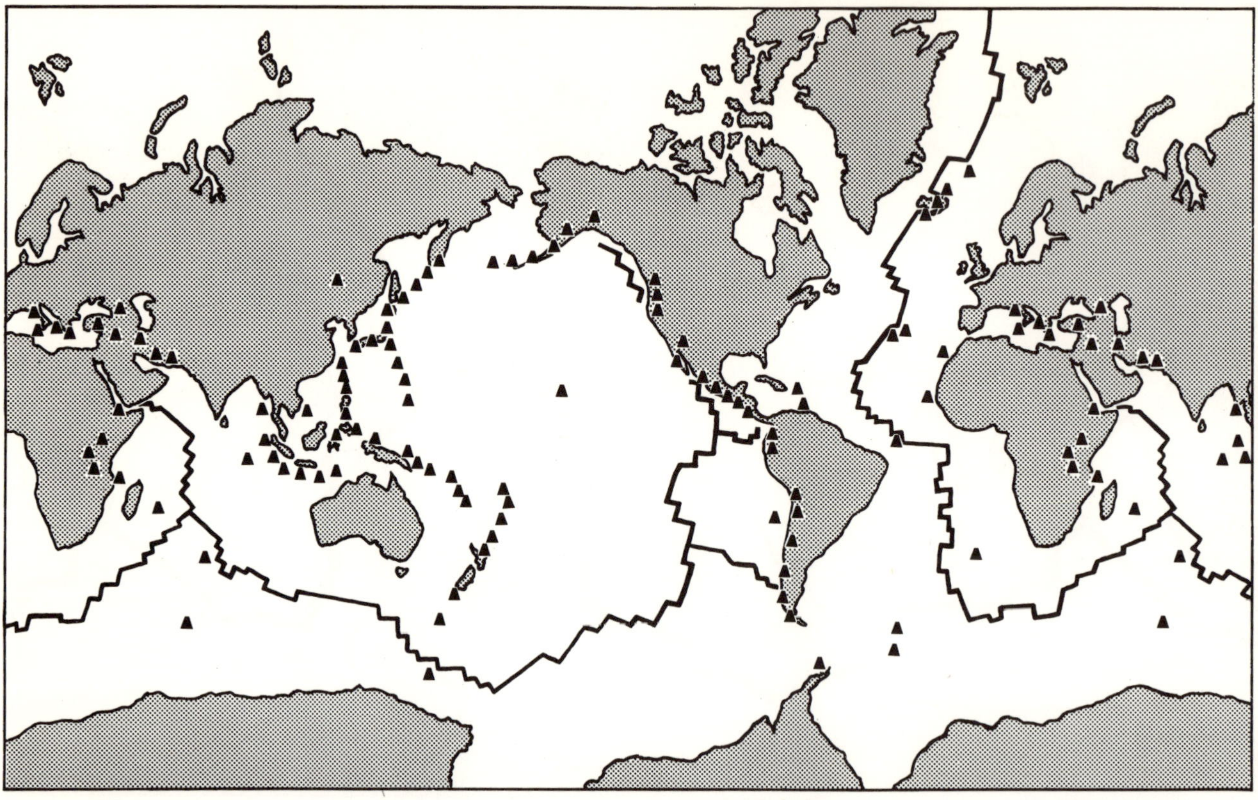

**Fig. 9.3.** The volcanoes of the world are primarily related to the edges of plates. Where plates converge, oceanic crust and sediments are dragged down into the mantle and melted down. The molten material rises to the surface where it erupts as volcanic lava. Volcanoes also form where plates pull apart and lava wells up the resulting cracks.

Other volcanoes occur in the middle of plates (e.g. Hawaii) and these are thought to be related to deep-seated "hot spots" in the mantle.

fissure openings this symmetry is not always present.

Not all oceanic ridges have axial rift valleys. For example, no rift valleys exist along the rapidly spreading East Pacific Rise and Pacific-Antarctic Ridge (east and south of New Zealand), where the new material generated along the ridge crest seems to merely separate and move down the sloping flanks.

Where spreading is slower, however, as from the Atlantic and Indian oceanic ridges, the plate edges slip upwards for hundreds of metres, forming precipitous scarps on either side of a narrow rift valley, before moving down the flanks (Fig. 8.2). The reason for this phenomenon is not at all clear.

Narrow belts of earthquakes are associated with the oceanic ridges (Fig. 9.1) and these show that the movements between the plates on either side of the ridges are only occurring at shallow depths, less than 30 km (19 miles), and that moreover such movements are limited to the ridge axes and the faults criss-crossing and offsetting the axes. Apart from this activity at the oceanic ridges (and similar but more pronounced activity at the plate edges, i.e. associated with island arcs and submarine trenches) the plates are virtually free of earthquakes and from this it is deduced that the oceanic lithosphere is moving as a rigid layer, without significant distortion to cause fracturing.

**Seamounts and guyots**

An enigmatic feature of oceanic plates is the presence of large volcanic islands and submerged, often flat-topped, seamounts that are usually situated in the middle of the oceanic plates, rather than along their edges.

The discovery of seamounts was yet another product of the Second World War. Harry Hess, geology professor at Princeton University (U.S.A.), served in the U.S. Navy during the war, and by the time of the island-hopping invasions of the Western Pacific had become commander of a troop transport. As the ship shuttled back and forth across the Pacific, carrying troops to landings in the Marianas Islands, the Philippines and Iwo Jima he kept the

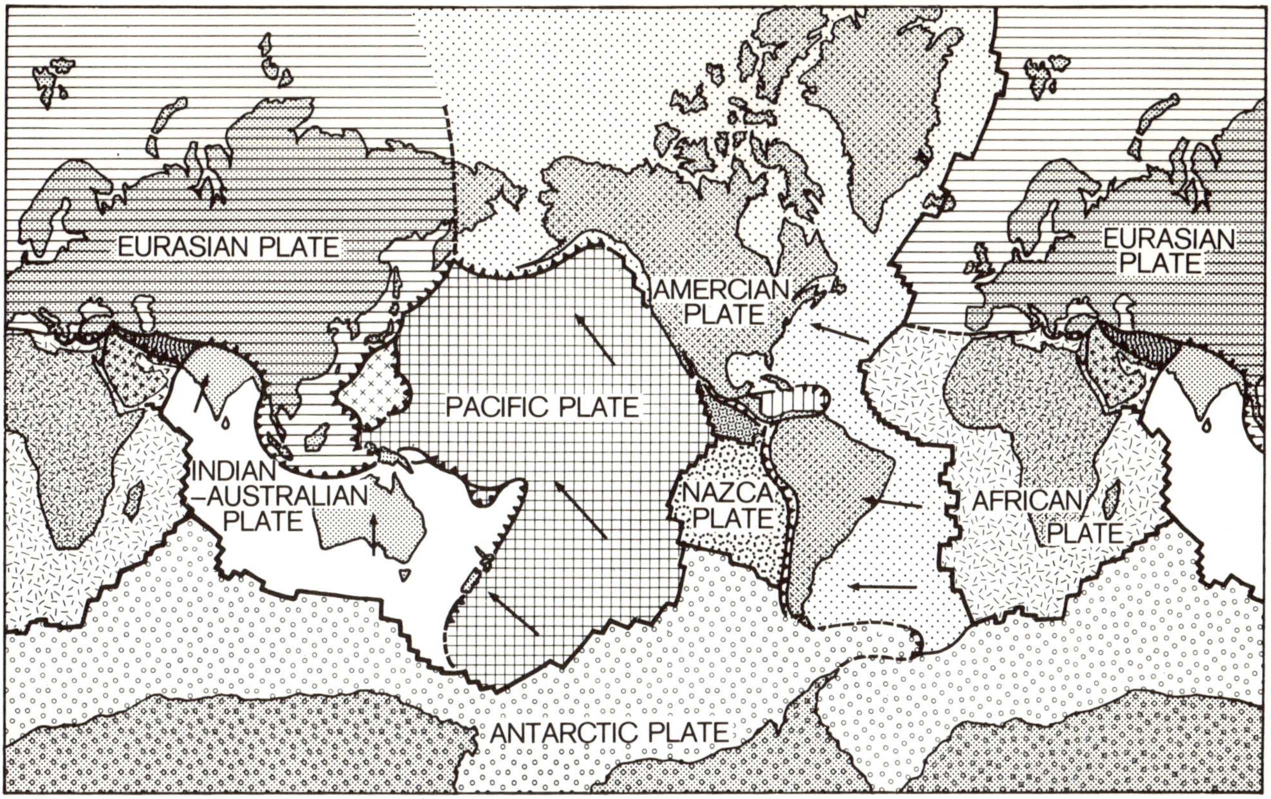


**Fig. 9.4.** The earth's crust is composed of at least 15 rigid virtually undistorted slabs or "plates" of lithosphere, seven of which occupy considerable areas of the globe and are shown on this map. Boundaries of plates are of four principal types: divergent or spreading zones, where plates are separating and new plate material is added; subduction zones, where plates converge and one plate is consumed; "collision zones", former subduction zones where continents riding on plates are colliding; transform faults, where two plates are simply gliding past one another, with no addition or destruction of plate material.

Almost all the earthquake, volcanic and mountain-building activity which marks the "active zones" of the earth's crust closely follows the boundaries of plates and is related to movements between them. As may be seen from the diagram, New Zealand straddles one such active zone, astride the boundary between the Indian-Australian plate and Pacific plate.

echo-sounder running continuously—having previously seen to it that his ship was fitted with the latest type of sounder, capable of recording deep sea depths as well as coastal shoals. This was rather a risky business as enemy submarines were capable of picking up the energy pulses and using them to home in on the ship. However, Hess could not bear the idea of all the good ship-time that oceanographically speaking was going to waste.

Hess's echo-sounder picked up many conical peaks on the sea floor that were undoubtedly small submarine volcanoes ("seamounts"). On a number of occasions however, he noted that the echo-sounder was showing that far beneath the ship, the sea floor had abruptly risen to a flat-topped summit resembling a volcano sliced off at the top (Fig. 9.5).

As the war drew to a close, Hess plotted the positions of these flat-topped seamounts and when the war was over, he reported the discovery of 160 of these "drowned ancient islands" of the Pacific basin. He named them "guyots" in honour of Arnold Guyot, Princeton's first geology professor. Now almost 500 guyots have been found in the Pacific, many of them scattered across the Pacific floor north of New Zealand, east of the Tonga-Kermadec Trench. A few guyots have also been found in the Atlantic.

In New Zealand's Northland a number of fossil seamounts and guyots, that formed on the sea floor in the Upper Cretaceous, some 70 million years ago, occur scattered around the countryside from Whangarei northwards. These volcanic blobs—called the Tangihua Volcanics—are composed of submarine lava flows and are surrounded by sea-floor sediments. Geologists interpret the area as a piece of 70-million-year-old sea floor that has been elevated by earth movements to form land. New Zealand is thus unique in this respect.

Hess found that most of the guyots on the modern Pacific floor rose some 3000 or 4000 m (10,000-13,000 ft) above the ocean floor, but rarely came closer than 1000 m (3500 ft) to sea level. Some were single, while others stood in groups, their flat tops all at the same height, as though they had all been planed off by waves when their peaks had stood above sea level. However, just to confuse matters, some of them, although close together, differed in height by as much as 300 m (1000 ft).

The unique flat summits of the guyots clearly suggest levelling by marine erosion at sea level some time in the geological past, followed by subsidence. The variation in height was a puzzle however, unless sea level had stood at a variety of heights. A major question was, do the guyots indicate that sea level was once some two kilometres lower? Or has the Pacific floor sunk by some two kilometres? Or have both remained stable while the Pacific acquired a two-kilometre-thick carpet of sediment which has displaced the water upwards by the same amount? Geologists pondered these questions for some years after Hess's discovery and some adventurous souls had the Pacific Ocean going up and down like a yo-yo, to plane off various groups of guyots. Others thought the guyots were extremely old, and dated back to a time when the Pacific was much shallower. In 1956 however, it was found that sediments dredged from the tops of some mid-Pacific guyots contained fossils of shallow-water marine animals that had lived in the Middle Cretaceous period, 100 million years ago. What could possibly have submerged these mountains to such great depths in so short a time?

The hypothesis of sea-floor spreading provides the best answer so far to the guyot problem. This hypothesis visualises many oceanic basalt volcanoes starting life as small cones on the crests of oceanic ridges, as a result of lava welling up along freshly created fractures. Then they continue to grow larger as they ride away on the moving sea floor.

A volcano may need as much as 10 million years in which to build up from the ocean floor and poke above sea level. By that time sea-floor spreading has probably moved it away from the sources of magma (molten volcanic rock) sited close to the crest of the mid-oceanic ridge and consequently activity of the volcano declines and it becomes dormant and then extinct. Once active eruptions cease, the volcano is eaten into by rivers and waves. It is likely that it may persist as an island for another 10-20 million years before being trimmed down below sea-level by erosion, where it remains as a flat-topped guyot (Fig. 9.6).

As the guyots ride the sea-floor conveyor belt down into the "jaw-crusher" subduction zones

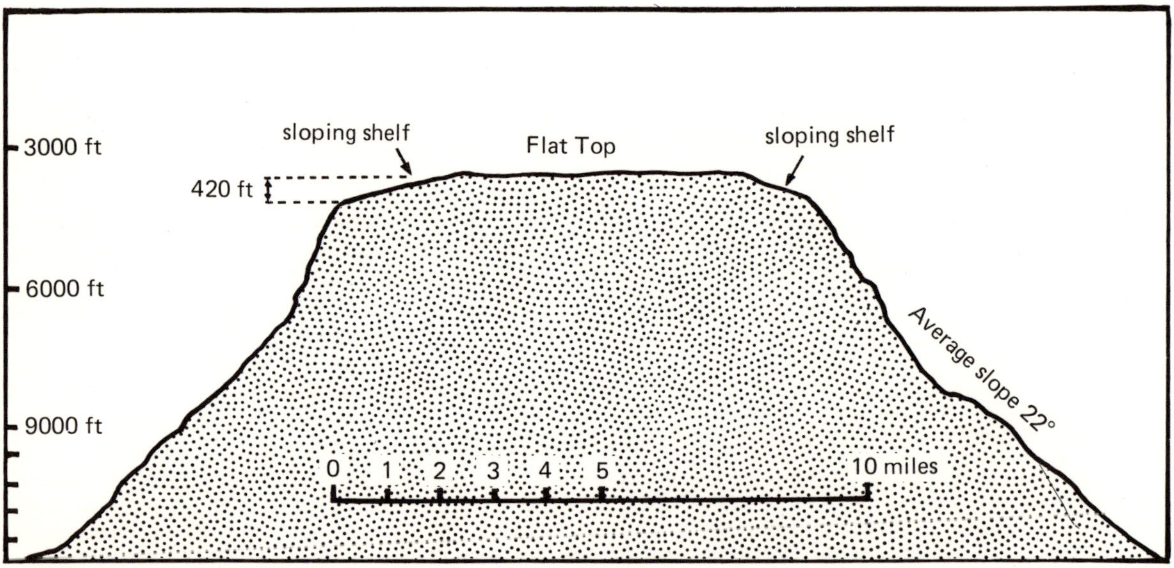


**Fig. 9.5.** The shape of a typical guyot, based on an echo-sounder trace.

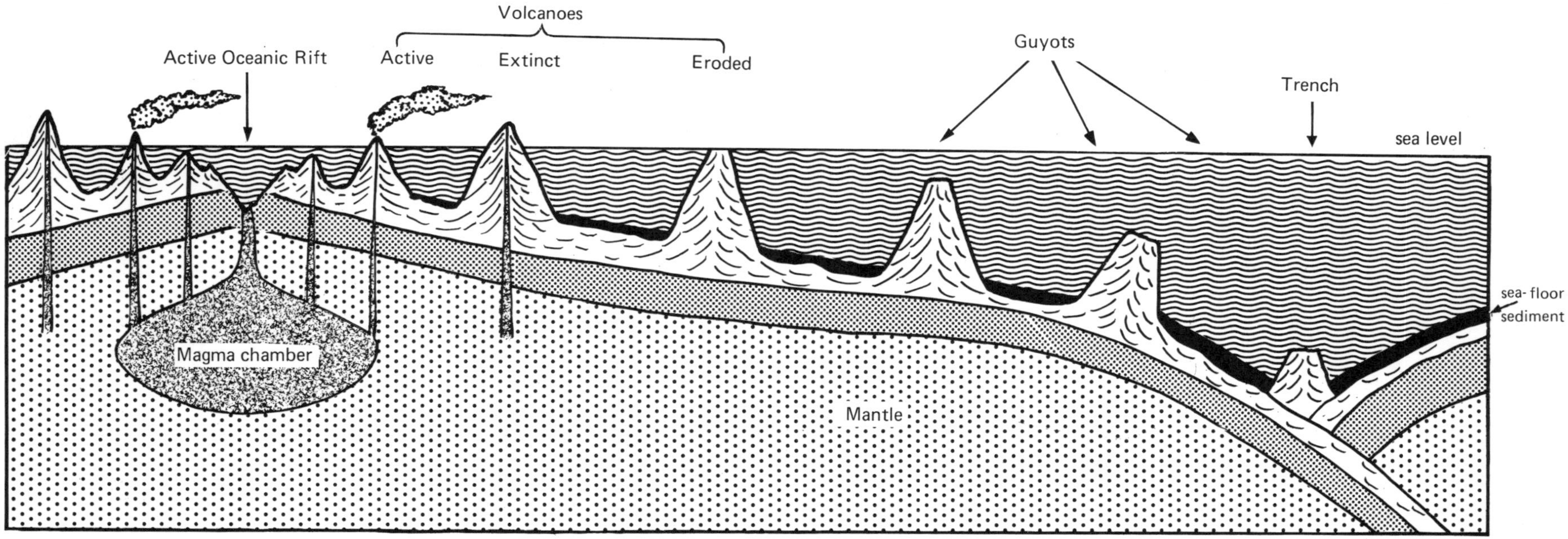


*Fig. 9.6*

**Fig. 9.6.** A composite diagram showing the growth, erosion, submersion and destruction of volcanoes on a spreading sea floor. A tilted guyot sits on the lip of the trench, while another, carried down into the trench, sits on its floor.

**Fig. 9.7.** Three stages in the formation of an atoll as a coral reef builds upwards on a sinking volcano.
A: Volcanic island with a fringing coral reef.
B: The volcano has ceased activity and erosion has reduced it to a skeleton. Meanwhile, sea-floor spreading has carried the volcano to greater depths, upward growth of the coral reef keeping pace with the slowly sinking volcanic platform.
C: The volcano has sunk out of sight, leaving a coral atoll enclosing a lagoon.

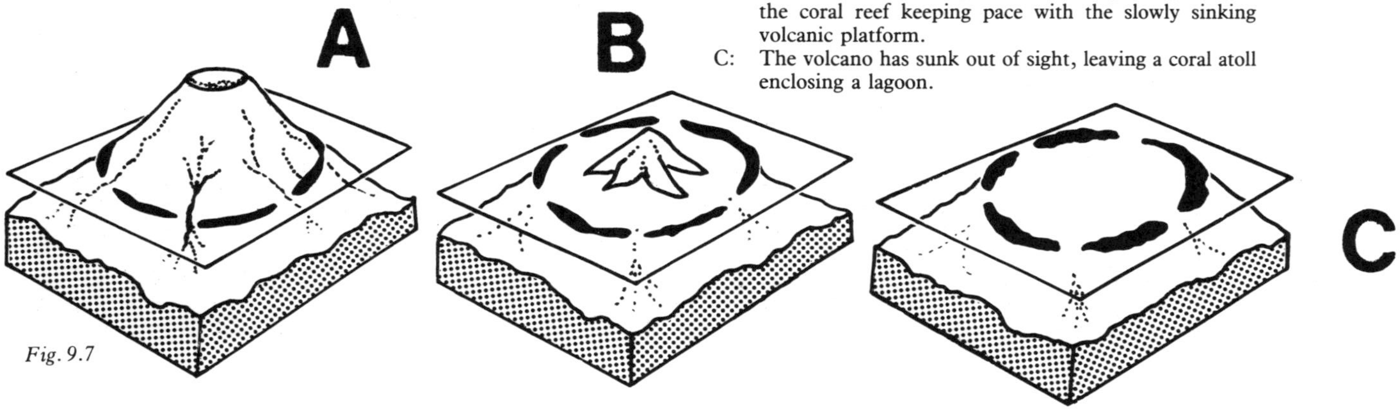


*Fig. 9.7*

they gradually get deeper and deeper. This is not wholly due to a simple dragging down of the crust into the subduction zone but rather is related to an actual sinking of the sea-floor foundation. When new sea floor is created at the oceanic ridges the newly solidified volcanic rock is still very hot and, as the tiny atomic particles within it are spaced widely apart, it occupies a greater volume than when cold. However, as the sea floor is conveyed further and further away from the oceanic ridge, it gets cooler and cooler (as indicated by heat-flow measurements on the sea floor), the atomic particles pack closer together, and its volume shrinks. Thus the sea floor gets steadily deeper with age, to either side of an actively spreading ridge, and this steady sinking of the sea floor will naturally affect the guyots, taking them down into deeper and deeper water.

Detailed oceanographic surveys in recent years have essentially confirmed the reality of this picture. Tilted guyots have been found angling down into some of the submarine trenches. One, north of New Zealand, on the seaward rim of the Tonga-Kermadec Trench, rises 8200 m (27,000 ft) from the ocean floor, coming to within 370 m (1200 ft) of sea level. It is one of the highest mountains in the world, but if anyone could see it, it would be leaning like the Tower of Pisa. Other guyots have ridden the moving carpet down into the trench and are sitting upright in the bottoms of trenches (Fig. 9.6).

Study of guyots and islands in the New Caledonia region, 1300 km (810 miles) north of New Zealand, has shown that often a bulge is developed in the oceanic crust, before it descends into the trench, just like a huge ruck in a table cloth. This giant bulge has raised and tilted those parts of New Caledonia and the adjacent Loyalty Islands closest to the New Hebrides Trench, as indicated by raised coral reefs and gravity anomalies.

**Sea-floor spreading and the odyssey of the green turtle**

A rather novel consequence of the creation of volcanic islands at oceanic ridges and their gradual rafting away and disappearance beneath the sea, as illustrated in Fig. 9.6, is that over geological time a group of Brazilian Green Turtles has become more and more detached from its ancestral breeding grounds.

For many years it has been known that the green turtles living and feeding on the coasts of Brazil leave the South American coast to swim some 2000 km (1300 miles) across the Atlantic to nest and breed on the beaches of Ascension Island in the central equatorial Atlantic, close to the Mid-Atlantic Ridge. The feat of navigation involved in this odyssey has puzzled zoologists for many years, but sea-floor spreading provides a possible answer.

The turtle's curious migration pattern has perhaps evolved as a result of the gradual separation of feeding and nesting sites by continental drift over millions of years.

Over 100 million years ago turtle populations travelled along the shores of a gulf linking northern South America and West Africa, as seen in Fig. 5.6.

Between 90 and 80 million years ago sea-floor spreading to either side of an active oceanic ridge had opened up a sea corridor extending southwards from the head of the gulf and separating Brazil and West Africa. Volcanic islands along the axis of the oceanic ridge were populated by turtles, who established a pattern of breeding on the islands and living and feeding on the Brazilian coast.

As sea-floor spreading continued, the islands originally on the ridge axis were rafted away, became inactive, and were submerged as seamounts and guyots (Fig. 9.6). They were progressively replaced, however, by new islands on the ridge. Thus as sea-floor spreading took place at a rate of about 2 cm (1 in) per year the turtles colonised successive islands as they appeared in place of others now submerged, having to migrate further and further, until today they are on the last and youngest of them (Ascension Island, created less than seven million years ago).

For all that we know, a number of other animal migration patterns that appear curious in terms of modern geography may relate back to ancestral patterns established when landmasses were either linked or were a lot closer than they are today. It may not be altogether idle speculation to say that the lemmings, for example, who hurl themselves off the coasts of Norway, are in fact trying to head for ancestral lands, now separated by many kilometres of open ocean.

**Atolls**

The relating of guyots and seamounts to the

hypothesis of sea-floor spreading has also helped throw some light on the origin of atolls, that in many ways are as odd as guyots. The coral atolls that start to dot the Pacific some 1300 km (810 miles) north of New Zealand consist of circlets of coral wholly or partially enclosing a lagoon. Some are little more than a strip of coral reef, but others are wide enough and sufficiently covered by debris to support vegetation.

On his celebrated voyage in the *Beagle* (1831-1836), Charles Darwin deduced that the various atolls he visited, and probably most others, were coral growths crowning the tops of sinking volcanoes. Darwin found that reef corals can live only in the upper levels of warm seas, roughly the top 60 m (200 ft). For maximum growth the water must be well aerated, warm (above 20°C), clear, clean and sunlit.

He concluded that eroded shelves around the Pacific volcanic islands make ideal foundations for corals that in this way could grow and eventually build fringing reefs. As the volcanic islands were reduced by erosion, and the ocean floor slowly subsided, the coral reefs built steadily upwards, to remain at sea level. Thus the initial fringing reef developed into a ring-like atoll (Fig. 9.7).

Darwin's interpretation of atoll formation was fully substantiated in 1952 when borings at Eniwetok Atoll (Marshall Islands) penetrated some 1500 m (5000 ft) of coral, perched on top of a submerged volcanic cone.

For many years up to the 1950s the presence of atolls in the Pacific, and Darwin's interpretation of them, had been held up as evidence in support of wholesale sinking of crustal rocks in the Pacific, or alternatively, of great fluctuations in sea level. However, like the guyots, atolls can now be better interpreted in terms of sea-floor spreading.

### Hot spots

Although as we have seen above, many oceanic volcanoes have been born on oceanic ridge crests, others have developed in the centre of oceanic plates. For example, a number of island chains can be seen in the middle of the Pacific Plate, bearing no apparent relationship to the East Pacific Rise.

The Hawaiian island group is an example, as it forms part of a long chain of islands, atolls and seamounts extending for 2500 km (1555 miles), from the large island of Hawaii on the south-east to the Emperor Seamounts at the north-west, extending to the very edge of the Pacific plate, before it descends down into the Kurile and Aleutian submarine trenches. Actively erupting volcanoes occur only on Hawaii, the youngest and most southerly island of the chain. The remaining islands of the Hawaiian group consist of extinct cones which are progressively older and more deeply eroded and dissected until, north-west of Kauai, they give way to a row of eroded submarine platforms, the stumps of former volcanic islands. Still further to the north-west lie the coral atolls of the Midway Group, consisting of reefs of coral growing up from stumps of still older volcanic islands, the lava from which has yielded dates of 15-20 million years. Beyond these is the submerged chain of the Emperor Seamounts dating back at least 75 million years (Figs 9.8-9.10).

Two other island chains in the Pacific, lying between Hawaii and New Zealand (Fig. 9.10), display a similar pattern of ages and also show parallel trends, all having a marked kink.

The idea that has been put forward to try to explain these chains of oceanic islands in the Pacific Ocean is that the Pacific plate has been moving in a north-west direction and has passed slowly over a deep permanent "hot spot", which has continuously erupted lavas and built one island after another along a zone of weakness in the plate. After moving beyond the zone of active eruptions, each island has been gradually eroded away and has then been submerged as it rides downward towards the trenches of the north-west Pacific, down which it disappears. Another probably less likely explanation that has been offered is that a prolific volcanic source migrated south-eastward beneath a fixed ocean floor, leaving a chain of islands in its wake.

### Continental rifts

Rift features such as the magnificent Rift Valleys of East Africa, the Rhine Graben of Germany and the Baikal Rift of Siberia have been interpreted as tensional (or "pull-apart") structures similar to oceanic rifts. Continental rifts occur in areas where the crust has been domed-up, and they are associated with numerous shallow earthquakes, and their valley floors transmit higher than normal amounts of heat. Unlike oceanic rifts, however, the walls of continental rifts are of ancient continental rocks and it looks as though they originate as giant tears in the earth's crust, related to stresses and strains within plates.

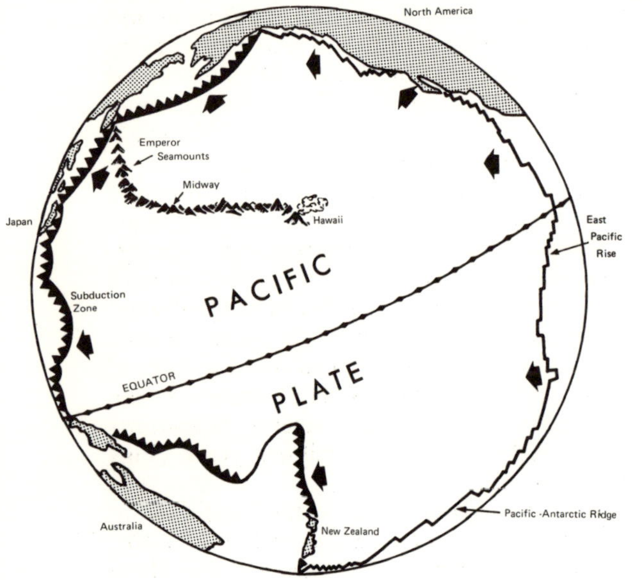


**Fig. 9.8.** The Pacific plate is being created at the East Pacific Rise and Pacific-Antarctic Ridge and is moving by sea-floor spreading towards the subduction zones in the Northern and Western Pacific, where it is being consumed. The broad arrows indicate the directions in which the plate is moving.

Hawaii is thought by some geologists to represent a weakness in the earth's mantle—a "hot spot"—up which volcanic lava is continuously welling. In the geological past the Pacific plate has moved across this hot spot, on its way to the trenches of the North-west Pacific.

The continuous volcanic activity of the hot spot has left a trail of extinct volcanic islands on the sea floor that are milestones marking the passage of the plate. The Hawaiian Islands grow older towards the north-west, beginning with Hawaii itself (Fig. 9.9). The age succession in the Hawaiian islands is continued into the Midway Group, ranging in age from 15.7 to 25 million years. Then follow other coral-capped volcanic seamounts and guyots that range up to 40 million years in age. These islands form a sharp bend in the otherwise linear chain of islands. This change in direction probably implies that the direction of plate movement changed between 40-20 million years ago, as the Pacific Plate skewed around on its axis.

A chain of seamounts continues northwards from the bend and extends down into the North Pacific Trench (cf. Fig. 9.6), where the Pacific plate is being consumed. The seamounts range in age up to 75 million years.

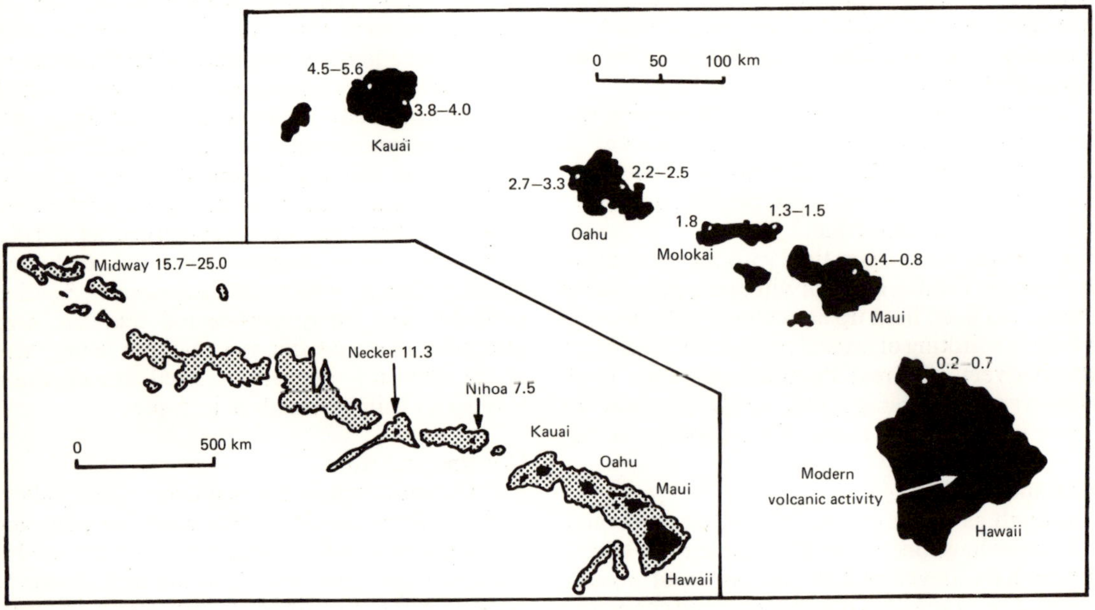


**Fig. 9.9.** The Hawaiian Islands form a linear chain of volcanoes extending for more than 500 km (300 miles). Individual islands show a progression in ages, ranging from youngest in the east to oldest in the west. Kilauea, an active volcano that produces voluminous basalt lavas, is situated at the south-eastern extremity of the island of Hawaii. Older extinct volcanoes make up the north-eastern part of the same island. The age progression continues to the north-west, to the oldest island in the Hawaiian group (upper half of diagram), and then continues along a submarine ridge to Midway (lower half of diagram). The age progression can be interpreted as indicating the tracks formed by the movement of the sea floor over stationary "hot spots" in the earth's mantle.

The islands are shown in solid black and dates of volcanic activity are given in millions of years. In the lower diagram submarine topography ranging down to 4000 m (13,000 ft) in depth is indicated by the dot pattern.

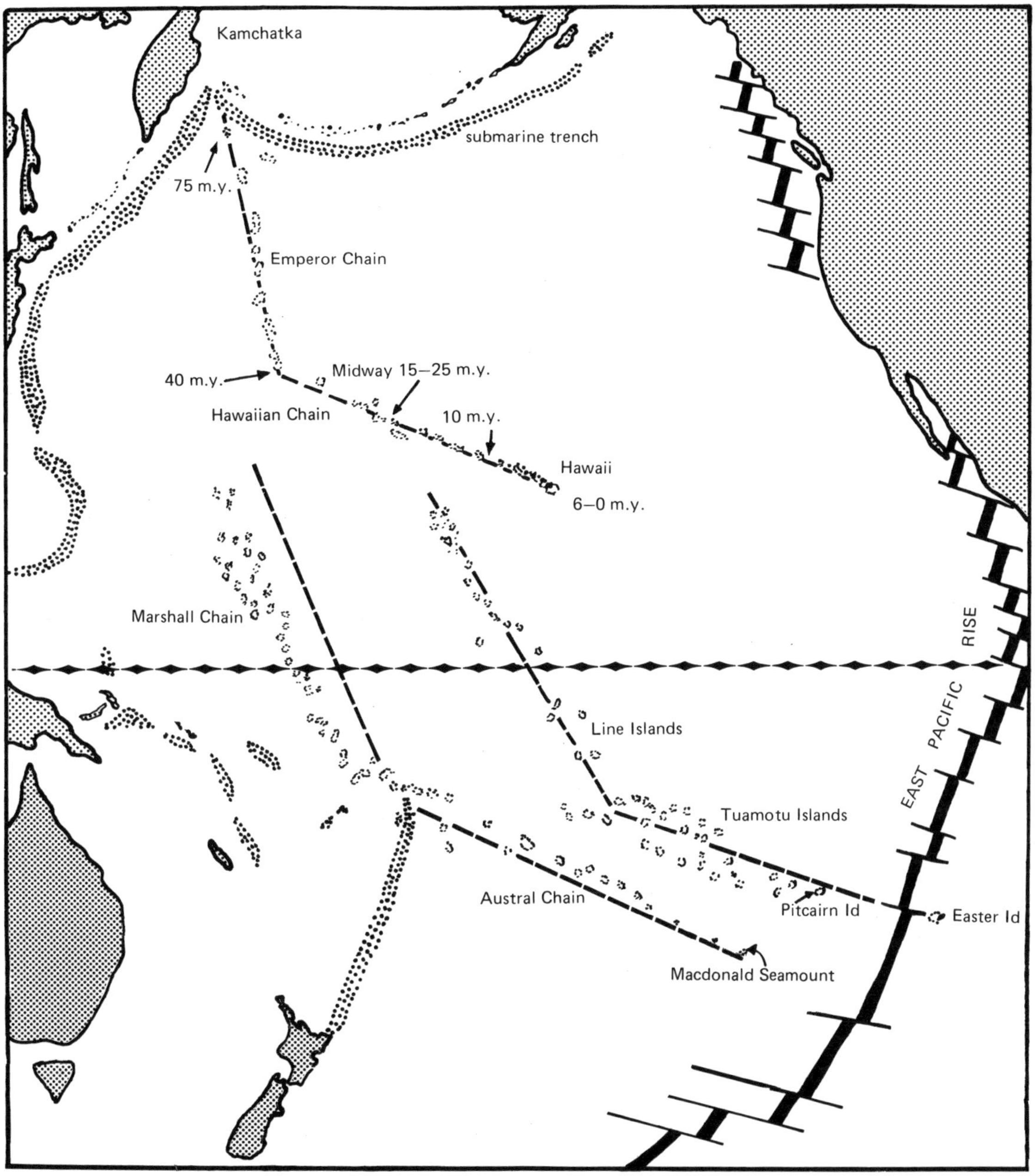


**Fig. 9.10.** The age progression seen in the Hawaiian island chain and its north-west continuation (Figs 9.8, 9.9) also appears to be parallelled in two other island chains between Hawaii and New Zealand.

The Tuamotu chain begins with Pitcairn Island (active between 0.9 and 0.4 million years ago) and the Austral chain with the active Macdonald Seamount.

Seamounts, guyots and atolls extend northward from the Tuamotu and Austral chains and, like the Hawaiian chain, form a marked kink, perhaps related to a change in direction of rotation of the Pacific plate, some 40-20 million years ago (Fig. 9.8).

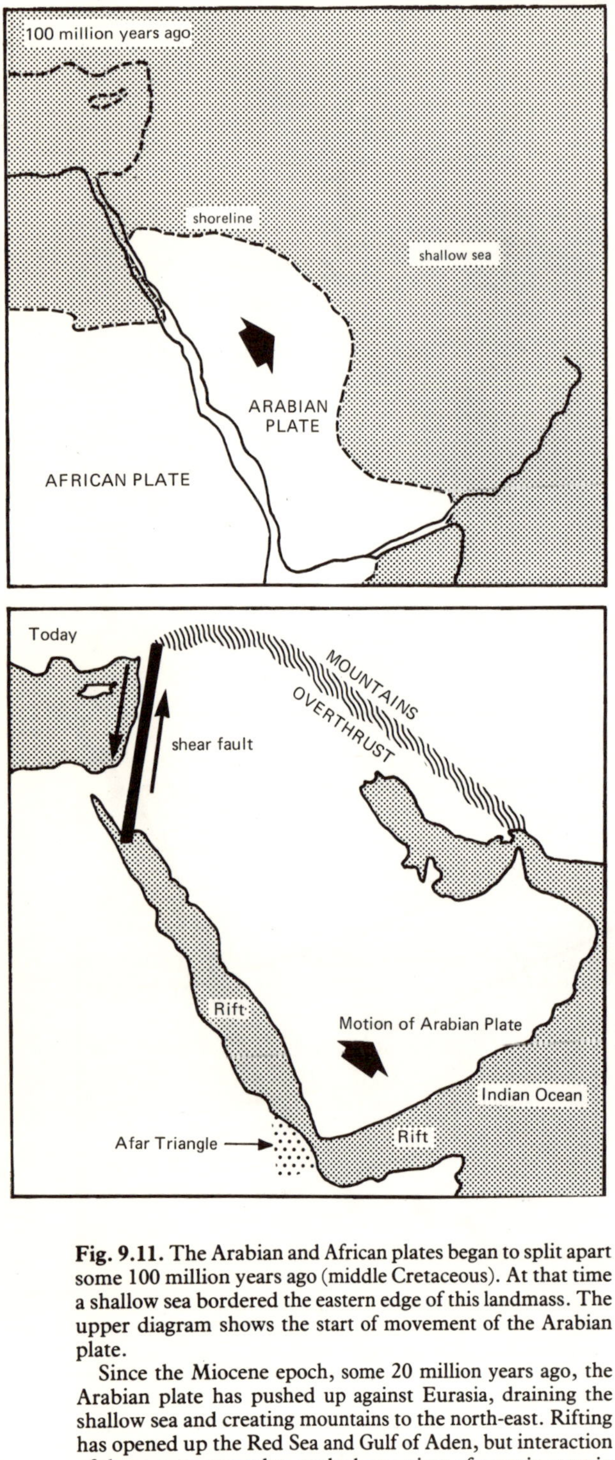

**Fig. 9.11.** The Arabian and African plates began to split apart some 100 million years ago (middle Cretaceous). At that time a shallow sea bordered the eastern edge of this landmass. The upper diagram shows the start of movement of the Arabian plate.

Since the Miocene epoch, some 20 million years ago, the Arabian plate has pushed up against Eurasia, draining the shallow sea and creating mountains to the north-east. Rifting has opened up the Red Sea and Gulf of Aden, but interaction of these movements has pushed up a piece of oceanic crust in the Afar Triangle (bottom diagram).

**Fig. 9.12.** A rift system extends from the Mozambique coast of East Africa some 9656 km (6000 miles) to Lebanon and Syria, where it dies out against the crumples of the Taurus and other mountains in Southern Turkey. In the vicinity of the Afar Triangle, opposite Aden (see Fig. 9.11), a branch of the rift system extends through the Gulf of Aden to link with the Indian Ocean's Carlsberg Ridge. In this diagram the rift zone is delineated by the clear areas and the black lines indicate the trends of some of the major fractures.

The East African lakes are shown in black.

The rift (or "Graben"—a German word for trough) within which the Rhine River lies for a stretch of 290 km (180 miles) between Basel in Switzerland and the castle-studded hills below Mainz is thought to have been an actively opening tear some 30 million years ago. It is about 40 km (25 miles) wide.

The Baikal rift of Siberia is younger—it was active some 10 million years ago. Lake Baikal that lies in this rift is so long and deep that it contains roughly one-fifth of the world's lake water.

The East African Rift Valleys are related to a modern rift system stretching 9656 km (6000 miles) from Lebanon and Syria in the north to Mozambique in the south, traversing *en route* a wide variety of continental rocks. It represents an incipient tear in the earth's crust, the result of herculean forces that are swinging the African continent around and northwards and are opening up the Red Sea as an embryonic ocean (Fig. 9.12).

The Red Sea appears to have started as a continental rift, rather like the East African Rift Valley. Now however it has opened out so much that new oceanic basalt is welling up along a slit-like central valley in its floor. Within some of the deepest depressions of this valley are pools of hot brines bearing various metal ores, as though ore deposits are here in the making (p. 385).

The Red Sea and Gulf of Aden are widening at about 2 cm (1 inch) per year, as Arabia moves north-eastward away from Africa. The rifting along the Red Sea continues via the Gulf of Aqaba, into Israel, Jordan and Lebanon, where it forms a longitudinal depression occupied by the Dead Sea, River Jordan, Sea of Galilee, Lake Tiberias, etc. North of Lebanon the rifting dies out against the massive rucks of sediment squeezed and distorted as the leading edge of Gondwana pushed its way into Laurasia (Fig. 9.11).

At the other end of the Red Sea, opposite Aden, the East African rift system encounters that of the Red Sea, while from the east enters an oceanic rift related to the Indian Ocean's Carlsberg Ridge. The three rift systems converge in an area called the Afar Triangle, lying chiefly within the boundaries of Ethiopia.

Wegener in his early discussion of continental drift singled out this triangle as being of particular interest. He made the observation that if the Afar Triangle is cut out, the opposite corner of Arabia fits perfectly into the gap. This fit has been dramatically confirmed by satellite photos.

The tremendous forces interacting at the junction between the three rift systems have squeezed up a triangular part of the floor of the Red Sea so that we can study it on dry land. The Afar Triangle is the result and, like Iceland sitting astride the Mid-Atlantic Ridge, it is a natural laboratory in which to see the birth pangs of new ocean crust.

The activity that is pushing the Red Sea apart is expressed in the landscape of the Afar Triangle as massive rifts, swarms of fractures, boiling springs and lava flows. A chain of seven volcanoes, all active, traverses the barren desert and often the topography is broken into a giant's staircase of large step-like blocks 30-105 m (100-350 ft) high, heaved up along fracture zones.

The spectacular 3220 km (2000 miles) East African Rift Valley began to form between 10-20 million years ago, starting in Ethiopia in the north, and slicing through Kenya and Tanzania on its inexorable journey south (Fig. 9.12).

In places the tension has simply stretched the crust, forming a depression, while in others the crust has snapped, causing huge chunks to sink between vertical fault lines. Volcanic activity was a natural concomitant of these massive crustal distortions and lava and ash deposits fill the valley and 30 active or partially active volcanoes are scattered along the rift. Large lakes lie in many of the depressions and in the Ice Age and earlier (2-5 million years ago), provided hospitable environments for ancient man, whose remains are now eagerly sought after by archaeologists—the associated volcanic material, readily dated by radioactive techniques, providing a precise timetable for man's evolution.

While the Red Sea and Gulf of Aden are widening at about 2 cm (1 in) per year, the East African Rift Valley is only opening at a rate of about 0.02 cm (0.01 in) per year and has spread no more than 15-25 km (10-15 miles) in the past 15-20 million years. If these movements continue, it is obviously going to be a long time before an embryonic ocean traverses East Africa—perhaps a slit-like sea, similar to the Red Sea today, may feature in the maps 50 million years hence (Fig. 9.30).

**Zones of convergence**

Old lithosphere is reduced in area or removed from the surface along zones (subduction zones) where two moving plates converge (Fig. 9.13). It is thought that what actually happens depends on whether both plate margins are oceanic, or whether one is oceanic and one continental, or whether both are continental.

If both converging plate margins are composed of oceanic crust, one will over-ride the other and the lower one will bend downward and be thrust or sink into the asthenosphere, forming a Benioff Zone. Friction between the two plates generates intense earthquake activity, the earthquake shocks extending to great depths. A submarine trench forms along the site of under-thrusting. As it descends, the under-thrusting plate carries oceanic crust down into the mantle, together with some of the sea-floor sediments. However, much of the low-density sediments are scraped off and crumpled up in the process. All remaining oceanic crust is partially melted or resorbed at depths between 100 and 300 km (65-190 miles), but the plate continues downwards to about 700 km (435 miles) before finally breaking up. Being less dense than the mantle, the molten crustal material rises towards the ocean floor, where much of it erupts as lava, and, together with the scraped-off sediments, builds up a chain of volcanic islands, or an "island arc", on the margin of the over-riding plate.

The molten rock that rises to the surface from subduction zones, and erupts as volcanoes, consists of melted-up oceanic (sima) material (from the consumed ocean floor) mixed with various proportions of continental (sial) material (from the consumed land-derived sediments originally lying on the sea floor). The proportions of sial added to the sima vary considerably, depending on local circumstances. If no sial is present, the resulting volcanic lava is a basalt, a basic rock (see p. 2), but as more and more sial is added, various "intermediate" rocks are produced, the best known of which is "andesite". New Zealand is the home of some notable andesite volcanoes. They are related to the Tonga-Kermadec subduction zone, which passes under the North Island (Figs 18.14-18.16). Sub-crustal melting of the down-going oceanic slab, and the subsequent mixture of sial and sima, has produced a number of andesite volcanoes, including Egmont, Ruapehu, Tongariro, Ngauruhoe and White Island.

If one plate margin is composed of continental crust, its buoyant sial will over-ride the less-buoyant sima of the adjacent oceanic crust. Once again a submarine trench will form at the line of under-thrusting, and the continental margin will be compressed into folded mountains capped by volcanoes (e.g. the west coast of South America) (Figs 9.14, 9.15).

If both plates are continental and therefore buoyant, profound sinking of either is unlikely. A number of things happen (Fig. 9.16):

(i) The oceanic strip that originally existed in between the two continents is closed and once the continental edges have met, the Benioff Zone becomes de-activated and disappears.
(ii) The two sialic masses become welded or "sutured" together, forming a permanently joined large continental area.
(iii) The collision rucks-up massive mountain ranges, incorporating sea-floor sediments and slices of oceanic material (i.e. the remnants of the strip of sea floor that originally existed between the two continents).
(iv) Thickening and under-thrusting of sial along the plate edges leads to buoyant uplift of mountains and plateaux.

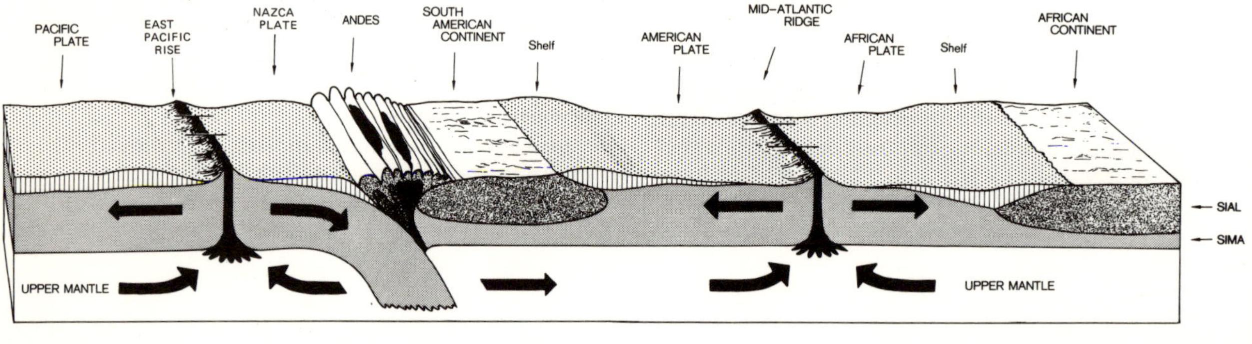


**Fig. 9.13.** Diagram to summarise the essential differences between zones of divergence (exemplified by the Mid-Atlantic Ridge and East Pacific Rise) and zones of convergence, exemplified by the western coastline of South America.

Volcanic magma is pictured as welling up in the rifts along the Mid-Atlantic Ridge and East Pacific Rise and also intruding into the Andes—in the last instance derived as a by-product of the melting-up of the descending plate.

*Fig. 9.14*

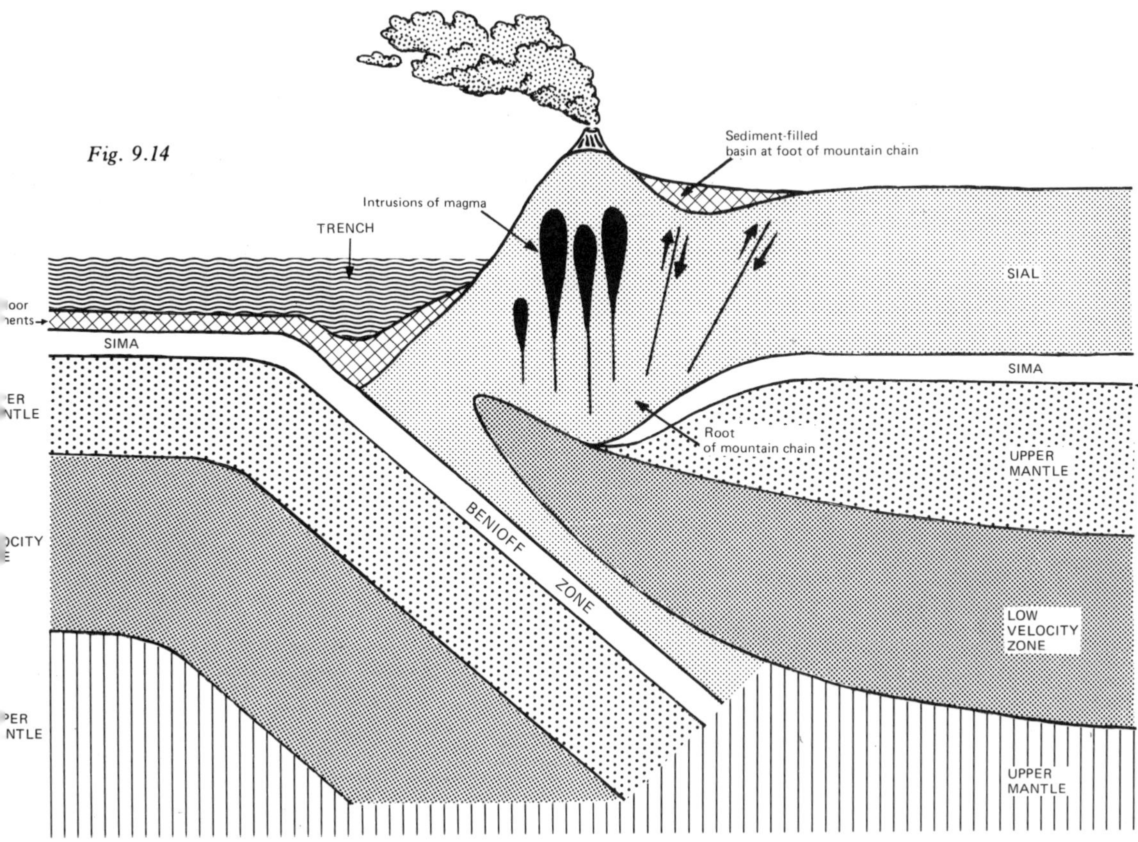


**Fig. 9.14.** Idealised cross-section through a subduction zone developed at the edge of a continent.

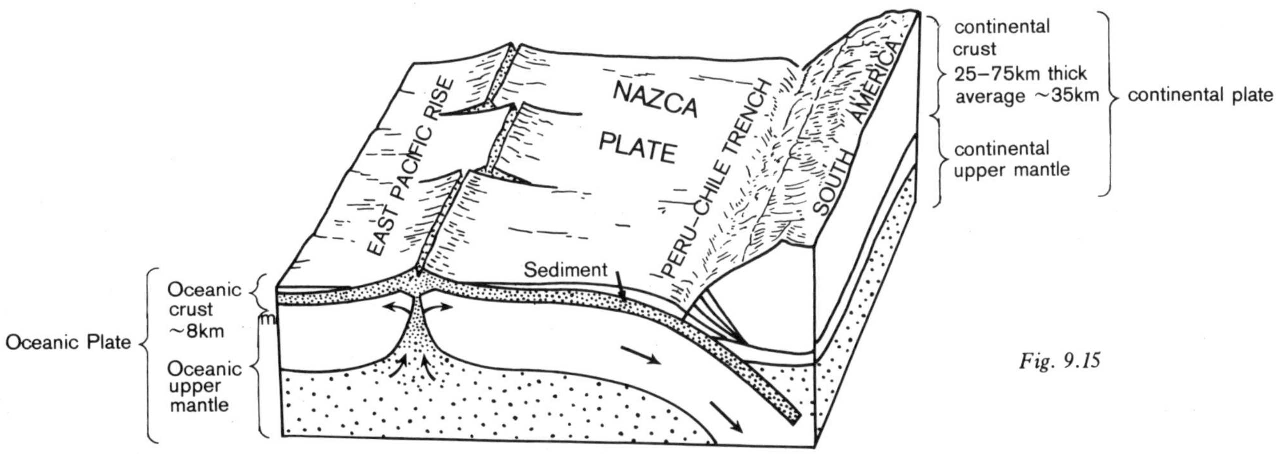


*Fig. 9.15*

**Fig. 9.15.** An actual example of the situation shown in Fig. 9.14.

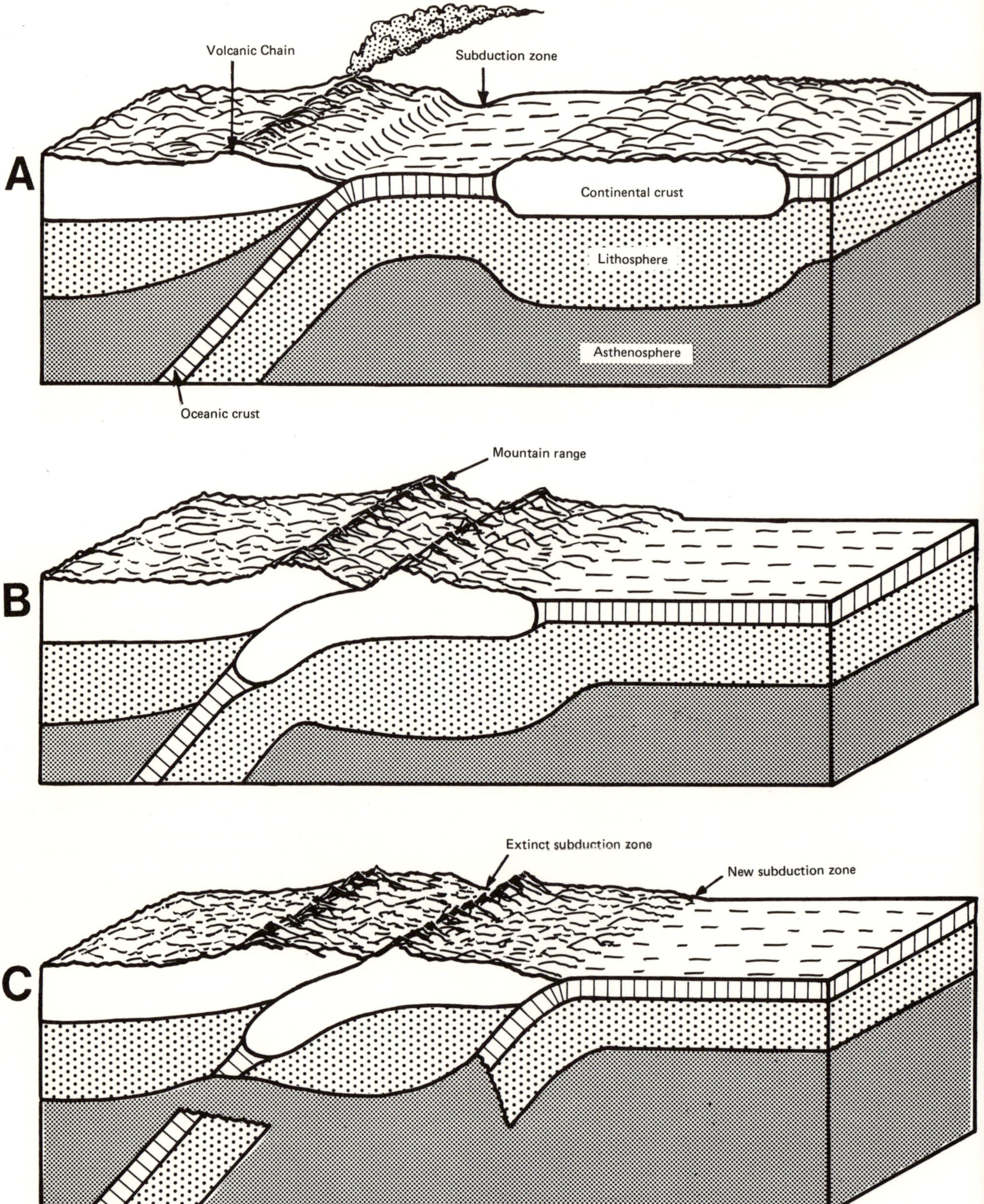

A
Volcanic Chain
Subduction zone
Continental crust
Lithosphere
Asthenosphere
Oceanic crust
B
Mountain range
C
Extinct subduction zone
New subduction zone

A prime example of a continent/continent collision is that of India and Asia, resulting in the creation of the Himalayan ranges and the uplift of the Tibetan Plateau. Another but more complicated example is the collision that occurred between the leading edges of Laurasia and Gondwanaland, that produced the various mountain chains fringing the modern Mediterranean (Fig. 9.17).

**Ophiolites**

As we have seen above, when two continents are moved together by sea-floor spreading, parts of the ocean floor that originally separated them, but have not yet disappeared down the subduction zone, become involved in the resulting collision. The tremendous shearing and tearing movements and massive squeezing that occur on the leading edges of the colliding continents may slice up the sea-floor strip and its overlying sediments and thrust them into adjacent rocks of the land (Fig. 9.18).

Similarly, the sliding of one plate beneath another may scrape off slices of both sediment and underlying crust and mantle and thrust them onto the leading edge of the over-riding plate (Fig. 9.18). This is all very well in theory, but just how do we identify slices of fossil oceanic crust and mantle that have been thrown up onto the land in this manner?

For many years geologists have been perplexed by sequences of very characteristic rocks that occur in elongated zones within some mountain systems. The rocks were grouped together as "ophiolites", coined from a Greek word meaning serpent or snake-stone. This description refers to the colour and texture of many of the rocks—shiny light green, with streaks of dark minerals, outlining scale-like patterns similar to those on a snake's body. The ophiolite rocks commonly consist of a sequence, sometimes jumbled together, but often quite clearly developed, starting on top with deep-water sediments, followed by submarine basalt lavas, and passing down into very heavy dense basic rocks (the so-called ultra-basic rocks), commonly with serpentine (a greenish rock, the name having the same origin as ophiolite).

**Fig. 9.16.** When two continents are rafted together by sea-floor spreading, the sea floor formerly present between them either disappears entirely down the subduction zone, or is rucked up and thrust onto the land (cf. Fig. 9.18).

The continents then make contact and a mountain range results, as young sedimentary material, originally deposited as an apron fringing the continents, is folded upwards under the compression exerted by the two plates (and also downward, forming the buoyant root of the mountain). Sea-floor material may be caught up in such movements. The arrival of light sialic material at the subduction zone effectively spells its end, as there is no way that the buoyant sial can be dragged down and consumed. Thus the tail-end of the old oceanic plate breaks off and continues its descent into the upper mantle. Meanwhile, the two continents become "welded" together. However, if compression continues, a new subduction zone will form on the oceanic edge of the continent, as in Fig. 9.21.

Up to a few years ago the ultra-basic rocks were assumed to have been erupted from deep within the earth and to have invaded the sediments in which they are found today at the very high temperatures typical of molten intrusions. However, the rocks around them showed no evidence of having been baked by such an invasion.

As more was learned about the structure beneath the sea floor—from seismic sounding, bottom sampling and drilling—it became clear that the oceanic crust was about 6 km (3.75 miles) thick and consisted of three distinct layers.

The uppermost layer, "Layer 1", is the carpet of unconsolidated sediment on the sea floor, in places up to a kilometre (¾ mile) thick. "Layer 2" is usually 1.5-2 km (1-1¼ mile) thick and consists of consolidated sedimentary rocks, such as limestones and shales mixed with great quantities of basalt lava. These basalts, erupted and rapidly chilled under water, form rounded lumps called "pillows". "Layer 3", about 5 km (3 miles) thick, extends down to the Mohorovicic Discontinuity (the Moho)—the boundary between crust and underlying mantle, marked by an abrupt increase in rock density.

The composition of Layer 3 is not known, as it is well beyond the reach of the *Glomar Challenger*'s drill. Geologists have surmised however that Layer 3 consists of heavy ultra-basic rocks like those exposed in the ophiolite sequences. Thus the ophiolite sequences of the world are thought by many geologists to represent an entire top-to-bottom cross-section of the oceanic crust—from the sediments, down through the pillow lavas, past the Moho and into the upper mantle (Fig. 9.19).

At the time of the Mohole "affair" (p. 89) some geologists pointed out that samples of the mantle were probably already lying unrecognised in geological rock collections around the world—there was no need at all to drill for them! But the sceptics said true mantle rock was so heavy that it

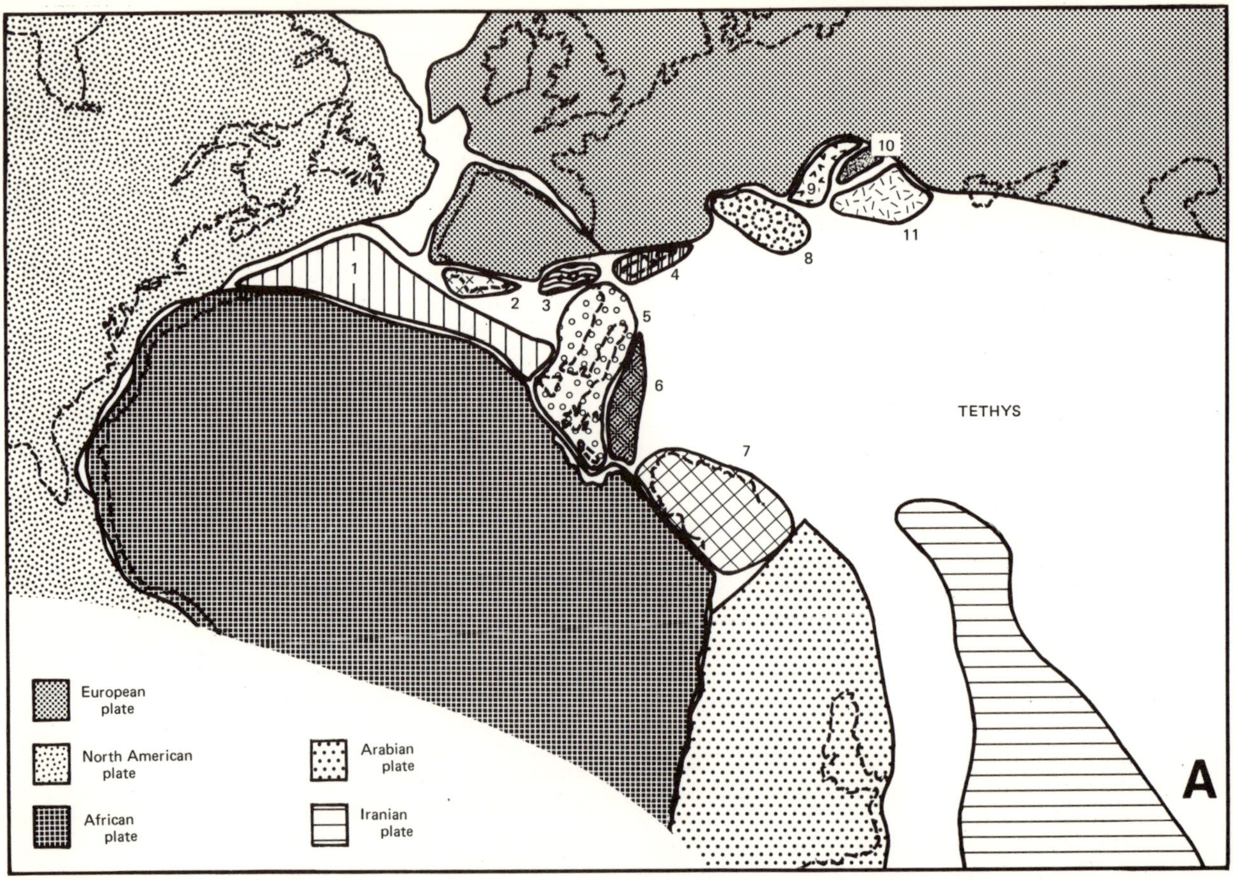

1
2
3
4
5
6
7
8
9
10
11
TETHYS
European plate
North American plate
African plate
Arabian plate
Iranian plate
A


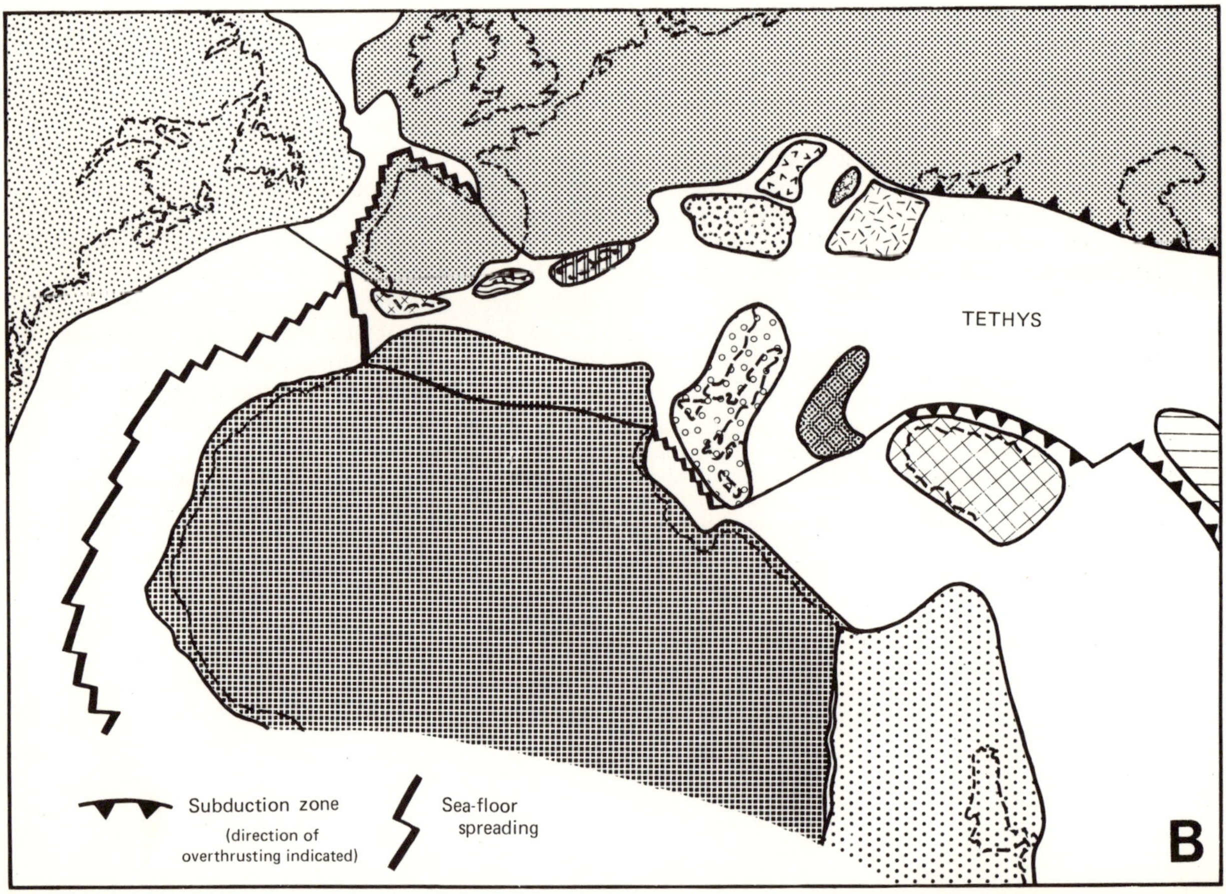

TETHYS
Subduction zone
(direction of overthrusting indicated)
Sea-floor spreading
B

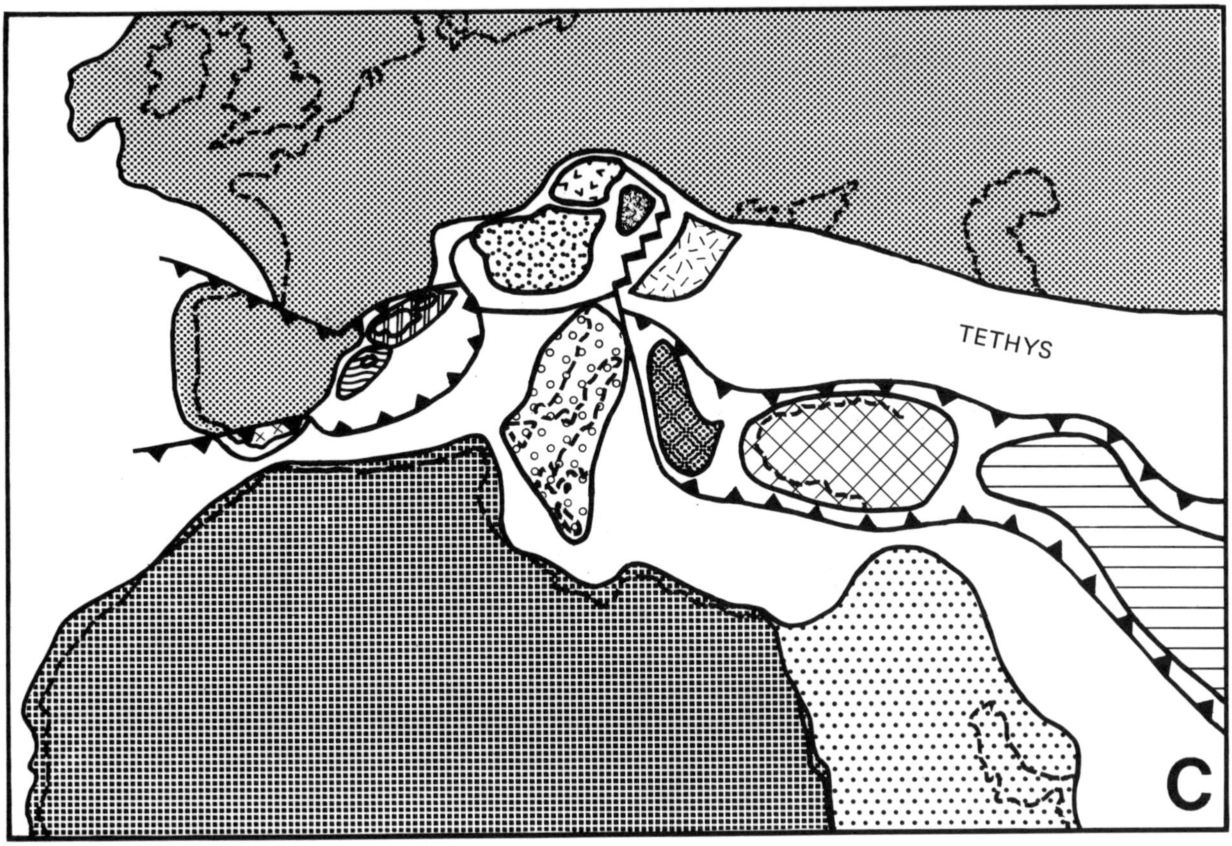


**Fig. 9.17.** The evolution of the Mediterranean Sea is the story of collision between the continental edges of Laurasia and Gondwanaland. As the two super-continents began to approach one another in the Mesozoic, some 200 million years ago, a number of fragments became detached, probably by the influence of convection currents flowing out from the continental edges. From the northern edge of Gondwanaland, a festoon of land fragments rotated northwards. These were the blocks that now largely make up the countries indicated (the numbers refer to the appropriate blocks in the diagrams):

1. North Africa
2. Betic Cordillera (southern Spain)
3. Balearic islands
4. Corsica and Sardinia
5. Italy, Sicily, Dinarides (Yugoslavia), Greece, Crete
6. Bulgaria, Serbia (Yugoslavia)
7. Turkey

Detached blocks also existed along the southern margin of Laurasia:

8. Carnic Alps (Austria)
9. Tatra Mountains (Czechoslovakia)
10. Hungary
11. Romania

Spain and Portugal rotated southwards, opening up the Bay of Biscay.

The resulting collisions produced the mountain chains that now ring much of the Mediterranean. Pieces of ocean floor that were caught in the gigantic earth movements were sliced up and thrust onto the land as ophiolites (see Figs 9.18, 9.19). Italy, a former part of Africa, has been driven violently into Switzerland, piling up, folding and squeezing material of the sea floor to make the Alps. Thus many Swiss alpine peaks (e.g. the Matterhorn) are composed of large sheets ("nappes") of African material thrust over a foundation of European granites etc.

A: Late Triassic (200 million years ago). A large sea, the Tethys, occupied much of the area that is now the Mediterranean.

B: Early Cretaceous (120 million years ago). Sea-floor spreading occurred in the Atlantic and between Africa and the Italian block. Subduction zones started to consume oceanic crust in the eastern Tethys.

C: Eocene (50 million years ago). Ever-increasing amounts of sea floor were consumed in subduction zones, the African-derived blocks were rafted northwards and the Tethys was steadily closed.

During Oligocene and Miocene times the jaws of the Africa-Europe vice gradually closed and the individual blocks were moved into their modern positions.

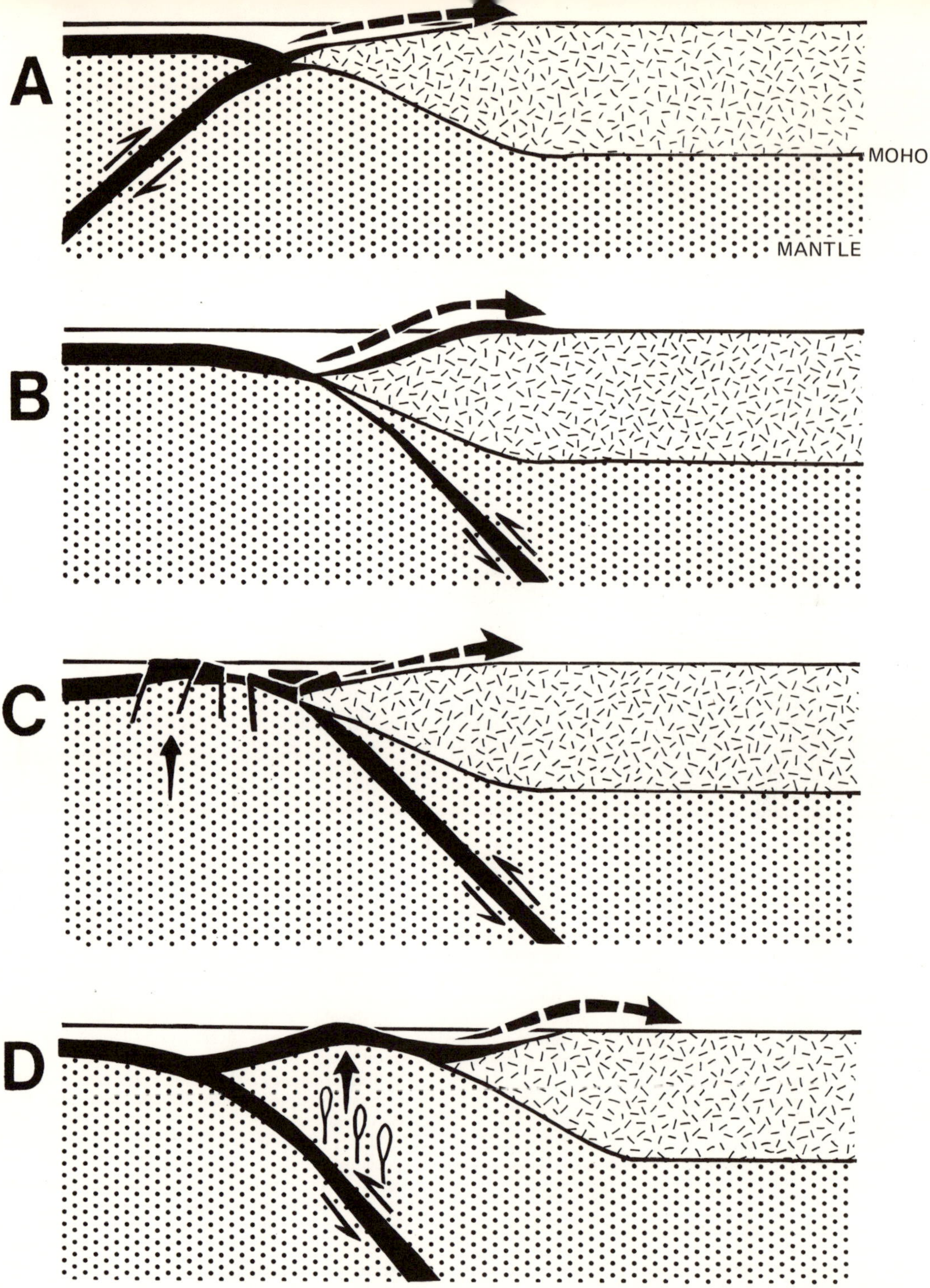


**Fig. 9.18.** These diagrams show four possible ways in which oceanic crust may be thrust up onto a continental margin as ophiolites. Continental crust is shown schematically by the speckled pattern and oceanic crust by solid black. The broken arrow shows in each case the direction of movement of detached slices of oceanic crust.

In diagrams A & B the edge of the continent acts rather like a chisel, paring off slices of oceanic crust as it descends into a subduction zone. In diagram C the chisel edge of the continent has bulged up the adjacent sea floor and split it into blocks, some of which may be thrust up onto the continental margin.

In diagram D, melting of material deep in the subduction zone has produced a marked crustal bulge, as molten magma forces its way towards the surface. Blocks of oceanic crust, derived from the bulge, may slide onto the continental margin.

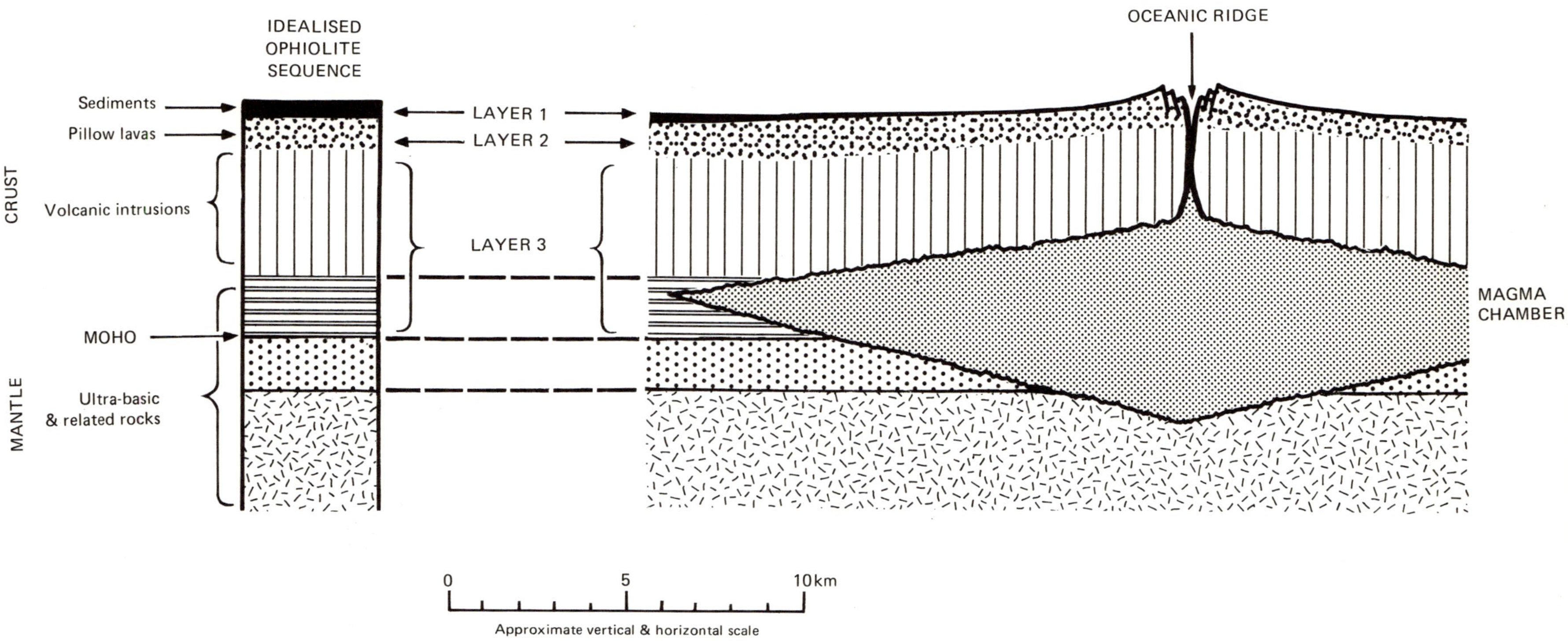


**Fig. 9.19.** The ophiolites that are found thrust up onto the continents are thought to represent slices of the oceanic crust, caught up in the earth movements accompanying plate collisions.

The sketch at the left compares the sequence of rock types found in a typical ophiolite sequence, such as that in New Zealand (Dun Mountain Ophiolite Belt), with that thought to be formed by sea-floor spreading to either side of active oceanic ridges.

would rarely, if ever, be pushed up to the surface, within reach of a geologist's hammer.

The plate tectonics theory suggested a mechanism for squeezing up slices of ocean floor and in 1971 it was proposed that ophiolite rocks occurring in the Troodos Massif, in the middle of Cyprus, were part of a section of oceanic crust and mantle.

The geology of Cyprus was found to be strikingly similar to that of Iceland and has all the characteristics of a mid-oceanic ridge, but dating back to Cretaceous times, some 80 million years ago. The deepest layers, exposed in the Troodos Massif, were formed on an active ridge in the now-vanished Tethys Sea, that existed between Gondwana and Laurasia (Fig. 19.9). As this part of the sea floor moved away from the ridge it became sprinkled with iron-rich muds that are exact parallels of the precipitates accumulating now, for example, on the East Pacific Rise. The only important oceanic feature that seems to be missing is the magnetic striping. As the leading edges of Gondwana (the African and Arabian plates) moved northwards to make contact with the Eurasian plate of Laurasia, the intervening Tethyan sea floor was squeezed and buckled and a fragment thrust up and shaped by erosion into today's island of Cyprus.

Typically, the ophiolites have been thrust up into mountain systems where two continents have collided, crushing a section of sea floor between them. Normally, where such an intervening sea floor is being squeezed between converging continental plates the oceanic plate goes down a trench into the mantle, where it is consumed (Fig. 9.16). However, some slices of the oceanic plate may be thrown up against the far side of the trench, or are thrust up in the final stages of collision, to become embedded in all kinds of scrapings and slumpings, and incorporated into the resulting mountains. The term "obduction", the converse of subduction, has been coined to describe the overall process. In this way mantle-type rocks are thrust while still cold into rocks that are very much younger.

Recognition that ophiolites mark ancient plate boundaries has led to a world-wide search for such rocks. Ophiolites have been found to occur in well-defined bands, as would be expected if they have developed along plate boundaries. They occur in the Himalayas, marking the collision zone between the Indian and Eurasian plates. Another ophiolite band extends along the axis of the Ural Mountains, defining the boundary between two plates that collided, and became welded together, some 300–250 million years ago. A well-known ophiolite band lies at the base of the Matterhorn, in the Swiss Alps, marking the junction between the southern rim of Europe and the northern tip of the African plate (which includes Sicily and Italy) (Fig. 9.17).

Turning to the New Zealand region, an ophiolite sequence very much like that of the Troodos Massif is present on Macquarie Island. Here, a virtually undeformed slice of ocean floor has been pushed up to form land. Both the Troodos Massif and Macquarie Island are very close to plate boundaries—Macquarie Island almost straddling the boundary between the Indian-Australian and Pacific plates. To the north of New Zealand, in New Caledonia, an ophiolite sheet has been thrust up and over a large part of the island (Fig. 16.29B).

Most other ophiolites are much more deformed, as they are often in the form of complexly folded sheets or slices of sea floor stacked on top of one another and thrust into contact with other rocks. Such a highly deformed zone of ophiolite rocks cuts through New Zealand, where it is called the Dun Mountain Ophiolite Belt (Fig. 9.20). On land it can be traced from Kaipara Harbour (Northland), through Pio Pio (near Te Kuiti) and D'Urville Island and then down to the Nelson region, where it is cut by the Alpine Fault. It can be picked up again in south-west Otago, on the other side of the fault, and traced in a sweeping arc through Mossburn and Otama to the south-east Otago coast. Some geologists maintain that the belt can be traced, by means of magnetic anomalies (as the rocks within the belt are highly magnetic), both to the north of New Zealand, some 600 km along the axis of the Norfolk Ridge, and to the south, some 1100 km south-eastwards across the Campbell Plateau (Fig. 9.20).

On land the New Zealand ophiolite belt is sometimes up to 3 km wide, and contains a great range of sediments, volcanic rocks, ultra-basic rocks and serpentine.

Dun Mountain, rising behind Nelson City, is prominent within the Nelson segment of the belt, although the belt itself can be traced from D'Urville Island in the north to Red Hill on the Wairau Valley, where it is cut by the Alpine Fault. Dunite, a very characteristic ultra-basic rock,

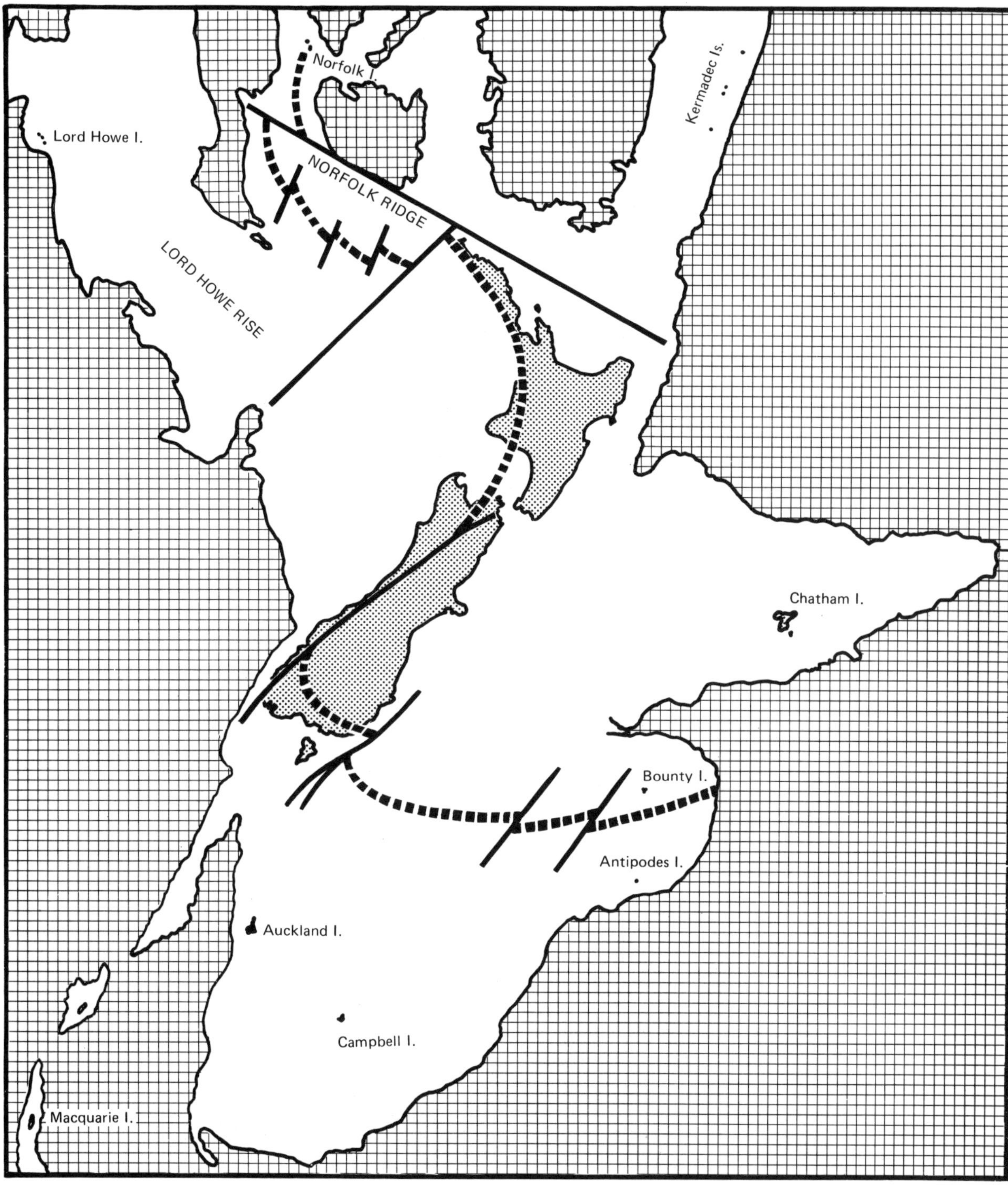


**Fig. 9.20.** The Dun Mountain Ophiolite Belt is the distorted remnant of a strip of ancient sea floor that was caught up in the earth movements that created a large landmass in the New Zealand region some 140-120 million years ago. On land it can be traced from Northland to Otago. Highly magnetic rocks that are part of the ophiolite belt can be traced across the sea floor north and south of New Zealand.

The Dun Mountain Ophiolite Belt is shown as the heavy broken line. Major faults are shown as continuous heavy lines. Sea floor deeper than 3000 m is indicated by the squared pattern.

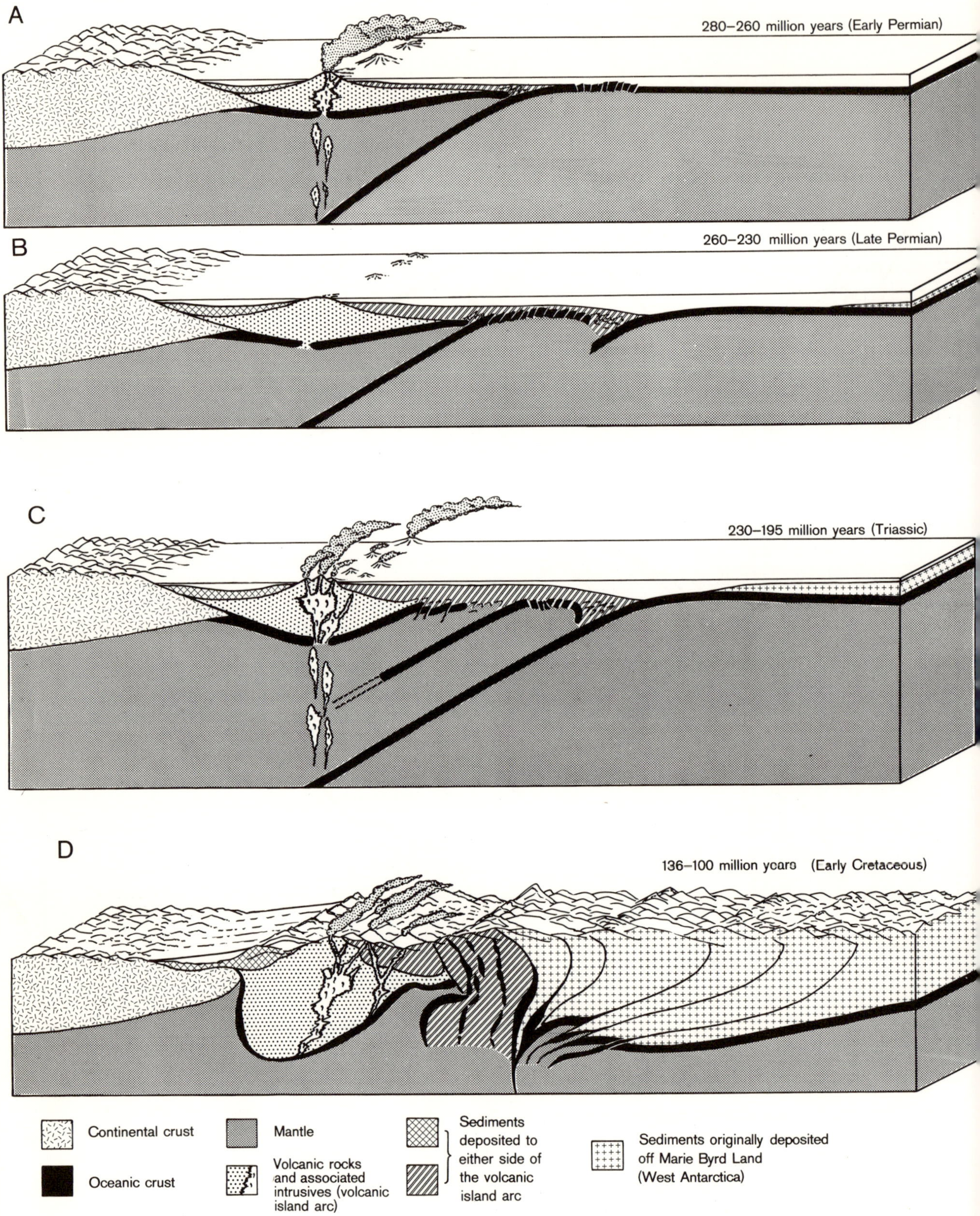
A
280–260 million years (Early Permian)
B
260–230 million years (Late Permian)
C
230–195 million years (Triassic)
D
136–100 million years (Early Cretaceous)
Continental crust
Oceanic crust
Mantle
Volcanic rocks and associated intrusives (volcanic island arc)
Sediments deposited to either side of the volcanic island arc
Sediments originally deposited off Marie Byrd Land (West Antarctica)

**Fig. 9.21.** Various models have been proposed to account for the geological features of the late Palaeozoic and Mesozoic of New Zealand and one of them is summarised in these diagrams.

In the early Permian, some 280 million years ago (diagram A), a volcanic island arc system, related to an active subduction zone, existed over much of New Zealand. Bulging of the plate that was being actively subducted at the time resulted in slices of sea floor and mantle being thrust into the sediments accumulating on the flank of the volcanic arc and in the submarine trench.

In the later Permian and Triassic (260-195 million years ago) (diagrams B, C), the subduction zone seen in diagram A became inactive and a new one formed to the east. Sea-floor spreading associated with this new subduction zone rafted sedimentary material that had accumulated off Antarctica into contact with the locally derived material. The resulting collision buckled up new land to form an ancestral New Zealand (diagram D).

was named in 1864 by the Austrian geologist Ferdinand von Hochstetter after Dun Mountain and since then the rock rodingite and the minerals hydrogrossular, awaruite and wairauite have been described from the belt's Nelson segment or its continuation in Southland and Otago.

Many of the rocks in the ophiolite belt have been sheared and contorted and the ultra-basic rocks themselves have often been altered by heat and superheated steam to serpentine. However, large relatively unaltered blocks weather out of the soft serpentine, to stand out as rubbly knobs or "knockers" that are a characteristic feature of the ophiolite belt's topography.

Minerals abound in the ophiolite belt and for this reason it is sometimes called the "Mineral Belt". Nickel and chrome minerals are abundant and in the early days of European settlement chromite was mined from Dun Mountain, the first tramway in New Zealand being constructed (1865) to transport it to Nelson. Iron, potash, magnesia and other elements are also plentiful. The chrome and iron minerals give an orange-red colour to the rocks and it is perhaps no coincidence that prominent reddish mountains mark the two ends of the ophiolite belt cut and displaced by the Alpine Fault: Red Hill in the Wairau Valley and Red Mountain in north-west Otago, near the Hollyford Valley.

During the Second World War, when supplies of rock phosphate were in very short supply, a scheme was envisaged to "fill-out" the little phosphate available with serpentine. It was then found that the "filler" was in fact a good fertiliser in its own right and that furthermore, when aerial topdressing was introduced, it made the fertiliser run freely out of the aircraft hoppers.

However, although the ophiolite belt is full of minerals, at times it is almost too full and when the concentration of chrome is too high, most plants cannot grow and these areas can be picked out as belts of bare rock or scrubby vegetation.

The New Zealand ophiolite belt is the remains of pieces of ocean floor and upper mantle dating back to Permian times (250 million years ago) that were caught in earth movements, leading to the creation of a large ancestral landmass on the site of modern New Zealand (Fig. 9.21) (p. 241). The distortion of the ophiolite belt can be used as a guide to the original shape of New Zealand (Figs 18.22-18.24).

### Transform faults

Sometimes the crustal plates merely slip past one another, without pulling apart or over-riding. This type of plate boundary involves neither the addition nor the destruction of lithosphere but instead consists of massive faulting, with large amounts of horizontal movement, resulting from the crushing and fracturing related to neighbouring plates sliding and grinding past one another (Fig. 9.22).

The San Andreas fault system in California (Fig. 9.23) has been formed along the line where the Pacific plate is sliding past the American plate. New Zealand's Alpine Fault and its branches in the North Island (often together referred to as the New Zealand Shear Belt) (Fig. 9.24) is related to sliding movements between the Indian-Australian and Pacific plates. The San Andreas Fault joins two parts of an oceanic ridge system, whereas the Alpine Fault joins two subduction zones (Fig. 18.7). As may be seen, both faults have a limited horizontal extent—they do not go on for ever, but are "transformed" into something else—in the two examples quoted either an oceanic ridge or subduction zone. Thus the name "transform fault" has been coined for these rather special faults.

The crests of oceanic ridges are cut by numerous transform faults and in these, active and non-active sections can be distinguished, the latter being "fossil" and relating back to previous sea-floor spreading movements.

### Evolution and structure of the continents

The great continental masses of the world have

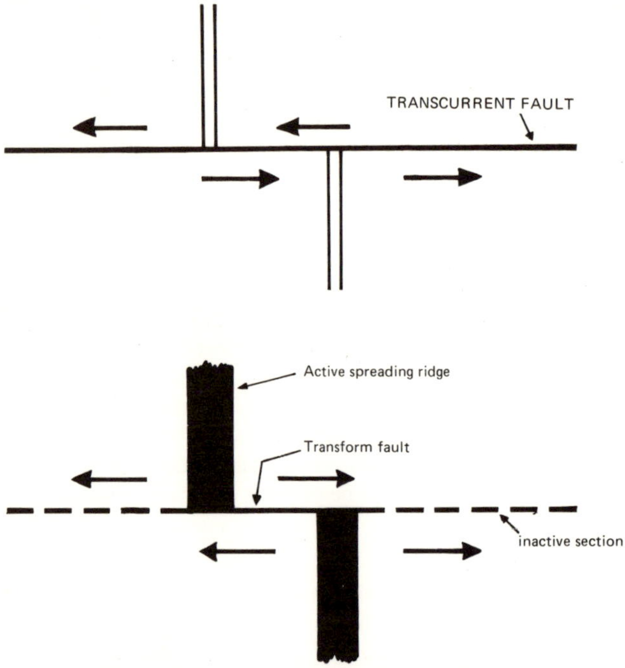


**Fig. 9.22.** In some regions the plates comprising the earth's surface may slip past each other, without very much distortion. They do so along lines of movement called transform faults. For a number of years "transcurrent faults" have been recognised—where movement is dominantly horizontal—as seen in the top diagram. As exploration of the sea floor proceeded it was found that actively spreading centres were criss-crossed with faults that, although they involved horizontal movements, died out to either side. These faults are termed transform faults. In the lower diagram only the section of the fault joining the two segments of the spreading ridge is active and therefore generates earthquake activity (Fig. 9.1).

vast structures known as "platforms" and "shields".

Platforms comprise two units: a rigid stable "basement" composed of old, highly deformed crystalline rocks, and a "cover" of younger, flat-lying or gently inclined strata deposited by the sea at various times on planed-off surfaces cut across the basement. This cover can be very thick in places and includes the extensive flat regions of the Great Plains of North America, the Russian Steppe and the Congo Basin.

Shields, on the other hand, are extensive outcrops of the basement kept clear of cover by persistent uplift of the crust. Shield areas predominate in large areas of both Canada and the U.S.S.R., for example.

Each continent is a combination of platforms and shields bordered or crossed by younger fold-mountains formed in the past 500 million years. At least four-fifths of the continental crust, exposed in shields or buried under sedimentary strata in platforms, is Precambrian basement older than 600 million years.

The structure of this old basement is extremely complicated as it has generally passed through a series of folding stages accompanied by melting or partial melting. Steeply dipping "hairpin"-like, folded structures are very common, although often in the very oldest crystalline rocks the folds are near-horizontal, as if flattened by the super-incumbent load. An association of rocks that characterises many ancient crystalline terrains in South Africa, India and Australia consists of belts of "greenstone" (basalt lava altered by heat and pressure) "swimming" in a sea of granite or altered granite (granite-gneiss). It is difficult to relate these rocks and structures to those seen in younger fold-mountain ranges. Some geologists maintain that the greenstone belts must be pieces of very ancient oceanic crust, but their location in granite is not easily explained. A curious feature of many old shield areas is that, despite the appearance of extremely ancient rocks at the surface, the continental crust beneath maintains a constant thickness.

Large areas of the continents are covered by layers of undeformed sedimentary rock accumulated in shallow seas which once flooded the continental surface. In the centres of some deep sedimentary basins the basement is depressed to more than 15 km (9.5 miles) below the surface. The cause of such persistent down-warping, lasting over hundreds of millions of years, is obscure. Most of the sedimentary cover has been deposited in the past 600 million years, but some is very ancient; the famous gold-bearing sediments of South Africa are more than 2000 million years old.

In the fold-mountain belts younger than 500 million years, the ancient crystalline basement is often reactivated and takes part in younger folded and over-thrust structures; it commonly builds large parts of the interior regions of younger fold-mountains, such as the Central Himalayas. In fact, in Africa and South America, extensive belts of re-activated crystalline basement are all that remain of some of these mountain ranges that have been almost completely eroded away. Until fairly recently it was not possible to identify older and younger parts of shields with any precision. Now

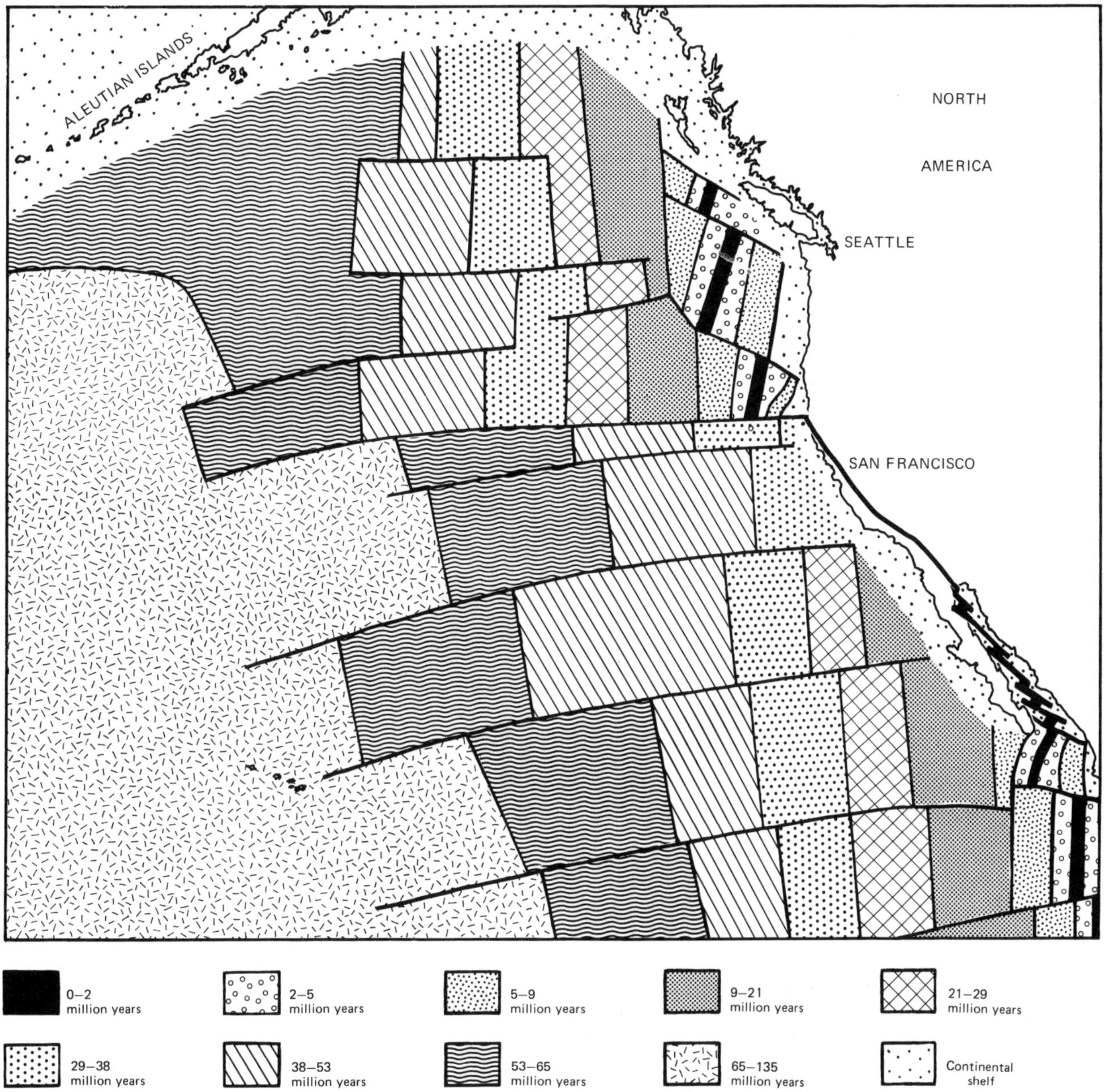


**Fig. 9.23.** Early in the 1950s ship-board magnetometer surveys established the presence of magnetic stripes on the sea floor off the west coast of North America. At first the stripes were regarded as virtually unintelligible, and as representing purely random variations in magnetism. Then in the 1960s characteristic magnetic signatures were recognised and these provided clues that led to the delineation of distinct blocks of magnetic patterns. This in turn led to the discovery that many of the patterns were displaced sideways along gigantic fractures, hundreds of kilometres in length, extending out into the Pacific as far as Hawaii and beyond.

The displaced patterns provide an indication of the amount of sea-floor spreading that has occurred in the North-east Pacific over the geological ages. New oceanic crust is being generated at the active ridge that runs into the Gulf of California, which is the northern continuation of the East Pacific Rise (Fig. 8.1). The San Andreas Fault provides a link through California, joining a spreading ridge extending towards Alaska. Spreading is opening up the Gulf of California and moving the North-east Pacific sea floor in the direction of the Aleutians, where it is consumed in a subduction zone. As it is created the new sea floor is broken into short segments separated by transform faults. The magnetic patterns of the individual segments can be correlated with time scales established on land and the ages thus obtained provide a series of "growth rings", rather like tree rings, for the sea floor.

with radioactive age determinations, the age pattern of a continent can be identified like the rings in a tree-trunk. The continents are seen to consist of ancient cores on to which are welded successively younger extensions. Fig 9.25 shows growth stages in the exposed and concealed basements of North America, though it should be emphasised that much of the rock composing younger growth stages is simply re-activated older basement. How has this pattern developed? The plate tectonics theorists imagine a number of small "micro-continents" bordered by ocean trenches colliding with each other and with oceanic island arcs to build up larger and larger units. Growth was not regular: world-wide interruptions marked by intense deformation and igneous activity occurred around 2700, 1800 and 1000 million years ago. After each interruption, mountain-building cycles tended to follow parallel, superimposed tracks, in places at right angles to previous patterns, until the next new pattern was established. A surprisingly large proportion of the continental crust was already in existence 2500 million years ago, much of it formed at very high temperatures and pressures unknown in mountain belts younger than 1500 million years.

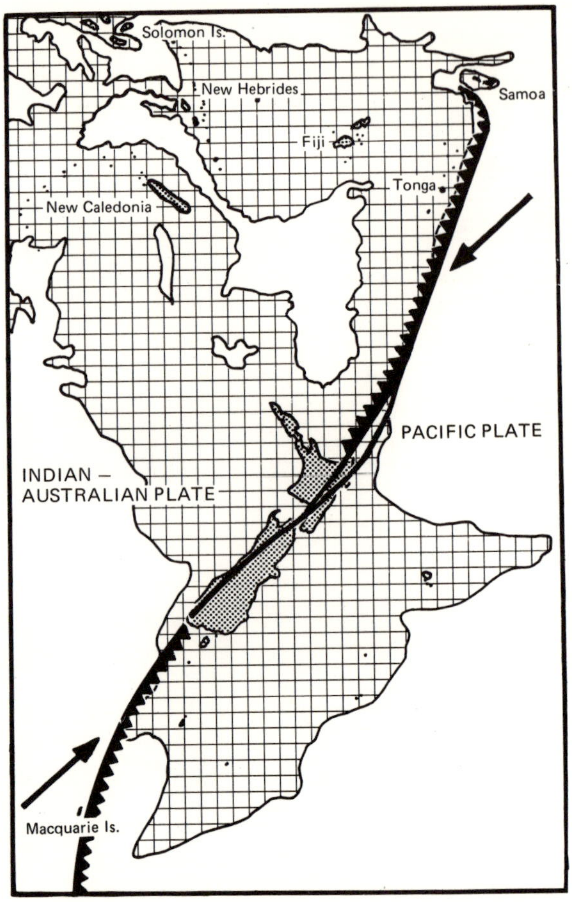


**Fig. 9.24.** New Zealand's Alpine Fault (Fig. 4.11) is a transform fault linking the Tonga-Kermadec Trench with the Macquarie Trench. In the North Island and the northern part of the South Island the Alpine Fault splits and forms a complex zone of individual faults (the New Zealand Shear Belt), that together pass into the Tonga-Kermadec Trench (cf. Fig. 18.17). The arrows show the direction of relative motion (vectors) between the Pacific and Indian-Australian plates and they lie on the plate edges being consumed down the trenches (toothed pattern). The trench system extending through the New Hebrides and Solomon Islands and the linkage via Fiji is not shown (but see Fig. 9.4). Sea floor ranging in depth from 0 to 1800 m (2000 fathoms) is shown by the squared pattern.

## The origin of mountain ranges

The powerful deformation and dislocation of continental crust that occur along narrow "mobile belts" subsequently uplifted as mountain ranges is called "orogenesis" (the "genesis of mountains"). The deformation and uplift affect abnormally thick piles of sedimentary and volcanic rock—up to 20 km (13 miles) thick—accumulated during several hundred million years in deep water along the length of the belt. The whole process of deposition, deformation and uplift is called the orogenic cycle.

For many years it was thought that orogenic belts evolved in transcontinental down-warps or "geosynclines" were located over down-sinking currents in the mantle. Deformation occurred when the continental crust beneath the down-warp melted and so lost its rigidity, allowing the sediments above to be squeezed. Uplift of the deformed belt as a mountain range followed when the currents stopped or reversed direction. Now a new theory of mountain-building has been advanced which takes plate tectonics as the underlying mechanism. The process starts with rifting of a continent and formation of an ocean

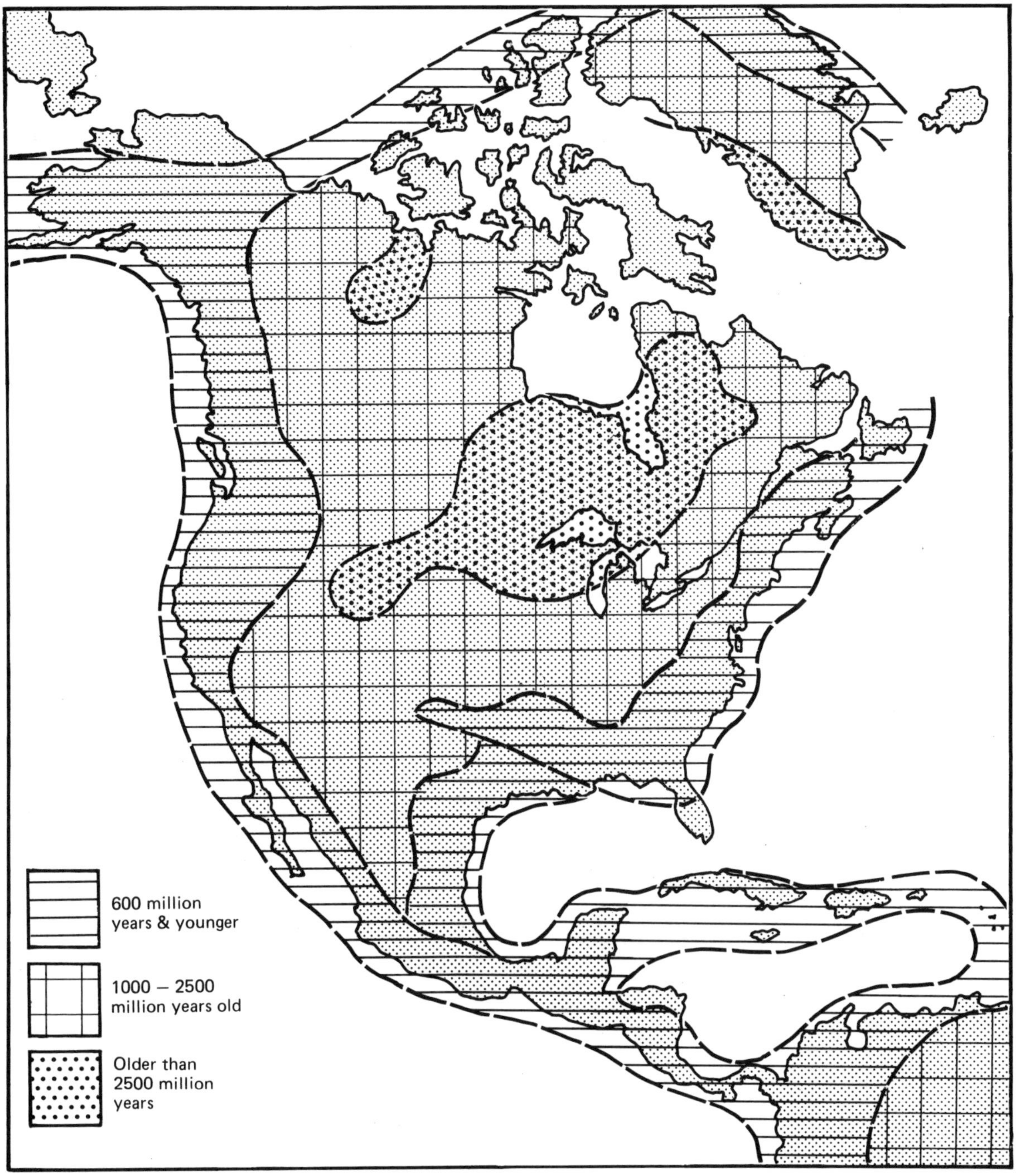


**Fig. 9.25.** The large continents of the world have been built up around very ancient cores (or "nuclei"), by the addition of belts of successively younger rocks. The belts have become "welded" onto the old nuclei by means of the process summarised in Fig. 9.16. North America, for example, has been built up around two ancient nuclei, the oldest rocks of which occur in Minnesota where they attain an age of 3600 million years.

An even older nucleus occurs in Greenland, where the oldest known rocks in the world occur, dated at 3800 million years.

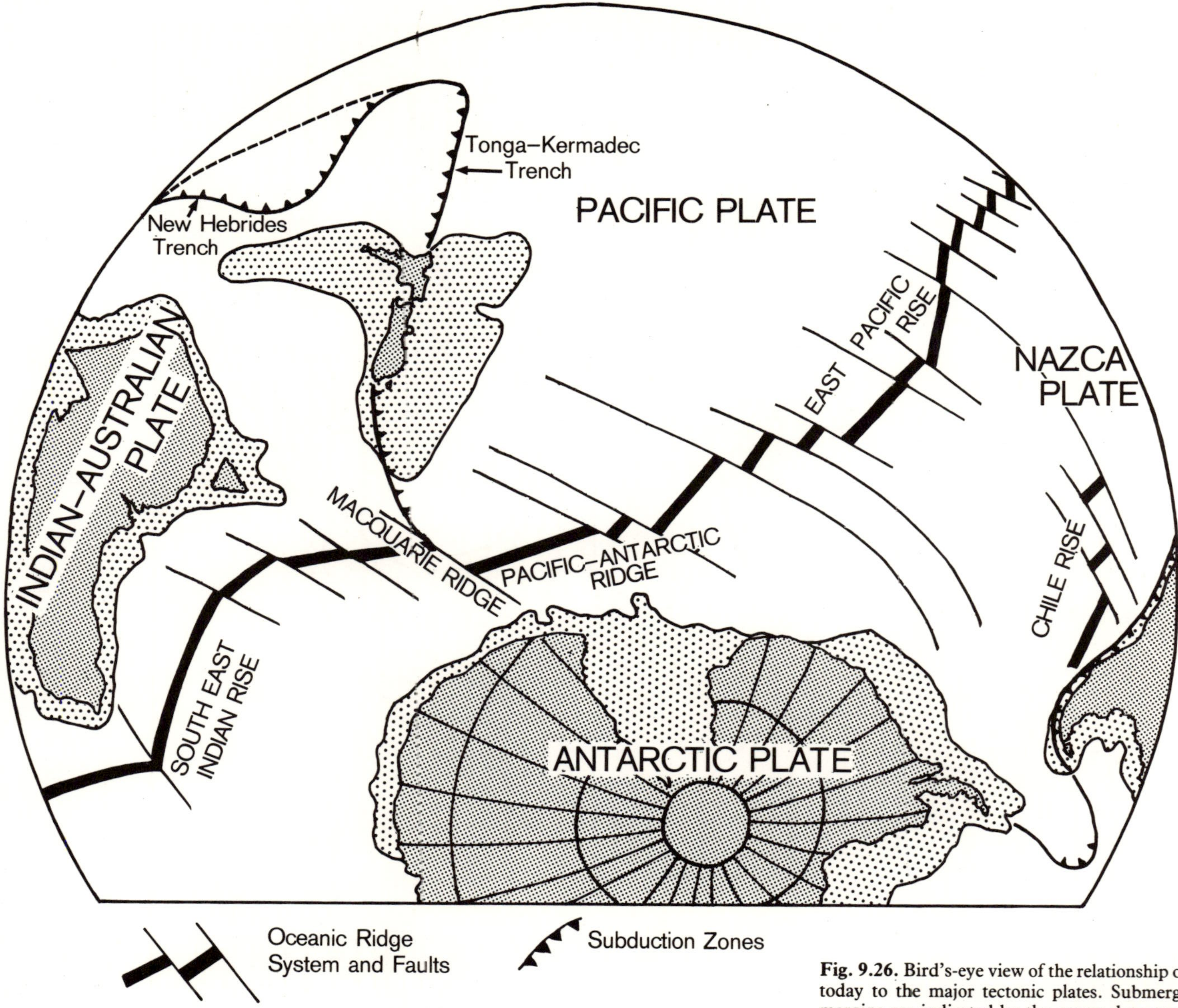


**Fig. 9.26.** Bird's-eye view of the relationship of New Zealand today to the major tectonic plates. Submerged continental margins are indicated by the coarse dot pattern.

by sea-floor spreading. Thick muds and silts accumulate on the continental slope alongside thinner limestones and sands on the continental shelves on both sides of the ocean. After perhaps one or two hundred million years, a trench forms on one side of the ocean and the oceanic crust slides under the continent. In our example, this is followed by consumption of the crust in the open ocean, leading to formation of a volcanic island arc. After further trenches and consumption zones have been established, the island arc is swept into the continent. Finally the ocean closes and the continents collide. The sediments at the continental margins and the volcanic rocks of the island arc are folded and over-thrust. Molten igneous rock, generated when the oceanic crust melts in the depths, rises into the crust. Uplift results when buoyant continental crust is thrust under adjacent continental crust.

### The New Zealand story

At this point we may pose the question: how does New Zealand fit into this global picture of moving plates? New Zealand has been referred to in various places in this chapter—with respect to subduction zones, Benioff zones, transform faults, andesite volcanoes, ophiolites etc. The reader may wonder how it all ties up.

The truth of the matter is that New Zealand has been different things at different times, as we shall see in Chapters 11-18.

The New Zealand region has been closely related to a convergent type of plate boundary for many millions of years—certainly since at least Carboniferous times, some 300 million years ago. Then New Zealand was an area on the ocean floor between the coasts of Eastern Australia and Antarctica. It was the site of an active subduction zone, and associated submarine trench and volcanic island arc. A Benioff Zone dipped westwards under the site of modern New Zealand. Active subduction and volcanic activity generated from the volcanic arc continued throughout the Permian, Triassic and Jurassic (280-136 million years), and substantial thicknesses of sediment were laid down on either side of the volcanic arc. All this time more and more sea floor was being conveyed to and consumed in the subduction zone, until eventually a mass of sediment that had originally been deposited off the coast of West Antarctica arrived at the subduction zone and, being buoyant continental material (sial), could not be consumed in the subduction zone. Thus it rode up and collided with the sediments that had accumulated in the submarine trench and on the flanks of the volcanic arc. Tremendous thrusting, shearing and buckling movements resulted and slices of sea floor and upper mantle were caught up and pushed into the contorted piles of sediment that were forced up to form land. These major earth movements, called the Rangitata Orogeny, are thought to have occurred mainly during the late Jurassic and early Cretaceous (140-120 million years ago). The ancestral New Zealand thus created—the first sizable landmass to exist here—extended northwards to at least New Caledonia and southwards to Campbell and Auckland islands.

The resulting accumulation of a contorted mass of sial "gummed up" the subduction zone, as it were, and it had to do a flip to the edge of the newly created land, similar to that illustrated in Fig. 9.16.

At about this time, sea-floor spreading started to occur between ancestral New Zealand and Australia, and the embryonic Tasman Sea came into being. The spreading ridge in the Tasman was active between 80 and 60 million years ago, and during this time ancestral New Zealand was rafted eastwards and pushed further and further over the Pacific floor, that progressively disappeared down the subduction zone lying along New Zealand's eastern margin.

Sea-floor spreading then started to occur between Australia and Antarctica, that up to then had been in close contact, and the Southern Ocean began to open. These movements first started about 55 million years ago, and are still continuing today. With the end of sea-floor spreading in the Tasman Sea some 60 million years ago, New Zealand became fixed in relation to Australia. In other words, it became part of the Indian-Australian plate, and as that plate pulled away from the Antarctic plate, and drifted northwards, so did New Zealand.

Thus today New Zealand straddles the boundary between the Pacific and Indian-Australian plates (Fig. 9.26). To the north of New Zealand the plate boundary is reasonably straightforward, with the Pacific oceanic floor being subducted under the edge of the Indian-Australian plate and the development of a submarine trench, Benioff Zone, volcanic island arc etc. (the Tonga-Kermadec system) (Fig. 9.27).

A similar subduction set-up exists to the south of New Zealand (the Macquarie system), complicated however by the fact that here the edge of the Pacific plate is apparently being pushed *over* the Indian-Australian plate. Thus all the features associated with subduction are developed as a mirror-image of the Tonga-Kermadec situation (Fig. 9.26).

New Zealand forms the link between the two—with the Tonga-Kermadec system dying out under the North Island, and the Macquarie system dying out off shore as it approaches Fiordland.

As may be expected, the geology of New Zealand reflects the complexities of being the go-between for the two systems—as it was squeezed, pushed and pulled and eventually torn apart. However, the Alpine Fault is one major feature that stands out amongst the maze of other related structures, and it, along with associated branch faults, forms a transform system, linking the two subduction systems. Thus throughout its long geological history New Zealand has played a variety of roles, but has never been very far away from an active plate margin of the convergent type.

**Fig. 9.27.** Schematic diagram summarising the deep-crustal processes affecting New Zealand.

As far as we know, a divergent plate margin (i.e. the one associated with oceanic rifting) has never traversed New Zealand, although it was very close to the scene of action when the Tasman Sea opened up by a process of rifting and sea-floor spreading between 80 and 55 million years ago.

## The future

The theory of plate tectonics has enabled us to look at the earth in a new light. By studying the magnetic "time pips" preserved on the sea floor we can now delineate vast areas of the oceans that have been formed comparatively recently (speaking in geological terms!) (Fig. 9.28). On the other hand, great areas of the world's major continents are immensely old and the land itself has been built up as a complex mosaic of strips welded on at various times.

Plate movements however are very slow in human terms. Greece, situated on one of the faster moving plates, has moved some 90-180 m (100-200 yards) since the time of Socrates (400 B.C.). Similarly, the Atlantic has opened up some 5-10 m (5-10 yards) since Columbus crossed it in A.D. 1492. The Atlantic is opening at a rate of 12-25 mm (½-1 in) per year, i.e. 12-25 km

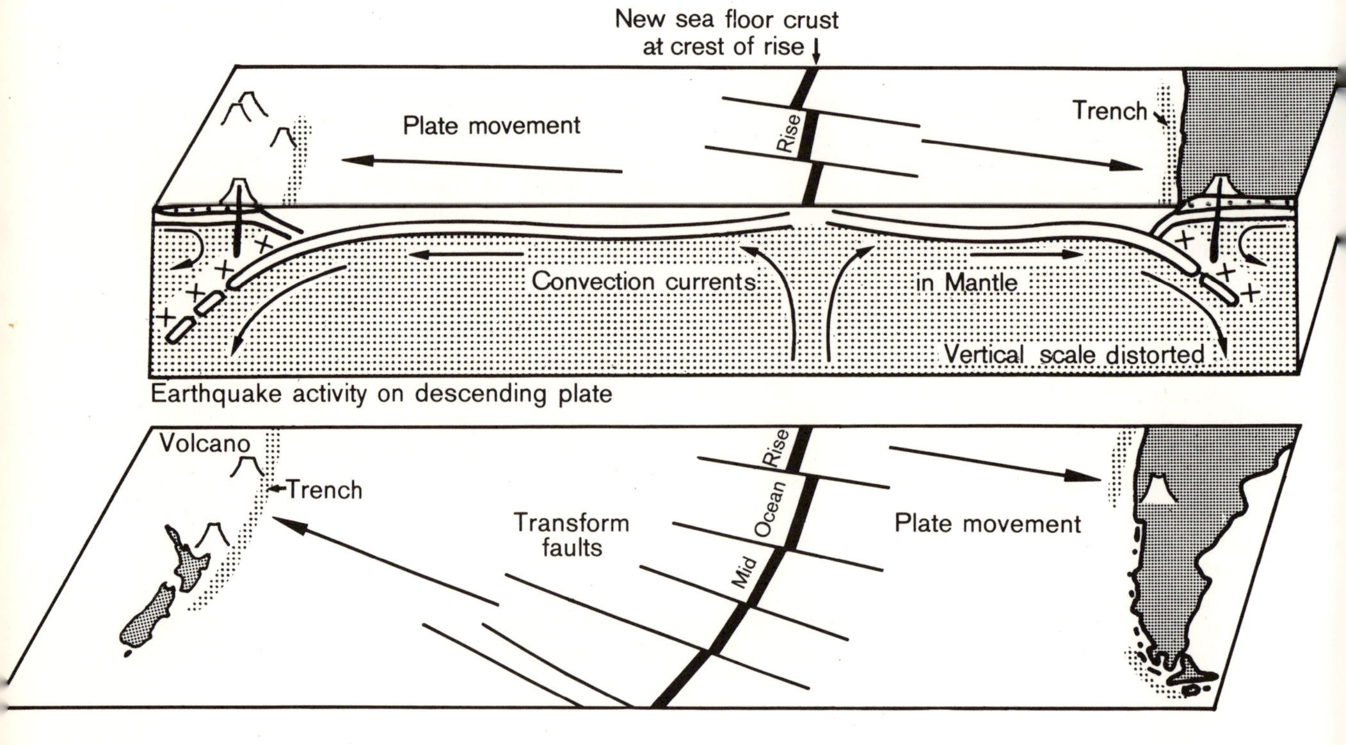

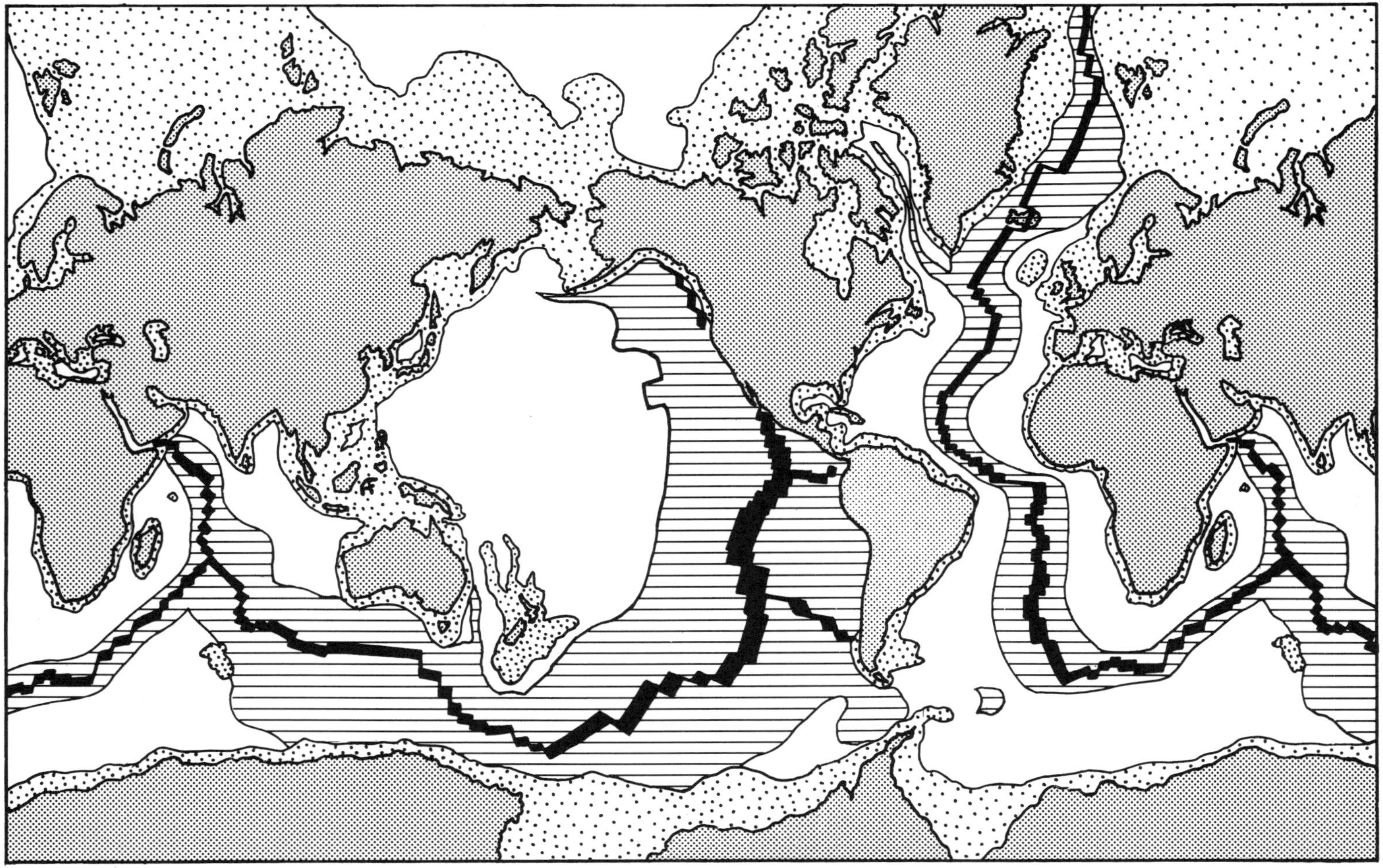

**Fig. 9.28.** Judging from magnetic studies and deep-sea drilling, vast areas of sea floor have been created in geologically "recent" times. This map shows the amount of sea floor that has been formed in the last 65 million years (horizontal lined areas). The actively spreading oceanic ridges are outlined in black and the submerged continental margins by a coarse dot pattern. The East Pacific Rise is creating new crust at a rate three times faster than that of the Mid-Atlantic Ridge, as can be seen by comparing the amount of material produced over the last 65 million years. Crust however is being almost equally actively consumed in subduction zones in the Western Pacific (Fig. 9.2), so that today there is no crust in the Pacific older than about 190 million years (Fig. 8.28).

(7.5-16 miles) per million years. The Pacific is opening at a rate of 35-90 mm (1½-3½ in) per year, i.e. 35-90 km (22-56 miles) per million years. In terms of our own bodily growth, the Pacific is opening up at the rate one's fingernails grow (0.10 mm per day) and the Atlantic at the rate one's toenails grow (0.05 mm per day). Thus as New Zealand (or at least some of it) drifts steadily northwards as part of the Indian-Australian plate, it will take 20 million years or more for her to start to bask in the tropical sun. Similarly, it will be the best part of 50-60 million years before Australia starts to nudge the Chinese mainland (Fig. 9.30).

In the meantime, to keep tabs on the rates of present-day plate movements, a global measuring project using special space satellites has been put into operation. In this project laser beams have been bounced off satellites orbiting around the earth. Distance measurements have been obtained from a global network (Fig. 9.29), in which each station is related to all others. These distances have been measured to an accuracy of about 100 mm (4 in). The measurements will then be repeated over a period of years, when it is hoped that the rate of the opening of the Atlantic, for example, can be measured directly. Technological improvements are expected to bring the accuracy down to 30 mm (1.25 in) or less and as some of the plates have been moving at about this rate annually in the past, and are presumably still doing so today, it is hoped that definitive results will be available within a few years.

In another related development, as part of the Apollo lunar landings, laser beam reflectors were positioned on the moon. These have been used to bounce laser beams between the earth and the moon and the resulting measurements will be used to gauge the rates of displacement of large plates such as the Pacific and North American plates.

Projections based on geologically recent sea-floor spreading rates, obtained via the magnetic polarity scale (Fig. 8.8), have allowed earth scientists to indulge in a bit of crystal-ball gazing. For example, a map of the world in 50 million years time has been visualised—that is if we haven't blown ourselves to bits by then!

A commonly accepted reconstruction of world geography at this time is shown in Fig. 9.30. As may be seen, some major changes are envisaged that will considerably alter world climatic patterns—as continents change their latitudes and ocean currents rearrange themselves in the wake of such changes.

**Fig. 9.29.** Diagram showing the network of stations used in measuring the width of the Atlantic by radio-astronomic methods.

The future of New Zealand, geologically speaking, is difficult to foretell. At the moment a major plate boundary bisects the country, and so in theory that part of New Zealand west of the Alpine Fault sits on the edge of the Indian-Australian plate and the remainder on the edge of the Pacific plate. The Indian-Australian plate is in the throes of drawing away from the Antarctic plate, a process that started about 55 million years ago. Thus part at least of New Zealand will presumably be dragged northwards along with the Indian-Australian plate.

However, various indications suggest that the modern New Zealand situation is rather transitional. The situation has probably been complicated by the arrival at the plate edge of that part of New Zealand now lying east of the Alpine Fault. Being sialic (and therefore buoyant), this segment of land will not go down any trench, and the scene may be set for a flip of the plate boundary eastwards, out into the Pacific, as in Fig. 9.16. If this in fact happens, all of the New Zealand landmass will drift steadily northwards, as part of the Indian-Australian plate. Alternatively, if the plate boundary remains in its present position, New Zealand may at some stage in the future find itself as two groups of islands as in Fig. 9.30. This no doubt will pose a few problems politically!

**Fig. 9.30.** Some 50 million years from now the world may look something like this. If in the future the world's plates continue their inexorable movements, the Atlantic and Indian oceans will grow at the expense of the Pacific (unless the world expands simultaneously). Today's Red Sea has widened out into a mini-ocean and the East African Rift System is the site of an embryonic strait. The northern drift of Africa has squeezed Spain and Portugal into the Bay of Biscay and virtually shut the Mediterranean. Australia and New Guinea have moved northwards to begin to nudge Eurasia, and New Zealand has similarly drifted in the same direction—although whether it will continue to be torn apart by the Alpine Fault, as shown in this diagram, or move as one piece on the edge of the Indian-Australian plate is not clear.

The part of California that is west of the San Andreas Fault (Fig. 4.12) will be severed from the rest of North America and begin drifting to the north-west. In about 10 million years Los Angeles will be opposite San Francisco, still fixed to the mainland. In about 60 million years Los Angeles will start to slide into the Aleutian Trench.

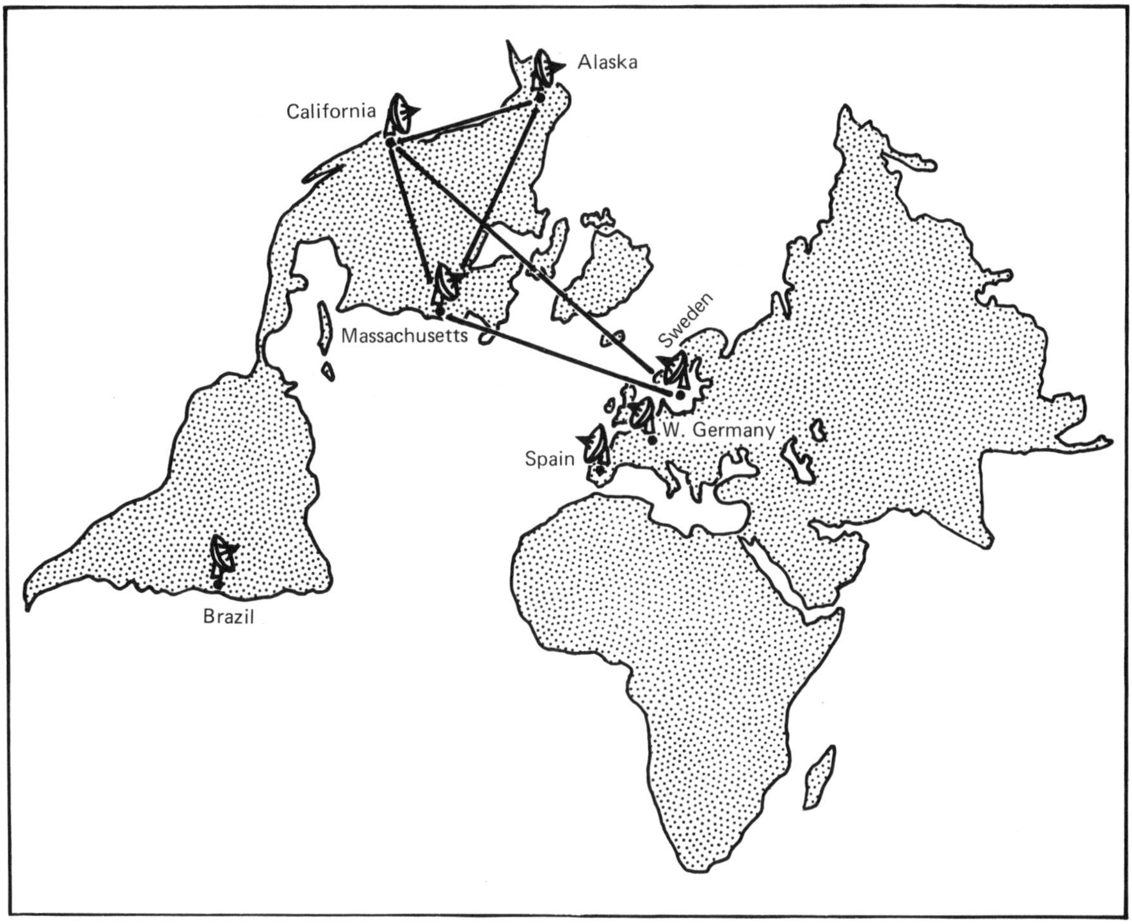


*Fig. 9.29*

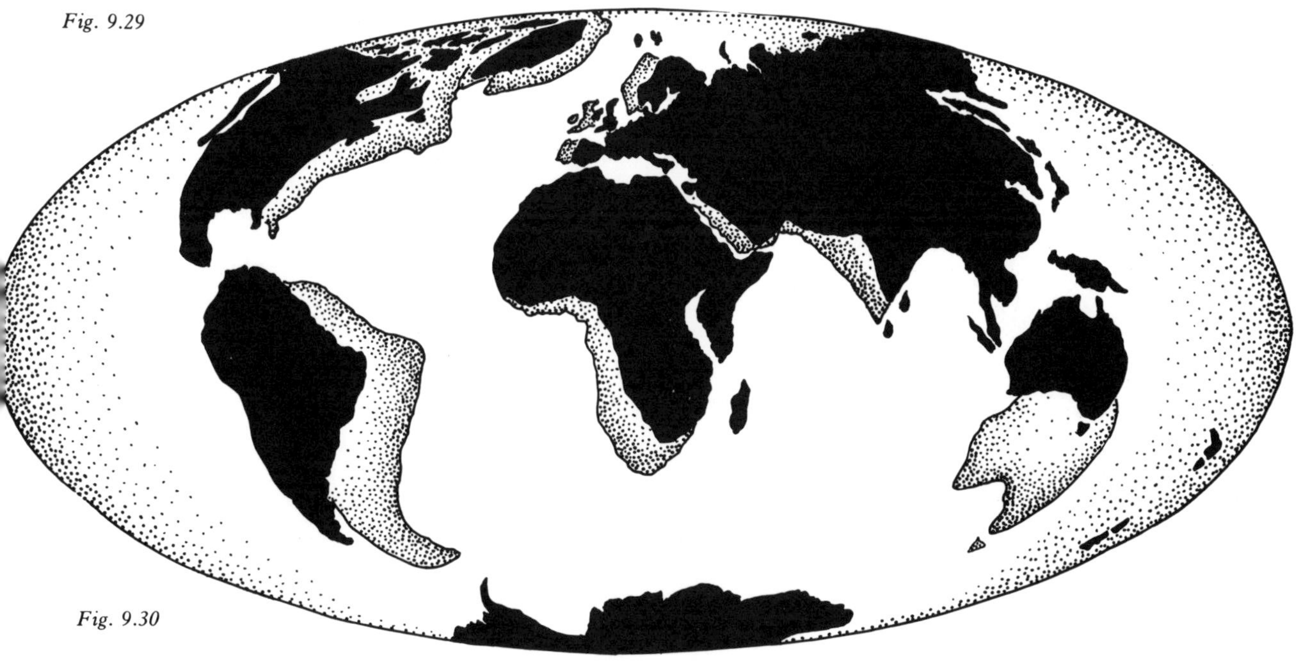

*Fig. 9.30*

CHAPTER TEN

# HOW IT ALL HAPPENS — OR SO WE THINK!

ALTHOUGH in Chapter 4 and other places we have touched briefly on the question of a mechanism for continental drift, we have tended to assume that movements *have* occurred, without worrying too much about how they happen.

In Alfred Wegener's time his theory of continental drift had come to grief on the problem of an ultimate cause—a mechanism of sufficient force to bring about the drifting of the continents. Wegener nonetheless thought it would only be a matter of time before a generally acceptable driving force was identified, because for him unshakeable evidence for continental drift could be seen on all sides.

The problem, of course, with speculation about mechanisms is that the action takes place tens and hundreds of kilometres below the earth's surface, concealed from our view. All that we see on the land and ocean are the superficial effects of such deep-seated processes. The field is therefore wide open to speculation—and controversy!

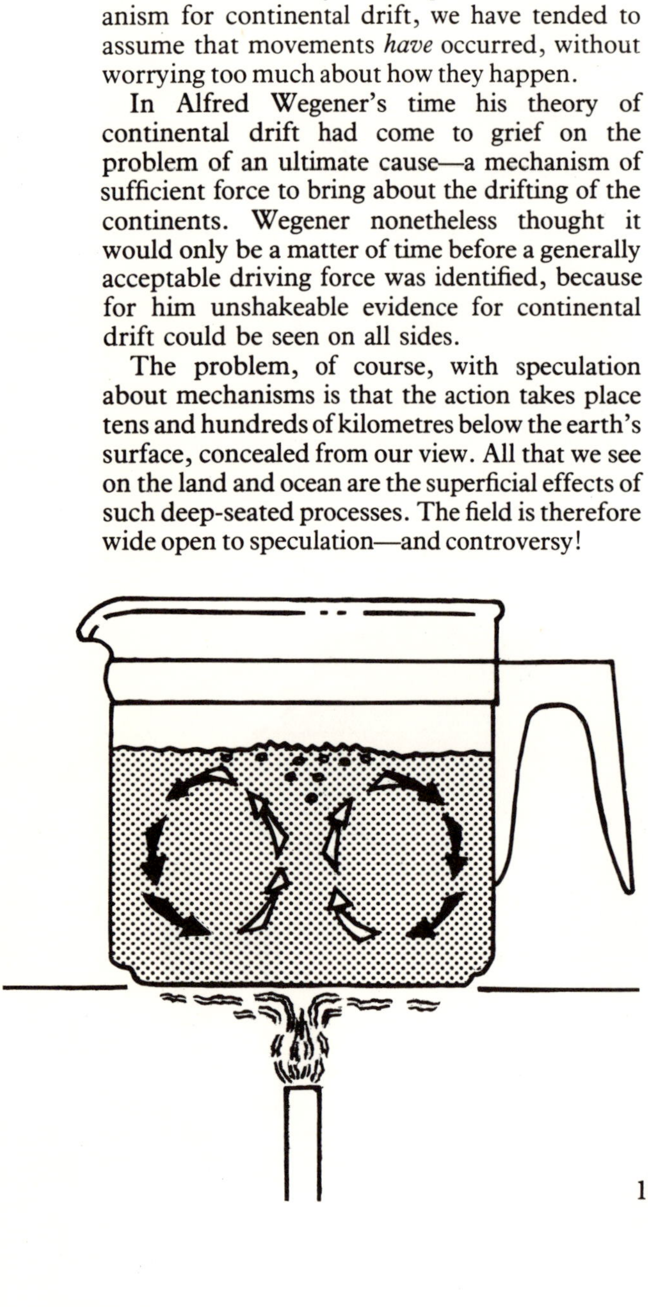

Nevertheless, over the years a vast store of information has been gleaned from study of earthquakes, rocks and volcanoes, and from chemical and physical laboratory research, and when taken together, these data point in certain well-defined directions and narrow the range of possibilities.

Most earth scientists have their pet theories to account for the why and how of continental drift but I think it is fair to say that nobody has yet come up with an overall scheme acceptable to everyone else.

A variety of factors interact to provide possible mechanisms for continental drift. They operate at a number of different levels, and in this chapter we will, for ease of reference, consider the various factors in order of scale, ranging from those operating world-wide down to those affecting individual plates. We must, however, remember that in nature all the various factors are inter-related and form a complex, interwoven system.

Most of the theories that have been put forward over the years to explain continental movements envisage the existence of convection currents, stirring the earth's material below the crust. This is a process by which hot material rises towards the earth's surface, spreads and cools and then sinks again. Such currents in the mantle may be the driving force for sea-floor spreading and for continental drift.

Convection is a well-known physical phenomenon. If a coffee pot is filled with water and placed onto a stove element to heat, the hot element transfers heat into the lower layers of water in the pot. As the bottom layers of water heat up, they become lighter and start to rise up towards the surface. Cold water then sinks down from the

**Fig. 10.1.** Convection occurs when water is heated in a coffee pot.

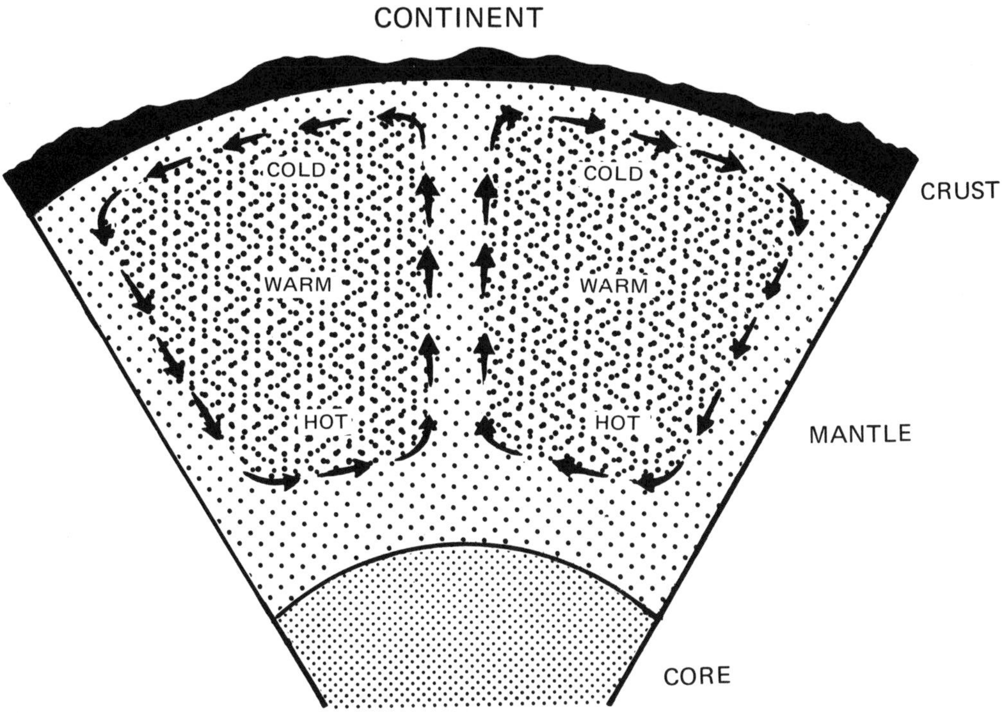


**Fig. 10.2.** Convection is envisaged as occurring in the earth's mantle, although whether the entire width of the mantle is involved is not known (cf. Fig. 10.16).

upper layers to take the place of the rising heated water. The newly arrived water is then in turn heated and eventually rises towards the surface (Fig. 10.1).

When the heated water in the coffee pot rises to the open upper surface, it loses a certain amount of heat into the atmosphere where it is visible as steam) and, as it cools, it gradually sinks to the bottom again, where it is heated once more.

Thus there is a steady transfer of heat from the bottom of the pot to the upper layers. If we add a dye to the water in the pot we can see that this transfer of heat generates a series of eddy-like currents (Fig. 10.1), some descending, others ascending; these flow circuits have been given the name "convection currents".

The term given to a pair of convection currents is a "convection cell". Thus a cell consists of two interacting currents—they converge, rise to the surface as virtually one stream of particles and then, after losing heat, split again and return to the depths in two separate flows (Fig. 10.2).

Before exploring the possible mechanisms, however, perhaps we should deal briefly with the origin of the earth and the solar system, as this leads us to a better understanding of the earth's architecture.

**How old is the solar system?**

The solar system, of which our earth is part, consists of planets, moons, asteroids, comets, meteorites, dust and gas, and a central body—the sun—about which all these revolve. This solar material circles the sun in the same direction, mostly spinning the same way and within the same plane, which suggests that it probably all formed together, at about the same time. If this is so, different parts of the solar system should have similar ages.

To study the ages of the various solar bodies we do need to have a sample in our hands, so that it can be analysed. This of course is rather a tall order and may never be accomplished. Even with our own planet, Earth, it has only been in the last few decades that a clear picture has emerged of the overall range of ages of rocks exposed on the surface. With the assistance of detailed field

*Fig. 10.3*

**Figs 10.3-10.5.** Meteorites are of great scientific importance as they provide us with a great range of samples of solar bodies that have disintegrated a long time ago. It has been estimated that ten to twenty million stellar fragments of various sizes enter our atmosphere every day. These fragments have been derived from explosions of celestial bodies out in deep space. The explosive disintegration of an asteroid, for example, would hurl vast amounts of material through space, which, coming within reach of the earth's gravitation, are swiftly drawn into our atmosphere. Almost all of the debris is probably burnt up by heat generated by the friction of its passage through the air—producing "shooting stars" and "falling stars". These heavenly rockets are called "meteors" from a Greek word meaning "things in the sky".

Some meteors are so large that they reach the earth before being completely burnt up. In such a case they burst shortly before reaching the ground, producing a shower of fragments—"meteorites". Large meteorites may explode noisily and excavate enormous craters such as the renowned Meteor Crater in Arizona. This feature is 1.25 km (¾ mile) across and 185 m (600 ft) deep, and it has been estimated that the meteorite responsible weighed about 60,000 tonnes and struck the earth at about 16 km (10 miles) per second.

**Fig. 10.3.** In January 1976 an iron meteorite weighing 49.5 kg (110 lb) was found by Mr Ted Dowie in a stream bed near Dunganville, Greymouth district, Westland. This photo shows Mr Dowie standing with the meteorite.

**Fig. 10.4.** This photo shows the discovery site of the Dunganville meteorite. To judge from pebbles stuck on its surface, the meteorite had evidently landed on a low terrace of stream gravels. Subsequently stream erosion had undermined the terrace edge and the meteorite had tumbled into the stream bed. The main mass of the meteorite and some 4 kg (10 lb) of fragments were found where the figures are.

**Fig. 10.5.** The Dunganville meteorite consists of a mass of nickel-iron alloy. Internally the alloy has separated into nickel-rich (30-60 percent nickel) and nickel-poor (4-7 percent nickel) portions arranged in interwoven bands parallel to the original crystal faces of the meteorite. These characteristic patterns ("Widmanstätten" patterns) show up when the meteorite is cut, polished and etched with acid, as in this photo, which is a cross-section, 33.3 cm (1 ft 1 in) in length, across the Dunganville meteorite.

The Widmanstatten pattern has been produced by infinitestimally slow cooling, such as that taking place within the core of an asteroid. This specimen is now on display in the Canterbury Museum.

sampling and radioactive dating (p. 66), it is now known that rocks on the earth range in age to almost 4000 million years (p. 147). These oldest rocks, however, are altered sedimentary rocks (sandstones, mudstones etc.) derived by erosion from pre-existing rocks, probably a lot older. Thus the original rocks of the earth's crust, crystallised from molten magma, may have formed some 4600 million years ago, and this is the generally accepted "age" of the earth.

Meteorites, derived from explosions of celestial bodies out in deep space, have also been a rich source of information on the possible range of ages within space (Figs 10.3-10.5).

Meteorites fall into three categories: "stones", "stony-irons" and "irons". The stones are exactly what they sound like—chunks of rock composed of silicate minerals similar to those found in terrestrial rocks. Meteoritic stones also have small amounts of a nickel-iron alloy, but sometimes this increases substantially so that there is half and half nickel-iron and silicate, to give "stony-irons". "Irons" on the other hand are pieces of nickel-iron alloy with only small

Fig. 10.4

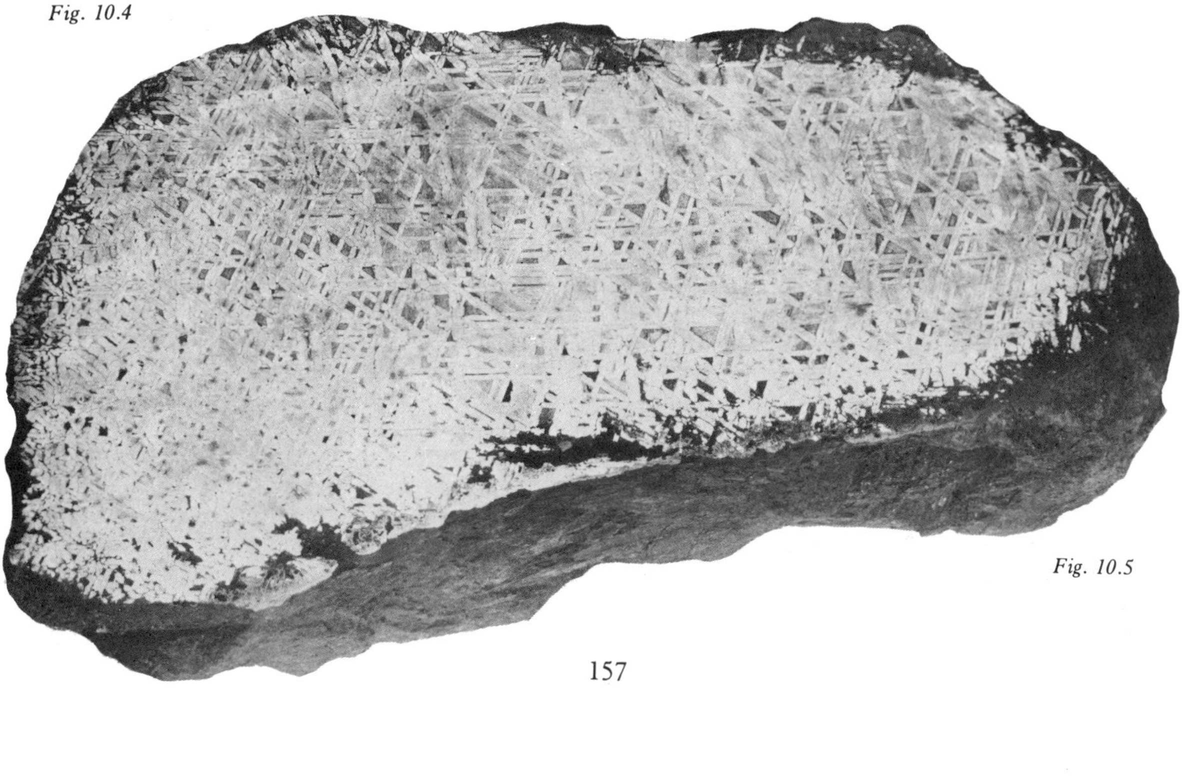

Fig. 10.5

amounts of silicate minerals. It is commonly believed that the three categories of meteorites correspond to the crust, mantle and core of the original stellar bodies.

Numerous ages have been obtained from meteorites of all three categories and these show that almost all of them were formed about 4600 million years ago.

In recent years, as successive spacecraft have brought back to earth their cargoes of immensely important samples of moon rocks, it has been shown that the ages of these rocks also fall consistently around 4600 million years.

Thus as there is close agreement in the ages of the solar material accessible to us—from the earth, meteorites and the moon—it is thought that the solar system formed some time around 4600 million years ago.

**In the beginning**

Three main theories have been put forward to account for the formation of the solar system. All involve condensation of material from clouds of gas and dust.

The first theory pictures a vast revolving cloud of gas and dust which gradually contracted under its own weight. This process of condensation made the centre of the cloud hot enough for nuclear reactions to set in, thus forming a new star, the sun. Condensation around other centres formed the planets and moons, but they may have been heated only to a point sufficient to melt their rocky interiors. Most scientists favour this theory, or variations of it, involving either slow accretion of cold stellar dust, or rapid accretion of already hot material. Even so, it is thought that the sun should have a faster speed of rotation than in fact it now has, if it had originated in this way.

The second theory envisages condensation of planets and moons from a vast streamer of gas and dust torn off the pre-existing sun by a star which grazed it as it passed. This theory however is considered by most scientists to be a remote possibility.

The third theory invokes a huge disc of gas and dust spun out from a vast and rapidly rotating sun. The solar bodies are then thought to have condensed from this disc, forming blobs which eventually consolidated. In terms of this theory only the large planets remote from the sun have retained large amounts of gas from the original condensation. The inner planets, on the other hand, have been stripped to their bare and rocky cores. This theory is new and as yet has not been tested.

An age of 4600 million years for the earth only dates the likely formation of the first rock. Many millions of years have to be allowed to pass for the primaeval earth to cool sufficiently for rock to solidify, rather like scum forming on the surface of jam during cooking.

As a round figure, a date of 5000 million years ago is often taken as the birth of the earth. At this time the earth was probably a ball of gas and interstellar debris. As time passed, the cosmic mixture gradually condensed. Some substances in the original cosmic mixture—especially the lighter substances—became separated off, and, escaping into the atmosphere, were lost to the mass which was to become the earth. The remaining assemblage of elements underwent further gravity separation, and layers were formed (p. 159). The layering of the earth, as revealed by earthquake studies (p. 21), is therefore interpreted by many scientists as indicating that at some point it passed through a molten stage.

Even if the earth had formed from a contracting cloud of cold interstellar dust (see above), it still must have become molten, at least internally, at some stage in its early history. This heating may have resulted from gravitational collapse and the decay of radioactive elements, producing heat. Only then could separation of the earth's layers have taken place.

The newly formed earth 4600 million years ago must have been very different from the earth we know today, or even that of 1000 million years ago. It may have been a much larger planet, not unlike Saturn, with a huge atmosphere of cosmic gas and a rocky core which was probably entirely molten. This primordial atmosphere was stripped away, possibly during a period of higher heat output by the sun. We know this from the very small content of rare gases (neon and xenon) in our present atmosphere compared with the cosmic mixture. Our present atmosphere and surface waters were derived entirely from the earth itself, that is, from the molten core of the primordial planet. Vast amounts of gas must have been given off from the semi-molten surface of the early earth. At the same time the primitive crust was forming. The first crust was probably of basaltic composition. As the crust repeatedly cracked and re-melted, a lighter granitic liquid would gradually have separated out, floating up to the surface. Further concentration of this granitic material into a thicker granitic (continental) crust may have had to await the commencement of erosion processes, caused by surface water condensed from the exhaled gases. Once under way, the processes of erosion would have provided a means of separating lighter (granitic) from heavier (basaltic) components. Granitic rocks would also have been produced by the re-melting of the sedimentary rocks.

Even after the formation of a totally solid crust and the first true oceans, intense volcanic activity went on, adding gases and water vapour to the atmosphere. These volcanic gases had a very low oxygen content and it is thought that for this reason the primordial atmosphere was very poor in oxygen. Oxygen was not added to the atmosphere in any volume until advanced forms of oxygen-producing plant life evolved, about 1900 million years ago.

As we have seen (p. 158), it is generally accepted that the earth originated either through slow accretion of cold stellar dust or rapid accretion of already hot material. In the former instance gravitational forces resulting in intense compression would progressively heat the interior as accretion continued. Thus the end product was probably the same—our earth was formed about 5000 million years ago with most of its mass at a high temperature, somewhere near 1000°C (1832°F).

**Differentiation**

Radioactive particles included in the accreted stellar material started to decay, generating heat. This new source of heat caused the internal temperatures of the earth to rise steadily. It is thought that perhaps some 4000 million years ago temperatures rose to the melting point of iron. The earth then probably underwent a profound reorganisation. As the particles of molten iron and other metals sank to the centre, a large part of the earth was converted to a molten state. This was the start of a separation process called differentiation, in which light material floats up to the top (in the case of the earth to form the crust), while heavy material sinks (to form the core). Differentiation, in effect gravity separation, can also occur on a small scale, as in lava flows, for example. In the case of the earth, differentiation converted the planet from a roughly homogeneous body, with the same kind of material at all depths, into a zoned or layered body with a series of concentric shells of differing composition and density.

Chemical affinity, however, was as important as gravity in determining where elements ended up in the differentation process. Uranium, for example, is one of the heaviest elements, yet it became concentrated in the outermost layers of the earth instead of sinking to the core. This is because uranium has a strong chemical attraction for oxygen, which is most abundant in the crust. Oxygen migrated upward because it forms lightweight and easily meltable compounds with calcium, sodium, potassium, aluminium and silicon. Thus, as differentiation of the earth progressed, there was a tendency for radioactive minerals to end up in combination with various components of the crust, and particularly the continental rocks.

As metallic particles sank and the core differentiated, large convection cells were set up in the overlying material. Over up-welling currents, where cooling was occurring, thin basalt crust began to form, like scum on hot jam (Fig. 10.6).

Where currents converged and descended, the thin rafts of basalt crust were dragged down and re-melted, producing further differentiation, separation of light components and generation of andesite magmas. Elongated andesite volcanic chains were thus created over the sites of descending currents and, as differentiation proceeded, andesite protocontinents were built up. As these protocontinents were located over the sites of down-going currents their position on the surface of the earth changed as the convection current pattern changed. In the earliest history of the earth, convection cells were undoubtedly less stable and more vigorous than they are today and the continents could have "skidded" all over the surface, like slag on top of molten iron in a blast furnace.

However, both in the blast furnace, and on the surface of the primaeval earth, the drifting masses of continental "slag" would sooner or later collide and accumulate to form larger masses (Fig. 10.6). These collisions may have been largely random but it is possible that some guidance may have come from the spinning forces or "Coriolis"

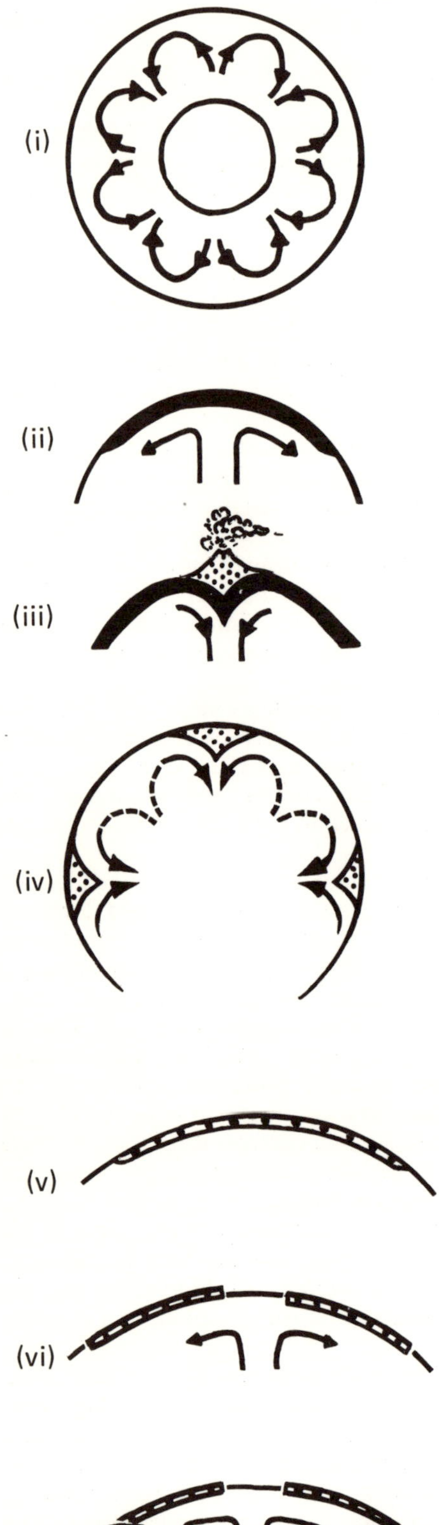


**Fig. 10.6.** Diagrams showing what may have happened in the early stages of development of the earth's continental crust. (i) About 4000-4500 million years ago the earth was a hot body, the heat resulting from the decay of radioactive minerals. The core had already differentiated but violent convection existed in the mantle.
(ii) Thin primaeval basalt crust forms over the site of upwelling convection currents.
(iii) Where convection currents descend, the primaeval basalt crust is remelted. This results in andesite magmas being created and erupted as elongated chains of volcanic islands along the axis of the downturning convection currents.
(iv) The andesite volcanic islands form the beginnings of continents. As these andesitic protocontinents are located over convective downwellings, their position on the surface of the earth changes as the convection current pattern changes.
(v) On the surface of the earth, the drifting masses of continental "slag" sooner or later collide and accumulate to form large continental masses.
(vi) If a convection upwell forms under it, the thin continental plate of dominantly andesitic composition will fracture and the pieces move apart, creating a rift zone. New basaltic (oceanic) crust is generated at the rift zone.
(vii) As the oceanic rift widens, the leading edges of the advancing continental margins encounter basaltic crust on an opposing convection current. A subduction zone forms and parts of the continental plate are melted, producing rocks of true granitic composition for the first time. With the passage of time, the initially andesitic continents gradually got more acidic until they became, as they are today, almost wholly granitic.

forces generated by the earth's rotation. The effect of the rotation would have been to deflect the rafts of continental "slag" towards either pole. Over a long period of time therefore there may have been a tendency for continental material to be guided towards the poles, to accumulate there as large super-continents.

If a convection current up-well formed beneath them, the thin primaeval continental plates of dominantly andesitic composition would fracture and the pieces move apart, leaving a widening rift zone. New basaltic (oceanic) crust would be generated in the rift zone. As the rift widened, the leading edges of the continental fragments thus torn apart would encounter basaltic crust on an opposing convection current. In the resulting over-riding and down-dragging of material (a subduction zone, see pp. 131–134), sea-floor sediment and parts of the continental crust would be melted, the consequent differentiation producing rocks of true granitic composition for the first time (Fig. 10.6). Thus plate tectonics as we understand it came into play, and with the passage of time, the initial andesitic continents gradually became more granitic, until they became almost totally so, as they are today.

Just when the pattern of continental and oceanic rocks was established we do not know. Many of the oldest rocks, more than 3000 million years old, are perfectly ordinary granitic rocks, derived by the heating up of sandstones etc. It is likely therefore that original differentiation of continental crust must have taken place very early on indeed, probably when the earth was a lot hotter and things happened rather faster than they do now. It seems then that the continents are the cumulative result, however indirect, of volcanic activity over thousands of millions of years, and the volume of continental material must be steadily increasing. On the other hand, most of the oceanic crust that is generated simply dives down under the continents and is re-absorbed into the mantle. Only a small percentage becomes involved in vulcanism and mountain-building activities and is added to the continental plates.

**Convection patterns**

Once the continental masses became established, they were moved around at the whim of changing patterns of convection currents. In both hemispheres continental fragments have interacted with neighbouring fragments—at times colliding and becoming welded or sutured into one mass—or at other times breaking up into individual pieces. As we study continental areas, we can see tell-tale signs of this process of interaction. Old weld lines can be seen: some 300 million years ago, in Carboniferous and Permian times, for example, eastern Siberia was welded onto the remainder of Russia, the welded seam being the Ural mountain belt of today. Study of these old weld lines and other features has shown that the break-up of Pangaea that started some 150 million years ago, and is still in progress, is only the most recent of a number of break-ups that have occurred in the geological past. Similarly, the conglomeration of continents into Pangaea is only the most recent of a number of such groupings (Fig. 10.7).

Some scientists maintain that the changing patterns may have been a response to fundamental changes in the deep internal structure of the earth—notably the size of the core. One idea is that as the core gradually grew in size, as a result of differentiation of the mantle and the sinking of metallic particles into the core, there would be enforced changes in the operation of convection cells.

At first, when the earth was very young, the convection pattern took its simplest form, with perhaps a single system of flow, or "convection cell", within the earth. All the cold material would thus sink in the same region. As it did so, lighter components would resist sinking and would form a "slag" on the surface. In this way a patch of lighter rock accumulated, while the heaviest material (chiefly iron and nickel) settled into the core.

Once the core became established, a two-cell convection system probably came into being, but with continuous growth of the core and consequent restriction of the width of the mantle, there was a change from two to three and then eventually four convection cells (Fig. 10.8). It has been suggested that the change from four to five cells was the event that initiated the break-up of Pangaea some 150 million years ago and the continental drifting that is still continuing today (p. 162).

Steady growth of the core would have also resulted in a gradual increase in the earth's inertia (i.e. the drag exerted by the spinning earth). As a consequence there would be an accompanying gradual increase in the length of the day. This is not as far-fetched as it may seem, as studies of growth rings in fossil corals suggest that 400 million years ago the day lasted only 22 hours. (Studies of modern corals have shown many of them to be very sensitive indicators of time, as they add lime to their skeletons only in daylight hours—and these studies can be applied to fossil corals.)

Another process that may have had some influence on continental break-up is the "blanketing" of heat by large continental masses. Large continental masses, such as existed at various times in the past, may have accumulated substantial quantities of heat underneath. This heat comes from two sources. First, heat is passing outwards from the hot interior regions of the earth. Second, heat is being generated within rocks, by the steady decay of radioactive minerals. As we have seen, radioactive minerals tend to accumulate in the lighter oxygen-rich rocks (i.e. granitic rocks).

Although radioactive minerals, even in continental rocks, are only present as "traces" (e.g. a few parts per million of the whole rock is

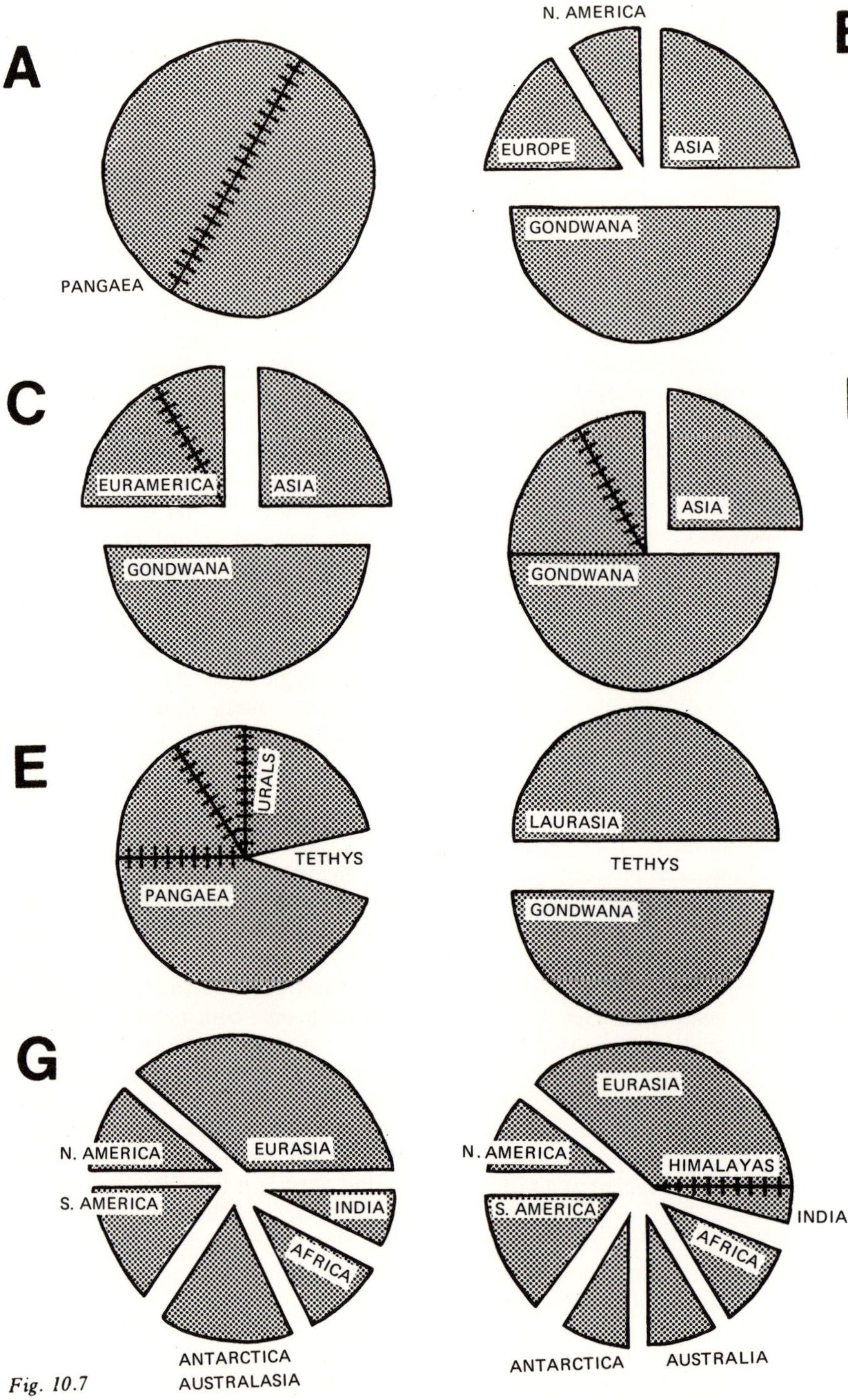
A
PANGAEA
B
N. AMERICA
EUROPE
ASIA
GONDWANA
C
EURAMERICA
ASIA
GONDWANA
D
ASIA
GONDWANA
E
URALS
TETHYS
PANGAEA
F
LAURASIA
TETHYS
GONDWANA
G
N. AMERICA
EURASIA
S. AMERICA
INDIA
AFRICA
ANTARCTICA
AUSTRALASIA
H
EURASIA
N. AMERICA
HIMALAYAS
S. AMERICA
INDIA
AFRICA
ANTARCTICA
AUSTRALIA

*Fig. 10.7*

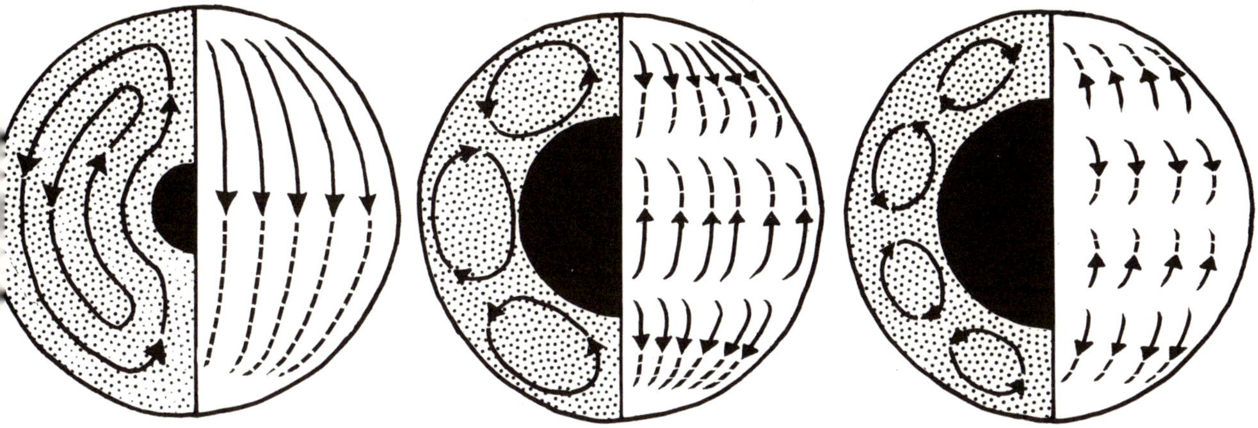

Fig. 10.8. Some scientists maintain that the earth's core has steadily increased in size throughout its history. This would have progressively decreased the width of mantle available for the operation of convection currents. Thus early in the earth's history large convection cells may have been generated (diagram to left). Then as the mantle's width became progressively restricted, the convection cells increased in number and decreased in size.

Thus a gradual change in the patterns of convection currents occurred throughout geological time. The surface effects of such changes are shown to the right of each diagram. Continents are carried to sites where convection currents converge.

Other scientists maintain that there has been steady expansion of the globe as a whole, as well as the core (cf. Fig. 10.9).

Fig. 10.7. These simplified diagrams summarise some of the continental groupings, break-ups and re-groupings that have occurred over the last 700 million years. The gaps represent oceans.

A late Precambrian super-continent (A), which some geologists believe may have existed 700 million years ago, may have been formed from previously separate continents. The Cambrian world (B), some 570 million years ago, consisted of four continents. On the other hand, the Devonian period (C), some 390 million years ago, was characterised by three continents, as by this time ancient Europe and North America had collided and become welded or sutured into one landmass (Euramerica). In the Carboniferous period (D), some 300 million years ago, Euramerica in turn became welded onto Gondwana. In the late Permian period (E), about 225 million years ago, Asia became welded to the remaining continents along the Ural mountain belt, and Pangaea was reformed, but with a gap corresponding to the Tethys Ocean. In Jurassic times (F), 195-136 million years ago, Laurasia and Gondwana were more or less separate.

By the late Cretaceous period (G) 70 million years ago, however, Gondwana had become highly fragmented and Laurasia partially so. The modern continental pattern (H) has India welded onto Eurasia.

fairly common), when they are taken together in a large mass, particularly if a super-continent (such as Pangaea) has been aggregated together, the amount of heat being generated by the minute nuclear reactions is in total quite significant.

The blanketing layer of continental material, often some 70-80 km (45-50 miles) thick, impedes the passage of heat out into the atmosphere. We thus have the situation in which a large continental mass can accumulate under it considerable quantities of heat.

This accumulation of heat will eventually start to migrate out to the edges of the continent, where it will be dissipated into the atmosphere. Transfer of heat in this way may generate current systems in the material underlying the crust, and in turn, over a long period of time, stresses and strains may be set up in the overlying crustal layers, so that eventually the super-continents break up and pieces begin to separate and move independently. In this way new convective systems may be generated and new patterns of continental arrangements.

**Expansion of the earth?**

Up to now we have assumed that the earth has remained the same size throughout its long history. However there are some scientists who believe that, quite apart from expansion of the core (see above), there has been slow and steady expansion of the entire globe.

If you take the present shapes of the continents and squeeze them all together, they fit together snugly on a globe one-half its present size. In this situation the present continental crust covers the entire surface of the earth, except for some narrow seaways. The fact that the continents do

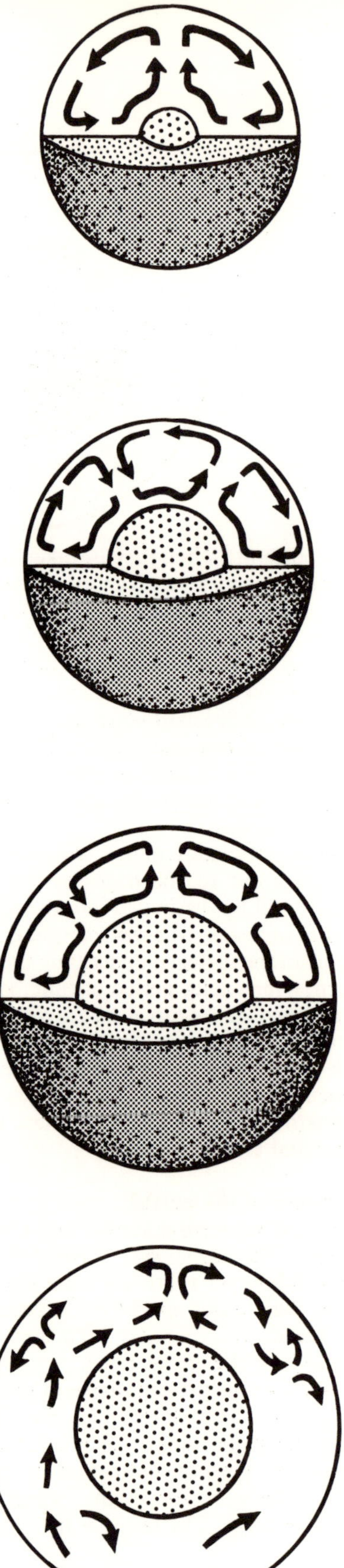

fit together on a smaller globe does seem to be more than a coincidence—and some geologists view this as a major piece of evidence for earth expansion.

The aggregate areas of the modern continents, if squeezed together, would cover a globe of 3500-4000 km (2200-2500 miles) radius, compared with the present-day globe of 6370 km (4000 miles) radius. If we assume that the present continental crust made up the entire surface of the globe some 4500 million years ago, the average rate of increase in global radius (to produce the oceans) is about 0.60 mm (0.025 in) per year. When we look at global geology we find that belts of folded rocks of an average age of 2750 million years can fit together symmetrically on a globe of a radius of 4400 km (2700 miles). This corresponds to an expansion rate of 0.75 mm per year (0.03 in). Similarly, belts of folded rocks of an average age of 650 million years, when fitted together indicate a global radius of 6000 km (3750 miles), corresponding to an expansion rate of 0.58 mm (0.02 in) per year.

We also find that during the past 600 million years progressively smaller continental areas have been covered with sea water at various times. If we assume that the total volume of sea water has remained unchanged, and the expanding ocean basins have progressively drained the continents of water, the average rate of expansion is 0.76 mm to 0.54 mm (0.03-0.02 in) per year.

Initially, when primaeval crust covered most of the earth and the core was just starting to form as a result of the precipitation of metals from the developing mantle, it is possible that the interior of the earth was stirred by one large convection cell. This may have been the situation some 4500 million years ago. Then as the core gradually expanded, and the entire earth as well (although probably not at the same rate), changes in the convection patterns followed, leading to creation

**Fig. 10.9.** Figure 10.8 illustrated the situation in which changes in the patterns of convection currents were generated by changes in the size of the earth's core. Some scientists maintain, however, that there was expansion of the globe, as well as expansion of the core. These diagrams show how these changes can influence the convective pattern. Past patterns are shown as regularly-shaped cells, but if the modern pattern, as shown in the bottom diagram, is anything to go by, they were probably very irregular and sited randomly across the globe.

of several smaller cells rather than one large one. Thus a two-cell pattern of convection currents (see p. 161) may have developed, and this may have been the situation about 2500 million years ago (Fig. 10.9).

As the earth expanded, however, and the core also, a three-cell convection pattern may have developed in the thinner mantle. This may have been some 1500 million years ago.

A four-cell convection pattern may then have been evident some 400-300 million years ago. The most recent shift, from four to five cells, possibly occurred 200-100 million years ago and may have been the event that initiated the present episode of continental drift (p. 162).

If changes in the size of both the globe and the core did in fact force changes in the deep circulation patterns of the earth in the way suggested, then it probably follows that over the length of geological time we have had migrations of zones of convergence and divergence. Thus sometimes a certain part of the globe would be the site of converging convection currents; crustal material would then be moved towards this area and the resulting collisions would weld individual pieces together. The current system might then change, the site of convergence migrate elsewhere, and in its place arise a zone of current divergence. Crustal material would then be torn apart and drifted for long distances.

The idea of an expanding earth has long been championed by the Tasmanian geologist S.W. Carey (Fig. 3.7) but, as Wegener found when he originally proposed his theory, the problem of a mechanism has been a major stumbling block to general acceptance of the expansion theory.

Various possible mechanisms to account for expansion have been suggested. Some ideas involve changes in the earth's gravity and the consequences of such changes on the globe. Cosmological studies, for example, suggest that the gravitational constant of the earth has been diminishing with time. This means that in the geological past the force of gravity was greater than it is today and, exerting a greater compressive force, held the earth's matter in a more condensed state; hence the globe was of a smaller radius. As the force of gravity gradually diminished as the present day approached, so the earth's radius steadily expanded.

Another by-product of diminished gravity may have been the conversion of various minerals to less dense forms, which would result in expansion. We know that many minerals can change into denser forms (pp. 167–172). At a given temperature these changes depend on pressure. If the force of gravity diminishes, so will the pressure at all depths, and minerals may therefore change to less dense phases.

It has also been calculated that gravity differentiation *within* the earth may have added to the general expansion of the globe.

You may say that the above suggestions are purely theoretical—as did the opponents of Wegener's theory. However, help may be at hand. The Apollo 15 mission placed a reflector on the moon which will return any light directed onto it back along the path it came. It is proposed to send a laser beam to the moon and observe its return at observatories in Honolulu, Tokyo and Canberra. The laser beam is reflected back from the lunar reflector in such a way that the distance can be measured very accurately. If these measurements are repeated for different positions of the moon, not only can the distances from the observatories to the moon be measured, but also their distances from each other, with an accuracy of a few centimetres. Over a number of years the measurements will determine the movements of plates on the earth's surface, but with a bit of luck they may also tell us whether any expansion is occurring at the same time.

Another imaginative experiment that is planned is to send a pulsed beam of neutrinos (i.e. highly accelerated particles) right through the earth from the high energy accelerator at Brookhaven (Chicago) to Cocos Island in the Indian Ocean. Neutrinos are the most elusive of all sub-atomic particles. They have no rest-mass, no charge, and so do not respond to magnetic fields. They can go through an atom, and even a nucleus, with rarely any interaction and so the bulk of them will pass right through the earth. The problem is focussing them into a beam. However when a beam is generated, it is thought that it should be possible to measure its length with an accuracy of a few centimetres. This again would, after a few years of repetition, prove whether or not the earth is expanding.

Finally, it may be possible to measure absolute gravity to such a degree of accuracy that some time in the future it may be possible to say something definite about whether or not gravity is diminishing with time.

Thus, bearing in mind the unfavourable reaction to Wegener's theory early this century, and its subsequent history, it may be best in the meantime to keep an open mind on the question of the expanding earth.

If we leave the matter of a mechanism to one side, the idea of an expanding earth has many attractions, particularly in reconstructions of past geography.

In 1976 a major paper was published in the prestigious *Philosophical Transactions of the Royal Society of London* in which H. G. Owen argued that the individual continents of Wegener's super-continent Pangaea have moved apart as a consequence of a 20 percent expansion of the earth's diameter since the early Jurassic, some 190 million years ago.

Without a doubt some of the reconstructions of past geographies, made on the present globe, pose awkward problems of continental overlap and misfit which have not been adequately explained away (although some have been; see p. 34). For example, many geologists have been worried by the loss of much of Central America in the fit depicted in Figs 5.6, 5.8, as it is known that extensive areas of old continental rocks occur there. Likewise, if one adopts the fit depicted in Figs 5.4, 5.7, West Antarctica does not run naturally into its obvious geological continuation in Patagonia. More disturbingly, a large gap is left west of Australia, which has led to the suggestion that this region (the Wharton Basin) was very old ocean. This idea however has now been disproved by the *Glomar Challenger*'s drill, which has demonstrated that the oceanic crust on the floor of the basin is as young as elsewhere. On the other hand, if India is fitted against Australia to close this gap, a corresponding hole is left in the western Indian Ocean. In other regions, gaps often separate areas which the local geological history would tie together.

If the continents are reassembled as in Fig. 5.8, again using a modern globe, a large Tethyan ocean results, which seems to gape awkwardly eastwards. Evidence from rocks and fossils indicates that at various periods throughout late Palaeozoic and Mesozoic times, shallow seas lapped onto the margins of the lands around the modern Mediterranean Sea, Indian Ocean and Western Pacific (i.e. flanking the Tethys in Fig. 5.8). These seas had the character of shallow seaways, rather than broad extensive oceans. Shallow-water marine creatures migrated along the shores of the shallow seas and at various times in geological history there was interchange across the Tethys, that obviously would not have been possible if a vast ocean had been present, as in Fig. 5.8.

Owen maintains, however, that many of these difficulties disappear when global expansion is taken into account. He believes the yawning Tethyan gap of Fig. 5.8 is merely a geometric artefact. As can be seen from Figs 8.39, 8.40, the marine animals mentioned above can migrate in shallow seas lapping onto the edges of the various landmasses that in the reconstruction are now brought closer together.

Another distinctly Southern Hemisphere problem with many of the reconstructions like that of Fig. 5.8 is that Australia, New Caledonia, New Zealand and West Antarctica are usually pictured as nudging the South Pole in mid-Mesozoic (i.e. Jurassic) times.

Although we can accept a high-latitude position for these countries in the Carboniferous and Permian, when we have unmistakable evidence of ice sheets on land, icebergs in the sea and cold-water marine life, we cannot accept such a geographic position in the Jurassic. As we will see in Chapter 12, the shallow-water marine creatures populating these regions in the Jurassic were all of warm-temperate type—i.e. not tropical, or even subtropical, but on the other hand not cool or cold.

A mid-latitude position (around 45°S Latitude) is thus indicated, and again, this may be obtained by closing up the Tethyan gap and moving Australia, New Zealand etc. correspondingly northwards. This is shown in Fig. 12.26, which is supported by palaeomagnetic evidence. Owen's reconstruction, Figs 8.39, 8.40, while easing the geographic problem somewhat, still has West Antarctica very close to the South Pole in the Jurassic. However, no doubt it will only be a matter of time before these problems of fitting reconstructions to the known geological framework are resolved.

**The fiery underworld**

One way of looking at an overall driving force for earth movements is to consider the earth as a giant heat engine, driven by radioactively generated heat plus the heat left over from the differentiation that took place early in the planet's history.

The earth's internal engine ultimately supplies the energy for the entire spectrum of geological processes—in the form of plate motions, earthquakes, mountain-making and volcanic eruptions.

Volcanoes and hot springs are obvious examples of the flow of heat from the interior of the earth to the surface. Over the years the heat flowing out from the earth has been monitored all over the globe. At first these measurements were restricted to the land but then in the early 1950s a device was perfected to probe the heat flow of the sea floor—quite an achievement from a bobbing boat in some miles of water. Some areas were found to have a greater heat flow than others—because they are close to volcanic activity, either modern or recent, or are sitting on top of great thicknesses of young wet sediments that transmit heat at a slower rate than old dry sediments.

It has been known for at least three hundred years that temperatures steadily increase as one progresses to deeper levels in a mine, and it was found that on the average there was an increase of 2° or 3°C (4-5°F) per 100m (300ft) of depth. The actual rate of increase (the "geothermal gradient") depended on the local rock conditions—e.g. young wet rocks or old dry rocks, or proximity to volcanic activity.

As mining technology improved, and men burrowed deeper and deeper into the earth, the sheer quantity of heat flowing out from the interior became very obvious. In South Africa gold-bearing sedimentary layers have been tilted at a 21° angle by past earth movements so that now they extend from the surface down to depths of about 4km (almost 2½ miles). These layers constitute the famed Witwatersrand Reefs, and in their quest for the precious metal, successive generations of miners have pushed tunnels progressively deeper, following the dipping gold-bearing layers. Today 50 mines tap an arc of reefs 500km (300 miles) long, and together yield more than 900 tonnes of gold a year, two-thirds of the world's production. As most of the South African rocks are very old and dry, they conduct heat very rapidly and the geothermal gradient is lower than the average—varying between 2 and 4°C (4.5° and 7.5°F) per 300m (1000ft) of depth.

Today the gold mines have reached 3500m (11,400ft) and at these depths the rock temperature is 54°C (129°F). It is envisaged that mining will eventually extend to about 4000m (13,000ft), where temperatures are expected to be 59°C (138°F) (Fig. 10.10).

Whilst an average geothermal gradient of 2° or 3°C (4-5°F) per 100m of depth holds true for those parts of the earth that we can reach with a thermometer—down to about 8km (5 miles), in deep boreholes—it doesn't mean that we can simply extrapolate the temperature gradient all the way to the earth's core at 2900km (1800 miles). If we did this it would mean that the core would be at a temperature of at least 25,000°C and that most of the earth would probably be molten. However, it is believed that most of the heat-generating radioactive minerals are concentrated in the outermost layers of the crustal rocks, where, as we have seen (p. 159), they have migrated along with the oxygen-bearing compounds.

Thus below this outermost "hot zone", the rate of temperature increase in relation to depth probably slackens markedly. The deeper layers of the earth may also transfer heat more efficiently than those nearer the surface—again helping to decrease the temperature gradient.

Earthquake studies and laboratory experiments have helped us form some idea of temperatures within the mantle and core. As we have seen in Chapter 4, earthquake waves tell us that the Low Velocity Layer, or asthenosphere (Fig. 4.7), is partially molten. We can therefore "fix" the temperature of this zone, say at a depth of 100km (60 miles), at about 1100 or 1200°C (2012-2192°F), which, allowing for the effects of pressure (which elevates the melting point of materials), is probably the temperature of incipient melting of basalts and other similar materials at this depth. (At the surface they are well and truly molten at these temperatures—but pressure considerably increases the point at which they melt.)

A bit further down in the mantle, at depths of 400km and 700km (250 and 450 miles), the character of earthquake waves suddenly alters in response to changes in mineral composition. These changes have been duplicated in the laboratory, using equipment capable of reaching the high pressures at those depths, and it is thought that the changes are related to a more dense packing of the atoms of olivine, a major mineral constituent of the mantle. This sudden densification, or "phase change", occurs when critical pressures and temperatures are reached. Using the laboratory and earthquake information it has been estimated that the phase changes have occurred at temperatures of about 1500°C (2732°F) at 400km (250 miles) and 1900°C (3452°F) at 700km (450 miles).

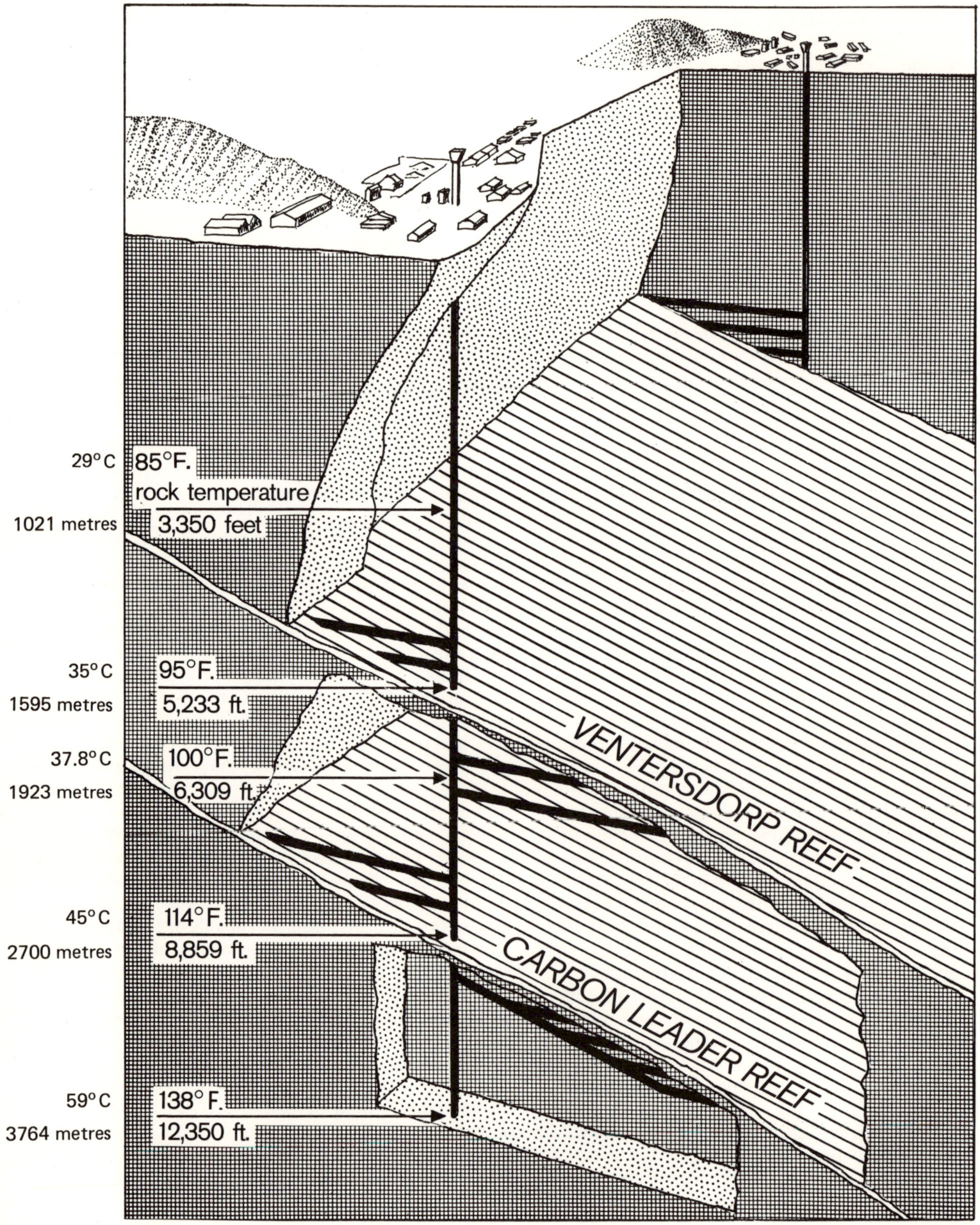
29°C
85°F.
rock temperature
1021 metres
3,350 feet
35°C
95°F.
1595 metres
5,233 ft.
37.8°C
100°F.
1923 metres
6,309 ft.
VENTERSDORP REEF
45°C
114°F.
2700 metres
8,859 ft.
CARBON LEADER REEF
59°C
138° F.
3764 metres
12,350 ft.

*Fig. 10.10*

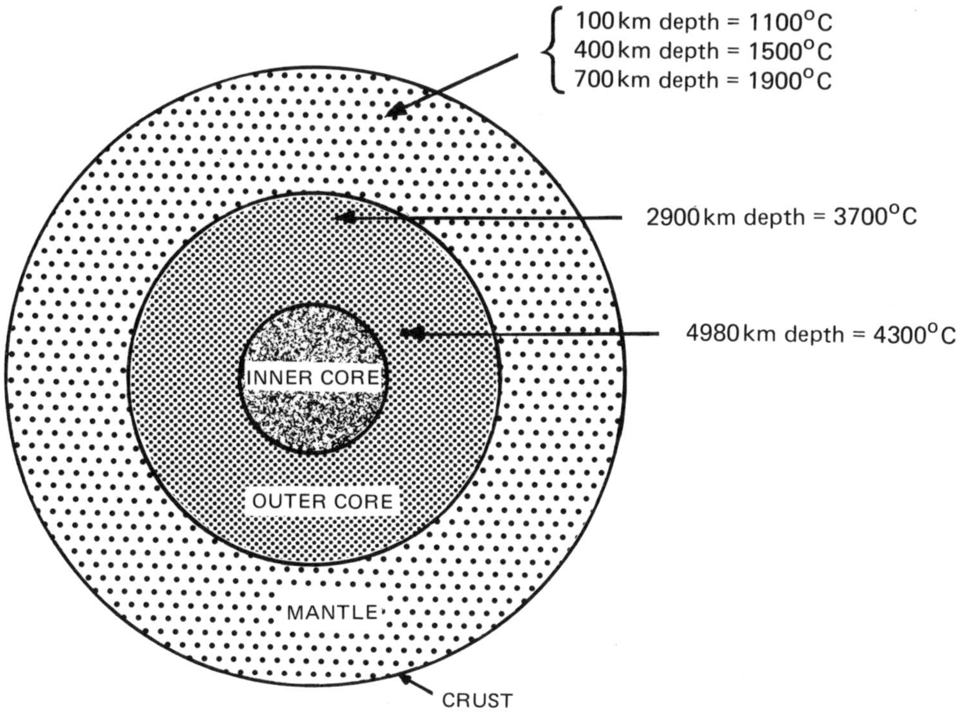


**Fig. 10.11.** By studying the patterns of earthquake waves as they travel through the earth, and the behaviour of rocks in laboratory experiments, it is possible to reach some conclusions as to the likely range of temperatures within the earth, as this diagram shows.

Earthquakes have also told us that the core is probably nickel-iron, molten in the outer part and solid in the inner part. Laboratory studies on the melting point of nickel-iron in relation to pressure indicate that temperatures in the molten outer part of the core are at about 3700°C (6724°F) and that the solid inner core, deeper than 5100 km (3200 miles), is probably at about 4300°C (7804°F) (Figs 10.11, 10.14).

In the outer core the pressures and temperatures are such that the melting point of nickel-iron is exceeded and hence it is in the molten state. However, the pressures are such in the inner core that the nickel-iron is held probably just below its melting point and therefore is solid. The two cores, both presumably of approximately the same material, thus illustrate nicely the delicate balance between pressure and melting points.

**The earth's mantle**

In previous chapters (e.g. pp. 21–24) we have touched on some of the properties of the mantle. Now we review these properties to consider how they may provide possible mechanisms for earth movements.

The mantle, accounting for 83 percent of the earth's volume and 67 percent of its mass, is a thick shell of hot rock separating the earth's metallic core from the cooler rock of the thin crust. Mantle rocks take up nearly half of the earth's radius,

**Fig. 10.10.** The Western Deep Levels mine, near Johannesburg, the deepest of the South African gold mines, taps two gold-bearing reefs (Ventersdorp Reef and Carbon Leader Reef), dipping down to the right. Shafts descend vertically and from them spread labyrinthine tunnels and chambers.

Heat from deep within the earth raises the temperature of the rocks. Hotter and hotter temperatures are encountered as the shafts go deeper. Rock can explode without warning in the relentless pressure, and scalding water may suddenly burst from hidden fissures. The high temperatures and humidity make hard physical work very difficult. Ventilation holds the key in such a brutal environment. However, because of the effect of compression, fresh air pulled down from the surface to provide ventilation becomes itself hotter and hotter the deeper it goes. Thus temperatures at the working faces have to be lowered by massive refrigeration plants, sited close by. At Western Deep Levels, a refrigeration plant has been installed that has a capacity equivalent to 3000 tonnes of ice at 0°C (32°F), and is designed to maintain the air temperature at the working faces at 29°C (85°F). Put another way, the cooling capacity is equivalent to that of one window air conditioner for every linear metre (3 ft) of working face.

**Fig. 10.12.** Plate tectonics in action, as exemplified by the South Atlantic Ocean. Plates of lithosphere, which includes the crust and part of the upper mantle, migrate laterally over the asthenosphere (Low Velocity Layer).

In the diagram the South American and African continents are riding on their respective plates.

Material from the asthenosphere rises below the crust of the Mid-Atlantic Ridge, melting to produce lava that is erupted to form new crust on the ocean floor. The Pacific plate is being thrust under South America and a subduction zone has been created, with associated ocean trench and volcanic activity.

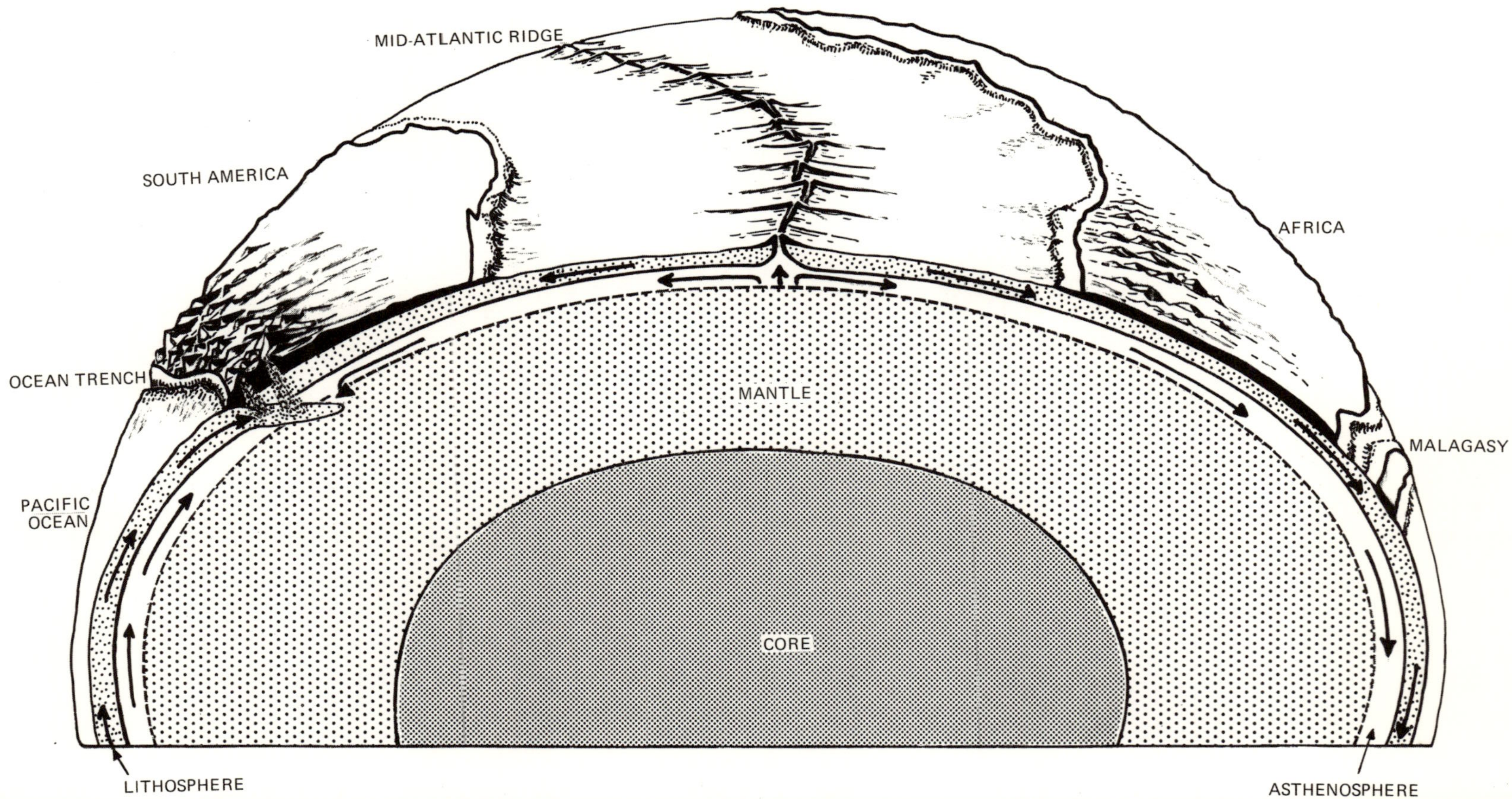

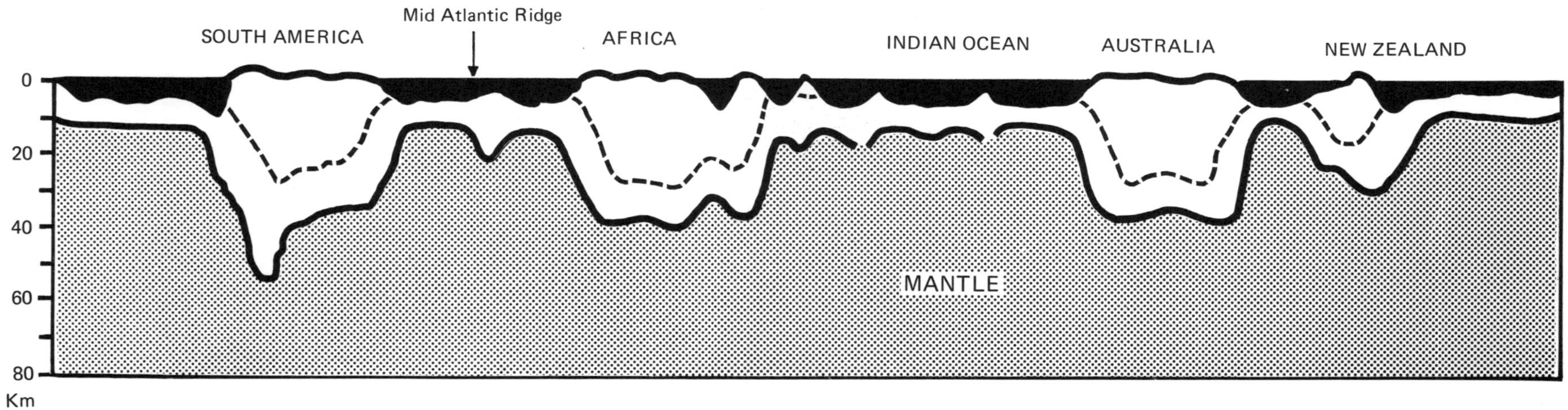


**Fig. 10.13.** The Conrad Discontinuity (dotted line) can be traced throughout continental areas and is thought by some scientists to represent the contact between the overlying granitic rocks (sial) and underlying basaltic rocks (sima).

extending from a depth of 35-45 km (22-28 miles) below the surface and continuing to a depth of some 2900 km (1800 miles) (Fig. 10.14).

The boundary between crust and mantle was first picked up from earthquake studies, as the speed of earthquake waves altered markedly at this junction. As we have seen on pp. 21, 89, 135, this boundary is termed the Mohorovicic Discontinuity, or "Moho". In turn, the mantle is separated from the outer core by another seismic discontinuity.

The sudden change in the speed of earthquake waves at the Moho can be interpreted in two ways—either as a phase change or as a difference in chemical composition.

A phase change involves a change in the physical properties of a substance without any change in composition. For example, both graphite and diamond are phases of pure carbon. Diamonds are formed at high pressures deep in the earth and, as their atoms are packed tightly together, the resulting stones are hard and dense. Graphite, on the other hand, is formed at low pressures, and the atoms are loosely packed, giving rise to a soft and less dense material. Most scientists, however, view the change at the Moho as being one related to chemical composition. The lower part of the crust is generally pictured as being composed of rocks of the sima—basalts and their coarse-grained equivalents known as gabbros. Then if, to judge from the seismic signals, the Moho represents a chemical change, the upper mantle must be composed of rocks whose minerals are denser and more rigid than those of basalts and gabbros (Fig. 10.14).

At the temperatures and pressures which must be present in the upper mantle the most likely contenders to give the appropriate seismic characters are ultrabasic rocks such as dunite and peridotite, which consist almost entirely of the magnesium and iron silicates olivine $(Mg, Fe)_2SiO_4$ and pyroxene $(Mg, Fe) SiO_3$.

Another rock that has the required seismic features and therefore may be found in the upper mantle is eclogite. Eclogite, while having the same composition as a basalt or gabbro, is very dense, as the atoms of its minerals are more closely packed. Laboratory experiments tell, however, that pressures equivalent to a depth of about 60 km (38 miles) are needed before eclogite will form. Thus, as the Moho under the oceans is only 11 km (7 miles) down, it is unlikely that eclogite occurs in any great quantity in the uppermost mantle, but probably is present at deeper levels, especially where compression occurs (p. 179).

Other pointers towards the composition of the upper mantle are obtained from study of volcanoes, ophiolites and meteorites.

Some volcanoes erupt lavas that have apparently been generated within the mantle. As these lavas move upwards towards the surface they carry pieces of rock torn from the walls of the deep-seated magma chamber and volcanic pipe. These rock fragments are usually peridotite, particularly in oceanic regions (where the mantle rocks are closer to the surface), and have presumably been derived from the upper mantle.

Some volcanic pipes have in the past penetrated very deeply into the mantle. Now uplift and erosion have exposed these deep-seated volcanic pipes and the material they contain.

The most celebrated volcanic pipes are those of southern Africa that were actively feeding volcanoes in the late Cretaceous, some 80 million years ago, but are now deeply eroded. These volcanic pipes are notable in that they are diamond-bearing and have therefore been worked to substantial depths, e.g. 1080 m (3520 ft) in the Kimberley Mine. The material filling the pipes is a peculiar rock called kimberlite (named after the mine), which is thought to be debris of ultra-basic rocks, greatly altered by superheated steam and other volcanic gases. Near the surface the kimberlite is thoroughly decomposed into a yellow mud, known to the miners as Yellow Ground. This passes down into blue clays, known as Blue Ground. Diamonds are scattered throughout the kimberlite, indicating that carbon fragments originally present have been subjected to intense pressure and temperature and have thus undergone a transformation, or phase change, to produce diamond. Experimental work, that, incidentally, has led to the commercial production of man-made diamonds, indicates that the temperatures and pressures required to produce diamonds are equivalent to a depth of about 150 km (94 miles) in the earth, where kimberlite presumably originates. As it travels upwards to penetrate the crust, the kimberlite carries up fragments of mantle material, and on study these prove to be ultrabasic rocks such as peridotite and eclogite.

As we have seen on p. 135, ophiolites occur in areas of the earth where through violent colliding

## Properties of the Earth's Layers

| | ZONES AND DISCONTINUITES | DEPTH TO BOUNDARIES (KM) | PERCENTAGE OF EARTH'S VOLUME | DENSITY | PRESSURE (KILOBARS) | NATURE OF REGIONS | POSSIBLE NATURE OF BOUNDARIES |
|---|---|---|---|---|---|---|---|
| LITHOSPHERE | CRUST (SIAL) | Continental average,33 | | 2·7 | | Heterogeneous Solid | |
| | Conrad | Mountain ranges, 65 | 1·55 | 2·8 | | Oceans basaltic | |
| | CRUST (SIMA) | Oceanic (from sea level), 10-11 | | | | Continents granitic | |
| | Mohorovicic | | | 2·9 | 9 | | Chemical-change from basalt above to peridotite below |
| | UPPER MANTLE | | | 3·3 | | Various types of peridotite | |
| | | ca.50 | | | | | |
| ASTHENOSPHERE | Low-Velocity Layer | | | | | Partly fused peridotite | |
| | | ca. 250 | | | | | |
| MESOSPHERE | UPPER MANTLE | | 82·25 | | | Peridotite with high density minerals | Phase change to higher density minerals |
| | | ca. 1000 | | | | | |
| | LOWER MANTLE | | | 4·3 | 270 | Peridotite with high density minerals | |
| | | | | 5·7 | 1 368 | | Change from silicate mantle to metallic core |
| | Core-Mantle Boundary | 2 900 | | 9·7 | | Liquid | |
| | OUTER CORE | | | 11·8 | 3180 | Probably nickel-iron alloy | Phase change from liquid to solid |
| | | 5 000 | | | | | |
| | INNER CORE | | 16·20 | ca. 14 | 3300 | Solid | |
| | CENTRE | 6371 | | ca. 16 | 3600 | Nickel-iron | |

**Fig. 10.14.** The various properties of the earth's layers are summarised in this table.

movements parts of the upper mantle have been brought close to the surface and subsequently laid bare by erosion. Outcrops of these rocks occur in various places in New Zealand and in all these instances, and in the occurrences overseas, peridotite is the dominant rock type present (p. 140).

Lastly, just as metallic meteorites can be used to throw some light on the possible composition of the earth's core (p. 158), so stony meteorites, supposedly representing the mantle of an exploded planet, can be used to indicate the composition of the earth's mantle. Almost invariably stony meteorites are of composition similar to that of peridotite.

Thus all in all it can be seen that the balance of the available evidence favours the contention that the upper mantle is largely of peridotite, with eclogite appearing in deeper zones.

Within the mantle, the velocity of earthquake waves increases with depth and it is thought that the rock is all basically the same—peridotite in various forms, depending on the prevailing temperatures and pressures.

Notable jumps (or "discontinuities") in seismic velocity occur within the lower mantle at depths of 400 and 700 km (250 and 440 miles) and these are thought to represent phase changes, where the main mineral components of the mantle are forced, by ever-increasing pressure related to depth, to re-adjust their crystal structure so that the atoms are more densely packed.

As we have seen on pp. 22–24, a Low Velocity Layer has been recognised in the mantle, in which earthquake waves markedly slow down. The zone of retarded velocity extends between 50 and 250 km (32-155 miles). There is still doubt whether there is a gradual diminution of wave velocity over the whole zone, or whether there are several very narrow layers within the zone in which seismic velocities are very much slower. These narrow layers, if they exist, seem to lie somewhere between 70 and 150 km (44-94 miles) below the surface.

It is thought that the low-velocity zone is a region of partial melting of the mantle's peridotite rock. It has been calculated that if only 5 percent of the peridotite melted to give liquid rock, this would be enough to retard the seismic wave velocities to match those observed in the Low Velocity Layer.

As has been stated on pp. 22–24, the Low Velocity Layer is the obvious site along which sliding movements of the earth's outer layers can occur. In this respect it is a far more important and significant layer than the Mohorovicic Discontinuity. The Moho is merely the boundary between the crust above and upper mantle below, and on seismic evidence both these layers are solid, and so not capable of movement.

Some scientists maintain, therefore, that the terms "crust" and "mantle" should be discarded in favour of terminology emphasising the importance of the Low Velocity Layer. They suggest that the outer part, consisting of both the crust and upper mantle above the Low Velocity Layer, should be termed the "lithosphere" (i.e. the "rock sphere"). The Low Velocity Layer, where movement between the outer and inner parts of the earth is most likely to occur, is termed the "asthenosphere" (i.e. the "weak sphere"), whereas the mantle below the Low Velocity Layer is labelled the "mesosphere" (or "centre sphere") (Figs 10.12, 10.14).

**The crust**

As we have seen (pp. 21, 89, 135), the Moho is now known to separate the rocks of the earth's crust from those of the mantle. The depth of the Moho, and hence the thickness of the earth's crust, has now been determined in most places around the globe. Under the ocean basins the crust is thin and averages about 5 km (3-4 miles) thick. It is probably entirely composed of basalt ("sima"). Under the continents however the crust varies between 20 and 65 km (13-41 miles) in thickness. For example, under the prairie country of North America and the steppes of Russia, the crust is between 30 and 35 km (19-22 miles) thick, whereas it thickens to 60-65 km (38-41 miles) under mountain ranges such as the Swiss Alps, Himalayas and Andes. In an exaggerated way the Moho reflects the topography of the earth's surface (Fig. 10.13).

Within the thick crust making up the continents can be recognised a seismic break called the Conrad Discontinuity (Fig. 10.14). Although it is by no means as distinct or as continuous as the Moho, it nonetheless divides the crust under the continents into lower and upper parts. The upper layer, which can be seen and examined in great detail, is the granitic or "sial" layer (p. 2). The lower layer, however, is more dense than the sial and is rather mysterious, as nobody really knows for certain what it is made of. At the edges of many continents it seems to be continuous with normal oceanic crust (i.e. "sima"), and in some areas (e.g. Lapland), where huge dislocations have brought the lower layer to the surface, it is of oceanic (i.e. basaltic) composition. However, where it is deeply buried under the continents, it may be a greenish-black metamorphic product of basalt known as amphibolite (i.e. a basalt changed into a denser phase by pressure and temperature). Thus it is possible that the sima continues under the continents (as is shown in Figs 8.36, 9.14, for example), but that in the deeper parts at least, it is probably changed by heat and pressure into denser minerals.

**A matter of mechanics**

The mechanics of continental drift is a topic wide open for speculation. It is fair to say that nothing is known for certain. However, some reasonably firm ideas have emerged that (at the moment) seem to tie in with what we know about the earth's materials and structure.

Although other mechanisms have been proposed, the deep-seated driving forces for continental drift can probably be best explained in terms of simple heating and cooling. As we have seen, vast amounts of heat are being continually transmitted through the earth. As not all regions of the earth are losing heat at the same rate (some,

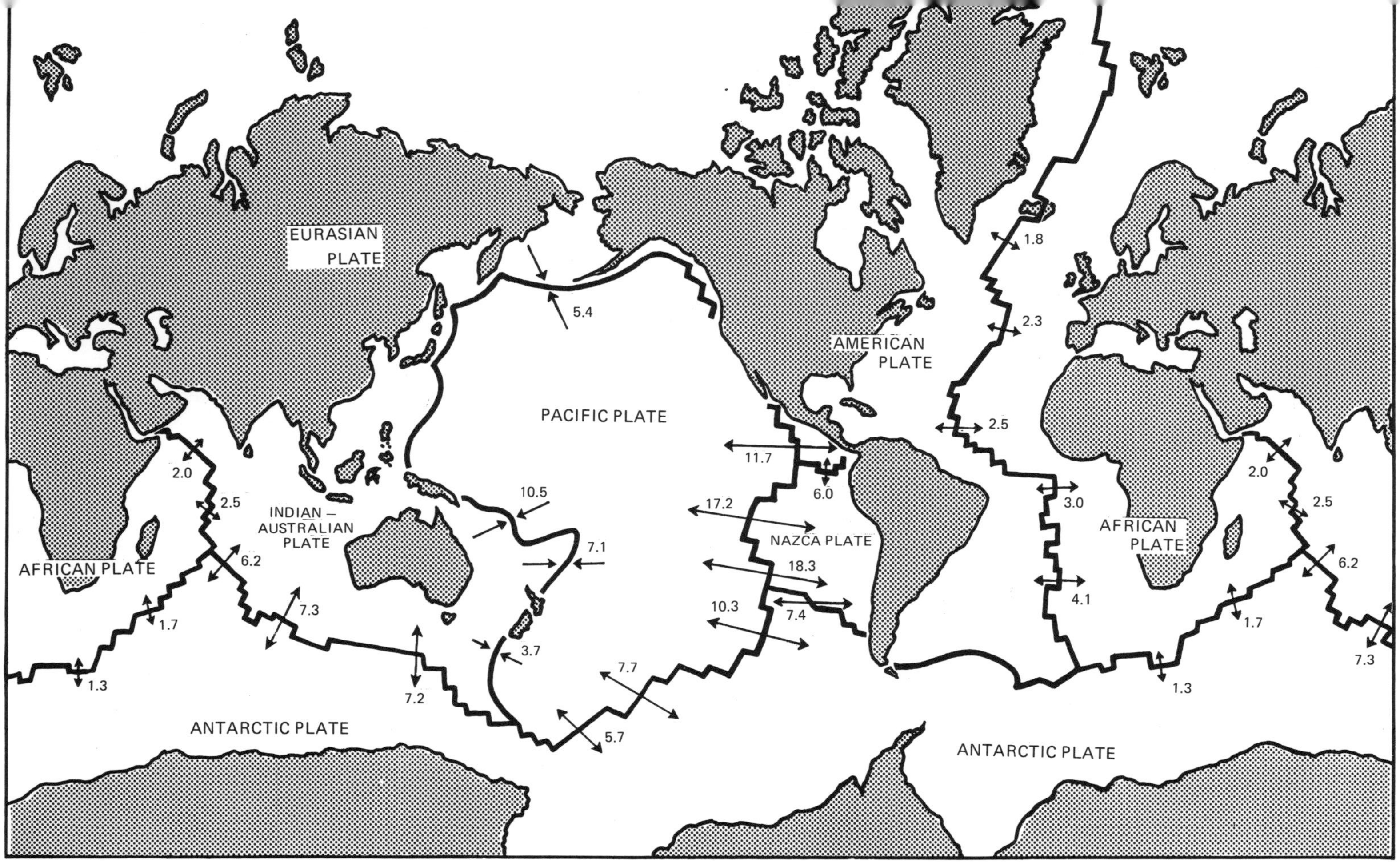


**Fig. 10.15.** Evidence for convection currents in the mantle, and some idea of the rate at which they move, is supplied by the motions of the dozen or so plates into which the earth's crust is divided. Some of the major plates are shown on this map (see Fig. 9.4 for other details). The lines with arrowheads pointing outward show where plates are moving apart at ridges. The numbers indicate the rate they are moving apart, expressed in terms of centimetres per year. The Indian-Australian, Pacific and Nazca plates are moving the most rapidly.

Lines with opposed arrowheads show where plates are moving towards each other, often at trenches. Plates can also slide past one another along transform faults such as New Zealand's Alpine Fault and California's San Andreas Fault (Figs 4.11, 4.12).

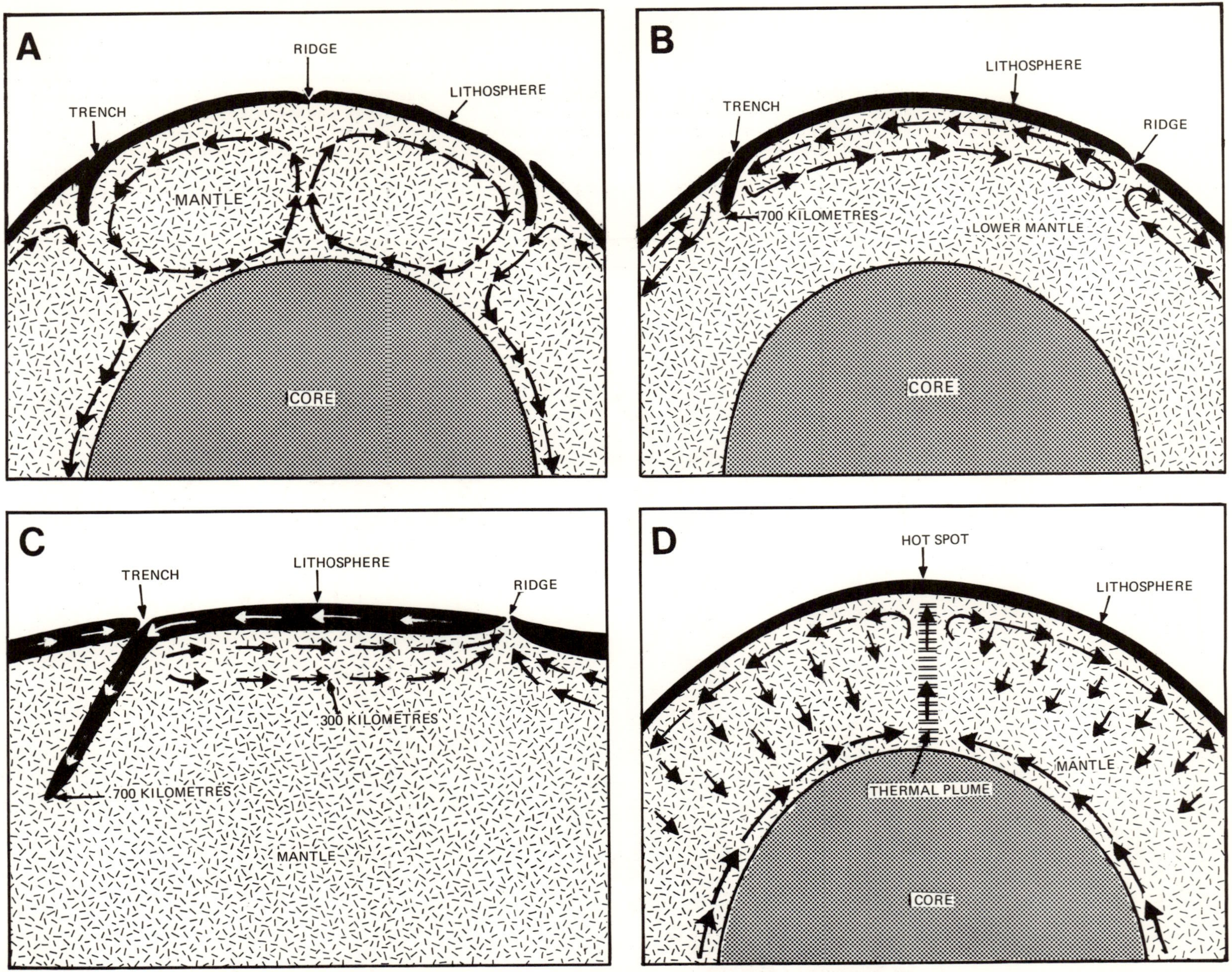
A
RIDGE
LITHOSPHERE
TRENCH
MANTLE
CORE
B
LITHOSPHERE
TRENCH
RIDGE
700 KILOMETRES
LOWER MANTLE
CORE
C
TRENCH
LITHOSPHERE
RIDGE
300 KILOMETRES
700 KILOMETRES
MANTLE
D
HOT SPOT
LITHOSPHERE
MANTLE
THERMAL PLUME
CORE

notably those associated with volcanic activity, are losing heat faster than others), a global system of convection currents, involving the upward flow of hot material and the downward flow of cool material is assumed.

During the first half of this century it was widely believed among geophysicists that convection could not occur in the solid rigid mantle. Recently, however, major advances in seismology and in experimental studies of the behaviour of rocks under high pressures and temperatures, some of which have been touched on in this book, have made convection currents the "front runner" among possible mechanisms for continental drift.

There is little doubt that the movement of such currents, if they exist, is exceedingly slow in human terms. The plates, and the continents that ride on them, drift at the rate of a few centimetres per year (Fig. 10.15). If we suppose the driving material in the mantle moves at a rate of 5 cm per year, this represents about 0.005 mm per hour. The tip of the hour hand of a typical household clock moves 5 cm per hour, but its movement is not readily detected by the human eye. Yet its speed is 10,000 times faster than the suggested movements in the mantle! Even so, a movement of 5 cm per year adds up significantly over geological time, measured in millions of years. For example, at a rate of 5 cm per year, rocks could move from the very base of the mantle up to the top in 58 million years, which is only a small fraction of the earth's probable age of 4600 million years.

Very slow convection movements in the mantle may be accomplished through the slow deformation of solid, yet hot, rocks. This is similar to the bending of heated steel which, after it has cooled, has great rigidity and strength. Experimental work shows that peridotite crystals, of which the mantle is probably largely composed, are capable of plastic flow when hot—by a process of repeated crystallisations, separated by liquid or semi-liquid phases allowing movements to occur. On these grounds, some geologists maintain convection movements can occur anywhere within the mantle, and that large convection cells extend through its entire thickness (Fig. 10.16A).

**Fig. 10.16.** Various models of convection currents have been proposed to explain plate movements. One model (A) contends that convection cells extend throughout the entire mantle. A second model (B) confines them to depths above a phase change in the mantle that is thought to occur at 700 km (435 miles). A third model (C) limits convection to the asthenosphere (Low Velocity Layer). The thermal-plume model (D) envisages all upward movement as occurring in a few thermal plumes, downward flow being accomplished by slow movements of the remainder of the mantle.

Many geologists, however, dispute the possibility of convection stirring the entire mantle, as they point out that descending crustal slabs go down to only 700 km (435 miles) before breaking up. Thus a school of thought has arisen that confines convection to the upper mantle above 700 km (where, as we have seen on p. 173, there is a possible phase change indicated by a seismic discontinuity) (Fig. 10.16B). Another school of thought pictures convection as occurring solely in the Low Velocity Layer or asthenosphere (Fig. 10.16C).

Yet another model for the plate-driving mechanism involves the development of thermal "plumes" in the mantle. According to this idea, all upward movement of mantle material is confined to about 20 plumes, each plume a few hundred kilometres in diameter, rising from the core-mantle boundary. The return flow is accomplished by a slow downward movement of the rest of the mantle (Fig. 10.16D). Where a plume reaches the lithosphere the flow becomes horizontal, spreading radially in all directions. A plume creates a "hot spot" with volcanic activity at the surface and it may cause an upward doming of the lithosphere. In such ways the plumes would cause movement of lithosphere plates. Some authorities have maintained that the kimberlite diamond pipes that reach deep into the earth (p. 172) have originated as thermal plumes in the geological past.

## Sliding plates

Regardless of the exact pattern of convection currents within the earth's interior (if they exist!), there is a steady transfer of heat from the core and mantle to the surface. There it spreads out and influences the various plates making up the surface of the earth.

Although various non-thermal mechanisms have been proposed to account for the movements of individual plates (e.g. the spinning, or Coriolis, force of the earth; the wobbling of the earth on its axis; expansion of the earth), as with the deep-seated processes it is still probably best to think of heat in the form of convection as providing the main driving force.

As we have seen (p. 120), the earth's outer layer, or lithosphere, is made up of seven major and at least eight minor plates (Fig. 9.3), that are like huge ice floes jostling together on an ocean. As with ice floes all the action occurs at the edges—plates crash and collide together and sometimes pull apart or over-ride one another. Sliding movements of the plates are undoubtedly made possible by the "lubricating" properties of the Low Velocity Layer. Heat, and its relationship to pressure, play a vital role in maintaining the unique properties of the Low Velocity Layer. Within this zone the rocks are very close to their melting point and in fact some five percent may be molten. Below the Low Velocity Layer the rocks are held at such high pressure that although they are hot, they remain solid. Above the Low Velocity Layer the rocks are largely cold and therefore rigid.

In terms of the convection current hypothesis, ascending (i.e. hot) currents rise from the mantle and as they do so, the cold rigid lithosphere above the Low Velocity Layer is split and moved sideways, to produce a divergent plate boundary. As the splitting (rifting) proceeds, localised relief of pressure occurs, the melting point of the rocks is exceeded and molten material moves upwards into the fresh gap between the plates.

Thus the rising convection currents are pictured as losing most of their heat in the rift zones (or divergent plate boundaries; see p. 121) of the world.

As they lose their heat the currents begin to sink, until eventually a zone of descending currents can be distinguished and plates are brought together (to form convergent plate boundaries; see p. 131). Processes such as subduction and continental collision occur at these sites.

The convection current story is, however, only part of the total picture. As with all natural processes, the forces at work are probably immensely complex. Along with convection, physical properties of rocks and differences in individual plates are other factors that may play significant roles in the interwoven processes that together add up to a driving force for plate movements.

The silicate rocks that make up the earth's crust and mantle respond to heat in a rather special way, because of two physical properties.

The first of these is the sensitivity of the silicate rocks to variations in temperature. At temperatures well below the melting point of the rocks (i.e. those normally at the earth's surface), they are hard and brittle, and when they are distorted in any way they may perhaps yield a bit, but then eventually break, splinter or shatter. However, when experiments are performed, and the rocks are distorted in the same way, but at progressively higher temperatures, it is found that the rock simply responds to the stress by what is called "ductile deformation" (or "creep"), rather as a steel bar bends once it has been heated.

The rocks are thus flowing in the solid state, much as glacier ice flows down a valley. This behaviour generally only becomes important when the rock is more than three-quarters of the way towards its melting temperature and even then the rates of flow are very small in human terms and are of the order of a few centimetres per year.

The ductility of rocks allows for the various buckling and flexing movements that accompany plate interactions. Ductility probably reaches its optimum development close to the Low Velocity Layer, so that although much slippage undoubtedly takes place within this zone, where partial melting has occurred, there is also a zone of bending in the upper mantle capable of absorbing substantial amounts of horizontal movement.

The second important characteristic of silicate rocks is their behaviour as conductors of heat. Compared with many everyday materials with which we are familiar (e.g. various metals—copper, iron, etc.), silicate rocks are poor conductors. Thus stone houses are cool in summer and warm in winter. If you place a stone in an oven for a number of hours, as country people did in Victorian times, for making a bed-warmer, you will find that after it is removed from the heat it will stay warm for a length of time which largely depends on its size and how long it was heated. In general, rock heats up very slowly and cools down very slowly. When the farmer carried the stone, wrapped in flannel, to his bed he was transferring heat by convection, i.e. he was transferring heat by the movement of hot material. Once placed in his bed the stone would transfer heat to the bed by conduction, rapidly at first while the rock was much hotter than the bed, and then more slowly when the temperature difference between them became less. The conductive properties of silicate rocks are made use of in a modern context in electric night-storage heaters, which use concrete blocks, rather than stones.

Another well-known physical effect of heating is that as objects such as rocks are heated, they expand in volume. This is because the tiny atoms making up the various constituents of the rock respond to heating by becoming more widely spaced. The rock thus expands slightly and becomes slightly less dense. When the rocks are cooled, the atoms pack closer together, the rock's volume contracts and its density increases. This property was used by the ancient Egyptians, Greeks and Romans in their stone-masonry, as rapid localised heating and cooling, involving expansion and sudden contraction, can be employed to crack large rocks to yield specific shapes.

Conduction in rocks has a vital part to play in the transfer of heat to the earth's surface. As plates of lithosphere move apart, there is an up-welling of material from below. Hot mantle material is moved rapidly upwards into the gap, along the oceanic rises. As cooling of this rising material is slow, it commonly arrives at the surface at temperatures above 1000°C (1832°F). Once on the surface however, heat is rapidly lost by conduction to the ocean floor, and cooling and solidification occur, and as a result fresh rigid rock is soon added to the edges of the adjoining plates. Then as movement of the plates continues, the fresh material fractures and hot material again arises from below. There is thus a great flow of heat at the crest of the oceanic ridge. The newly injected hot rock material produces a doming to either side of the crestline of the ridge. As the new rock material is moved to either side of the ridge by the sea-floor-spreading "conveyor belt" (p. 105), it progressively loses heat by conduction, contracts and increases in density (as we have seen above), and begins to sink. This leads to a gradual sinking of the ocean floor, the amount of sinking being in direct proportion to its age. Thus, as was found from sea-floor drilling by *Glomar Challenger*, the ocean gets progressively deeper and older away from spreading centres (p. 101). Some geologists have suggested that the "sliding" of plates "downhill" off the flanks of oceanic ridges may provide at least some of the energy required for plate movements. The injection of lava at the ridge crests may also play some part in forcing plates apart.

Other geologists point to the density changes that occur as a plate is moved away from its birthplace on an oceanic ridge. As plates are moved to either side of the ridge, and lose heat by conduction, they cool, contract, and become more dense. The sinking of old, cold and heavy rocks at plate edges, like the "sliding downhill" mentioned above, may also help in the overall driving mechanism for plate movement, by physically "pulling" plates down into the mantle.

As it sinks, the old slab may be over-ridden by a younger and hotter (and hence lighter) plate and be actually *pushed* down into the mantle as well (so that a subduction zone forms). However it is thought that the strength of rocks is such that any "pushing" or "pulling" is rather limited.

Phase changes (p. 172) of rocks within the plates may also help in the subduction process. As the plate edge is drawn down into the mantle, the basalt and gabbro of the sima (density 2.9 or 3.0) are subjected to intense compression. The rocks respond to such compression by transforming themselves into a denser phase or variety. The resulting highly compressed rock is called eclogite, the density of which is 3.4. The heavy eclogite may act as a "sinker" and, like the other factors mentioned above, may contribute to the subduction process.

Regardless of the exact nature of its driving force, there is no doubt that subduction is a very efficient process for swallowing up vast areas of ocean floor for, as the *Glomar Challenger*'s drill has shown, all the ocean floor older than Jurassic (i.e. older than 195 million years ago) has disappeared down trenches (p. 101) and been melted up in the mantle and "recycled" again.

The downgoing slab of oceanic lithosphere bends as a flexured beam and penetrates the mantle at an angle to the horizontal of between 30° and 70°. As the slab descends into the mantle it forms an elongated, inclined cold tongue extending to depths of some 700 km (435 miles), if we are justified in linking the deepest earthquakes with the existence of the slab. Initially the slab carries its own temperatures virtually unchanged into the depths of the mantle (as indicated by the temperature lines on Fig. 10.17), but it is gradually heated up; and this is the result of a number of interacting factors.

First, heat simply flows by conduction (see p. 178) into the cold descending slab from the surrounding warm mantle. As the conductivity of rock increases with temperature (cf. the bed-warmer stone, p. 178), the conductive heating becomes more efficient with increasing depth.

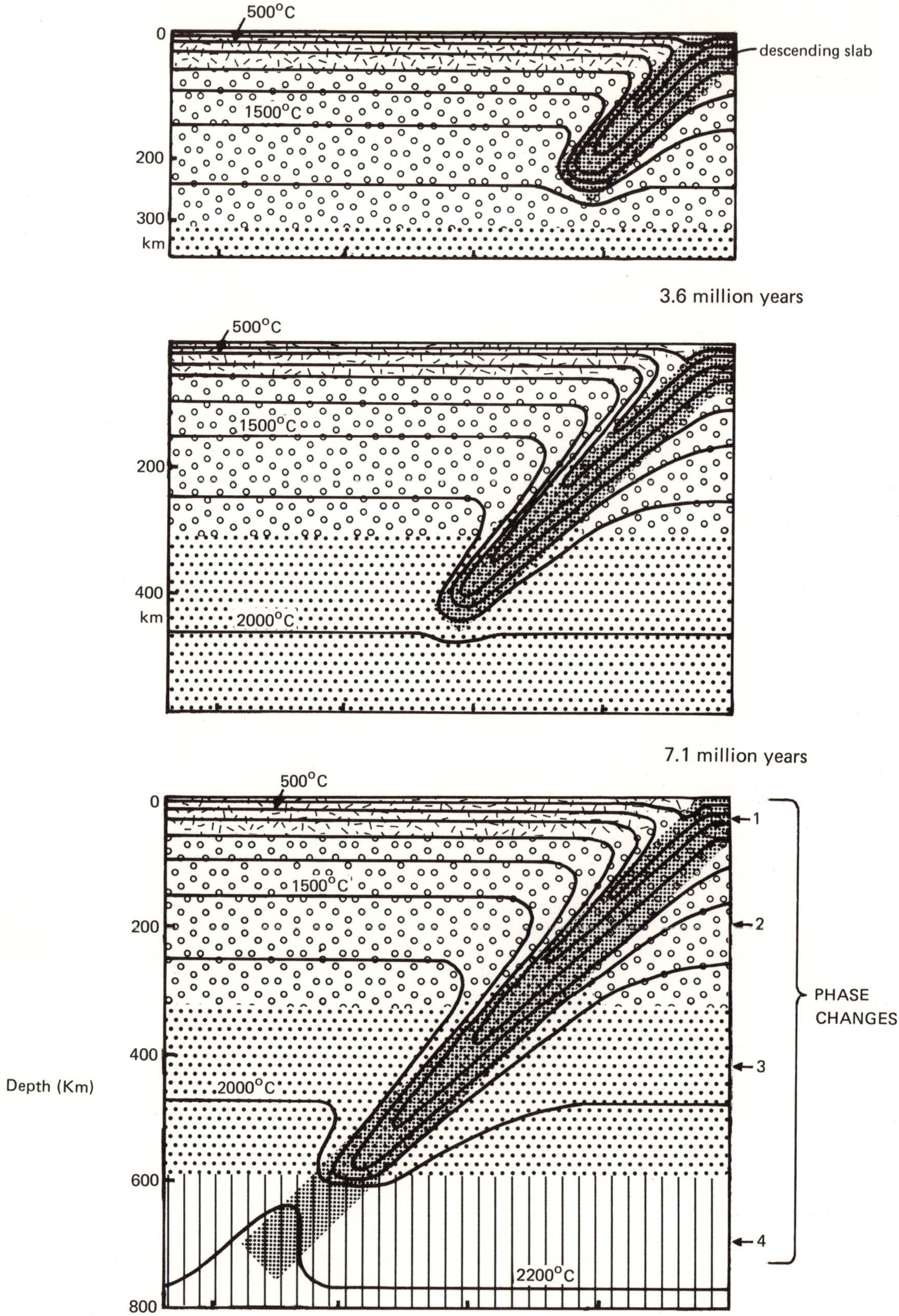

500°C
0
descending slab
1500°C
200
300
km
3.6 million years
500°C
1500°C
200
400
km
2000°C
7.1 million years
500°C
0
1
1500°C
200
2
PHASE CHANGES
400
3
Depth (Km)
2000°C
600
4
2200°C
800
800
600
400
12.4 million years
Distance (Km)


*Fig. 10.17*

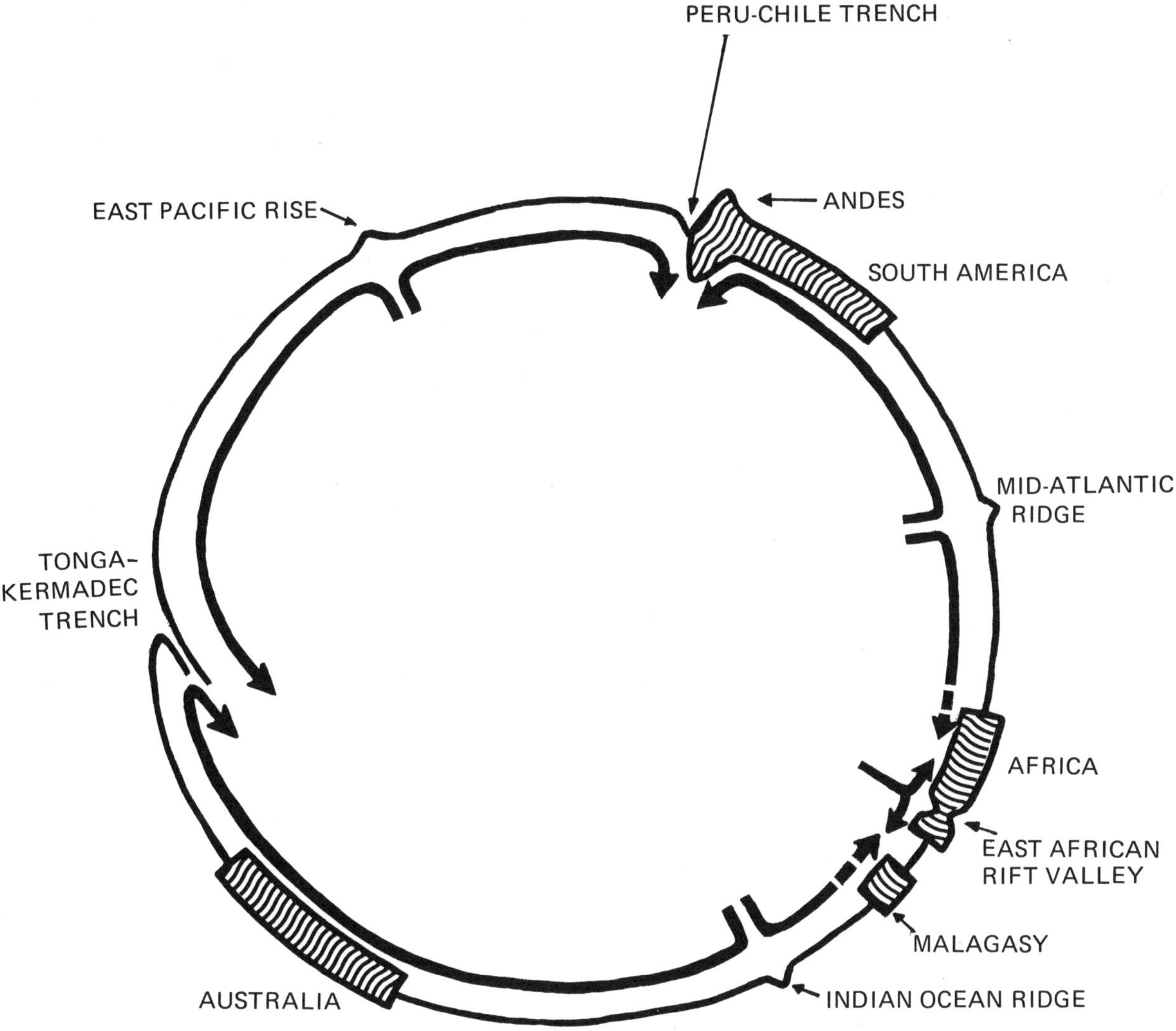


**Fig. 10.18.** A slice through the earth at 20°S Latitude shows that there is not a simple relationship between convergent plate junctions, where, in terms of the convection theory, convection currents are descending, and divergent junctions, where currents are ascending. Thus the underlying pattern of convection currents, if it exists, is probably exceedingly complex.

---

**Fig. 10.17.** The likely fate of a descending slab of lithosphere has been studied by means of computer models.

The three diagrams depict a slab subducting at an angle of 45° and at a rate of 8cm (3¼in) per year. Temperature contours are shown both within the crust and mantle and the descending slab.

Mineral phase changes, induced by increased pressure, are thought to normally occur at depths of 70, 300 and 600-700km (i.e. 44, 187, 373-435 miles) (shown by the patterns). In the descending slab, however, the first two phase changes occur at shallower depths, because of the slab's lower temperature. The transformation of minerals to denser forms provides a source of heat and may also act as a "sinker". As the slab is heated it undergoes assimilation into the surrounding mantle and by a depth of 700km (435 miles) has lost its original identity.

Second, as the slab descends it is subjected to increasing pressure, which introduces heat of compression.

Third, the slab is heated by the decay of radioactive minerals of uranium, thorium and potassium (p. 159), which are present throughout the crust and add heat at a constant rate to the descending slab.

Fourth, heat is released when the minerals originally present in the slab change to denser phases (e.g. eclogite, p. 172), as they are subjected to higher and higher pressures during descent.

Finally, heat is generated by the friction and

shearing stresses that occur at the boundaries between the downgoing slab and the surrounding mantle.

All these heat sources, particularly the first and fourth, contribute to a gradual warming-through of the slab. Thus as it descends, more and more of the slab is dissipated, to become part of the mantle. The slab becomes progressively thinner until it eventually breaks up and loses its identity altogether. At this stage, somewhere about 700 km (435 miles) deep, or less, the lithospheric plate becomes completely "assimilated" or absorbed by the mantle.

Although it is simple enough to say that subduction zones are the sites of descending convection currents and that oceanic ridges, continental rifts etc. are the sites of ascending currents, it is difficult to make any regular pattern out of the modern global situation which appears to be completely random (Fig. 10.18). Some people use the irregularity of this pattern as evidence against the existence of convection currents. However, in view of the complexities of the earth's internal structure it is probably a bit much to expect a simple regular pattern of convection cells, similar to that existing in our coffee pot of Fig. 10.1. It must also be remembered that any existing pattern is being further complicated by the continuous migration of the plates across the surface of the earth.

To conclude this chapter, it may be said that some kind of convection within the mantle—involving either the entire mantle, or only the Low Velocity Layer—probably provides an overall mechanism for plate movement. At the same time however, all the other factors mentioned above—spinning and wobbling of the earth, global expansion, density differences, gravitational sliding, magma injection, phase changes etc.—also make a contribution, along with many other factors that we are not fully aware of, or do not even know about!

CHAPTER ELEVEN

# EARLY NEW ZEALAND

As we have seen, continents like Australia have been built up around nuclei or centres of very old Precambrian rocks ("shield" rocks, older than 570 million years; see pp. 143–147). Most parts of New Zealand, on the other hand, have no obvious "core" of Precambrian rocks. Indeed, for many years it was thought that New Zealand had been created entirely from young rocks and its status—whether it was truly "continental" or somewhere intermediate between "oceanic" and "continental"—was in some doubt. Geophysical studies, however, have shown that New Zealand is truly continental in structure.

Recent datings have shown that Precambrian rocks occur as small patches along the West Coast of the South Island (Fig. 11.1). Other patches, however, long thought of as Precambrian, have recently been shown to be of Ordovician age.

Up to some three or four years ago it was thought that Silurian and Carboniferous rocks were missing in New Zealand, but we now know that all

**Fig. 11.1.** On the West Coast of the South Island occur outcrops of Charleston Gneiss, a metamorphic rock created by heat and pressure in mid-Palaeozoic time (about 400 million years ago) from sediments originally laid down in the Precambrian (older than 570 million years ago).

In this view of the West Coast at Charleston, banded gneiss outcrops appear in the foreground. The mouth of the Nile River, in the middle distance, is bordered by a ridge of Oligocene limestone, laid down by the sea some 37-25 million years ago. The granite mass of the northern Paparoa Range forms the background.

the geological periods are represented in New Zealand—although, particularly in the Palaeozoic, sometimes only by mere fragments of earth history.

Although our Palaeozoic history is fragmentary we do know that New Zealand has not always been on its lonesome out in the South-west Pacific. For all of the Palaeozoic, and for most of the Mesozoic also, the piece of the earth's crust that is now New Zealand was lying at the edge of the super-continent Gondwanaland, somewhere off the coasts of Victoria, Australia and Victoria Land, Antarctica.

**Fig. 11.2.** A series of expeditions to South-eastern Australia, New Zealand and Victoria Land (Antarctica), sponsored by UNESCO, have established the existence of close links between these countries in the early Palaeozoic. In the Cambrian, Ordovician and Silurian (570-390 million years ago) Tasmania is believed to have been squeezed between the Adelaide region and northern Victoria Land. New Zealand, on the other hand, was roughly opposite Victoria (south-east Australia). Diagram A illustrates the inferred situation, taking that of 400 million years ago as an example. Belts of similar rocks and fossils can be traced through the four landmasses, that were then closely adjacent.

The relationships changed in the early Devonian, about 380 million years ago, when some geologists believe the block containing Tasmania, Antarctica and New Zealand slid sideways, as in diagram B. The once continuous belts of diagram A were displaced to the south-east. Earth movements in the middle and late Devonian disrupted the fairly simple relationships seen in diagrams A and B and a new belt of structures (the Lachlan Fold Belt) was superimposed across the earlier geological patterns (diagram C). In all diagrams New Zealand is shown with movement along the Alpine Fault reversed (Fig. 4.11).

### Australasian jig-saw puzzle

The first chapter of New Zealand's long and varied geological history that we can distinguish clearly is in the periods of time geologists call the Cambrian, Ordovician and Silurian, between about 570 and 390 million years ago. Studies have shown that very close similarities exist between the rocks and fossils laid down at this time in New Zealand and those laid down in Victoria (Australia); so much so that the same names for fossils and rocks can be used in both countries. Less close similarities exist, however, between the rocks laid down at this time in Victoria on one hand, and South Australia and Tasmania on the other, in spite of the fact that they are now close neighbours.

It was realised that the similarities could be matched by sliding New Zealand and Tasmania sideways, towards the west, in relation to the

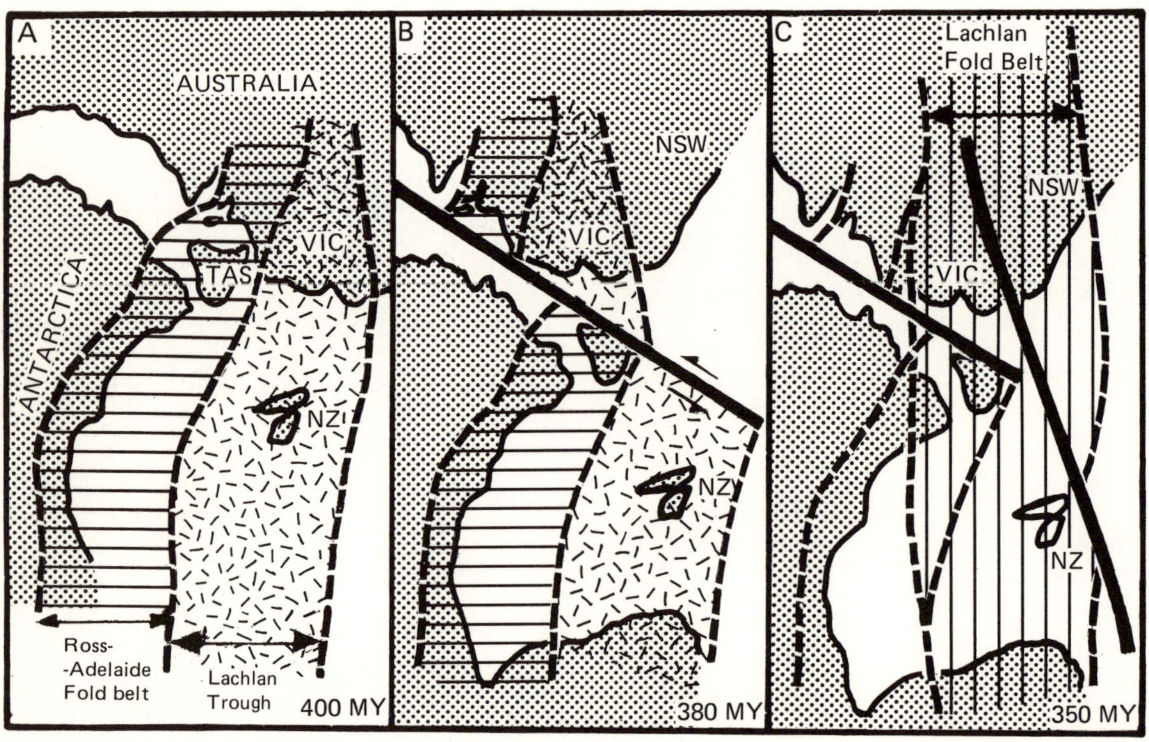

Australian mainland. Such a manoeuvre placed Tasmania nearer to South Australia and New Zealand opposite Victoria. The various Cambrian, Ordovician and Silurian rocks and fossils now matched perfectly (Fig. 11.2). The land masses probably occupied these positions at the very beginning of Palaeozoic time, some 570 million years ago.

Another problem was whether, if New Zealand and Tasmania could match up respectively with Victoria and South Australia, an equally good match was possible with Victoria Land, Antarctica. For if the fitting of the jig-saw puzzle had been done correctly (as in Fig. 11.2), this part of the Antarctic coast would at this time have been nudging South Australia. To test this idea a joint New Zealand-Australian expedition was mounted in 1974 with the specific aim of studying the Cambrian, Ordovician and Silurian rocks and fossils in Victoria Land, Antarctica. This expedition found that the northern Victoria Land rocks and fossils had close similarities with those of Tasmania. Thus the reconstruction shown in Fig. 11.2 was well and truly confirmed.

In the Devonian (390-345 million years ago), however, a marked change occurred, and the rocks and fossils that were laid down in New Zealand at that time lost their Victorian relationships and instead gained links with New South Wales. Similarly, Tasmania lost its ties with South Australia and gained links with Victoria. This marked change can perhaps be correlated with a sliding movement of Tasmania and New Zealand towards the east, along a major fault line passing through the modern Bass Strait, so that Tasmania was now opposite Victoria and New Zealand opposite New South Wales (Fig. 11.2). Except for these sliding movements of Tasmania and New Zealand, it is thought that New Zealand stayed at the edge of Gondwanaland, wedged between New South Wales, Tasmania and

**Fig. 11.3.** The Palaeozoic country of North-west Nelson. The peak in the foreground is Mt Patriarch (1687 m; 5535 ft). The bluffs and summit of Patriarch are composed of early Ordovician limestone, laid down in the sea about 480 million years ago. Below the bluffs are shales of earliest Ordovician age, passing downwards into rocks of late Cambrian age (500-515 million years ago). The rocks underlying the landscape beyond Mt Patriarch (i.e. to the west) are early Palaeozoic sediments and granites. The massive slips scarring the hills in the background date back to the Murchison Earthquake of 1929.

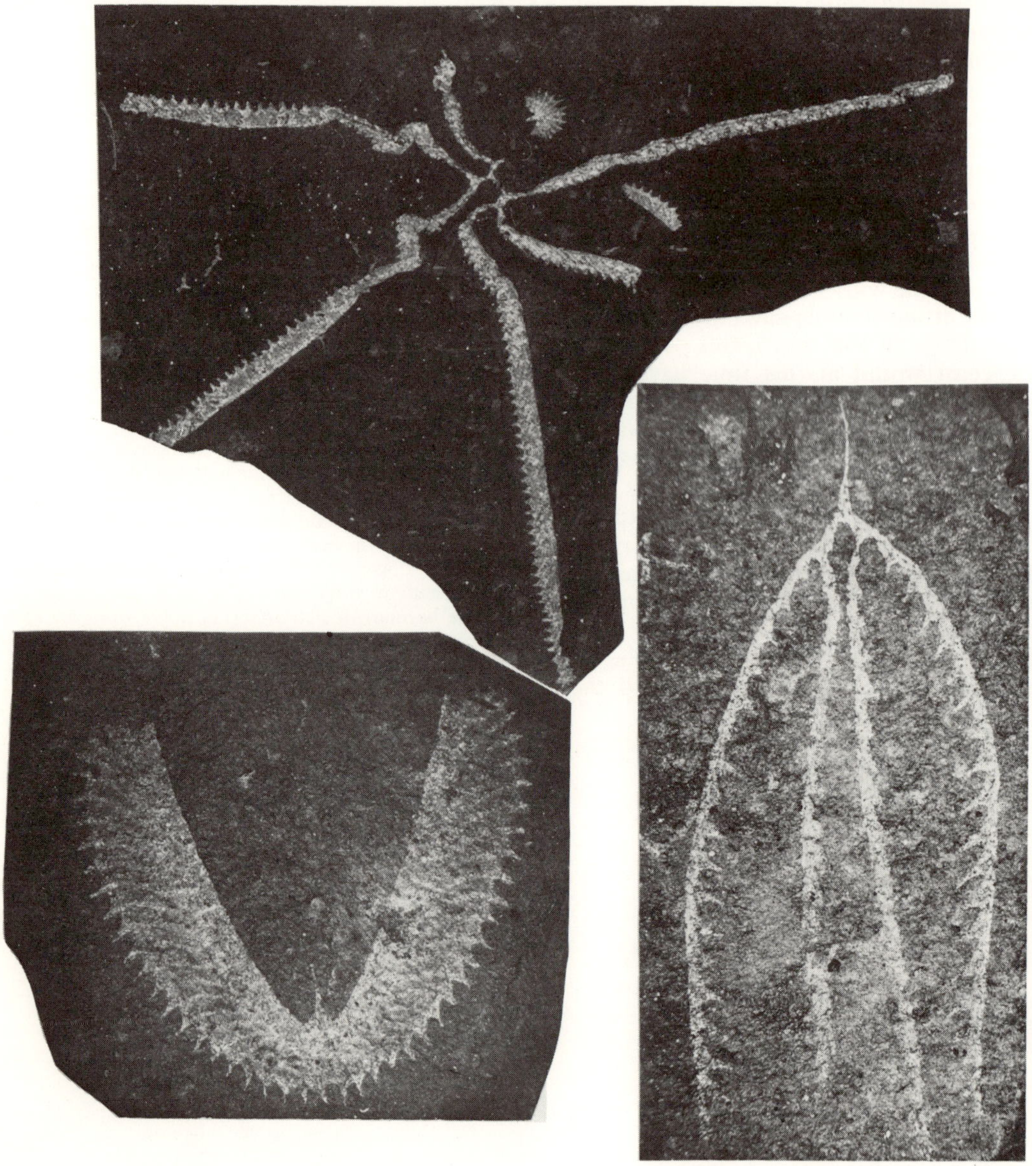

**Fig. 11.4.** Fossils called graptolites provide clues to former links between New Zealand and other countries in the early Palaeozoic.

They appeared in the middle Cambrian, 540 million years ago, reached their acme in the late Cambrian, Ordovician and Silurian and died out in the late Carboniferous, 300 million years ago.

Graptolites were floating colonies of animals, looking rather like seaweed, but are thought to be ancestral forms of primitive back-bone-bearing animals (Protochordates).

The three specimens illustrated all come from N.W. Nelson and are about 480 million years old (early Ordovician). The uppermost specimen, *Dichograptus*, is 73 mm wide (2¾ in); the specimen lower left, *Isograptus*, is 20 mm wide (¾ in) and the specimen lower right, *Tetragraptus*, is 8 mm wide (⅓ in).

The teeth-like projections each held a little animal—rather like a coral polyp—and the individuals were joined together by a nerve cord (notochord).

Antarctica, for about the next 200 million years, until Gondwanaland started to break up in the middle Mesozoic, some 150-100 million years ago.

During the Lower and Middle Palaeozoic, when Tasmania and New Zealand were on the move, the piece of crust that was eventually to become New Zealand was largely covered by sea. Throughout Cambrian, Ordovician and Silurian times (570-390 million years ago), the Australian and Antarctic continents were being actively eroded and the sedimentary products dumped in the New Zealand area in an off-shore trough-like geosyncline (Fig. 11.3) (p. 146). By the early Devonian (390-370 million years ago), however, the rate of erosion of the landmasses of Antarctica and Australia had largely slackened and materials such as shales, sandstones, limestones, etc. were being laid down in the sea.

Bizarre fossils such as graptolites and trilobites (Figs 11.4, 11.5) provide information on Palaeozoic links between the South-west Pacific countries and elsewhere. During the early Devonian a lamp shell (brachiopod), *Pleurothyrella*, lived in the shallow seas that covered New Zealand (Fig. 11.6). This shell also lived at the same time in seas lapping onto South Africa, Bolivia and Antarctica, leading us to believe that there was a shallow-water migration route around the margin of Gondwanaland available to these and other animals.

Extensive earth movements occurred in eastern Australia in the middle Devonian (370-359 million years ago) and similar movements—called the Tuhua movements—created large areas of land in New Zealand, out of areas formerly occupied by sea, and therefore leaving a gap in the marine sedimentary record. Meanwhile in Antarctica the sea had also been pushed off the land and extensive lake deposits accumulated, with rich fish remains.

To judge from the animals that lived in our seas at the time (the remains of which occur in rocks as fossils), New Zealand had a climate essentially similar to that of today—i.e. not too hot and not too cold.

The landmass created by the Tuhua earth movements was progressively whittled away by the effects of river and stream erosion until eventually, in the Carboniferous period, 345-280 million years ago, sea again covered most of what is now New Zealand.

This sea however had a vastly different character from that covering New Zealand before

**Fig. 11.5.** The study of trilobites also provides information on past links to New Zealand. Trilobites get their name from the three(tri)-lobed body and in some respects were rather like large marine slaters (wood-lice). They appeared in the early Cambrian, 570 million years ago, reached their acme in the Ordovician, and became extinct in the late Permian, 230 million years ago. The upper photo is of a specimen of *Ogygites*, distorted by earth movement, from Paturau River, Nelson. The specimen is 35 mm (1½ in) wide. The lower drawing is of *Incaia*, also from the Paturau River, 28 mm (1¼ in) wide. Both trilobites are of late Ordovician age, 450 million years ago.

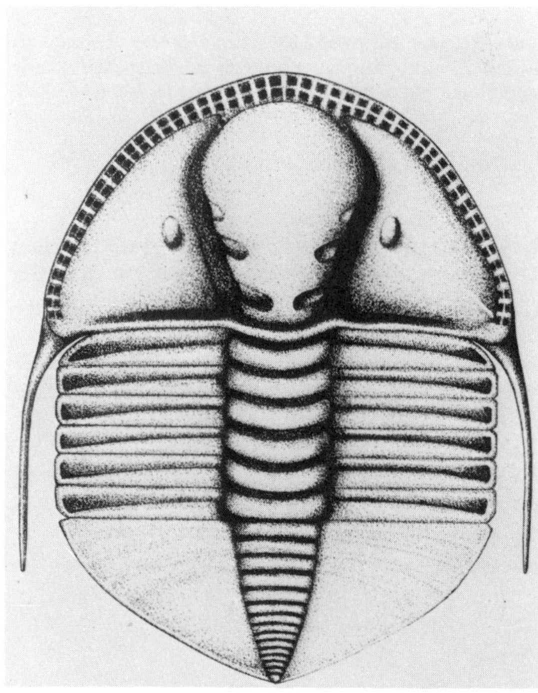

**Fig. 11.6.** In the Devonian, 390-345 million years ago, lampshells (Brachiopods) were shared by many of the southern Gondwana countries, indicating the existence of shallow-water marine links at this time.

*Pleurothyrella* (upper left) is known from South Africa, South America and Antarctica, as well as New Zealand. Its shell is 17 mm (¾ in) at maximum diameter. *Acrospirifer* (upper right and bottom) are winged brachiopods of southern affinities. The specimen upper right is 70 mm (2¾ in) in length and the one at bottom is 120 mm (4¾ in) in length. All three specimens are from Reefton.

the Tuhua movements. The sea bordered rugged newly made land to the west, in Eastern Australia and Eastern Antarctica. This land was mountainous, with many active volcanoes, and was being constantly shaken by earthquakes. Rivers and streams were actively eroding the raw slopes of the new land and volcanoes were pouring out ash and lava onto both land and sea.

Off-shore from this land vast quantities of gravel, sand, mud and volcanic debris were accumulating on the sea floor. The great weight of this sedimentary material buckled downwards the crust in the New Zealand area, producing in the sea floor a rapidly subsiding trough-like depression called the New Zealand Geosyncline (Fig. 11.7).

Fragments of the leading edge of the old Palaeozoic landmass that formed a westward margin to the New Zealand Geosyncline, which were left behind on the edge of New Zealand when the Tasman Sea opened, can be seen in Fiordland, coastal Westland and north-west Nelson (Figs 11.1, 11.3).

The New Zealand Geosycline was much larger in all respects than the early Palaeozoic geosyncline—it extended northwards at least as far as New Caledonia and as far south as Campbell and Auckland islands. In width it extended at least from the seas off Westland across to those east of the Chatham Islands. It was in existence for some 200 million years, between Carboniferous and late Jurassic times, and during this interval it was the dumping-ground for immense thicknesses of sedimentary material, derived from the erosion of Gondwanaland (Fig. 11.8).

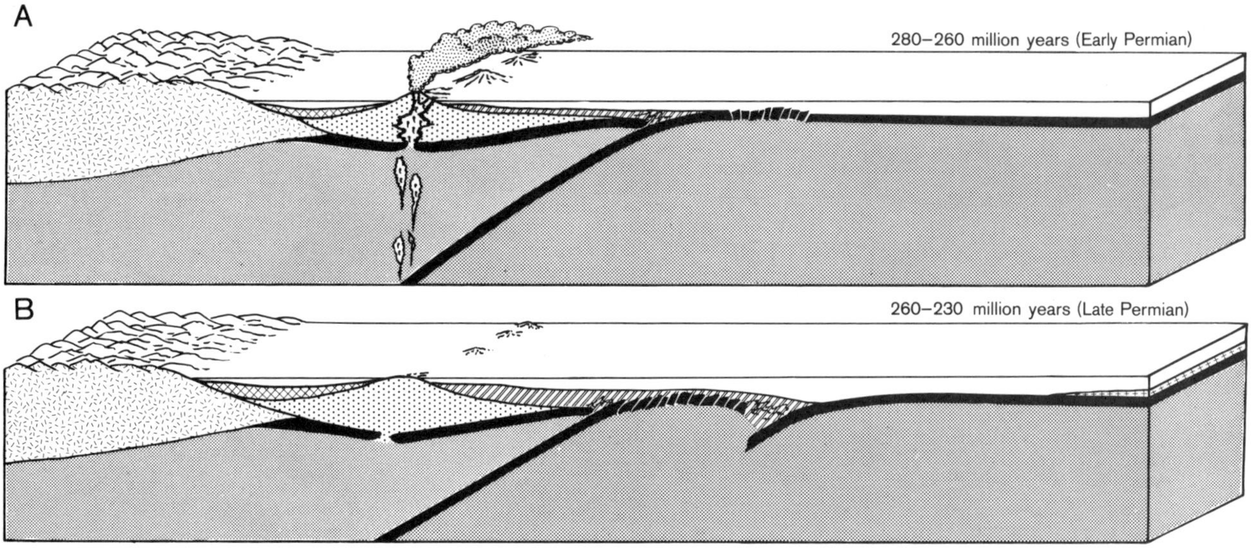


*Fig. 11.7*

**Fig. 11.7.** In the Permian the New Zealand Geosyncline may well have had this appearance. The continents of Antarctica and Australia and the early Palaeozoic part of the South Island of New Zealand were being actively overthrust across the floor of the ancient Pacific Ocean, and a subduction zone was formed along the continental margin (diagram A). Melting of the subducted slab generated an oceanic island arc, with active volcanoes. Sediment derived from the continents was laid down between the continental shoreline and the island arc, whilst volcanic-derived material accumulated to either side of the arc.

In the late Permian (diagram B) the subduction zone was disrupted by a change in plate movements and a new subduction zone formed. Meanwhile, sea-floor spreading steadily rafted sedimentary material derived from Marie Byrd Land (Antarctica)—seen on the right side of diagram B.

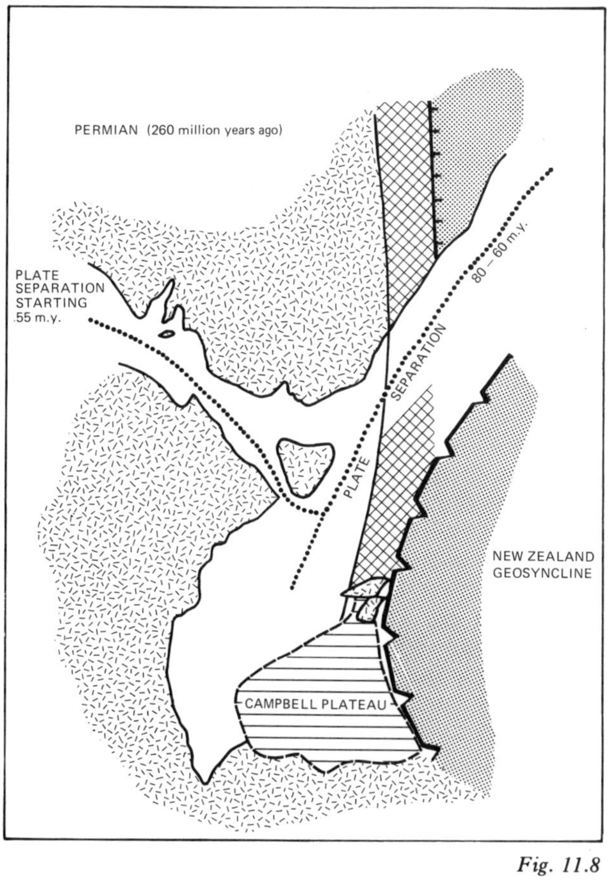


*Fig. 11.8*

**Fig. 11.8.** The links established in the Devonian between New South Wales and the early Palaeozoic part of New Zealand (Fig. 11.2C) continued into Permian times and the rocks and fossils of both countries were very similar.

In this diagram the New Zealand Geosyncline is shown lying off-shore from the margin of Australia, Antarctica and the early Palaeozoic part of New Zealand. The dotted lines show the lines of separation that occurred in the Tasman Sea (80-60 million years ago) and Southern Ocean (starting 55 million years ago).

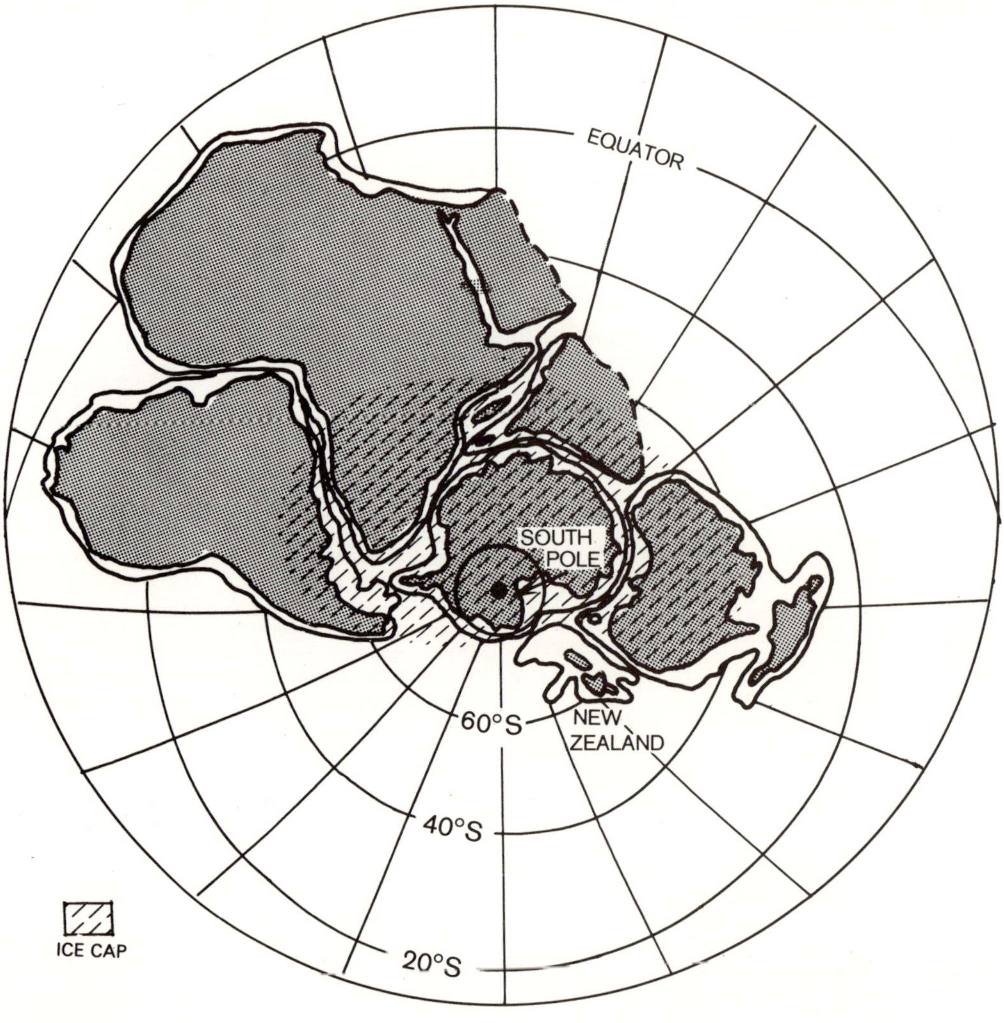


**Fig. 11.9.** In the Carboniferous and Permian, 345-230 million years ago, the super-continent Gondwanaland was straddling the South Pole and experiencing a series of glacial and interglacial periods. During the glacials a large ice cap built up on South America, South Africa, India, Australia and Antarctica. New Zealand, largely under the sea at the time, was marginal to the ice sheet.

During the interglacials climate returned to normal or warmer-than-normal (e.g. coral reefs existed in New Zealand: cf. Figs 11.10, 11.11).

As the crust subsided, large fractures appeared up which welled molten rock (magma), to be erupted on the sea floor, producing quantities of lava and fragmentary volcanic material. Chains of volcanic islands dotted the surface of the sea. However, as far as is known, little substantial land, apart from volcanic archipelagoes, existed in the New Zealand region throughout much of the history of the New Zealand Geosyncline (cf. Figs 13.1-13.3).

Momentous things were also happening on Gondwanaland as a whole.

**Fig. 11.10.** The volcanic islands that dotted the seas of the New Zealand Geosyncline during the Permian (cf. Figs 11.7, 13.1) erupted masses of lavas both on the sea floor and on the islands themselves. The top photo shows part of a pile of submarine lava flows exposed in cliffs at Marble Bay, Whangaroa Harbour, Northland. The height of the exposure is 3m (10ft).

At Marble Bay the submarine lavas have flowed around some limey sea-floor sediments containing reef corals and minute warm-water creatures (foraminifera). The presence of these tropical animals indicates that the sediments were laid down during one of the interglacial periods. The lime-bearing sediments have been recrystallised by the heat and pressure of the lavas to form marble, although recognisable fossils are still present (Fig. 11.11).

The middle photo shows a large bed of marble and the bottom photo a small marble layer incorporated in the lava.

## New Zealand on ice

Prior to the Carboniferous, some 345 million years ago, most of Gondwanaland straddled the temperate climate zones of the Southern Hemisphere. For example, in the Cambrian, Ordovician and Silurian, the equatorial region extended through Central Asia, North America and Eastern Antarctica. To judge from palaeomagnetic measurements, however, throughout this time world-wide drifting movements had pushed Gondwanaland closer and closer to the South Pole. As a result of this steady southwards drift, there began in the Carboniferous a series of ice ages (or glaciations) (Fig. 17.12). Ice sheets covered large areas of Gondwanaland, and the scraping and scouring actions of the glaciers left tell-tale evidence in many regions (Figs 5.9-5.11, 11.9). The unmistakable implications this evidence had for past climates had originally impressed Wegener, du Toit and other stalwarts of the old continental drift theory, and even in the darkest days of the theory the fact of Permo-Carboniferous glaciation provided indisputable proof that the continents *had been* in vastly different positions in the past (pp. 35–38).

A characteristic of glacial periods is that there is an alternation between cold and warm periods (called respectively "glacials" and "interglacials"). The precise reason for this to-ing and fro-ing of climate is not known, but, like most things affecting this earth of ours, it is probably a combination of many things, such as fluctuations in the sun's heat, cyclic build-up and melting of ice sheets, etc. (pp. 336–341).

Anyway, in the Gondwana countries, including

**Fig. 11.11.** A group of Permian corals from Marble Bay, Whangaroa Harbour (Fig. 11.10) and a cross-section through the skeleton of a single coral animal. The coral slab is 100 mm (4 in) wide, and the coral specimen 20 mm (¾ in) in diameter.

Although Gondwanaland experienced a number of glacial periods during Carboniferous and Permian times (Fig. 11.9), these were separated by interglacial phases—times when climate returned to normal or warmer-than-normal. During the Permian glacials cold-temperate organisms such as *Glossopteris* (Figs 5.11-5.13) lived in New Zealand, whereas during the interglacials climate improved to allow reef corals (such as those illustrated) to thrive.

Australia and New Zealand, we can see in the animal remains preserved in rocks laid down in the sea throughout the Carboniferous and Permian (345-230 million years ago) clear signs of changing climates. Animals that inhabited cold water are found to alternate with warm water animals. In New Zealand, for example, Permian corals are found, sometimes aggregated together as reefs (as at Marble Bay, Whangaroa Harbour, North Auckland: Figs 11.10, 11.11)—providing clear evidence of the former existence of tropical waters, as modern reef corals cannot live in waters colder than 22-25°C (72-77°F). Equally, some of the Permian shellfish found in other parts of New Zealand have very thick shells, and this factor, combined with other information (notably oxygen content, see p. 214), tells us that these creatures lived in cold waters, as modern shellfish living in cold waters tend to grow thick, massive shells. Thus, in our fossil record, we can distinguish a series of cold and warm water populations, corresponding, it is thought, to fluctuating glacial and interglacial climates.

On land huge snow fields accumulated in the areas of high country and, as the snow compacted to become hard glacier ice, ice sheets formed and under the influence of gravity began to move across the lowlands, scouring and scraping rock surfaces (to produce "glacial pavements"), and transporting great quantities of debris, both on top and within the ice sheet (glacial moraine) and dragged along underneath (glacial till).

Thus glacial pavements, of Carboniferous and Permian age, often overlain by deposits of till and/or moraine, typify many regions in India, South Africa, South America, Australia and Antarctica (Figs 5.9-5.11).

Continuous ice sheets nonetheless did not cover the entire landscape and the ice-free areas adjacent to the ice sheets were populated by hardy plants, rather like our modern alpine and sub-alpine shrubs. These plants, called *Glossopteris*, had large robust spatula-shaped leaves, with prominent veins (Figs 5.11-5.13). *Glossopteris* occurs in rocks of Permian age in the Gondwana countries, often in close association with glacial deposits, but is completely unknown from Laurasia. The presence of *Glossopteris* in India, South Africa, Australia, South America and Antarctica has been known for many years and was remarked on by Wegener and du Toit. The first *Glossopteris* specimens from Antarctica were collected by the ill-fated Scott expedition (p. 39).

*Glossopteris* however, appeared to be notably

absent from New Zealand and as the theory of continental drift steadily attained respectability, and people began to think seriously of where New Zealand fitted into the jig-saw puzzle, the absence became more and more a glaring anomaly, so much so that some people began to wonder whether New Zealand had in fact ever been part of Gondwanaland.

**As work progressed on the New Zealand Permian rocks the reason for the absence became apparent. It was found that all of the New Zealand Permian rocks were of marine origin and so had preserved in them only the remains of marine creatures. The chances of finding the remains of land plants in our rocks were therefore rather slender.**

One isolated leaf fragment was, however, located in Southland in 1970 (Fig. 5.13). This leaf had probably drifted off-shore from the main landmass of Gondwanaland and on becoming waterlogged had settled to the sea floor, there to be incorporated in sands and muds. This fossil's discovery proved conclusively that New Zealand had originally been part of Gondwanaland, although as we now know, it was then only a small piece of the sea floor off Australia and Antarctica (Figs 13.14, 18.6).

## The Age of Reptiles

In the next period of geological time, the Triassic, the same global movements continued that had twisted and rotated Gondwanaland in the Carboniferous and Permian. By the start of Triassic times however, some 230 million years ago, the movements, having swung Gondwanaland right across the South Pole in the Carboniferous and Permian, were starting to rotate it northwards again, into warmer climes. Thus the super-continent gradually lost its ice sheets, and numerous reptiles roamed widely across its surface. These reptiles were the ancestors of the dinosaurs, turtles, lizards, crocodiles, flying reptiles and birds that appeared later in the Mesozoic.

At first, in the early Triassic, 230-215 million years ago, comparatively few reptiles were able to migrate between the two super-continents, and Laurasia and Gondwanaland had distinct reptile populations. Laurasia's reptiles, for example, gave rise to the dinosaurs, while Gondwanaland's reptiles were mammal-like in appearance, with skulls and bodies similar to those of wolves or foxes. (They have been called "reptilian wolves" and may in fact have been hunters like their mammal namesakes.) (See Fig. 11.12.).

In the middle Triassic (215-205 million years ago), however, the lines between Gondwanaland and Laurasia became progressively blurred, and this process continued into the late Triassic (205-195 million years ago), when apparently many of the land reptiles were able to roam far and wide across both super-continents (Fig. 11.13).

**Fig. 11.12.** Two reptiles that lived in southern Gondwanaland in the Triassic were *Cynognathus* (top), about the size of a large wolf (i.e. 1.8 m, 6 ft long), and *Lystrosaurus* (bottom), about the size of a large sheep (i.e. 1.2 m, 4 ft long), but very heavily built.

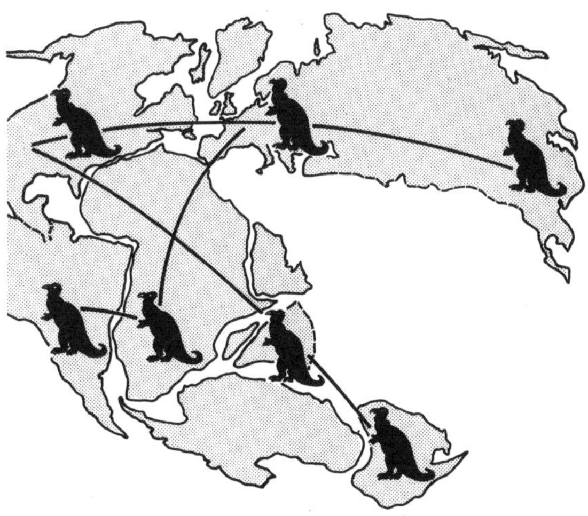

**Fig. 11.13.** In the late Triassic (205-195 million years ago) large land reptiles (dinosaurs and their relatives) were able to roam virtually the length and breadth of Gondwanaland and also migrated into southern Laurasia—thus indicating the existence of good land links and a reasonably uniform climate.

One such widespread group was the rhynchocephalian (i.e. "beak-headed") reptiles that provided the ancestors of the New Zealand tuatara lizard (*Sphenodon*). The progressive breaking down of the distinctions between Laurasia and Gondwanaland as Triassic time progressed may be interpreted to mean that the climates of both super-continents gradually became the same—as Gondwanaland drifted further and further away from the South Pole. So in late Triassic times cold-blooded reptiles could travel extensively across both super-continents.

As we have seen, the occurrences of *Glossopteris* tie together the various landmasses of Gondwanaland in the Permian. A similar linking job can be done in the Triassic by the plant *Dicroidium* which, like *Glossopteris*, is found in all Gondwana countries, including New Zealand (Fig. 5.16).

Reptiles also provide links. As we have seen in the Permian, *Mesosaurus*, an alligator-like reptile, is found only in Southern Brazil and South Africa (Figs 5.1, 5.16). Similarly in the Triassic a wolf-like reptile, *Cynognathus*, is found only in Argentina and South Africa. This reptile had a skull similar to that of a dog (hence the name, *cyno* = dog, *gnathus* = jaws) and its teeth were adapted to a carnivorous diet. The body was slender and supple, and the feet suited for rapid running. Some of these reptiles must have been rather wolf-like or fox-like in both habits and appearance (Fig. 11.12).

Another reptile that occurs with *Cynognathus* in South Africa is *Lystrosaurus* (Figs 5.14, 11.12). *Lystrosaurus* was a heavily built animal, about the size of a large sheep (i.e. 1.2-1.5 m or 4-5 ft in length) with a barrel-like body, stout stocky legs, broad feet and a short tail. The head is very distinctive, with beak-like jaws and a pair of large tusks. *Lystrosaurus* has long been known from South Africa, India and Western China (Sinkiang and Shansi), where it is an important Triassic fossil. Its discovery in Antarctica in 1970 provided a major palaeontological link, binding together the Triassic Gondwanaland (Figs 5.15, 5.16).

Before the finding of *Lystrosaurus* in Antarctica, its presence in the other countries—South Africa, India, Western China—could be readily explained in terms of the modern world map. The reptile could simply walk up through the African continent, across Asia Minor and into India and Western China. Such a wide geographic distribution would not be at all unusual as the leopard, for example, not so long ago ranged from the Cape of Good Hope, through Africa, across the Middle East and Asia and well into Siberia.

*Lystrosaurus*, however, could not have walked into Antarctica without some help from continental drift. For although it was a lover of lakes and rivers and may have been like a miniature hippopotamus, it was nonetheless essentially a land animal, and with its heavy body would have sunk like a stone in deep water. It could no more cross great stretches of ocean than can a hippopotamus today. It had to move from one region to another by a land route, albeit one that was well-watered. Thus the *Lystrosaurus* story adds another string to the Gondwanaland bow (see also p. 43).

A side issue is the presence of *Lystrosaurus* in Western China, the only occurrence outside Gondwanaland. One interpretation could be that *Lystrosaurus* was able to wander far beyond Gondwanaland, to populate the easternmost reaches of Laurasia. Another interpretation, however, could be that China and South-east Asia were originally parts of Gondwanaland and have since been welded onto the Eurasian block. If such were the case, *Lystrosaurus* and its companions were members of a widely distributed Gondwana population.

As may be seen from Fig. 5.15, *Lystrosaurus* is absent from South America. All other evidence indicates that South America and Africa were intimately connected at the beginning of Mesozoic history. We may ask, is this absence real? Perhaps it is, and *Lystrosaurus* and its companions were barred from South America by some factor unknown to us. On the other hand, the absence may only be apparent and may be one of the many gaps in our knowledge of past life. One day *Lystrosaurus* might well turn up in some as yet unexplored part of South America and then our picture will be complete.

**Triassic New Zealand (230-195 million years ago)**

Throughout Triassic time New Zealand was still largely under the sea. The New Zealand Geosyncline dominated the region, and great thicknesses of material eroded off Gondwanaland, together with volcanic debris contributed by volcanoes both on land and in the sea, were dumped on the subsiding ocean floor (Fig. 13.2).

Earth movements continued at intervals, causing massive submarine sliding of loosely compacted materials that had earlier been laid down on the sea floor. Large masses of chaotically jumbled layers, seen in the rocks today, provide us with dramatic evidence of such submarine sliding (Fig. 13.13).

Small islands may periodically have been created by such earth movements and evidence for these is seen in the rock record (Fig. 13.2). Such islands, and those of volcanic origin, were, however, apparently ephemeral and soon eroded away.

Climate in the Australasian region throughout the Triassic was temperate, i.e. similar to that of today, and perhaps slightly warmer. On the land reptiles roamed far and wide across Gondwanaland and Laurasia.

In the sea many creatures became widespread, particularly those floating or swimming in the upper layers of the sea and so able to travel across vast areas of deep ocean. Early ammonites, ancestors of the modern *Nautilus* (Figs 12.1-12.4), which could swim (by means of long tentacles and by jetting out streams of water) and float (by virtue of air chambers), became distributed throughout almost every Triassic ocean, and New Zealand had identical forms to those in South-east Asia, Southern Europe, and North America.

But on the other hand, many animals that were common in the Triassic seas of South-east Asia, for example, did not make it to New Zealand. These animals lived mainly on the sea floor and on rocks in shallow coastal waters and the probable reason they were absent from New Zealand is that at that time the seas covering the region were cut off from other surrounding oceans by extensive areas of deep water. The absence of shallow water routes into the South-west Pacific meant that only floating or swimming creatures could come to New Zealand; all others were excluded.

The sea-floor animals that were already present in New Zealand seas developed independently, therefore, of the other types populating other oceans, and a number of special or "endemic" races appeared, some quite bizarre (Figs 11.14, 11.15), restricted to the South-west Pacific (New

**Figs 11.14, 11.15.** In the mid and late Triassic, 215-195 million years ago, the shallow coastal seas then covering New Zealand and New Caledonia were populated by groups of bizzare winged brachiopods, endemic to the region. Representatives of some of these are shown in these two illustrations.
Fig. 11.14 shows two specimens of *Rastelligera*, from the late Triassic of Nelson. The upper shell is 72 mm (2¾ in) wide; the lower shell 65 mm (2½ in) wide.
Fig. 11.15 shows two views of *Clavigera*, characteristic of the latest Triassic, about 197-195 million years ago. These specimens, from Marokopa, Waikato coast, are 60 mm (2¼ in).

*Fig. 11.14*

*Fig. 11.15*

Zealand and New Caledonia). Isolation of New Zealand seas continued into the early Jurassic, 195-172 million years ago.

The Jurassic is often called the "Age of the Ammonites". These squid-like creatures with intricately coiled shells became extremely common in Jurassic seas (Figs 12.5-12.15). Many of them, using a combination of swimming and floating, could cross vast oceans and were therefore independent of coastlines. Others, however, hovered close to the sea floor and lived on fish and other animals inhabiting the submarine plains.

In the early Jurassic the free-swimming and floating ammonites were able to migrate to New Zealand, but the bottom-dwelling ones, limited to shallow coastal waters, were excluded. Thus we had the same geographic situation as in the Triassic—the New Zealand seas cut off from others by extensive areas of deep ocean. As we will see however (p. 245), conditions changed markedly in the middle and late Jurassic. New seaways opened up, particularly to the north, into the Indo-Pacific region, allowing great numbers of shallow water marine creatures to migrate to New Zealand.

CHAPTER TWELVE

# THE FOSSIL STORY

ALL AREAS of New Zealand have been covered by the sea at some time in past geological ages. A great variety of animals and plants lived in the New Zealand seas of the past, and as they died their bodies settled to the sea bottom, there to be entombed in the gradually accumulating layers of sand, silt and clay. In most instances the creature's soft parts soon disintegrated and disappeared, leaving only the hard parts (e.g. the shells of shellfish, the skeletons of fish) as fossils in the sediment layers.

Often, after the passing of many millions of years, and the accumulation of great thicknesses of sedimentary layers on the sea bed, movements related to processes deep within the earth's crust buckle and push up the sea-floor layers to form land. The new-born land is immediately cut into by rivers, streams and ocean waves, so that fossil-bearing layers, once part of the sea floor, become exposed in hillsides, gorges and sea cliffs. Frequently the once-horizontal sedimentary layers have been tilted to various angles or sometimes even turned upside-down. The layers are often dislocated by large cracks, along which movement has occurred ("faults").

Today, therefore, we can find fossils in the rocks on the tops of many of our high hills and mountains, many hundreds of metres above sea level. The fact that we can find the remains of sea creatures on mountain tops does not necessarily mean that the landscape has been drowned by the sea to that height. All that it means is that the particular rocks (and fossils) were laid down on the sea floor many millions of years ago and then pushed up to their present height by earth movements.

The fossils that we find now exposed in various parts of New Zealand tell us a lot about the history and geography of New Zealand in the geological past.

Some animals and plants have existed virtually unchanged throughout the aeons of geological time. *Lingula*, a lamp-shell (or brachiopod), is one shell that has lived continuously from the Cambrian period, 570 million years ago, through to the present day.

Most animals and plants, however, tend to change as they progress through generation after generation. Some of the changes that appear can be used to advantage by the plant or animal, enabling it to compete more effectively in its own little world and hence to gain more food, grow bigger and stronger, live longer and reproduce more actively. Some changes, on the other hand, may leave the animal or plant handicapped in some way, or prone to disease, and so in Nature's ruthless way, these organisms are very soon "weeded out". This is Darwin's "survival of the fittest" in the natural world, or to give it his term, "natural selection".

Anyway, the crux of the matter from the geological point of view is that throughout time plants and animals have been changing, and if we had a time machine and could travel back into time and trace a particular animal, for example, we would see it gradually changing from one form to another, with all sorts of transitional varieties, so that it would be difficult to say definitely at any point that type A had changed into type B. In practice, however, we very rarely see a perfect transition from one fossil type into another fossil type, although we can see from a comparison of their features that one is the ancestor of the other. This is because the fossil records or family trees that have been preserved for us in the rocks are so very fragmentary.

Only a minute fraction has been preserved of all the animals and plants that lived in the past. The creatures may be eaten by some scavenger soon after their death and their hard parts smashed into unrecognisable fragments by the scavenger's teeth. Or swift sea-floor currents may scour away the freshly entombed fossils, eroding and fracturing the shells or skeletons.

Even if the remains of the creatures become safely enclosed in the sea-floor layers, other factors may conspire to destroy them. The weight of successive layers of sediment may push the fossil-bearing beds deeper and deeper into the earth's crust, into hotter and hotter regions. Under these conditions, hot acid and alkaline fluids originating from molten rock below may seep up through the rocks and dissolve away all the fossils, but sometimes, if we are lucky, impressions are left in the surrounding rock.

In addition, the various tiny mineral flakes in the rocks surrounding the fossils may begin to melt, growing to form new minerals that can survive at high temperatures and pressure, and in the process obliterating any fossils present.

Then the rocks containing the fossils must be pushed up above the sea to form land and erosion must occur, sufficient to expose the particular layers but not so drastic as to remove them completely. Finally scientists must find the fossils and recognise their significance.

It is thus not surprising that our fossil record is full of holes—and our record of life on the land is in even worse shape than that of life in the sea. This is because, apart from entombment in lake or river deposits, or in caves and fissures, the possibilities for preservation of land creatures are extremely limited. Usually their remains are literally "scattered to the four winds" (pp. 245–246).

**The parade of life through the ages**

By piecing together the fragmentary record we have been able to establish an "order of succession", or chronology, of animal and plant life. We can, for example, recognise an age of "ancient life" or "Palaeozoic" (*palaios* = "ancient"), when most animals and plants were of extinct and often bizarre type. Then we can establish an age of "middle life" or "Mesozoic" (*mesos* = "middle"), midway between the old and the new. Then finally we have an age of "recent life" (Cenozoic, sometimes spelled Kainozoic or Cainozoic, derived from the Greek *kainos* or *cenos* meaning "recent"), during which we see the progressive appearance of animals and plants that can be recognised immediately as being like modern creatures (pp. 65–66).

Within these three ages there are finer and finer sub-divisions, again all based on changes in animal and plant life, and these changes can be used to identify, or label, the rocks in which they occur.

As we have seen on p. 67, however, a fossil in a rock will not give us an age in terms of millions of years. It will only tell us that the particular rock is older or younger than some other rock. We need to turn to radioactivity to help us to date the rocks directly. This method depends on the fact that certain minerals are internally unstable and, over a period of time, gradually change into other more stable minerals. This change can be very rapid, measured in fractions of a second, or very slow, measured in millions of years (pp. 66–67).

The minerals that are changing at very slow rates can be used to give some idea of how long the rock containing them has been part of the earth's crust. We can work out the rate at which the mineral is changing and by finding out the amounts of stable compared with unstable minerals can make an estimate of the time over which the change has been occurring.

Radioactive minerals often occur in close association with fossils, and the dates obtained from the minerals can be applied to the fossils. Thus when we say a fossil is "12 million years old", it is a shorthand way of saying that somewhere somebody has found that rocks with these or related fossils contain radioactive minerals that have been decaying for 12 million years.

Thus the study of fossils can provide us with a history of New Zealand, complete with dates. Fossils can also tell us a lot about past geographies of New Zealand.

**Ammonites: ancient sea snails**

Many fossils can be used to give us some idea of where New Zealand was in relation to the equator and South Pole in times past. As with modern sea life, certain creatures like warm tropical seas—for example, reef corals—while others prefer the cold polar seas. Often quite ordinary shellfish are very useful indicators of past ages and climates of New Zealand. Some of them live with coral reefs and are thus clearly tropical. Others have heavy ribbed shells and live in populations that have little variety (i.e. low diversity) which suggests that the waters they inhabited were fairly cool.

Our ancient seas, however, were populated by many bizarre creatures and some of these can give us even more detailed information about past climates. Animals called ammonites and belemnites were conspicuous members of the marine life in Mesozoic seas in many parts of the world, including New Zealand. In some areas of New

Zealand belemnites and ammonites were more common than the various types of shellfish living at that time.

Both groups are members of a class of animals called the Cephalopoda (a combination of words meaning "head" and "foot"). The name alludes to the fact that the molluscan "foot" (the plough-like fleshy projection that bivalves such as the toheroa use to crawl or burrow on the sea floor) is in these animals united with the head and forms a ring of tentacles (often called "arms") around the mouth. Another part of the foot forms a funnel, through which the animal can squirt out a jet of water (if necessary mixed with "ink") and thus propel itself rapidly backward to escape danger.

The earliest cephalopods, in the Palaeozoic era (570-230 million years ago), had a long narrow conical shell, the interior of which was divided into a series of chambers. At the open end of the cone there was a large chamber (body chamber) in which the animal itself was accommodated, with its head and tentacles emerging. The other chambers were filled with gas and fluids, by varying the relative proportions of which the animal could regulate its weight in relation to that of the sea water, and so adjust its level and position in the ocean, just like a submarine.

As geological time went by, forms evolved in which the conical chambered shell became coiled into a spiral, and this type still survives as the beautiful living animal called the Pearly Nautilus, an inhabitant of the tropical seas around New Caledonia, Fiji, Solomon Islands,

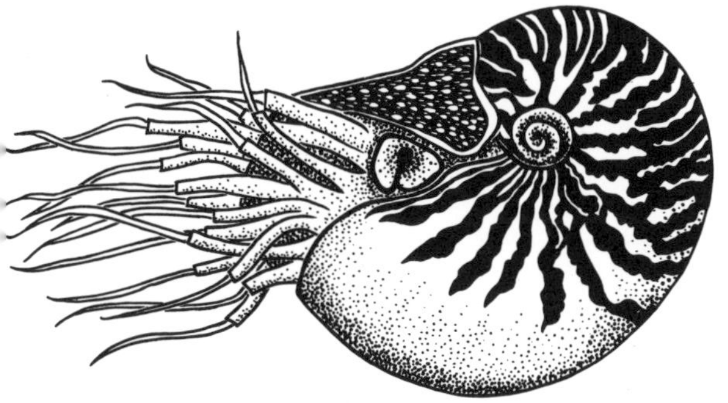

**Fig. 12.1.** The modern Pearly Nautilus is found in tropical waters in the vicinity of New Caledonia, the New Hebrides, Solomon Islands and the Philippines. Empty shells, can, however, drift as far as Japan, Australia and East Africa. The *Nautilus* floats by means of a chambered shell filled with a mixture of gas and fluid. It can alter its position in the sea by varying the proportions of the mixture in the chambers.

**Fig. 12.2.** The living *Nautilus* usually has a camouflage pattern of stripes to break up its outline in the water (top photo). The shell is made of layers of mother-of-pearl that can be highly polished (bottom photo). The shells illustrated measure 110 mm (4.2 in) (top photo) and 160 mm (6.5 in) (bottom photo) in width.

New Guinea, Philippines and S.E. Asia (Figs 12.1-12.4).

The living *Nautilus* floats about in the sea with its coiled part uppermost. The animal itself lives in a large chamber at the lower end of the shell, with the squid-like head and tentacles protruding, but able to be retracted if circumstances demand. As the animal outgrows its living chamber, it pulls itself forward by means of large muscles attached to the shell wall and builds behind itself a new wall, dividing off a chamber.

The chambers however are not dead structures, but are lined with tissue, and nerves and blood vessels pass from chamber to chamber via a tunnel-like hole (siphuncle) in the chamber wall. The *Nautilus* can therefore vary the proportions of

**Fig. 12.3.** The *Nautilus* body lives in a chamber at the end of its shell, attached by muscles to its walls (see lines of muscle attachment in photo). As the body grows it pulls itself forward, extending the chamber as it does so, and sealing off the chamber's back part by secreting a new wall. Thus in time a coiled chambered shell is produced, as is seen in this longitudinal section through a *Nautilus* shell.

The chambers are lined with living tissues and nerves and blood vessels link each one, passing through the chamber walls via a tunnel-like necked hole (siphuncle).

The chambers are used in the same way as a submarine uses its tanks—the *Nautilus* being able to adjust the gas and fluids in the chambers and thus rise or sink in the water. The shell illustrated is 160 mm (6.5 in) in width.

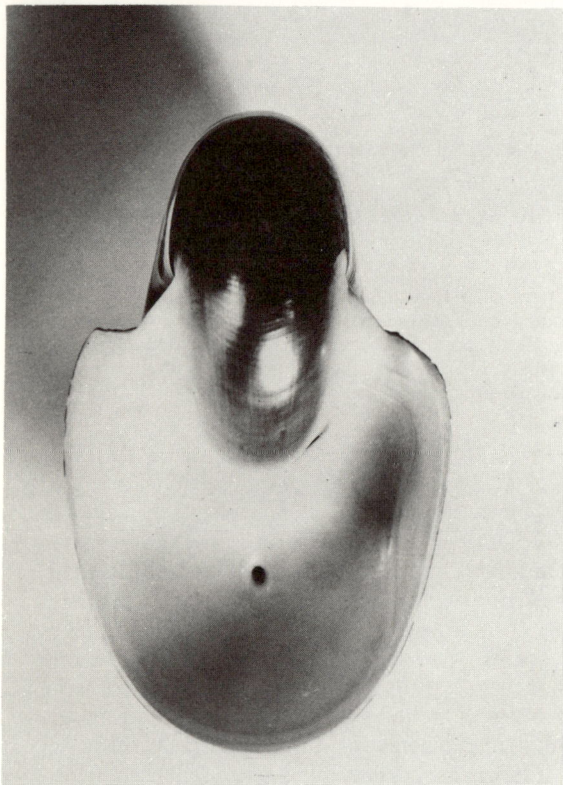

**Fig. 12.4.** A view of the end chamber wall of a *Nautilus* shell, showing the siphuncle passing into the next chamber (cf. Fig. 12.3). The shell illustrated is 110 mm (4.2 in) in height.

gas and fluid in the chambers and alter its level and attitude in the sea. It can either float on the surface of the sea, with its tentacles spread out to snare unsuspecting fish, or it can sink to deeper levels.

The ammonites were built like the *Nautilus* and most had a coiled shell too (Figs 12.5-12.15). Some details are different however, one of the most important being that the walls between successive chambers were not shaped as in *Nautilus*. In *Nautilus* the chamber wall forms a simple curve (Fig. 12.3). In ammonites, however, the chamber walls became very complexly folded and in some, the suture line (the pattern traced by the chamber wall as it joins the external shell) has the superficial appearance of branching, fern-like structures (Figs 12.14, 12.15). It is thought that this complex folding of the chamber walls provided support to the shell, allowing the ammonite to penetrate into deep or turbulent water without being crushed.

The outer side of the shell, smooth in *Nautilus*, is often richly ornamented in the ammonites—with branched or unbranched ribbing, knobs, spikes—and some forms have a rib-like keel along the outer edge of the shell. Some ammonites resemble the spiral horn of a large ram and the whole group has been named after the Egyptian deity Ammon, who had the head of a ram. Abundant ammonites in the cliffs of Yorkshire—to the local people appearing to be like coiled up snakes turned to stone—led to a legend identifying them with a plague of snakes quelled by St Hilda.

In some instances ammonite specimens have the original iridescent mother-of-pearl layer preserved on the shell, and these are remarkable for their beauty.

Most ammonites were probably good swimmers, but their great diversity of form suggests that they lived in a wide variety of habitats. Some

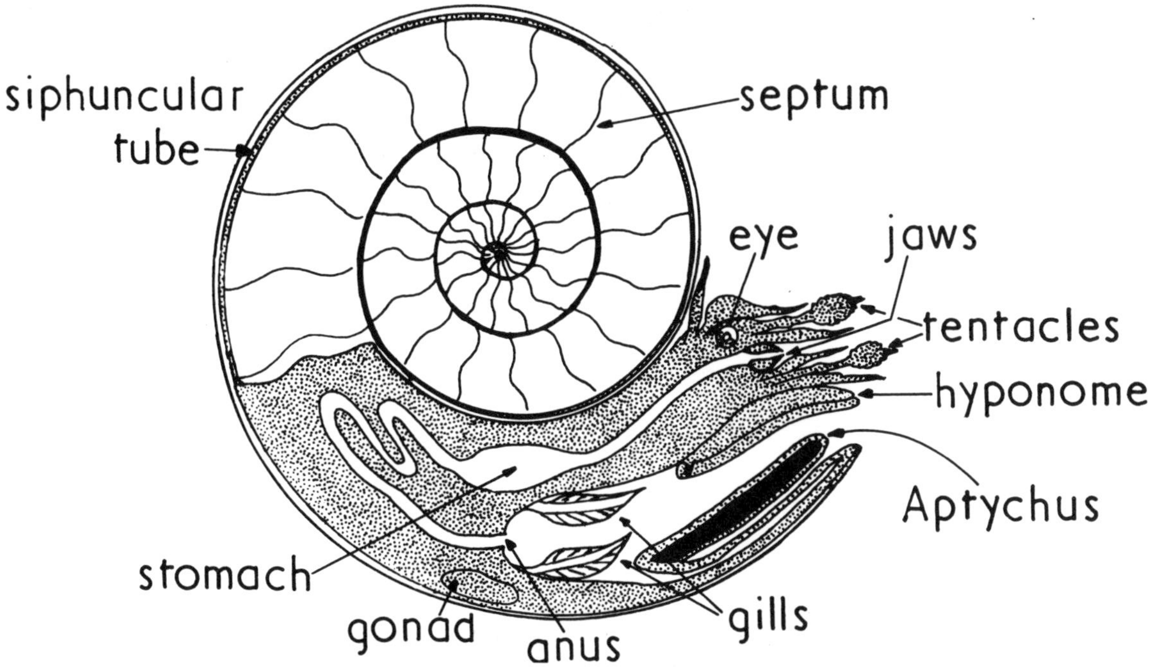


**Fig. 12.5.** A longitudinal section of an ammonite. In ammonites the siphuncle was on the lower (ventral) part of the chamber wall (septum), rather than in the middle as in the *Nautilus* (cf. Fig. 12.4). Also as in *Nautilus*, the mouth had a parrot-like beak and a funnel-like structure (hyponome) was present in the gill cavity through which the animal could pump water and jet-propel itself through the sea.

However, unlike the *Nautilus*, most ammonites had a plate (Aptychus) or paired plates, that could close off the entrance to the shell once the animal had retracted (rather like the "cats-eye" in periwinkle shellfish).

may have spent most of their life on or close to the sea floor; others may have floated in the sea, like *Nautilus*, and still others—the forms characterised by smooth streamlined shells—may have been energetic swimmers.

Ammonites were probably preyed upon by other sea dwellers. They could protect themselves by retracting the head and tentacles into the living chamber of the shell and closing off the open end by means of a plate, or two plates, similar to the periwinkle's "cat's eye". Nonetheless ammonite shells have been found that bear traces of injury scars, where pieces of the shell, particularly close to the head, have been sliced out, presumably by biting predators. Large sharks abounded in the Mesozoic seas and probably were not averse to ammonite meals.

Mosasaurs (sea monitors) were also most common in Mesozoic seas (Fig. 12.16). They were great marine lizards, very closely related to present-day monitors, but attaining lengths of up to 14-15 m (45-50 ft). The mosasaurs were probably the most rapacious predators of the Mesozoic seas, hunting various types of swimming animals, and at least some were in the habit of hunting ammonites. One ammonite shell that has been studied bears the puncture marks from sixteen successive bites, as the mosasaur repeatedly bit across the shell, trying to pull the animal out of its shell in order to swallow it.

Most ammonites are between 50-200 mm (about 2-8 in) in diameter, but giant forms are known. New Zealand's largest ammonite, from rocks south of Kawhia Harbour (in the Waikato) 140 million years old, is only a fragment, about one-third of the original shell, which was 1.42 m (almost 5 ft) in diameter. The largest ammonite known in the world is 2.55 m (over 8 ft) in diameter, and when uncoiled would attain a length of 10.7 m (35 ft).

**Belemnites**

Belemnites, thought to have been ancestors of present-day squids and cuttlefish, flourished in the seas around New Zealand 190 to 65 million

**Fig. 12.6.** Polished cross-section of an ammonite from Kawhia Harbour, Waikato coast (50 mm, 2 in, in width).

The coiled chambered part of the shell (cf. Fig. 12.3) is clearly visible, as is the undivided body chamber, formerly occupied by the animal. The body chamber is filled with sediment and limey material now hardened into rock and lime crystals. Fine sediment has also percolated via the siphuncle (cf. Fig. 12.4) into the chambers.

**Fig. 12.7.** Ammonites can be preserved "in the round", as it were, with the original shell preserved, the shape undistorted and with fine sediment filling the chambers. This specimen is an example. It is called *Aulacosphinctoides*, and is from the latest Jurassic, some 137 million years ago. It was found at Kawhia Harbour on the Waikato coast. The ammonite measures 80 mm (3 in) at its greatest diameter.

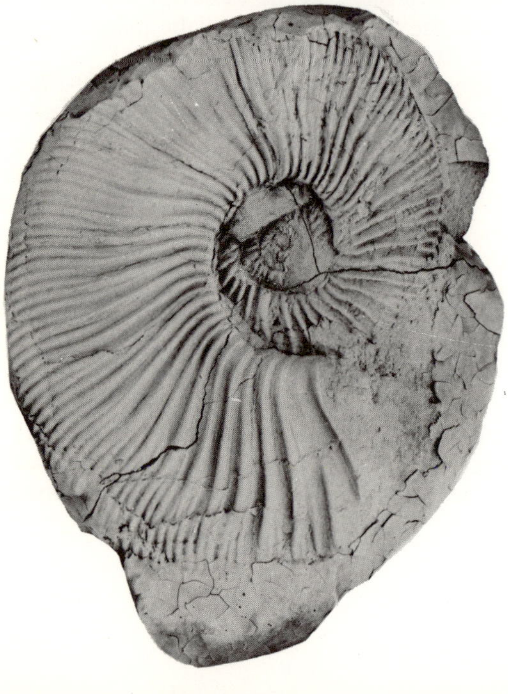

**Fig. 12.8.** Sometimes acid waters percolating through the rocks leach away the shell material, leaving only the impression of the ammonite in the surrounding rock. This is the impression of a *Kossmatia* from the late Jurassic (140 million years ago), found at Kawhia Harbour, Waikato coast. The impression measures 83 mm (3¼ in) at its greatest diameter.

**Fig. 12.9.** New Zealand's largest ammonite was found in February 1977 by Mrs Jean Gyles, of Khandallah, Wellington, a member of the New Zealand Rock and Mineral Club.

At first only the white-coloured area at the top of the specimen was showing in the cliff-face, but the remainder was revealed after a three-day "dig" by Geological Survey staff, followed by extensive laboratory treatment.

This photograph shows Mrs Gyles with the ammonite. It was found 5 km (3 miles) south of Kawhia Harbour, Waikato coast, in rocks of late Jurassic age (about 142 million years ago).

**Fig. 12.10.** The specimen illustrated in Fig. 12.9 represents only one-third of the entire ammonite shell. The remaining two-thirds of the fossil was recovered in 1978 after two weeks of blasting and excavating. The photograph shows the complete ammonite after reconstruction and restoration. The specimen is 1.42 m (almost 5 ft) in diameter. In life a squid-like head and long tentacles protruded out of the open end of the shell (cf. Fig. 12.13).

MUSEUM F. NATURK.
MÜNSTER i W.
W. GERMANY
TOP

**Fig. 12.11.** Another large New Zealand ammonite, also from Kawhia Harbour, is more complete than that of Fig. 12.10 and furthermore is attached to a large piece of fossil wood. The maximum diameter of this specimen is 320mm (12½ in).

years ago (in Jurassic and Cretaceous times).

They had a squid-like body but differed from present-day squids in that they had a special type of hard shelly internal skeleton, divided into three parts (Figs 12.17-12.23). One part was towards the front of the animal and consisted of a thin shoe-horn-like shell (pro-ostracum). The second was a chambered cone (phragmocone) extending backwards from the pro-ostracum, which fitted into a deep conical pit in the front end of the third part of the skeleton. The third portion is known as the guard. It is a bullet- or cigar-shaped solid calcareous structure which extended into the belemnite's tail and is often the largest of the three parts and the most commonly preserved.

Belemnite guards occur in great abundance in certain areas of England and Europe and were well known to ancient peoples. The word "belemnite" is very old and the first written reference to it was in the treatise *De Natura Fossilium*, published in 1546 by the "Father of Mineralogy", Georgius Agricola. The word itself came from the Greek *belemnon*, meaning a dart. The ancient Greeks themselves, however, called them by the name *lynkourion*, and the Romans called them *lapis lyncis*—both expressions meaning "lynx-stone", and based on the idea that they were the excrement of wild cats. Around the shores of Kawhia Harbour, New Zealand, Maori children used to gather belemnites to play with them and they were called *rokekanae*, or excrement of the mullet, which they thought was in the habit of leaping out of the water and leaving belemnites behind on the shore. In mediaeval times belemnites were believed to be the toe and finger nails of devils drowned in Noah's Flood. Later and more plausible interpretations were that belemnites were stalactites produced underground, or thunderbolts fallen from the sky ("thunder-stones"). In the seventeenth century the Sicilian naturalist and painter Agostino Scilla concluded that belemnites must be some kind of shell formed by an unknown type of mollusc. His inference we now know to have been correct, for at the beginning of the nineteenth century the French naturalists Cuvier and Lamarck were able to prove that they were the remains of extinct mollusca related to squids and cuttlefish. More recently, in the late nineteenth and early twentieth century, specimens have been found in England and Germany with guard, phragmocone and pro-ostracum intact and with impressions of the soft parts, including the arms (with rows of hooklets) and ink sac, outlined in the surrounding rock.

Study of the fossils that preserve impressions of flesh show that belemnites looked and presumably acted like the modern squid. The body was torpedo-shaped and was surrounded by a fleshy mantle spreading out into two lateral fins that kept the body right side up and could be moved to guide it upward or downward.

There was a constricted "neck" behind the head and the eyes, like those of the squid, were extremely efficient. Ten tentacles surrounded the mouth and were set with rows of strongly curved, sharp hooks (some of which are known from New Zealand rocks), for grasping and holding soft-bodied prey. Like most other cephalopods, belemnites probably swam by taking water into the mantle cavity and forcing it out as a jet through the funnel. When the funnel pointed forward, the belemnite moved backward; when it was pointed to the rear, the belemnite moved forward. Alarm would cause the animal to release a thick dark liquid from an ink sac that opened into the cavity and through the funnel. The ink spread in an opaque cloud, confusing the belemnite's predator.

An ink sac like that found in modern squids is sometimes discovered in well-preserved fossil belemnites. In at least one case the pigment was put back into solution and a drawing made of the fossil with the very ink the living animal secreted some 175 million years ago.

The pro-ostracum of the belemnite is not often preserved and probably served as an internal

**Fig. 12.12.** The largest ammonite known in the world was found in 1895 in Westphalia, Germany. Its name is *Pachydiscus suppenradensis* and it lived in late Cretaceous times, about 75 million years ago.

This photo shows a fibreglass cast of the giant specimen being unpacked at the Los Angeles County Museum of Natural History, where it is now on permanent display. The specimen is 1.9m (6ft 5in) in diameter. Part of the body chamber is missing, and if this is taken into account the diameter is increased to 2.55 m (over 8 ft).

**Fig. 12.13.** The reconstructed size of the largest New Zealand ammonite (1.42 m, almost 5 ft) (Figs 12.9, 12.10) (top), compared with that of the world's largest (2.55 m, over 8 ft) (Fig. 12.12) (bottom).

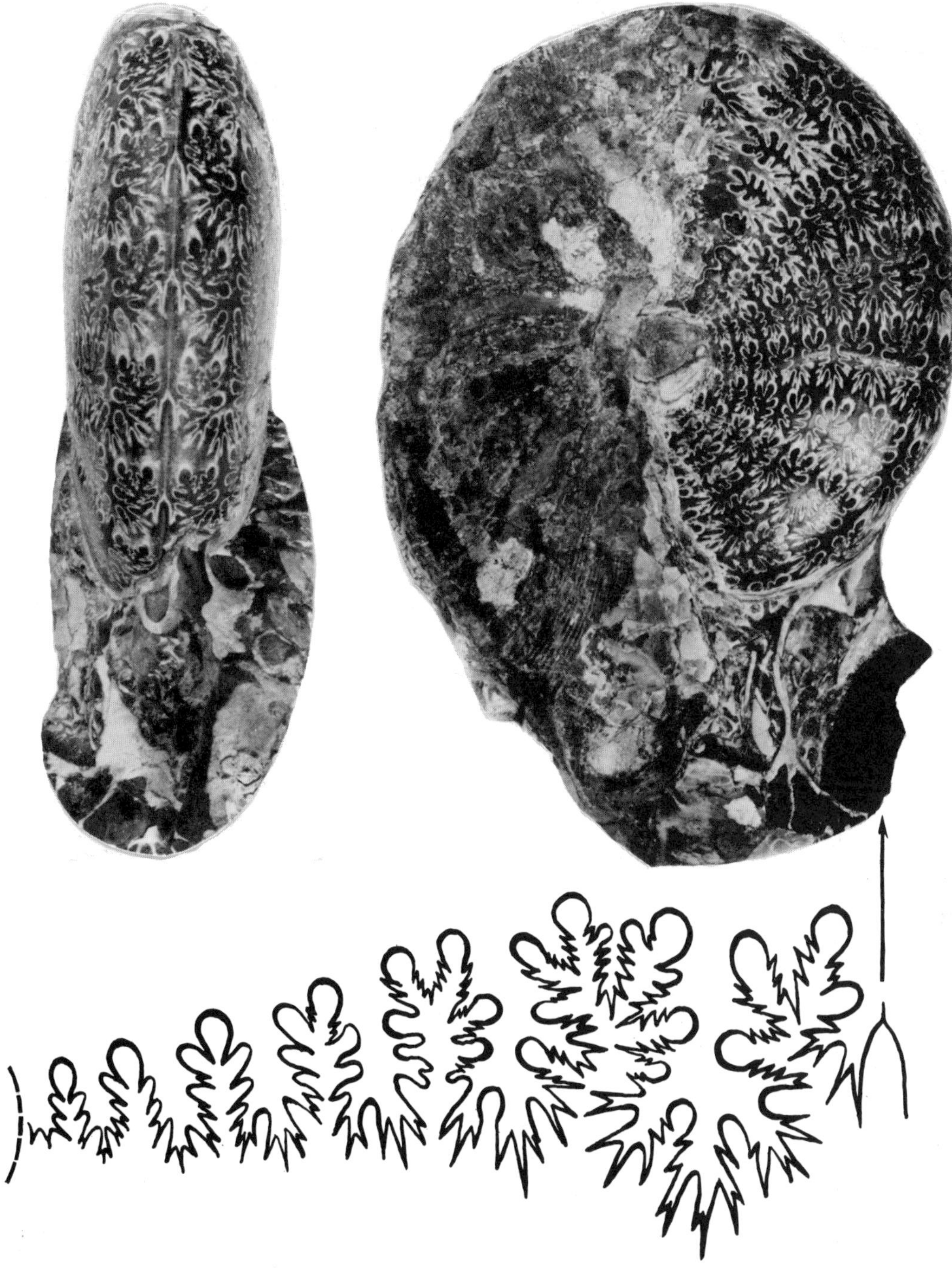

**Fig. 12.14.** *Nautilus* has very gently curved chamber walls, whereas in most ammonites they have been intricately folded. The complex pattern revealed when the shell is removed and the sediment-filled chambers are left is called the "suture".

The top photo shows an ammonite *(Phylloceras)* from Kawhia Harbour, of late Jurassic age, with part of the shell removed, to show the suture. The ammonite measures 85 mm (3½ in) at its greatest diameter.

The bottom diagram shows a drawing of the suture, extending from the mid-line of the shell (indicated by the arrow) to the inside of the chamber wall (indicated by the curved dashed line).

Fig. 12.15

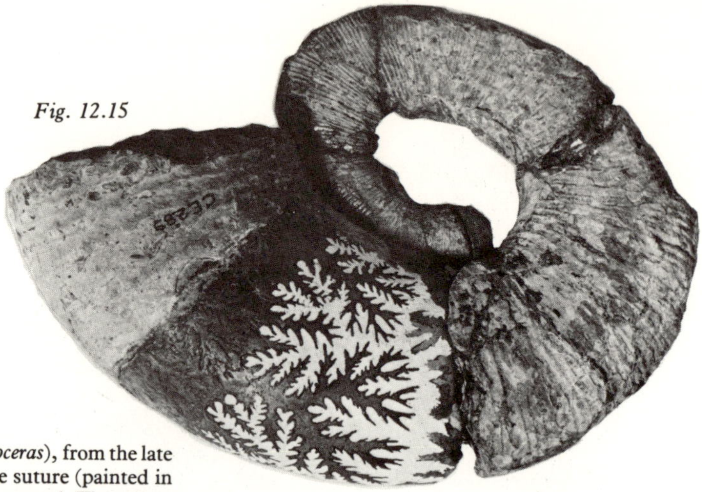

**Fig. 12.15.** This is a larger ammonite (*Lytoceras*), from the late Jurassic of Kawhia Harbour, showing the suture (painted in white) revealed when part of the shell is removed. The suture is quite different to that of Fig. 12.14 and in a larger shell is more widely spaced. The ammonite measures 170 mm (6¾ in) across.

**Fig. 12.16.** Mosasaurs flourished in New Zealand seas in the Cretaceous. They were giant marine lizards, very closely related to present-day monitors. However whereas today's monitors vary in length from 0.25 m (10 in) to 3 m (10 ft) (as in the famous Komodo "dragon"), the Mosasaurs grew still larger, up to 13-15 m (45-50 ft). The mosasaurs were the most rapacious predators of Cretaceous seas and probably hunted various types of fish—but also had a taste for ammonites.

support for the forward portion of the body and an attachment for muscles.

The chambered phragmocone, sometimes found still joined to the guard, but more often as an isolated piece (Fig. 12.23), probably performed the same function as the chambered shell in the ammonites and *Nautilus*—as a buoyancy apparatus, helping the animal to suspend its weight in the water without undue muscular effort.

The guard served as a counterweight, balancing the main part of the body, including the head and arms, located on the opposite side of the buoyant phragmocone. It probably also provided protection for the delicate tip of the phragmocone. The guard is composed of fibres of lime

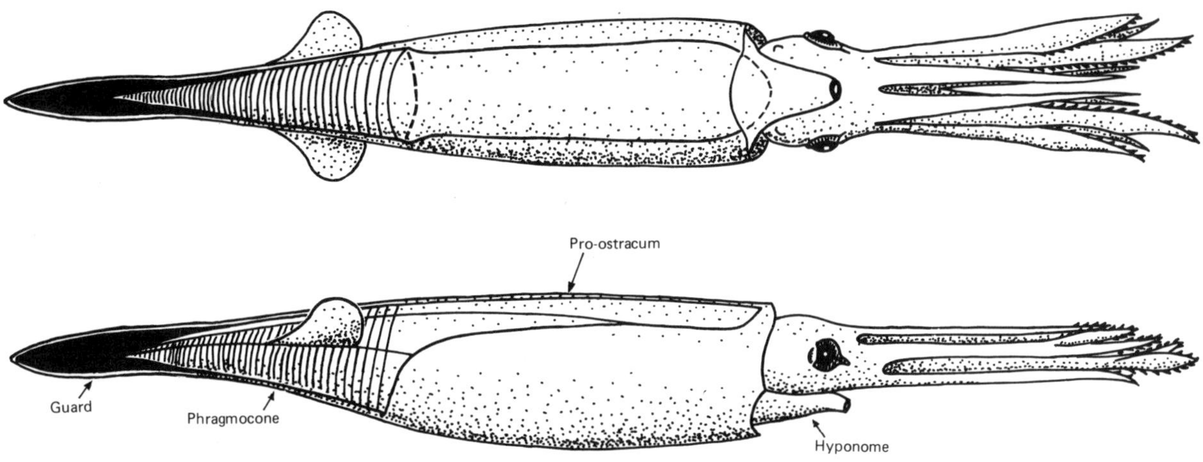


**Fig. 12.17.** Belemnites had a squid-like body, but with a hard shelly internal skeleton, divided into three parts.

These diagrams depict underside (top diagram) and side (bottom diagram) views of a belemnite. The main body part of the belemnite has been depicted as though it is transparent—to show the three elements of the internal skeleton: the solid guard (black) at the tail, the phragmocone (a series of air chambers) and the flattened cuttle bone-like pro-ostracum. The chambered phragmocone corresponds to the chambered spiral shell of the *Nautilus* and ammonites (Figs. 12.1-12.6), and presumably served the same function i.e. to assist the animal in positioning itself in the water. A funnel (or hyponome) projects from the gill chamber and belemnites were able to make quick reverse dashes, particularly in emergencies, by squirting water out of the funnel, like modern squids. Should a collision occur, the compact guard would protect the delicate phragmocone, but fractures and dislocations in guards, healed over by later growth layers, record traffic accidents in the Mesozoic seas.

After death the guard often separates from the remainder of the body (Fig. 12.19) and so is frequently found on its own. The length of the guard is about ⅛-1/10 of the total body length.

In belemnites the length of the guard ranges from 457 mm (18 in) to 37 mm (1½ in). In terms of total body length the largest known guard probably came from a belemnite squid 4.6 m (15 ft) long.

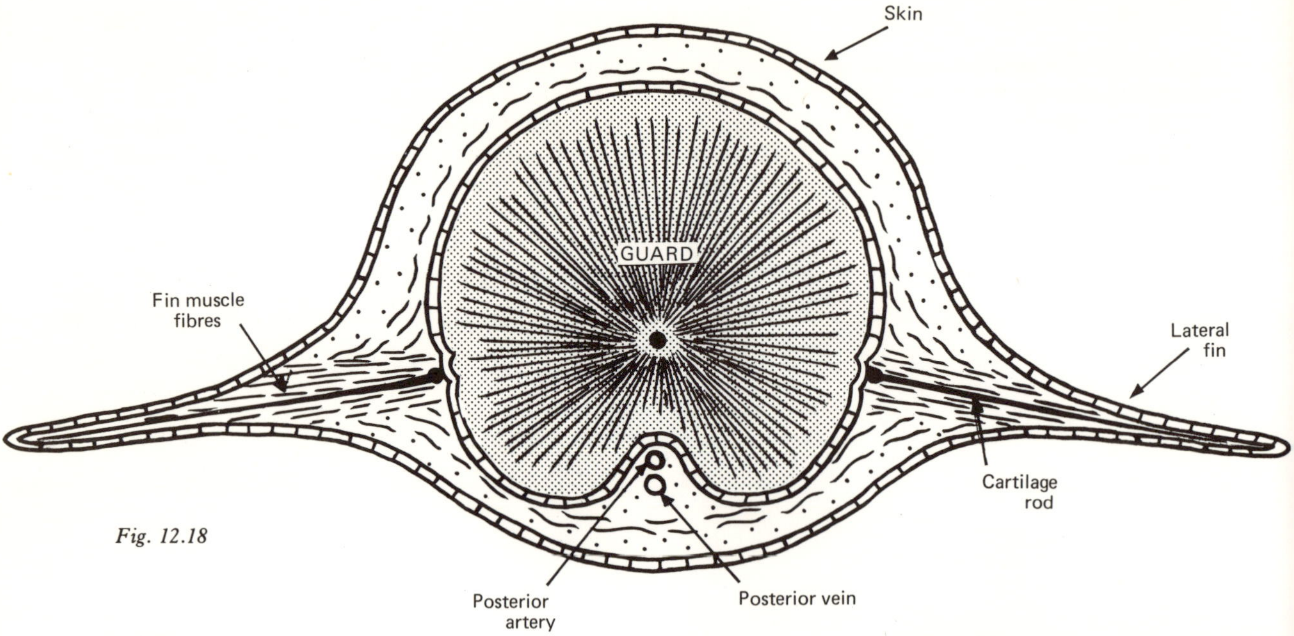


*Fig. 12.18*

**Fig. 12.18.** Cross-section through a belemnite guard and the surrounding soft tissues. Large blood vessels run along the underside of the squid-like body, sending out branches to the fins (Fig. 12.17). Often the blood vessels lie along a groove in the ventral surface of the guard (Fig. 12.19).

carbonate (calcite), orientated at right angles to the surface. As the belemnite grew, extra layers were added to the guard so that there was also a concentric layered structure, like growth rings of a tree trunk.

Like present-day squids, belemnites probably preyed upon small fish and crustacea and in turn were preyed upon by larger fish. Cuts, scratches and punctures on the surface of belemnite guards have been attributed to attacks by predators, and it has been suggested that the principal enemies of belemnites were fish (particularly sharks), ichthyosaurs and mosasaurs. Over 250 belemnite guards were found in the stomach of a large fossil shark—it would be interesting to know whether death was due to indigestion!

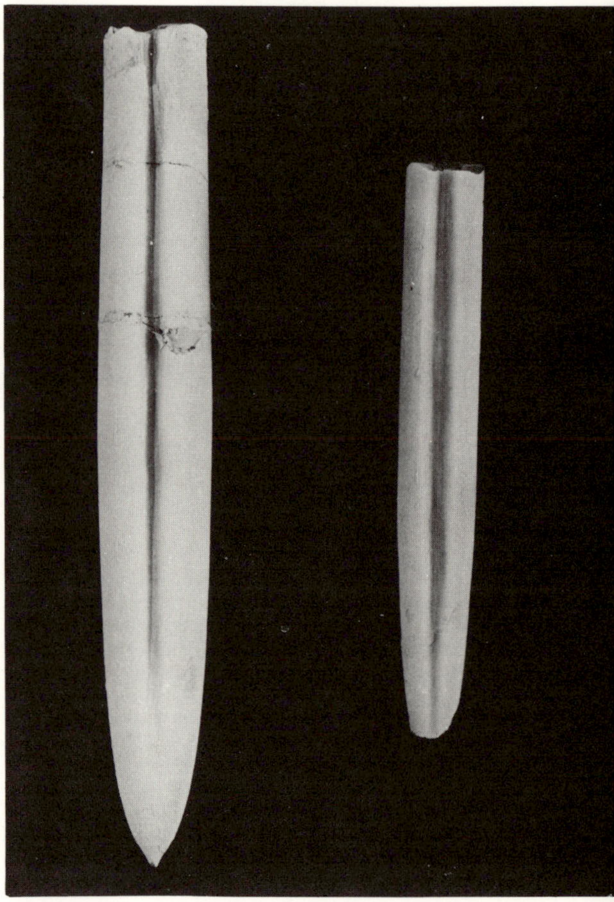

**Fig. 12.19.** Bullet-shaped guards, the tail-bones of extinct squids (belemnites). The guard on the left is New Zealand's largest specimen (114 mm; 4.5 in long), from Waikato Heads; that on the right (76 mm; 3 in) is from Kawhia Harbour. Both came from late Jurassic rocks, laid down in the sea about 140 million years ago. The larger guard (on the left) came from a belemnite squid about 1.2 m (4 ft) long.

*Fig. 12.19*

**Fig. 12.20.** A section cut across a belemnite guard 12.7mm wide (0.5in), from Kawhia Harbour, showing the way the shelly material has been added in a series of layers, like tree rings.

Temperature analyses (Fig. 12.25) suggest that the major layering, clearly visible in the photo, is annual. If this is correct the guard illustrated is three years old.

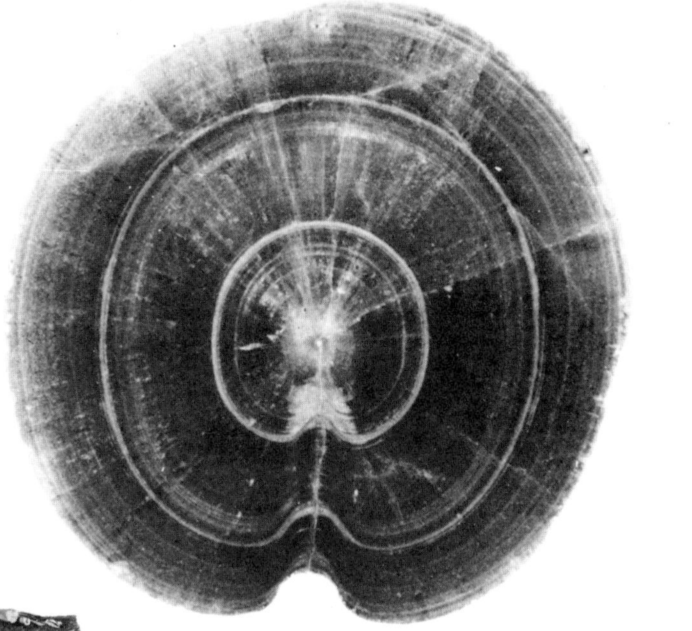

*Fig. 12.20*

*Fig. 12.21*

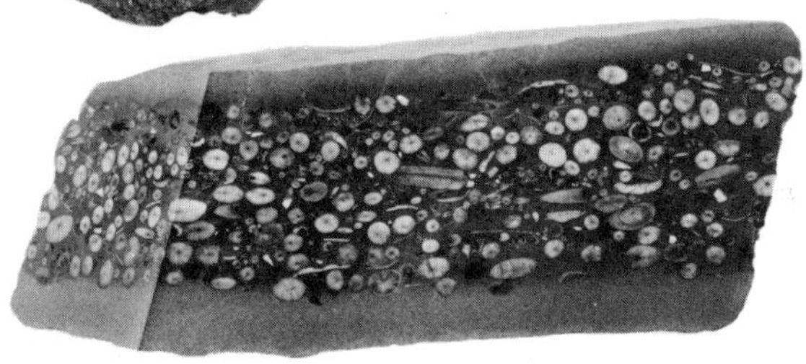

When the belemnites died and sank to the sea bottom, their flesh decayed, and usually all that survived to become fossils in the enclosing rock were the hard guards. The total body length of the original animal, from the tail to the tip of the tentacles, was probably nine or ten times that of the guard.

The largest belemnite guard known comes from Indonesia and is 457mm (18in) long. The animal was probably about 4.6m (15ft) from the tail to the tip of the tentacles. The largest from New Zealand is a 113mm (4½in) specimen from Waikato Heads, probably from a 1.2m (4ft) long animal. Other New Zealand belemnite guards range down to 40mm (1½in) long, representing animals about 0.4m (1½ft) long.

Thus the belemnites were comparable in size to

**Fig. 12.21.** Belemnite guards often occur in discrete beds (sometimes called "belemnite battlefields"). Often these beds have been created by current action sweeping along the seafloor, gathering up loose guards and depositing them in the one site.

Another possibility is that the guards are the remnants of mating swarms, as it is thought that belemnites, like modern squids, came close inshore to mate and died once the eggs were laid and fertilised.

The specimens illustrated—naturally weathered (top) and cut and polished (bottom)—come from a belemnite bed of late Cretaceous age (70 million years) outcropping on the North Canterbury coast near Haumuri (Amuri) Bluff. The top slab measures 440mm (17½in) in length and the bottom slab 300mm (12in).

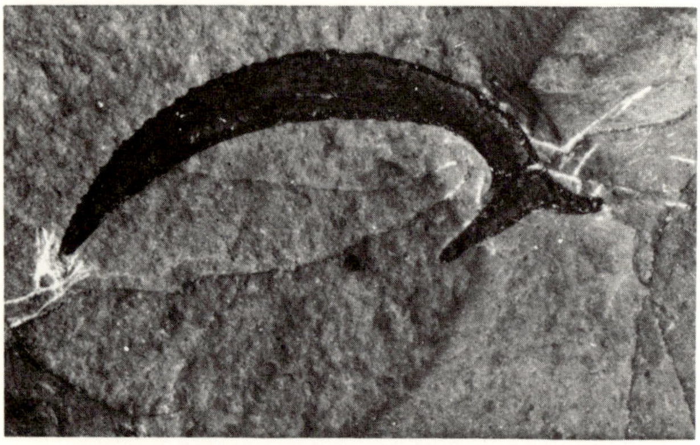

**Fig. 12.22.** Isolated hooklets, separated from the arms of belemnites (cf. Fig. 12.17), have been found in the late Jurassic beds of Kawhia Harbour (Waikato Coast). This specimen is 22 mm (¾ in) in length.

**Fig. 12.23.** Isolated phragmocones occur in some belemnite beds and these specimens are from the late Jurassic of Kawhia Harbour. They measure 50 mm (2 in) (lefthand specimen) and 43 mm (1¾ in) (righthand specimen) in length. Note the walls dividing off the chambers (cf. Fig. 12.17).

the small or medium-sized squids that populate many of our coastal seas today. Giant squids are found in deep waters (some have been washed up on beaches close to Cook Strait), may reach a length of 18 m (60 ft) and are the largest known invertebrate animal (Fig. 12.24).

Belemnite guards are found in many parts of New Zealand, and are used as "identity tags" to recognise rock strata in widely separated areas—notably Waikato Heads, Kawhia Harbour, Wairarapa, Haumuri Bluff (N. Canterbury), Brighton (near Dunedin) and the Hokonui Hills (Southland).

Ammonites and belemnites are some of the marine creatures that can give us clues about the geographies of the past.

In the Jurassic (195-136 million years ago), for example, we can recognise certain groups of ammonites and belemnites that appear to have lived fairly close to the North Pole as it then was, as they are known only from northern places like Siberia, Greenland, Spitzbergen, Northern Europe and Northern Canada. They have been given the name "Boreal", signifying their northern distribution. Other ammonites and belemnites lived in the seas that covered parts of Central and Southern Europe and these creatures at times could spread to South America, India, Africa, South-east Asia, Indonesia and even to New Zealand and Australia. As these ammonites and belemnites lived essentially in the Tethys and various interconnected seas, they have been called "Tethyan" (Fig. 12.26).

As a glance at the globe will show, we should expect to have in the Southern Hemisphere an equivalent of the Northern Hemisphere Boreal life, living in the seas peripheral to the South Pole.

For many years however, as exploration and study proceeded, no Southern Hemisphere equivalents of the Boreal fossils were found. The final nail in the coffin, as it were, was provided by expeditions to West Antarctica in 1964, that discovered Jurassic belemnites very similar to those known in New Zealand. Therefore it was evident that in Jurassic times Tethyan animals could penetrate as far south as 75°S Latitude (in terms of modern geography), and the likelihood of finding Boreal-type faunas in Southern Hemisphere high latitudes was very small indeed.

We know from studies of Tethyan ammonites and belemnites that in the Mediterranean region they were associated with reef corals and other obviously tropical animals. The Boreal ones on the other hand had no such associations. Thus it was suspected that Boreal creatures were generally inhabitants of cool or cold water and Tethyan of temperate or warm water.

At this point chemistry came to our aid.

**Fig. 12.24.** Although dwarfed in size by the Blue Whale (36 m; 120 ft), the largest living vertebrate, giant squids are the largest known invertebrate animal, reaching a total length of 18 m (60 ft). One giant squid was washed up, still alive, on Wellington's south coast, at Island Bay beach, on 6 June 1880. This drawing is exactly to scale and shows the recorder of the incident, Mr T. W. Kirk, measuring with a metre rule (3 ft).

The squid had a 3 m (10 ft) body, a 3.3 m (11 ft) circumference and tentacles 7.6 m (25 ft) in length. The largest known belemnite probably had a total length (body and tentacles) of 4.6 m (15 ft). The largest known ammonite (Fig. 12.12) had an uncoiled length of 10.6 m (35 ft).

Another small squid was washed up at nearby Lyall Bay on 23 May 1879. It had a body length of 3.4 m (11 ft 1 in) and arms 1.2 m (4 ft) long. Lyall Bay was the scene of yet another stranding in October 1880. Then a squid with a comparatively short body (2.3 m; 7 ft 9 in) but with incredibly long arms (15.0 m; 49 ft 3 in) was washed up.

Other large squids have been washed up on the east coast of New Zealand between Waimarama in the north and Kaikoura in the south.

All these specimens have probably been derived from the deep waters of the Hikurangi Trough, that lies near the coastlines of these regions.

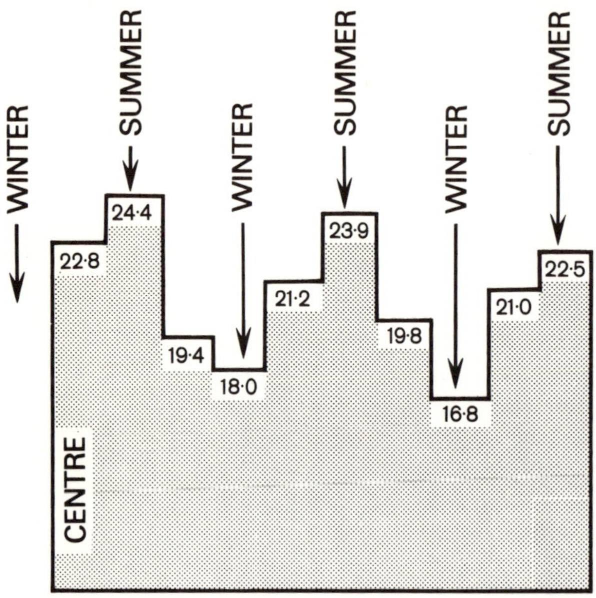


**Fig. 12.25.** The lime laid down in the layers of a belemnite guard preserves a record of the temperatures in the sea at that time. This is because the proportions of two oxygen varieties (or "isotopes"), that occur in the lime vary with the sea-water temperature.

This graph shows the temperatures obtained by sampling the oxygen isotopes of the guard illustrated in Fig. 12.20, extending from the centre of the guard to the outside. The graph shows that the belemnite lived through three winters and died in the third summer. It also indicates that the major layering of the guard, visible in the photo, is probably annual.

## Fossilised temperatures

The ammonites and belemnites and other sea creatures lived in and "breathed" sea water that contained dissolved oxygen of two main varieties (or "isotopes"), oxygen 16 and oxygen 18, which occur in proportions related to the temperature of the sea water. Thus in any one area the proportions change with the seasons, or over longer periods if there is a prolonged cooling or warming trend.

Lime (calcium carbonate), containing the oxygen isotopes, was extracted from the sea water by marine life such as shellfish, ammonites and belemnites, and used to build their shells. The ammonites and most shellfish built fairly thin shells that after death did not usually withstand the stresses and strains of burial in the sands and muds—so that the original structure became altered in various ways. Some of the more robust forms, however, were more resistant and are therefore better preserved (e.g. thick oyster shells). On the other hand, the bullet-like belemnite guard is a solid bone-like structure more resistant to chemical and physical changes.

As we have seen (p. 210), the belemnite gradually added shell material in successive layers to its guard during its lifetime so that a cross-section of a guard shows a series of concentric growth layers, like tree rings. Each layer nestles into and protects adjoining layers, so that if any harmful chemicals succeed in penetrating the guard, they are limited in the extent of the changes they can cause in the shell layers. The oxygen isotope content of the limey material comprising each growth layer depends on the temperature of the sea water at the time the layer formed. If a layer was laid down during the winter, for example, it would have the winter sea temperatures recorded within it. Therefore, if the belemnite guard is well preserved, each growth layer preserves a record of the temperature changes of the ancient seas. Temperature fluctuations of as little as half a degree may be detected and measured.

Detailed temperature analyses of individual belemnite guards, growth layer by growth layer, give the summer and winter temperatures of the seas in which the belemnite lived and the climatic fluctuations thus revealed tell us whether the belemnites lived in tropical seas, subtropical and so on. Also, by counting the number of seasons the belemnite's age can be obtained. Many of the New Zealand belemnites seem to have lived for three or four years, an age span roughly comparable with that of many of the squids living in our coastal waters today (Fig. 12.25).

Oxygen analyses of Boreal belemnites showed that, as had been suspected, they were inhabitants of coolish (but not frigid) water. As far as we know from the fossil evidence, no polar ice caps mantled the North and South Poles in the Jurassic, and so the polar seas were probably cool rather than cold. The oxygen analyses confirm that this was indeed the case.

Oxygen analyses of Tethyan belemnites, on the other hand, showed that those from the Mediterranean and southern Europe that are found in association with reef coral, lived in seas with temperatures equivalent to the modern tropics. Other Tethyan belemnites, including those from New Zealand and Australia, were shown to have lived in subtropical or temperate seas, rather like those between New Zealand and

New Caledonia today. The Tethyan belemnites from Antarctica were found to have the same temperature ranges also.

Thus a picture emerged showing that in Jurassic times countries like Scandinavia, Greenland, northern Europe and the U.S.S.R., northern Canada etc., were close to the North Pole and the seas that lapped onto them at various times were rather coolish in temperature (geographers would use the term "cool-temperate")—somewhat like the waters that now surround New Zealand's sub-Antarctic islands—Campbell and Auckland islands.

The remainder of the landmasses were in areas of either tropical or subtropical waters. By tracing the occurrences of Jurassic reef coral and other tropical animals we can delineate a belt corresponding to the Jurassic tropics—and find that it is slightly north of and at an angle to the present equator (Fig. 12.26). To either side of the tropical belt we can recognise belts in both Northern and Southern Hemispheres corresponding to the subtropical-warm temperate of zones of today. In the Southern Hemisphere, however, subtropical or warm temperate fossils—our Tethyan fossils—were, as we have seen, evidently able to live quite happily in Australia and New Zealand and then to spread to Antarctica and Patagonia. This distribution does not make sense when we have a look at the modern globe, unless we have subtropical and warm temperate waters extending to the South Pole. We know however from the Northern Hemisphere that cool-water Boreal animals lived in seas around the North Pole at that time. So if all things were equal we should be able to find similar cool-water animals in the Southern Hemisphere. However things were evidently not equal, as, in spite of diligent searching, only warm water Tethyan forms have been dis covered—even in Antarctica.

We know from palaeomagnetism that Laurasian countries (Greenland, Scandinavia, northern Europe, the U.S.S.R., northern Canada etc.) were grouped around the North Pole in the Jurassic—so therefore it is quite predictable that the animals living in their seas at the time would have been types with a preference for cool water.

Palaeomagnetism also tells us that in the Jurassic India, Africa, Australia, New Zealand, South America and Antarctica were all squeezed together as the Southern Hemisphere supercontinent Gondwanaland. Some of the latitudinal "fixes" are, however, rather vague and some people have placed Gondwanaland fairly close to the South Pole, to match Laurasia in the north.

These reconstructions would have New Zealand, Australia, Antarctica and South America nudging the South Pole in the Jurassic. This is not in accord with our reading of the record preserved in the rocks in New Zealand and elsewhere in the Southern Hemisphere. We can see quite clearly from our fossils that New Zealand (and much of Gondwanaland) was close to the South Pole in the Permian. But in the Triassic and Jurassic there is every indication that the seas covering New Zealand were subtropical or warm temperate, not cool or cold. Temperate forests grew in New Zealand, Antarctica and Patagonia at this time, as indicated by the presence of fossil forests, coal beds and layers containing plentiful leaves, seeds and pollen.

There is perhaps no way of avoiding the conclusion that New Zealand, Australia, Antarctica, South America, and in fact probably all of Gondwanaland, were some distance away from the South Pole for most of Triassic and Jurassic time. The Jurassic South Pole was probably just a spot in the midst of a large southern ocean (Fig. 12.26).

On the areas of Gondwanaland that were land, large dinosaurs lumbered around and many of them managed to traverse much of Gondwanaland and at least the southern (i.e. warmer) part of Laurasia. As they were cold-blooded creatures, dependent on the sun's warmth to provide their body heat, presumably they kept very much to warmer areas and their presence in Antarctica is perhaps another point in favour of the idea that Antarctica has not always been the frigid continent it now is.

Dinosaurs however are not yet known from New Zealand—probably because we were mainly under the sea for much of Mesozoic time. Our seas however did have large swimming reptiles—dinosaurs that had taken to the water. These were the long-necked plesiosaurs and the short-necked dolphin-like ichthyosaurs. Although we lack the marvellous skeletons of some overseas countries, isolated bones, often from the heavy backbone or the thick paddle-like flippers, have been found in the Hokonui Hills (Southland), Nugget Point (Otago), Upper Rangitata and Waipara (Canterbury), Haumuri Bluff (North Canterbury), Wairoa Gorge (Nelson), South Wellington Coast,

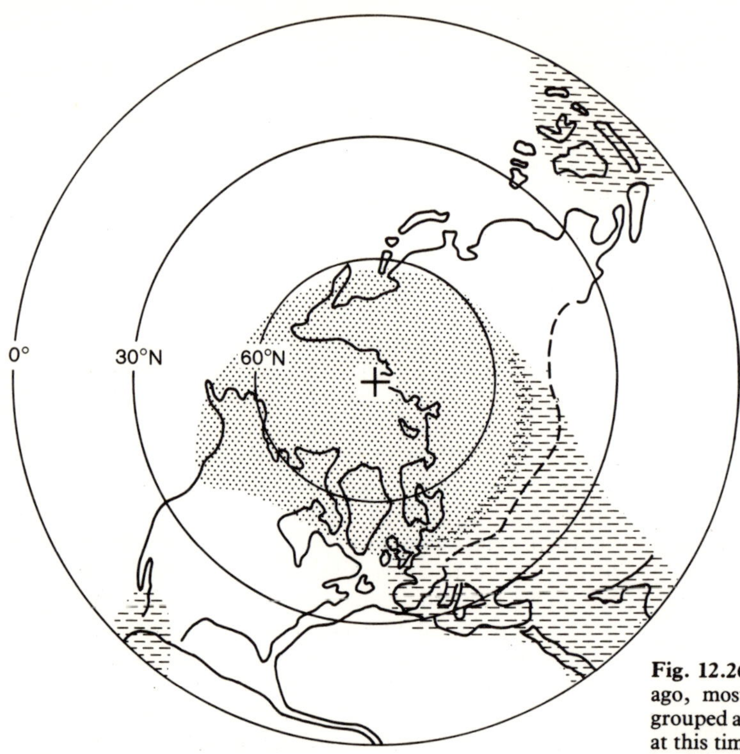


**Fig. 12.26.** In the late Jurassic, about 162-136 million years ago, most of the Northern Hemisphere landmasses were grouped around the North Pole, as in the upper diagram, and at this time the seas lapping onto these lands were populated by "Boreal" creatures that preferred cool-temperate conditions.

Other lands straddled the equator of the time, and many of them were covered by an extensive sea, the Tethys. Reef corals and other warmth-loving creatures thrived in the tropical waters of the Tethys, while other "Tethyan" animals spread out to populate seas in sub-tropical and warm-temperate regions.

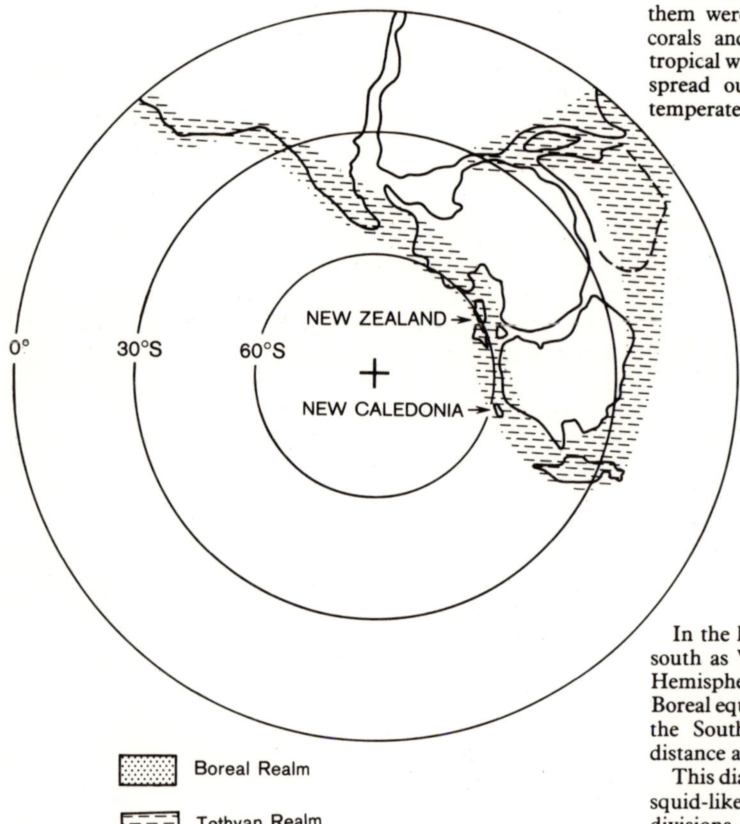


In the late Jurassic Tethyan marine animals ranged as far south as West Antarctica, and equivalents of the Northern Hemisphere Boreal animals are not apparent. This absence of Boreal equivalents is interpreted as indicating that at this time the Southern Hemisphere countries were grouped some distance away from the South Pole (as in the lower diagram).

This diagram shows how the world-wide distribution of the squid-like belemnites can be divided into two major subdivisions (or "realms")—Boreal and Tethyan—related to cool-temperate (Boreal) and tropical-subtropical-warm temperate (Tethyan) sea water temperatures. Other animal and plant groups show similar distribution patterns.

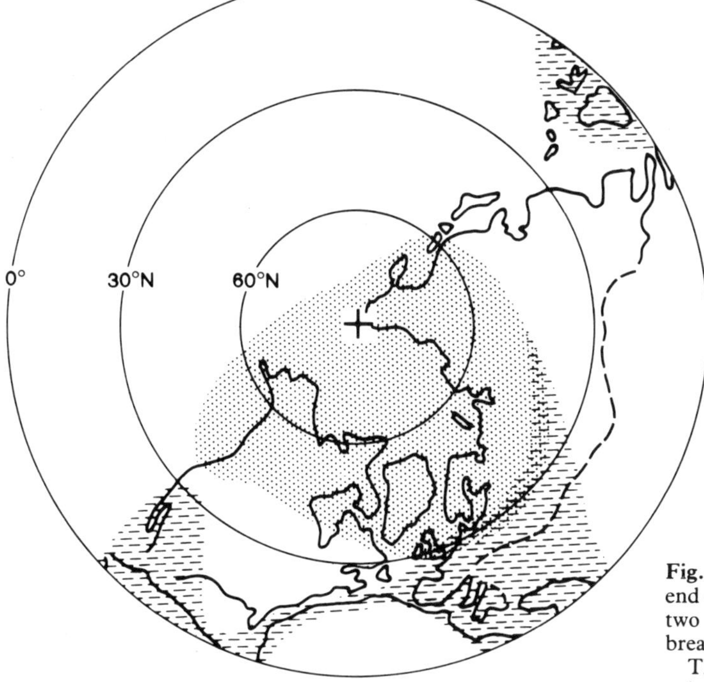


**Fig. 12.27.** Major changes in world geography occurred at the end of the Jurassic, and extended into the Cretaceous, as the two super-continents Laurasia and Gondwana started to break up.

These changes were reflected in the distribution of the animals and plants living at the time. As may be seen from a comparison with Fig. 12.26, for the first time since the Permian, distinctive southern or "Austral" organisms can be recognised in the Southern Hemisphere.

The diagrams summarise the situation in early Cretaceous times, 136-100 million years ago, when cool-temperate Boreal organisms populated the seas around the North Pole (at this stage the Northern Hemisphere landmasses had barely started to split apart).

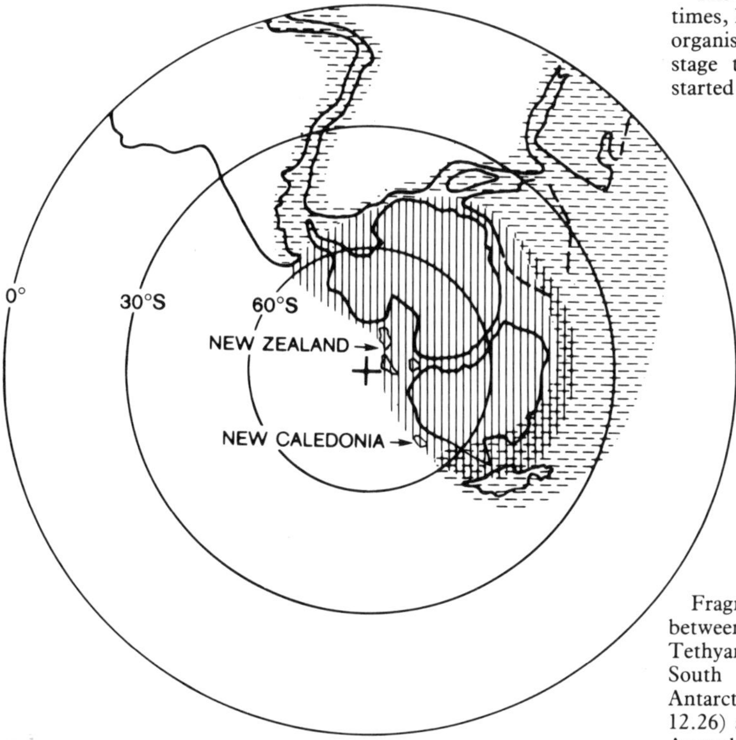


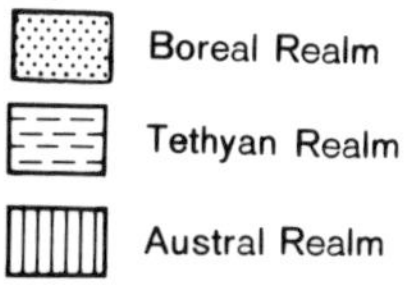


Fragmentation of Gondwanaland had opened up a split between South America and Africa, allowing warm-water Tethyan organisms to migrate along the shores of the infant South Atlantic Ocean. Movements of Australasia and Antarctica had brought them closer to the South Pole (cf. Fig. 12.26) and the seas of these countries were populated by Austral organisms, that, like the Boreal ones, preferred to live in cool-temperate seas.

Although, as in Fig. 12.26, the diagrams illustrate the distribution of the major groups of squid-like belemnites, comparable patterns are seen in other animals and plants.

Austral faunas and floras became even more marked as Australasia and Antarctica drifted further south in late Cretaceous and early Tertiary times (cf. Figs 15.1-15.5).

Tararua Range and Mangapakeha (east of Masterton). New Zealand may have the world's largest ichthyosaur, as vertebrae 450 mm (17.7 in) in diameter and ribs 1060 mm (42 in) long have been found in the Upper Rangitata—although unfortunately nothing more of the skeleton. If these measurements are reliable indications, an animal some 30 m (98 ft) long may have lived in Canterbury seas.

At the end of the Jurassic there is a break in the fossil story preserved in our rocks. It looks as though no marine rocks were laid down in the New Zealand area for some 26 million years, covering a fair chunk of early Cretaceous time. This probably means that land occupied most of the area that is now New Zealand and that it took many millions of years for erosion to eat away the land to allow the sea to transgress again.

When the New Zealand fossil story starts again in the early Cretaceous some 110 million years ago, the animals living in our seas are not the subtropical Tethyan types of before (Fig. 12.27). Instead the animals are cool-water types and oxygen analyses of belemnites and ammonites show that the seas of the time corresponded in temperature to those now found between Stewart Island and Auckland and Campbell islands. Similar animals—called "Austral"—also lived at this time in seas covering Australia, Antarctica and Patagonia. So, therefore, for the first time since the Permian, we have living in Gondwanaland seas animals equivalent to the cool-water Boreal animals of the Northern Hemisphere. Although other factors, such as changing world climate or oceanic currents, may also have influenced this marked swing from Tethyan to Austral animals in our New Zealand fossil record, a favoured interpretation is that the change is related to continental drift.

A lot was happening to Gondwanaland at this time. Splits were starting to appear—the Tasman and the Atlantic were opening and landmasses were moving and rotating. Palaeomagnetic measurements tell us that Australia, New Zealand, Antarctica and South America were gradually swung southwards during the breaking-up movements. As they moved closer and closer to the South Pole, their shores were deserted by the warm currents that had flowed around them in the Jurassic and Triassic, and cool currents were introduced in their place. Thus the warm water Tethyan animals moved away from New Zealand and Australia and instead are found at this time, for example, in India, South-east Asia and New Guinea (Fig. 12.27).

Cool-water Austral animals moved into the places vacated by the Tethyan animals and so in the Cretaceous we find that we have a completely new set of fossils, with marked southern affinities. Austral animals lived in New Zealand and Australia for all of the Cretaceous, and oxygen analyses show that cool waters continued to wash New Zealand shores throughout this time. In the early Tertiary, however, tropical and subtropical immigrants, signalling warming seas, start to make their appearance, and the Austral animals gradually disappear. This warming of New Zealand's seas may reflect a gradual reversal of the drift of New Zealand towards the south, and, with the opening of the sector of the Southern Ocean between Australia and Antarctica, starting 55 million years ago, Australia and New Zealand started to separate from Antarctica and began the long journey northwards back into subtropical realms. By contrast Antarctica drifted steadily southwards and from this time on the fossil story of Antarctica, on the one hand, and New Zealand-Australia on the other, that up to now had so much in common, began to take divergent paths.

# A LANDSAT VIEW OF NEW ZEALAND

The photos presented in this folio of plates were taken by the Landsat satellites operated by the U.S. National Aeronautics and Space Administration and have been supplied by courtesy of the Department of Lands and Survey, in co-operation with the Physics and Engineering Laboratory, Department of Scientific and Industrial Research.

Landsat 1 was launched on 23 July 1972 and is now inactive. Landsat 2 was launched on 22 January 1975 and Landsat 3 on 5 March 1978. The satellites are positioned to circle the Earth in orbits passing close to both North and South Poles at an altitude of about 900km (560 miles); the tracks they follow are such that the same Earth scene is covered once every nine days. The satellites carry instruments capable of electronically recording on magnetic tape a continuous earth scene 185km (115 miles) in width as it passes beneath the satellite.

The basic instrument carried by the Landsat satellites is a multi spectral scanner, a highly sensitive light meter which responds to and records the various types of light reflecting from objects on the Earth's surface. Just as light from the sun can be split into various colour components (the light spectrum) by using a prism (or in the atmosphere by falling raindrops acting as tiny prisms and together producing a rainbow), so the sunlight reflected from objects has similar components. However, the components of the reflected light vary considerably with the object from which they are being reflected. Most objects tend to absorb various components from the full colour spectrum of sunlight they receive and reflect other components that give the object its characteristic colour. Green vegetation, for example, absorbs colours in the red, yellow and blue parts of the spectrum and reflects green colours—and thus appears green. A white rock, such as limestone, reflects all colours of the spectrum, whereas a black rock, like basalt, absorbs all colours and reflects none. More specific differences are also apparent and individual types of vegetation, or healthy or diseased plants, or types of rock, differ markedly in the nature of the light they reflect. For this reason the satellite photos are of great importance in a wide range of mapping projects.

The multi spectral scanner records the nature of the reflected light coming from individual areas of the Earth's surface in four spectral bands: Band 4 (0.5–0.6 micrometres wavelength, in the green-yellow part of the spectrum); Band 5 (0.6–0.7 micrometers wavelength, in the orange-red part of the spectrum); Band 6 (0.7–0.8 micrometers wavelength, in the deep red part of the spectrum); Band 7 (0.9–1.1 micrometers wavelength, in the near infra-red part of the spectrum). The bands can be printed separately or in combination, to bring out details of rocks, natural and cultivated vegetation, water masses, settlements and so on.

The information from the scanners can be transmitted simultaneously to the ground, or stored in tape recorders for transmission when the satellite is within the range of a ground station. Receiving stations are sited in the U.S.A., Canada, Italy and Brazil and others are being constructed elsewhere, including Australia. On the ground the information from the scanner is electronically transferred onto film and an image built up line by line, like a TV picture. The resulting products are photographic negatives or prints of sequences of earth scenes, each individual scene covering 34,000sq.km (185 × 185km).

The photos of the earth's surface obtained by the Landsat satellites are used for a variety of mapping and monitoring purposes. They are used to map types of forest, pasture and field crops; in the field of regional planning they provide information on land use, settlement patterns and roading systems.

Snowfield and glacier mapping and monitoring, for use in meteorological studies, are greatly simplified by their use. Computer processing of selected portions of the spectrum allows the mapping of sediment patterns in lakes and along sea coasts and the monitoring of their seasonal variations, for use in erosion studies. Landsat photos provide information on the exact location and size of offshore islands and reefs, the positions of ships and the distribution of sea ice, icebergs, etc. In the field of environmental monitoring they are used to show the effects of catastrophes such as bushfires, droughts, floods and outbreaks of plant diseases or insect pest plagues. Changes related to timber cutting, strip mining, pasture cropping and fluctuations in reservoir levels can also be monitored.

In geology major rock structures extending over long distances may be mapped and an appreciation gained of their relationship to other features—although geologists must still climb hills and clamber up stream beds to collect the all-important rock and fossil samples!

# PLATE 1

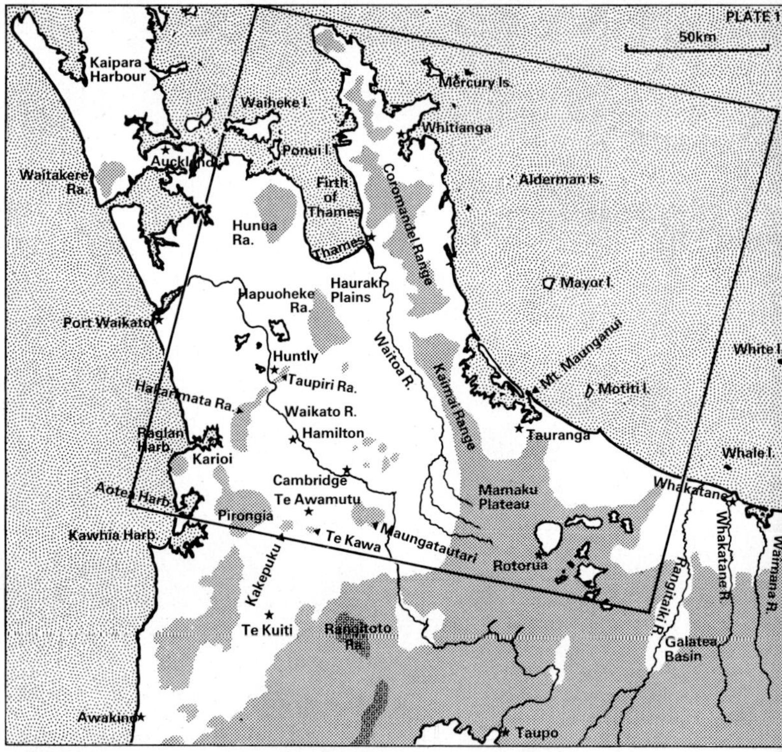


This region and the one covered by Plate 2 spans some of the most active volcanic areas in New Zealand, both today (Rotorua—Taupo region) and in the recent geological past, extending from Miocene times (Waikato coast and Coromandel Peninsula).

The rocks forming the geological foundation of the entire area have been formed from sediments, chiefly muds and sands, but sometimes also lavas and tuffs resulting from submarine volcanic eruptions which were originally laid down on the sea floor in the New Zealand Geosyncline (p. 221; (Figs 11.7, 11.8, 13.1–13.3). These rocks are exposed in northern Coromandel, Waiheke Island, Ponui Island, the Hunua and Hapuohoke Ranges and hills south and east of Te Awamutu. The Firth of Thames occupies a trough-like depression formed in these rocks by the sinking of a large block to depths of 1,500m (5,000ft) below sea level. Scattered occurrences of fossils show that all these rocks were laid down in late Jurrasic times, some 150–140 million years ago.

At the same time as sediments were being laid down in the New Zealand Geosyncline, other sediments were being deposited on the areas of continental shelf flanking the old Gondwana landmass (Figs 13.1–13.7). These rocks, of greater variety and more fossiliferous than those of the New Zealand Geosyncline, were later bent by mountain-building forces into a giant U-shaped fold (or syncline). In this part of New Zealand rocks of both Triassic and Jurassic age were folded in this way and are found along the Waikato and northern Taranaki coasts, extending from Port Waikato to Awakino. They are particularly well exposed around the shores of Raglan, Aotea and Kawhia Harbours. Drillhole evidence suggests they also continue under Taranaki and probably southwards to Nelson (Fig. 13.8; Plate 8).

In the region covered by the photograph these Mesozoic shelf rocks underlie and also outcrop extensively in a broad belt extending parallel to the coast and, along with younger rocks, form a western edge to the Waikato lowland, part of which is clearly visible in this photo as the Hakarimata and Taupiri Ranges. To reach the sea the Waikato River cuts through this rim of older rocks at Taupiri gorge.

The Mesozoic rocks of this area, like those of the rest of New Zealand, were pushed up out of the sea to form land by mountain-building movements that geologists call the Rangitata Orogeny (Fig. 13.23; pp. 241). These movements lasted from about 140–110 million years ago and were followed by 70–80 million years of comparative tranquility, during which the old land was gradually worn down and in many areas was again invaded by the sea (Figs 16.2, 16.32). During the latter stages of this down-wearing phase, in middle and late Eocene times, 47–38 million years ago (Figs 16.10, 16.11), extensive coal swamps developed in low-lying areas on the surface of this old land; these were to give rise to coal deposits now mined in the Huntly district and elsewhere in the Waikato.

The long period of quiescence was broken in Miocene times by the eruption of chains of volcanoes along the length and breadth of the Coromandel Range. It is believed that this volcanic activity was related to the initiation and propagation through New Zealand of the boundary between the Indian-Australian and Pacific plates (Fig. 18.6).

The Coromandel line of volcanic activity was preceded, in late Oligocene-middle Miocene times, by other lines of volcanoes extending through western Auckland (including the Waitakere Range) 28–15 million years ago, and eastern Northland 18–15 million years ago (Figs 16.33, 16.36).

The Coromandel volcanoes built up massive lava flows, tuff beds and intrusions over the period between middle Miocene and late Pliocene (15–3 million years ago). As the volcanic activity declined the area became peppered with hydrothermal areas, similar to those of the Taupo-Rotorua area today. This hydrothermal phase continued until middle Pleistocene time (750,000 years ago) and the colourful silica deposits of this phase (including agate and jasper), now exposed by erosion, are highly sought-after by rock hounds.

Mineral-bearing solutions that accompanied the volcanic and hydrothermal activity laid down deposits of gold, silver, lead, zinc and copper that have been the basis of mining activities both in the past and on a smaller scale at present.

Most of the old volcanic cones have now been deeply eroded and their original forms are not recognisable as such. However, individual lava flows and intrusions of volcanic rock stand up as rugged knobs in the landscape.

Volcanic cones dating back to the dying stages of Coromandel volcanism, in late Pliocene-middle Pleistocene, are found on the Mercury and Alderman Islands and adjoining parts of Coromandel Peninsula.

Outbreaks of volcanic activity also occurred behind the main Coromandel chain. These consisted mainly of lines of individual volcanic cones—one extending from Karioi in the west through Pirongia, Kakepuku and Te Kawa to Maungatautari in the east. Another line of cones had a trend parallel to the Firth of Thames and extended from Waiheke Island in the north to a group of three cones (Kiwitahi, Maungakawa and Te Tapui) close to Cambridge in the south.

As the Coromandel volcanoes became extinct, volcanic activity shifted eastwards and southwards to Whitianga (6–3 million years ago), Tauranga (3–0.75 million years ago—for example, Mt Maunganui, Mayor Island, Motiti Island) and then to the Rotorua–Taupo region, the site of modern volcanism. This migration of volcanism, like that in the Oligocene and Miocene, probably reflects changes in the direction and orientation through New Zealand of the boundary between the Indian-Australian and Pacific plates (Fig. 18.6).

The present volcanic arc extends from Tonga in the north through the Kermadec Islands and White Island to the volcanoes of

Tongariro National Park in the south (Fig. 18.11). Activity in this arc commenced about 0.75 million years ago and has been virtually continuous ever since (*cf.* Plate 2).

Some 15,000 cubic km of volcanic material has been erupted from vents within the Taupo-Rotorua zone. Although much has contributed to the filling of the trough-like structure underlying the Taupo-Rotorua region (Plate 2), substantial amounts have fanned out in all directions. In the area covered by this photo, for example, sheets of the volcanic rock ignimbrite cap the Kaimai Range and Mamaku Plateau (*cf.* Plate 2).

Pumice material from many of the eruptions covered vast areas of the central North Island (Fig. 18.18) and was also transported by the Waikato River and other rivers of the region to fill up low-lying areas. The Waikato basin, Firth of Thames, Hauraki Plains, Galatea basin, Rangitaiki Plain and the Tauranga-Whakatane coastline are examples of features that have been filled in this way.

# PLATE 2

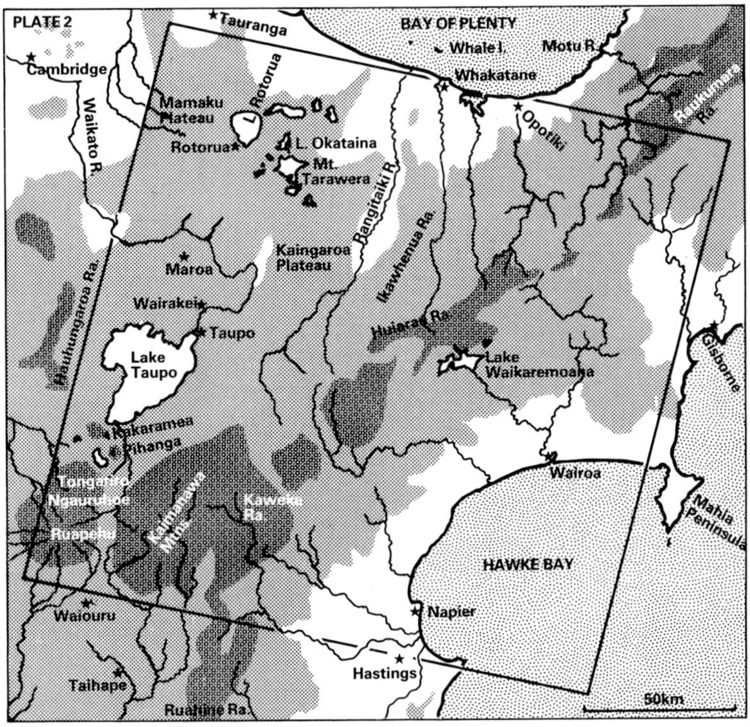


Like that covered by Plate 1, the entire region shown in this photo is underlain by argillites and greywacke sandstones of the New Zealand Geosyncline (p. 221). Fossils occurring in these rocks in the Hauhungaroa Range (beyond the left edge of the photo) and in the Huiarau and Ikawhenua Ranges show they are probably largely of late Jurassic age (about 150–140 million years ago). These ancient rocks were crumpled, folded and uplifted to form land during the earth movements of the Rangitata Orogeny, 140–110 million years ago (p. 241; Figs 13.23, 13.24).

This ancestral New Zealand landmass then underwent some 70–80 million years of erosion by rivers, streams and the ocean, until eventually large parts of it were worn down to below sea level (Fig. 16.2). The area of the photo was largely under the sea for most of early Cenozoic time (65–25 million years ago) and during this phase marine sediments were deposited across the worn-down landscape of ancient rocks (Figs 16.8, 16.10, 16.11, 16.13, 16.32, 16.33). Although later erosion and earth movements have removed or buried most of the sediments laid down at this time, large areas of middle Miocene rocks (17–12 million years ago) are preserved around Lake Waikaremoana. Towards the end of the Ice Age (or Pleistocene period), some 80,000 years ago, a giant landslide involving these rocks blocked a large valley, forming a natural dam. Lake Waikaremoana owes its origin to the water thus impounded.

Earth movements of the Kaikoura Orogeny, beginning some 12 million years ago, pushed up the ancient New Zealand Geosyncline rocks to form the modern ranges (p. 314). Some of the faults along which this movement has occurred are visible in the photo as SW-NE trending lines extending along the ranges. These faults are part of a large fault system extending from Marlborough to the Bay of Plenty coast (*cf.* Plates 3, 4). To either side of the ranges the same movements pushed down the ancient rocks to form the Hawke's Bay and Rotorua-Taupo depressions. Sea occupied the Hawke's Bay depression until well into the Ice Age (Fig. 17.21), so that some 9200m (30,000ft) of Miocene, Pliocene and early Pleistocene sediments, and thinner layers of older sediments, underlie most areas of the Hawke's Bay plains seen in this photo.

Early in the Ice Age, about 2 million years ago or slightly earlier, extensive volcanism broke out in the trough-like Rotorua-Taupo depression. Tongariro, Pihanga and Kakaramea volcanoes were built up in the period 2–1.5 million years ago (Fig. 17.17). Activity then shifted to Ruapehu, which commenced erupting 750,000 years ago, and then moved to Ngauruhoe some 2,500 years ago. Both Ruapehu and Ngauruhoe are intermittently active today (p. 346; Figs 18.20, 18.21). The lavas and ash showers of these five major volcanoes and other minor cones filled the southern end of the Rotorua-Taupo depression. Elsewhere, however, vast ignimbrite eruptions occurred, the last major outburst being in A.D. 130, but with smaller eruptions in A.D. 500, A.D. 700 and A.D. 1020 (p. 410). These eruptions of fiery clouds of extremely hot volcanic material, ejected with great force from fissures in the earth, spread out across the central part of the North Island, reaching southern suburbs of Auckland city, coastal Waikato, inland Hawke's Bay, and the flanks of Tongariro. Material from successive ignimbrite eruptions fills the central and northern parts of the Rotorua-Taupo depression, often to depths of 3,000–4,000m (10,000–15,000ft). Many of the exotic forests of the Central North Island, such as the Kaingaroa Forest, are sited on the plateaux formed by such ignimbrite sheets.

Many of these huge ignimbrite eruptions are believed to have originated from Lake Taupo, which is considered to be at least in part a giant volcanic crater excavated by explosive eruptions. There is evidence, however, that subsidence has also played a part in its formation—presumably the result of collapse of the crust after vast amounts of sub-crustal material had been melted and erupted. Lake Rotorua and the city of Rotorua are sited in another basin-shaped collapsed area (or caldera) and the Mamaku Plateau extending to the west and north is capped by an ignimbrite sheet up to 180m (600ft) thick which was erupted from within its boundaries. Smaller eruptive centres are present at Okataina and Maroa.

Other more localised volcanic activity occurred at the same time as the ignimbrite eruptions, and 163 individual volcanic vents have been mapped in the Rotorua-Taupo area. Mt Tarawera erupted in 1886, killing 153 people and spreading a mantle of ash and debris over the surrounding farmland and forest. Lake Rotomahana occupies a crater formed by explosive activity during this eruption. Hot springs and geysers are common throughout the Rotorua-Taupo area today and a geothermal power station is sited at Wairakei.

The Rotorua-Taupo volcanic zone continues north-eastwards as a linear belt extending through Whale Island, White Island (Fig. 18.19) and the Kermadec Islands to Tonga and along with the accompanying earthquakes, faulting and mountain building is the surface expression of the boundary between the Indian-Australian and Pacific plates (Figs 18.6, 18.7, 18.17).

# PLATE 3 SOUTH TARANAKI—WESTERN WELLINGTON

# PLATE 3

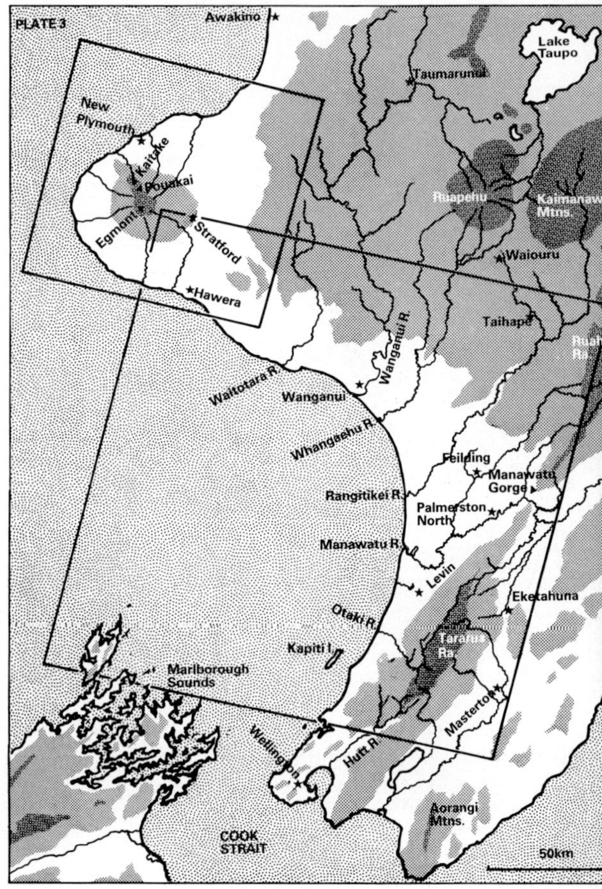


The Tararua and Ruahine Ranges, and their continuations to north and south, are composed of the New Zealand Geosyncline rocks (mainly argillite and greywacke sandstone) also encountered in Plates 1 and 2. Fossils of late Triassic age (about 200 million years ago) are known from these rocks in the western and central regions of both ranges and of late Jurassic age (150–140 million years ago) from the eastern flank of the Tararua Range, near Eketahuna. These ancient rocks have been pushed up along fault lines (which impart a SW-NE trending grain to the landscape of the ranges and often provide a sharp boundary between mountains and plains) so that now they stand above the surrounding lowlands (*cf.* Fig. 16.2). It is generally believed, however, that such rocks, or their South Island counterparts, also continue under virtually the entire region covered by the photo, although in some areas (for instance, near Wanganui) they lie at depths of 4,500m (15,000ft) below the surface. To either side of the ranges the ancient rocks are overlain by sandstones, siltstones and mudstones deposited by the sea in late Tertiary and early Pleistocene times (5 million years-600,000 years ago).

Like those elsewhere, the old New Zealand Geosyncline rocks were folded and pushed up by the earth movements of the Rangitata Orogeny (140–110 million years ago) to form an ancestral New Zealand landmass. This land was then gradually worn down by erosion and eventually inundated by the sea. By Oligocene times (30–25 million years ago) much of the area now occupied by modern New Zealand had been submerged (*cf.* Figs 16.13–16.32).

From middle Miocene times (12 million years ago) onwards earth movements accompanying the Kaikoura Orogeny, or mountain-building phase, gradually brought more and more land above sea level (Figs 16.35, 16.36). However, the coastal area adjacent to Wanganui did not rise out of the sea until the end of early Pleistocene time, about 600,000 years ago, and was one of the last major areas of modern New Zealand to be reclaimed from the sea in this way (Fig. 16.36). At about the same time, or slightly earlier, the Tararua and Ruahine Ranges began to be upthrust and have risen to their present heights by means of individual movements spread out over late Pliocene and Quaternary time, but more especially over the last 1½–1 million years (pp. 346–347).

The last parts of the embryonic ranges to remain under the sea were the southwest Wellington region adjacent to Cook Strait, the Manawatu Gorge region, and the Kuripapango district of inland Hawke's Bay. For most of late Pliocene times (5–2 million years ago) the Manawatu Gorge was the site of a marine strait, passing across the rising ranges, and the approximate limit of the rocks laid down in this strait are shown in the photo by the break in the vegetation on the Tararua and Ruahine Ranges to north and south of the present Manawatu Gorge.

During the glacial episodes of the Pleistocene or Ice Age period (2 million years-10,000 years ago) world-wide sea levels dropped substantially as water in the form of ice and snow accumulated on the land and so was withheld from the ocean. Glacial sea levels reached their lowest point, 183–238m (600–775ft) below modern sea level 17,000–14,000 years ago. At this time much of the area that is now sea in this photo was dry land and Cook Strait ended in a blind gulf to the west of Kapiti Island (Fig. 17.2).

Another Pleistocene development in this region was the construction of a chain of volcanoes in western Taranaki, of which Egmont (seen to the left of the photo surrounded by a circle of rain forest, the outer boundary of which is that of Egmont National Park) is the most recent example. Early in the Pleistocene, about 1.75 million years ago, volcanoes existed at the Sugar Loaves and Paritutu, on the outskirts of New Plymouth. About 500,000 years ago volcanic activity shifted to Kaitake, and then about 250,000 years ago to Pouakai (Fig. 17.16). These volcanoes are now deeply eroded, and in the case of the Sugar Loaves and Paritutu only stumps of resistant intrusive rocks remain.

Egmont itself began to erupt about 70,000 years ago and has built up a cone of classic proportions so that it is often called the Mount Fuji of New Zealand (Fig. 1.6). The last major lava flows from Egmont were erupted about 3,000 years ago but small outbursts, mainly of ash, occurred in A.D. 1500, A.D. 1655 and A.D. 1755.

# PLATE 4

Most of the rocks visible in this photo date back to the time of the New Zealand Geosyncline (*cf.* Plates 1–3), when they were originally laid down on the sea floor as sands and muds, but with occasional intercalations of lava and ash, derived from underwater volcanic eruptions. Fossil occurrences indicate that the rocks comprising the Marlborough Sounds and the terrain north of Wairau Valley are of Permian age (280–230 million years ago), whereas those elsewhere are largely of late Triassic age (200–195 million years ago), although probably extending into the Jurassic (190–135 million years ago) in coastal Marlborough.

Earth movements of the Rangitata Orogeny, 140–110 million years ago, crumpled and folded these ancient rocks and uplifted them to form an ancestral New Zealand landmass. A long period of erosion ensured, lasting in many parts of New Zealand for the next 70–80 million years, during which the ancestral land was worn down and large areas progressively flooded by the sea. Although most of the marine sediments laid down at that time were stripped by later erosion remnants of middle and late Cretaceous age (100–65 million years ago), some are preserved along the Marlborough coast and in the Inland and Seaward Kaikoura Ranges (Figs 16.4, 16.6, 16.9).

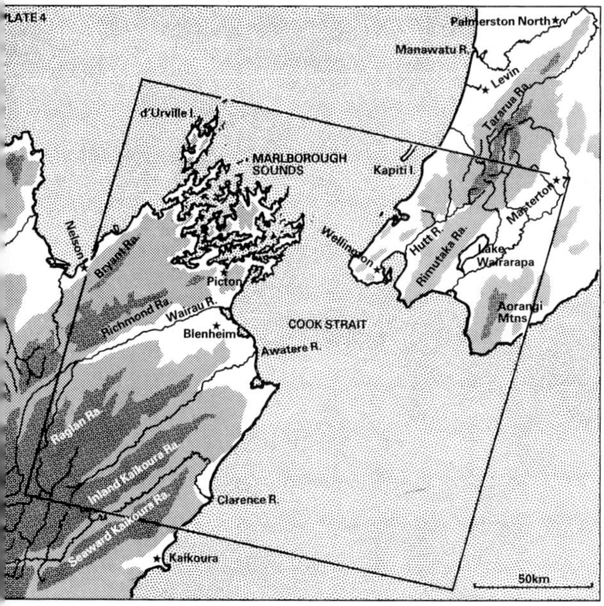


The long period of down-wearing and submergence was followed by another phase of earth movements (Kaikoura Orogeny), which began about 12 million years ago and continues today. These renewed movements twisted and distorted the ancient New Zealand Geosyncline rocks as well as the younger covering sediments and created extensive zones of fault lines (*cf.* Plates 2, 3). Movements along these faults wrenched apart and pushed up or down large blocks of terrain. Cook Strait probably owes its origin to a cross-cutting fault system lying between Marlborough and Wellington.

At the same time, however, movement along the Wairau Fault (following the Wairau Valley) has progressively moved the Nelson-Marlborough Sounds block northeastwards, imparting an S-shape to Cook Strait. Such movements were also accompanied by down-warping and the Marlborough Sounds has been formed by the submergence of a rugged landscape of hills and valleys. Kapiti Island, seen in Plate 3, is believed to lie on the submerged leading edge of the Marlborough Sounds block.

A number of major fault lines are clearly visible in this photo—traversing the Wairau, Awatere, and Clarence Valleys—and the Hope-Kaikoura Fault runs on a parallel trend to the south of the photo (seen in Plate 4), intersecting the coastline between Kaikoura and the mouth of the Clarence River (Fig. 18.17). All these faults are branches of the Alpine Fault, responsible for the uplift of the Southern Alps and for moving N.W. Nelson 450km (280 miles) north-eastwards from its original position off Fiordland (Figs 4.11, 18.17). The Alpine Fault is considered to be a transform fault, forming the boundary between the Indian-Australian Plate and Pacific Plate (p. 357; Figs 9.24, 18.7).

Uplifting and tilting movements associated with the Marlborough faults have created a series of ranges (notably the Richmond, Spencer-St Arnaud-Raglan, Inland Kaikoura and Seaward Kaikoura Ranges), separated by broad valleys. All the Marlborough faults are believed to cross Cook Strait and to continue north-eastwards as the Pukerua, Owhariu, Wellington, West Wairarapa and East Wairarapa faults—responsible for major features of the Wellington landscape such as the Hutt Valley and Wairarapa depressions, and the Rimutaka, Tararua and Aorangi Ranges. A zone of faults continues the same trend north-eastwards to eventually intersect the Bay of Plenty coastline near Whakatane (some of these faults are seen in Plates 2 and 3).

# PLATE 5

The old argillite and greywacke sandstone rocks of the New Zealand Geosyncline seen in Plate 4 continue into northern and central Canterbury. They outcrop in the snow-capped ranges of the Southern Alps (Figs 13.9–13.11) and their foothills, and also in the Lowry Peaks Range and Cheviot area. Fossils of late Triassic age (200–195 million years ago) occur in the rocks of the Southern Alps and of late Jurassic age (150–140 million years ago) in the Lowry Peaks Range and Cheviot area. The same rocks continue eastwards under the Canterbury Plains and Banks Peninsula, and are believed to extend along the Chatham Rise east of Chatham Islands (Fig. 13.14).

The faults of the Marlborough region (Plate 4) also continue into north Canterbury (Fig. 18.17). The Hope-Kaikoura Fault traverses the upper left part of the photo and a short length of the Clarence Fault is also visible close to the upper left edge. Both faults, together with the Wairau and Awatere Faults, are branches of the Alpine Fault, which also traverses the upper left corner of the photo. A prominent feature of the Hope-Kaikoura Fault is the Hanmer Plain, occupying a topographic depression formed by warping along the fault line.

Banks Peninsula is the eroded stump of a large volcanic island that began erupting off the Canterbury coast some 10 million years ago (Figs 16.35, 16.36). The volcanic island was composed of twin basalt volcanoes, the dead craters of which, hollowed out by erosion and breached on one side by the sea, now form Lyttelton and Akaroa Harbours.

During the glacial episodes of the Ice Age, between about 2 million years and 10,000 years ago, extensive glaciation occurred in the Southern Alps and active frost erosion elsewhere (Fig. 17.2). Masses of rock debris were transported by rivers and streams, to be deposited in the sea between the alpine foothills and the Banks Peninsula island. This material augmented that already accumulated in earlier times as a result of uplift and erosion of the Alps. In time a vast gravel plain fringing the alpine foothills was built and was gradually extended seawards so that eventually Banks Peninsula island became tied to the mainland, and the modern Canterbury Plains were formed (Fig. 17.5). The latest phase in this natural land-tying process has been accumulation of gravel and sand, under the influence of coastal currents, to cut off a large bay of the sea, now Lake Ellesmere. In time Lake Ellesmere will be filled in naturally, to provide another land link to Banks Peninsula.

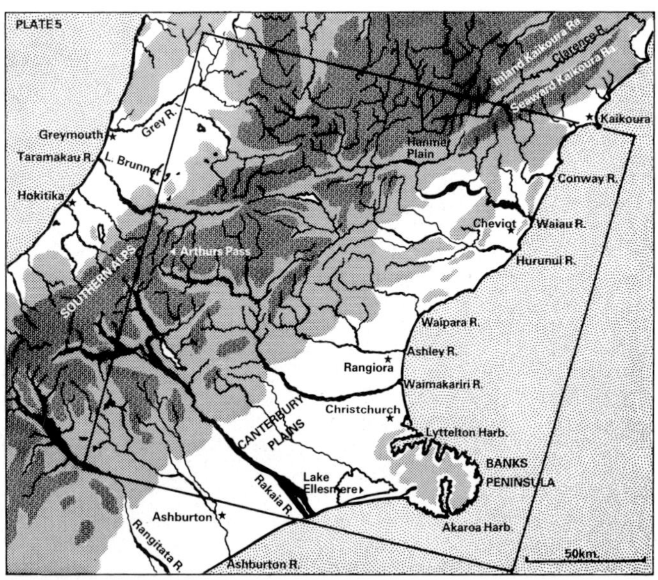

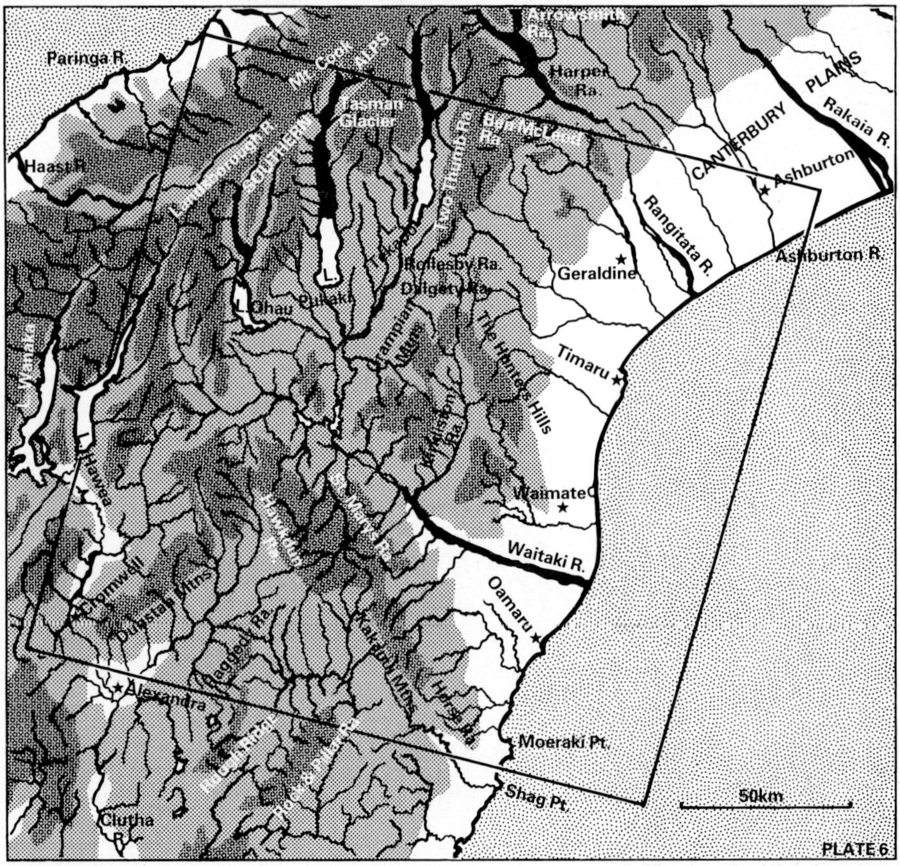


# PLATE 6

Many of the rocks seen in Plate 5 continue southwards into this region of South Canterbury and North Otago. Like the remainder of Canterbury, the argillite and greywacke sandstone rocks of the New Zealand Geosyncline are thought to underlie the entire region—although in many areas, especially near the coast, they are covered by younger rocks. Fossils of late Triassic age (200–195 million years ago) occur in these rocks in the Hawkdun and Horse Range and in the ranges flanking Lakes Ohau, Pukakai and Tekapo. Fossils of Permian age (280–230 million years ago) occur in the rocks of the ranges between the Waitaki and Ashburton Rivers (for example, Kirkliston Range, Grampian Mountains, the Hunters Hills, Rollesby and Dalgety Ranges and, north of the area covered by the photo, the Ben McLeod, Harper and Arrowsmith Ranges). South Canterbury is notable, however, for having remnants of the rocks deposited in the New Zealand Geosyncline at the very beginning of its development—in Carboniferous times (345–280 million years ago). These old rocks occur in Kakahu Hill, west of Geraldine.

In some areas the rocks of the old New Zealand Geosyncline have been altered by heat and pressure. While strips of these altered rocks occur in some of the south Canterbury ranges, they occur more extensively in the hilly topography south of the Waitaki (for example, Dunstan Mountains, Raggedy Range, Rough Ridge) and form much of the landscape of central and coastal Otago extending south from the area covered by the photo (see Plate 7). The altered rocks, called schists, have been changed (or metamorphosed) by heat and pressure deep within the earth's crust (p. 236; Figs 13.8–13.20). They are thought to represent the lower parts of the vast thickness of sediments which originally accumulated in the New Zealand Geosyncline (Figs 11.7, 13.23). These sediments were solidified into rocks, deeply buried, and pushed down to depths where heat and pressure changed the minerals originally present in the sediments into other mineral types stable at higher temperatures and pressures (pp. 234–236).

The deep burial and exposure to heat and pressure also provided conditions for formation of metallic ores so that the Otago schists are notable for deposits of scheelite (the ore of tungsten), antimony and gold. Although originally occurring as veins in the schists these ores have often been eroded out of the rocks and then, after having been concentrated in the river channels, laid down as placer deposits in old river gravels.

The Otago schists continue north-westwards until they are cut and displaced by the Alpine Fault (p. 26; Plate 8). Horizontal movements along this major fault line, the largest in New Zealand, have slid the schists and other adjoining rocks some 450km (280 miles) to the northeastern tip of the South Island, where they form the landscape of the Marlborough Sounds (Plate 4). The same ore deposits as those in Otago (i.e. scheelite, antimony and gold) also occur in the Marlborough schists.

South of the Waitaki River the old schist rocks have been broken into a series of large blocks, creating a basin and range landscape—with rib-like ranges of the old rocks projecting above and separated by gravel-filled basins (Fig. 16.38).

During the glacial episodes of the Ice Age (between 2 million and 10,000 years ago) snow and ice accumulated on the peaks and in the valleys of the Southern Alps and huge glaciers pushed out into the surrounding lowlands (Fig. 17.2). Great quantities of rock debris were transported by the glaciers and in turn by rivers draining from them, resulting in the formation of extensive gravel plains throughout Canterbury (Fig. 17.5) and Otago, and the filling of the basins between individual blocks of hills and ranges.

Heaps of moraine, dumped by the glaciers during their greatest extension, form natural dams impounding lakes such as Ohau, Pukaki and Tekapo. The Ice Age rivers and streams also transported great quantities of gravel and sand out to the coast, where it was added to the shoreline, resulting in steady outgrowth of the Canterbury Plains during Quaternary time. The influence of a large river like the Waitaki on this building-out process is seen clearly in the photo. The amount of debris brought down by this river has been such that a large rounded promontory has been built on the coastline at its point of discharge.

# PLATE 7

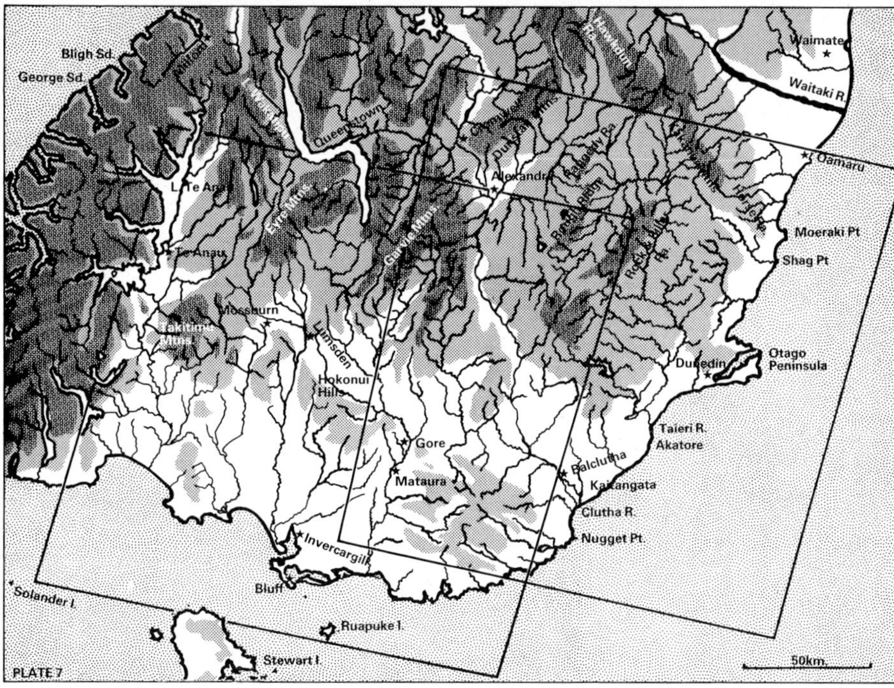


The greywacke sandstone and argillite rocks of the New Zealand Geosyncline that are such prominent features of most of the regions covered by the previous photos are very much in the minority in this region of Otago. A small strip of these rocks is present in the Horse Range and an isolated block at Akatore, on the coast south of the mouth of the Taieri River. The rocks of both these areas contain fossils of late Triassic age (200–190 million years ago).

Otherwise, the northern half of the photo is dominated by schist rocks, formed from the argillites and greywacke sandstones by the effects of heat and pressure whilst buried deep within the earth's crust (Figs 13.18–13.20). These schist rocks are the southward continuations of those seen in the previous photo (Plate 6).

During the earth movements of the Rangitata Orogeny, or mountain building phase, some 140–110 million years ago, the sediments of the New Zealand Geosyncline (including deposits that are now argillites, greywacke sandstones and schists; Plates 1–6), were folded and pushed up to form an ancestral New Zealand landmass, much larger than modern New Zealand. In southern New Zealand this ancestral land extended eastwards to beyond the Chatham Islands and southwards to the edge of the Campbell Plateau, beyond Auckland and Campbell Islands (Figs 13.14, 15.6). During the following 70–80 million years erosion by rivers, streams and the sea gradually lowered the ancestral land until eventually it was reduced to a low-lying plain, or "peneplain" (Fig. 16.2). By middle and late Oligocene times (30–25 million years ago) most of modern New Zealand (Fig. 16.13) had been reduced to a peneplain—even including areas now occupied by some of our most rugged topography (as in N.W. Nelson, the Southern Alps and Fiordland). The wearing-down of the old landmass in this way was followed by gradual encroachment of the sea. Thus layers of young sediments were laid across the peneplain surface.

In Otago, development of the peneplain across the schist rocks had largely been accomplished by late Cretaceous times (75–65 million years ago) and sediments of this age and of the early Tertiary (65–37 million years ago) were laid down on its surface. However, most of these sediments were stripped off by erosion in late Tertiary and Quaternary times, so that the "fossil" peneplain was "exhumed" and now forms the rolling landscape of central Otago. Remnants of the formerly overlying late Cretaceous and early Tertiary sediments are still preserved along the Otago coast and in some inland depressions. Many of these deposits contain coal-bearing beds (for example, the Kaitangata coalfield).

The former continuity of the exhumed peneplain surface was broken by earth movements of the Kaikoura Orogeny, beginning 12 million years ago. In north Otago these movements broke the peneplain into a series of raised blocks, separated by depressed "basin" areas, filled with Ice Age gravels—and created the "basin and range" topography typical of this region (*cf.* Plate 6; Fig. 16.38). The movements also downwarped the peneplain towards the coast, where it disappears under remnants of the covering deposits.

Also involved in the earth movements of the Rangitata Orogeny were sediments originally laid down on the continental shelf flanking the New Zealand Geosyncline. These sediments, the equivalents of the Triassic and Jurassic rocks of the Taranaki and Waikato coasts (*cf.* Plate 1; pp. 221–227), but also including flanking Permian rocks, were folded into a giant, U-shaped fold by the mountain building forces of the Rangitata Orogeny. This U-fold (or syncline) was cut across by the erosion that created the late Cretaceous-early Tertiary peneplain and, along with many parts of New Zealand, was subsequently submerged by the sea. Later erosion in the last 12 million years has, however, re-exposed the syncline and etched-out its inner structure. Thus the rock layers standing on edge along the flanks of the U-fold of the Southland Syncline now show up as ribs of rock forming prominent escarpments extending north-west from Nugget Point on the Otago coast for 150km (95 miles) to near Mossburn in N.W. Southland. Such rocks continue, although not so spectacularly, to meet the Alpine Fault at Pyke River (near Lake McKerrow, W. Fiordland) where they are displaced by horizontal movements along the fault for a distance of 450km (280 miles) towards the north-east, so that they reappear in the Nelson region (Fig. 4.11).

As may be seen in the photo, the ribs of rock of the Southland Syncline are abruptly cut off at the modern coastline, but geophysical and oceanographic investigations have shown that they continue under-sea across the Campbell Plateau, the submerged continuation of the New Zealand landmass (Figs 13.14, 18.4).

Although Otago and Southland are today geologically quiet, and far removed from the area of present volcanic activity, this was not so in the past and throughout most of Miocene times the Dunedin area was the scene of extensive volcanism (Fig. 16.35).

Three major episodes of eruptions occurred between 15 and 9 million years ago and intrusions of volcanic rock, lava flows and beds of volcanic ash and tuff dating back to this time make up the entire Otago Peninsula area, and extend short distances along the coastline to north and south.

# PLATE 7   OTAGO AND EASTERN SOUTHLAND

# PLATE 8

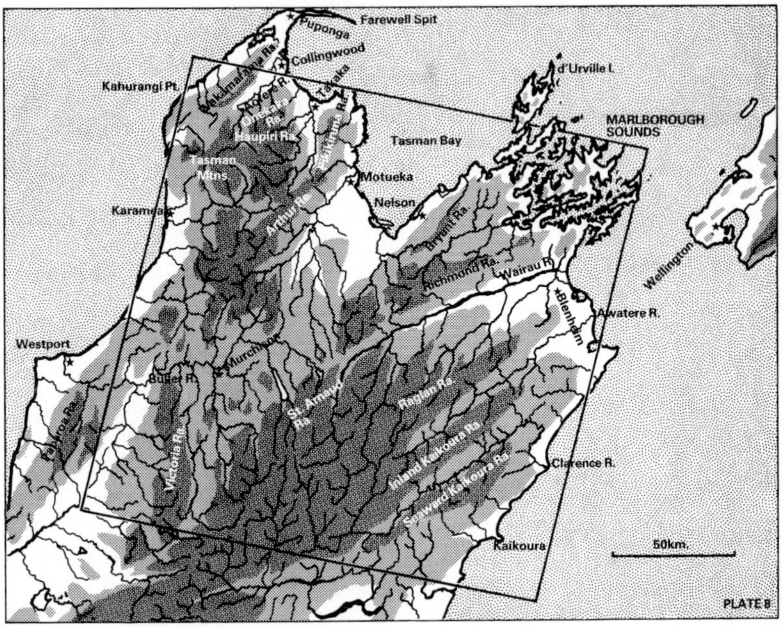


As we have seen in Plate 7, the Palaeozoic and Mesozoic rocks of Otago and Southland can be traced westwards and north-westwards until, in the vicinity of Lake McKerrow in western Fiordland, they are cut and displaced by the Alpine Fault. Horizontal movements along this fault, spread out over millions of years, have literally torn the South Island diagonally apart—sliding rocks originally lying to the west of Fiordland for a distance of 450km (280 miles) north-eastwards, so that today they form the north-west part of the South Island (Fig. 4.11). Some geologists maintain that these movements date back to 140 million years ago, whereas others believe them to have occurred over the last 25 or 12 million years.

Whatever the exact timing of the movements, it is nonetheless a striking fact of New Zealand geology that rocks that were once continuous are today found at opposite ends of the South Island, separated by the Alpine Fault. These rocks can be matched precisely by using their range of unique and contrasting characters in the same way as lines of print can be used to match torn pieces of newspaper. Thus the rocks of N.W. Nelson, west of the Moutere-Waimea depression, are equivalents of those of Fiordland; the Triassic and Permian rocks forming the hills behind Nelson, and extending north-east and south-west, are equivalents of those of the Southland Syncline; the Marlborough Schists are equivalents of the Otago Schists (Fig. 4.11).

The ancient rocks of N.W. Nelson are thought to be a remnant of the old Gondwana continent that originally flanked New Zealand to the west. In Palaeozoic and early Mesozoic times Australia, Antarctica and the older rocks of the Lord Howe Rise were all one landmass and formed the eastern margin of Gondwanaland (Figs 11.9, 13.14). For most of this period the area that is now modern New Zealand was largely a subsiding part of the sea floor (the New Zealand Geosyncline) lying immediately offshore from these land-masses (Figs 11.7, 11.8).

The opening of the Tasman Sea, between about 110 and 60 million years ago, split off a strip of rocks from the leading edge of the old Gondwana continent, and as the embryonic Tasman opened they were left behind to provide a foundation for the modern Lord Howe Rise and to form a western fringe to the modern New Zealand landmass (Figs 11.2, 11.8). Today these old Gondwana rocks occur in N.W. Nelson as three strips of sediments extending in a north-south direction through the Tasman Mountains, Haupiri Range and Pikikiruna Range. They contain fossils of Cambrian, Ordovician and Silurian age (570–390 million years ago) and, together with an isolated patch of Ordovician rocks at Cape Providence, S.W. Fiordland, comprise the oldest fossil-bearing rocks in New Zealand.

These old sediments have been deformed and altered by at least three episodes of New Zealand-wide mountain building movements or orogenies: Tuhua Orogeny 420–300 million years ago; Rangitata Orogeny 140–110 million years ago; Kaikoura Orogeny, extending from 12 million years ago to the present day. They are also intruded and altered by many granites dating back to these times of major earth movements. Mineral-bearing solutions associated with these granites introduced a great variety of ore deposits, including gold, iron, lead, copper, silver, zinc and molybdenum.

Oil prospecting drillholes show that the Palaeozoic rocks of N.W. Nelson extend offshore in a westwards direction and also northwards as least as far as opposite the Waikato coast (Fig. 13.8). These submerged rocks probably extend even further westwards to link with those forming the geological foundation of the more ancient parts of the Lord Howe Rise (Fig. 13.14 B).

Like most other areas of New Zealand, N.W. Nelson was worn down virtually to sea level during 70–80 million years of erosion in Cretaceous and early Tertiary time (*cf.* Plate 7; Figs 16.2, 16.13). A low-lying landscape or peneplain was thus formed, on the surface of which coal-bearing and other terrestrial sediments accumulated that now outcrop extensively in the region between Kahurangi Point and Puponga. In Oligocene times (37–25 million years ago) the peneplain surface was flooded by the sea and marine sediments of this age, notably limestones, were deposited. Later erosion associated with the uplift of N.W. Nelson to form the present rugged terrain has stripped away most of these covering sediments, but remnants are preserved along the coast between Kahurangi Point and Cape Farewell, Aorere Valley and the Onekaka-Takaka-Tarakohe area. The peneplain surface, stripped of marine sediments, caps the Onekaka Range and remnants are present on summits in the Wakamarama and Haupiri Ranges. Elsewhere the peneplain surface has been largely destroyed by later uplift and erosion.

Mountain building movements of the Kaikoura Orogeny have broken the N.W. Nelson region into a series of massive blocks, now carved by erosion into mountain ranges. Intervening areas, depressed by faulting, folding or tilting, are marked by valleys such as those of the Aorere and Takaka Rivers and the Waimea-Moutere Plain.

During the glacial phases of the Ice Age, 2 million-10,000 years ago, massive glaciers formed in the high country of N.W. Nelson and carved broad U-shaped valleys (Fig. 17.2). As the glaciers and the effects of the harsh climate ate into the landscape, great quantities of rock debris were transported to coastal area by rivers and streams. During the glacials, world sea levels were greatly lowered, sometimes to 183–238m (600–775ft) below modern sea level, and at these times the shoreline was situated, in the N.W. Nelson area, for example, in the western approaches of Cook Strait (Fig. 17.2). Broad gravel plains were built out to these shorelines and were submerged by the sea as it rose to its present level after the end of the Ice Age. Remnants of the gravel deposits built up at this time are preserved, however, in the Moutere Hills and in various coastal areas. These gravel deposits and the underlying coastal rock formations provide an abundant supply of marine sediment and spits, bars and banks built from such material are a prominent feature of the coastline. Farewell Spit, 25km (15 miles) in length, is a notable example of such an accumulation built up by the interaction of coastal currents originating from the Karamea coast on the west and Golden Bay on the east.

CHAPTER THIRTEEN

# THE NEW ZEALAND LAND

THE VARIOUS changes that took place in the marine life living in New Zealand Mesozoic seas, outlined in the last chapter, were accompanied by far-reaching changes in the character of the sea basin (the New Zealand Geosyncline) occupying the site of modern New Zealand. The New Zealand Geosyncline, the successor to similar but probably smaller structures in the early Palaeozoic (see Chapter 11), began its life in the Carboniferous, some 345-280 million years ago. Then for some 200 million years, until late Jurassic times, it was the dumping-ground for masses of sedimentary material, eroded off the flanking Gondwanaland continents (Figs 13.1-13.3).

In Carboniferous times newly created mountainous terrain existed in eastern Australia and its extension into the Tasman, the modern Lord Howe Rise. Similarly, new raw mountains had been created in eastern and western Antarctica. The steep crumbling slopes provided great quantities of coarse rock debris that was quickly eroded away by rivers and streams and transported out to sea, to eventually end up in the New Zealand Geosyncline.

Also at this time active volcanoes were liberally sprinkled amongst the mountain peaks of eastern Australia and to some extent Antarctica too, and these contributed masses of material—blocky lavas, powdery ash and other volcanic debris—to the New Zealand Geosyncline. Volcanic islands also dotted the seas of the geosyncline, depositing ash and lavas directly onto the sea bed (Figs 13.1-13.3).

**Continental shelf and slope**

The rocks that were laid down during the time of the New Zealand Geosyncline now form the rocky mountainous backbones of both North and South Islands and also underlie substantial areas to each side of the main ranges. If we look at these rocks, we can separate them into three distinct groups. First, we can recognise layers of rocks that were originally laid down in shallow coastal waters, on the continental shelves flanking the New Zealand Geosyncline to the west, immediatly off-shore from Australia, for example. These rocks are sandstones, siltstones and argillites; originally they were sands, silts and muds lying on the sea floor of the continental shelf (Fig. 13.4). As any fisherman knows, marine life is at its most abundant on the warm, sunny continental shelves and so it is that often these ancient continental shelf rocks are crammed with fossil shellfish, the remnants of colonies similar to modern pipi and oyster beds.

Fossil plant material is also very abundant—leaves, woody stems etc., as well as finely mashed-up plant debris that, when incorporated into sea-floor muds and reduced to carbon by the rock-forming processes, gives a very dark colour to the rock.

Also present amongst these rocks are beds of fine volcanic ash, often white or creamy in colour, but sometimes reddish or greenish (if iron is present) (Fig. 13.5). This ash, ejected from nearby volcanoes, either on the continental landmass or on volcanic archipelagos, drifted across the sea and, settling down to the ocean floor, formed discrete layers of pure volcanic ash. Such waterlain deposits of ash are called tuff. Some parts of the Permian, Triassic and Jurassic strata contain almost a half by proportion of volcanic tuff beds and good examples can be seen in the Hokonui Hills, Taringatura Hills and Takitimu Mountains of Southland, the Nuggets coast of south Otago, and the Kawhia and Marokopa coasts of the Waikato.

When we have a look at these old shelf sediments in detail, we find clear evidence that the sea floor currents originally responsible for their deposition had come from the west. For example, rocks of Triassic and Jurassic age cropping out along the Kawhia and Marokopa coasts

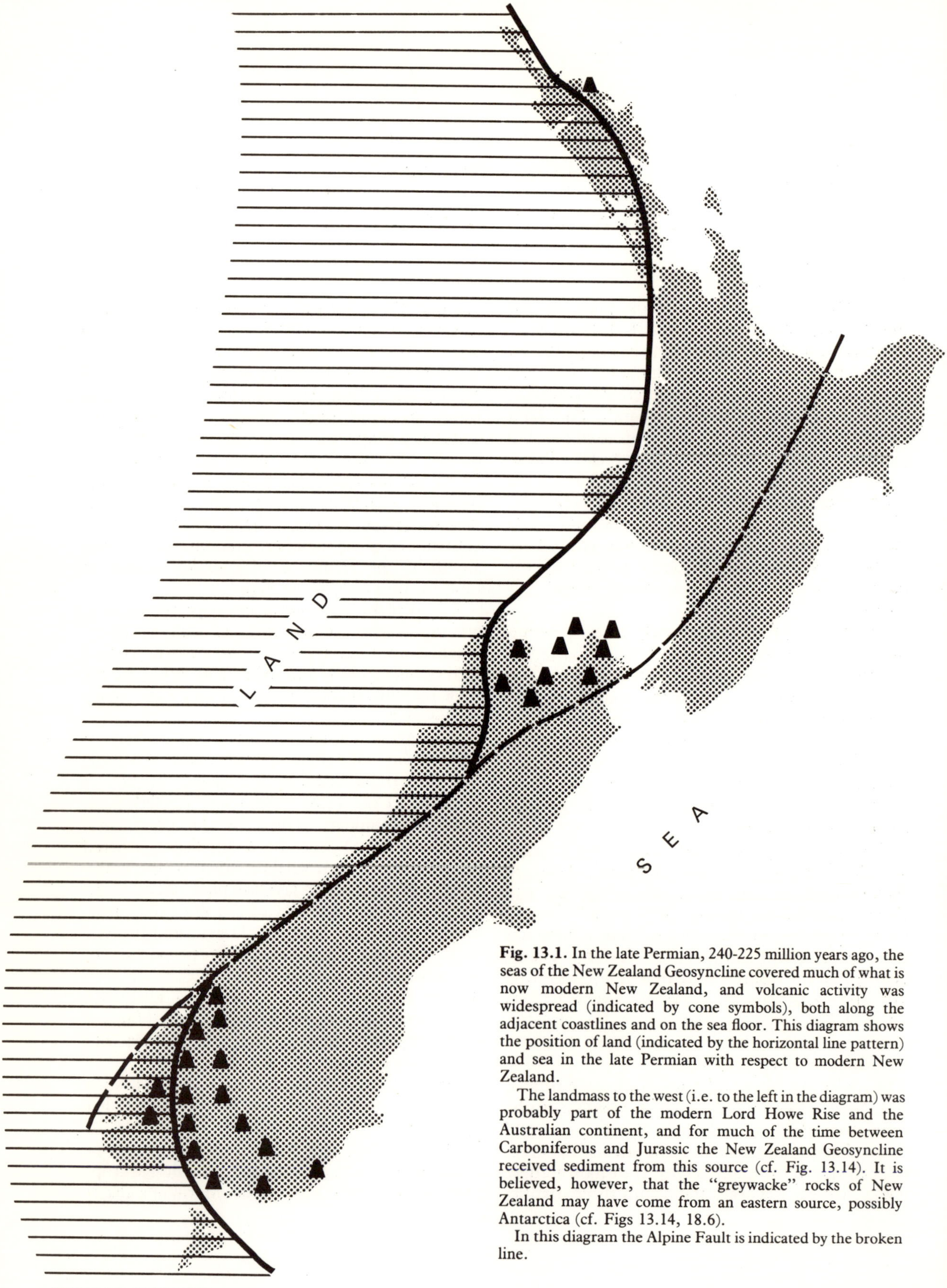


**Fig. 13.1.** In the late Permian, 240-225 million years ago, the seas of the New Zealand Geosyncline covered much of what is now modern New Zealand, and volcanic activity was widespread (indicated by cone symbols), both along the adjacent coastlines and on the sea floor. This diagram shows the position of land (indicated by the horizontal line pattern) and sea in the late Permian with respect to modern New Zealand.

The landmass to the west (i.e. to the left in the diagram) was probably part of the modern Lord Howe Rise and the Australian continent, and for much of the time between Carboniferous and Jurassic the New Zealand Geosyncline received sediment from this source (cf. Fig. 13.14). It is believed, however, that the "greywacke" rocks of New Zealand may have come from an eastern source, possibly Antarctica (cf. Figs 13.14, 18.6).

In this diagram the Alpine Fault is indicated by the broken line.

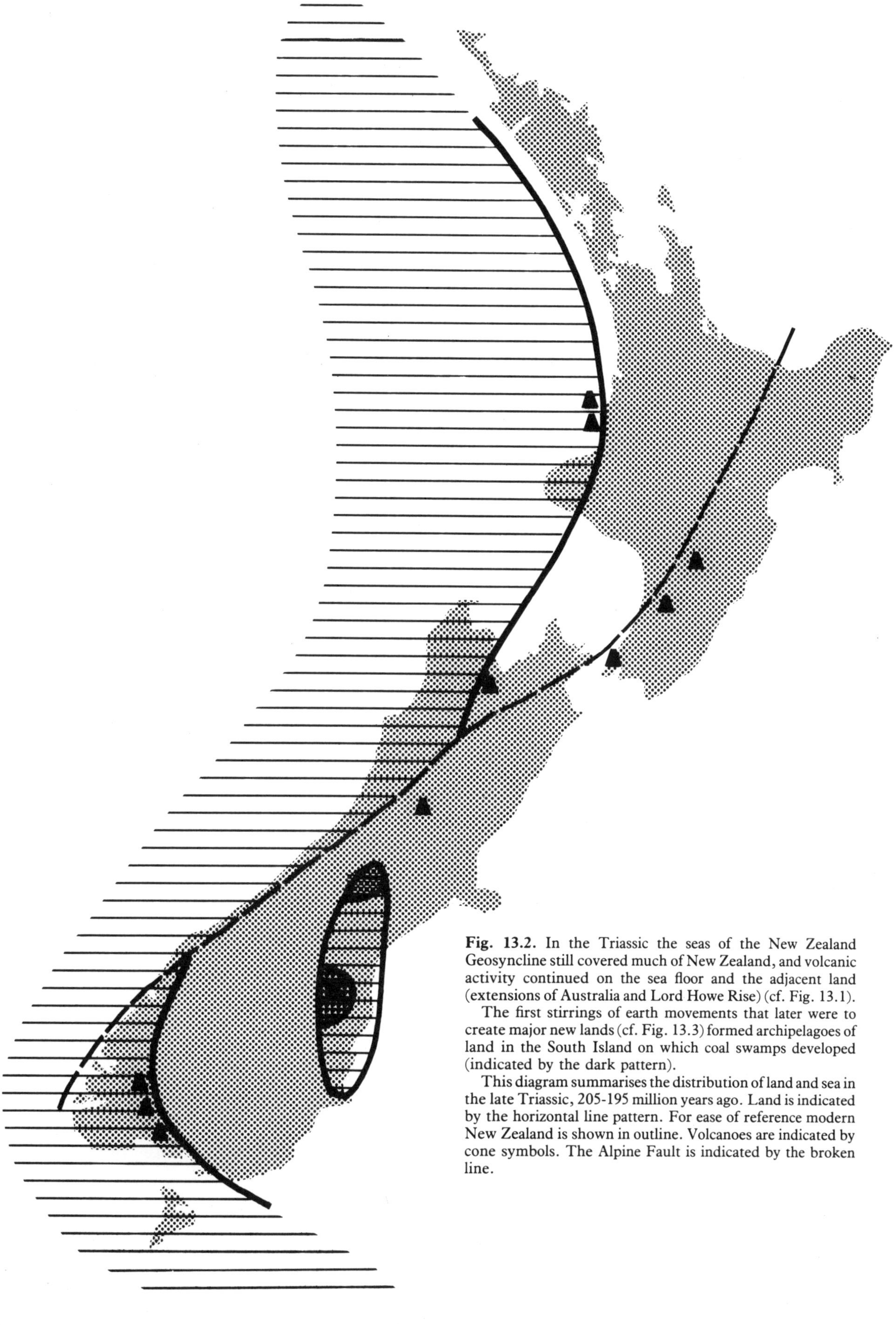

**Fig. 13.2.** In the Triassic the seas of the New Zealand Geosyncline still covered much of New Zealand, and volcanic activity continued on the sea floor and the adjacent land (extensions of Australia and Lord Howe Rise) (cf. Fig. 13.1).

The first stirrings of earth movements that later were to create major new lands (cf. Fig. 13.3) formed archipelagoes of land in the South Island on which coal swamps developed (indicated by the dark pattern).

This diagram summarises the distribution of land and sea in the late Triassic, 205-195 million years ago. Land is indicated by the horizontal line pattern. For ease of reference modern New Zealand is shown in outline. Volcanoes are indicated by cone symbols. The Alpine Fault is indicated by the broken line.

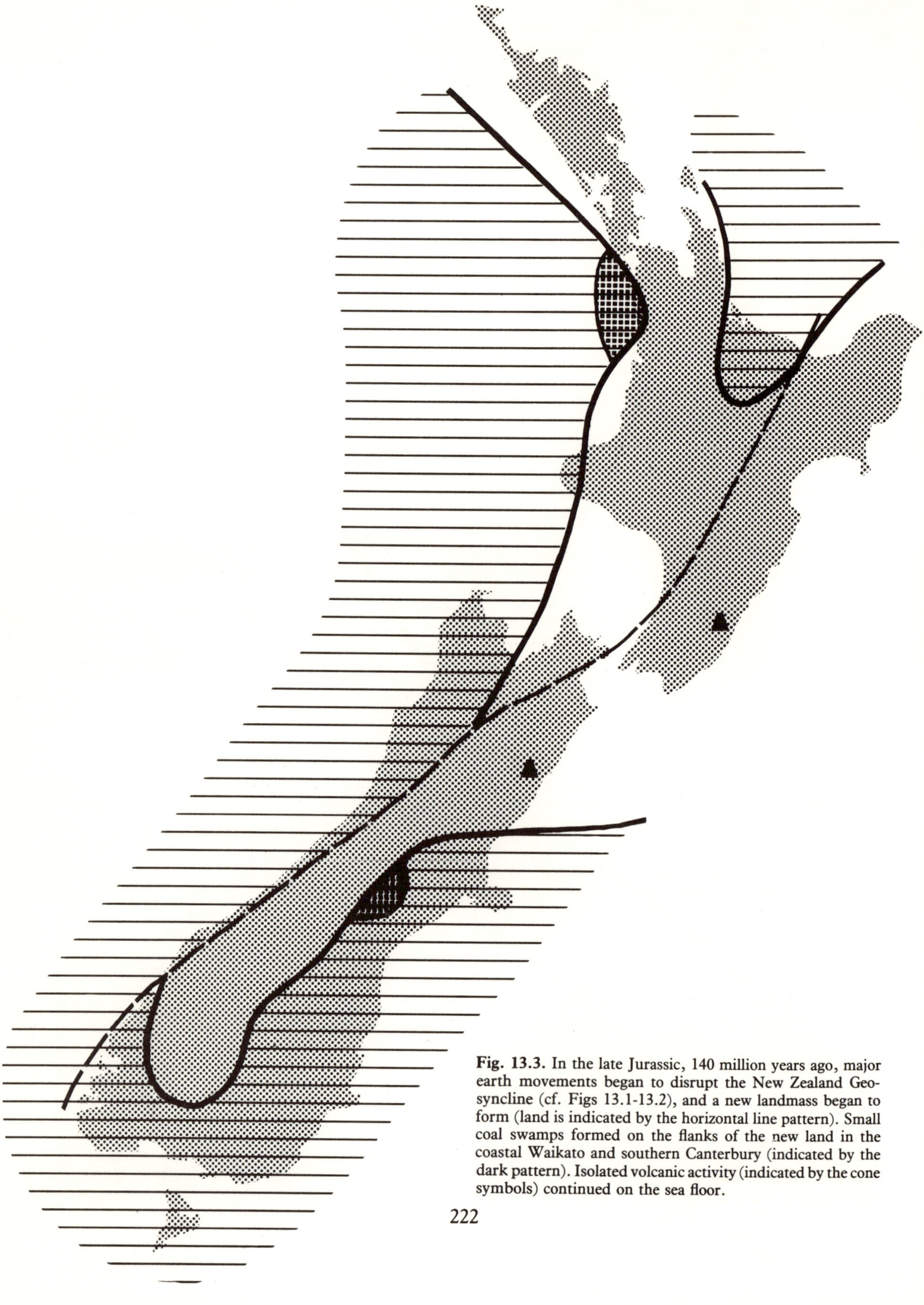

**Fig. 13.3.** In the late Jurassic, 140 million years ago, major earth movements began to disrupt the New Zealand Geosyncline (cf. Figs 13.1-13.2), and a new landmass began to form (land is indicated by the horizontal line pattern). Small coal swamps formed on the flanks of the new land in the coastal Waikato and southern Canterbury (indicated by the dark pattern). Isolated volcanic activity (indicated by the cone symbols) continued on the sea floor.

of the Waikato are full of large boulders, gravel, tree trunks, etc. (Fig. 13.6). If these same rocks are traced inland however—towards Hamilton and Te Awamutu—the large-sized materials (boulders, gravel, logs, etc.) are soon left behind and the materials making up the rocks get finer, i.e. silts and muds progressively appear. Thus it is quite clear that the material was coming from the west, off-shore from the present New Zealand coast (Fig. 13.7).

It does not necessarily follow, however, that everything came from as far away as Australia and the Lord Howe Rise. In times past, a large chunk of South Island rocks probably extended as a belt parallel to the present west coast of the North Island, and may have also been a source of erosion debris.

This old landmass has been traced with the help of oil prospecting wells put down into the sea bed in the North Taranaki Bight, South Taranaki Bight and Cook Strait (Fig. 13.8). Together these have provided us with enough information to be able to trace the rocks of north-west Nelson northwards out into the Tasman Sea. How far the rocks extend westwards is anyone's guess at the moment, but nonetheless it is quite clear that the West Wellington, Taranaki and Waikato coasts were once fringed by a belt of old rocks and that they have since been eaten back by erosion, and submerged beneath the present continental shelf.

**Fig. 13.4.** These rocks at the mouth of the Marokopa River, on the Waikato coast, were once silts and sands lying on the continental shelf of the New Zealand Geosyncline in the late Triassic, some 205-195 million years ago (cf. Fig. 13.2). Now they have been hardened and compacted to form beds of siltstone and sandstone and have been tilted by later earth movements.

Eastwards of the rocks that were laid down on the continental shelf of the New Zealand Geosyncline we can often find, particularly in the northern half of the North Island, a second belt of rocks, characterised by a great richness of volcanic material. These sediments had originally been laid down on the outer parts of the shelf or on the continental slope, inclining down into the abyssal depths. Although a lot of volcanic material is also present in the shelf sediments, its presence is often masked by the great quantities of other sedimentary material (gravel, sand, silt). However, because of the presence of light crystal material and trapped air and gas bubbles, volcanic material can be transported further off-shore than many other sediments. Thus the proportion of volcanic material steadily increases as we move eastwards in the Hamilton region, for example, indicating that all the gravel, sand, silt etc. had been left behind in regions closer to the old coastline and that only lighter materials (in this instance volcanic materials) were transported further off-shore.

Still further eastwards we encounter a third type of rock. This rock consists of thin layers of dark argillite and thick layers of grey sandstone (Fig. 13.9). Often the two rocks are together referred to as "greywacke", although strictly speaking the word (from the German *Grauwacke* = grey stone) should be only applied to the sandstone. The argillites are often very dark and coal-like in appearance (and have been mistaken for coal) and were originally fine muds, rich in mashed-up plant material, which, when converted to carbon by the rock-making processes, is responsible for the black colour. The sandstones, on the other hand, are dove-grey in colour when fresh and very tough. A split face sparkles with tiny crystals. These greywacke sandstones are very rich in quartz (silica or silicon dioxide)—which show up in fresh rocks as tiny clear diamond-like crystals—and feldspar (sodium and/or calcium aluminium silicate)—which can be seen as whitish crystals in the fresh rock. A varying amount of volcanic material is also present, and as this is rich in iron, it "rusts" (or "weathers") on exposure to the elements and so the "weathered" rock has a "rusty" appearance.

The argillite and greywacke sandstones make up much of the main ranges of both North and South Islands and are thought to underlie many of the younger rocks elsewhere (Figs 13.10-13.13).

For many years it was thought that the argillites and greywacke sandstones had been laid down even further off-shore in the New Zealand Geosyncline—at the bottom of the continental

*Fig. 13.5*

*Fig. 13.5*

**Fig. 13.5.** The light-coloured beds in this outcrop were originally volcanic ash layers ("tuff") laid down on the sea floor, derived from eruptions on the adjacent land and volcanic islands (cf. Figs 13.1-13.3). Later settling and compacting movements have distorted and displaced the originally horizontal ash beds. The upper photo shows faults displacing the beds. The lower photo shows small-scale buckling of a single ash bed.

Late Jurassic, Kawhia Harbour.

**Fig. 13.6.** Extensive beds of boulders and gravel occur on the Waikato coast, near Marokopa. These were derived from a land lying to the west, where there is now the Tasman Sea.

Before the opening of the Tasman, the Lord Howe Rise and Australia were immediately westwards of New Zealand and shed sedimentary material into the New Zealand Geosyncline (Figs 13.1-13.3). The granite gravel illustrated in this photo is of late Triassic age (about 205 million years ago) and had such a westwards source—although a slice of South Island rocks also extended off-shore at this time (cf. Fig. 13.8) and probably provided some sediments.

slope or in the abyssal depths of the ocean (Fig. 5.5). Although this explanation may still apply in some instances, modern research has favoured the idea that the New Zealand greywacke rocks (often shown on geological maps as "Torlesse", a name coined from Mount Torlesse in the Canterbury foothills of the Southern Alps) have come from a source separate from that of the other rocks of the New Zealand Geosyncline. Whereas the other rocks had come from the west—from eastern Australia, the Lord Howe Rise and a western extension of the old rocks of New Zealand (now eroded)—the greywackes are thought to have come from Marie Byrd Land, Western Antarctica. This part of Antarctica has extensive outcrops of granitic rocks (largely composed of quartz and feldspar), the erosion of which would account for the richness of the New Zealand greywackes in these minerals.

If we squeeze New Zealand back into its original position, between Australia and Antarctica (Fig. 13.14), we can see how this interpretation has come about, although at first sight it sounds impossible. Greywacke rocks similar to those of mainland New Zealand occur in the vicinity of the Bounty Islands, which lie at the eastern edge of the Campbell Plateau, the segment of continental crust that extends south-east from New Zealand (Fig. 18.1). Reconstructions of the New Zealand region (Fig. 13.14)

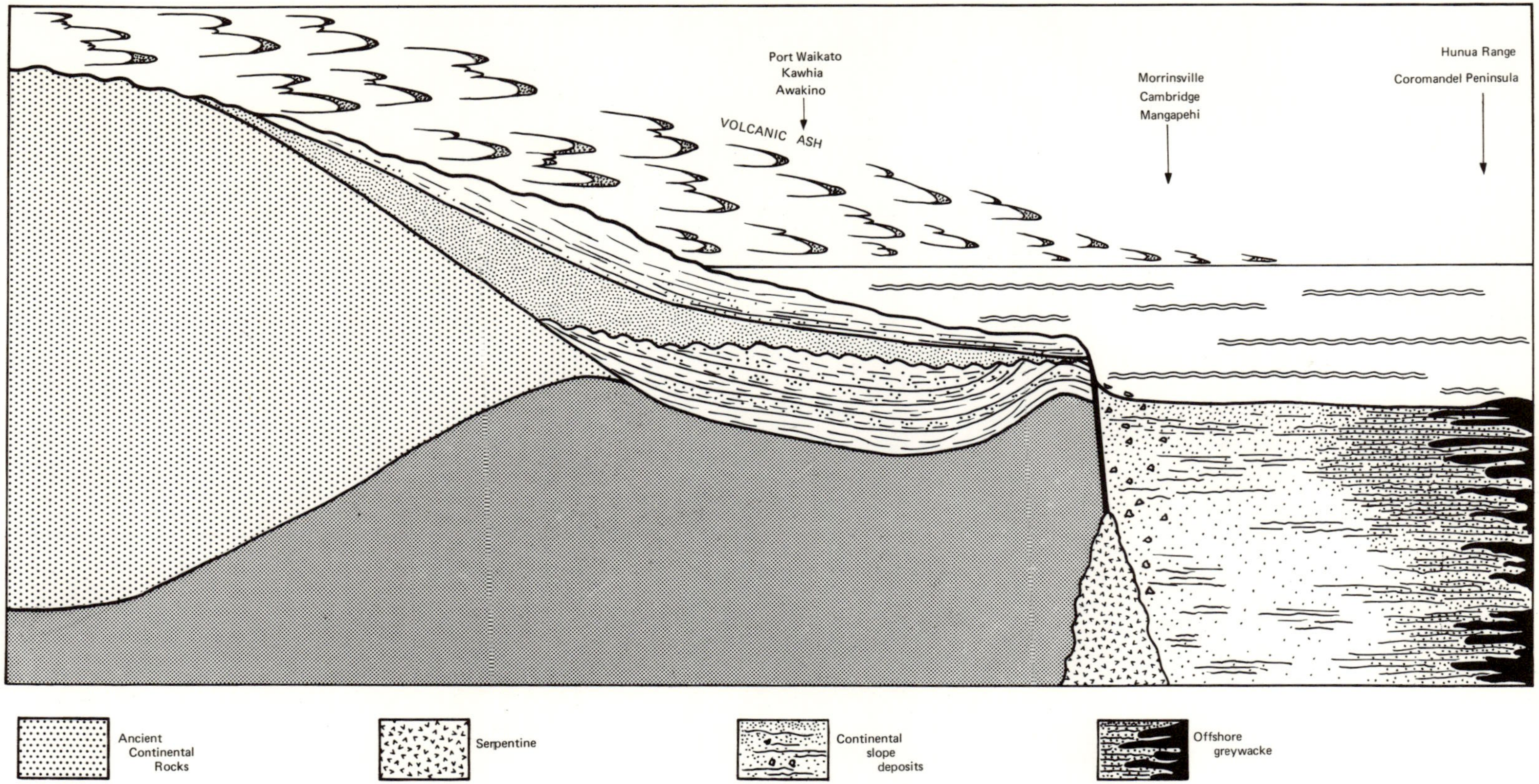


**Fig. 13.7.** During late Triassic and Jurassic time (205-136 million years ago), that part of the New Zealand Geosycline traversing what is now southern Auckland and the Waikato received sediment and volcanic material from the west, derived from the ancient continental rocks of the Lord Howe Rise and Australia (although a slice of old South Island rocks may have also extended off-shore, cf. Fig. 13.8).

Today as we pass eastwards away from the Waikato-south Auckland coast we encounter rocks that at the time of the New Zealand Geosyncline were laid on the inner and outer parts of the continental shelf, and in the deeper water beyond.

The edge of the old continental shelf is marked today by the Waipa Fault, that extends as a north-south line on the eastern side of the coastal hills lying between Ngaruawahia and Raglan.

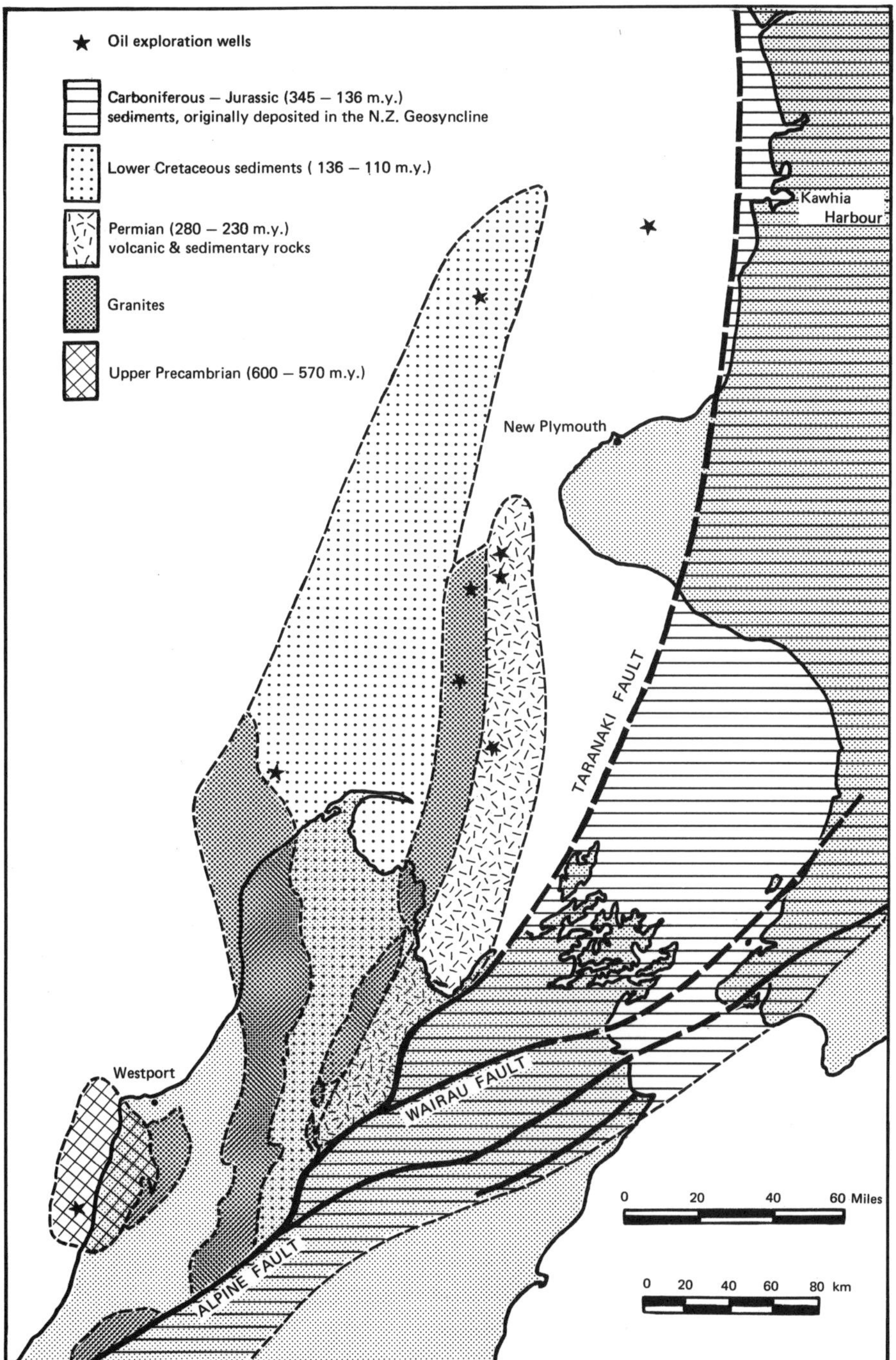


**Fig. 13.8.** Oil exploration drill holes in the sea floor off northern Taranaki and in the western approaches to Cook Strait have revealed the presence of belts of South Island rocks, extending northwards to the region off-shore from the Waikato coast.

These rocks may have formed a fringe to the landmasses of the Lord Howe Rise and Australia, that edged New Zealand to the west during the geological ages before creation of the Tasman Sea.

*Fig. 13.9*

**Fig. 13.9.** Alternating beds of Torlesse argillite and sandstone ("greywacke"). Mount Cook National Park.

*Fig. 13.10*

**Fig. 13.10.** The Torlesse greywacke rocks make up most of the main ranges of both North and South Islands of New Zealand, and are believed to underlie many of the younger rocks to either side of the ranges.

Earth movements during the Rangitata Orogeny folded the Torlesse greywackes into intricate patterns. Enormous folds can be distinguished, such as this one on the western face of Häckel Peak (2940 m, 9649 ft) in the Malte Brun Range, Southern Alps. The visible face extends vertically for some 1050 m (3500 ft).

**Fig. 13.11.** In other areas of the Southern Alps the tops of folds have been eroded off, leaving the layers of rock standing up on end, like a pack of cards. Mt Cook National Park.

**Figs 13.12, 13.13.** Although at least the major folding of the Torlesse rocks was accomplished during the Rangitata Orogeny some of the minor complicated folds may have formed during deposition, when masses of semi-consolidated layers of mud and sand became unstable (perhaps as a result of an earthquake, or current action accompanying a storm) and slumped, producing an underwater landslide.
Fig. 13.12 shows folding (shaped like an "S" on its side), at Pencarrow, eastern entrance to Wellington Harbour. The piece of driftwood to the right foreground is about 1 m (3 ft) long.
Fig. 13.13 shows complicated folding in the Ngauranga Gorge, Wellington.

*Fig. 13.12*

*Fig. 13.13*

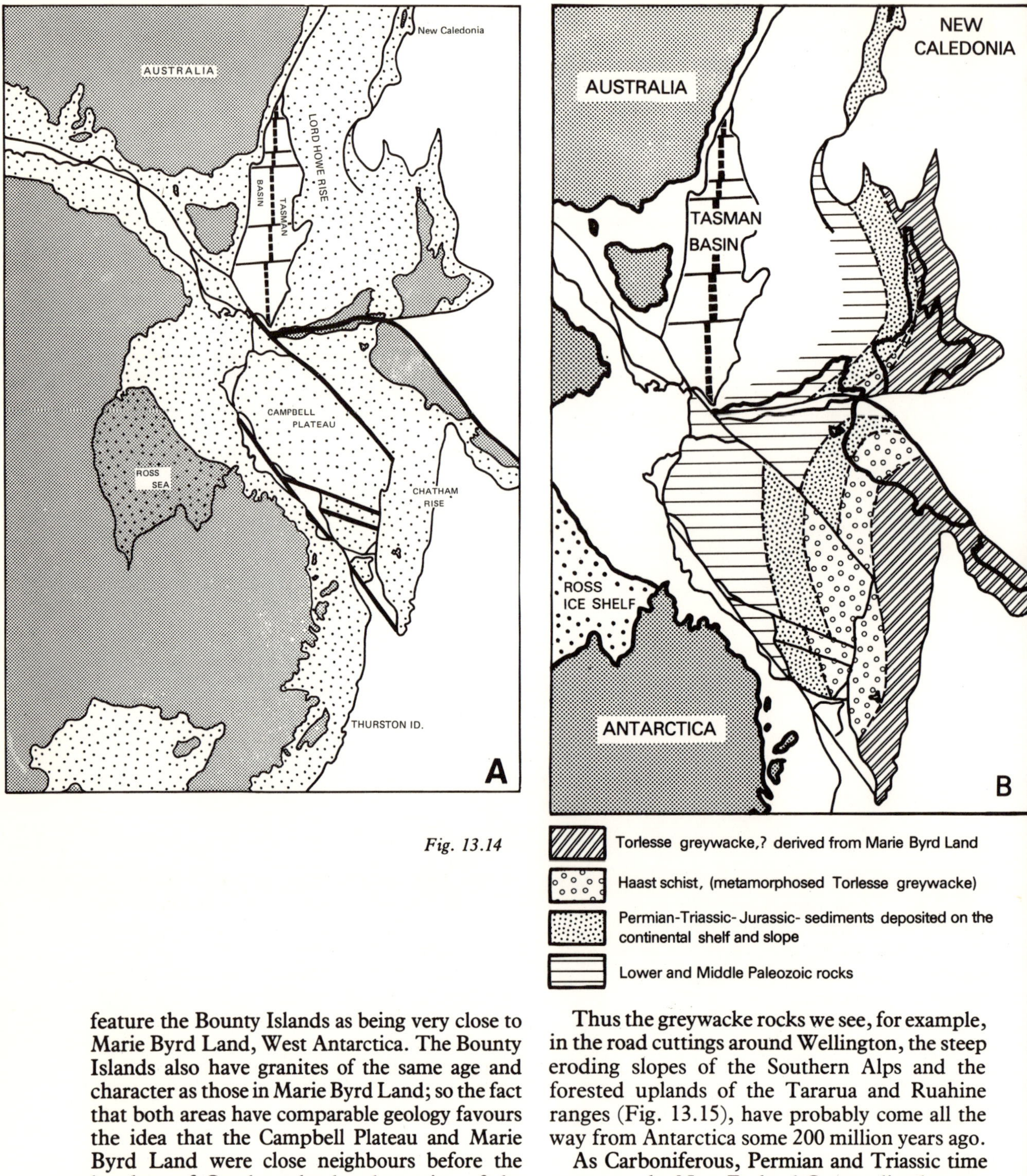


*Fig. 13.14*

feature the Bounty Islands as being very close to Marie Byrd Land, West Antarctica. The Bounty Islands also have granites of the same age and character as those in Marie Byrd Land; so the fact that both areas have comparable geology favours the idea that the Campbell Plateau and Marie Byrd Land were close neighbours before the breakup of Gondwanaland and creation of the intervening South-western Pacific Ocean basin.

Thus the greywacke rocks we see, for example, in the road cuttings around Wellington, the steep eroding slopes of the Southern Alps and the forested uplands of the Tararua and Ruahine ranges (Fig. 13.15), have probably come all the way from Antarctica some 200 million years ago.

As Carboniferous, Permian and Triassic time went on, the New Zealand Geosyncline became gradually filled with material—coming from

**Fig. 13.14.** It is believed that the Torlesse greywacke rocks of New Zealand came originally from the vicinity of Marie Byrd Land, West Antarctica.

The geology of the various sub-Antarctic islands (Campbell, Auckland, Snares, Bounty, Antipodes etc.), combined with information on sea-floor drilling, dredging and magnetism, indicates that the geology of the South Island of New Zealand continues to the edge of the Campbell Plateau. The Southern Ocean, now flanking the Campbell Plateau, began to open up some 80 million years ago. Thus today the New Zealand landmass is separated from the presumed source of the greywackes by a broad stretch of ocean.

This reconstruction shows the Chatham Rise and Campbell Plateau set against Marie Byrd Land, the presumed source of the Torlesse rocks and their metamorphosed equivalents, the Haast schists.

To arrive at the reconstruction, representing the South-west Pacific about 80 million years ago, the sea-floor spreading process has been reversed, using ocean magnetic patterns as a guide, and the magnetics and bathymetry have been matched by computer. On the other hand, the New Zealand block has been regarded as a flexible unit and various distortions have been introduced. The parts of New Zealand on either side of the Alpine Fault have been reversed back along the fault and rotated. A large shift of 1000 km (630 miles) is proposed on the Alpine Fault, together with an earlier (pre-80 million years) shift in the opposite direction, to link the displaced ends of the New Zealand Geosyncline.

The Bounty Trough, between the Chatham Rise and the Campbell Plateau, has been closed and so also has the Solander Trough, south of Fiordland. On the other side of New Zealand, the Tasman Basin and New Caledonia Basin have been closed.

The pie-segment-shaped gap between the two halves of New Zealand is thought to represent the amount of Cretaceous-Cenozoic subduction. However, this may prove to be a geometric artefact and it is likely that it would disappear if possible expansion of the earth is taken into consideration (i.e. if the reconstruction was plotted on a smaller globe, e.g. 80 percent of its present diameter, cf. Fig. 15.1).

Although most geologists now accept that New Zealand, New Caledonia, Australia and Antarctica probably fitted together something like this there is still a difference of opinion regarding the geometry of the fit, the amount of distortion undergone by New Zealand and the amount and timing of movement along the Alpine Fault (cf. Fig. 18.6).

Diagram A indicates the fit along the 2000 m (7000 ft) submarine contour. Landmasses are indicated by the dark pattern, continental shelves by the large dots. Major faults are shown by the heavy lines.

Sea-floor spreading is pictured as actively taking place on either side of an oceanic ridge in the embryonic Tasman Sea.

Diagram B shows how it is thought the rocks of the New Zealand land continue under the sea along the Lord Howe Rise and Norfolk Ridge and across the Campbell Plateau.

north-west, west and south-east (Figs 13.1-13.3).

By late Triassic time, about 200 million years ago, there were signs that parts of the geosyncline were starting to fill up completely as at times islands were appearing, although they were soon removed by erosion.

Other processes were occurring out of sight in the depths of the huge sediment pile filling the geosyncline.

### Inside the New Zealand Geosyncline

The sediments accumulating in the New Zealand Geosyncline were heavy with water, and as their thickness increased, so the underlying rocks of the earth's crust buckled downwards with the weight. This downbuckling in turn made more room for further sediments to accumulate in the geosyncline. The process continued throughout the Permian and Triassic, and into at least the early half of the Jurassic period, together spanning about 129 million years. The total amount of sediment accumulated during this time is difficult to estimate, as it varied from place to place and period to period. However, *at least* 27,000 m (90,000 ft) accumulated and this represents a rate of formation of about 0.25 mm (0.01 in) per year. This rate is, of course, only an average for the whole time. In some regions the rate may have gone up to several millimetres per year for a while and then slackened off.

**Fig. 13.15.** The sediments that accumulated on the continental shelf of the western edge of the New Zealand Geosyncline are now found as belts of rocks in both North and South Islands. They are folded into a large trough-like structure (called a "syncline").

The formerly continuous belts have been disrupted by the horizontal movement of 450 km (280 miles) along the Alpine Fault.

In both islands the Torlesse greywacke rocks lie east of the shelf rocks, but in some areas (e.g. Kaimanawa Ranges in North Island; Otago and S. Canterbury in South Island) heat and pressure have changed them and other rocks into schists and gneisses ("metamorphic" rocks).

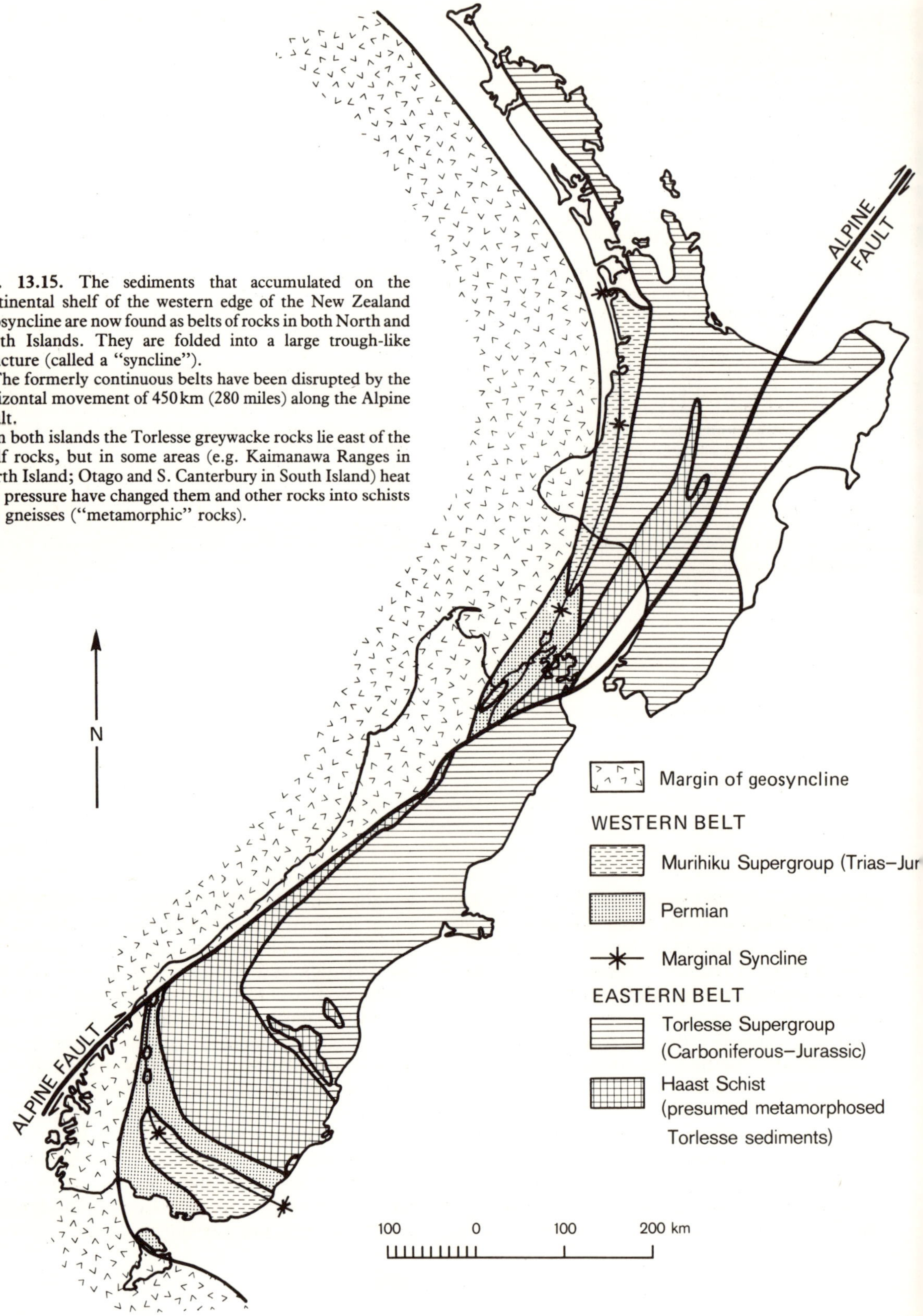

**Fig. 13.16.** Joints in Torlesse greywacke sandstone. Mt Cook National Park.

**Fig. 13.17.** Quartz veins in Torlesse greywacke sandstone. Mt Cook National Park.

*Fig. 13.16*

*Fig. 13.17*

Downbuckling of the earth's crust under the weight of the sediments in the New Zealand Geosyncline carried the sediments filling the lower levels of the geosyncline deeper and deeper into the earth's crust.

As we have seen (p. 167), temperature goes up about 2°-3°C (4-5°F) for every 100 m (300 ft) we go down into the earth's crust. This means that the lower sediments in a geosyncline are carried down into regions of progressively higher temperatures. The higher temperatures begin to dry the sediments out, by driving water up into higher, cooler, regions of the geosyncline. As the rocks get hotter and hotter, so the temperature of the water being driven off increases. As the water gets hotter, so its reactivity increases and as it migrates through the sediments it dissolves various constituents from the sediments—lime, silica, iron, manganese etc.

Apart from the effects of heating, the sediments filling the lower levels of the geosyncline are buried the deepest. Burial by 3 km (2 miles) of overlying sediment means that the rock is subjected to pressures of the order of 800 tonnes per 0.09 square metre (800 tons per square foot) and to temperatures in the region of 100°C (212°F).

Burial to various depths substantially alters the original sediment, and by having a look at the resulting rock we can arrive at an estimate of the depth to which it has been buried. These estimates are based on mineral and rock research, supplemented by experiments in which sediments are subjected to pressures and temperatures likely to be found at various depths in the earth's crust. This experimental work is of course limited in scope, as there are no machines capable of reproducing the extremely high pressures and temperatures existing very deep in the earth's crust. Nonetheless, conditions at depths of a few kilometres can be created successfully in the laboratory, and changes produced in various types of sediment studied and compared with similar changes in the natural rocks.

As the soft sediments originally deposited on the sea bed are buried deeper and deeper, more and more water is squeezed out from between the individual mineral particles. The particles (of sand, mud, etc.) are pushed together and the sediment is compacted into a rock. Various

*Fig. 13.18*

**Fig. 13.18.** A specimen of schist from Central Otago. The specimen is 260 mm (10¼ in) wide.

**Fig. 13.19.** An outcrop of schist in the Shotover River, Central Otago, showing kink-folds produced by intense contraction along the layering of the schist (a result of lateral compression of the rocks).

**Fig. 13.20.** Schists underlie some of the most spectacular country of the South Island and their mineral layering often dominates the landscape. This photo is of schists (foreground) at the northern end of the Olivine Range, N.W. Otago. The Cascade River flows from left to right in the middle distance.

*Fig. 13.19*

changes appear in the rock, depending on how deeply it has been buried. After burial of about 1.5 km (1 mile) the rocks are not quite hard enough to need blasting; rippers, bulldozers and carryalls can tackle most of them. After burial to a depth of about 4 km (2½ miles), the rocks are compacted to such an extent that blasting is necessary to excavate them. Some of these rocks can be used as a poor type of road metal. After burial to a depth of about 12 km (7½ miles), most rocks are hard enough to be used as concrete aggregate and road metal. After about 16 km (10 miles) of burial, the intense pressure and high temperatures tend to completely break down

*Fig. 13.20*

the original sediment and a new type of rock (a "metamorphic" rock), with new minerals, is formed.

The intense pressures found deep in the earth's crust produce cracks and fissures in the rocks. These are called "joints" and they provide lines of weakness along which the rock will tend to break. Quarrymen and stonemasons make use of such joints when splitting slabs of rock (Fig. 13.16).

After burial to a depth of some 3 km (2 miles), the rock is jointed in a rectangular pattern, parallel to the layers (or strata) of the rock. But after burial to a depth of about 6.5 km (4 miles), the rocks are jointed in a very complex and closely spaced pattern, both parallel and oblique to the rock layers.

As has been mentioned, water is progressively squeezed out from the sediments as they are buried deeper and deeper. At comparatively shallow depths the water is probably not very hot and any reaction occurring between the water and surrounding rocks mainly affects the lime minerals, the most sensitive rock minerals. These are dissolved by water and deposited again in the joints, which tend to be the main migration routes for the water. The joints thus become filled with new minerals ("secondary minerals") and these form "veins" in the rock. The lime minerals that are formed at comparatively shallow depths, between 8-16 km (5-10 miles), all belong to the group of hydrated calcium-aluminium-silicates known as zeolites. They are usually white and soft and can be readily scratched with a knife.

Below a depth of about 16 km (10 miles), the water circulating through the rocks is probably very hot and very reactive. Silica, which is very common in all sediments, is dissolved from the rocks at this depth and deposited along the joints to form quartz veins (Fig. 13.17). These veins, like the zeolite veins, are white, but because quartz is a harder material, they cannot be scratched with a knife.

If the rocks in the lower levels of the geosyncline are buried even more deeply, perhaps to a depth of 24-32 km (15-20 miles), the increased temperature and pressure changes or "metamorphoses" the rocks. In this process many of the original minerals break down into their constituent atoms and these atoms are re-grouped to form new mineral compounds ("metamorphic" minerals), stable under conditions of high temperature and pressure.

At the same time, the high temperatures tend to soften the rock and it tends to behave like a viscous liquid. In this condition the rock can flow in response to local stresses and strains within the crust and as it does so it can fold itself into fantastic shapes. The rock formed by this first phase of metamorphism is called a schist (Figs 13.18-13.20).

Even deeper burial produces further regrouping of atoms and appearance of new minerals; the rock formed is a gneiss. Some of these new minerals can grow to a large size in the rock. For example, garnet crystals (iron aluminum silicate) formed in this way can each measure 63 × 63 × 50 mm (2.5 × 2.5 × 2 in) (Fig. 13.21). If burial of the rock continues, wholesale melting takes place and granite, an "igneous" rock (i.e. once-molten) is formed.

Rocks formed as a result of the deep burial of sediments filling the New Zealand Geosyncline cover large areas of the South Island: schist in the Marlborough, Canterbury and Otago regions and gneisses and granites in Fiordland and Westland. An isolated patch occurs in the Kaimanawa Ranges, in the North Island (Fig. 13.22).

The sediments filling the New Zealand Geosyncline, like sediments off our modern coasts, contained minute amounts of metallic elements (e.g. gold, mercury, antimony and tungsten), more or less evenly dispersed through them. These metallic elements had been eroded from the rocks, especially the granites and volcanic rocks, of the old landmass.

It has been mentioned already that, as the sediments in the geosyncline were pushed deeper and deeper into the crust, so the water originally present in them was driven upwards towards the higher, cooler, levels of the geosyncline. Along with this water went the metallic elements. Depending on the depth of burial this migrating water was able to react with the sediments and remove various minerals: zeolites at depths of less than 16 km (10 miles), quartz below 16 km.

At the same time as quartz is removed, so are many of the metallic elements in the rocks. This process concentrates the metals and in the higher, cooler, parts of the geosyncline the percolating water leaves behind metallic deposits in cavities and fissures, often, in the case of gold, associated with quartz. Thus many of the rocks that were originally deep in the New Zealand Geosyncline contain a variety of metals—for example gold in the greywackes and schists, antimony and tungsten in the schist and gneisses—or have been the source of ores that have been eroded out and concentrated in rivers and gravels (e.g. alluvial gold and scheelite). The search for these metals was the basis of the gold rushes of last century and various other mining operations that have continued up to the present day.

**Mountain building**

There is obviously a limit to the life of a geosyncline, as sinking of crust and accumulation of vast quantities of sediment cannot go on for

**Fig. 13.21.** A gneiss from Fiordland with large garnet crystals developed. The specimen is 190 mm (7½ in) wide.

ever. Sooner or later forces within the earth will combine to restore the balance and the geosyncline and its contained sediments will be pushed up above the sea (Fig. 13.23).

This uplifting process is helped by the various internal changes within the geosyncline. Thus it is thought that the drying-out of the sediments in the lower levels of the New Zealand Geosyncline, and their alteration by heat and pressure (described above), will tend to lighten the rocks, and a light area or "root" will form at the base of the geosyncline. This light root will have a natural tendency to rise towards the earth's surface. Eventually, as more and more of the rocks are lightened in this way, the forces acting to push the geosyncline deeper are neutralised and the light root will tend to rise under the remainder of the geosyncline and will help push the sediment-filled trough above the sea to form a new mountainous landmass.

In the past geologists thought that the buoyant rise of sedimentary material played a major part in the uplift of geosynclines. It is now known, however, that throughout the geological ages geosynclines have been sited at plate boundaries and that their evolution is intimately linked to individual plate movements.

The rise and fall of geosynclines is therefore now seen as part of the overall plate tectonic story and the development of internal buoyancy is relegated to a relatively minor role.

When viewed regionally, in the context of the entire South-west Pacific, the New Zealand Geosyncline is visualised as having originally developed along the boundary of the ancestors of the modern Indian-Australian, Pacific and Antarctic plates.

The history of this boundary, over some 1000 km (700 miles) in length and extending in time from Carboniferous to Jurassic (some 200 million years) was probably exceedingly complex. However, after studying the intricacies of the rocks presently exposed we can arrive at some idea, albeit highly generalised, of the various plate movements that influenced the development of the New Zealand Geosyncline.

Initially, in the Carboniferous and Permian, the New Zealand Geosyncline was situated in a gap that existed between the Australian and Antarctic continents, sitting on their respective plates. A mid-oceanic rise generated new sea floor between the two continents and subsequently a westerly-dipping subduction zone was developed along the arc now occupied by New Zealand and New Caledonia. At the site of this subduction zone the sea floor was dragged down to form a submarine trench and in Nelson and elsewhere

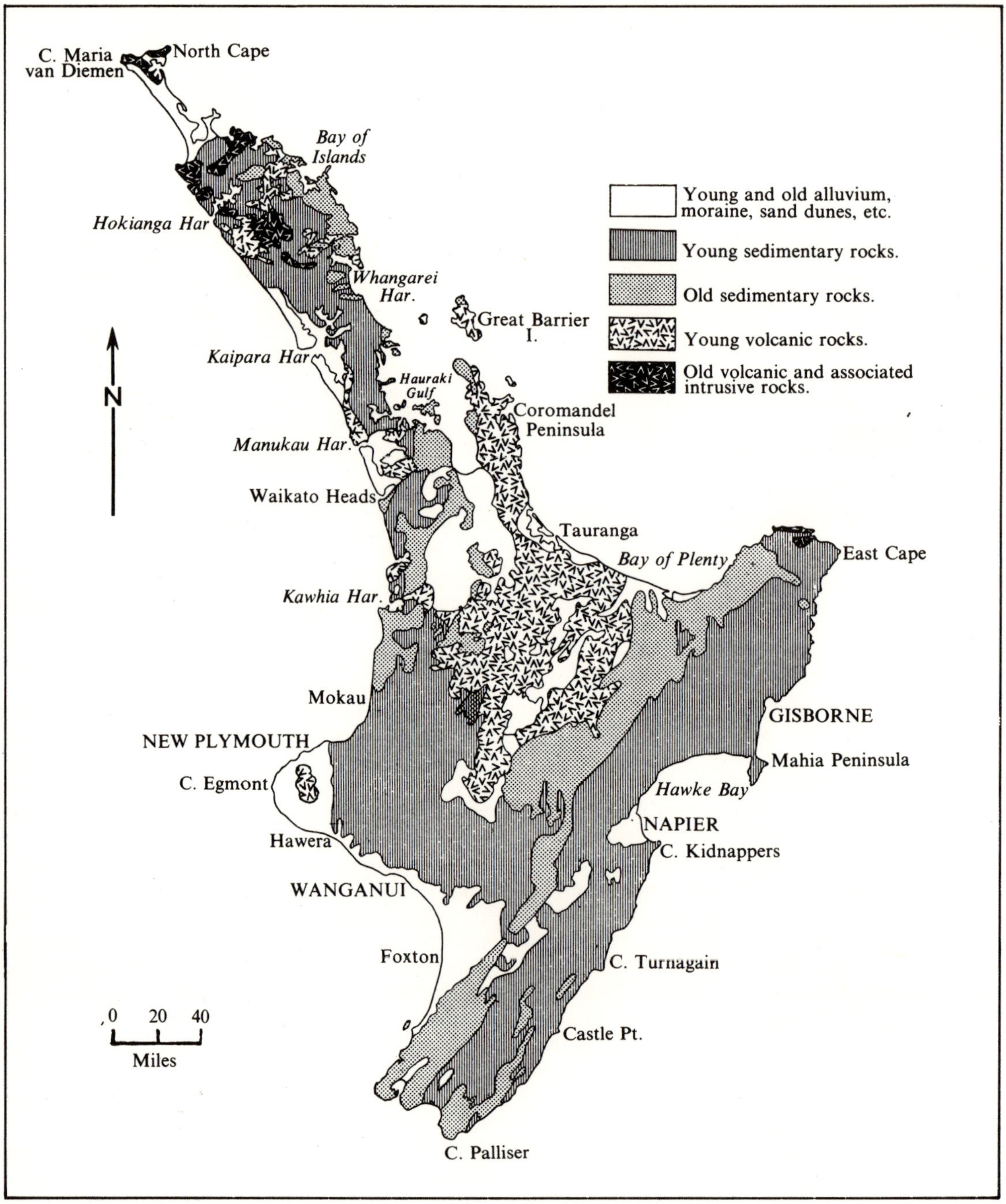


**Fig. 13.22.** Simplified geological maps of New Zealand. The rocks have been grouped into broad categories and their approximate equivalence with the Geological Time Scale is as follows:

Young alluvium etc.: Holocene to mid-Pleistocene
Young sedimentary rocks: early Pleistocene to late Cretaceous
Young volcanic rocks: Cenozoic
Old sedimentary rocks: middle Cretaceous to ? Precambrian
Metamorphic rocks: Mesozoic to ? Precambrian
Granitic rocks: Mesozoic to ? Precambrian

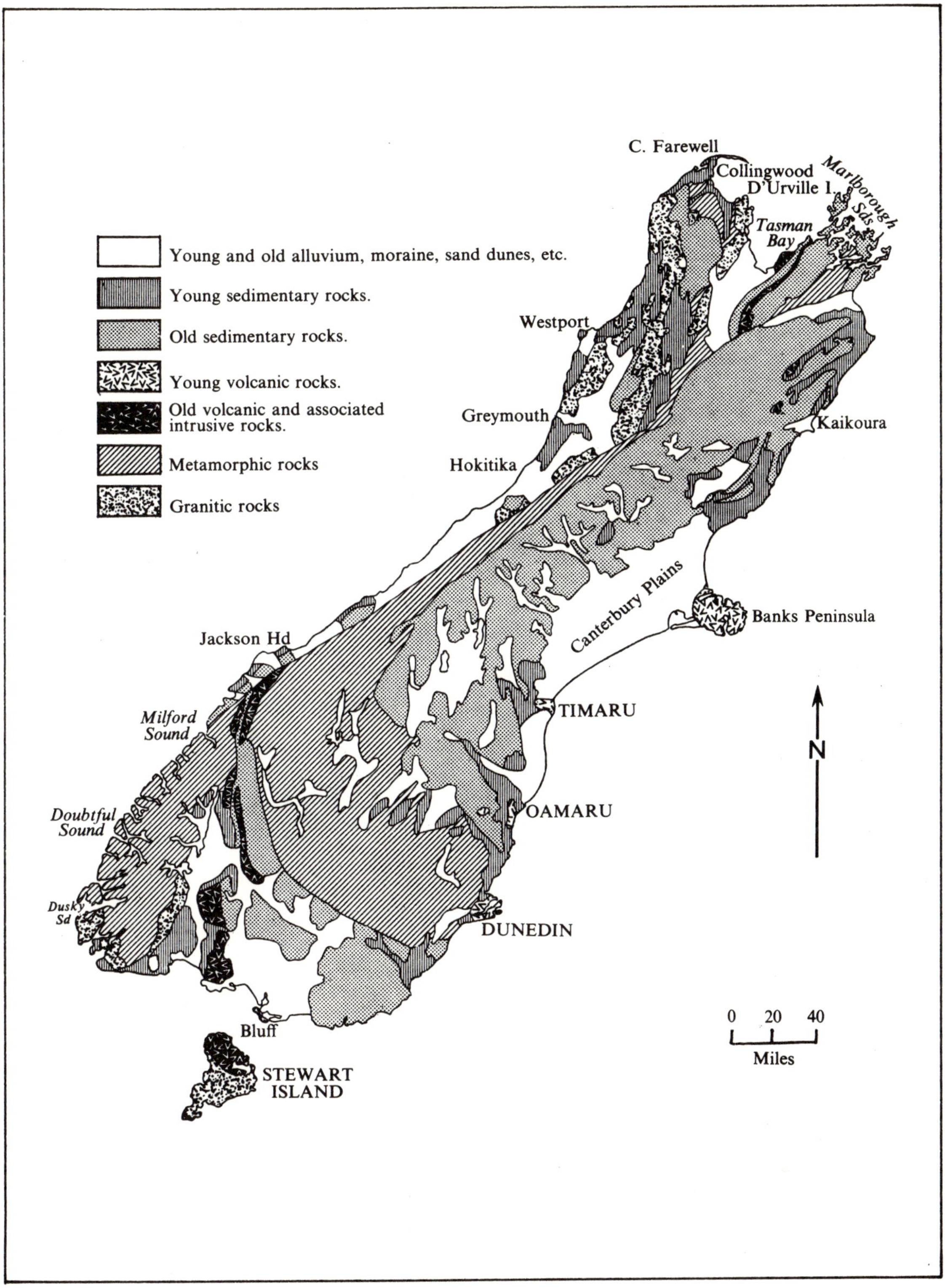
Young and old alluvium, moraine, sand dunes, etc.
Young sedimentary rocks.
Old sedimentary rocks.
Young volcanic rocks.
Old volcanic and associated intrusive rocks.
Metamorphic rocks
Granitic rocks
C. Farewell
Collingwood
D'Urville I.
Marlborough Sds
Tasman Bay
Westport
Greymouth
Hokitika
Kaikoura
Canterbury Plains
Banks Peninsula
Jackson Hd
Milford Sound
TIMARU
N
Doubtful Sound
OAMARU
Dusky Sd
DUNEDIN
Bluff
STEWART ISLAND
0 20 40
Miles

rocks have been identified that are thought to have been deposited in such an environment. As crustal melting occurred, a volcanic arc was built up parallel to the submarine trench. Thus in the Carboniferous and Permian we had all the features of a classic convergent plate margin (Fig. 8.36) present in New Zealand (Fig. 11.7). The volcanic rocks extruded at this time as part of the volcanic arc are now seen, for example, in Nelson as the "Brook Street Volcanics".

As we have seen, sediment accumulated on the fringes of both Australian and Antarctic continents and to either side of the volcanic arc. As seafloor spreading progressed and more and more

**Fig. 13.23.** These two diagrams, a continuation of those in Fig. 11.7, give some idea of the processes leading to termination of the New Zealand Geosyncline and generation of the Rangitata Orogeny.

In diagram A the Permian subduction zone has died out (see Fig. 11.7) and a new one has been initiated on its oceanic side.

Volcanic arcs and thick sedimentary accumulations occupy the site of modern New Zealand, flanked to the west (left) by the ancient continental areas of the Lord Howe Rise and Australia.

To the right (east) of diagram A sediments laid on the sea floor off Marie Byrd Land are being steadily rafted towards the left (west), as ocean floor is steadily consumed in the subduction zone. These sediments are thought to be ancestral to the modern Torlesse greywackes and Haast schists of New Zealand.

It is believed that throughout the Jurassic more and more Torlesse/Haast sediment arrived at the subduction zone. This sediment, being continental-derived material, and therefore light, could not be consumed by the subduction zone, and so piled up and collided with the island arc and associated western-derived material.

At this point in time (late Jurassic-early Cretaceous) the subduction zone died out and the two masses of sediment (from western and eastern sources) were folded and squeezed together in the paroxysm of the Rangitata Orogeny.

Diagram B gives an interpretation of the scale and nature of the movements involved.

In the later Cretaceous, some 100-80 million years ago, the ancient continental rocks of Lord Howe Rise and Australia split away, to leave New Zealand as a segment of newly-created continental rocks.

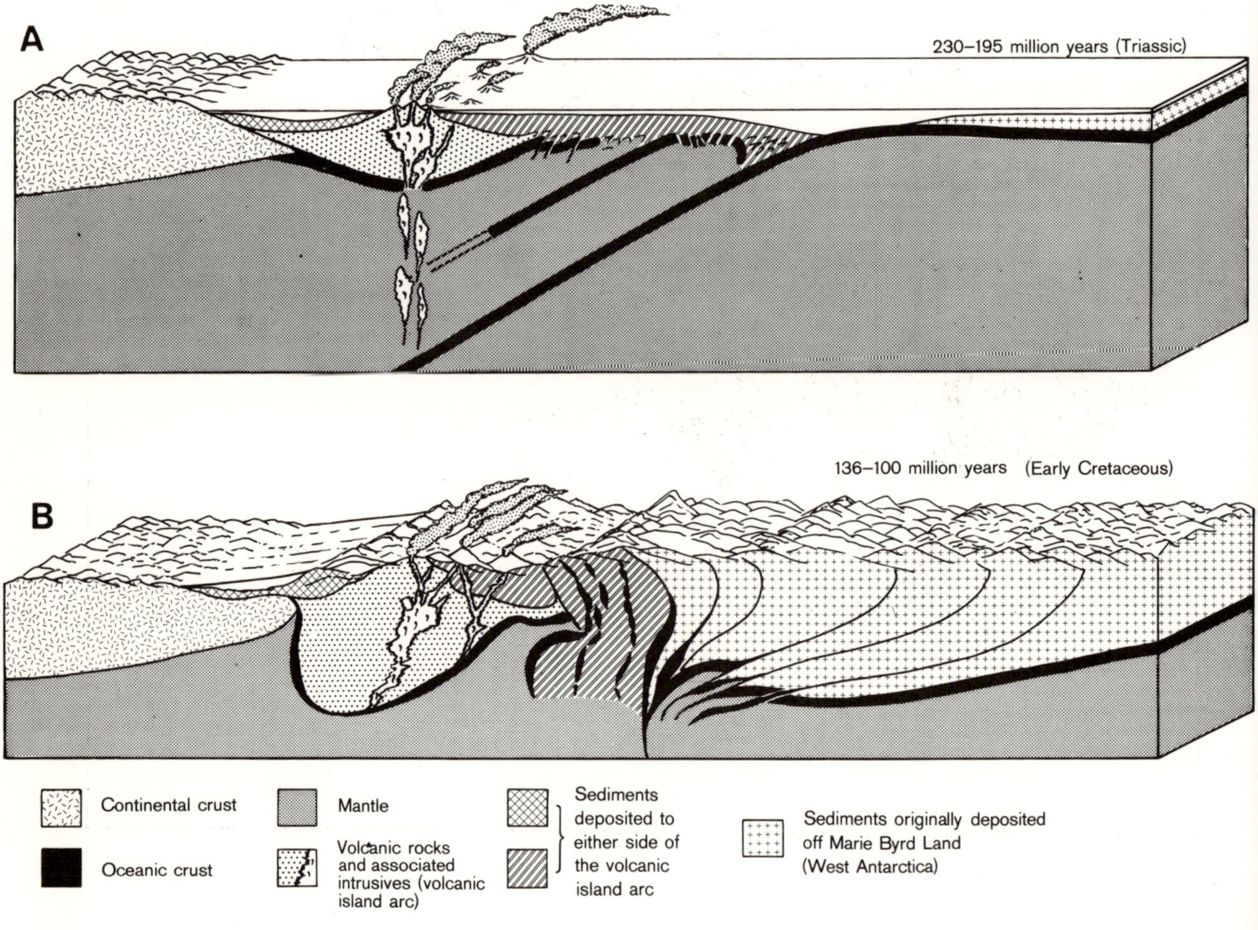

of the intervening ocean floor between Australia and Antarctica disappeared down the subduction zone, the continents and their fringing sediments were gradually rafted together. The two areas may have been separated by anywhere between 700-7000 km (450-4500 miles) at first, but were brought together at a rate of 1 to 10 cm (½-4 in) per year extending over some 70 million years. Thus in time, probably in the Triassic and Jurassic, the sediments that had originally been shed off Antarctica (and now form the New Zealand greywacke rocks) were rafted by sea-floor spreading across to the very edge of the subduction zone. There they encountered the sediments that had been derived from a westerly source (i.e. Australia—Lord Howe Rise), as well as those from the volcanic arc and submarine trench (Figs 13.23, 16.29). Being essentially sialic in composition, and therefore light, both lots of sediments were skimmed off the ocean floor as it descended into the subduction zone. The resulting clogging of the subduction zone with light sialic material, unable to be disposed of by re-melting in the mantle, caused the zone to side-step to one side, towards the ancient ocean, as seen in Fig. 13.23.

This disruption of the subduction zone is thought to have occurred in Triassic times, 195-230 million years ago, and, apart from everything else, greatly complicated the structure of the New Zealand Geosyncline. On top of all this, the collisions between the sedimentary masses produced a jumbled mass of rocks, with the greywackes, of eastern (Antarctic) origin, jammed up against rocks of western origin (Australia and ancient volcanic arc).

Also, as geological time moved on into the Jurassic, the Australian and Antarctic continents edged closer and closer together. The contents of the New Zealand Geosyncline were thus squeezed as if caught between the jaws of a giant vice. The relentless pressure rucked and buckled the sediments and pushed them up to become land, thus bringing an end to the New Zealand Geosyncline phase. The phase of major earth movements, or "orogeny", is called the Rangitata Orogeny, as the ancestral Southern Alps were formed at this time. During these mountain-building movements segments of the ancient sea floor were caught up, torn into slices and thrust into and pushed over sediments that became part of the new landmasses. This gave rise to the ophiolites we can see in a number of areas throughout New Zealand (pp. 135–143).

The landmass created by the Rangitata Orogeny was not built, as it were, overnight. Rather, the known effects of historic earthquakes suggest that it was accomplished by means of a series of small upward movements, of perhaps 3-5 m (10-15 ft) at a time, spaced out at intervals of 100 or more years. These small movements, extending over millions of years, raised the pile of sediments originally filling the geosyncline high above the sea. Thus sediments originally laid down on the sea floor, and containing fossilised seashells, can be seen in what have become hills and mountains many hundreds of metres high.

The reader will probably realise that when fossil seashells are seen on mountain tops, some distance away from the nearest sea, it does not necessarily mean that the sea once covered the land to that height, but rather that the sea was at about its present position and the mountains were pushed up from beneath the sea. The severe forces necessary to bring about this uplift tend to break and bend the pile of sediments which thus become "faulted" and "folded". In some instances the once flat sedimentary layers may become completely overturned, with the oldest uppermost, thus presenting mute testimony to the tremendous subterranean forces that have been at work.

During Permian times volcanic islands may have dotted the seas covering the New Zealand Geosyncline, but they were essentially ephemeral (Fig. 13.24). More substantial land appeared in the late Triassic, some 200 million years ago, but again it was in the form of isolated islands (although largely of sedimentary not volcanic origin), did not last very long and was soon reduced by erosion and disappeared beneath the waves. The bits and pieces of land that were formed at this time may have been related to the disruption of the subduction zone, and its side-stepping, that were occurring in the late Triassic (Fig. 13.23).

By the middle Jurassic, some 160 million years ago, substantial amounts of land had appeared, for example, in Southland (Fig. 13.24). It is at this time that we have the first clear evidence that the earth's crust, bowed down under the great weight of the pile of sediment, was starting to rebound into its original position, and marked shallowing occurred in places as widely separated as New Caledonia and the Waikato coast of New Zealand. From then onwards the sediments

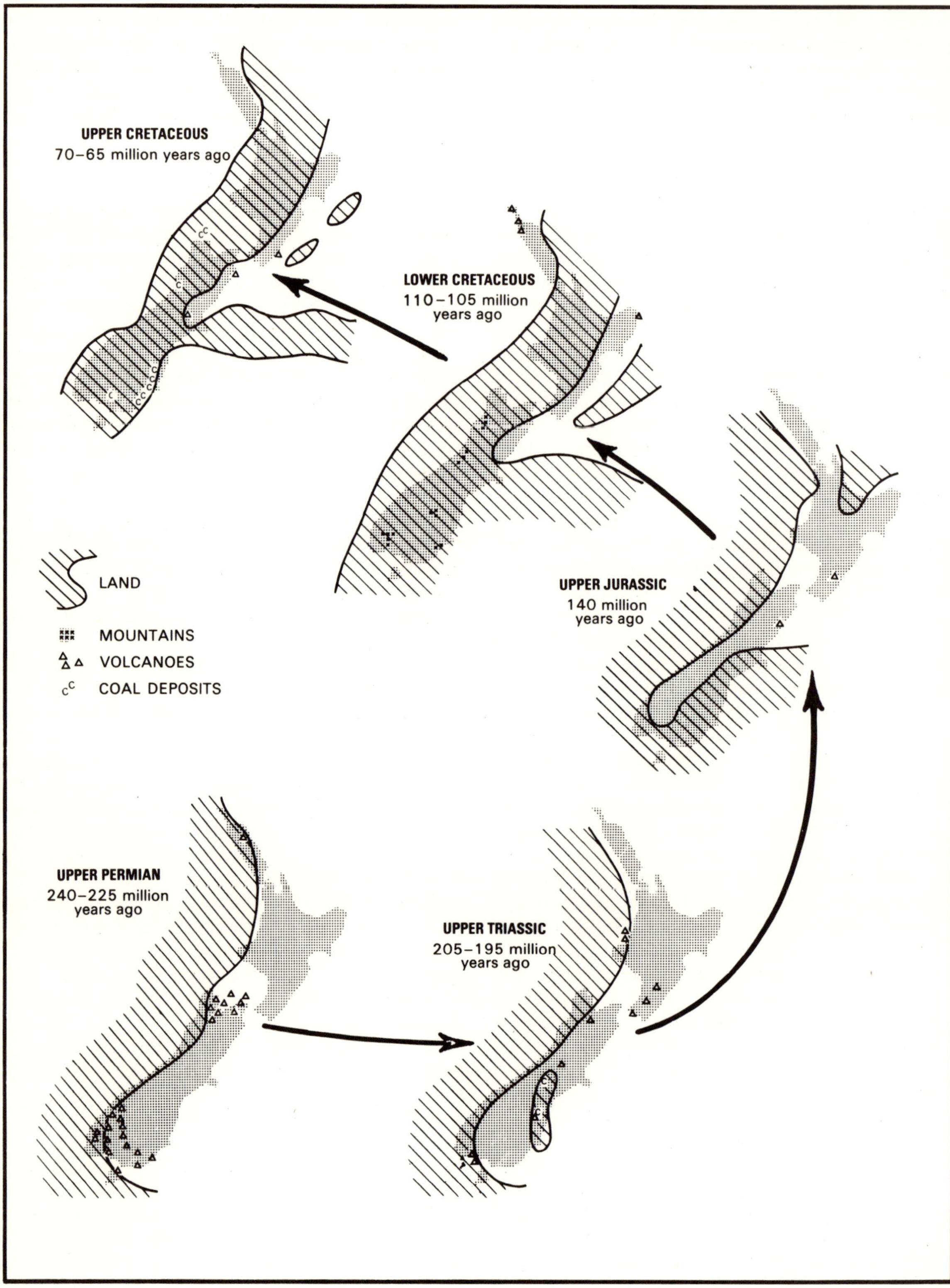
UPPER CRETACEOUS
70–65 million years ago
LOWER CRETACEOUS
110–105 million
years ago
UPPER JURASSIC
140 million
years ago
LAND
MOUNTAINS
VOLCANOES
COAL DEPOSITS
UPPER PERMIAN
240–225 million
years ago
UPPER TRIASSIC
205–195 million
years ago

filling the New Zealand Geosyncline were squeezed and folded upwards and pushed above the sea to form a new mountainous landmass. It is at about this time that the western-derived and eastern-derived sediments (i.e. those originally shed from Australia and Antarctica) were probably starting to collide, after having been rafted together by sea-floor spreading (Fig. 13.23, 16.29).

We can see from the geological record that throughout the middle and late Jurassic, increasing amounts of land poked up above the sea, as a result of the combined effects of sea-floor rafting, vice-like squeezing (between Australia and Antarctica) and buoyant uplift. Then between 140 and 110 million years ago (latest Jurassic and early Cretaceous), the rising land probably occupied all of the area that is now New Zealand, and quite a bit beyond also. How far this new landmass stretched is difficult to say exactly. It is quite likely that there was continuous land south to Campbell and Auckland islands and to the edge of the Campbell Plateau. Land also may have extended south-eastwards to the Bounty Islands and Bounty Platform and eastwards to Chatham Islands.

In a north and north-west direction much of the Lord Howe Rise may have been land, and fingers of land may have extended to New Caledonia and perhaps some distance towards New Guinea. All the areas just mentioned are thought to have been part of the New Zealand Geosyncline and therefore presumably shared in the earth movements that brought the geosyncline to a close, creating an ancestral New Zealand very much bigger than life-size. As far as we know, this was the first large landmass to appear in the South-west Pacific region and from it were carved, very much later on, the modern New Zealand and New Caledonia.

**Fig. 13.24.** Fluctuations of the New Zealand shoreline during Permian-Cretaceous times.

During this phase of development the New Zealand Geosyncline, flanking Australia, was gradually filled and gave rise at the end of the Jurassic to a new landmass, ancestral to modern New Zealand.

The distribution of land is shown by the diagonal line pattern and the coastline by the heavy line.

To assist in orientation each diagram is super-imposed on a shaded outline of the modern New Zealand landmass, but it must be emphasised that New Zealand as we now know it was not then in existence.

CHAPTER FOURTEEN

# MOAS, KIWIS AND TUATARAS

IN MESOZOIC times, before creation of the ancestral New Zealand landmass (i.e. prior to about 160 million years ago), the New Zealand seas, and presumably the ephemeral islands dotting them, were rather isolated and, as we have seen (p. 195), marine life (and undoubtedly land life also) developed special ("endemic") varieties, restricted to the New Zealand Geosyncline. Although climatic differences may have played a role at times, it is nonetheless thought that the major factor responsible for the isolation was the presence of extensive areas of deep water surrounding the New Zealand Geosyncline. As we have seen in Chapter 12, most marine animals need shallow sea to live and breed in, and extensive areas of deep sea are very effective barriers to their movements.

Although we have no idea of what animals and plants lived on the islands that dotted the waters

**Fig. 14.1.** Fossil coelacanths (pronounced "see-la-kanth") and relatives have been found in 370-million-year-old rocks in Scotland, Norway and Canada. These lobe-finned fish, or related forms, are thought to be ancestral to the four-legged animals that eventually crawled out of the water to populate the land, giving rise to the amphibia and later the reptiles. Formerly, coelacanths had been thought to have become extinct along with the dinosaurs, at the close of the Mesozoic era, 65 million years ago. However, in 1938 one was found alive off the South African coast and another in 1952 off the Comoro islands, north of Malagasy. Since then some 85 coelacanths have been pulled out of the Indian Ocean, all from near the Comoros.

All the coelacanths caught have been generally about 1.5-1.8 m (5-6 ft) in length, heavy (often 58 kg, 130 lbs), dripping with oil, and covered with massive steel-grey bony scales flecked with white. A double tail is present and fleshy lobed fins, fringed with stout hollow spines (coelacanth means "hollow spine").

This drawing is of *Latimeria chalumnae*, the first coelacanth found alive, discovered in 1938.

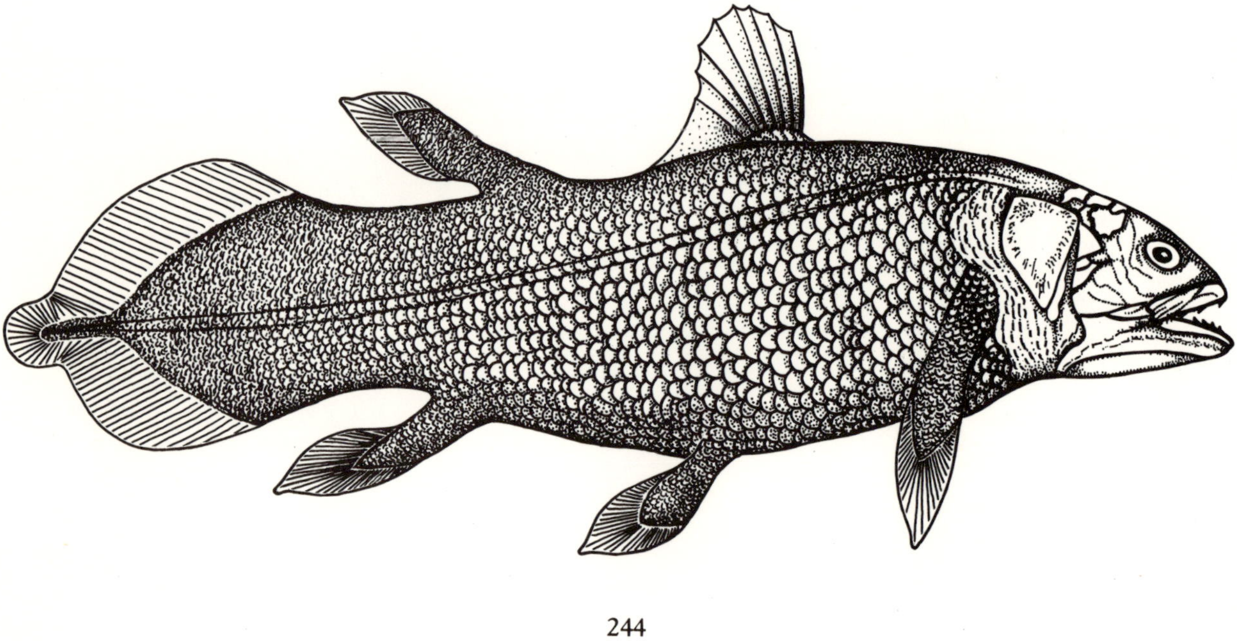

of the New Zealand Geosyncline, it is quite likely that they, like the marine creatures, had also developed "endemic" varieties in the isolation of the New Zealand region.

The earth-moving forces that created the ancestral New Zealand landmass out of the New Zealand Geosyncline pushed out long fingers of land, particularly to the north-west, along the trends of the modern Lord Howe, New Caledonia and Norfolk ridges and to the south-east, across the modern Campbell Plateau towards Antarctica. These new areas of land were edged with shallow continental shelves.

The record of marine fossil life tells us that floods of new marine creatures arrived in New Zealand and New Caledonia at about this time—in the middle and late Jurassic, 160–135 million years ago. The new immigrants were shellfish, ammonites, belemnites etc. related to, or the same as, those populating seas that then covered Iran, Iraq, the Himalayas, South-east Asia, Indonesia and New Guinea. From this we can deduce that the new landmass and its extensions north-west had provided very good shallow-water routes along which marine creatures could travel from the Indo-Pacific region.

Judging from fossil occurrences in Antarctica and South America, it is also deduced that, after travelling to New Zealand, many of the marine creatures were able to continue via equally good migration routes into Western Antarctica and into the Antarctic Peninsula, Tierra del Fuego, Patagonia and Chile. Migration in the reverse direction (i.e. travel to New Zealand via South America and Antarctica) was also theoretically possible, although all our evidence, at least as far as the marine animals is concerned, points towards migration to New Zealand from the Indo-Pacific rather than from South America. As migration of marine creatures along a coastline is often at the whim of currents it is quite likely that the pattern of coastal currents then existing favoured the movement of marine creatures from the Indo-Pacific to New Zealand and thence via Antarctica to South America and not *vice versa*.

As such good shallow-water routes were available, it follows that good land routes may also have existed. It was not necessary, however, to have continuous land to form a "bridge". Rather, strings of closely-spaced islands, like those of eastern Indonesia today, may have provided a perfectly adequate route for many land creatures. As we have seen in Chapter 6 (pp. 49–58), we should not under-rate the effectiveness of chance dispersal. Many land animals can survive quite lengthy voyages on floating trees, in the aftermath of a storm, and some, if they have to, can swim considerable distances. The main thing, however, is that at about the time of the establishment of the ancestral New Zealand landmass, 160-110 million years ago, we had, for the one and only time in our entire history, optimal conditions for land creatures to either walk or island-hop into New Zealand. The sad thing is that we have very little record of past land life in New Zealand. This is because of the vagaries of fossil preservation.

**The chances of fossil preservation**

If creatures are living in the sea, the chances are fairly good that eventually they will be preserved in some way or other. A sea creature when it dies sinks to the bottom and starts to decay. If this creature has landed in an area in which muds and sands are accumulating at a rapid rate, it is quickly buried, and as the various bacteria, fungi etc. causing decomposition require oxygen for their activities, burial puts an immediate stop to decay, and the creature is preserved in entirety. If, however, the dead creature lies on the sea floor for some time, scavengers may consume it, or break it into pieces. Alternatively, if decay runs its full course, the creature's flesh is reduced to a soft pulp that is eventually winnowed away by currents. Only the hard parts of the body remain and these may also disintegrate under the attack of acids and alkalis generated as by-products of the decay process. The disarticulated hard parts, if they survive, may, however, be eventually buried and thus preserved (pp. 197–198).

In time the ocean-floor sediments are raised above sea level to form land, rivers and streams cut gorges, and, if we are lucky, geologists will come along and find the preserved creatures—often now "petrified" or "turned to stone" by the addition of various minerals that soaked into every pore and crevice of the creature's body. To the geologist it often does not matter whether the fossil is perfect or not. Pieces of some fossils are so distinctive that they are instantly recognisable (at least to the palaeontologist's eye!).

Although you may think that the preservation of marine creatures is a chancy business, the dice are even more loaded against land creatures. If, for instance, a cow dies out in a paddock, decay rapidly sets in and the body begins to disintegrate. Scavengers may hasten the disintegration. Very soon all that is left are some sun-bleached bones lying scattered on the surface, and in time even these will decay, under the combined effects of sunlight's ultraviolet rays, acidic rain and the various natural acids and alkalis in the soil. Often very soon nothing is left.

Even if the cow's body, or parts of it, are swept into a nearby river or stream, the effects of a tumbling journey down the river bed strewn with stones and boulders soon reduce the body to unrecognisable fragments, particularly if the bones have already become powdery through exposure to sun and rain.

All in all, therefore, the chances are not very good of preserving a sample of the land creatures that lived in an area at any particular time.

At times in the past dinosaurs and moas, for example, have been preserved in ancient lake, swamp or river deposits. Also, creatures have sometimes fallen down pot-holes or other cracks in the earth and have been preserved. Others have been preserved in caves. Compared with the chances of preservation in the marine environment, however, these types of preservation all require special circumstances. Large parts of North America, for example, were swampy river plains at the time (Jurassic and Cretaceous) when ancestral New Zealand was created. Meandering rivers, continually shifting, formed large areas of quicksand and swamps, suitable for trapping many of the huge dinosaurs that roamed the land at that time. Many animals were thus preserved in deposits on the land, and, as little erosion or earth movement has occurred since then, the deposits are still in their original position.

The ancestral New Zealand land undoubtedly had creatures roaming its surface and many were probably preserved in river, swamp and lake deposits. As we will see, however (p. 285), this primaeval land was gradually whittled away by rivers, streams and the sea, so that by about 30 million years ago most of it was back below the ocean again. Thus any land deposits dating back to the Mesozoic and early Tertiary have probably been destroyed.

Even in geologically recent times, in the last 20 million years, when land again started to appear in the New Zealand area, it has been continually shaken and disturbed to such an extent that the few land deposits now surviving have only done so because they occupied subsiding swampy areas that continued their downward movements over a long period of time.

The major point of all this is that while areas like central North America, with a long history of earth stability, have a rich record of land life, going back to the Mesozoic and beyond, New Zealand, going up and down like a yo-yo, has a very poor land record. In fact, the very few deposits of land fossils we have are all very young—Miocene, Pliocene, Pleistocene (i.e. spanning the last 25 million years)—and are of little help in giving us an idea of our Mesozoic land life. However, by using information from other countries with good records of fossil land life, and from pollen analysis (to give us an idea of the vegetation), we have been able to build up a picture, albeit rather sketchy.

**The succession of animal life**

Before human beings arrived on the scene at least 1000 years ago, New Zealand was a land with a very "unbalanced" flora and fauna. Many plant and animal groups, widespread elsewhere in the world, were absent from primaeval New Zealand. This was because New Zealand had been cut off from the rest of the world during the development of these animals and plants.

As we go back through geological time, we can trace, through the study of fossil life, a steady progression of animals and plants. As time passed the original flora and fauna living in a particular area were progressively replaced by more advanced forms that were probably better equipped to survive in the environment. In the animal kingdom, for example, we can see an overall progression from life in the sea to life on the land. The sea provided a cosy almost insulated environment, whereas the land, with its variety of often harsh habitats, represented a great challenge to the early animals. Thus in the Precambrian, Cambrian, Ordovician and Silurian, beyond about 390 million years in geological time, much of the development in the animal kingdom, as far as we know, took place in the sea. Various fish types evolved and then in the Devonian, 390-345 million years ago, appeared fish like the living fossil, the "coelacanth", whose discovery in waters off Madagascar caused a scientific sensation in the 1950s (Fig. 14.1). These fish had stout fleshy fins, that look well on

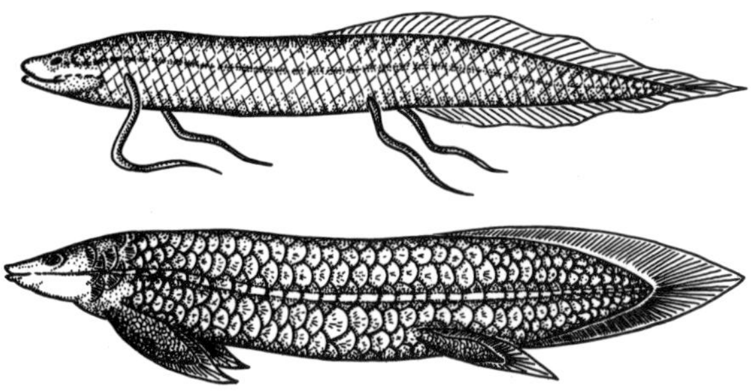

**Fig. 14.2.** True lungs first appeared in groups of Devonian fish (390-345 million years ago), ancestral to the modern lungfish. The lungfish today, virtually unchanged from those fossil pioneers, look a bit like eels with massive jaws and rope-like or flattened fins. They live in swampy areas in Africa, Australia and South America and when the dry season comes they dig into the mud and estivate (a type of hibernation). As the mud dries rock hard they are imprisoned and lie fasting and breathing until the rains wash them free, sometimes many months later.

The diagrams are of the African Lungfish (*Protopterus*) top, and the Queensland Lungfish (*Neoceratodus*) bottom. Typical lengths are 60 cm (2 ft) and 1.5 m (5 ft) respectively.

the way to becoming structures that could turn into legs—so much so that the coelacanth has been nicknamed "Old Four Legs". Some of these fish may also have had functional lungs rather like those of the modern lungfish now found in Africa, South America and Queensland (Fig. 14.2). The Queensland lungfish can survive severe droughts by living in stagnant pools, gulping fresh air from the surface, while other fish die in the foul water. The African and South American lungfish go even further—they can burrow into the mud of a drying-up pond and form a cocoon linked to the surface via a vent. Curled up in its cocoon the lungfish uses its lungs to breathe air through the vent. When things improve and the water returns, the lungfish wakes up and starts swimming again. It is claimed that dry cocoons have been sent through the post from one scientist to another and that the dozing fish suffered no ill effects, swimming around quite happily as soon as the cocoon had been soaked off.

Thus the presence in these early fish of stout fleshy fins, capable of developing into legs, and a functional lung, led to the eventual development of the amphibians, able to live on both land and water, but still dependent on the water for their reproductive cycle.

A factor that undoubtedly helped nudge the coelacanth and relatives out of the water was that, to judge from fossil occurrences, they populated lakes and rivers dotting parts of Laurasia subject to periodic droughts. It was probably this inhospitable environment that forced the lobe-finned giants to crawl in search of water.

The reptiles, that first appeared in the middle Carboniferous, some 300 million years ago, and presumably evolved from amphibia, were the first animals to cut the ties to water, and could freely live and reproduce on the land. The reptiles were covered with massively thickened scales, developed from those of their fish and amphibian ancestors, to protect them from predators, but also to insulate them from the rigours of the terrestrial environment. They were cold-blooded and although their scaly covering did help to conserve bodily heat, their activity was governed by the temperature of the environment. Cool temperatures would slow them down so that they were capable of only sluggish movements.

The reptiles had their heyday in the Permian, Triassic, Jurassic and Cretaceous (280-65 million years ago) and in many countries large dinosaurs ruled the land. Some reptiles developed paddle-like limbs and a streamlined shape and returned to the sea. Examples are the long-necked plesiosaurs and the dolphin-like ichthyosaurs, both of which are found as fossils in New Zealand. Other reptiles developed large sail-like flaps of skin extending between their greatly elongated fingers and their sides. Some had a wing-spread of 15 m (50 ft). These could glide in the air, but it is doubtful whether they could actually fly by flapping their wings. Some reptiles, however, formed feathers from their reptilian scales, and developed light hollow bones and a large keel-like chest bone to provide attachment for powerful wing muscles. These feathered reptiles, with scale-clad legs (still present in modern birds, as a hint of their reptilian ancestry), reptilian teeth and clawed wings (both lost in modern birds), first appeared in the late Jurassic, about 150-140 million years ago.

## The first New Zealanders

By Jurassic times, therefore, amphibians and reptiles were roaming the land and the first birds had just evolved from reptilian stock. Higher animals, such as mammals, had not yet appeared (they evolved in the late Cretaceous and Eocene; pp. 262–265).

Thus when the ancestral New Zealand landmass came into being, in the late Jurassic and early Cretaceous, 140-110 million years ago, and probably for the one and only time had extensive land links (north-west to the Indo-Pacific and South-east Asia; south-east to Antarctica and South America), the land animals that then migrated to New Zealand were representatives of the amphibia, reptiles, birds and other organisms then in existence and living in Gondwanaland.

These early arrivals have been drastically thinned over the geological ages, but even today we can pin-point certain groups in our land fauna whose ancestors probably migrated to New Zealand at a time when it was part of Gondwanaland.

Some of these animals are now found only in New Zealand (i.e. are endemic), and have their roots deep in the geological past. They are "living fossils" because they are the only survivors of primitive groups that evolved in late Palaeozoic and early Mesozoic times and subsequently gave rise to, and were eventually replaced by, more advanced groups unable to reach New Zealand because of its isolation. These "archaic" groups include, for example, the native frog (*Leiopelma*) and the tuatara (*Sphenodon*).

Other New Zealand native animals, again of very ancient lineages, and with life habits that make them unable to cross any extensive stretch of salt water, belong to groups that are today only found scattered across remnants of the old Gondwanaland. These creatures, showing "southern" affinities (i.e. with relatives in southern South America, sub-Antarctic islands, Australia, and South Africa), date back to the time when Gondwanaland was one continent, and include many birds, the *Peripatus* and some native earthworms, insects, snails, slugs and freshwater animals.

**Fig. 14.3.** Three species of New Zealand native frog are known. All are very tiny, none being more than 40 mm (1⅝ in) long. In the past *Leiopelma* was fairly widespread, but its modern distribution is patchy. *Leiopelma hochstetteri*, the most common species, is found quite widely in Auckland province. *Leiopelma archeyi* is known only from Coromandel Peninsula. The third species, however, *Leiopelma hamiltoni* (illustrated here), is perhaps one of the rarest frogs in the world. For many years the only known locality was a small 0.25 hectare (about ½ acre) heap of stones near the summit of Stephens Island, north of the Marlborough Sounds, but it is now known to occur in a 15 hectare (40 acre) patch of forest on nearby Maud Island.

Thus because of its early separation from Gondwanaland, and subsequent isolation, New Zealand has more than its fair share of unusual animals and some of the more notable of these are dealt with in the following pages.

**A frog without a tadpole**

Charles Darwin in his *On the Origin of Species* remarked that frogs are absent from truly oceanic islands, i.e. those that have not had any land links in the geological past. This is because frogs are extremely sensitive to exposure to sea water. The possibility of their being able to "raft" from one place to another (on floating logs etc.) is thus very remote. The New Zealand native frogs are therefore prime candidates for inclusion amongst the first New Zealanders and certainly needed continuous land, albeit well-watered, for their journey to New Zealand. Three species of native frog are known in New Zealand, all belonging to the genus *Leiopelma* (Fig. 14.3). They are the most primitive living frogs in the world and are virtually the same as frogs that are known as fossils from rocks of late Jurassic age.

As may be expected, they have a number of primitive features, amongst which is the presence of tail-wagging muscles (although they have no tail!). They have no ear-drums or vocal sacs and the absence of the latter means that they can only manage a rather un-frog-like thin high-pitched squeak.

One of the most fascinating things, however, about the native frogs is their life history. Unlike most frogs, they do not have a free-swimming tadpole. All development takes place within a gelatinous capsule derived from the egg originally laid by the female. It is because of this life history that *Leiopelma* is not tied to pools and streams, as most frogs are, and can inhabit a wide range of environments. *Leiopelma* also lacks any webbing between the toes of its hind feet. Webbed feet would be of little use to frogs that often never see enough water to swim in!

**Old Beak Head**

The tuatara, *Sphenodon punctatus* (Fig. 14.4), is the only surviving member of the reptile order Rhynchocephalia ("beak-headed"). The rhynchocephalians first appeared in the early Triassic, some 220 million years ago, and became relatively abundant in the later Triassic and Jurassic, fossils of this age being known from East Africa, North

**Fig. 14.4.** The tuatara (*Sphenodon punctatus*) is the sole survivor of the once diverse reptile order Rhynchocephalia. Members of this order were abundant and widespread in the Mesozoic but elsewhere died out along with the dinosaurs, some 65 million years ago.

An adult male may grow to 700 mm (28 in) in length and weigh up to 1000 g (36 oz).

In geologically recent times, probably up to several hundred years ago, tuataras lived on the New Zealand mainland and ranged from North Cape to Bluff. Now however they are restricted to off-shore islands in the Auckland-Coromandel-Bay of Plenty region and Cook Strait.

and South America and Europe. A few small members of the group occur in early Cretaceous rocks in South America and Europe, but these are the last known in the fossil record, and it is possible most became extinct at the end of the Cretaceous, some 65 million years ago.

The tuatara is almost identical in structure to a rhynchocephalian, *Homoeosaurus*, of the late Jurassic, some 140 million years ago. Small wonder that the tuatara is often referred to as a "living fossil" and is regarded as being one of the most interesting reptiles in the world.

Although the tuatara is lizard-like in appearance, it has features, many of them primitive, which clearly separate it from the lizards. The skull, ribs and backbone all have primitive features that hark back to the "stem reptiles" of the Permian and Triassic (i.e. the group from which all reptiles are presumed to have descended). However, perhaps the most notable feature is that a vestigial "third eye" is present on the top of the head in young tuataras, but becomes covered over and is invisible in adults.

**Fig. 14.5.** It is believed that the ancestors of the flightless ratite birds (moas and kiwis of New Zealand, ostriches of Africa, elephant birds of Madagascar, rheas of South America, emus of Australia, cassowaries of New Guinea and Australia) walked to the various southern landmasses some time in the Jurassic, as reconstructed in this diagram. At this time good land links were available and the lands themselves were grouped some distance away from the South Pole—allowing migration to occur in temperate climes.

*Fig. 14.5*

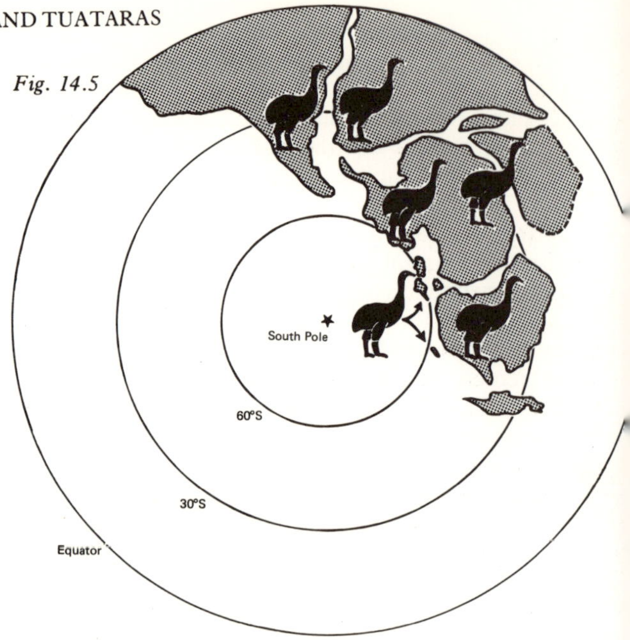


## "Southern" animals

In some other groups of land animals Southern Africa, South America, Australia and New Zealand share certain "archaic" forms, all needing continuous land for migration, and it is fairly clear that they probably originally spread to these countries before the break-up of Gondwanaland. These "southern" animals include, for example, flightless birds, the caterpillar-like Onychophora (e.g. *Peripatus*), certain primitive earthworms, freshwater mussels, freshwater crayfish, and some insects and spiders. Some of these long-time inhabitants of New Zealand are treated in greater detail below.

## Ratite birds

New Zealand is well-known as the home of the kiwis and in the very recent past, the moas. Both are members of a group of flightless birds called the ratites (Suborder Ratiti), all characterised by the absence of a keel on the breast bone, to which the large wing muscles are attached in birds that fly (hence the origin of the name, derived from the Latin *ratis*: a raft or flat-bottomed boat, referring to the fact that ratite birds have a large flattened keel-less breast bone).

*Fig. 14.6*

**Fig. 14.6.** The largest moas (e.g. *Dinornis maximus*, *Dinornis giganteus*) attained a height of some 3.0-3.6 m (10-12 ft), and weighed some 235 kg (520 lb), although others were much smaller and ranged down to a size not much bigger than a large turkey.

*Dinornis* had a small head, a blunt beak with a down-curve like a grubber, powerful thighs and stout legs, no wings and no tail. The feathers were similar to those of the emu.

Moas and kiwis are very closely related. The kiwi however retains tiny rudimentary wings, whereas the moa had gone even further towards flightlessness and had completely lost the wing bones.

The original of this photo was reproduced in the *Otago Witness* in 1903. The reconstructed Moa was placed in a natural setting in the Dunedin Public Gardens. The hunters are from left: Te Rangi Hiroa (Sir Peter Buck), Hemi Papakakura and Tutere Wirepa.

Besides the moas and kiwis of New Zealand, the ratites include the ostriches of Africa (and in the past also of Europe and Asia), the extinct elephant birds of Madagascar, rheas of southern South America, emus of Australia and cassowaries of New Guinea and Australia.

Once it was thought that the ratites as a group had diverged from the ancestors of other birds, before birds had acquired the power of flight. It is now recognised, however, that ratites show so many structural features of flying birds that they are considered to have arisen from flying ancestors by degeneration of wings and breast muscles as they increased in size and began to depend on their running ability to escape enemies. The various ratite birds, although they are now widely separated geographically, are at the same time so similar that they must already have been flightless before being dispersed among the southern continents.

Although some living ratites can swim for very short distances, they cannot cross ocean barriers. Thus they were probably distributed before the break-up of Gondwanaland, at a time when continuous land links were available.

**Fig. 14.7.** The Brown or Common Kiwi (*Apteryx australis*), New Zealand's national emblem.

Everything points, therefore, to the probability that the ancestors of moas and kiwis walked into New Zealand before it became isolated from the other Gondwana countries (Fig. 14.5).

Other birds, with normal powers of flight, probably accompanied the ratites on their migration to New Zealand. Three families of endemic New Zealand birds may have such an ancestry: the New Zealand wrens (including the rifleman and rock wren), the native thrushes (piopio) and the wattlebirds (huia, saddleback and kokako).

Other birds may also have come at this time, or perhaps slightly later, when the steadily widening distance between New Zealand and the remainder of Gondwanaland could still be crossed by birds that were capable of island-hopping or being blown by storm winds. After migrating to New Zealand, these birds, in the absence of marsupial and mammal predators, lost their power of flight, or had it greatly curtailed. In this category are the New Zealand rails (takahe, weka, pukeko and several extinct forms), a large flightless goose (*Cnemiornis*, now extinct), flightless parrots (e.g. the kakapo and several extinct forms) and the flightless ducks, now extinct, that lived on Auckland Island and Campbell Island.

Flightlessness and large size in birds tend to go together and the New Zealand moas were a prime example of this, although it must be remembered that not all of them were giants (as in Fig. 14.6) and some were about the size of a large turkey.

Studies have shown that moas and kiwis are very closely related to one another, and rather more distantly related to the other ratites. This indicates they have been together for a very long time and isolated from the remainder of the ratite birds for an equally long time.

It is thought that the ancestors of the ratites originated in South America. The tinamous that live today in some parts of Central and South America are thought to have split off from the ratite group at about this time. Subsequently, the ancestral ratites lost their powers of flight and soon after this ratite groups became fairly widely distributed across Gondwanaland. One early group made its way from South America, across West Antarctica and into New Zealand, there to eventually produce the moas and kiwis. Fairly soon after this another group, living in the area of Africa and Madagascar, gave rise to the ancestors of the elephant birds, that became isolated on Madagascar when it separated from the remainder of Gondwanaland sometime in the Cretaceous, about 70 million years ago.

Following this, another group of ancestral ratites dispersed to Australia via East Antarctica. There they gave rise to the emus and cassowaries. Finally the rheas and ostriches developed from a common ancestor present in both South America and Africa.

**Moas**

Two groupings of moas can be recognised. The Dinornithidae, or Greater Moas, had some eight species and ranged up to 3.0-3.6 m (10-12 ft) in height.

The largest moa, *Dinornis maximus* (Fig. 14.6), stood higher than any living mammal except the giraffe and elephant, and grew leg bones more massive than those of a draught-horse. The Greater Moas needed an enormous amount of food to sustain their great bulk. Most of them probably grazed grassland plains, shearing off the grasses with their massive beaks as neatly as an old-time harvester with his scythe. *Dinornis maximus* probably required as much grass per day as a bullock, and to help his digestion, carried around gizzard stones, often 200-500 in number, and up to 101 mm (4 in) in diameter, together weighing some 2.2-2.7 kg (5-6 lb). In ancient New Zealand the grassland moas thus filled the role now occupied by grazing mammals (sheep, goats, cattle etc.).

The Anomalopterygidae, or Lesser Moas, had some 16 species and tended to be stouter and shorter birds with more rounded skulls. Some of the Lesser Moas were 1.2-1.5 m (4-5 ft) high, with bandy, massive legs supporting a squat barrel-like body. Others were about the size of a large turkey. Many of the smaller moas lived in the forests, where they browsed on the twigs and leaves of many shrubs (e.g. *Carmichaelia*) and low-growing trees, such as ngaio (*Myoporum*) and the berries and fruits of trees (e.g. the podocarps, many of which have fleshy fruits; see p. 259).

The earliest Polynesian inhabitants of New Zealand, the Moa Hunters, who arrived sometime before A.D. 1000, used the moas for food, their bones for implements and ornaments and their eggs for water bottles. This hunting pressure soon led to extinction of the large moas, and it is generally thought that most had disappeared by about A.D. 1500. The smaller bush moas, however, may have lingered on in remote parts of the South Island until early in the nineteenth century.

**Kiwis**

The kiwis of today are in a way miniature versions of the extinct moas (Fig. 14.7). Moas and kiwis are, in spite of their size difference, very closely related. Moas, however, had completely lost their wing bones, whereas the kiwis still retain tiny wings, under which they place their bills while asleep, like any normal bird. It is thought that the ancestors of kiwis were much larger birds, like small moas, and grew smaller as they became specialised for nocturnal life in dense rain forests. There are three species of kiwi: the Brown or Common Kiwi (*Apteryx australis*), the Little Grey or Little Spotted Kiwi (*Apteryx oweni*) and the Roa or Great Spotted Kiwi (*Apteryx haasti*).

Kiwis are about the same size as domestic fowls. The Great Spotted Kiwi, the largest, can weigh up to 3700 g (8 lb 2 oz), while the Little Spotted Kiwi, the smallest, weighs about 1100 g (2 lb 7 oz).

All have long slender bills with the nostrils near the tip. Rudimentary wings, about 35-50 mm (1.5-2.0 in) long, and ending in a claw, are hidden beneath the feathers. A tail is absent and one-third of the body weight is occupied by powerful muscular legs.

**Extinct giant woodhens**

*Aptornis*, a giant type of woodhen (i.e. similar to the modern weka), lived in New Zealand up to just before the arrival of Europeans (its remains have been found in Polynesian ovens). Like the weka, the *Aptornis* belongs to the rail family of birds, that also includes the takahe and pukeko.

*Aptornis* stood about 1 m (3 ft) high, and had massive leg bones reminiscent of a small moa. It had a large head and an enormous bill that together were as large as the skull of *Pachyornis*, the second largest moa.

The powerful beak was probably used, like that of the takahe, for deftly chopping off clumps of tussock grass.

Although unable to fly, unlike the moas *Aptornis* retained the bones of a tiny wing, that in life was probably completely concealed beneath the feathers.

Although the rails as a family were probably

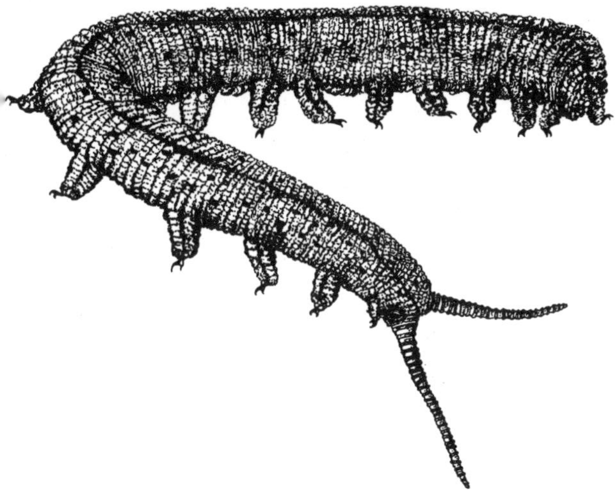

**Fig. 14.8.** The New Zealand *Peripatus* looks like a squat velvet-covered caterpillar-like animal about 50 mm (2 in) in length. It has stumpy clawed legs and a pair of soft flexible antennae with a small beady eye at the base of each.

*Peripatus* is widely distributed in the cool moist New Zealand rain forest and during daylight can be found under rocks and logs.

better able to get around than the ratites, for example, and may have indulged in a bit of island-hopping to get to New Zealand, nonetheless they are, or were, shared by many of the southern countries, and like the ratites, once they became isolated by continental drift, gave rise to many interesting flightless varieties.

New Zealand's *Aptornis*, for example, is very closely related to the cagou (*Rhynochetos*), a flightless rail that is the national emblem of New Caledonia (Fig. 16.30). As New Zealand and New Caledonia are now separated by some 1500 km (930 miles) of open ocean, the close affinity shown by the *Aptornis* and cagou must date back a very long way. However, as they are more advanced forms than the ratites (e.g. moas and kiwis) they may have been slightly later immigrants into the South-west Pacific.

### New Zealand's "missing link"

The New Zealand *Peripatus* (Fig. 14.8) is a member of a very ancient group of animals (the Phylum Onychophora—or "claw bearers") which have existed from the middle of the Cambrian period, some 550 million years ago. All have a body divided, rather like that of an earthworm, into many segments (14-44), which cannot, however, be seen from the outside. Each segment bears a pair of small stumpy legs with two tiny claws.

From the time of its discovery, *Peripatus* has been regarded as something of a biological oddity. Because of its unusual combination of anatomical characteristics, its affinities with other invertebrate groups have in the past been rather confused. Now, however, it has been firmly placed in a phylum all by itself and is generally recognised as the "missing link" between the segmented worms (Phylum Annelida) and the jointed-legged invertebrates such as crayfish, woodlice etc. (Phylum Arthropoda).

The New Zealand *Peripatus* is a small greyish-green velvet-skinned caterpillar-like animal commonly found under logs and rocks in the bush. It is a nocturnal hunter and feeds on a wide range of small animals such as springtails, termites and insect larvae. *Peripatus* uses a unique method of food capture. When faced with a likely meal, it raises its head and squirts a stream of slime, that is emitted from glands on either side of the mouth. On contact with the air the slime thickens and hardens, becoming incredibly strong, falling over the victim like a net. The *Peripatus* then moves in and by means of its jaws cuts a neat hole in the body wall and sucks the prey dry. The jets of sticky fluid also come in handy for defence.

The New Zealand *Peripatus* is most closely related to species found in South Africa and Australia, but others are known from South and Central America, Malaysia, India and New Guinea. This distribution pattern probably dates back to the time when Gondwanaland was a continuous continent and land-bound creatures such as *Peripatus* were able to move about freely.

### Native earthworms

Even the lowly earthworm is yet another remnant of the original Gondwanaland terrestrial fauna. Not, however, the ordinary worms in your garden—these have been introduced—but rather the less common native worms, including the milk worm (or Maori Worm), which can exceed 300 mm (12 in) in length, and the spectacular North Auckland worm *Spenceriella gigantea*. The latter worm, New Zealand's largest, reaches a

**Fig. 14.9.** The giant weta of Little Barrier Island (*Deinacrida heteracantha*)—the biggest insect endemic to New Zealand—can reach a body length of 82 mm (3.2 in) and weigh up to 71 g (2½ oz).

The wetas are considered to be representatives of a very ancient group of insects that migrated to New Zealand sometime in the Mesozoic era some 200-140 million years ago, when it was still part of Gondwanaland, and have changed very little since that time.

In the primaeval New Zealand forest the wetas developed the habits and life styles of rats and mice and for this reason have been called "insect rodents".

length of 1400 mm (4 ft 7 in) and a thickness of 11 mm (almost ½ in). Its burrows are often 20 mm (greater than ¾ in) in diameter and have been found still continuing downward at a depth of 3.5 m (11 ft 6 in).

Some 173 native worm species still survive, mainly confined to areas of native forest. Although some of these species are related to Australian worms, many of them also have relatives in southern South America, the Falklands, southern South Africa, and New Caledonia as well as Australia. Although most earthworms can live for some time in fresh water, most are unable to pass even a narrow arm of salt water.

Worm cocoons are laid below the surface of the earth and are normally inaccessible to birds and other animals and it is unlikely that birds would transport either worms or cocoons in mud caked on their legs. Thus, as with the other animals we have already mentioned, we again have to seek the origin of such a distribution pattern back in geological time, when Gondwanaland was in existence and terrestrial creatures could move freely across continuous land. This means that many of our native worms, particularly those with southern affinities, have been long-time inhabitants of New Zealand.

**An insect rodent**

A striking feature of the New Zealand insect fauna is the considerable number and variety of large flightless ground-surface insects. Also obvious is the number of these that have relatives, if any, only in those countries that were once part of southern Gondwanaland, e.g. southern South America, Australia etc. Thus the insects in this category were formerly able to migrate across the southern parts of Gondwanaland and, once isolated in New Zealand, were able to develop in the absence of mammal predators.

Perhaps the most notable of New Zealand's flightless insects are the wetas (*Deinacrida*, *Hemideina*) (Fig. 14.9). Wetas, rather like giant flightless crickets, have a long fossil history and related forms have been found in rocks of Triassic age (about 200 million years old) from Queensland. Thus the ancestors of the wetas may well have migrated to New Zealand at about this time or later, in the Jurassic, when, as we have seen (p. 245), land routes were probably at their best.

In the absence of terrestrial mammals, the wetas developed habits and life styles similar to those of many small rodents. Thus wetas, like mice and rats, hole up during the day and emerge at night to roam the forest floor. As they feed on vegetation and forest floor debris, they produce rodent-like droppings. For this reason the rat-sized giant weta has been called an "insect rodent". Large wetas were once fairly widespread in the North Island lowlands, but became extinct throughout most of the area in the early days of European settlement, probably with the establishment and spread of European rats (which indicates that the wetas and rats occupy, and compete for, the same ecological "niche").

**Freshwater life**

At first sight the inhabitants of our lakes, rivers and streams—the freshwater fishes, mussels and crayfish ("crawlies")—may appear to be good candidates for inclusion as long-time residents of New Zealand, as all would not survive very long, if at all, in the sea. Studies suggest, however, that most of the New Zealand representatives of these groups have probably been derived from saltwater ancestors. Nonetheless, some "southern" affinities are detectable, notably in the fish *Galaxias* (the young of which are the New Zealand delicacy "whitebait") and lampreys and some of the mussels. Thus although trans-oceanic migration has occurred to some extent, there is also a background of ancestral links dating back to the time of Gondwanaland.

**Spiders, snails and slugs**

Animals as diverse as spiders, snails and slugs also show traces of their Gondwanaland heritage. Most of the native harvestmen and spiders of New Zealand show southern affinities, although

**Fig. 14.10.** The native land snail, *Paryphanta*. Upper side (top), lower side (bottom). The shell is 50 mm (2 in) across.

there is an overlay of affinities reflecting later trans-Tasman dispersal (perhaps via drifting vegetation or wind dispersal of spiders "ballooning" on parachutes of web). Even more marked southern affinities are shown, however, by the very characteristic native slugs (often olive-green in colour, with a veined body) and the large native land snails.

A number of large native snails are found in New Zealand and their glossy shells are much prized by collectors. One group, including *Paryphanta* (Fig. 14.10) and *Wainuia*, have coiled shells up to 110mm (4½in) in diameter, and studies suggest that they belong to the oldest family of land snails, having originated in the late Palaeozoic or early Mesozoic. Thus these land snails probably came to New Zealand at about the same time as the native frog and tuatara and have survived in isolation, whereas elsewhere the original ancestral lineages have given way to others that have produced the ubiquitous European Snail (*Helix aspersa*), the bane of many a gardener's life.

**Why no dinosaurs?**

As we have seen (pp. 193–194), the giant land reptiles known as dinosaurs spread to many lands in the Mesozoic and late Palaeozoic eras, some 280-65 million years ago. Fossil dinosaurs dating back to these eras are known from Australia, but yet none are known from New Zealand.

Giant marine reptiles (ichthyosaurs, plesiosaurs etc.), contemporaries of the dinosaurs, are known from New Zealand, but this merely testifies to the excellence of our marine record. If the land links in the South-west Pacific were good enough in the Jurassic and early Cretaceous to allow the ancestors of the moas, kiwis and tuataras to walk to New Zealand, what stopped the dinosaurs?

However, the absence of fossil dinosaurs in New Zealand may be more apparent than real. As we have seen, the record of our ancient land life, except for plants, is almost non-existent. For example, we know full well that ancestral tuataras probably lived in New Zealand in Mesozoic times, but yet the only fossil tuataras known in New Zealand are in deposits of Pleistocene age that formed two million years ago. The absence of fossil dinosaurs from our records is therefore no

**Fig. 14.11.** *Agathis australis*, the New Zealand kauri. "Tanemahuta", 52m (170ft) high. Waipoua Forest, Northland.

Kauri trees can also be seen in the New Caledonian forest—a country that up to some 55-37 million years ago (Paleocene-Eocene) had very close links to New Zealand (see p. 297).

The ancestors of the kauri probably migrated to the South-west Pacific in Triassic times, 230-195 million years ago.

**Figs 14.12-14.14.** The ancestors of the podocarps probably migrated to New Zealand in the early Jurassic, some 195-172 million years ago.

14.12 Totara (Tuatapere, Southland)

14.13 Kahikatea (Westland)

14.14 Rimu (regenerating rimu, Westland)

*Fig. 14.11*

*Fig. 14.12*

guarantee that they were indeed absent from New Zealand.

If on the other hand the absence of dinosaurs is genuine, this may indicate that at this time the climate of New Zealand was a bit too chilly for their taste (at this time New Zealand was further south than Australia; see Fig. 12.27). Another possibility is that the original migration route over which the moas, kiwis and tuataras reached New Zealand was not continuous land, but a series of closely spaced island stepping-stones, so that certain animals, such as the lumbering and often heavily armoured dinosaurs, were excluded.

### The succession of plant life

A similar evolutionary progression to that of the animals (pp. 246–247) had also occurred in the plants.

The first organised living things were bacteria that probably appeared on the scene more than 2000 million years ago. Then around about 2000 million years ago algae appeared, the first oxygen-producing creatures. These were followed, probably in late Precambrian or Cambrian times, 1000-600 million years ago, by plants ancestral to the modern fungi, mosses and liverworts. These early plants had relatively simple bodies, and it was not until the late Silurian, about 400 million years ago, that the first plants with true stems (the ancestors of the modern horsetails and club mosses) appeared.

Then in the Devonian, some 390 million years ago, appeared the first plants with true leaves, the ancestors of the modern ferns. At this time large areas of the land were probably colonised by plants. Some of the ancient horsetails grew to 10-12 m (30-40 ft), whereas today they barely reach 1 m (3 ft).

By Carboniferous times (345 million years ago) tree-sized horsetails, club mosses and ferns had reached the zenith of their development. Horsetails and club mosses reached a height of 21-33 m (70-110 ft) and some tree ferns grew 10.5-16.5 m (35-55 ft). The debris of this luxuriant flora became the basis of the deposits of black coal now mined extensively in Europe and North America (hence the basis for the name "Carboniferous" = carbon-bearing).

The ferns, horsetails, club mosses etc. all reproduce by means of spores. In the Carboniferous, however, appeared the first seed-bearing plants—the conifers or cone-bearers, ancestors of the modern pines and cycads. The development of seed was an event of great significance in the plant kingdom as it allowed vegetation to spread

*Fig. 14.13*

*Fig. 14.14*

**Figs 14.15-14.16.** The tree ferns and other ferns so characteristic of New Zealand forests are of archaic varieties that came to the South-west Pacific in Triassic or earlier times. Many of New Zealand's tree ferns are shared with New Caledonia, which was closely linked with New Zealand until about 55-37 million years ago (Paleocene-Eocene).

beyond the limited breeding ground of damp earth (necessary for spore growth). For in the conifers male cells are formed in pollen, which is blown by the wind to the female cells in another cone. After fertilisation takes place, the seeds are nourished in the female cone until ripe, and then released, often explosively, so that they are fired for some distance. Since they carry with them their own supply of food, seeds have considerably more chance of survival than do spores.

Coal swamp vegetation continued into the Permian and the coal deposits of India and Australia, for example, were laid down at this time. Some of the early conifers grew to a large size in the coal swamps and were up to 44m (145ft) high. Then in the Triassic, some 230-195 million years ago, a group of conifers, the araucarian pines, appeared that gave rise, for example, to the modern kauri and Norfolk pine (Fig. 14.11). These were followed, in the early Jurassic (195-172 million years ago), by another group of specialised conifers, called podocarps, ancestral to many of the New Zealand forest trees, such as totara, miro, matai, kahikatea, rimu, tanekaha, toatoa etc. (Figs 14.12-14.14).

The pines, spruces and araucarian pines all bear fully developed cones, but in the podocarps the cones are greatly reduced, often to merely a single scale. The seed, however, is often surrounded by a fleshy cover, or is borne on a fleshy stalk, hence the name *Podocarpus*, which means "seed with a foot".

The next, and final, step of plant evolution took place in the middle Cretaceous, about 100 million years ago, when the first flowering plants appeared. They were not only the highest forms of plant life but the most diversified and widespread (pp. 265–267).

**Relics of Gondwanaland forest**

Evidence from fossils tells us that the early ferns, araucarian pines and podocarps spread widely throughout Gondwanaland and Laurasia and in many areas may have been the main elements of the forest.

However, in many regions they later gave way to the more advanced flowering plants, that started to appear from the middle Cretaceous onwards. But in New Zealand the podocarps, kauri and many of our ferns, including the magnificent tree ferns, have survived as relics, or "living fossils", because of the land's long-standing isolation (Figs 14.11-14.16).

The ancestors of the ferns and kauri, with origins in the Triassic and late Palaeozoic, may have colonised New Zealand at the same time as the tuatara and New Zealand frog. The podocarps, however, as they evolved somewhat later (Jurassic) and have a poor dispersal ability (their heavy seeds sink quickly in water), may be contemporaries of the moas and kiwis, spreading from Gondwanaland to New Zealand via the improved land links developed in the late Jurassic and early Cretaceous.

Thus we can see that the heritage of New Zealand's link with Gondwanaland in the geological ages past is still apparent today in both plant and animal kingdoms.

CHAPTER FIFTEEN

# MAMMALS AND THE BIRTH OF THE TASMAN SEA

THE *Glomar Challenger*'s drill and magnetic studies have shown that new sea floor started to form in the Tasman Sea about 80 million years ago. This does not necessarily mean, however, that the Tasman was created at this time exactly.

It is thought that an episode of continental drift is preceded by a period of rifting, or breaking apart, of the two landmasses that are later to separate. This occurs some time before substantial areas of new sea floor are created, and an embryonic ocean begins to form.

Thus the splitting-off of New Zealand from Australia probably started with a rifting phase, during which an extensive zone of cracking and splintering appeared along a belt between the two

**Figs 15.1-15.5.** These five diagrams summarise stages in the development of the Southern Ocean and the break-up of Gondwanaland during the Mesozoic and Cenozoic. The reconstructions are based on the theory of an expanding earth (pp. 163–166). All the landmasses have been drawn in their modern form, to aid identification.

The sequence of maps serves to illustrate the influence of changing world geography on animal and plant migrations in the Mesozoic and Cenozoic.

As comparison with other reconstructions will reveal (cf. Figs 13.14, 18.6), although the relationships of continents to one another are reasonably well established, the actual latitudinal and longitudinal positions are less definite. For example, it is thought that on the basis of the fossil flora and fauna, and oxygen isotopes, the southern landmasses illustrated in Fig. 15.1 were situated in middle latitudes (e.g. 45°-50°S Latitude) in the mid and late Jurassic (cf. Fig. 12.26).

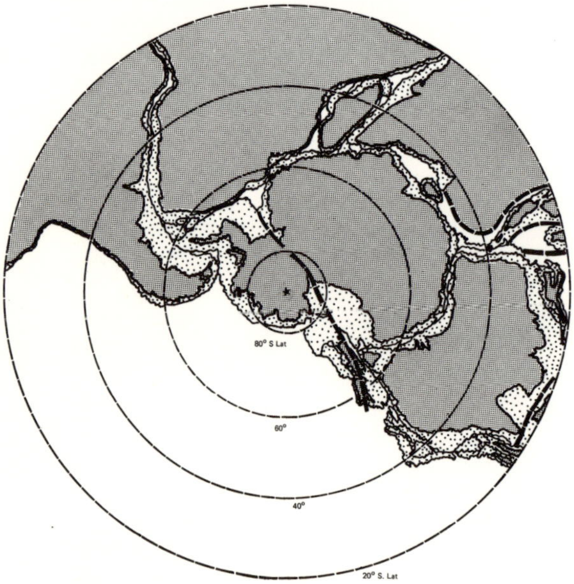


**Fig. 15.1.** The Southern Hemisphere landmasses 180 million years ago (early Jurassic). Based on a global diameter 80 percent of the modern value. Plant and animal migrations were possible across the width and breadth of southern Gondwanaland. New Zealand's moas, kiwis, podocarps etc. probably arrived at about this time (pp. 248–259).

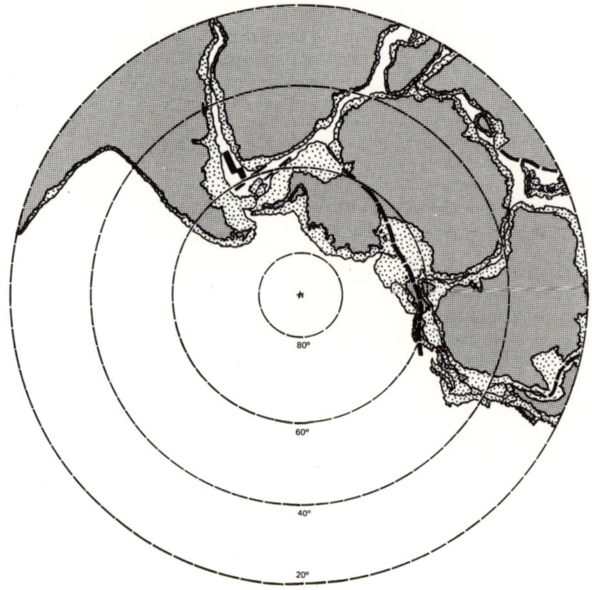


**Fig. 15.2.** The Southern Hemisphere landmasses 120 million years ago (early Cretaceous). Diameter of the earth is 87 percent of the modern globe. Sea-floor spreading has commenced in the South Atlantic, between South Africa and South America, and accompanying rifting movements are occurring between West Antarctica and South Africa.

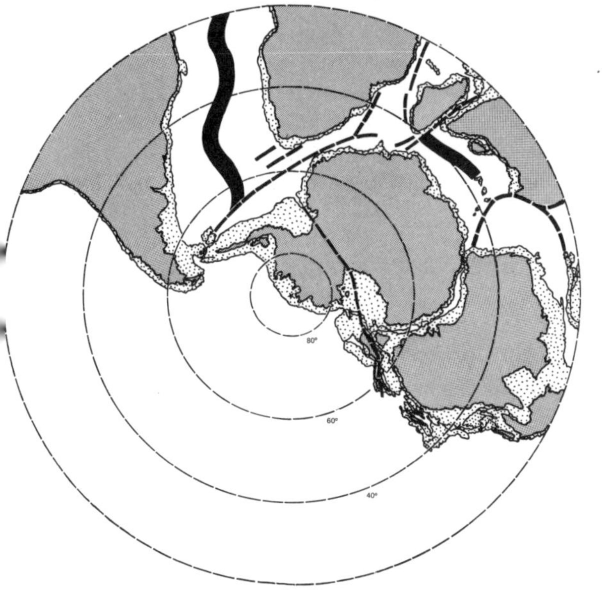

Fig. 15.3. The Southern Hemisphere landmasses 90 million years ago (late Cretaceous). Diameter of the earth is 90 percent of the modern globe. Sea-floor spreading has been under way in the South Atlantic Ocean for 30 million years (cf. Fig. 15.2). Continued rifting has separated South Africa, Malagasy, India and Australia. Sea-floor spreading has commenced in the Indian Ocean, between India and Antarctica.

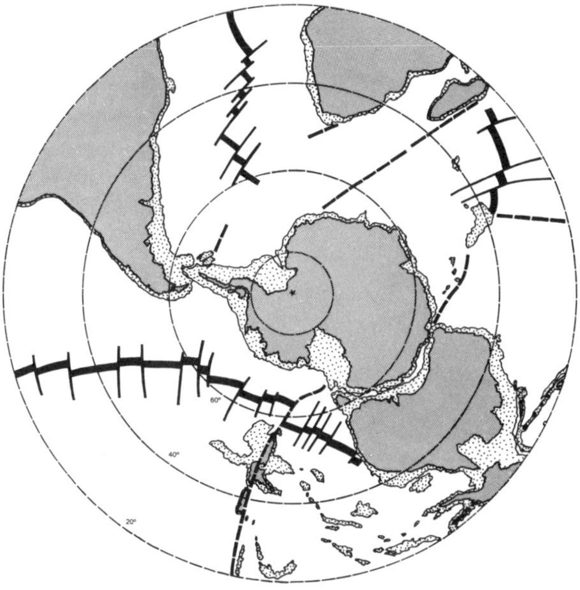

Fig. 15.4. The Southern Hemisphere landmasses 60 million years ago (Paleocene). Diameter of the earth is 93 percent of the modern globe. Sea-floor spreading is actively under way in all the major oceans, except the segment of the Southern Ocean between Antarctica and Australia.

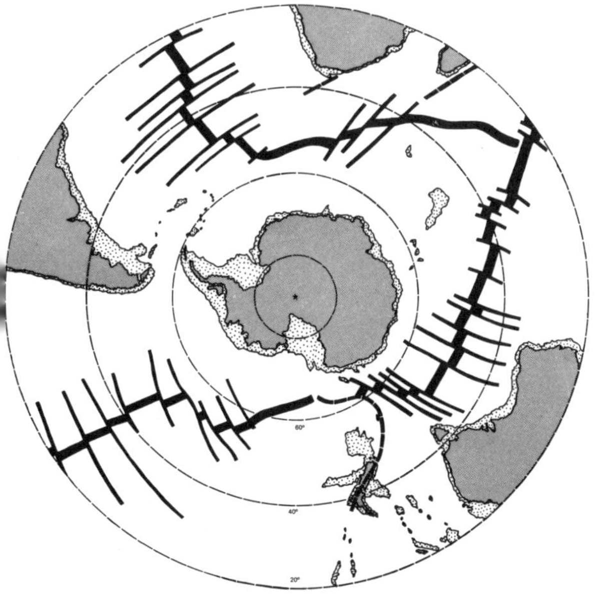

Fig. 15.5. The Southern Hemisphere landmasses today, with active spreading ridges in the Pacific, Indian and South Atlantic Oceans.

continents. Sinking movements probably also accompanied these events, leading to flooding by the sea. As tension continued, and the continents moved further away from each other, the rift zone was completely pulled apart, exposing the underlying sima. A large tensional crack thus developed in the crust, along which melting occurred, magma rose to the surface, a spreading ridge formed, and new sea floor with magnetic stripes began to take shape. It is thought that the date of 80 million years ago, obtained from the magnetism of the Tasman Sea, probably refers to the beginning of this last phase, i.e. that of actual sea-floor production (Figs 15.1-15.5). However, as may be appreciated, this does not necessarily date the cutting of any land links between Australia and New Zealand—it is merely a minimum age and therefore all we can say is that probably there were no land links *after* this date.

To judge from the geological record preserved on both sides of the Tasman, rifting between Australia and New Zealand probably began in the upper part of the early Cretaceous, about 110-105 million years ago, after which fracturing, subsidence and inundation probably cut any land links. Thus those animals and plants needing continuous land for migration probably found their way barred between Australia and New Zealand at about this time (Fig. 15.6).

There was a period, however, probably between about 110 and 80 million years ago, and perhaps a bit later, when the newly created gap between Australia and New Zealand could still be crossed by creatures capable of island-hopping or rafting across oceans, providing they were not too wide and stormy (Figs 15.1-15.5).

By the middle part of the Paleocene, however, some 60 million years ago, magnetic studies indicate that the Tasman Sea had opened up to its present width and from this time onwards any trans-Tasman immigration probably had to span some 1850 km (1200 miles) of open ocean (Fig. 15.4).

As creation of new sea floor drew gradually to a close in the Tasman, the focus of sea-floor spreading shifted towards the south-west, to between Australia and Antarctica, and as a result a wedge of new ocean floor began to appear between these two continents some 55 million years ago.

**Snakes alive!**

While the ancestors of our moas and kiwis were walking to New Zealand from the rest of Gondwanaland via the Jurassic land links, other animals were appearing elsewhere in the world.

For example, the reptilian stock gave rise to snakes. Although, as with most terrestrial creatures, we are again dogged by a poor fossil record, it is generally thought that the snakes evolved from lizards, and that they first appeared on the south-eastern margin of Laurasia (i.e. today's Central Asia) in the late Cretaceous, about 100-65 million years ago. It is believed that these first snakes successfully colonised a great diversity of environments (as modern ones do), and soon spread throughout Laurasia.

At this time Laurasia and Gondwanaland were still virtually intact (as in Fig. 11.13), and little rifting had as yet occurred. Thus the newly evolved snakes were able to migrate via land connections into Gondwanaland. Snakes were consequently able to spread, for example, from Central Asia, into South-east Asia and then via Indonesia into New Guinea and Australia. A route into Australia from South America, via Antarctica, may also have been possible at this time, although there are indications that this was a cool-temperate route (pp. 215–220), and for this reason may have been unattractive to cold-blooded creatures such as snakes.

Thus towards the end of the Cretaceous, or perhaps in the Paleocene, some 65-60 million years ago, snakes probably arrived in Australia. However, as we have seen, by this time New Zealand had lost its land links to the rest of Gondwanaland, and sea-floor spreading, and creation of ocean, had been proceeding in the Tasman Sea since at least 80 million years ago. New Zealand and New Caledonia (which at that time formed essentially the one landmass) thus missed out on snakes—much to the relief of New Zealanders and New Caledonians today!

**The rise of mammals**

At about the time the first snakes were appearing in Central Asia, tiny ancestral mammals were running across the plains of North America. Mammals, like birds, had developed a warm-blooded circulation system and could therefore cope with a great range of environments. Cold weather did not restrict their activities as it did those of reptiles. They could thus "compete" with the environment and, because of this, and also as many were very efficient predators, they soon gained the upper hand. The reptile's method of reproduction is by laying eggs (as the turtle does) and, as the eggs are generally left unattended, mortality is high. The early mammals probably had quite a taste for dinosaur eggs and this may have been one of the factors that hastened the demise of the dinosaurs.

Mammals also have a vastly superior reproduction system. The word mammal is coined from the mammary glands, or milk-producing organs that all mammals use to feed their young. These glands ensure that there is a ready supply of easily digested milk for the young mammal and, as a by-product, engender the "mothering instinct" necessary to care for and protect the young.

It is thought that the first mammals to appear were the ancestors of the modern platypus and echidna (or spiny anteater), collectively called the monotremes. These animals share many anatomical features with the reptiles: the females lay eggs as birds and reptiles do, but unlike these, have hair and suckle their young after hatching.

The appearance of the first monotremes was probably soon followed by that of the ancestral marsupials (or pouched mammals). The marsupials show an advance over the monotremes in their method of reproduction and in many other ways are that much further removed from their reptilian ancestors.

Young marsupials are born at a very immature stage, and crawl from the birth canal along a well-defined parting in the mother's fur, leading to the "marsupium" or pouch. Once inside the pouch, the young attach themselves to mammary glands and remain there until fully developed. Well-known marsupials are the kangaroo, opossum, koala and wombat.

Somewhat later in geological time, probably during the Paleocene or Eocene (65-37 million years ago), more advanced mammals appeared. These are called "placentals" because all of them have a placenta, a complex feeding structure, within them. In placentals the young are kept within the female's body for a considerable time and are born at a stage when they are quite advanced in development. Whilst within the female's body the young are enclosed in a uterus and are fed through the placenta, a pad-like structure of blood vessels providing a link and filter between the mother's and infant's blood systems. Dissolved foods pass through the placenta into the infant, and the infant's waste products pass via the placenta into the mother's blood system. When the infant is born it can often fend for itself immediately after birth (as do lambs, calves and foals), but is fed by the mother's mammary glands for some time. The human infant, because of the great development of the brain, has to be born at an immature stage (otherwise the large head could not pass through the mother's birth canal) and thus requires long parental care.

Soon after their appearance in North America in late Cretaceous times the marsupials became very widespread and were a very successful and diverse group, producing a wide range of forms, filling many ecological niches and today ranging in size from that of a mouse to a large dog (marsupial wolves), although elephant-sized forms are known as fossils.

Marsupials migrated to South America, from North America, in late Cretaceous and early Tertiary times and are thought to have reached Australia at about the same time. The route the marsupials took to reach Australia is not known—it could have been via Asia or Antarctica. Unfortunately, however, we encounter a "blank" in our geological record in both these areas, as they lack fossil marsupials. Nonetheless, the Antarctic route is generally favoured, as we know that Antarctica and Australia were largely in contact up to some 55 million years ago (Figs 16.14, 16.15, 18.6), whereas the history of links across the Indo-Pacific region is rather obscure.

Whatever the route taken, it is clear that links to Australia had been broken by the time placental mammals had appeared and were beginning to spread, set on a course to conquer the world. It is thought that the placentals originated in Eurasia some time at the start of the Tertiary period—for this reason often termed the "Age of the Mammals".

Marsupials were thus isolated in Australia and, free from other mammalian competition, evolved rapidly and took up many ecological niches, just as placentals did on other continents.

Australia, with its unique fauna, remained isolated throughout Tertiary times, but as the gap between it and Antarctica widened, and Australia drifted northwards closer to Asia (Figs 16.14–16.19), the chances of rafting or island-hopping improved. Thus in the Oligocene some bats and later, in the Pliocene, members of the mouse family, were able to gain entry to Australia, probably via Asia. During the early Pleistocene period, about two million years ago, the ancestors of the Aborigines migrated to Australia from South-east Asia, aided by the low sea levels of the glacial periods that exposed and made accessible vast areas of continental shelf and provided stepping stones across the Indonesian region (p. 328). Later, the racially distinct Melanesians migrated from the same source, but did not reach Australia.

The first Aborigines brought the dingo, while the European settlers introduced domestic and other placental mammals. Over the years the predations of these animals have steadily decreased the populations of the various Australian marsupials. Nonetheless, a natural boundary between the Australian and Oriental faunas, dating back to the Tertiary isolation, remains obvious even today (see pp. 294–304).

To sum up, we can see that the natural distribution (as distinct from human-induced introductions) of animals such as snakes and mammals was very much controlled by the continental movements under way at the time they

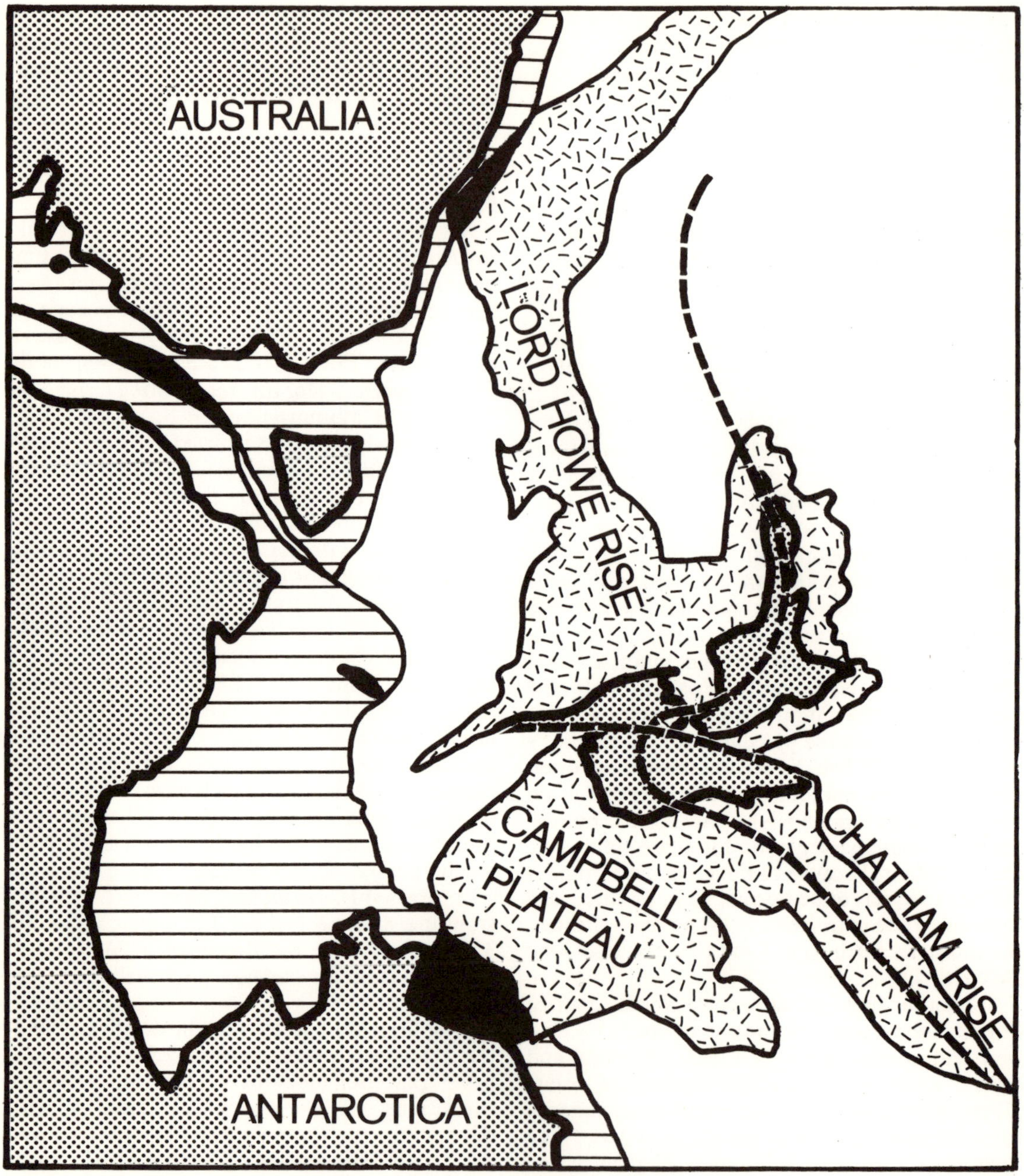


were spreading throughout the world. As we have seen, a lot of things were starting to happen in Gondwanaland in Cretaceous and early Tertiary times. Splitting movements were starting to occur that, as time went on, were to eventually break Gondwanaland into fragments. India was starting to pull away, to embark on its long journey northwards. A split was appearing between Africa and South America, and sea-floor spreading was occurring to either side of a newly formed oceanic ridge. Similar splitting movements were also beginning to affect New Zealand and Australia and by the time the snakes and marsupials reached Australia, they found

**Fig. 15.6.** Sea-floor spreading, accompanied by creation of new oceanic crust, began between Australia and the Lord Howe Rise about 80 million years ago. Between 80-60 million years ago the Tasman Sea opened up to its full width and New Zealand (and probably New Caledonia) were left isolated in the South-west Pacific. The episode of sea-floor spreading was preceded, however, by rifting and subsidence, during which any land links were probably broken. Thus the terrestrial migration routes between Australia and New Zealand, that probably existed in the Jurassic, were probably cut early in the Cretaceous, some 110 million years ago.

This map illustrates the reconstruction obtained if some 450 km (280 miles) of horizontal movement is reversed on New Zealand's Alpine Fault, Australia and Antarctica moved together, and the New Zealand-Lord Howe block moved without distortion into a flanking position. Some overlaps occur, as indicated by the black areas. The Tasman Sea opened up between the Lord Howe Rise and Australia (cf. Fig. 15.4).

Other reconstructions (cf. Figs 13.14, 18.6, 18.22) introduce postulated fault movements and distortions of the various crustal blocks to achieve a better "fit" of the jig-saw puzzle.

their passage blocked by extensive stretches of the infant Tasman Sea and, as far as we know, they did not ever manage to migrate to New Zealand.

As the snakes and marsupials were populating Australia, the first placentals were starting to appear in Central Asia. The placentals, predators *par excellence*, soon replaced the less efficient marsupials in Asia and Europe and gradually moved southwards into South-east Asia. However, by the time they were poised to follow the marsupials into Australia all available routes had been disrupted and Australia was isolated.

Thus by a quirk of timing of the earth movements that were starting to split Gondwanaland at that time, New Zealand was left isolated, bearing on its surface a remnant of ancient Mesozoic plants and animals. Australia, on the other hand, was left with a remnant of the once-numerous marsupial faunas, that elsewhere in the world had disappeared under the onslaught of the placentals. Australia and New Zealand are thus in many ways biologically unique.

Before its discovery by the human race, New Zealand's fauna was "unbalanced" biologically because of isolation dating back to Mesozoic times. The only native land mammal was a bat, which had presumably flown in across the infant Tasman Sea. As we have seen, many niches in the environment, elsewhere filled by mammals, were in New Zealand occupied by flightless birds, large insects and other non-mammalian forms.

In the birds, for example, the absence of predators encouraged flightlessness and nesting on the ground—features that were to prove the downfall of many species once carnivorous mammals were introduced. Races of giants developed, such as the moas. Some bird groups expanded and developed endemic families, like the New Zealand wattle birds and wrens.

As New Zealand has been so remote for so long, mammals could not arrive casually, e.g. on floating logs. Polynesians, storm-blown in canoes, were probably the first strictly terrestrial mammals to set foot in New Zealand, possibly more than 1000 years ago, and were followed later, traditionally in A.D. 1350, by better-organised Polynesian voyagers bringing dogs and Polynesian rats.

Because of in-breeding, extinction has often been the lot of island communities and in New Zealand a lot of "thinning" has undoubtedly gone on—weeding out the animal and plant pioneers that originally travelled to New Zealand in the Mesozoic—so that considerable gaps now remain.

Another thinning-out also occurred in the "Ice Age" or Pleistocene period (2 million-10,000 years ago), when New Zealand experienced a series of cold episodes. Our isolated island situation meant that warmth-loving organisms could not escape northwards when temperatures fell, so each cold episode brought repeated decimation (pp. 342–344).

**The rise of the flowering plants**

As New Zealand drifted away from Australia, it also carried on its landscape populations of ancient ferns, araucarian pines and podocarps that were then abundant in Gondwanaland (p. 259). At this time, in middle Cretaceous, some 100 million years ago, new plants were appearing that were to achieve the acme of plant development. These plants, that were to develop into the flowering plants of today, are called the angiosperms. Angiosperms have been enormously successful, as, like mammals, they have great powers of reproduction and survival. Non-woody perennial types can survive winter by dying down to their roots in autumn and sprouting new shoots in spring. Trees discard their leaves (and use them as a vehicle for discarding their waste products) and become dormant. But the flower is the most important angiosperm structure, and is

*Fig. 15.7*

**Fig. 15.7.** The New Zealand forests of Southern Beech (*Nothofagus*) are very similar to those found in southern South America—even to the extent of having the same plant and insect parasites inhabiting the bark of the trees. Maruia Saddle, Nelson.

unique to angiosperms. The flower is reproductively far more efficient than the separate male and female cones of the conifers. Usually containing both pollen (male) and ovules (female), the flower may be pollinated either by wind, or by insects it attracts (by colour, perfume or nectar), or even by itself.

In the middle and late Cretaceous angiosperms evolved and spread rapidly throughout the world. They were probably moving into Australia at about the same time as the marsupials and snakes and no doubt used the same land routes—via South-east Asia or Antarctica. Like the snakes and marsupials, the angiosperms found the infant Tasman Sea blocking their passage to New Zealand. Nonetheless, judging from our records of fossil pollen, New Zealand did receive some of the early angiosperms very soon after they first appeared—in the mid-Cretaceous, 100 million years ago. It is clear, therefore, that many of the angiosperms were well-equipped for dispersal across open ocean and to them the Tasman presented little difficulty. We know, however, that some angiosperms were not so well-endowed, but nevertheless they also managed to get to New Zealand. The Southern Beech, a major constituent of many of our modern forests, is in this latter category.

## Southern Beech

The New Zealand beech species belong to the

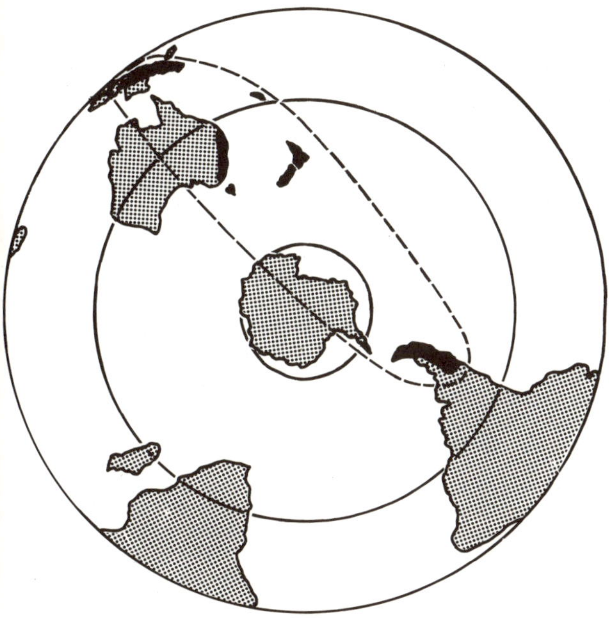

**Fig. 15.8.** Today the Southern Beech (*Nothofagus*) is found in New Guinea, New Caledonia, eastern Australia, Tasmania, New Zealand and southern South America. In the past it grew in Western Australia and West Antarctica. Such distribution (occurrences in black, boundary of distribution indicated by the broken line) led to speculation on how the plant spread, as it is not readily dispersed by wind, and the seeds die rapidly after exposure to salt water. *Nothofagus* undoubtedly needed continuous land for its dispersal, indicating that very good southern land links were available in the middle Cretaceous (100 million years ago), when the Southern Beeches first appeared (cf. Fig. 15.9).

**Fig. 15.9.** In the early and middle Cretaceous (110-90 million years ago) the southern landmasses were closely grouped in cool-temperate latitudes, enabling the Southern Beeches and other plants and animals to migrate and achieve a wide distribution. The reconstruction used is one derived from palaeomagnetic studies.

genus *Nothofagus* ("false beech") of the family Fagaceae which also includes the well-known Northern Hemisphere *Fagus* (the true beech), as well as the oak and chestnut. *Nothofagus* species are restricted to the Southern Hemisphere and so are often known as the "southern beeches". Species occur in New Guinea, New Caledonia, eastern Australia, Tasmania, southern South America and New Zealand. Fossil wood and pollen suggest that in the past southern beeches populated Antarctica and Western Australia. There is no record of them, however, in South Africa or Madagascar (Figs 15.7, 15.8).

Fossil evidence suggests that the beeches colonised these southern countries in the middle and late Cretaceous and for many years—in fact, since 1847 when Joseph Dalton Hooker (who had been a naturalist on the southern cruises of H.M.S. *Erebus* and H.M.S. *Terror* and was later a renowned Director of the Royal Botanic Garden at Kew) started publishing descriptions of many southern plants in his great *Flora Antarctica* and *Flora Novae Zealandiae*—there has been endless speculation as to how such a distribution could have been achieved. This is because *Nothofagus* is very poorly equipped for trans-oceanic travel. Its seeds sink very quickly in water, and die if exposed to salt water. Therefore the seeds could not float across the ocean, nor could they be transported on floating logs, as splashing by salt water would kill them. *Nothofagus* seeds also cannot withstand freezing, so transport by means of floating ice is not possible. The seeds are not attractive to birds, making bird transport unlikely. Finally, *Nothofagus* is not easily spread by wind, as exemplified by the fact that it has not re-established itself, after the deforestation of the Ice Age, on Stewart Island (32 km, 20 miles, off New Zealand) nor on Great Barrier Island, a mere 18 km (11 miles) from the nearest seed source.

Thus *Nothofagus* undoubtedly needed virtually continuous land for its distribution and, as a consequence, many of the early botanists postulated extensive land bridges, connecting the southern lands now forested with *Nothofagus*. Continental drift, however, has provided a mechanism to achieve such dispersal without the need to cross oceans. It has also solved the apparent enigma of fossil *Nothofagus* in Antarctica in the Cretaceous and early Tertiary, for, as indicated in Fig. 15.9, Antarctica was in more temperate latitudes at this time.

A conclusion we can draw from the *Nothofagus* story (Fig. 15.10) is that although land links between Australia and New Zealand may well have broken by the time these plants arrived on the scene, New Zealand retained fairly good land links to the south (to Antarctica and South America) for some considerable time after the Tasman started to open.

Other plants, also with poor dispersal ability, accompanied *Nothofagus* on its southerly route. They included members of the Magnolia, Protea and Fuchsia families.

Today, representatives of the Magnolia family in New Zealand include tawa, pukatea and horopito, but others of the family are found elsewhere in the Southern Hemisphere, on other fragments of Gondwanaland.

The family Proteaceae is now only represented in New Zealand by the rewarewa (*Knightia excelsa*) and toru (*Persoonia toru*). But fossil evidence indicates that before the Ice Age the family was numerous and diversified in New Zealand, particularly in the Eocene, during which time it was also present in Antarctica. The New Zealand Proteaceae have very close links with Australia, New Caledonia and southern

**Fig. 15.10.** Occurrences of fossil pollen enable geologists to chronicle the establishment of New Zealand's beech forests. The photo is of a single fossil beech pollen grain extracted from rocks in Marlborough laid down some 100 million years ago. The pollen grain measures 40/1000 mm in diameter.

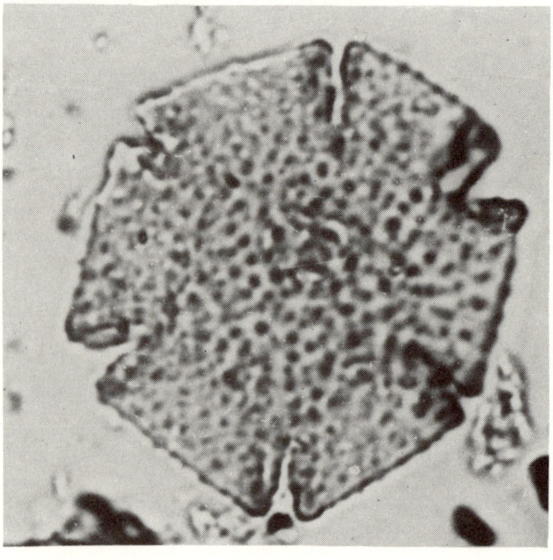

South America, but less close links to South Africa.

The colourful *Fuchsia* that graces many of our gardens today and is a conspicuous shrub in our forests also arrived via the same southern route and has a good fossil record in New Zealand. Modern *Fuchsia* species are distributed in the southern Pacific and Central and South America.

The effectiveness of these southern land links at the close of the Mesozoic times is strikingly demonstrated by the fact that *Nothofagus* forests in New Zealand and southern South America have the same parasitic fungus plants, moss plants and flightless sucking bugs inhabiting their bark—suggesting that one continuous southern forest probably linked the two countries in these times.

As may be expected, the forested landmass was bounded by coastlines offering excellent shallow-water migration routes and, as we have seen (p. 220), throughout most of Cretaceous time New Zealand received waves of immigrant coastal marine creatures with strong southern affinities (the Austral faunas), attesting to the effectiveness of such routes. Oxygen isotope temperature results from these Austral faunas indicate, furthermore, that the route was cool-temperate in character (pp. 214–220).

Thus in summary we can say that two land links—a northern warm-temperate, or sub-tropical, and a southern, cool-temperate—joined New Zealand to other Gondwana countries and were available for animal and plant migrations in Jurassic and to some extent Cretaceous times. Between them these links made unique contributions to the New Zealand flora and fauna and the timing of their opening and closing was a crucial factor in allowing some animals and plants in, but excluding others.

All available evidence suggests that northern land links had virtually been broken some time in the early and middle Cretaceous (probably by about 110 million years ago, at the latest) but that good southern land links, via the Campbell Plateau and Marie Byrd Land (Western Antarctica) persisted long enough into the Cretaceous to allow pioneer angiosperms to reach New Zealand.

To judge from New Zealand's marine fossil record we continued to receive many shallow-water marine immigrants from Antarctica and Patagonia right through to the end of the Cretaceous, some 65 million years ago, indicating that shallow-water links were still present. But it looks as though actual land links had disappeared before this, as marsupials had spread to South America at about this time and, as far as we know, none used this southern route to get into New Zealand, although they could probably get across or around Antarctica to Australia, which was still firmly joined to Antarctica at this time.

Magnetic stripes preserved on the sea floor between the edge of the Campbell Plateau and Antarctica indicate that creation of new sea floor was occurring in this region some 80 million years ago (Fig. 18.5). Thus if we allow time for a certain amount of rifting and subsidence before this sea-floor spreading phase, some cutting of southern land links may possibly have started about 100-80 million years ago.

Shallow-water marine links, as distinct from land links, between New Zealand, West Antarctica and southern South America continued into the early Tertiary; in the Paleocene and early Eocene, 65-50 million years ago, these countries were still sharing many coastal marine animals. However, some of the newcomers to the flora and fauna of these times came to New Zealand across the newly formed Tasman Sea and had northern (Australian and Malayo-Pacific) affinities.

These northern affinities became more dominant towards the close of Eocene times, between about 50 and 37 million years ago, while Austral affinities declined, and from this it can be inferred that the shallow-water links between southern South America and New Zealand had probably broken some time in the late Eocene. New Zealand was thus left on its lonesome, surrounded by deep seas, and all later faunal and floral imigration to its shores was restricted to those groups capable of trans-oceanic dispersal.

CHAPTER SIXTEEN

# CHANGES IN LAND AND SEA

AS THE ancestral New Zealand landmass moved steadily away from Australia and Antarctica, and became more and more isolated geographically, events and processes were occurring on its surface that profoundly altered the landscape.

An extensive landmass, with steep slopes and mountainous terrain in many areas, had been created by the series of earth movements (Rangitata Orogeny) that gradually brought the New Zealand Geosyncline phase to a close between 136 and 110 million years ago. The new land stretched at least as far as New Caledonia in the north, to Campbell, Bounty and Auckland islands in the south, to the Chatham Islands in the east and probably some distance along the Lord Howe Rise to the west (Fig. 18.1).

The newly created land was very soon attacked by rivers and streams, and waves nibbled at its edges. Erosion didn't have its own way entirely, however, as along the West Coast of the South Island and probably elsewhere, earth movements associated with the rifting movements preceding

**Fig. 16.1.** The massive bluff at Hawks Crag, in the Lower Buller Gorge, Westland, and other neighbouring outcrops, provide evidence that in the early Cretaceous, 110 million years ago, rift-valley-like features developed along the West Coast of New Zealand's South Island, as rifting movements began between Australia and New Zealand. These rifts were the forerunners of major splits that culminated in the formation of the Tasman Sea. The rocks of Hawks Crag and other nearby areas comprise coarse gravels and rock debris derived from the fresh steep scarps bounding the newly-formed rift valleys.

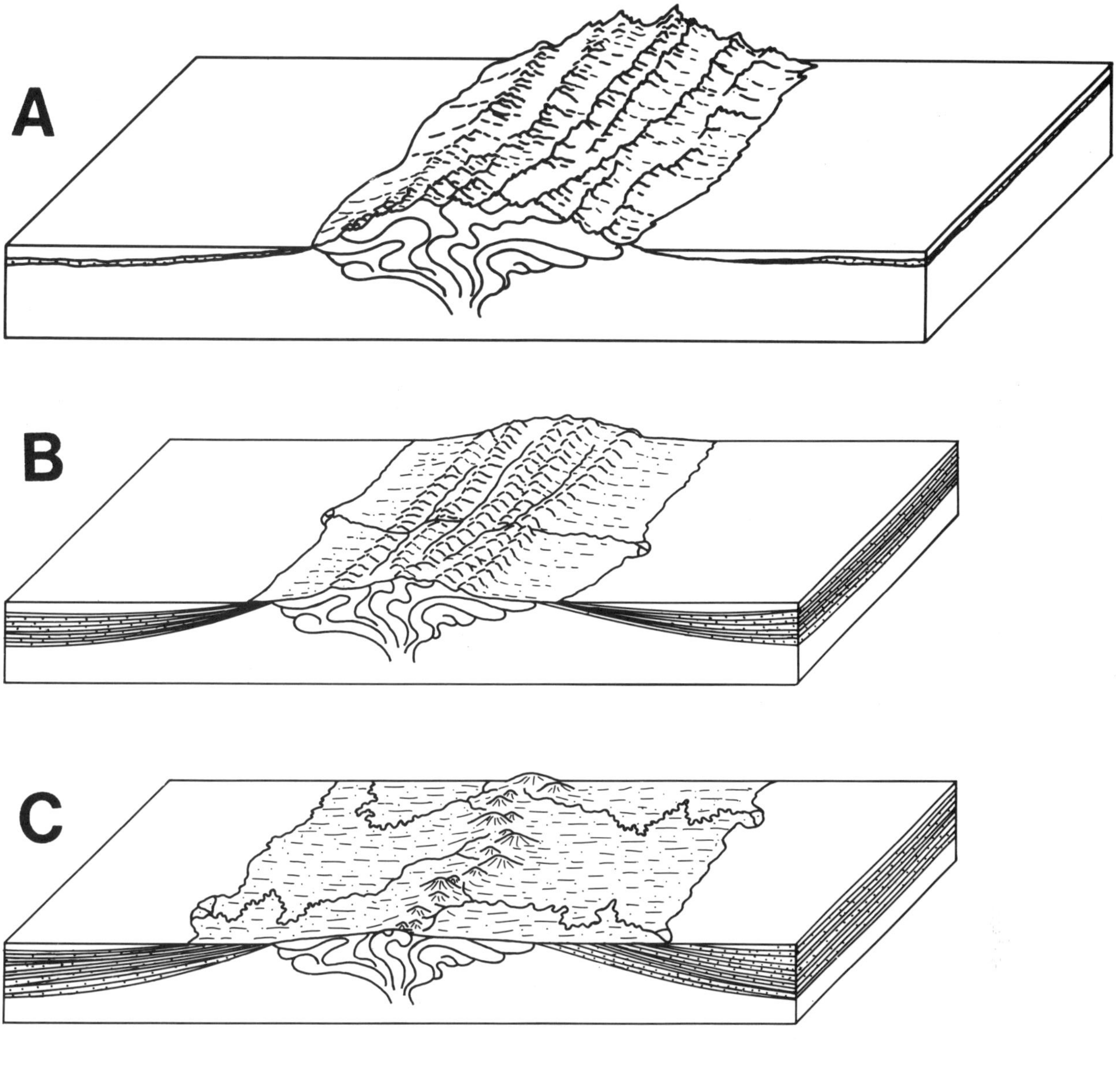


**Fig. 16.2.** The earth movements of the Rangitata Orogeny (primarily those occurring in the late Jurassic and early Cretaceous, 162-110 million years ago) created a new landmass, much larger than modern New Zealand, that extended northwards towards New Caledonia and southwards to Campbell and Auckland islands (top diagram).

This new landmass was attacked by erosion (rivers, streams and the sea) and progressively levelled, the products of such erosion being deposited in the sea to either side (middle diagram). By middle and late Cretaceous time (100-65 million years ago), some parts of New Zealand had been worn down almost to sea level—a peneplain being formed (bottom diagram).

Other parts, however, were not worn down to the peneplain stage until the Paleocene or Eocene (65-37 million years ago).

the opening of the Tasman created new rugged terrain. Thus we can see evidence that in the early Cretaceous, about 110 million years ago, rift-valley-like features, bordered by massive scarps, formed along the West Coast into which poured masses of gravel and other rock debris (samples of which can be seen at Hawks Crag in the Lower Buller Gorge; Fig. 16.1).

Soon, however, the forces of erosion regained the upper hand and large areas of the new land were progressively lowered in elevation and eaten away by erosion until in late Cretaceous, between about 100 and 65 million years ago, substantial

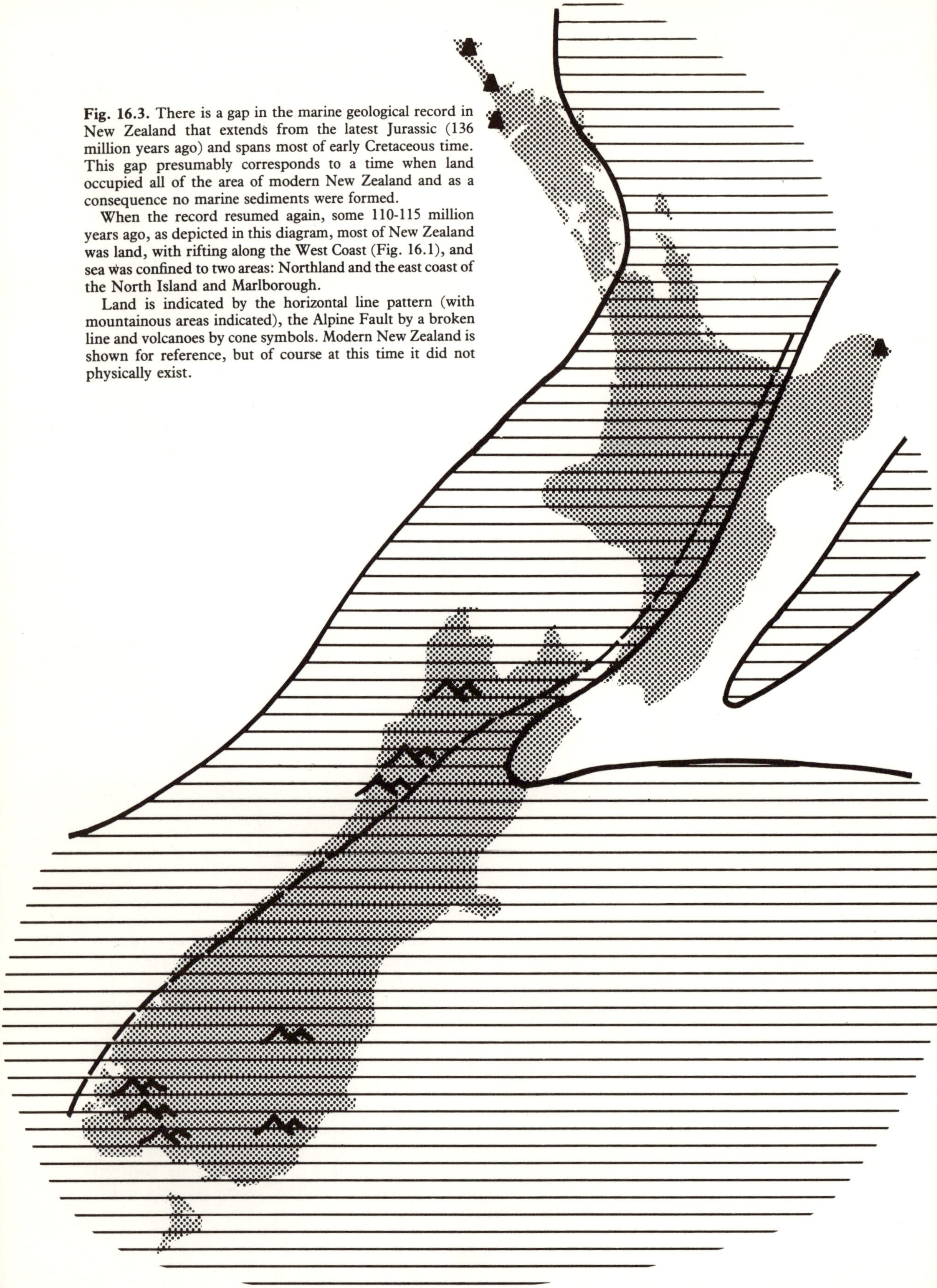

**Fig. 16.3.** There is a gap in the marine geological record in New Zealand that extends from the latest Jurassic (136 million years ago) and spans most of early Cretaceous time. This gap presumably corresponds to a time when land occupied all of the area of modern New Zealand and as a consequence no marine sediments were formed.

When the record resumed again, some 110-115 million years ago, as depicted in this diagram, most of New Zealand was land, with rifting along the West Coast (Fig. 16.1), and sea was confined to two areas: Northland and the east coast of the North Island and Marlborough.

Land is indicated by the horizontal line pattern (with mountainous areas indicated), the Alpine Fault by a broken line and volcanoes by cone symbols. Modern New Zealand is shown for reference, but of course at this time it did not physically exist.

Fig. 16.4. Siltstones and sandstones of mid-Cretaceous age exposed in the Wharfe Gorge, Clarence Valley, Marlborough.

parts had disappeared beneath the sea (Fig. 16.2). We know, for example, that towards the end of the Cretaceous much of the Lord Howe Rise that earlier, in the Jurassic and early Cretaceous, had probably provided some of the land links traversed by the moas and kiwis, had been inundated by the sea. Sea also covered major parts of coastal New Zealand in the east—Marlborough, Wairarapa, Hawkes Bay, East Cape, Northland (Figs 16.3, 16.4).

By the close of the Cretaceous, 65 million years ago, many of those parts of New Zealand not already under the sea had been worn down almost to sea level (Fig. 16.2). Coal swamps developed in some low-lying areas, notably in the western approaches to Cook Strait (as revealed by off-shore oil prospecting drill holes), the Greymouth area and coastal South Otago (Figs 16.3, 16.5, 16.8).

### The New Zealand peneplain

The end-product of the erosion of a landmass is called a peneplain (Fig. 16.2). As some parts of the original ancestral New Zealand landmass took longer to erode down than others because of harder rocks, distance from a coastline etc., the New Zealand peneplain was developed at different times in different parts of the country. Thus in coastal eastern New Zealand (Marlborough, Wairarapa, Hawkes Bay, East Cape, Northland etc.) the land had been worn down to a peneplain by at least late Cretaceous time, 100-65 million years ago. In other parts of the country, however, the peneplain did not develop until the Paleocene or Eocene (65-37 million years ago) (Figs 16.5, 16.8).

All this time while the ancestral landmass was being worn away (some 40 to 75 million years, depending on the part of the country), the sediment brought down from the eroding land by rivers and streams was being dumped in the sea to either side of the land, to form flanking geosynclines (Fig. 16.2).

On the land the greywacke rocks of which the landscape was largely composed were gradually broken down into their mineral components. Many of the minerals (feldspars and micas) in the greywacke rock are relatively soft and readily decompose to form muds. But quartz, present in both the rock itself and as veins filling cracks and crevices, is a very hard mineral and resists erosion. Thus, as millions of years of erosion passed by, the softer minerals were reduced to muds and carried out to sea by rivers and streams. But while some of the quartz was carried out to sea as sand, a lot of it remained as white-coloured gravel and sand in low-lying areas of the landscape, on valley floors and river beds. These quartz deposits, when buried by later rocks, became consolidated as the individual grains packed down together. Then sometimes water percolating through the deposits brought in further quartz in solution, and this was deposited in the gaps between individual grains. In this way quartz grains became cemented together, and the resulting hard flinty light-coloured rock is called a quartzite. Quartz deposits and quartzites are known from many areas of New Zealand (e.g. Otago, Southland, Westland, Canterbury, N.W. Nelson, Western Wellington), and all date back to the time the peneplain was forming. Similar deposits on the Chatham Islands and Campbell Island indicate that the peneplain also extended well beyond the modern New Zealand shores.

At the same time, a certain amount of quartz was dissolved by rivers and streams and trans-

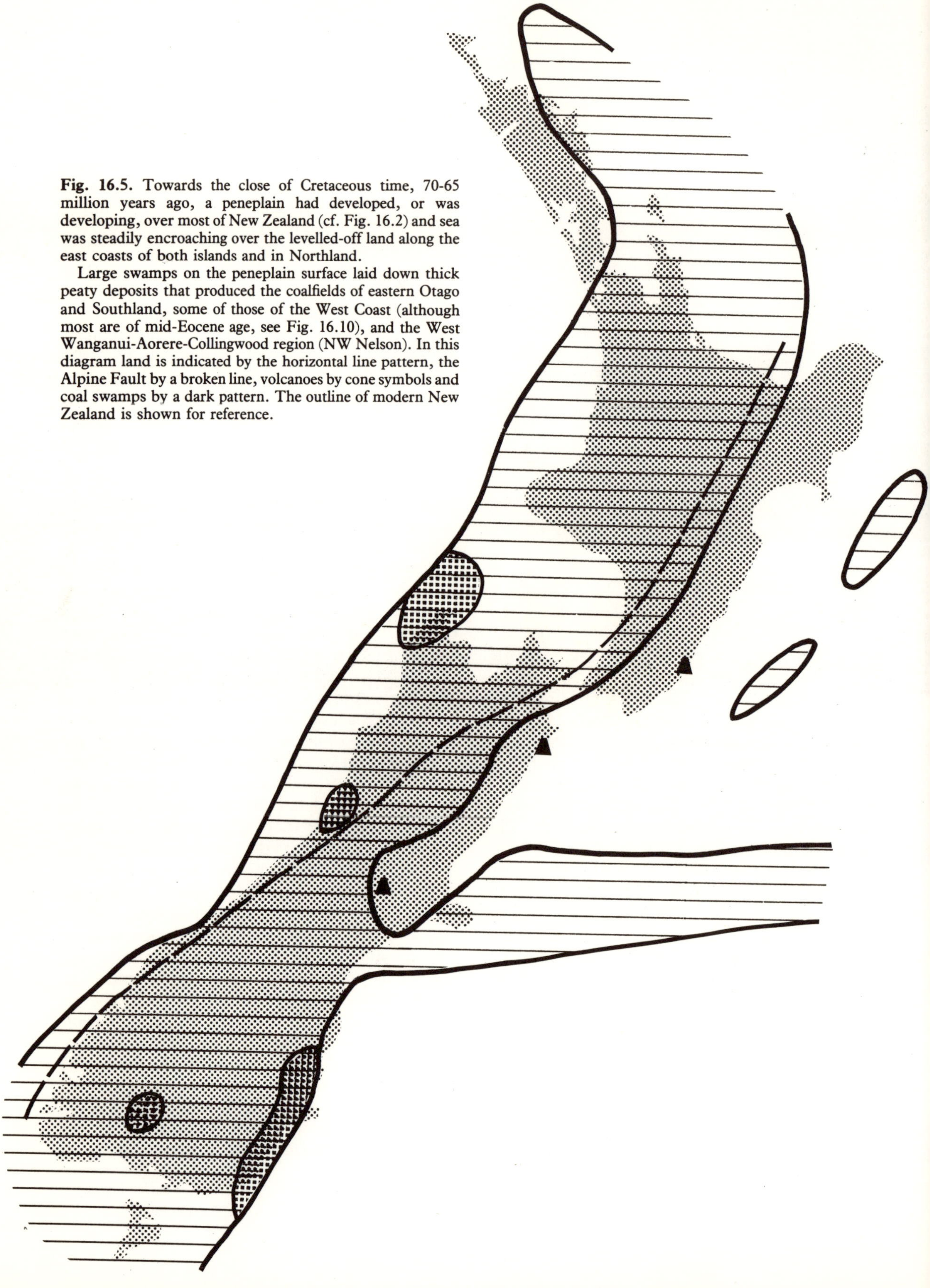

**Fig. 16.5.** Towards the close of Cretaceous time, 70-65 million years ago, a peneplain had developed, or was developing, over most of New Zealand (cf. Fig. 16.2) and sea was steadily encroaching over the levelled-off land along the east coasts of both islands and in Northland.

Large swamps on the peneplain surface laid down thick peaty deposits that produced the coalfields of eastern Otago and Southland, some of those of the West Coast (although most are of mid-Eocene age, see Fig. 16.10), and the West Wanganui-Aorere-Collingwood region (NW Nelson). In this diagram land is indicated by the horizontal line pattern, the Alpine Fault by a broken line, volcanoes by cone symbols and coal swamps by a dark pattern. The outline of modern New Zealand is shown for reference.

**Fig. 16.6.** This view, deep in the "Cretaceous country" of inland Marlborough, spans a large part of the Cretaceous record of New Zealand.

The hills in the left foreground are composed of mid-Cretaceous sediments (110-80 million years old), that pass upwards (towards the right) into late Cretaceous (80-65 million years old). These are in turn overlain by greensands that traverse the Cretaceous-Tertiary boundary (65 million years ago) and pass upwards into the Amuri Limestone (65-55 million years ago) capping the prominent steep-faced ridges in the right foreground and middle distance (the Chalk Range).

Mt Tapuaenuku in the background (2885 m; 9465 ft) consists of mid-Cretaceous volcanic and sedimentary rocks, on the other side of the Clarence Fault, that runs behind the line of Amuri Limestone ridges.

ported out to sea, there to be incorporated in sea-floor muds and sands. Thus many of the sediments then being laid down in the sea offshore from the developing peneplain were rich in quartz, and such sediments are common all along the eastern part of New Zealand: Marlborough, Wairarapa, Hawkes Bay, Gisborne, East Cape and Northland.

Iron and calcium were also being leached from the landmass while it was being worn down. The iron compounds were laid down on the sea floor to form greensands, and the calcium compounds to form limestones, limey muds and sands. So all in all this time in New Zealand's geological history was marked by a great development of "out-of-the-ordinary" sediments instead of the rather monotonous mudstones and sandstones that had dominated the New Zealand Geosyncline (e.g. Fig. 16.6).

As we passed from the Mesozoic into the Tertiary, therefore, large areas of New Zealand

Fig. 16.7

**Fig. 16.7.** Many plants and animals became extinct at the end of Cretaceous times, 65 million years ago. The large ribbed clams, *Inoceramus*, that are distinctive fossils of Jurassic and Cretaceous marine sediments in New Zealand disappeared at this time, as did ammonites and belemnites. A Jurassic *Inoceramus* 150 million years old is on the left and a Cretaceous one 100 million years old is on the right.

**Fig. 16.8.** In the Paleocene, 65-55 million years ago, the old New Zealand landmass was worn down still further and the peneplain (Fig. 16.2) became very widespread.

The sea continued to inundate the margins of the peneplain—extending further westwards on the east coasts of both islands, and for the first time since the end of the Jurassic (Fig. 13.3), started to encroach on the west coasts.

Large swamps occurred on low-lying parts of the peneplain in what is now the western approaches to Cook Strait. Coals formed from these swamps and later ones (mid-Eocene; see Fig. 16.10) were found in oil-prospecting drill holes off-shore from Taranaki and Nelson, and it has been suggested that the oil and gas of the Kapuni and Maui fields originated as a by-product of the coal-forming processes.

In this diagram land is indicated by the horizontal line pattern, the Alpine Fault by a broken line, volcanoes by cone symbols and coal swamps by a dark pattern. Modern New Zealand is shown in outline, for reference.

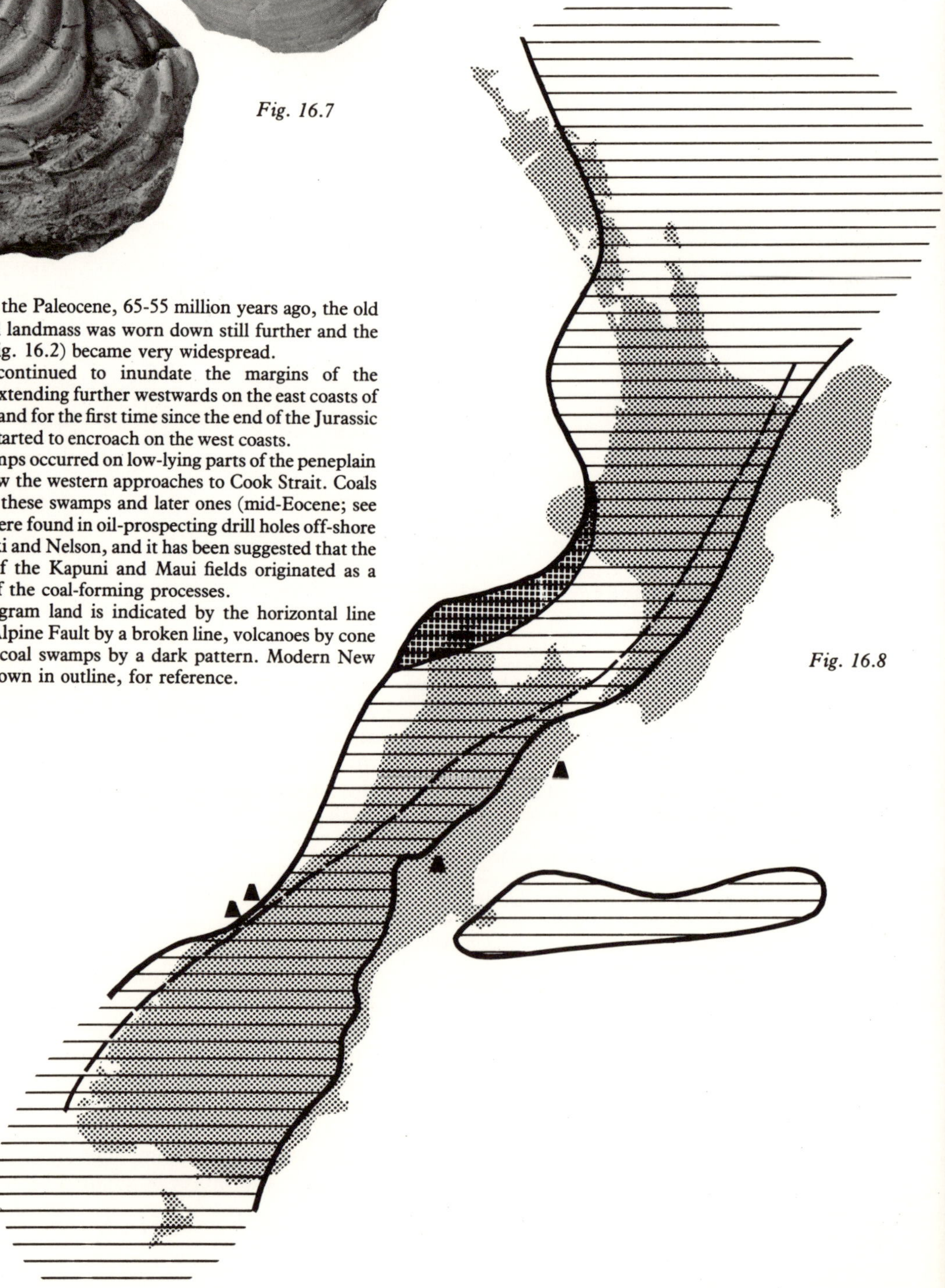

Fig. 16.8

**Fig. 16.9.** The Cretaceous-Tertiary sequence is continuous in a number of areas in New Zealand, notably in Marlborough and Hawkes Bay. The upper photo shows the rocks exposed at Woodside Creek, Marlborough. The boundary is situated half way up the cliff face.

The lower photo is in the Chalk Range, Inland Kaikoura Range, Marlborough (cf. Fig. 16.7). The ridge is capped by Amuri Limestone and the Cretaceous-Tertiary boundary is situated in the strata immediately underlying.

had been worn down to the peneplain state, or were well on the way towards it, and other parts had been worn down below sea level. Opening of the Tasman Sea stopped in the middle of the Paleocene, 60 million years ago, but opening of the South-west Pacific Ocean south and east of New Zealand (i.e. between the Campbell Plateau and West Antarctica), that had started about 87 million years ago, has continued steadily right through to the present day.

With the cessation of movements in the Tasman the scene of sea-floor spreading shifted to the area between Australia and Antarctica. As with the separation of New Zealand from Australia, the phase of actual creation of new sea floor was probably preceded by a splitting (or rifting) phase, involving fragmentation and subsidence of once-continuous land and cutting off of migration routes. This initial splitting may have been taking place at about the start of Tertiary time. Anyway, sea-floor magnetic stripes tell us that by at least late Paleocene times, 55 million years ago, new sea floor was being created to either side of the infant South-east Indian oceanic rise, and the consequent sea-floor spreading was beginning to raft Australia northwards and Antarctica southwards.

As was mentioned in Chapter 12, throughout the Cretaceous the seas around New Zealand and Australia were populated by cool-temperate Austral animals, and the inference is that both countries were then slightly more southerly in latitude than they are today.

During the Cretaceous large parts of Gondwanaland were being affected by splitting movements. The Atlantic, Indian and Pacific oceans and Tasman Sea were all opening. Movements were therefore occurring all around Australia and Antarctica, and it is likely that some southwards rotation took place, perhaps as a result of pivoting among the newly developed spreading centres in the Tasman Sea, Pacific and Indian oceans.

As a result of these rotating movements, ancestral New Zealand probably ended up somewhere in the sub-Antarctic region, with a climate rather like modern Campbell and Auckland islands, and it is noteworthy that towards the close of the Cretaceous the Austral affinities of our marine fossils become very strong indeed, perhaps suggesting that at this time the southern drift of New Zealand had reached its maximum extent.

Ancestral New Zealand continued to receive Austral marine animals well into middle Paleocene times, 60 million years ago, but in contrast to the peak of Austral immigration in the late Cretaceous, the Austral influence gradually waned and gave way to Australian, Indonesian and Malaysian organisms migrating to New Zealand across the infant Tasman Sea.

This gradual change may be interpreted to mean that ancestral New Zealand, after having been rotated progressively southwards during Cretaceous times, was gradually rotated northwards again, as sea-floor spreading got underway between Australia and Antarctica. Changing climates and ocean currents may also have played a part too, but at present we have no way of pin-pointing these effects, if indeed they existed.

**Mass extinctions**

In New Zealand, and elsewhere in the world, there is a major change in the fossil record at the boundary between the Mesozoic and Cenozoic, 65 million years ago.

The often bizarre creatures that were such a feature of the Mesozoic—the dinosaurs and their marine relatives, the ichthyosaurs and plesiosaurs that swam through New Zealand seas, and the ammonites and belemnites, as well as many specialised clams (e.g. *Inoceramus*; Fig. 16.7), corals and plankton—all disappeared forever.

As far as we know, all these extinctions were world-wide and occurred simultaneously. Numerous scientists have pondered on the reasons for the massive extinctions. It has been suggested, for example, that the dinosaurs died of chronic constipation, as they were adapted to a diet of the resinous giant ferns, clubmosses, horsetails and other plants of the Mesozoic, that were replaced at about the close of the Cretaceous by the more advanced, and perhaps less laxative, angiosperms (p. 265). Then, there is the idea that the dinosaur eggs, left unattended, provided easy meals for the small mammals that were then beginning to become quite numerous. Another, rather sad, suggestion is that as climate cooled, the cold-blooded dinosaurs were all too big to find any suitable cranny in which to hibernate (smaller dinosaurs having died out earlier in the Cretaceous).

Extinctions occurred, however, in the depths

of the seas, as well as on the land, and affected a great variety of plants and animals. Thus no single explanation suffices by itself and the extinctions were undoubtedly the result of a complex combination of circumstances.

The transition from Mesozoic to Cenozoic, 65 million years ago, was marked by major changes throughout the world. In many respects the Mesozoic was marked by relative calm, and the South-west Pacific was one of the few regions to experience the turmoil of mountain building during this time.

Also, because of the various groupings of the continents, a fair proportion of the world enjoyed tropical, or subtropical or warm-temperate conditions. All this started to change, however, as the Cretaceous drew to a close.

To judge from the records preserved in the rocks a whole host of things were happening at once, all generating major changes in the environment.

In both Northern and Southern Hemispheres continents were beginning to move apart, and this led to changes in ocean basins and climates (re-routing of ocean currents, wind-systems, etc.; positioning of more land close to the poles etc.). At the same time earth movements, linked to the various continental break-ups, were occurring in many regions. These, together with changes in ocean basins, may have had far-reaching environmental consequences. Falling sea levels (resulting from either uplift of the land, or deepening of the ocean basins) may have disturbed many habitats on both land and continental shelf. Rising sea levels, on the other hand, would have diminished many terrestrial habitats and forced changes onto shallow-water organisms.

As seas transgressed, nutrients from the land dwindled, plankton productivity in the sea declined and marine food chains were disturbed. Active mountain building also had the same effect—smothering the terrestrial and marine environments with coarse gravel and sand and disturbing the supply of nutrients.

Volcanic activity accompanying earth movements may have laid waste, or poisoned, many environments and the volcanic dust thrown up into the atmosphere may have seriously diminished the amount of the sun's energy received on earth, leading to a cooling of climate and a decrease in the chemical processes (photosynthesis) in plants.

Extra-terrestrial effects may also have had a part to play. A supernova close to the earth, or a giant solar flare, may have led to an increase in high-energy radiation, exposing organisms both on land and in the sea to the possibility of genetic damage (leading to mutations etc.).

The increased radiation may also have boosted the production of nitrogen oxides (by combining nitrogen and oxygen) in the atmosphere, thus leading to a chemical "smog" that would filter out the sunlight, cooling the earth and adversely affecting plant growth.

Evidence suggests also that the Mesozoic-Cenozoic boundary coincides with a major polarity reversal (pp. 80–85). If, as a consequence of the polarity change, the earth's magnetism was lost for a time and the various layers in the atmosphere that protect us from cosmic radiation (and are held in place by the magnetic field) were temporarily thinned, or lost altogether, then organisms on the earth would be exposed to serious radiation damage.

From this sample of the various possibilities it can be seen that chains of causes and effects may have been set up ("chain reactions") that may have led eventually to major extinctions.

While it is true that a whole host of extinctions occurred at the Mesozoic-Cenozoic boundary, many groups did in fact survive. In these latter groups it is nonetheless noticeable that a great number of evolutionary changes (related to possible genetic effects?) occurred on or near the boundary. Many of the animals and plants that survived the "crisis" (if we can use this term) had just recently evolved (e.g. the mammals and angiosperms), had greater flexibility and reproductive potential than the dinosaurs, for example, and so were able to cope with the major environmental changes that everyone agrees undoubtedly occurred.

A similar "crisis" for organisms occurred at the Palaeozoic-Mesozoic boundary, 230 million years ago, and this change, like the Mesozoic-Cenozoic one, is also very evident in the New Zealand fossil record. However, because it is that much older, and its story therefore less clear in the rocks, we cannot be at all certain whether it too coincided with a similar variety of terrestrial and cosmic factors—but the odds are that it probably did.

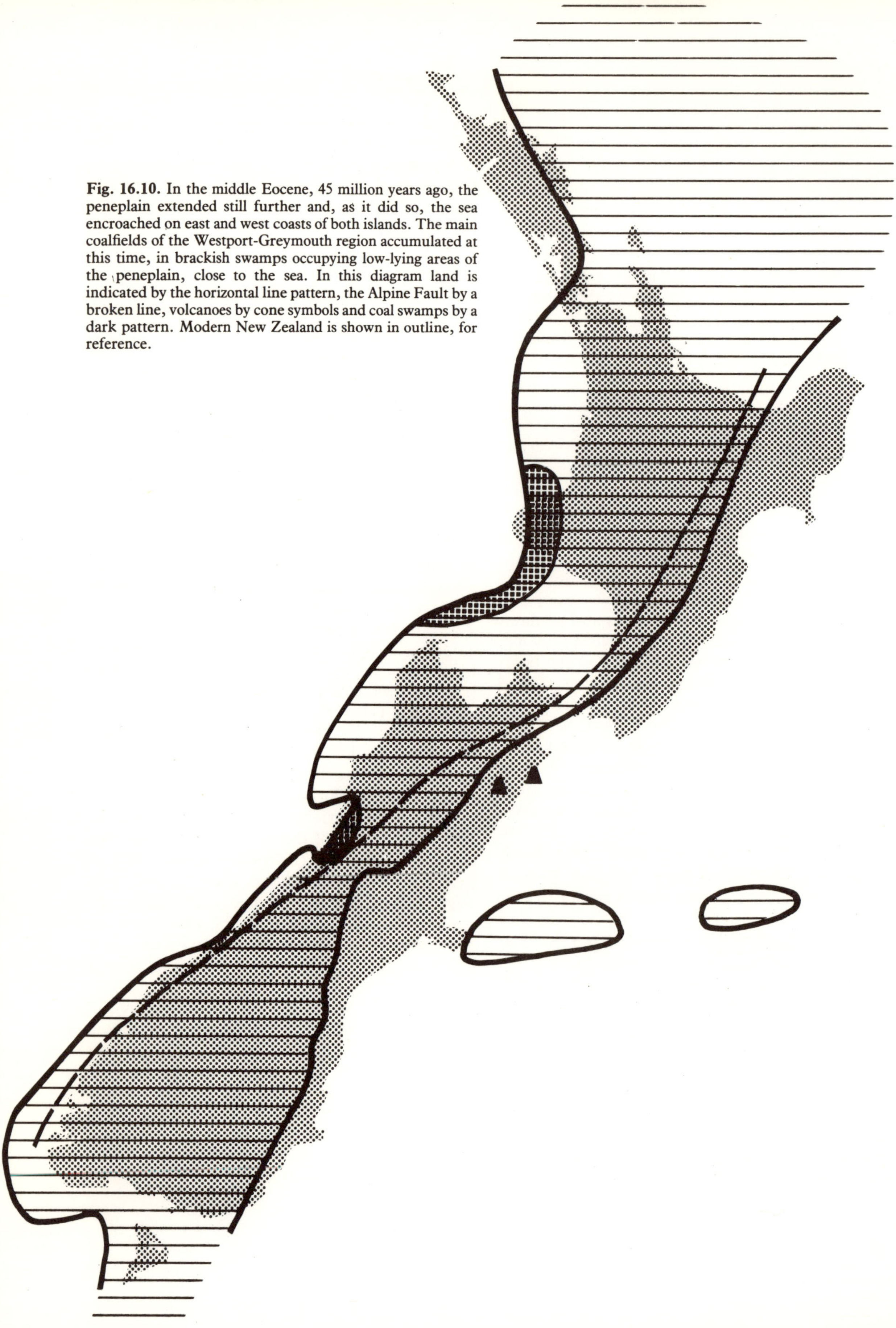

**Fig. 16.10.** In the middle Eocene, 45 million years ago, the peneplain extended still further and, as it did so, the sea encroached on east and west coasts of both islands. The main coalfields of the Westport-Greymouth region accumulated at this time, in brackish swamps occupying low-lying areas of the peneplain, close to the sea. In this diagram land is indicated by the horizontal line pattern, the Alpine Fault by a broken line, volcanoes by cone symbols and coal swamps by a dark pattern. Modern New Zealand is shown in outline, for reference.

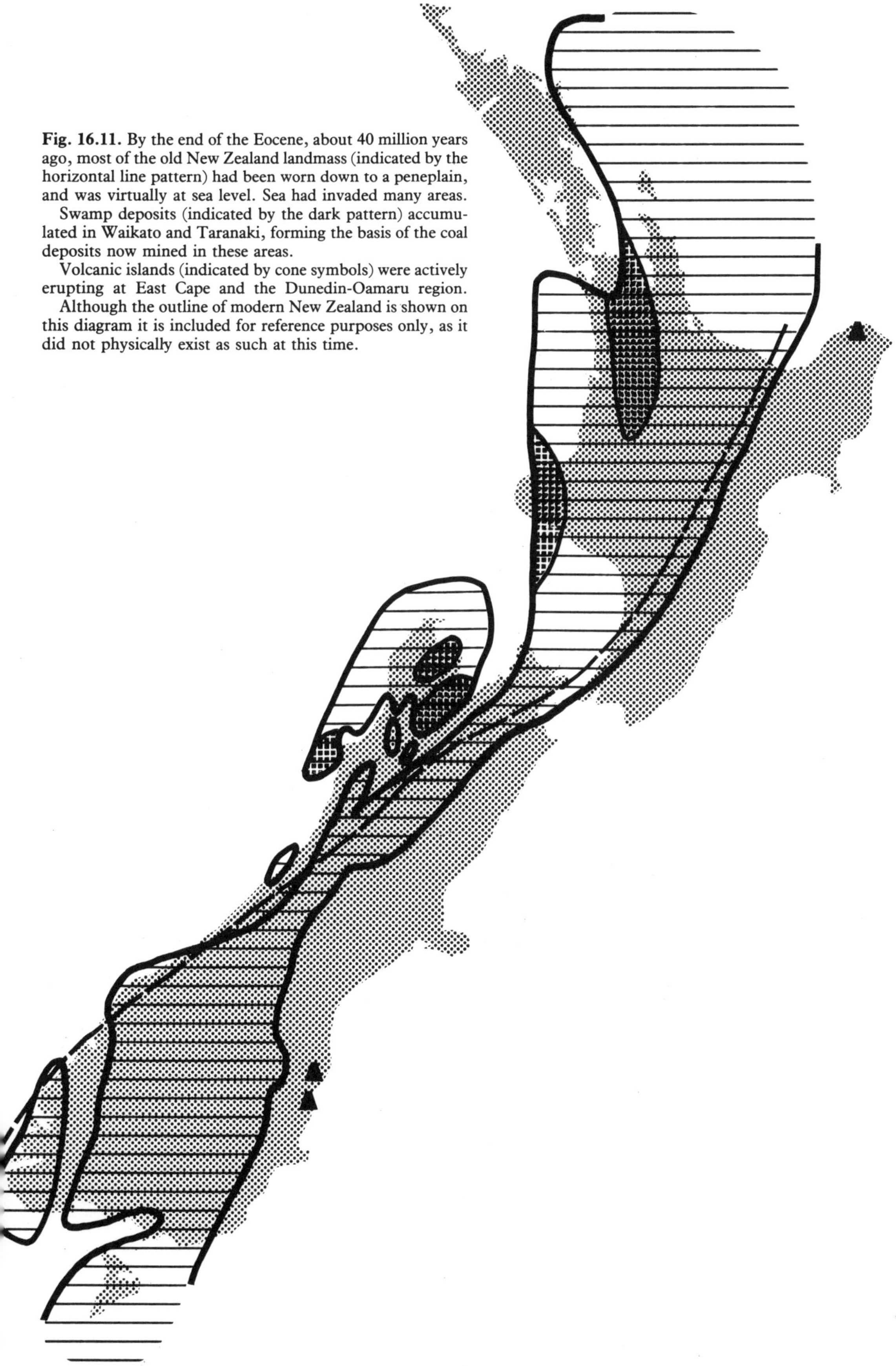

**Fig. 16.11.** By the end of the Eocene, about 40 million years ago, most of the old New Zealand landmass (indicated by the horizontal line pattern) had been worn down to a peneplain, and was virtually at sea level. Sea had invaded many areas.

Swamp deposits (indicated by the dark pattern) accumulated in Waikato and Taranaki, forming the basis of the coal deposits now mined in these areas.

Volcanic islands (indicated by cone symbols) were actively erupting at East Cape and the Dunedin-Oamaru region.

Although the outline of modern New Zealand is shown on this diagram it is included for reference purposes only, as it did not physically exist as such at this time.

The Ice Age, or Pleistocene period, that spanned from about 2 million years to 10,000 years ago, was yet another crisis point for many animals and plants, the aftermath of which is still very evident in New Zealand's native flora and fauna.

Although like the other crises the Ice Age was the result of a combination of factors, major climatic deterioration, and its interacting consequences, played a pre-eminent role (pp. 342).

**Coal, oil and natural gas**

To judge from oxygen isotope analyses of fossil shellfish (pp. 214–220), and the waning influence of southern (Austral) migration, sea-water temperatures around ancestral New Zealand gradually rose throughout Paleocene times (65-53 million years ago), to reach a maximum in the Eocene. Meanwhile on land, erosion steadily lowered the landscape and peneplain conditions became widespread. As the land was worn down, so the sea transgressed progressively further, and by the end of the Paleocene, 53 million years ago, sea had covered large parts of Northland and the east coasts of both North and South Islands (Figs 16.8, 16.9). Also, sea had lapped over the areas now occupied by the modern Chatham Islands, Campbell and Auckland islands and the Lord Howe Rise.

In the Eocene, 53-37 million years ago, peneplain conditions extended still further across the land, but with the added influence of a tropical or subtropical climate (Figs 16.10, 16.11).

It is a well-known fact in chemistry that the adding of heat tends to speed up chemical reactions. Thus under cold conditions chemical changes occur slowly, whereas they speed up considerably as more and more heat is applied. In the landscape all the various exceedingly complex chemical reactions that together are acting upon the rocks to break them down (the process we call "weathering") are greatly accelerated under tropical and subtropical conditions. The great heating of exposed rocks that occurs under a hot climate also helps the weathering process, by breaking and splintering rock and exposing numerous surfaces to the effects of chemical reactions.

Thus under humid tropical conditions great thicknesses of weathered material, mainly clays, accumulate on the landscape and geologists know from experience in tropical areas that often you look in vain for a decent rock outcrop and frequently the only way to see the underlying strata is to excavate trenches or to drill.

Throughout Eocene time the New Zealand landmass experienced intensive tropical weathering and great areas of rolling densely forested plain developed. Extensive swamps developed in low-lying hollows in the landscape—particularly in the Greymouth region, southern Taranaki and the part that is now Cook Strait between NW Nelson and Taranaki.

The swamp lands were clothed in luxuriant tropical vegetation and, as time passed, thicknesses of plant material accumulated in the subsiding swamp basins. Towards the end of Eocene times sea flooded in over the swamps and adjoining regions and many thousands of metres of marine sediments were laid down on top of the swamp deposits. In time heat and pressure gradually changed the woody plant material into the coal that is now mined in the Greymouth area.

The original swamps, particularly those in southern Taranaki and western Cook Strait, also had other constituents apart from woody material. Many plants have varying amounts of vegetable oils and waxes in their stems and leaves and, if the swamp was very wet, and especially if it was periodically invaded by the sea, numerous algae, also rich in such oils and waxes, may have flourished. These algae, on dying, became incorporated in the swamp deposits, and were eventually deeply buried and subjected to the same heat and pressure that changed the woody material into coal.

The heat and pressure in the depths of the crust altered the vegetable material to produce a mixture of heavy and light petroleum oils and gas, the proportions of which depend on the composition of the source material. Some algae are very rich in waxy compounds. When these are altered by the natural processes within the crust, they produce a heavy viscous crude oil, which, separated in an oil refinery, gives a rich yield of "heavy" products such as tar, bitumen and waxes, but a lesser yield of "light" products such as petrol and kerosene.

Some of the heavy crude oil obtained from oilfields (e.g. the Schoonebeck field in Holland), whilst it will flow at depths below the crust, because it is under great pressure, will virtually congeal once it reaches the surface, and it is necessary to install heaters in the well-head pumps, pipelines and storage tanks to keep it fluid.

The crude oil from other fields may be a "light" oil, free-flowing, and, as it contains a high proportion of "light" oils, it can be used virtually in the crude state, if necessary. Depending on the source material, varying amounts of natural gas are present, and this drives the oil to the surface when it is tapped by a drill hole.

The vast swamps that spread across the Paleocene and Eocene landscape of Taranaki, and the Cook Strait region between N.W. Nelson and southern Taranaki, became the source of the gas and oil of the Maui off-shore petroleum field. The Maui field and the smaller field at Kapuni, southern Taranaki, are primarily gas fields, but oil condenses from the gas as it is drawn off the wells. This condensate obtained from Maui and Kapuni is a very light grade of oil and would be quite capable of being used in a car, without being refined. For maximum efficiency, however, modern petrol engines need a carefully graded fuel and although in theory Maui and Kapuni oil would drive your car, it would be wiser to buy some from the nearest petrol pump!

Both the Maui and Kapuni fields were formed at the same time (Paleocene-Eocene) and their origin is intimately linked with the coal beds. Chemical analyses have shown that the natural gas in both fields has been given off during the coal-forming processes and has been trapped in adjacent porous rocks (elsewhere in the world oil and gas deposits are usually associated with marine sediments and probably originated by natural alteration of the remains of waxy and oily marine life, e.g. algae).

As the Eocene wore on, the land became progressively lower and in late Eocene, 40-37 million years ago, extensive swamps formed in Northland, Waikato, Southland and along the West Coast of the South Island (Fig. 16.11). The coal that formed in these swamps is the basis of the coal fields in these areas today. Thus, because of our vastly different geological history, the main New Zealand coalfields are of Paleocene and Eocene age, rather than Carboniferous, as in Europe and North America.

**Eocene immigrants**

The warm Eocene seas lapping over much of ancestral New Zealand encouraged many tropical and subtropical marine animals to spread southwards from Malaysia, Indonesia and eastern Australian waters. Reef corals, for example, flourished around New Zealand coasts.

The marine creatures that migrated southwards into New Zealand waters during the Paleocene-Eocene and later times were, however, all of types that could become widely distributed across open ocean—either because the adults could float or swim, or because the eggs or larvae had this ability. There is no evidence that continuous shallow-water links similar to those of the late Mesozoic existed between New Zealand and neighbouring countries in the Tertiary.

Oxygen isotope determinations show that sea-water temperatures reached a peak in mid-late Eocene times, when conditions were probably almost fully tropical. This peak was followed however by a fairly sharp decline in temperatures and by the end of the Eocene the first penguins were starting to arrive in New Zealand (see below), along with whales. Falling temperatures carried on into the Oligocene, but nonetheless the continuing presence of Malayo-Pacific creatures in the sea and a warmish type of vegetation on the land indicate that temperatures still remained slightly warmer than at present.

**Fossil penguins**

As Gondwanaland started to break up in the Cretaceous, and countries such as New Zealand and Australia started to rotate and drift closer to the South Pole (and in the process provided homes for cool-temperate Austral plants and animals, p. 220), the shores of southern New Zealand began to be populated by penguins.

The first penguins appeared in the late Eocene, about 43 million years ago, at a time when climate had cooled off appreciably and the first ice fields were starting to form in Antarctica. Penguins of this age are known from southern Australia, southern New Zealand and Seymour Island (Antarctic Peninsula) (Fig. 16.12).

Fossil penguins are known from other post-Eocene geological ages in Argentina, Chile, Seymour Island, Antarctica, and southern South Africa as well as New Zealand and Australia. New Zealand, however, has the world's most extensive record of fossil penguins, in terms of both range of geological age and variety of species

**Fig. 16.12.** The first penguins appeared at the end of the Eocene period, about 43 million years ago, at a time when climate had cooled appreciably and the first ice fields were starting to form in Antarctica.

Penguins of about this age are known from Australia (Victoria and South Australia), southern New Zealand, Seymour Island (Antarctic Peninsula) and southern South America. These occurrences are shown by penguin symbols on the diagram.

As may be seen from Figs 12.27, 15.1-15.5, all these countries had moved steadily to southern positions in late Cretaceous-early Tertiary times, as Gondwanaland broke up, and the Atlantic and Indian oceans developed.

This reconstruction, based on palaeomagnetism, shows the late Eocene positions of the continents, with the South Pole indicated by a star. Beginnings of ice sheets are pictured on Antarctica (circle symbols) and to judge from ice-rafted sediments encountered in deep-sea drilling, these sheets may have given rise to icebergs in the opening South Atlantic. Overall temperatures, however, were not unduly frigid in New Zealand, southern Australia, Antarctica and southern South America, and probably varied between warm and cool-temperate. Cool-temperate forests of southern beech and other trees were reasonably widespread (indicated by the tree symbols).

and it also includes the first to be discovered and the largest species.

New Zealand fossil penguins range in age from late Eocene through to the Pleistocene, spanning some 43 million years. The first fossil penguin known to science was found by a local Maori at Kakanui (near Oamaru) and given to Dr W. B. D. Mantell (a pioneer New Zealand scientist, associated with early discoveries of moa and takahe fossils), who later forwarded it to Dr T. H. Huxley of the British Museum, who published its scientific description in 1859.

The late Eocene penguins of New Zealand, while including some that probably ranged up to the size of the modern Emperor Penguins, included a giant species, the largest so far discovered. This penguin, rejoicing in the name *Pachydyptes ponderosus*, was found at Oamaru and described in 1930 by Dr W. R. B. Oliver, later Director of the Dominion (now National) Museum. Modern Emperor Penguins have a

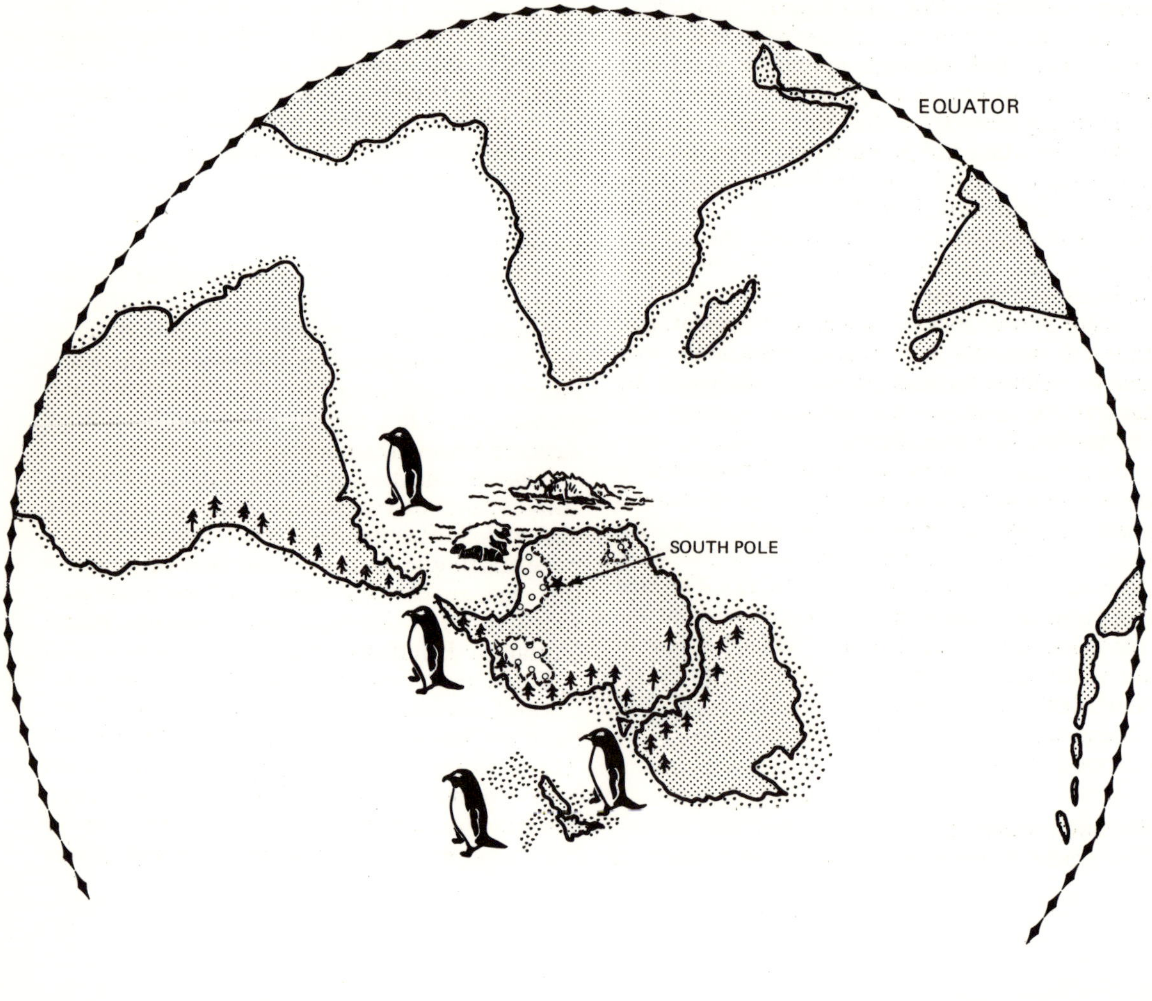

standing height of about 1 m (3 ft), and weigh some 30 kg (66 lb), whereas the Oamaru giant probably stood 1.62 m (5½ ft) and weighed in at a massive 100 kg (221 lb).

The fossil occurrences of penguins fall within their modern geographic range. Today penguins are found in southern New Zealand, southern Australia, the sub-Antarctic and Antarctic islands, Antarctica, Argentina, Chile, and southern South Africa, but the largest concentrations and greatest numbers of species occur in cold-temperate, sub-Antarctic and Antarctic seas.

The northernmost penguins live along the coasts of Chile and Peru and close to the equator at the Galapagos Islands, but this distribution is undoubtedly related to the cold Humboldt Current sweeping north along the South American coast.

Otherwise, small penguins range up to about 42°S Latitude (they live under houses in Eastbourne and Petone, Wellington Harbour), but a stray Emperor Penguin reached Invercargill in April 1967.

Penguins "fly" rather than "swim" in the water and have made the radical change from flying in a light medium (air), to flying in a dense medium (sea water). The penguins have retained large flying muscles in their chest and all the bones of a flying wing and their flippers are thus modified flying wings—for flight in water rather than air.

Diving petrels, like penguins, also fly underwater but are capable of aerial flight as well. This enables them to fly out to sea, and once there, to pursue food by underwater flight (as distinct from diving, in the usual sense of the word).

It has been suggested that diving petrels may have provided the ancestral stock for penguins—but guillemots are also strong contenders for kinship. However, at some stage penguins lost the powers of flight and it is thought that within modern families of diving birds capable of flight in both air and water there is a definite weight limit, probably about one kilogram, beyond which excessive wing loading and loss of manoeuvrability in flight makes flying difficult.

But yet, a body size larger than one kilogram facilitates sustained diving and allows birds wider scope in hunting. Therefore it is possible that the penguins were petrel-like birds that sometime in the Eocene opted for underwater expertise (hunting squids, fish etc.) and became flightless. As we have seen, this development was probably related to the splitting of Gondwanaland, the steady drift of the southern countries towards the South Pole and the consequent creation of a range of cool-temperate and colder habitats.

### New Zealand submerged

The Oligocene period, 37-25 million years ago, saw ancestral New Zealand virtually submerged by the sea. Some two-thirds of the area of modern New Zealand was covered by sea at this time and the lowland coal-forming swamps that had been in existence for most of Paleocene and Eocene times were buried under great thicknesses of marine sediments (Fig. 16.13). The remaining remnants of land were reduced to an elongated archipelago and isolated islands. Volcanic islands were active in the seas covering southern Westland, Dunedin, Canterbury and East Cape (Fig. 16.13).

In some groups of animals and plants isolation on scattered islands may have led to development of distinctive varieties. It is believed that some of the large native land snails, first isolated in this way at this time, may have evolved into species that, even today, with a greatly enlarged landmass, still retain their separate identities.

On the other hand, Oligocene times may have presented many problems of survival for the plants and animals living in New Zealand, as no doubt the sea then covered many formerly occupied habitats, and the amount of suitable land had been markedly reduced. Thus many extinctions of land organisms may have occurred at this time—thinning out still further the original plants and animals that had arrived in Mesozoic times.

The coastline of the Oligocene New Zealand archipelago was populated by many East Australian and Malayo-Pacific marine organisms, and oxygen isotope analyses of their shells indicate that although sea temperatures were still warm, they were lower than those of the Eocene.

### Debut of the Circum-Antarctic Current

Today many facets of life in the Southern Hemisphere are influenced by a major oceanic current, the Circum-Antarctic Current, that circles clockwise around Antarctica, passing just

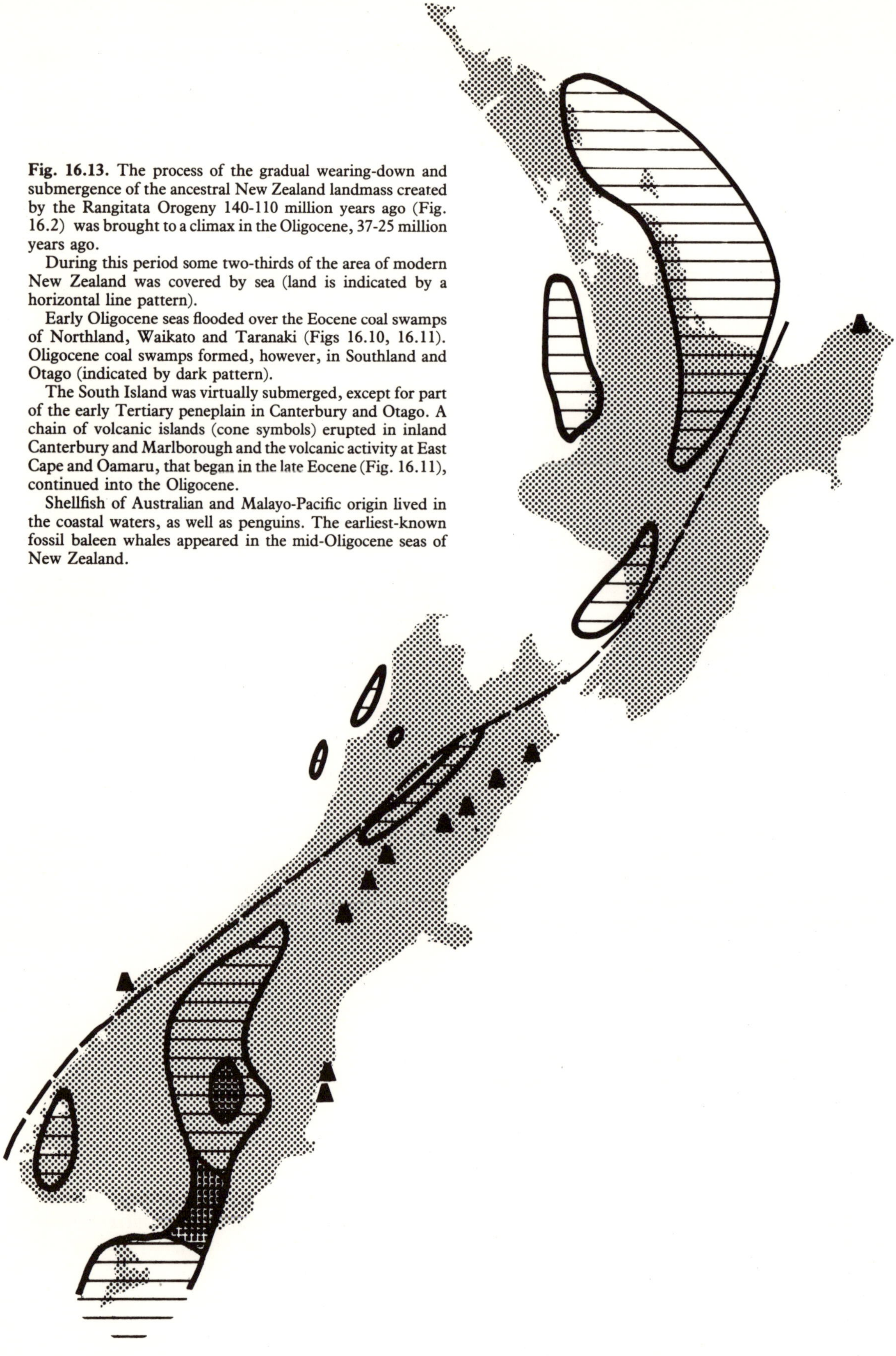

**Fig. 16.13.** The process of the gradual wearing-down and submergence of the ancestral New Zealand landmass created by the Rangitata Orogeny 140-110 million years ago (Fig. 16.2) was brought to a climax in the Oligocene, 37-25 million years ago.

During this period some two-thirds of the area of modern New Zealand was covered by sea (land is indicated by a horizontal line pattern).

Early Oligocene seas flooded over the Eocene coal swamps of Northland, Waikato and Taranaki (Figs 16.10, 16.11). Oligocene coal swamps formed, however, in Southland and Otago (indicated by dark pattern).

The South Island was virtually submerged, except for part of the early Tertiary peneplain in Canterbury and Otago. A chain of volcanic islands (cone symbols) erupted in inland Canterbury and Marlborough and the volcanic activity at East Cape and Oamaru, that began in the late Eocene (Fig. 16.11), continued into the Oligocene.

Shellfish of Australian and Malayo-Pacific origin lived in the coastal waters, as well as penguins. The earliest-known fossil baleen whales appeared in the mid-Oligocene seas of New Zealand.

to the south of Australia and New Zealand. In New Zealand, for example, it exerts a major influence on our weather.

Cold sea water is denser, and heavier, than warm sea water. Thus there is a "pushing-out" of cold water from the poles and, as this happens, other slightly warmer water flows in from lower latitudes to take its place, in turn cooling, sinking and flowing out.

Continual movement of water therefore occurs to and from the poles. In the Southern Hemisphere cold water pushes out from the Antarctic shelf regions, notably those of the Ross and Weddell seas. As it moves northwards substantial quantities of this water are diverted into the Circum-Antarctic Current. A proportion continues, however, along the sea floor, towards the equator, but rises to the surface in some areas, at boundaries with other currents, or near undersea ridges, the nutrient-laden water forming the basis for major fisheries and, as a by-product, the guano industry (as in Chile and Peru).

The Circum-Antarctic Current today transports more than 200 million cubic metres of water per second, probably the largest volume transport of any ocean current. It circulates completely around the Antarctic continent, unimpeded by any landmasses, and is the only current that makes major contributions to all the world's major oceans (except the Arctic Ocean).

Its plankton-laden waters make considerable contributions to Southern Hemisphere fisheries and it has a significant effect on oceanic circulation and everything that derives from this—meteorology, climate, animal and plant distributions etc. Development of the Circum-Antarctic Current had to wait, however, until continental drift had opened up enough ocean between the various Gondwana landmasses.

As we have seen, creation of ocean between the edge of Campbell Plateau and Antarctica's Marie Byrd Land began in the late Cretaceous, 80 million years ago (Fig. 18.6), and Africa started to pull away from Antarctica at about this time or later. Sea-floor spreading, starting some 65 million years ago in the Drake Passage, opened up a gap between the Antarctic Peninsula and southern South America. Thus by Paleocene times, 65-53 million years ago, Australia and Antarctica were the only southern landmasses still joined together and hindering free circum-Antarctic oceanic circulation (Fig. 16.14).

Drilling by the *Glomar Challenger* in the sea floor south of Australia has documented the various developments that laid the foundations for the Circum-Antarctic Current. Sea-floor spreading between Australia and Antarctica began about 55 million years ago and the two continents drifted apart at a rate of about 40-50 mm per year (1½-2 in).

At first a shallow sea stirred by weak oceanic currents existed between Australia and Antarctica. These conditions continued until almost the end of the Eocene, between 40 and 37 million years ago, when enough of a gap had been opened to allow more open-sea conditions to occur. Initiation of the Circum-Antarctic Current was not possible however until Tasmania, and more critically, the South Tasman Rise, its submarine continuation (proved by the *Glomar Challenger*'s drill to be continental) had cleared the northern tip of Antarctica's Victoria Land (Fig. 16.15). By latest Eocene time, although some marine life was able to get from the South-east Indian Ocean into the South-west Pacific Ocean, across the South Tasman Rise, it was evident that only a very shallow marine connection had been established.

During the early Oligocene the South Tasman Rise still had not moved sufficiently north to allow free current movement, although open-sea conditions were apparent to either side (Fig. 16.16). However in the middle and late Oligocene, about 30-25 million years ago, enough of a gap had opened between the South Tasman Rise and Antarctica to allow the Circum-Antarctic Current to sweep through (Figs 16.17-16.19). And "sweep" it did—for the *Glomar Challenger*'s drill has revealed that the newly created current immediately started to scour the sea floor and carry away masses of sediment. Such scouring has been detected by drilling in all the southern oceans—the Atlantic, Pacific, Indian and Antarctic. Sea creatures also began to take advantage of the swift-flowing current, to migrate to otherwise inaccessible areas.

Drilling results suggest that apart from some minor changes the deep-sea oceanic circulation south of New Zealand has remained essentially the same since the late Oligocene, so it may be said that initiation of the Circum-Antarctic Current marked the first move towards development of modern climatic patterns in the Southern Hemisphere.

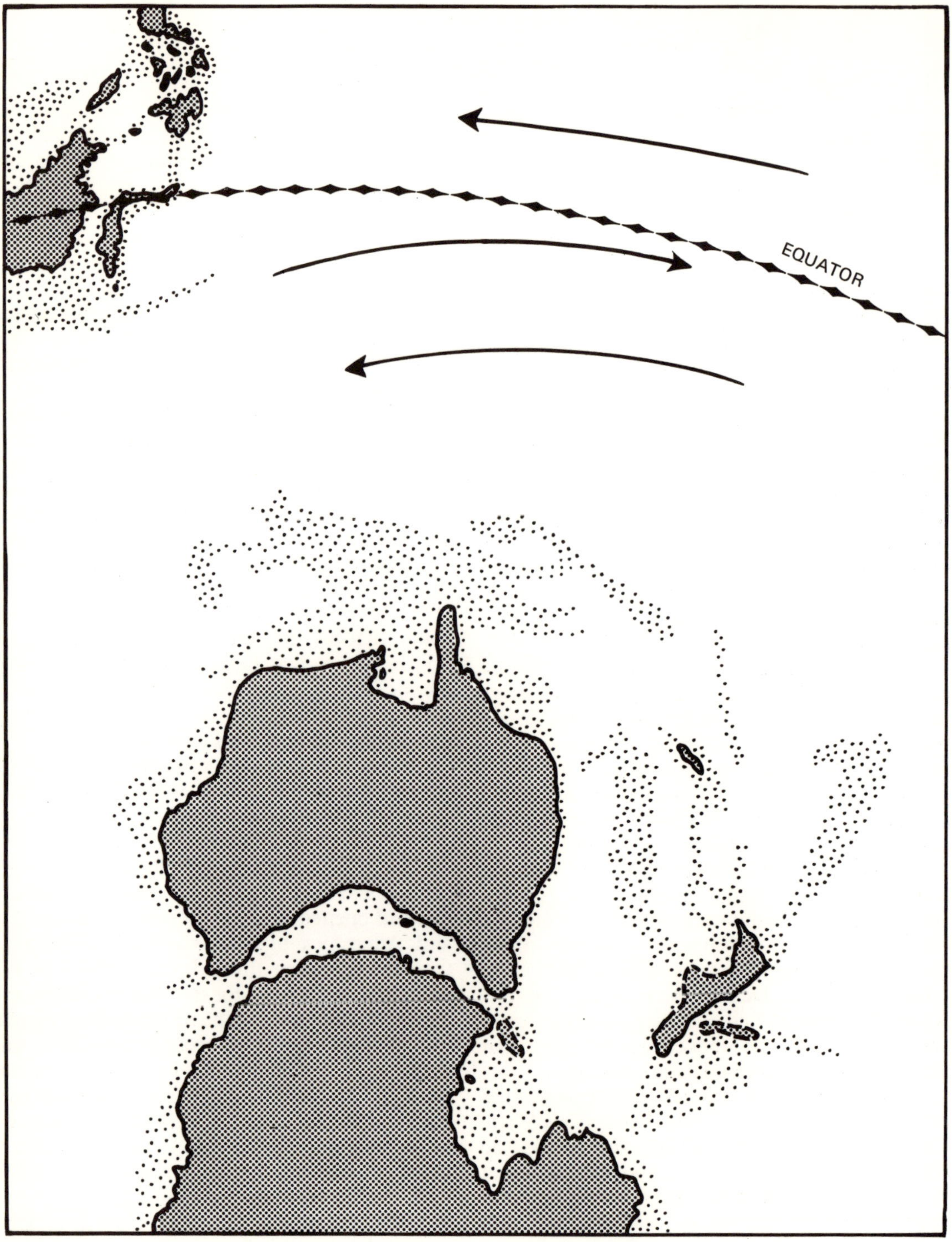


**Fig. 16.14.** A reconstruction of the South-west Pacific for middle Paleocene, 58 million years ago. Between 80 and 60 million years ago (late Cretaceous-early Paleocene), sea-floor spreading had opened up the Tasman Sea to its full width. At the time of this reconstruction sea-floor spreading was dying out in the Tasman, but was still continuing between the edge of the Campbell Plateau and Marie Byrd Land, Antarctica. Australia was still firmly attached to Eastern Antarctica. An equatorial-subtropical circulation was the dominant feature of the oceanic current pattern of the time. Off-shoots from these currents, penetrating into the South-west Pacific, provided transport to New Zealand for island-hopping marine creatures of northern Australian and Malayo-Pacific origin.

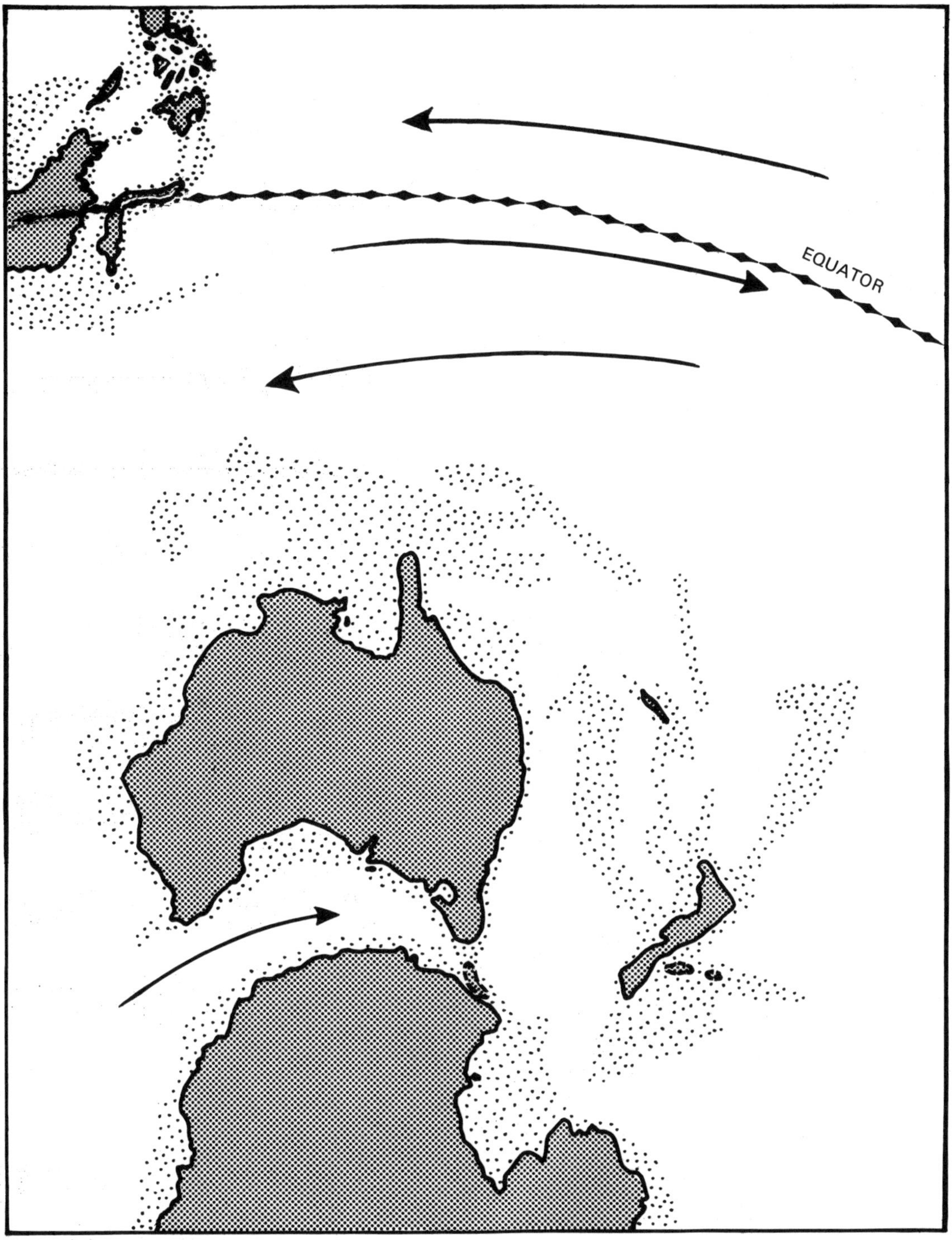


**Fig. 16.15.** By middle Eocene times, 47 million years ago, sea-floor spreading between Australia and Antarctica had been under way for some 3 million years. Such movements had opened up a slight gap between the two continents.

Minor oceanic currents were active in the western segment of the widening gap.

The oceanic circulation of the South-west Pacific was however still dominated by an equatorial-subtropical pattern, with southwest-trending off-shoots facilitating the migration of northern Australian and Malayo-Pacific marine immigrants.

In this figure, and Figs 16.16-16.19, the directions of major oceanic currents are indicated by bold arrows. Continental margins (submerged extensions of the continents) are indicated by coarse dot patterns.

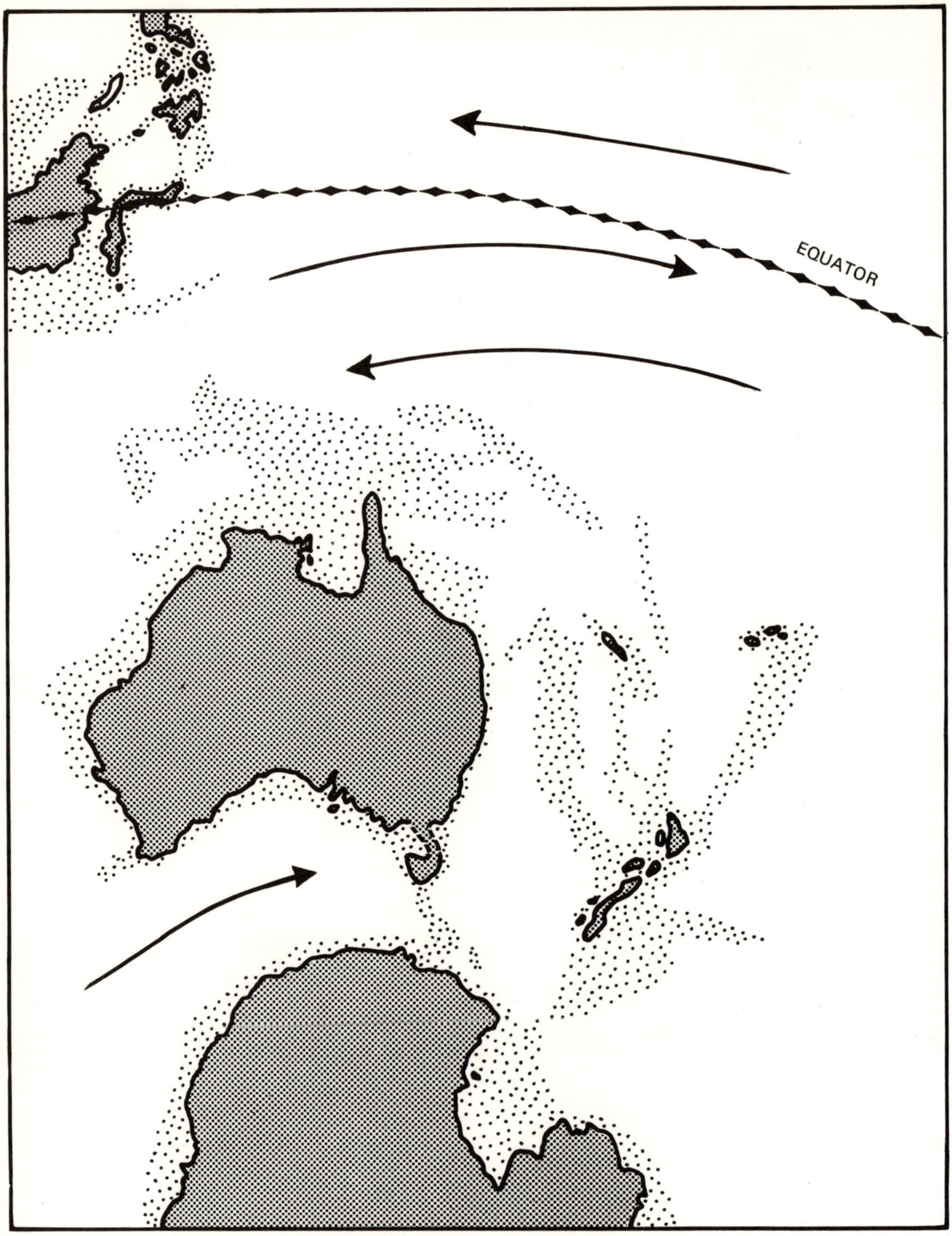


**Fig. 16.16.** Early Oligocene, 35 million years ago.

The gradually widening sea gap between Australia and Antarctica was now stirred by an independent current system, but little of it penetrated to New Zealand, as Tasmania was still linked to Victoria and the South Tasman Rise (south of Tasmania) was still linked to Antarctica.

New Zealand continued to receive marine immigrants from northern Australia and the Malayo-Pacific region.

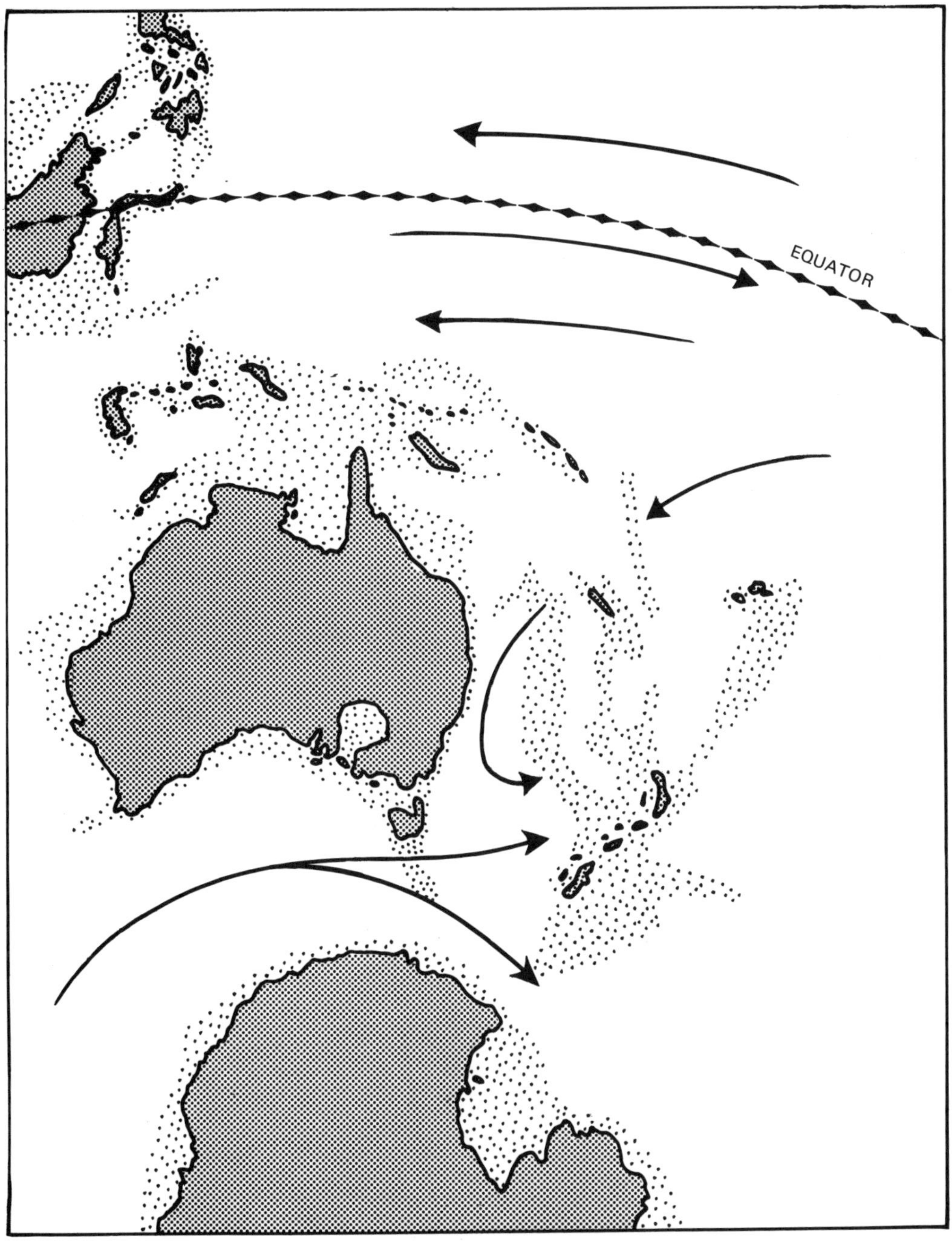


**Fig. 16.17.** Middle and late Oligocene, 30-25 million years ago.

Sea-floor spreading between Australia and Antarctica had opened up enough of a gap to allow the beginnings of the Circum-Antarctic Current to sweep through.

On the other side of the Pacific, similar sea-floor spreading movements between the Antarctic Peninsula and southern South America had opened up the Drake Passage.

The opening of these gaps, together with the meteorological conditions associated with the gradual build-up of ice on Antarctica, allowed establishment of the globe-encircling Circum-Antarctic Current.

New Zealand began to receive influxes of "southern" marine immigrants, as well as those derived from northern (Malayo-Pacific) sources. The earliest known baleen whales (that feed on plankton) appeared in New Zealand at this time and it has been suggested that initiation of the plankton-rich Circum-Antarctic Current may have stimulated their development.

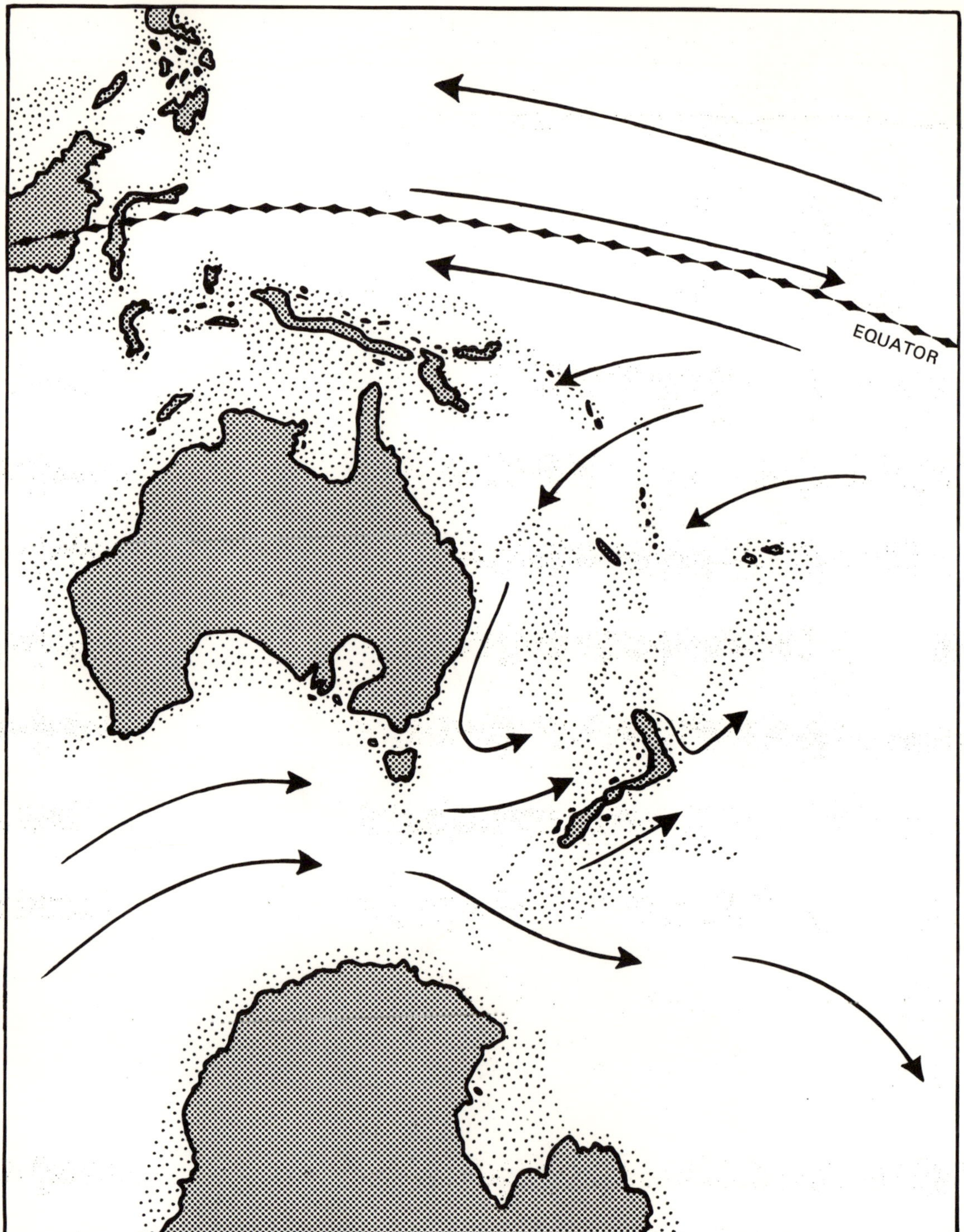


**Fig. 16.18.** As the oceanic gap widened between Antarctica and Australia, and Antarctica's infant ice cap had an ever-increasing effect on the meteorological conditions of the Southern Pacific, the Circum-Antarctic Current became firmly established.

In the middle Miocene, about 13 million years ago, depicted in this reconstruction, warm currents from the north brought tropical plants and animals to New Zealand. At the same time, however, there were steady contributions from the Circum-Antarctic Current. It is believed that some of the tropical newcomers, after migrating southwards to Australia and New Zealand, were transported by the Circum-Antarctic Current to other southern lands.

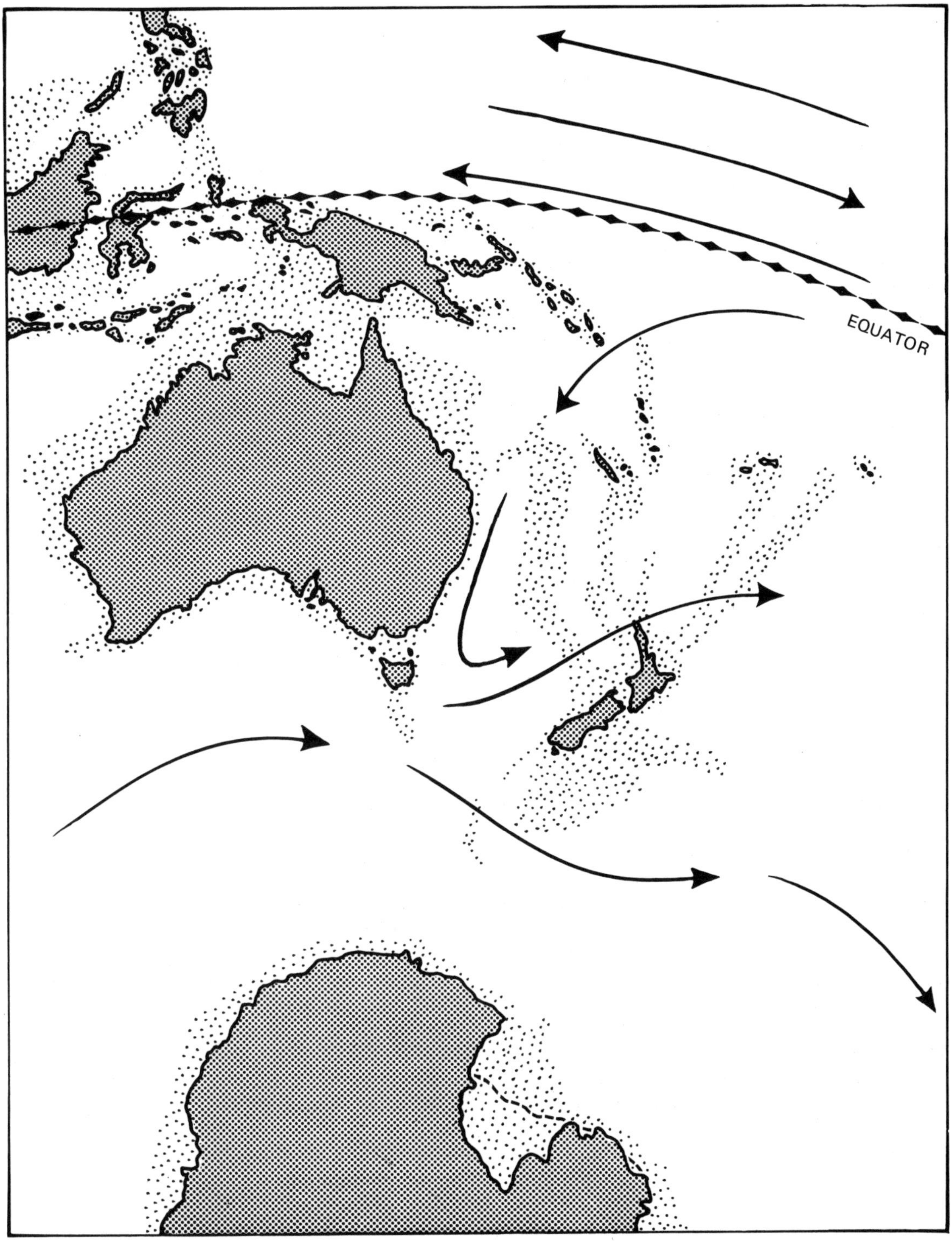


**Fig. 16.19.** At the present time, as depicted in this diagram, the Circum-Antarctic Current dominates the oceanic circulation of the South Pacific and New Zealand receives (and also contributes; see Fig. 6.11) many "southern" creatures. The warm East Australian Current influences western New Zealand.

**The drift of Australia**

As we have seen, before it broke away from Antarctica, some 55 million years ago, Australia shared many plants and animals, both terrestrial and marine, with South America and West Antarctica, testifying to the closeness of the land links (Figs 16.20, 16.21). These links, however, were gradually broken as rifting and subsidence, followed by sea-floor spreading and generation of oceanic crust, widened the gap and isolated Australia, with its distinctive flora and fauna (Figs 16.22-16.27).

To judge from palaeomagnetism, in the early Tertiary, before it started to drift northwards, most of Australia was south of about 28°S Latitude (approximately the present latitude of Brisbane). Then as the Southern Ocean opened, Australia drifted northwards some 15° of latitude. New Guinea, which today defines the leading edge of the Australian plate, was largely submerged during much of the early Tertiary. However it began to be uplifted in the late Oligocene (30-25 million years ago) and these movements have continued to the present day.

The uplift of New Guinea added a broad fringe of land to the leading edge of the Australian continent. This uplift, coupled with the steady northern drift of Australia as a whole, meant that both countries became more and more accessible to migration from Asia.

Judging from the influxes of Indo-Pacific (Tethyan) immigrants that came into New Zealand and Australia in the middle and late Jurassic, shallow-water routes, and perhaps also land routes, were available between Australasia and South-east Asia during these times. Such routes, however, largely disappeared in the Cretaceous, and the continental movements heralding the break-up of Gondwanaland rotated Australia and New Zealand southwards, opening up an oceanic gap to the north of Australia.

The steady northwards drift of Australia, as it pulled away from Antarctica, coupled with the uplift of New Guinea, gradually narrowed the gap between Australia and Asia and it is thought that sometime during the Miocene, 25-5 million years ago, the Australian plate was largely in contact with South-east Asia and exchange of animals and plants was starting to occur. However, it was evidently very much an island-hopping exercise, as it is today, although low sea levels during the glacial periods may have helped considerably (p. 330). Thus the migration route from this direction, unlike the one that formerly existed via Antarctica, was a very selective one and filtered out many intending immigrants. The effectiveness of this filter is illustrated by the fact that not one Australian marsupial gets past Wallace's Line (Fig. 16.28) and no native placental mammals have reached Australia and New Guinea, except for bats, that were obviously able to fly across the intervening ocean gap.

Even today a marked boundary exists between Oriental faunas and floras on one hand and Australian on the other. Oriental faunas and floras can be traced from Pakistan and India southwards through South-east Asia into Indonesia. Australian faunas and floras extend from the Australian continent into New Guinea and eastern Indonesian islands such as Timor, Celebes and the Moluccas. The boundary between the two groups is known as "Wallace's Line", after Alfred Russel Wallace, a contemporary of Charles Darwin (p. 51). The line marks one of the most striking faunal and floral boundaries of the world. Wallace drew a line between Bali and Lombok, east of Java, stating that the "strait is here 15 miles [24 km] wide, so that in two hours we may pass from one great division of the earth to another, differing as essentially in their animal life as does Europe from America". From Bali the line runs northwards, between Borneo and Celebes and between the Philippines and the Sangi and Talaud islands (Fig. 16.28).

Wallace's Line probably marked a major dividing line between the South-east Asian landmasses on one hand, and those linked to Australia on the other, during the time in the early Tertiary when a considerable gap existed between the two. During this time Sumatra, Borneo, Java etc., lying on the Asian continental shelf, were repeatedly joined to the Asian mainland and received influxes of plants and animals. Migration of such plants and animals beyond Wallace's Line was, however, impeded by various ocean barriers but, as the northward drift of Australia and New Guinea progressively narrowed the gap between them and Asia, crossings of the line occurred in the late Tertiary and Quaternary so that now the Philippines, Celebes, Moluccas, Kei, Timor, Flores and Lombok form a transitional zone, as these islands

**Fig. 16.20A.** Australia and Antarctica were locked together as part of the ancient continent of Gondwanaland until about 55 million years ago.

The black dots mark the positions of earthquakes that have occurred in the past 10 years on the South-east Indian Rise, a segment of the world-encircling system of mid-oceanic spreading ridges (Fig. 9.2). Their location coincides with the primitive cleavage line between Australia and Antarctica as seen when the two continents are rotated towards each other by an equal amount along the "latitude" lines centred on the axis of rotation (arrow).

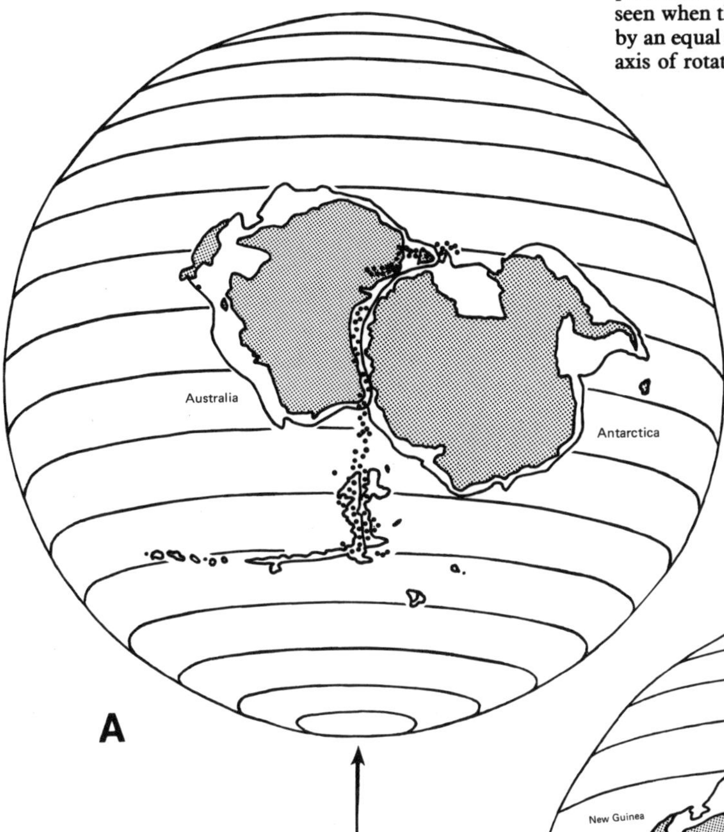


A

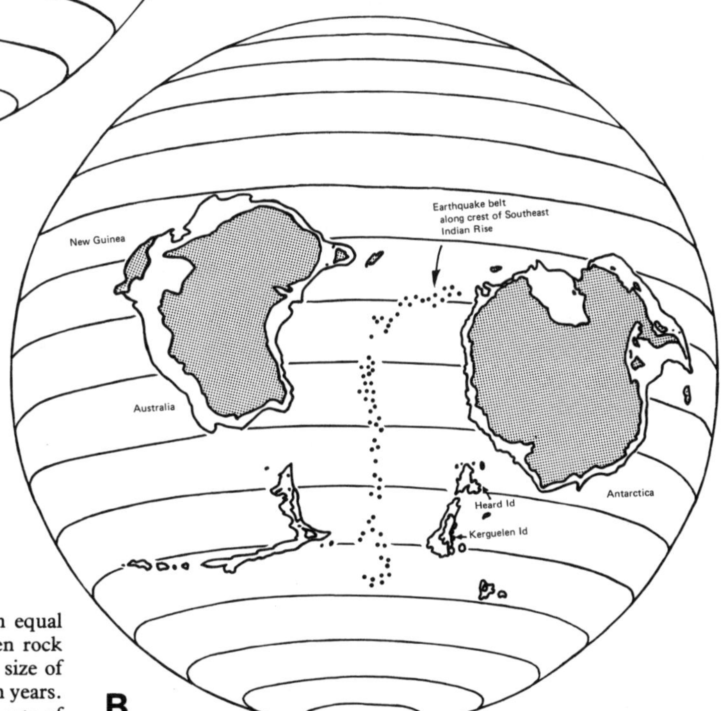


B

**Fig. 16.20B.** Today Australia and Antarctica are an equal distance from the still-active spreading ridge. Molten rock flowing into splits along the ridge has increased the size of each plate by an equal amount over the past 55 million years. The two continents are still being pushed apart at a rate of about 40-50mm (1½-2in) per year.

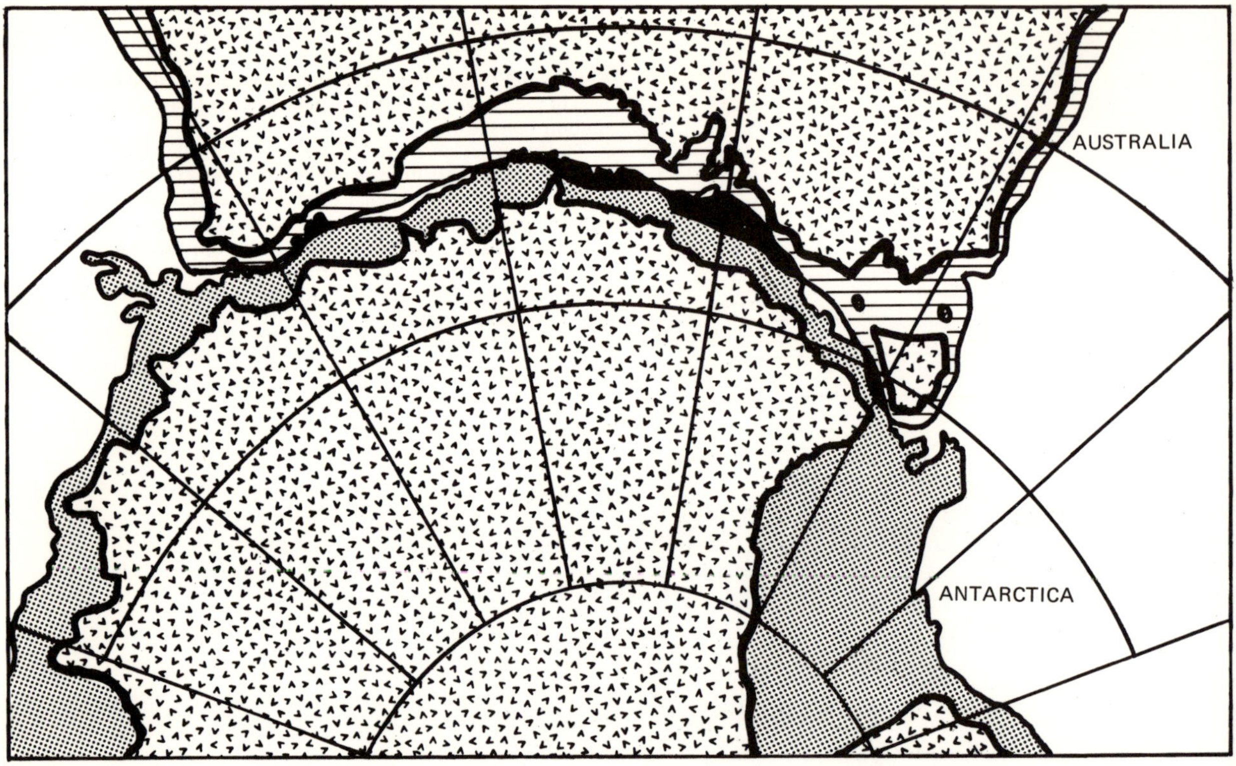


**Fig. 16.21.** Even today the geometric fit between Australia and Antarctica is surprisingly good, considering they have been apart for at least 55 million years.

In this diagram the Australian continental shelf is indicated by horizontal lines and the Antarctic shelf by dots. The black areas indicate overlaps.

have received by means of chance dispersal both Oriental and Australian elements (Fig. 16.28).

The Australian desert and semi-desert developed in post-Eocene times as the continent moved into the "horse-latitudes", the worldwide belt of low rainfall at the edge of the tropics.

In the future, as Australia continues its northward drift, the present climatic belts will gradually shift, no doubt causing major upheavals to human existence. Thus the cool-temperate rain forest now in its south-eastern corner (Victoria and Tasmania) will give way to subtropical and then eventually tropical rain forest. Similarly, the areas of Australia that now have a Mediterranean-type climate (in the south-west and south) will become desert, while the present desert area will become moist and encourage the growth of open woodland and savannah. The vast central areas of wattle (*Acacia*) shrublands will change to savannah and then to rain forest (Fig. 19.19).

The Aussies of the future may have to learn to cope with some major climatic changes!

**Antarctica**

As we have seen, Antarctica served as a route for both land and marine migrants throughout much of Jurassic and Cretaceous time (p. 220). For example, in mid-Cretaceous times, 100 million years ago, the links between southern lands were such that virtually continuous forests of the ancestors of the Southern Beech (*Nothofagus*) probably eventually stretched from New Zealand, across what is now the Campbell Plateau, and then via Antarctica, the Antarctic Peninsula and Scotia Arc to southern South America (pp. 267–269).

Beech forests similar to those of southern New Zealand continued to cover much of Antarctica during Paleocene-Oligocene time. Now, however, they were isolated, as sea-floor spreading had begun between the Campbell Plateau and Marie Byrd Land some 80 million years ago and in the Drake Passage, south of Patagonia, at about the same time.

As climate deteriorated, ice fields formed on

the mountain tops, and glaciers reached out down the valleys to the sea. To judge from drilling by the *Glomar Challenger* off-shore from Antarctica, ice first started to lie on the Antarctic mountain tops in the early Eocene, 52 million years ago. Then at the Eocene-Oligocene boundary, 37 million years ago, a major cool period occurred and tongues of ice reached the Antarctic coast. However, it was not until the Middle Miocene, 10-13 million years ago, that the first major ice caps accumulated—initially in East Antarctica, but later in West Antarctica.

At first the beech forest could probably live alongside the glaciers and ice fields, as it does in Patagonia and southern New Zealand, but eventually further chilling would lead to massive growth of the ice fields, and retreat of the plant life that, with nowhere to retreat to (links to New Zealand and possibly South America having been broken), became eventually extinct.

*Glomar Challenger*'s drill shows that by 5–4 million years, in Pliocene times, the land of Antarctica had virtually been submerged under a thick ice cap, writing *finis* to any forest life (p. 320).

**New Zealand and New Caledonia: the parting of the ways**

From almost the very beginning of geological time New Zealand and New Caledonia have been "as one", geologically speaking, and therefore probably very close geographically.

The oldest rocks in New Caledonia, of Permian age (280-230 million years ago), are identical in both type and fossil content to those in New Zealand. Extremely close similarities continue throughout the Triassic and Jurassic (230-136 million years ago) and at this time, as in the Permian, New Zealand and New Caledonia were actively involved in the New Zealand Geosyncline. Slight regional differences show up, however, and it looks as though New Caledonia was more involved in volcanic activity than New Zealand and probably straddled the volcanic island arc, developed adjacent to the subduction zone extending into New Zealand (Fig. 16.29A). The various earth movements of the Rangitata Orogeny (136-110 million years ago), that brought the New Zealand Geosyncline to a close, affected both New Zealand and New Caledonia. Cretaceous rocks, deposited on the eroded and planed-off landmass created by the Rangitata Orogeny, are also virtually identical in both countries. Close similarities continue into the Paleocene and Eocene, when both countries were low-lying, deeply weathered and eroded, and quartz-rich sediments accumulated off shore and coal beds in low-lying depressions on the land. However, towards the end of Eocene times, about 37 million years ago, came the parting of the ways for New Caledonia and New Zealand.

Drilling by the *Glomar Challenger* has shown that the South Fiji Basin, north of New Zealand, opened up between 37-27 million years ago and these and other movements probably rifted New Zealand and New Caledonia apart and broke their continuity. The intensity and scale of these tearing and rifting movements is indicated by the fact that during the Oligocene (37-25 million years ago) a vast piece of ancient oceanic crust was thrust up over New Caledonia (Fig. 16.29B) and in northern New Zealand, north of Auckland, entire chunks of country were turned upside down, thrust over each other and jumbled around into a chaotic mess.

New Caledonia also shared in New Zealand's primitive terrestrial flora and fauna (Chapter 14) and although, as in New Zealand, "thinning out" has occurred over geological time, geographic isolation has ensured that many interesting relics remain.

Thus it is uncanny to go some 1300km (800 miles) across the seas to New Caledonia and there to see a country with so many faunal and floral similarities to New Zealand—a country with a primitive flightless bird as a national symbol (the Cagou, *Rhynochetos*; Fig. 16.30), and a forest with many New Zealand types (tree ferns, podocarps, *Nothofagus* etc.). Like New Zealand, New Caledonia lacks land snakes, marsupials, and fossil records of land dinosaurs and its only mammal is a native bat, that had presumably winged its way across the infant Tasman early in the Tertiary.

However, unlike New Zealand, it lacks indigenous amphibians, primary freshwater fish and freshwater mussels, even as fossils. But although lacking spectacular primitive animals such as New Zealand's tuatara, *Leiopelma* etc., New Caledonia more than makes up for it in the plant kingdom. It teems with primitive plants and probably no richer, nor more peculiar, archaic and endemic relics of plant life can be

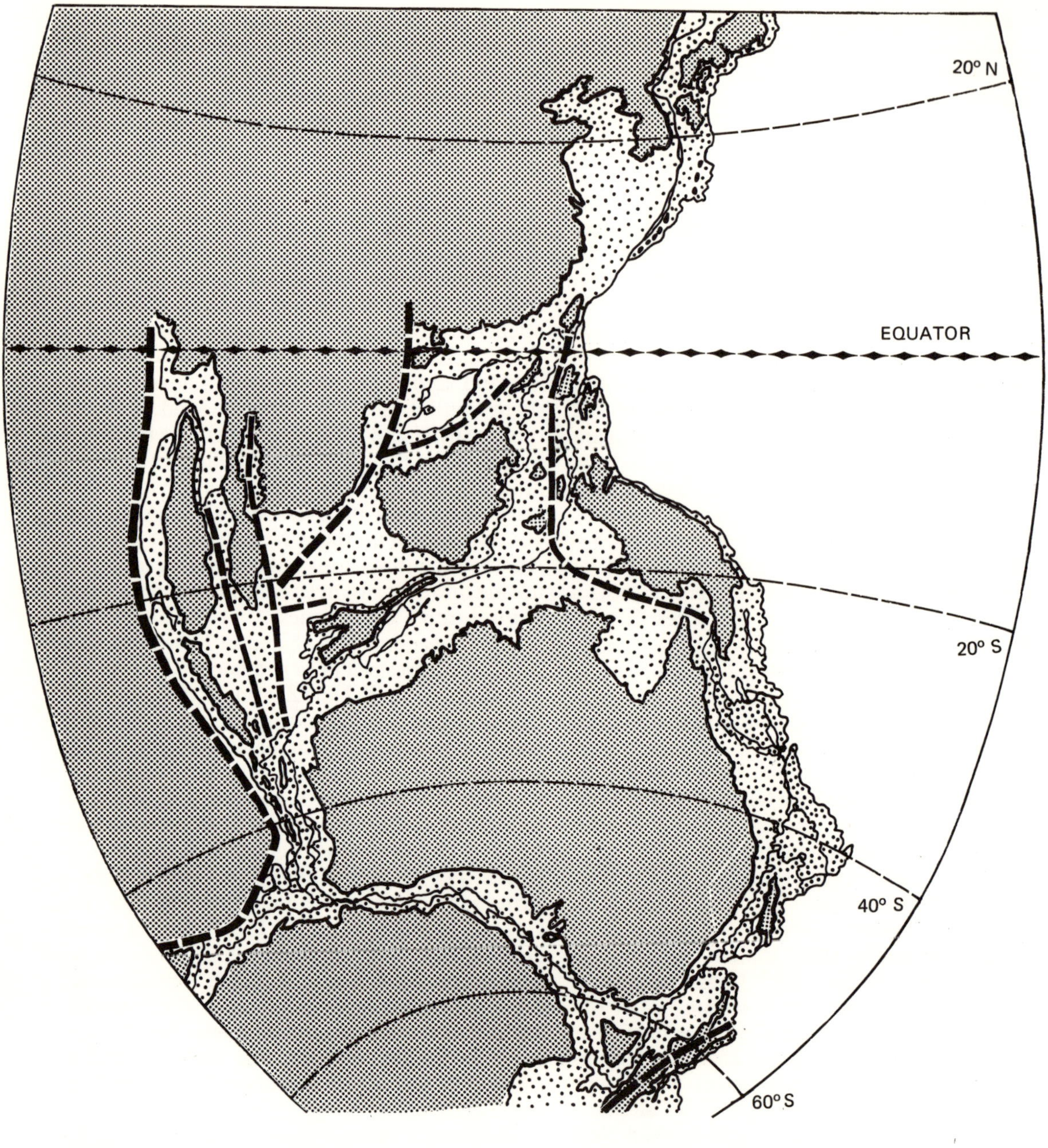


**Figs 16.22-16.27.** These reconstructions, based on the theory of an expanding earth, illustrate the changing fortunes of floral and faunal links between Australasia and South-east Asia during the past 180 million years.

Continental shelves are shown in a light dot pattern, zones of movement by medium broken lines and sea-floor spreading centres by thick lines with arrows showing directions of spreading.

Fig. 16.22. Australasia and South-east Asia 180 million years ago (early Jurassic). The diameter of the earth is 80 percent of its present diameter. Free migration probably occurred between Australasia and South-east Asia. It is likely that West Antarctica, Australia and New Zealand may have had more northerly latitudes than depicted in this diagram (cf. Fig. 12.26).

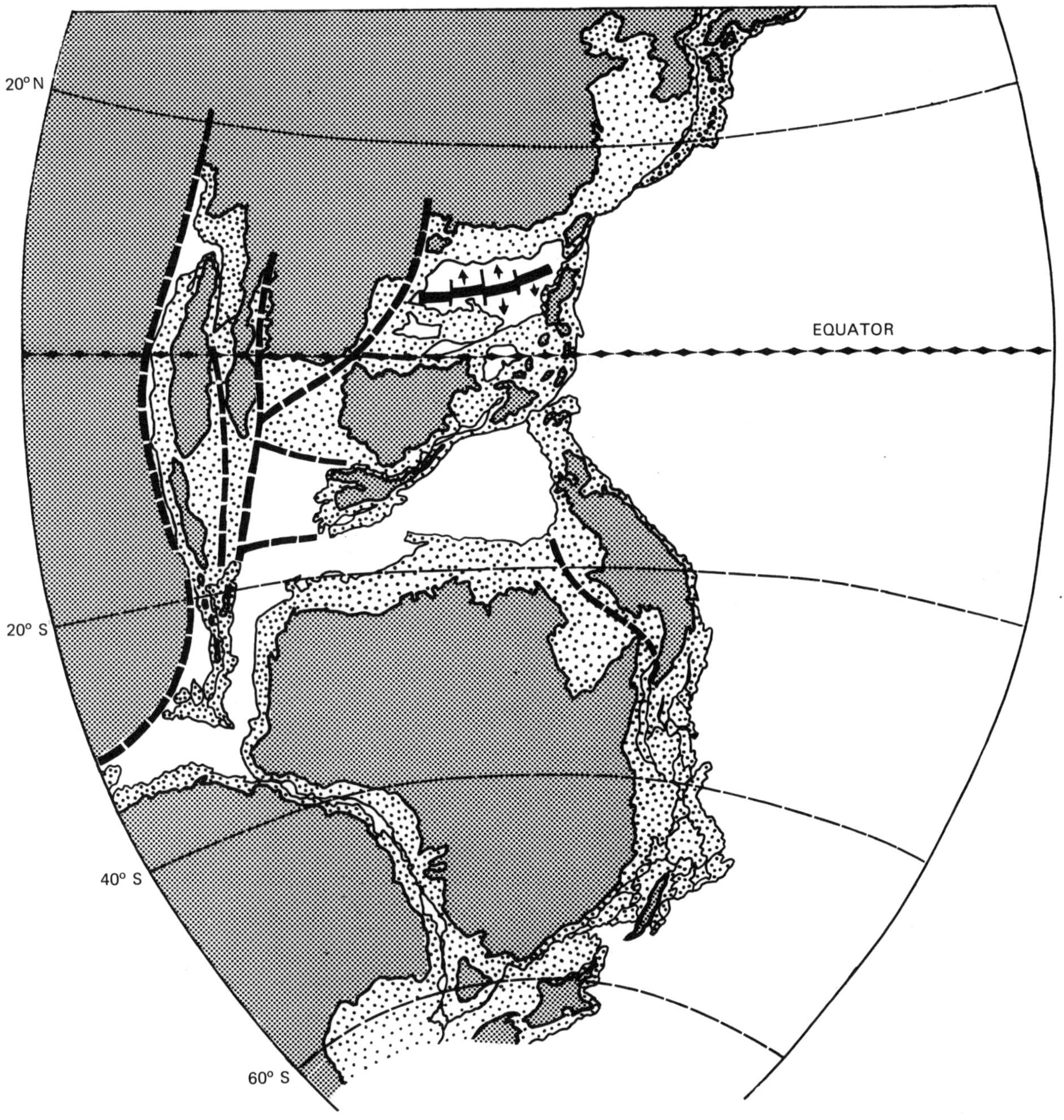


**Fig. 16.23.** Australasia and South-east Asia 120 million years ago (early Cretaceous). The earth's diameter is 87 percent of its present value. Sea-floor spreading began in the South China Sea. Australasia rotated southwards, loosening its links to the north.

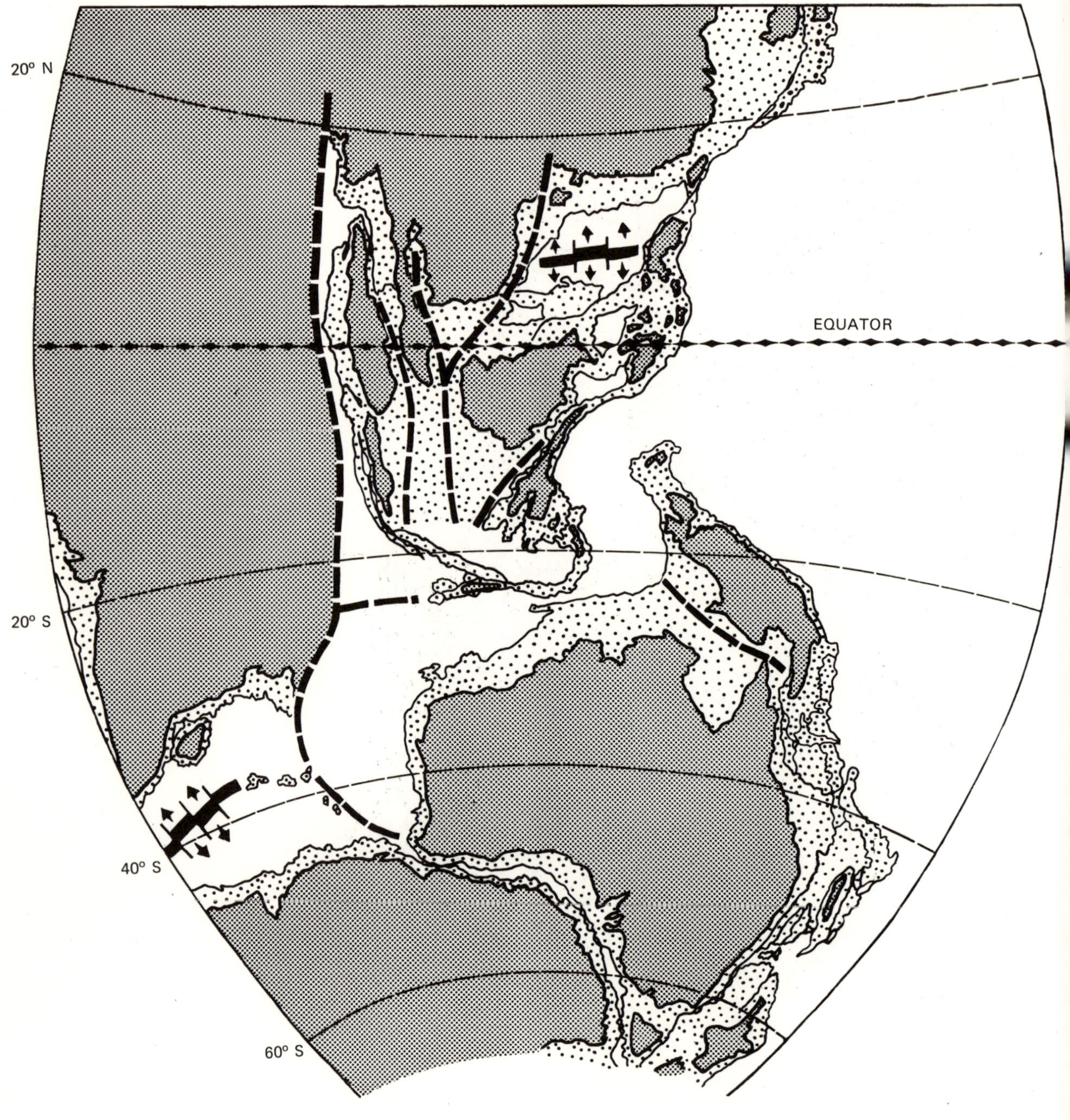


**Fig. 16.24.** Australasia and South-east Asia 90 million years ago (mid-Cretaceous). The diameter of the earth is 90 percent of its present diameter. Australasia had lost its land links to the north and a moat-like oceanic gap separated it from South-east Asia.

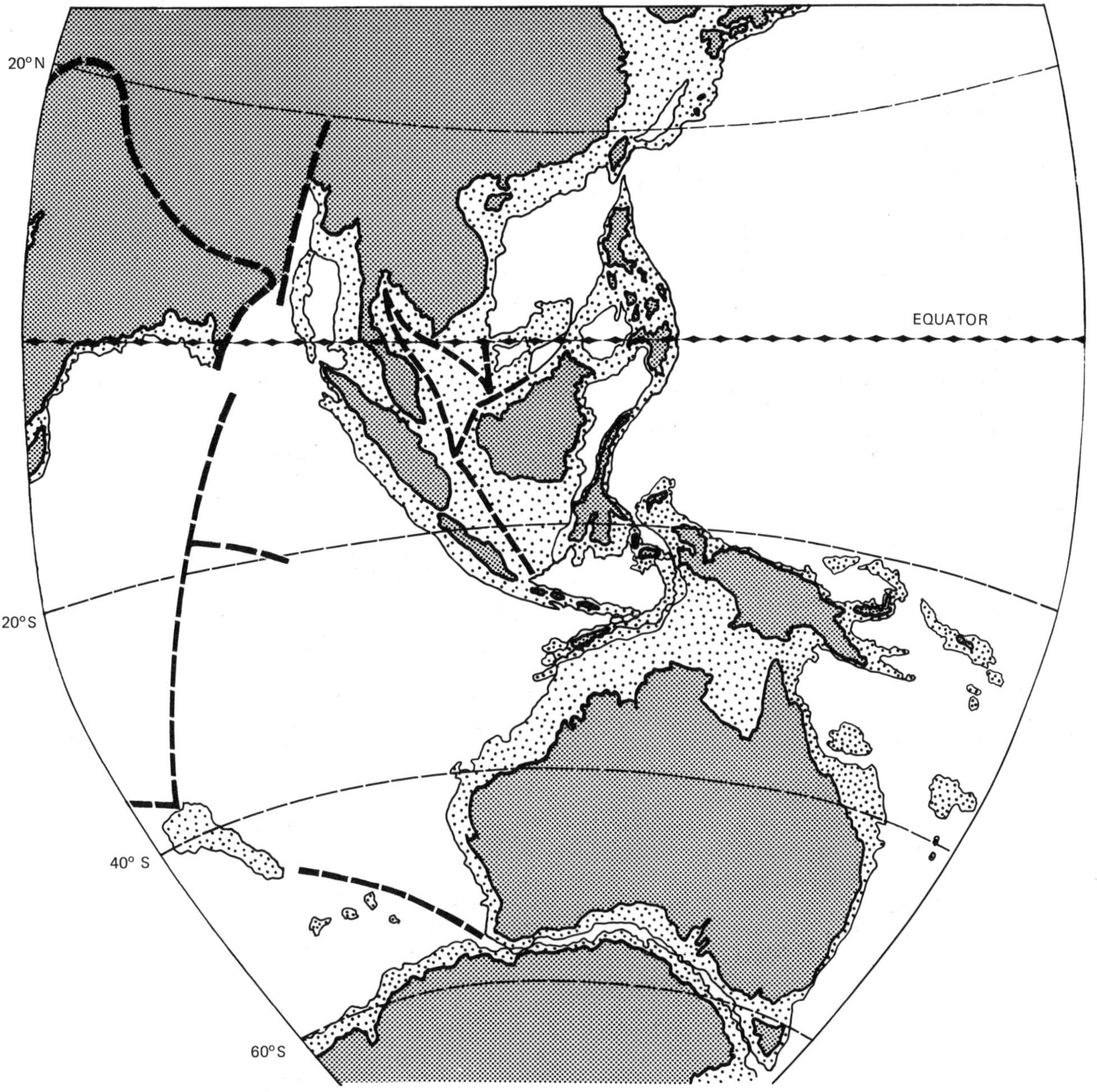


**Fig. 16.25.** India, South-east Asia and Australasia 60 million years ago (Paleocene). The earth's diameter is 93 percent of its current value. Ephemeral land links via the island arcs of Indonesia may have been possible at times between Australia and South-east Asia. The Tasman Sea had opened up and New Zealand had now cut all terrestrial links with Australia.

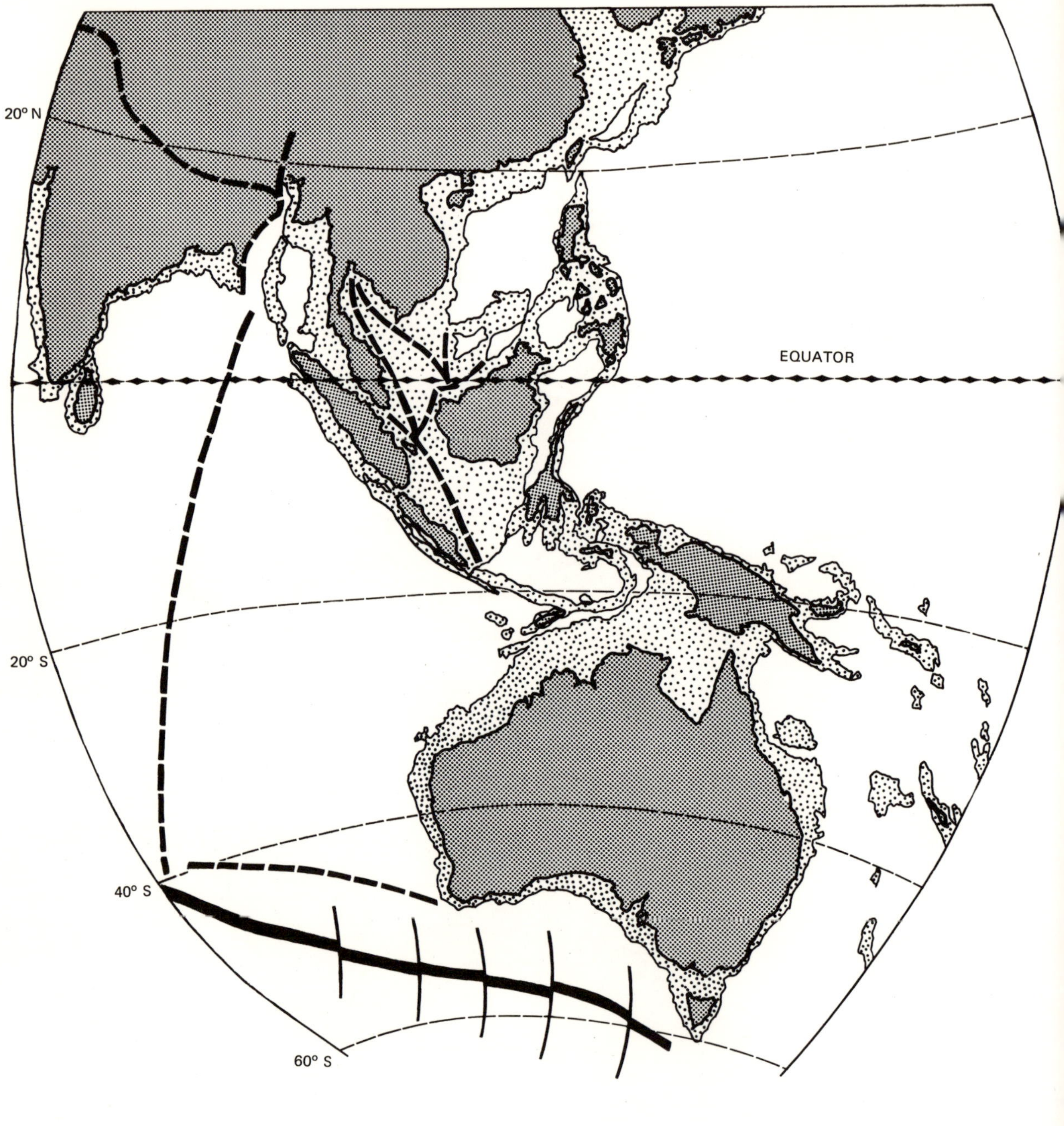


**Fig. 16.26.** India, South-east Asia and Australasia 30 million years ago. Diameter of earth is 97 percent of modern value. Ephemeral land links via the Indonesian islands may have been possible at some intervals between Australia and South-east Asia. Australia had separated from Antarctica.

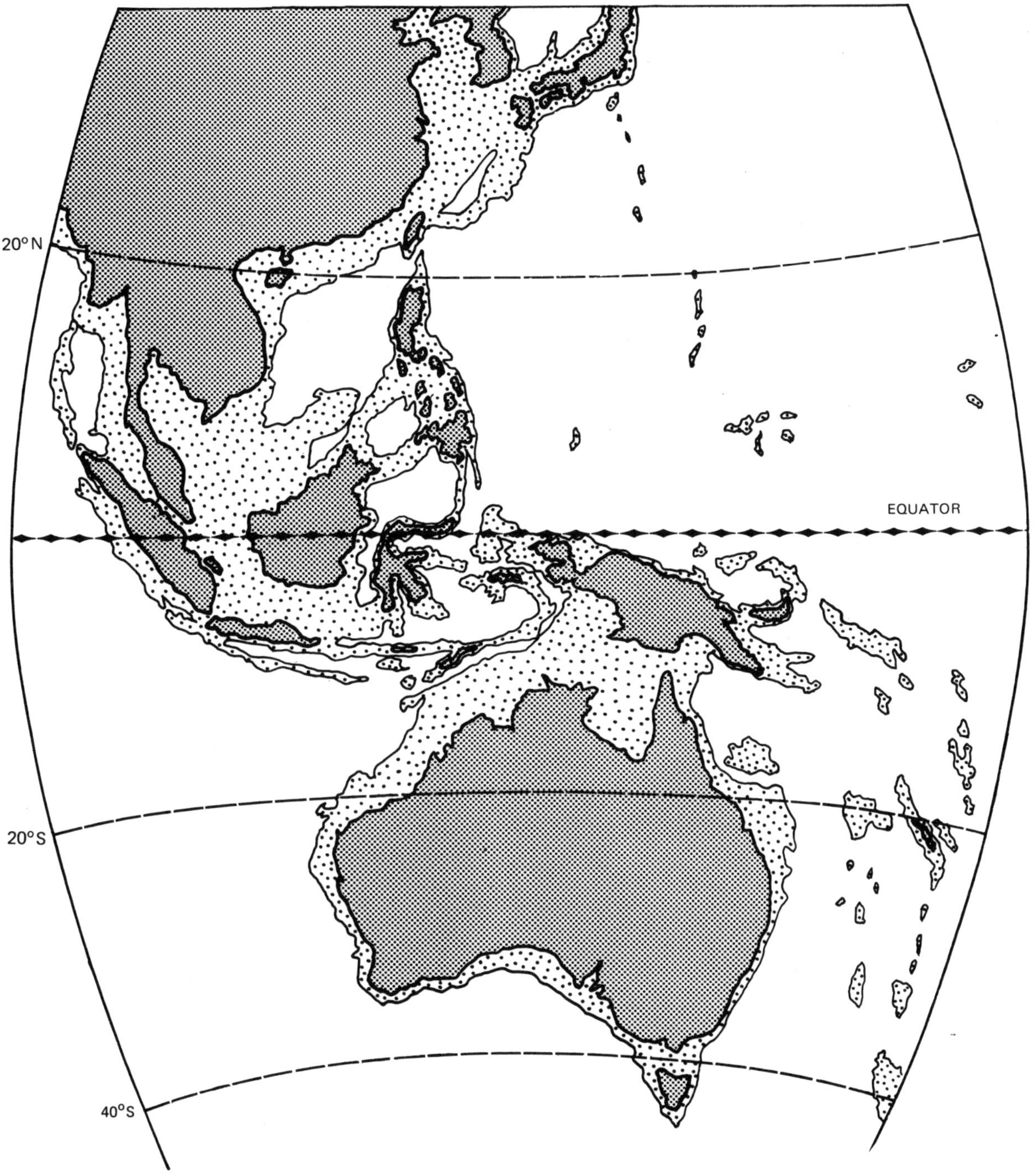


**Fig. 16.27.** South-east Asia and Australasia as seen today.

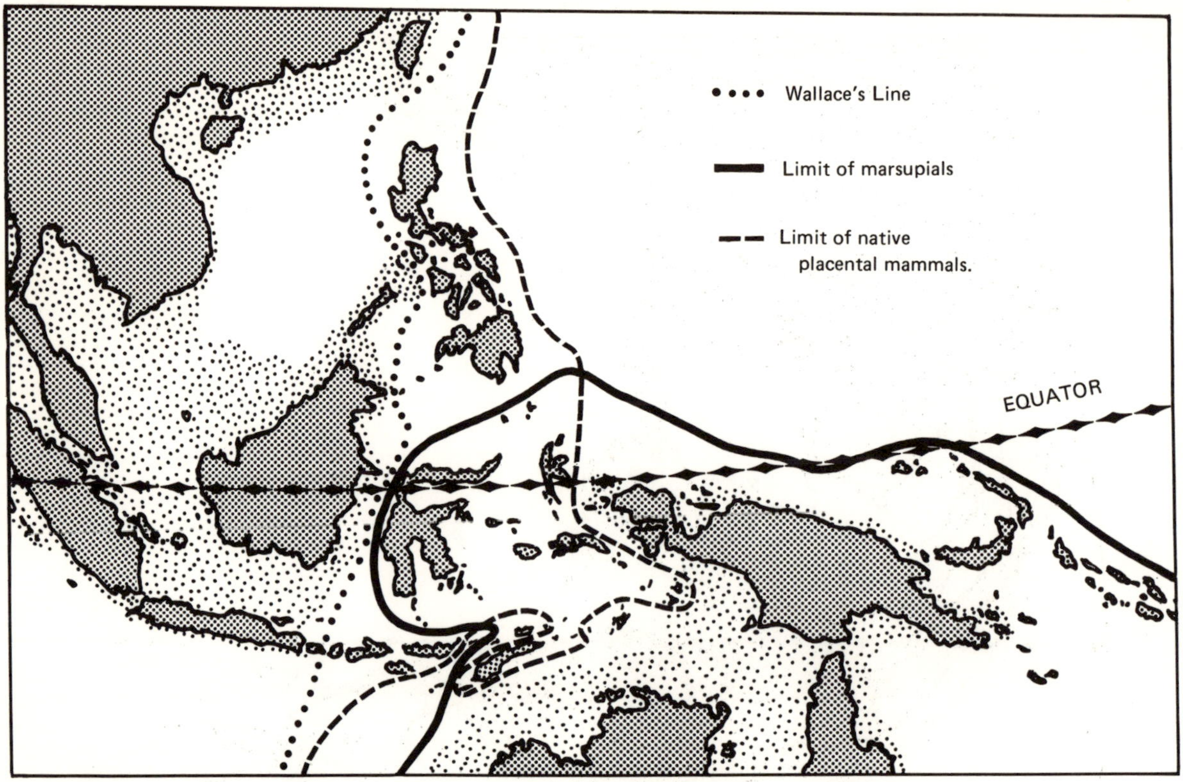


**Fig. 16.28.** The changing geography of South-east Asia in the geological past, as illustrated in Figs 16.22-16.27, had a decisive influence on animal and plant migrations.

From late Cretaceous time onward (Figs 16.24-16.27) the Asiatic and Australian faunas faced one another across an oceanic "moat". During this period earth movements provided ephemeral island links that at times allowed partial bridging of the moat.

These links allowed the carnivorous placental mammals that colonised the Asiatic mainland in late Cretaceous and early Tertiary to penetrate as far as the Moluccas (broken line), but only bats reached Australasia. In the absence of placentals, marsupials survived in Australia and New Guinea, but managed to co-exist with them in the Moluccas (heavy line).

Other Australian animals, chiefly birds, used the island stepping stones in a reverse direction and intermingled with Asiatic types in the Sunda islands, but were unable to spread further than Wallace's Line, lying between Lombok and Bali.

Thus there is a transitional zone between New Guinea and Wallace's Line in which Australian and Asiatic animals intermingle, as a result of temporary land links in the geological past, from the directions of both Australia and Asia.

In this diagram the continental shelf, extending to depths of 200 m (650 ft) is shown by the medium dot pattern.

found elsewhere in the world compressed into such a small area (one-seventh the size of New Zealand's North Island) (Fig. 16.31).

## Coral reefs and coconuts

The comparative tranquility of the Paleocene, Eocene and Oligocene (65-25 million years ago), when the ancestral New Zealand landmass was gradually worn away and shallow seas flooded in, was ended in the Miocene (25-5 million years ago), and from this time onwards the New Zealand islands were the scene of restless activity (Figs 16.32, 16.33). Patterns of folds, welts and troughs developed under the influence of earth movements. Changes in geography were frequent as troughs sank rapidly and welts rose in complementary fashion. Segments of land moved up and down as a series of short interfingering or branching folds. Large parts of New Zealand had the form of an ever-changing archipelago. It almost seems as though a kind of writhing of this part of the Pacific margin went on over most of the later Tertiary.

In the early Miocene, 25-15 million years ago, although there were many signs that the restless land had begun to stir, there is no evidence of truly mountainous terrain. However, a chain of active volcanic islands extended from Auckland to Hokianga (Fig. 16.33).

Tropical seas lapped around northern New Zealand at this time and reef-building corals similar to those now found on the Great Barrier Reef (Queensland) lived around the Northland and Auckland coasts and extended south as far as East Cape. It has been estimated that these seas had a seasonal range of surface temperature some 5–7°C higher than today, corresponding to a range of 18–27°C (64.5–80.6°F). The early Miocene marked the high point of Tertiary climate and of the influx of warmth-loving immigrants it brought from the Indo-Pacific areas to the north.

As we found in the Mesozoic, our knowledge of fossil plants and animals on the land is rather sketchy. We know, however, from fossil pollen studies that palms were particularly abundant at this time over the entire length and breadth of New Zealand. A bed of fossil coconuts, dating back to this time, is known from Cooper's Beach (Mangonui, Northland) (Fig. 16.34) and isolated specimens from the East Cape area.

In the late Miocene, 15-5 million years ago, earth movements were continuing and short-lived islands and fingers of land were appearing (Fig. 16.35). By the end of the Miocene a long narrow-gutted land had been formed and on it some areas of rugged landscape were giving rise to boulder beds in Wairarapa, Marlborough and Murchison. Volcanoes were active on land in the Coromandel Peninsula and around Dunedin, and volcanic islands were erupting on the site of the modern Banks Peninsula.

When we look at the animals that had lived in New Zealand seas in the late Miocene we see that many of the warmth-loving types that had migrated to New Zealand in the early Miocene had disappeared and this, combined with oxygen isotope determinations, tells us that the seas were cooling down, although they were still warmer than those of the present day.

During this time we have the first influxes of creatures that have used the West Wind Drift and Circum-Antarctic Current as a means of dispersal (Figs 6.11, 6.12).

**Riding the west wind**

Establishment of the oceanic Circum-Antarctic Current in the middle and late Oligocene (30-25 million years ago) brought into being a whole host of associated meteorological phenomena that have a decisive influence on our weather.

The most notable of these meteorological phenomena is the West Wind Drift—prevailing westerly winds that encircle the globe at latitudes between 40°S Latitude and 60°S Latitude, and give rise to the "Roaring Forties", "Furious Fifties" and "Screaming Sixties". These winds are so powerful and constant that they can blow a giant weather balloon from New Zealand across to South America in a mere five days and then around the globe eight times in 102 days, repeatedly crossing all of the southern continents and most islands (p. 56, Fig. 6.12), and can drive surface water along at comparably fast rates, producing a surface oceanic current to match the deeper-flowing Circum-Antarctic Current.

The transporting efficiency of such a wind and current system is so high that, once initiated, many animals and plants started to use it as a means of crossing the ocean after Antarctica ceased to be available as a stepping stone (cf. Fig. 6.11).

Thus from the Miocene onwards New Zealand began to gain a new lot of southern colonists. These, however, were rather different from those we received in the Cretaceous and early Tertiary—the Austral immigrants (p. 269). The Austral plants and animals had travelled across and around the then closely connected southern lands and were all of types that needed continuous land, or continuous shallow water, or closely adjoining islands for dispersal. The riders of the West Wind Drift, on the other hand, termed Neo-Austral (i.e."New-Austral"), were all types capable of crossing major water barriers. As from Miocene times onward there is no doubt that the distribution of land and sea in the Southern Hemisphere was virtually as it is now, one is forced to the conclusion that these organisms were capable of long-distance dispersal over oceanic gaps of up to 6000 km (3700 miles).

Possible mechanisms for such long-distance travel include birds, ocean currents and winds. Many birds, such as giant petrels, shearwaters and albatrosses, migrate widely in southern latitudes, with the assistance of the West Wind Drift. They may carry seeds encased in mud around their legs, hooked seeds attached to their feathers, or fleshy seeds in their stomachs. It has been suggested that Macquarie Island gained its

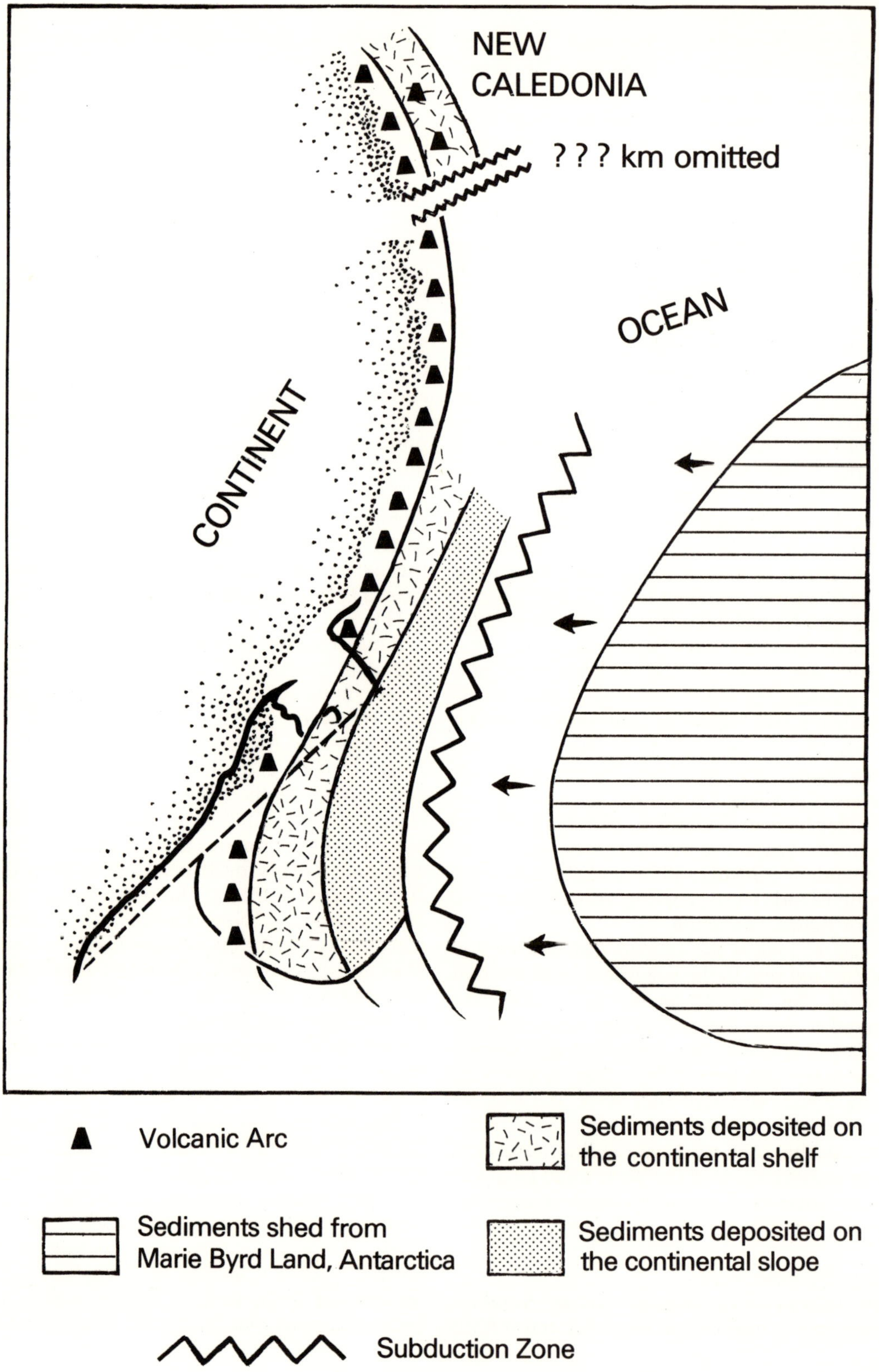
NEW
CALEDONIA
? ? ? km omitted
OCEAN
CONTINENT
Volcanic Arc
Sediments deposited on the continental shelf
Sediments shed from Marie Byrd Land, Antarctica
Sediments deposited on the continental slope
Subduction Zone

**Fig. 16.29A.** During the Permian, Triassic and Jurassic (280-136 million years ago) virtually identical rocks and fossils were laid down in New Caledonia and New Zealand and the two regions were part of a continuous trough of deposition, (New Zealand Geosyncline) and were probably also geographically very close together.

The New Zealand Geosyncline bordered a large landmass (Australia and the Lord Howe Rise; cf. Fig. 13.14) and was in turn flanked by a subduction zone (cf. Fig. 11.7). Sediments shed from the Lord Howe-Australia landmass progressively filled the western flank of the New Zealand Geosyncline and belts of sediments laid down on the continental shelf and continental slope (Fig. 13.15) can be distinguished.

Volcanic island arcs, related to the down-dipping subduction zone, formed extensive archipelagos in the New Zealand Geosyncline. Compared with New Zealand, New Caledonia had a greater share of such volcanic material.

As subduction continued a mass of sediment derived from Marie Byrd Land was rafted across from Antarctica (Torlesse greywacke and Haast schist, cf. Fig. 13.14).

The collision of these latter sediments with the westerly-derived material already laid down in the New Zealand Geosyncline generated the Rangitata Orogeny (between 136-110 million years ago), and produced a primaeval New Zealand continent.

This diagram is a sketch showing the geological relationships of New Caledonia and New Zealand just prior to the Rangitata Orogeny.

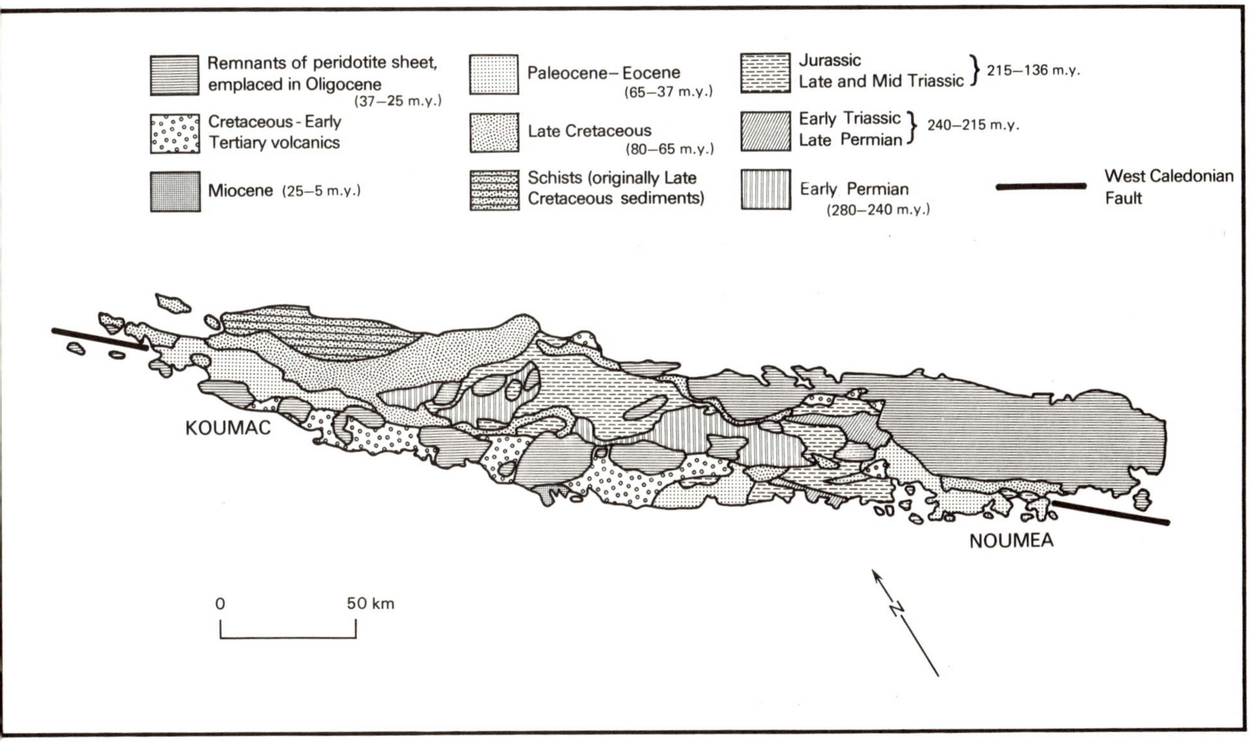


**Fig. 16.29B.** Later on, in Tertiary times, New Caledonia was involved in earth movements along an active plate boundary, probably the ancestor of the one now seen in the New Hebrides (Fig. 9.18). This intense crustal activity culminated in the Oligocene (37-25 million years ago), when enormous sheets of ocean crust and upper mantle material (ophiolites) were thrust over the entire island (cf. Fig. 9.18). Erosion since the Oligocene has either removed completely, or substantially thinned, the ophiolite sheets. Large blobs still remain, however, as seen in this geological map, and their mineral-rich layers form the basis of New Caledonia's wealth. Like New Zealand (Fig. 4.11) New Caledonia has been torn in half longitudinally by a transcurrent fault (West Caledonian Fault). Unlike New Zealand's Alpine Fault, however, the West Caledonian Fault is not active today (the Alpine Fault is a modern plate boundary).

*Fig. 16.30*

*Fig. 16.31A*

*Fig. 16.31A*

**Fig. 16.30.** The cagou (*Rhynochetos jubatus*), the emblem of New Caledonia. This primitive flightless bird is thought to be a remnant of the original Gondwanaland fauna. However, as it is a more advanced form than the ratites, (i.e. moas and kiwis) for example, it is thought to have been a slightly later immigrant to the South-west Pacific.

The cagou is closely related to New Zealand's *Aptornis*, an extinct giant rail (a family that also includes the weka, pukeko and takahe) (pp. 252–253). The cagou is about the size of a large weka or large domestic hen.

**Fig. 16.31.** New Caledonia is packed with numerous peculiar archaic plants. Although some of the plants are shared with New Zealand, many are restricted to New Caledonia. Notable amongst the latter are araucaria pines that are distinctive features of the landscape in many parts of the island. The columnar pine and other similar varieties are often planted in groves around Melanesian villages. Fig. 16.31A shows columnar pines (*Araucaria montana*) growing on the slopes of Koniambo massif, northern New Caledonia.

Candelabra pines (*Araucaria rulei*) grow on rugged terrain and Fig. 16.31B shows some of these trees on Boulinda massif (left) and Tiebaghi massif (right).

plants chiefly in this way. Ocean currents, driven by the West Wind Drift, can carry floating material between the southern continents at about 13 km (8-9 miles) a day. Floating or swimming animals, and plants with buoyant fruits and coastal habitats, are prime candidates for this type of dispersal and it is believed that plants such as *Hebe* and *Sophora* migrated between New Zealand, southern Chile and Gough Island in this way.

*Fig. 16.31B*

*Fig. 16.31B*

The high winds of the "Roaring Forties" may in their own right distribute plants and animals. Many plants have light seeds or spores, or plumed seeds, and some spiders, for example, create web parachutes by which they can "balloon" in the wind.

At the time the West Wind Drift was starting to get into its stride (i.e. early Miocene), New Zealand was also receiving many plant and animal colonists from across the seas to the north, chiefly from the Malayo-Pacific region, and some of these, after populating New Zealand, rode the West Wind Drift and by this means spread to other southern lands (Fig. 6.11).

**Circum-Pacific mobile belt**

During the late Miocene, and particularly the Pliocene, major earth movements, that in most instances have continued to modern times, were occurring all around the Pacific. A circum-Pacific "mobile belt" or "ring of fire" was produced—characterised by newly formed mountains, earthquakes and volcanoes—and, as the reader may appreciate, we in New Zealand are very much part of this belt. Some spectacular mountain chains were created—the mountainous backbones of Japan and New Zealand, for example, and the Rockies and Andes of the Americas and their continuations through Central America.

Before the Pliocene, North and South America had been cut off from each other and broad marine straits traversed much of what is now Central America. This set the scene for a dramatic faunal change.

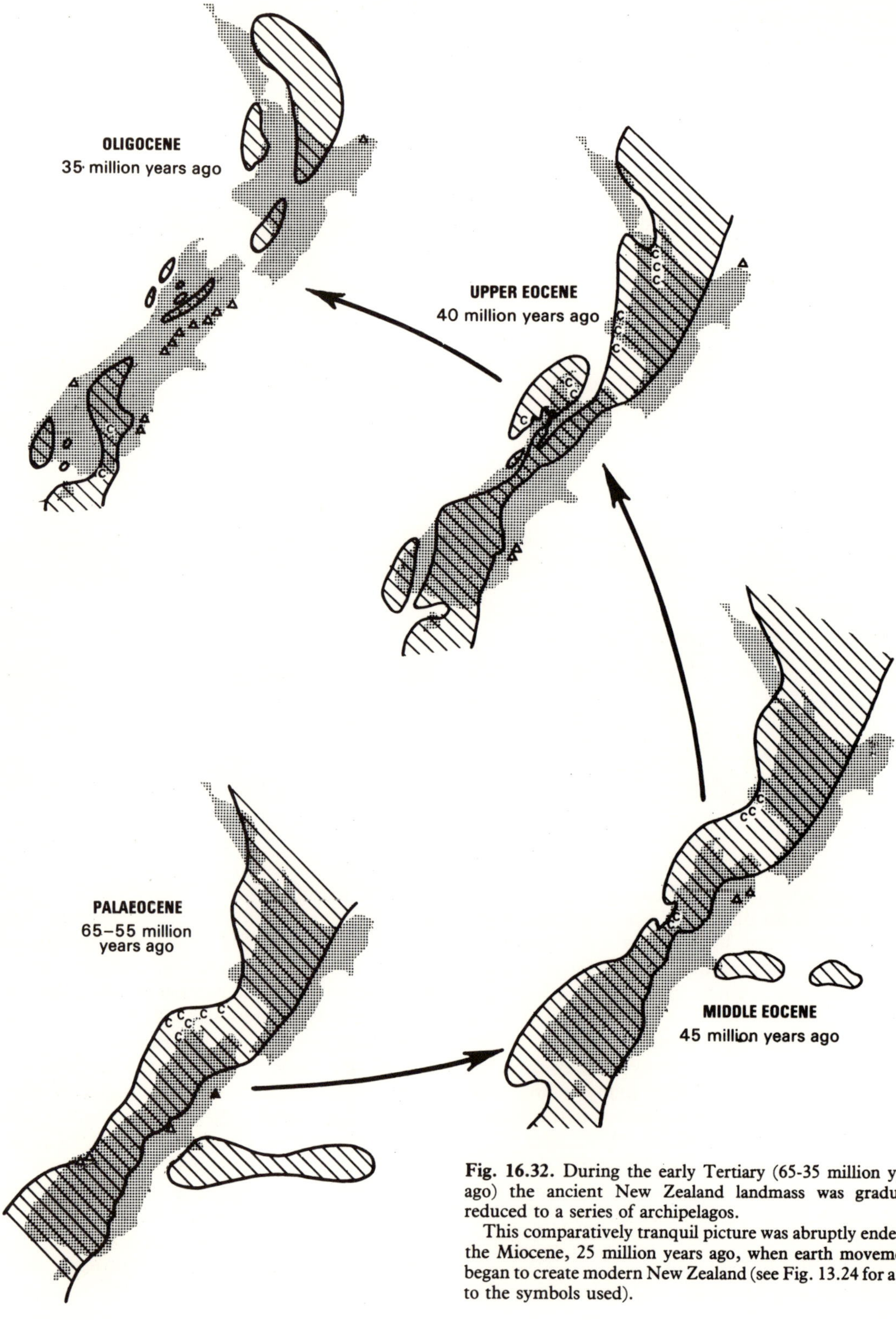


**Fig. 16.32.** During the early Tertiary (65-35 million years ago) the ancient New Zealand landmass was gradually reduced to a series of archipelagos.

This comparatively tranquil picture was abruptly ended in the Miocene, 25 million years ago, when earth movements began to create modern New Zealand (see Fig. 13.24 for a key to the symbols used).

**Fig. 16.33.** In the early Miocene, 25-18 million years ago, the long episode of peneplanation and submergence of the ancient New Zealand landmass was brought to a close (cf. Figs 16.5, 16.8, 16.10, 16.11, 16.13). New land began to appear in both North and South Islands (land is indicated by the horizontal line pattern). Swampy river deltas in the North Taranaki area laid down deposits that were later to become the Mokau coals (indicated by a dark pattern).

A rising ridge in Northland and Auckland was flanked by a chain of andesite volcanoes (cone symbols) and the first volcanic eruptions began in the Coromandel Peninsula.

Apart from volcanoes, however, the early Miocene land was not yet mountainous.

During the early Miocene warm-water plants and animals colonised most parts of New Zealand, and coral reefs grew around the coasts as far south as East Cape.

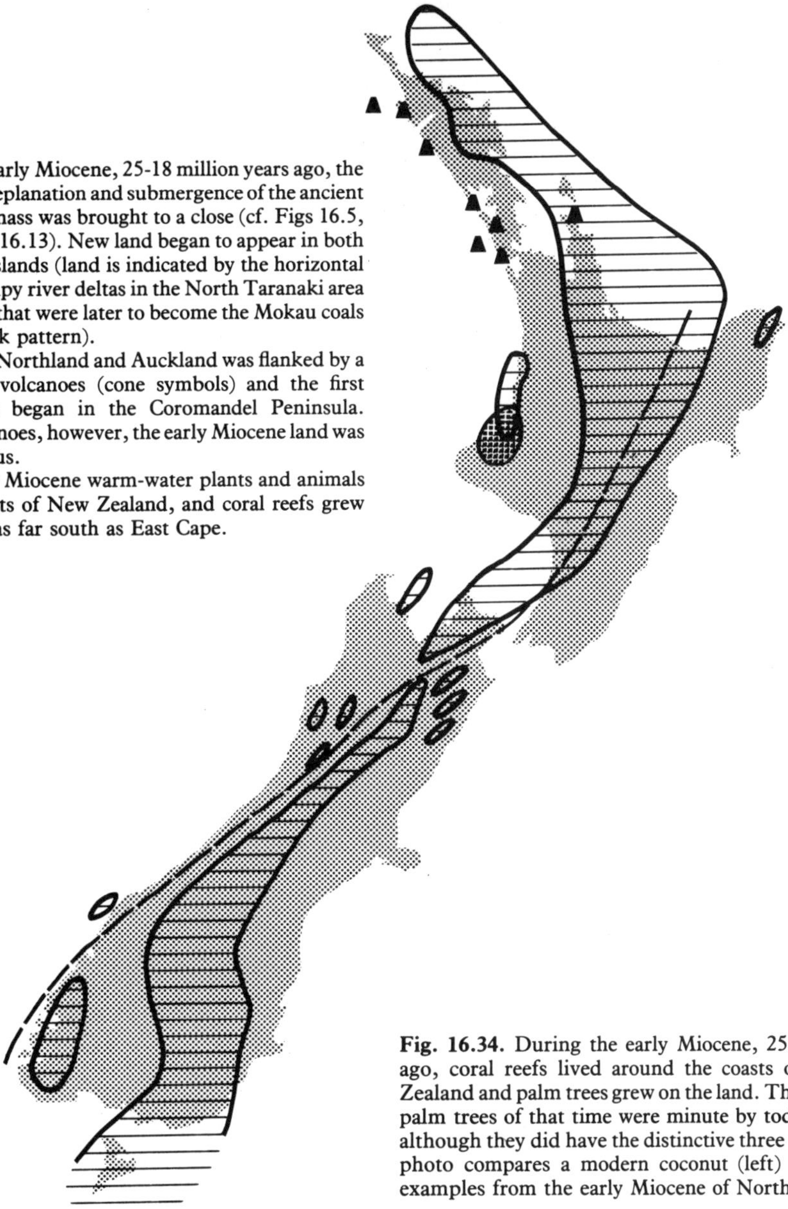

*Fig. 16.33*

**Fig. 16.34.** During the early Miocene, 25-18 million years ago, coral reefs lived around the coasts of northern New Zealand and palm trees grew on the land. The coconuts on the palm trees of that time were minute by today's standards—although they did have the distinctive three "eyes"—and this photo compares a modern coconut (left) with some fossil examples from the early Miocene of Northland.

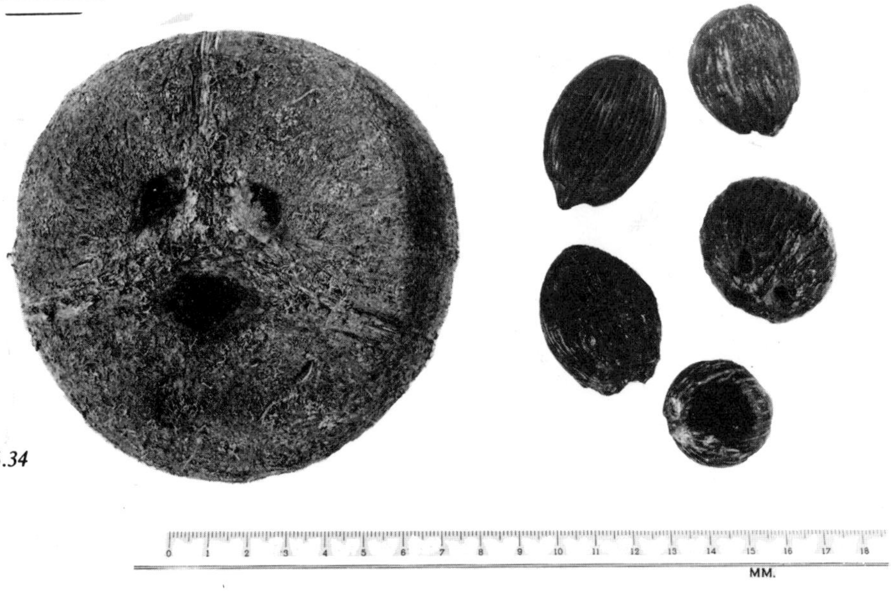


*Fig. 16.34*

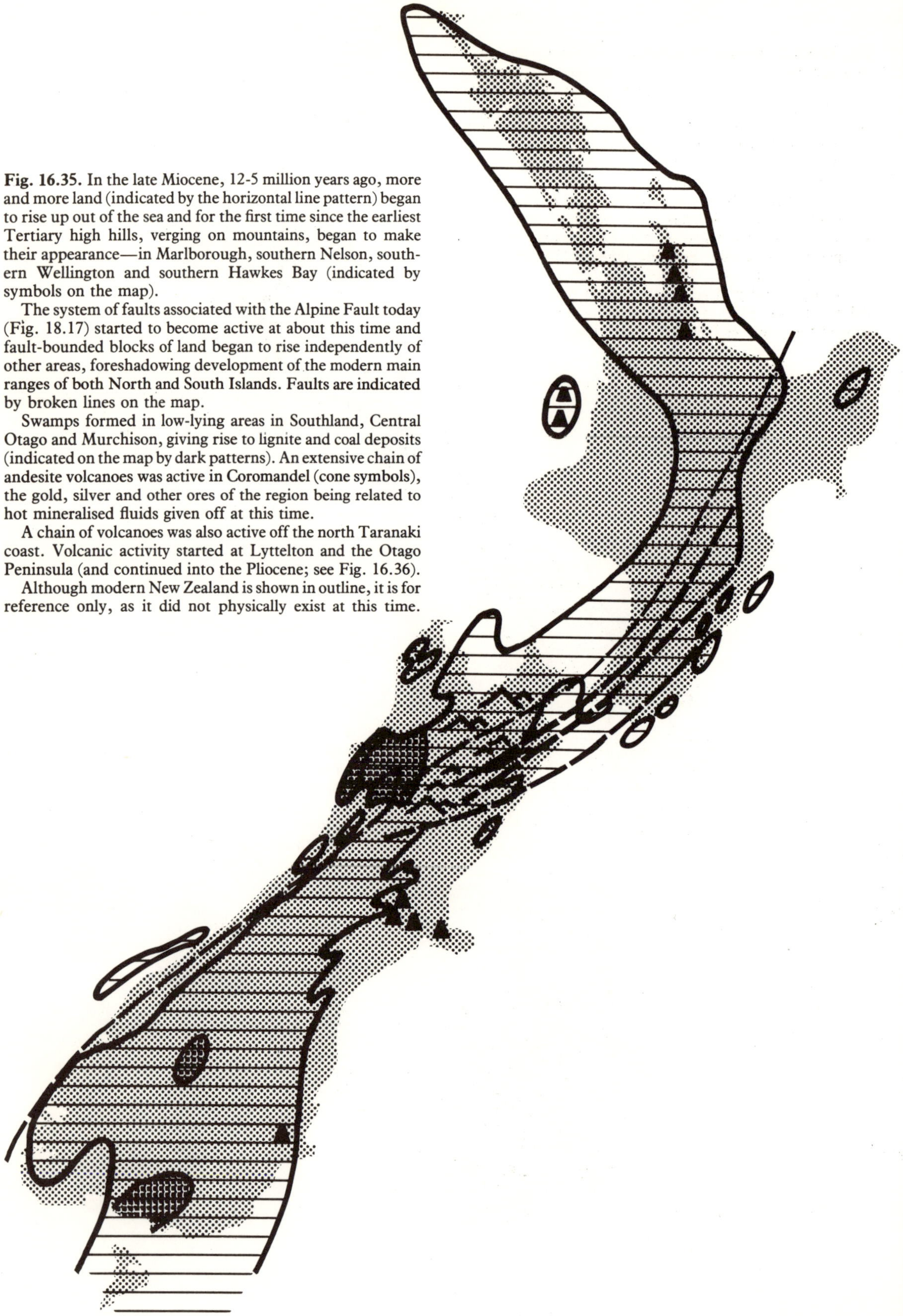

**Fig. 16.35.** In the late Miocene, 12-5 million years ago, more and more land (indicated by the horizontal line pattern) began to rise up out of the sea and for the first time since the earliest Tertiary high hills, verging on mountains, began to make their appearance—in Marlborough, southern Nelson, southern Wellington and southern Hawkes Bay (indicated by symbols on the map).

The system of faults associated with the Alpine Fault today (Fig. 18.17) started to become active at about this time and fault-bounded blocks of land began to rise independently of other areas, foreshadowing development of the modern main ranges of both North and South Islands. Faults are indicated by broken lines on the map.

Swamps formed in low-lying areas in Southland, Central Otago and Murchison, giving rise to lignite and coal deposits (indicated on the map by dark patterns). An extensive chain of andesite volcanoes was active in Coromandel (cone symbols), the gold, silver and other ores of the region being related to hot mineralised fluids given off at this time.

A chain of volcanoes was also active off the north Taranaki coast. Volcanic activity started at Lyttelton and the Otago Peninsula (and continued into the Pliocene; see Fig. 16.36).

Although modern New Zealand is shown in outline, it is for reference only, as it did not physically exist at this time.

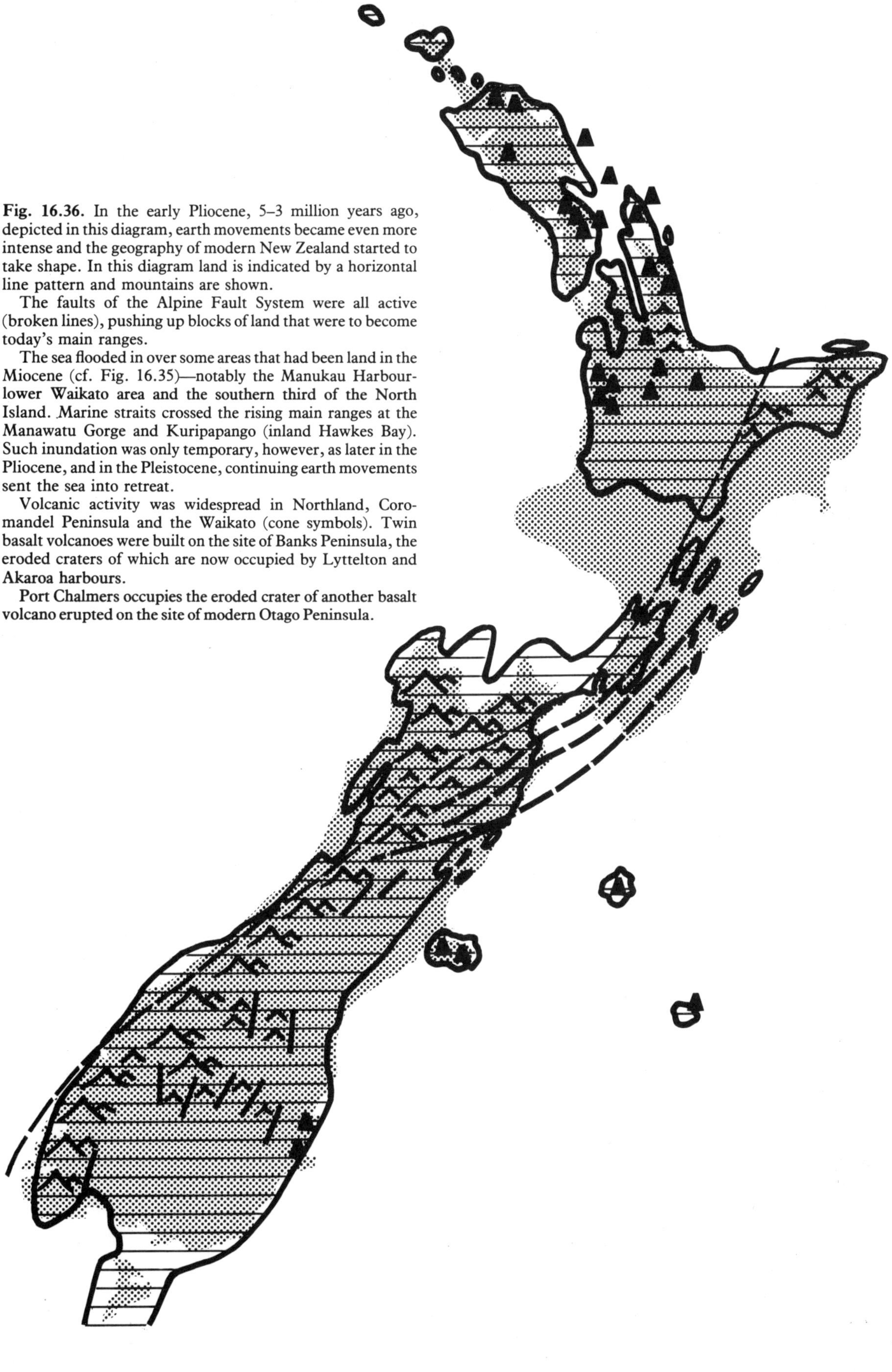

**Fig. 16.36.** In the early Pliocene, 5–3 million years ago, depicted in this diagram, earth movements became even more intense and the geography of modern New Zealand started to take shape. In this diagram land is indicated by a horizontal line pattern and mountains are shown.

The faults of the Alpine Fault System were all active (broken lines), pushing up blocks of land that were to become today's main ranges.

The sea flooded in over some areas that had been land in the Miocene (cf. Fig. 16.35)—notably the Manukau Harbour-lower Waikato area and the southern third of the North Island. Marine straits crossed the rising main ranges at the Manawatu Gorge and Kuripapango (inland Hawkes Bay). Such inundation was only temporary, however, as later in the Pliocene, and in the Pleistocene, continuing earth movements sent the sea into retreat.

Volcanic activity was widespread in Northland, Coromandel Peninsula and the Waikato (cone symbols). Twin basalt volcanoes were built on the site of Banks Peninsula, the eroded craters of which are now occupied by Lyttelton and Akaroa harbours.

Port Chalmers occupies the eroded crater of another basalt volcano erupted on the site of modern Otago Peninsula.

As we have seen (pp. 262–265), marsupials were progressively overwhelmed by the advance of placentals throughout most of the world (with the notable exception of Australia and New Guinea) in the early Tertiary and soon all became extinct in Europe and Asia, for example, and most in North America. Placentals were excluded, however, from South America for most of the Tertiary period—for about 60 million years.

Some time in the late Cretaceous marsupials had originally entered South America from North America via tenuous links across what is now Central America, then loosely linked chains of ephemeral volcanic islands. Throughout most of the Tertiary, however, these links had disappeared, so that South America remained completely isolated from other landmasses. During this time of isolation the early marsupials already established there evolved in their own peculiar fashion. For example, flesh-eating marsupials evolved, paralleling carnivorous placentals such as sabre-toothed cats and hyenas in their appearance. Likewise, curious giant marsupial ungulates filled various ecological niches, producing an amazing variety of forms, with superficial resemblances to horses, camels, rhinos and elephants.

Some 3–2 million years ago, as part of the Pliocene circum-Pacific earth movements, the Central American isthmus came into being and overland migration became possible between North and South America. A few South American animals like the opossum and armadillo moved into North America. In the opposite direction, however, many more North American animals (including for example the puma and llama) moved into South America, where they quickly took the places of the peculiar Tertiary marsupials, so that now these are only known as fossils in museum displays.

**Fig. 16.37.** The Southern Alps and other main ranges of New Zealand were pushed up by a series of earth movements grouped as the Kaikoura Orogeny. The movements began in Miocene times, 25-5 million years ago, and are still continuing today.

The Southern Alps have been pushed up by vertical movements along the Alpine Fault and this satellite view shows the line of the fault (indicated by arrows) sharply truncating the alpine topography.

The view extends from the Grey River (top, north) to Jackson Head (bottom, south).

### The Kaikoura Orogeny

As if in concert with the other countries in the circum-Pacific mobile belt, the Pliocene period in New Zealand (5-2 million years ago) saw a great increase in earth movements as the Kaikoura Orogeny, or mountain-building phase, got into full stride (Fig. 16.36). The orogeny is labelled after the mountains of the same name and these and the other ranges of New Zealand started to be formed at about this time. The earth movements were most acute in a belt trending from East Cape obliquely south-westward through the Southern Alps to Fiordland, and decreased towards the more stable districts of Northland and Otago-Stewart Island to either side. Kaikoura Orogeny movements are continuing today (Fig. 16.37).

The first stirrings of the Kaikoura Orogeny were in the Miocene, when thick layers of gravel, boulders and scree (Great Marlborough Conglomerate) were deposited in valleys such as the Clarence Valley, after having been eroded off the infant Inland and Seaward Kaikouras, then starting to be pushed up to either side.

Movements of the Kaikoura Orogeny produced characteristic "basin and range" topography in many parts of New Zealand, as some blocks of country were pushed up faster than others. Thus in Nelson and northern Westland the Murchison, Grey-Inangahua and Nelson

depressions lie between uplifted blocks. Similarly in Central Otago, Kaikoura Orogeny movements have pushed up a series of ranges e.g. the Rock and Pillar Range (1450 m; 4500 ft) and Pisa Range (1960 m; 6000 ft), each separated by depressions (Fig. 16.38). This part of the country is also notable for its vast stretches of peneplain landscape (p. 273), exhumed from beneath its former cover of marine sediments, and it is this surface that has been broken to produce the basins and ranges of today. The same peneplain, now thoroughly dissected, formerly arched across Fiordland, indicating uplift of the order of at least 1800-2100 m (6000-7000 ft).

**Fig. 16.38.** The earth movements of the Kaikoura Orogeny produced "basin and range" topography in many parts of New Zealand, as some blocks of country were pushed up faster than others. In Central Otago, for example, various fault-bounded blocks have broken up the old peneplain surface dating back to early Tertiary times, so that now there are a series of flat-topped ranges, separated by depressions.

The Alpine Fault, that had been initiated in the Rangitata Orogeny (pp. 236–243), became rejuvenated in the Kaikoura Orogeny. On the other hand, the various faults that branch off the Alpine Fault and traverse Marlborough and Wellington were probably formed entirely in the Kaikoura Orogeny. The Southern Alps were uplifted by vertical movement along the Alpine Fault in late Pliocene to modern times. An uplift rate of 10.7 m (35 ft) per 1000 years is known for the last 10,000 years (i.e. since the end of the Ice Age) from the Alpine Fault near Paringa River, Westland. Such a rate, continued for 3 million years, would provide 24,730 m (90,000 ft) of uplift.

It is believed, however, that the tempo of the Kaikoura Orogeny has been increasing towards the present day and that the observed post-Ice Age rates should not be extrapolated too far back into time. Nonetheless, uplift of the Southern Alps has been substantial and rapid—certainly of the order of at least 18,600 m (60,000 ft), judging from rocks altered by heat and pressure at great depths and now exposed at the surface.

The rate of uplift along the Alpine Fault has been so rapid that a very impressive steeply dipping "alpine front" has been produced on the West Coast side, as erosion has been unable to cut the slope back to a lower and more normal angle (Fig. 16.39). The Southern Alps, like much other New Zealand mountain scenery, have been cut out of the old peneplain surface, as shown by the fact that most peaks come up to about the same height ("summit accordance"), indicating a common plane of origin.

Horizontal movements also occurred along the Alpine Fault during the Kaikoura Orogeny and observed rates of horizontal movement in post-glacial time (i.e. last 10,000 years) average out at 3-4.6 m (10-15 ft) per 1000 years. As with the vertical uplift rates, however, it may not be wise to extrapolate these rates too far back into geological time. Kaikoura Orogeny movements along the Alpine Fault, both vertical and horizontal, have produced a broad zone of crushed, sheared and shattered rock some 300 m (1000 ft) wide, and other equally wide zones, but dating back to other earlier movements (as they are now naturally cemented by minerals), can be distinguished.

Uplift of the North Island ranges was at a slower rate than that of the Southern Alps and most did not reach their full heights until the Pleistocene.

As the ranges were rising, world sea level was rising also, probably in response to changing

**Fig. 16.39.** Uplift of the Southern Alps along the Alpine Fault has been geologically rapid and as a result an impressive "alpine front" has been formed on the West Coast side. The movements along the Alpine Fault, both horizontal and vertical, have produced a broad zone of crushed and shattered rock. Such weakened material has been eroded out by rivers and streams, resulting in formation of a well-defined longitudinal depression along the fault line.

The approximate line of the Alpine Fault is indicated. The view is inland from Kumara, looking north east. River flats of the Taramakau River are in the foreground, Lake Poerua is in mid distance and Inchbonnie is at the lower right edge of the photo.

ocean basins, in turn a reflection of continental drift and sea-floor spreading. Thus the late Pliocene saw many areas of New Zealand submerged for the very last time. The Southern Alps were high enough at that time to stand above the waters, but much of the North Island was inundated and marine straits crossed sags in the then quite low main ranges at the sites of the modern Manawatu Gorge and Kuripapango (Hawkes Bay). As the ranges rose to their full height in Pleistocene times the late Pliocene marine sediments were carried up to heights ranging to 1220m (4000ft).

Cook Strait first came into being during the late Pliocene marine transgression—the sea then flooding in over a low-lying transverse break in New Zealand's mountainous backbone (probably the site of a cross-cutting fault system). Later falls of sea level, however, during the Pleistocene ice ages, turned most of Cook Strait into dry land, except for a blind gulf-like arm of the sea.

In the far north, beyond Ahipara, the coastline now occupied by the Ninety Mile Beach was covered by the Pliocene sea and sea also flowed across the Auckland Isthmus, from the Hauráki Gulf to Manukau Harbour.

During the Pliocene volcanoes were active at Whangaroa, Northland, the Coromandel Range, and along the west Taranaki and King Country coastline at Whareorino and Orangiwharo. Eruptions at Lyttelton continued into the Pliocene. Development of the volcanic centres in the Taupo-Rotorua region was, however, subsequent to Pliocene times.

The steady cooling of climate that had started in late Miocene continued throughout the Pliocene and temperatures started to approach those of the present day, although in central New Zealand Pliocene sea temperatures were nowhere cooler than those of modern Northland.

The progressive loss of warm-water groups that had first started in the late Miocene continued throughout the Pliocene as temperatures steadily dropped.

The gaps thus created in our marine fauna were only partly filled by newcomers. Ancestors of many groups still living appeared, and some of them are dominant organisms of our shores. A number of the Pliocene marine immigrants are of Australian or northern origin, but others are clearly cool-water groups, distributed by the West Wind Drift.

Towards the end of the Pliocene there is evidence that a number of climatic fluctuations occurred in the New Zealand region, foreshadowing, as it were, the major ups and downs of climate during the next period, the Pleistocene, when the whole world was to feel the effects of successive severe ice ages.

As Antarctica drifted slowly southwards, moving at a rate of about 40-50mm (1½-2in) per year, it progressively came to straddle the South Pole. This southward movement had a number of far-reaching repercussions and eventually led to a revolution in world climatic patterns.

**Trans-Tasman migration**

As the southern continents moved closer to their present positions, and the West Wind Drift and associated oceanic and wind currents became established, so the effectiveness of trans-Tasman migration increased.

As we have seen, Australia has been a prolific source of New Zealand animals and plants from the very beginning of geological time. However, the nature of the Australian-New Zealand connection operating at any particular time has controlled the type of immigration. In the Mesozoic and early Tertiary, for example, the nature of migration routes between Australia and New Zealand ranged from terrestrial, or virtually terrestrial in late Jurassic (p. 269), to shallow-water marine in Cretaceous, to trans-oceanic in early Tertiary. As a consequence, the animals and plants able to migrate along these routes at particular times have varied from those needing continuous land for dispersal, through those capable of island-hopping or migration along continental shelves, to those able to travel across deep water.

Since the opening of the Tasman, 80-60 million years ago, New Zealand has been the recipient of trans-oceanic immigration from Australia and points north. Once the West Wind Drift got into its stride, however, from late Oligocene times onwards, increasing numbers of Australian plants and animals hitched rides across to New Zealand. Many of the immigrants used New Zealand as a stepping stone and continued to ride the West Wind Drift to the Chathams, Easter Island, South America and beyond (e.g. Fig. 6.11).

Others arrived in New Zealand and established themselves, only to be wiped out by the frigid

**Fig. 16.40.** It is known from fossil pollen occurrences that the typical Australian plants *Casuarina* (she-oak), *Acacia* (wattle) and *Eucalyptus* (blue gum) were well established in New Zealand before the Ice Age (Pleistocene period). They were, however, wiped out by the cold glacial climates and never re-established themselves naturally. This photo is an electron microscope photo of a single *Acacia* pollen grain. The grain measures 50/1000 mm in diameter.

climates of the Ice Age (as unlike Australia, New Zealand has no retreat route northwards along which warmth-loving plants and animals could move to escape the rigours of the glacial climate). Thus typical Australian plants such as *Eucalyptus* (blue gum), *Casuarina* (she-oak) and *Acacia* (wattle), to judge from the excellence of their fossil record (Fig. 16.40), were formerly well-established in New Zealand, but were wiped out during the Ice Age (they have since been introduced artificially).

Other immigrants arrived in New Zealand but for various reasons did not establish breeding populations. This point was brought home to zoologists during the drilling off-shore from Taranaki that preceded development of the Maui gas field. During this time entomologists operated insect traps on the drilling platforms and were amazed at the numbers of Australian moths and butterflies that arrived as exhausted stragglers in the wake of trans-Tasman gales. However of these only a fraction have managed to establish themselves on the New Zealand mainland.

Some successful natural colonisations nonetheless result. In the plants, *Epilobium, Celmisia* and *Veronica* are notable examples—but equally notable absentees (until introduced artificially) were the wattles (*Acacia*) and blue gums (*Eucalyptus*), each with over 500 species in Australia, and *Hakea* and *Grevillia*, each with over 100 species. Amongst the insects, Blue Moon Butterflies and various Australian moths regularly arrive on New Zealand's west coasts. The beautiful Monarch Butterfly is another Australian immigrant.

Many birds have successfully crossed the Tasman in the last 150 years. These include the waxeye, royal spoonbill, welcome swallow, grey teal, white-faced heron, black-fronted dotterel, Australian coot and spurwing plover. The New Zealand fantail and kingfisher, while similar to Australian forms, are slightly different and for these changes to have occurred, have probably been in New Zealand for some time.

Of these natural colonists the smallest is the waxeye (or silvereye) (*Zosterops lateralis*), now found throughout New Zealand from sea level to 1000m (3500ft). The spread of these lively and attractive colonists was well-documented by early naturalists. The first invading waxeyes reached southern Westland or Southland in the 1830s. By 1856 they had become established in Canterbury and Nelson, then in 1862 they crossed Cook Strait, reached Wanganui in 1863, Auckland in 1865 and the Bay of Islands by 1867.

The spread of the waxeye did not stop there, however, and by 1904 it had reached Norfolk Island and the Chatham Islands about 1861. By some strange chance, however, the Kermadec Islands were either missed or proved unsuitable. In 1874 waxeyes were found on lonely inhospitable Campbell Island and in 1904 Scott's Antarctic expedition found them to be the most abundant bird on the Auckland Islands.

As we have seen, as the Tasman widened and New Zealand moved further and further away, the chances of land animals successfully crossing the widening sea-barrier became less. Thus while there are over 80 rodent species in Australia, none reached New Zealand. Similarly, Australia has 57 species of bat, whereas New Zealand has only two. One of the New Zealand species, the Short-tailed Bat (*Mystacina tuberculata*), is unique to this country and has no close relatives elsewhere in the world (Fig. 16.41). It has therefore been in New Zealand for a long time and may have crossed the widening Tasman early in the Tertiary.

The other species, the Long-tailed Bat

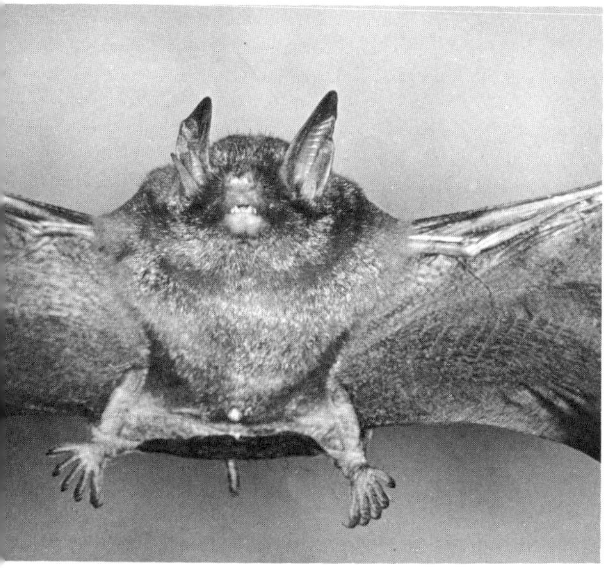

**Fig. 16.41.** The Short-tailed Bat, *Mystacina tuberculata* (upper photo), is unique to New Zealand and it is believed that it crossed the Tasman Sea early in the Tertiary, before Australia and New Zealand had reached the full width of their modern separation. Note the short tail clearly visible in this photo.

The long-tailed Bat, *Chalinolobus tuberculatus* (lower photo), is of Australian origin and probably reached New Zealand sometime in the Ice Age (2 million-10,000 years ago), after flying across the Tasman Sea. The long tail, edged to either side with a flying membrane, is clearly visible in the photo.

(*Chalinolobus tuberculatus*), is closely related to a species in North-eastern Australia and New Caledonia (Fig. 16.41). Its ancestors were probably able to fly to New Zealand (and New Caledonia), with the help of storm-generated winds, perhaps sometime in the Ice Age, and even today some large Australian bats have occasionally been blown across the Tasman to New Zealand.

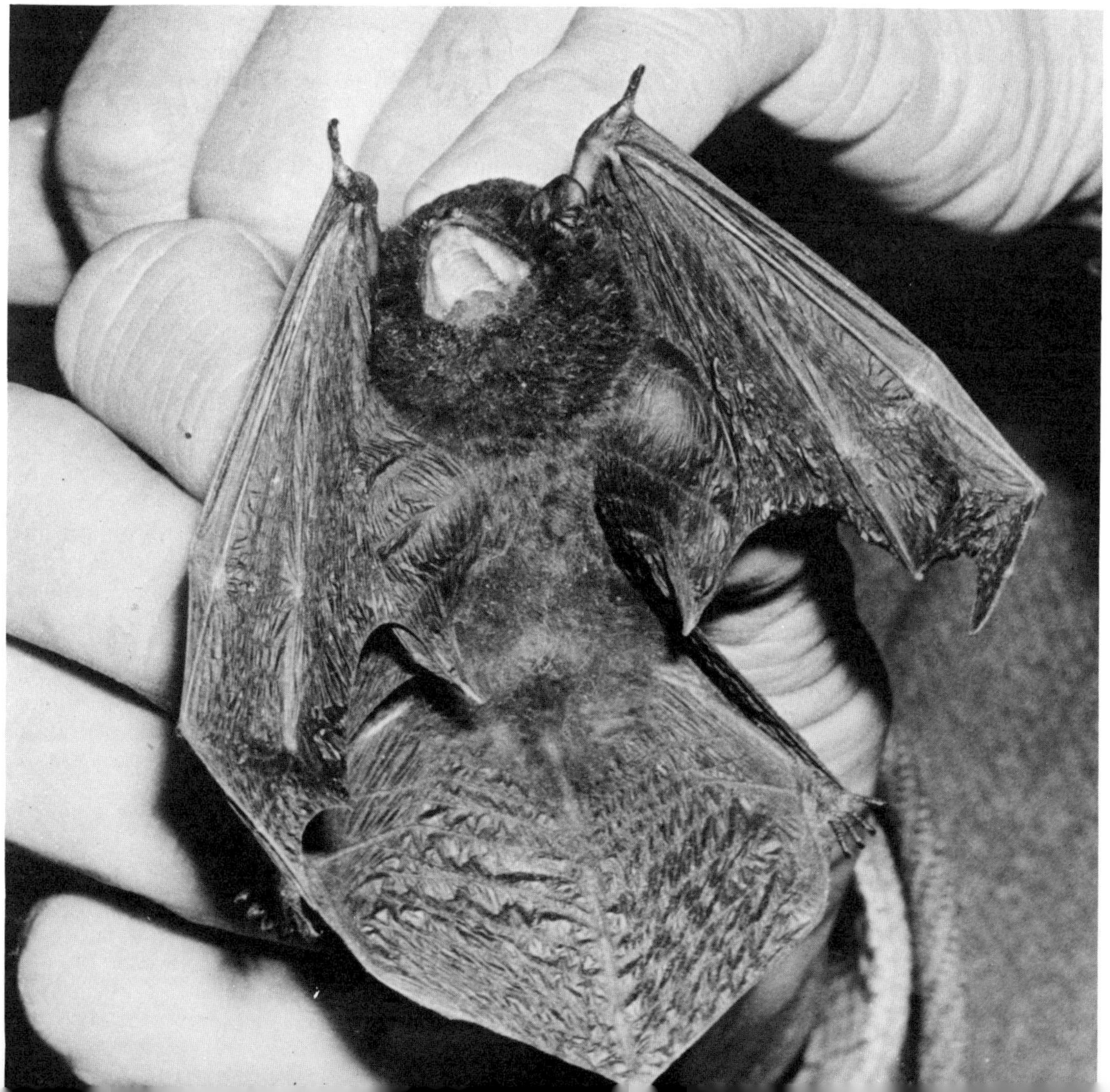

CHAPTER SEVENTEEN

# ICE-AGE MOULDING OF THE LAND

AS mentioned earlier, the activities of the deep-sea drilling ship *Glomar Challenger* in the oceans around Antarctica, particularly those south of New Zealand and Australia, have documented the gradual build-up of ice on Antarctica during the Cenozoic era (i.e. over the last 65 million years). This drilling was accomplished in spite of high seas, snow squalls, howling gales and the occasional near-miss with icebergs.

The activities of glaciers on land produce very characteristic deposits. These deposits accumulate as glacial moraine, glacial silts etc., but if the glaciation is occurring near a coastline debris is carried out to sea on floating ice, rather like loaded sand barges. When the floating ice eventually melts the ice-rafted debris sinks to the bottom and becomes incorporated into the ocean sediments.

The drill holes put down into the sea bed by the *Glomar Challenger* were able to sample this debris derived from glaciers, and by studying it and the associated remains of extinct marine life (and the variations in oxygen isotopes, see p. 214), geologists have been able to reconstruct the history of the build-up of the Antarctic ice sheet, as the continent drifted closer and closer to the South Pole.

The first signs of ice in Antarctica appeared in the early Eocene, about 52 million years ago, but all indications are that at this time Antarctica had a reasonably temperate climate, although glaciers were starting to fill many of the valleys at high elevations.

In the early Oligocene, about 35 million years ago, we have the first indication of a major cold period, and at this time individual glacial tongues reached out from the mountain highlands, to push their way down to the sea.

The glaciers that existed on Antarctica were, however, right through to middle Miocene times, about 15 million years ago, merely filling pre-existing valleys, formerly occupied by rivers and streams, and it was not until middle Miocene, and more particularly, late Miocene and early Pliocene (15-5 million years ago), that an actual ice sheet started to develop. In this process, so much snow and ice accumulated that the pre-existing valleys became completely filled with glacial ice and eventually ice spilled over and covered the land between the valleys. In this way extensive ice sheets were formed, completely submerging the land. Coalescing of the individual ice sheets led to formation of the modern Antarctic ice cap. Thus Antarctica today is a continent with a topography of mountains, plains, hills and valleys, completely covered by ice, often 200m (660ft) thick, but attaining a maximum of 4300m (14,500ft).

The gradual build-up of the Antarctic ice cap over the last 50 or so million years had wide repercussions. New Zealand, having been close to Antarctica over much of its history, has felt particularly strongly the effects of the build-up of the Antarctic ice cap, and these have left their mark in New Zealand rocks and fossils.

New Zealand geologists therefore have been able to track changes in New Zealand paralleling those in Antarctica. One effect has been a gradual extinction of warmth-loving organisms as we trace our fossil history back over the last 10 or 20 million years. The tropical and subtropical life that teemed in our seas during early Miocene times, for example, gradually declined until in the Pliocene there were only isolated remnants left. At this time most of our organisms had reached us from the direction of Australia or via the West Wind Drift. Then between two million and 10,000 years ago the Pleistocene period or "Ice Age" dealt the *coup de grace* to many of our plants and animals.

### Glacials and interglacials

Pictures conjured up by the popular press about

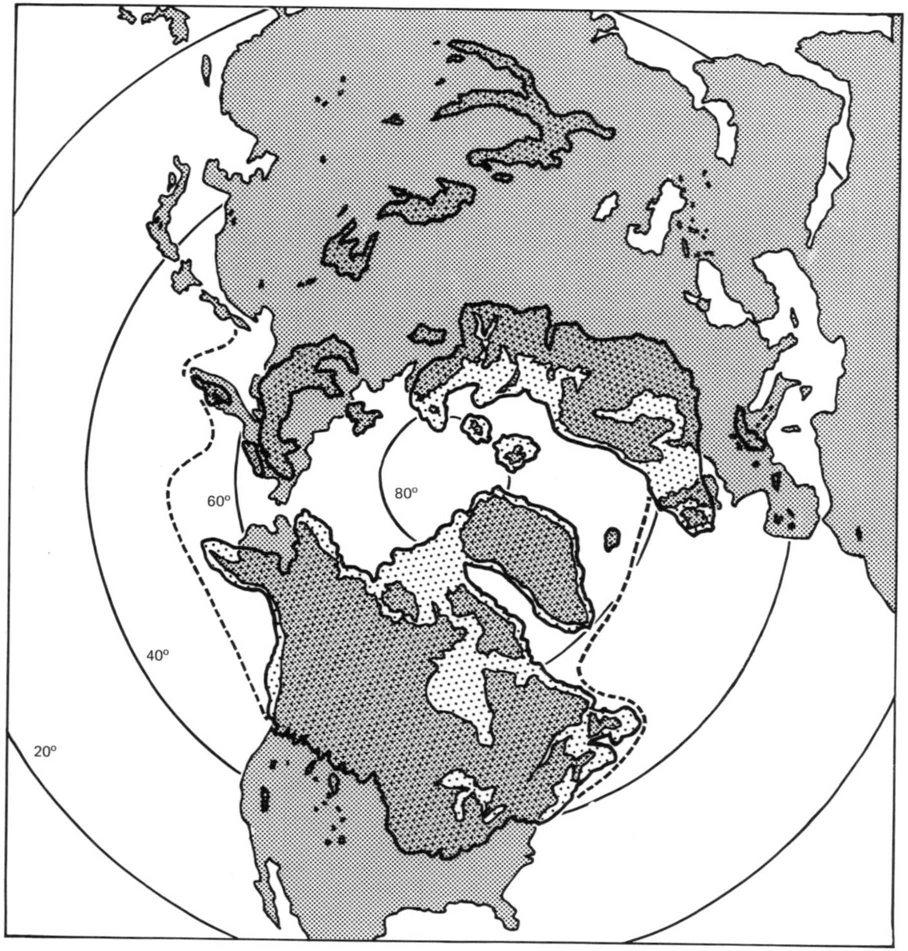


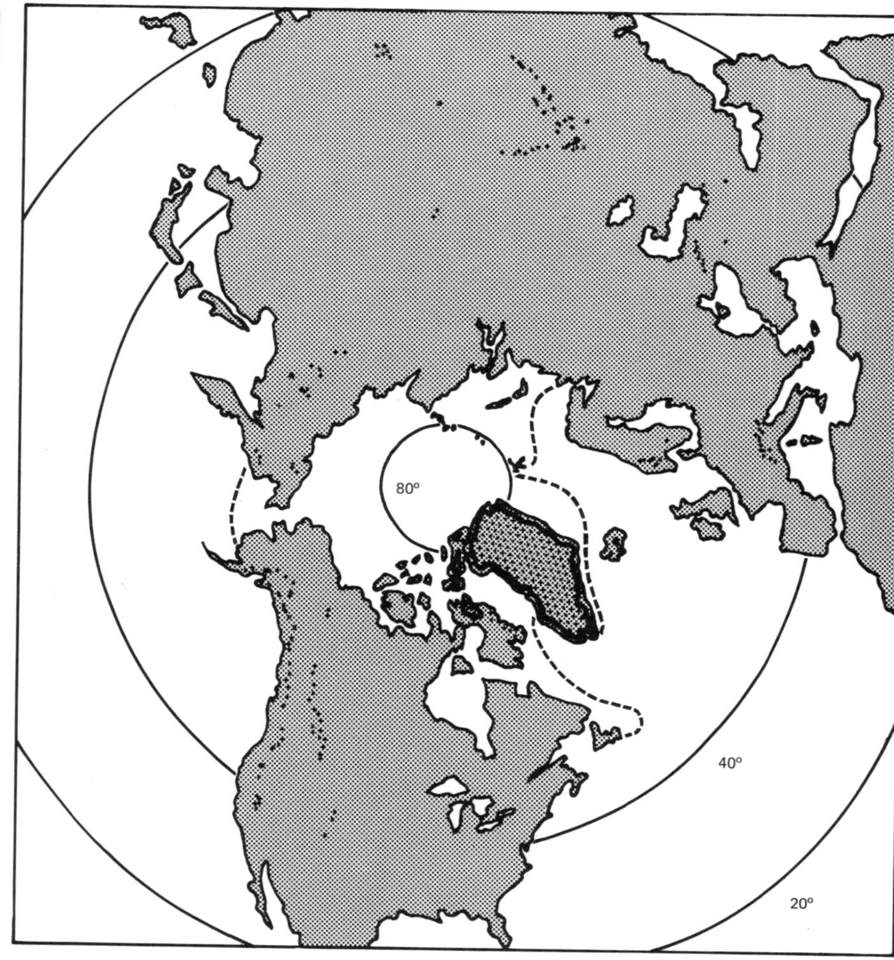


**Fig. 17.1.** In the last glacial phase ice covered about 27 percent of the earth's land surface, compared with about 10 percent today.

The left-hand map shows the distribution of ice 20,000 years ago (towards the end of the last glacial phase) in the Northern Hemisphere and, the right-hand map that of the present day. The areas of ice caps and large glaciers are indicated by the dark pattern. The approximate outer limit of sea ice at its annual maximum is shown by the heavy broken lines.

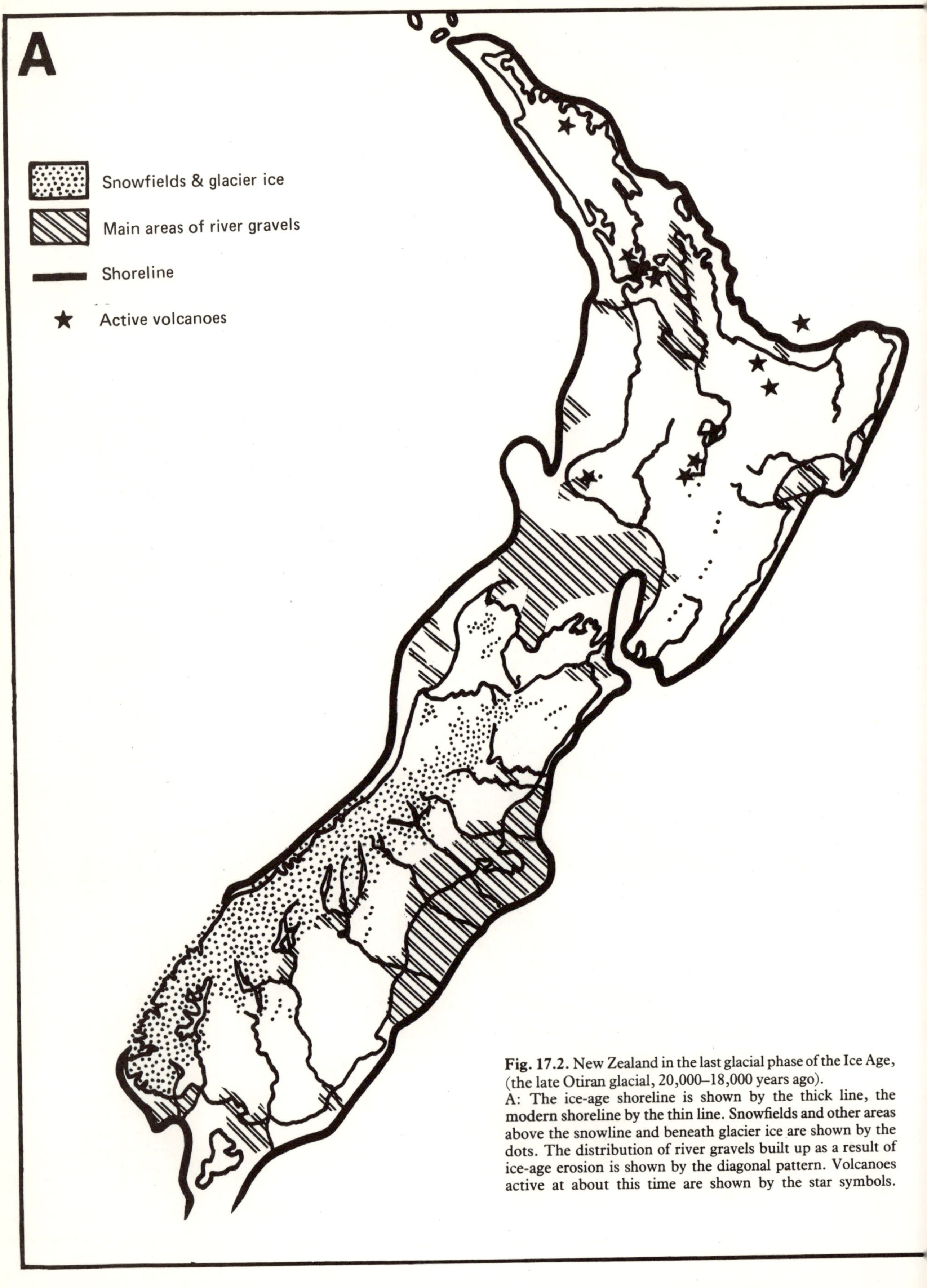


**Fig. 17.2.** New Zealand in the last glacial phase of the Ice Age, (the late Otiran glacial, 20,000–18,000 years ago).
A: The ice-age shoreline is shown by the thick line, the modern shoreline by the thin line. Snowfields and other areas above the snowline and beneath glacier ice are shown by the dots. The distribution of river gravels built up as a result of ice-age erosion is shown by the diagonal pattern. Volcanoes active at about this time are shown by the star symbols.

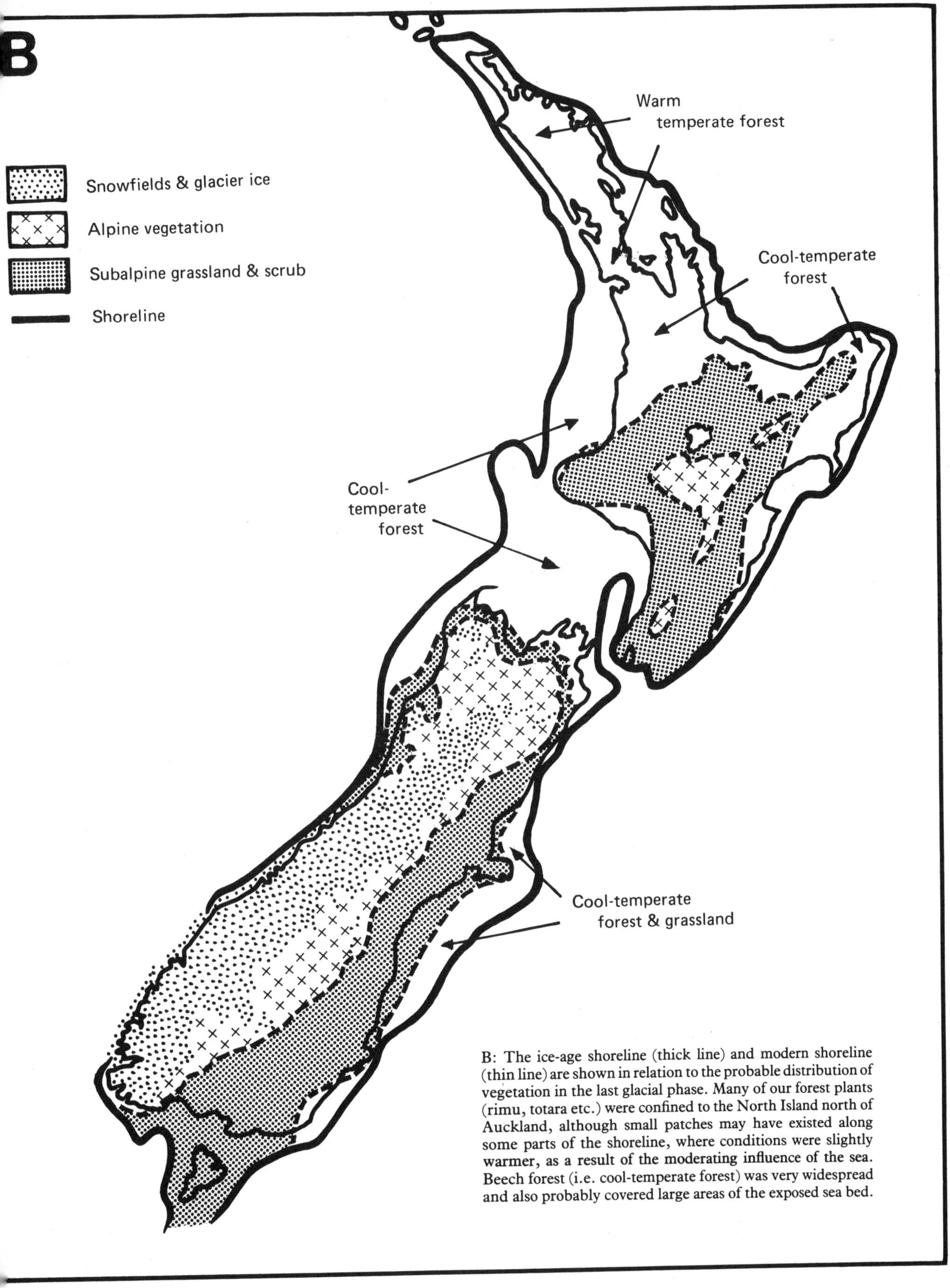


B: The ice-age shoreline (thick line) and modern shoreline (thin line) are shown in relation to the probable distribution of vegetation in the last glacial phase. Many of our forest plants (rimu, totara etc.) were confined to the North Island north of Auckland, although small patches may have existed along some parts of the shoreline, where conditions were slightly warmer, as a result of the moderating influence of the sea. Beech forest (i.e. cool-temperate forest) was very widespread and also probably covered large areas of the exposed sea bed.

the Ice Age, or Pleistocene period, always emphasise icy wastes, woolly mammoths and so on, but this is only part of the story. The term "Ice Age", although convenient, is in fact a misnomer. During the Pleistocene the world experienced a series of cold periods, called "glacials", separated by periods of normal or sometimes warmer climate called "interglacials". During some of the interglacials climate in the North Island of New Zealand became tropical, as indicated by the presence of tropical soils.

Ideas on the number of glacials and interglacials during the Pleistocene have very recently undergone a dramatic change. Since 1909 it had been largely accepted, almost as a scientific "fact", that there had been four glacials, separated by long warm interglacials. However, almost exactly as had happened with continental drift, as evidence accumulated over the last decade (particularly that from deep-sea drilling, along with magnetic, fossil and oxygen isotope studies) this idea had crumbled a little at the corners. Now the whole edifice has been swept

**Fig. 17.3.** Thick Ice-Age glaciers flowing off the Fiordland mountains scoured the landscape to west and east. The ice filled existing valleys and often scoured them out to great depths (depending on the thickness and therefore weight of ice), and forming vertical-sided U-shaped valleys. In some areas tongues and sheets of ice over-topped the valleys and invaded the intervening mountainous terrain, planing and smoothing-off the landscape.

Fig. 17.3A (top left) is a view of glaciated terrain west of Lake Te Anau. Takahe Valley with Lake Orbell is in right middle distance, Point Burn in left middle distance. The Murchison Mountains are in the background. The broad U-shaped valleys were occupied by Ice-Age glaciers, flowing into the Te Anau basin. The mountain peaks, projecting above the ice, were sharpened by severe frost action.

Fig. 17.3B (bottom left) is a view of glaciated terrain in the Darran Mountains, south east of Milford. The Gulliver River is in the foreground and Gertrude Saddle and Barrier Peak are in the background. Huge Ice Age glaciers flowed down the valleys of this landscape, discharging into the sea via glaciers filling the west coast fiords of Fiordland.

away and it is known that there were perhaps 20 glacials in the last two million years, separated by brief warm interglacials, each lasting only a fleeting 10,000 years.

The difference in mean annual temperature between a glacial and interglacial is thought to be some 14°C. These temperature drops had a greater effect in the middle latitudes of the world than at the poles and the equator. Where a region is already all ice, further cooling has little obvious effect and similarly, where the climate is already tropical, a small drop in temperature may affect vegetation and perhaps rainfall, but little else. But dramatic changes occurred in middle latitudes.

In the Northern Hemisphere extensive ice sheets formed on the land. An ice sheet centred on Scandinavia pushed southwards across Holland, Germany, Poland and European U.S.S.R., extending as far south as a line drawn through Amsterdam, Dresden, Krakow, Lvov, Kiev and Kharkov. In Britain ice pushed down from Scotland, the Lake District, Wales and Central Ireland to cover all the British Isles except that part of England south of a line drawn between London and Bristol. Both the Irish Sea and North Sea were filled with ice. In North America ice covered all of Canada and pushed southwards roughly to a line drawn through New York, Indianapolis and Vancouver. Some 20,000 years ago ice covered about 27 percent of the earth's land surface, compared with about 10 percent today (Fig. 17.1).

**Fig. 17.4.** During the glacial phases of the Ice Age sheets of ice spilled over from the glacier-filled valleys leading from the flanks of Fiordland and the Southern Alps. These moving sheets of ice scoured and sculpted the underlying landscape and here near Lake Wakatipu softer bands in the underlying schist rocks have been gouged out, producing parallel grooves.

The effects of glacial climates were not quite so severe in New Zealand, as most areas were not very far away from the moderating influence of the sea. Auckland's climate probably changed to that of Dunedin and Wellington experienced a frost climate similar to that of northern Norway today. Mounts Egmont, Ruapehu, Ngauruhoe and Tongariro had ice caps and small glaciers filled the upper reaches of some valleys in the higher parts of the Tararua and Ruahine ranges. Ice covered the higher parts of north-west Nelson and the St. Arnaud Range. Extensive glaciers formed in the Southern Alps and pushed out to

**Fig. 17.5.** Ice-Age glaciers occupying valleys in the Southern Alps transported great quantities of rock material down to the lowlands. Rivers and streams transported this material further, spreading it out as broad gravel plains (cf. Fig. 17.2A). The Canterbury Plains, fringing the Southern Alps to the east, have been built out in this way and have joined the former volcanic island of Banks Peninsula to the rest of the South Island.

This view looks across the Canterbury Plains westwards to the Rakaia Gorge.

west and east to reach the sea along the West Coast and spread along the inland flanks of the Canterbury Plains. A thick ice sheet developed in Fiordland and pushed out westwards to the sea and eastwards towards Central Otago (Figs 17.2-17.6).

In general, other parts of New Zealand south of a line passing approximately through the Waikato and the Bay of Plenty had a cold climate, with varying degrees of frost. In these areas it was not quite cold enough, nor possibly wet enough, for actual ice sheets to accumulate on the surface of the land. However, the cold frosty climate was sufficiently severe to discourage the growth of forest and many areas were either completely bare or thinly covered with alpine-type vegetation and certain hardy grasses, rushes and sedges. Forest was restricted to certain "refuges": areas in coastal North Otago, Canterbury, Nelson and Marlborough, well away from the mountains, and in the northern part of the North Island, coastal Hawkes Bay and East Coast and north of the Waikato. The dropping of sea level that accompanied the glacial periods (see below) uncovered strips of land, immediately adjacent to the sea (and hence with a milder climate) and now submerged, that also provided temporary refuges for forest plants (Fig. 17.2).

During each interglacial period, when the climate improved dramatically, perhaps even

reaching subtropical conditions, the forest extended out from these refuges, probably to re-establish itself over much of New Zealand, only to be pushed back again by the onset of another glacial period. Similarly, in the last 10,000 years, since the end of the Ice Age, our present forest vegetation, along with its accompanying fauna, has staged a "comeback" from these same refuges.

Water circulates freely in the present-day atmosphere. Heat from the sun draws up water vapour from the surface of the sea (as well as from lakes etc.). The moisture-laden air then passes on to the land and as higher country is encountered, the air is forced to rise and the moisture condenses to form clouds which in turn release rain on to the land. The rain runs off the land into streams and rivers and finally most of the water is returned to the ocean and other sources from which it came. But this cycle is disrupted during a glacial period. Then, moisture falls on the land not as rain, but as snow and ice, which

**Fig. 17.6.** Although large areas of the southern part of the North Island had a severe frost climate in the Ice Age, and the tree-line in some regions may have fallen below modern sea level, actual glaciers were confined to a few high-level valleys in the Tararua and Ruahine ranges. This view is of one such valley, Park Valley, in the Tararua Range, showing the characteristic glacially-scoured U-shaped profile, little modified by river erosion.

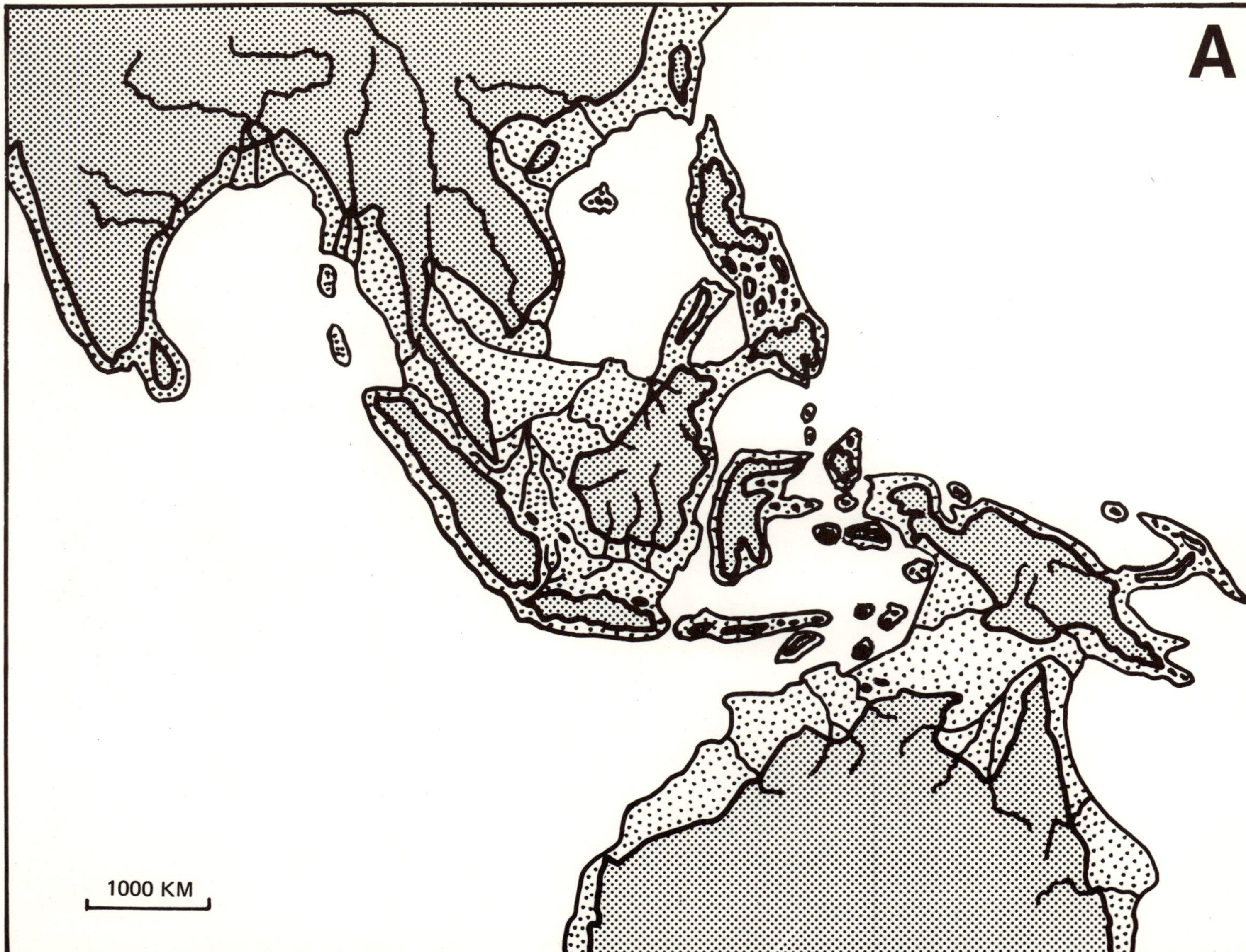
A
1000 KM

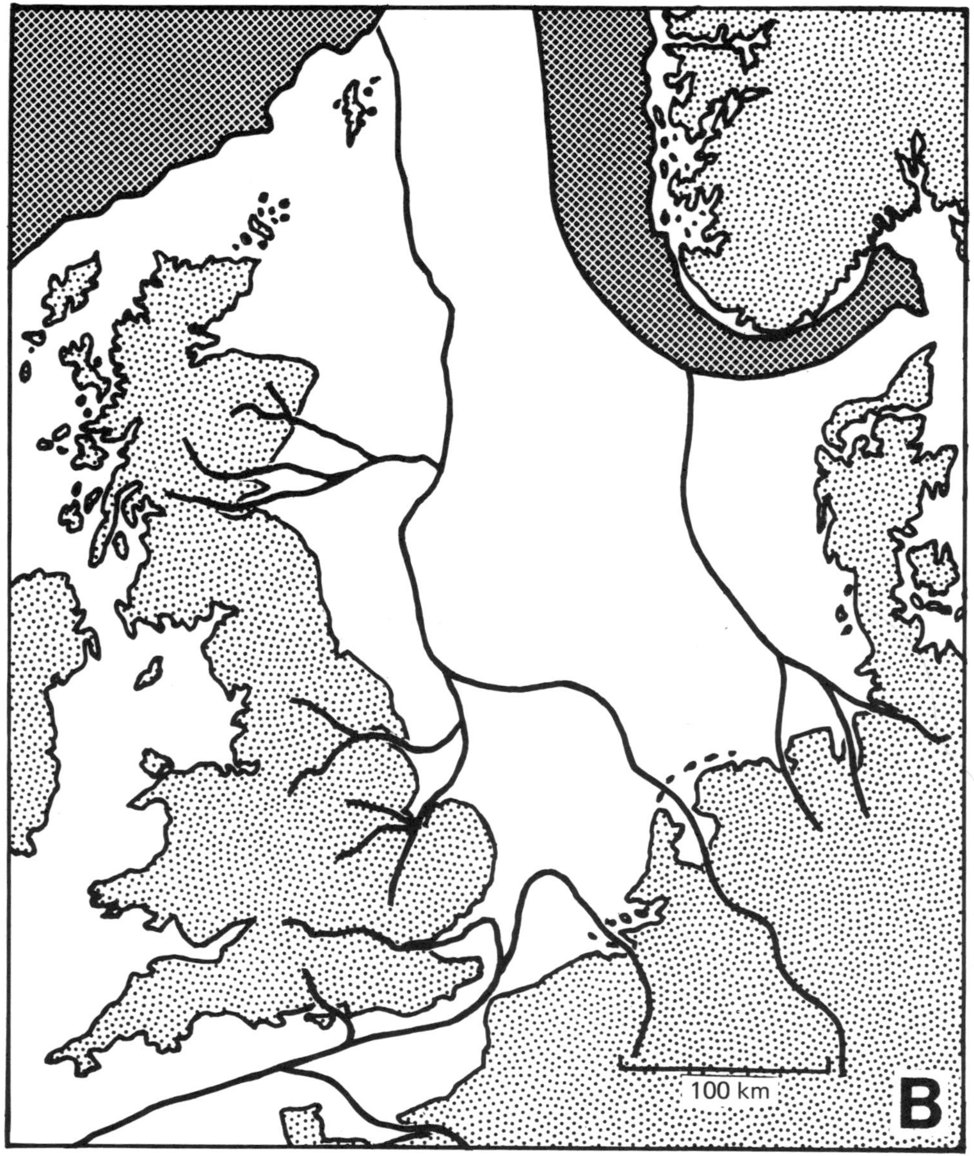


**Fig. 17.7.** During the glacial phases of the Ice Age so much water was locked up on the land in the form of ice, snow and vast inland lakes that the world's oceans fell to very low levels—reaching 183-238 m (600-775 ft) at times. Many of the myriad islands in Indonesia were thus connected by land, as shown in the upper diagram (A), allowing a certain amount of interchange between the Australian and Asiatic faunas and floras. These interchanges reinforced similar earlier migrations via ephemeral island arcs that had existed at various times in the Tertiary (Figs 16.22-16.28).

Carnivorous placental mammals, however, did not penetrate to Australia and New Guinea, and those Australian animals with better-than-average dispersal abilities (chiefly birds) did not get further than the Lombok Strait (Wallace's Line), between the islands of Bali and Lombok, that even in times of low ice-age sea levels remained an oceanic barrier.

During glacial low sea levels the North Sea, Irish Sea, English Channel and Western Approaches of the Atlantic were dry land, as shown in the lower diagram (B), and the only sea close by was beyond the Hebrides and Shetlands and in a trough close to the Norwegian coast. Faunas and floras were thus able to interchange across the exposed sea floor, as in Indonesia (diagram A). The major rivers of Western Europe and Great Britain extended out to the glacial shoreline, which as shown in this diagram is drawn at the 200 m (660 ft) isobath.

accumulate to form permanent ice sheets. Thus the build-up of ice sheets on the land during the Ice Age was done at the expense of the world's oceans. As the ice caps increased in extent and thickness so sea level steadily dropped. During the last glacial period, at a time 30,000–20,000 years ago, world sea level was 105–135m (350–450ft) lower than it is now, but between 17,000 and 14,000 years ago it dropped even further, to reach 183-238m (600-775ft) below modern sea level. During the glaciation which preceded the last, and perhaps in the one before that, respectively some 150,000 and 350,000 years ago, sea level fell to reach some 170m (550ft) below modern sea level.

These drops in sea level were world-wide and drastically altered the shapes of many of the present landmasses. Britain, for example, was joined to France, Belgium, Holland and Germany across what is now the North Sea. The myriad islands in Indonesia were all connected by land (Fig. 17.7). Around the New Zealand coast, the retreating sea exposed broad areas of sea floor (Fig. 17.2). In the Auckland region the entire area now covered by the Hauraki Gulf was land, and land continued to beyond Great Barrier Island. The area now covered by Hawke Bay was land. The North and South Islands were joined across Cook Strait and Stewart Island was linked across Foveaux Strait to Southland (Fig. 17.2).

Retreat of the sea had a subtle effect on the climate of the time, by altering the pattern of currents in the sea. Also, it meant that the oceanic climate of a place now near the sea was changed to a much more continental one.

### Ice in retreat

Retreat of the ice in New Zealand is thought to have started about 14,000 years ago, although in many parts of the world ice sheets lingered on until 10,000 years ago, so that this latter date is often taken as the "official" end of the last glacial (Fig. 17.8). At about this time the sea rose from its low glacial levels (135-183m; 450-600ft) to reach about 75m (250ft) below its modern height. Then in the period between 10,000 and 5500 years ago the sea rose at a rate of about 1m (3ft) per century, until about 5500 years ago it

**Fig. 17.8.** A sketch of the climatic changes that occurred over the last 85,000 years. The Last Glaciation spanned from about 75,000 to 10,000 years ago.

The Post-glacial Climatic Optimum, when average world temperatures were 1–3°C (2–5°F) higher than at present, ranged from 5500-4000 years ago.

The Little Ice Age, a cold phase that produced the lowest temperatures and greatest extensions of ice on land and sea since the glaciations, lasted from about A.D. 1550 to A.D. 1800.

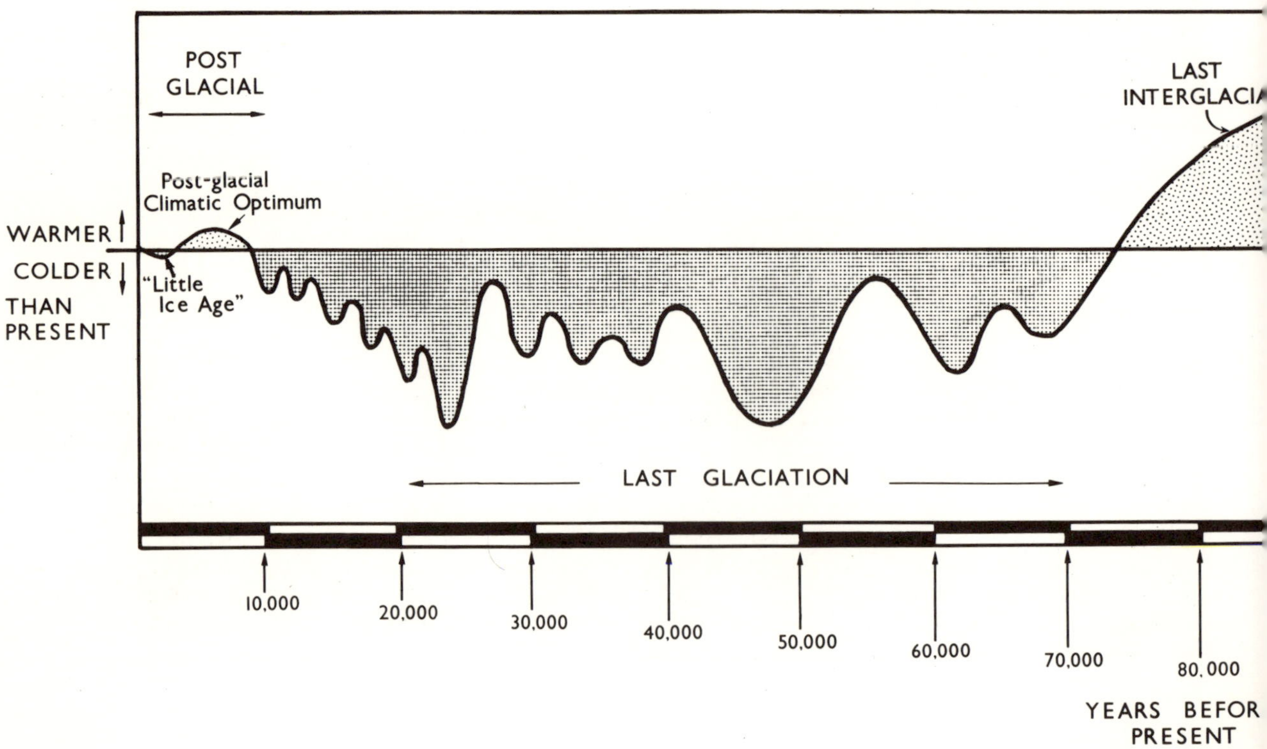

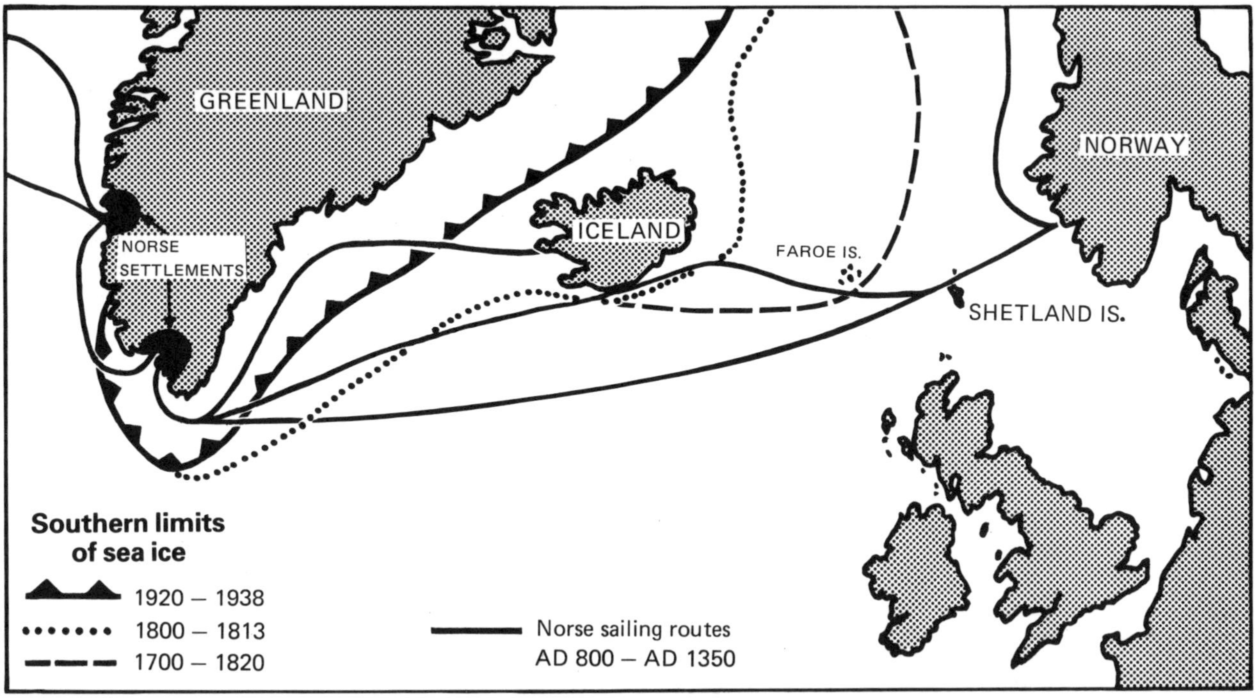


**Fig. 17.9.** A continuation of warm conditions after the Post-glacial Climatic Optimum enabled the Vikings to make numerous voyages across the ice-free North Atlantic Ocean.

Vikings settled both Iceland (in A.D. 870) and Greenland (in A.D. 986). They then used Greenland as a stepping stone to establish colonies on the coasts of Labrador and Newfoundland in about A.D. 1000.

However, after about A.D. 1200 the climate gradually deteriorated as the world began to pass into the "Little Ice Age" (cf. Fig. 17.8). The increasing cold, bringing with it sea ice and storminess in the North Atlantic, forced the Vikings to abandon their settlements in Greenland and Newfoundland.

The Little Ice Age reached a maximum between A.D. 1740 and 1750 and it was during this interval that the edge of the Arctic sea ice reached the Faroe Islands, 400 km (250 miles) north of Scotland.

This diagram shows the Viking sailing routes, between A.D. 800 and 1350, and the southern limits of sea ice in A.D. 1700-1938.

had reached a level about that of the present. The last remnants of the ice sheet covering Scandinavia had melted by 7000 years ago and those of the North American ice sheet by 5500 years ago.

Sea level then continued to rise, until in the period between 5500 and 4000 years ago it rose to a height 2-3 m (7-10 ft) above present sea level, although some people put the height at 3.6-4.6 m (12-15 ft). After this, the sea dropped to its present level, but 270-420 years ago it temporarily dropped a further 0.6 m (2 ft).

Studies of fossil plants and animals have chronicled the climatic variations accompanying the sea level changes and these indicate that some time before the time of the high sea level, about 7000 years ago or perhaps slightly earlier, the climate of the world began to become warmer than that of the present day, and from about 7000 to about 4000 years ago (about 5000-2000 B.C.) the world experienced a warm phase known as the "Post-glacial Climatic Optimum". Some researchers believe that the warming-up reached a peak about 5500 years ago (about 3500 B.C., at about the beginning of Egyptian culture). The Post-glacial Climatic Optimum petered out some time during the Middle Kingdom phase of Egyptian culture (2133-1786 B.C.).

At the height of the Climatic Optimum the average world temperatures were 1–3°C higher than at present, and many warmth-loving plants and animals extended their ranges.

These warm conditions led to thinning of the Greenland and Antarctic ice caps. The extra water produced in this way added at least 2-3 m (7-10 ft) to the sea level of the world, although exactly how much is open to debate.

It has been estimated that in the period between 20,000 and 5000 years ago summer temperatures in Denmark rose by some 9.5°C—and this serves to underline the spectacular improvement in the climate. We know however that the ice caps did not completely disappear, as it has been calculated that if all the ice in the modern world should melt, sea level would rise by some 65 m (215 ft).

The warm conditions of the Climatic Optimum were followed by a return to conditions similar to those of the present. The ice caps on Greenland and Antarctica were built up again and the sea level fell to its modern level. But climate continued to fluctuate like a yo-yo, and 1620-770 years ago (i.e. A.D. 350-1200) there was another spell of warm climate, with the warmest interval between A.D. 800-1000. During this time it has been estimated that summer temperatures in Alaska were some three degrees higher than those of today, vineyards were quite abundant in many parts of England, and the sea level was 0.5 m (1.5 ft) higher than at present (because of the water added to the oceans from the thinning of the world's ice caps).

The warmer conditions meant that Arctic sea ice virtually disappeared from the seas around Iceland and southern Greenland, enabling the Vikings to make numerous voyages of discovery, colonisation and trade in the North Atlantic Ocean (Fig. 17.9). Vikings settled both Iceland and Greenland (in A.D. 870 and A.D. 986 respectively) and voyaged via Greenland to North America to establish colonies on the coasts of Labrador and Newfoundland in about A.D. 1000, some 500 years before the arrival of other Europeans in the New World.

The Viking colony in Greenland eventually grew to 4000 people, but during the twelfth century the Norsemen began returning to Europe and by A.D. 1410 they had completely abandoned Greenland. For years historians have debated the cause of their mysterious exit. Were the Vikings driven out by hostile natives? Did excessive interbreeding cause genetic deterioration of the tough Norse stock? Now it looks as though there is a simpler explanation: the mild weather the Vikings originally encountered in Greenland gradually changed and became too harsh even for their hardy tastes. Plant fossils, in addition to recent information from deep cores taken in the Greenland and Antarctic ice caps, tell us that after A.D. 1200 the climate gradually deteriorated, until 420-170 years ago (A.D. 1550-1800) the world passed through a phase known as "The Little Ice Age" (Fig. 17.8). This period produced the lowest temperatures and the greatest extensions of ice on land and sea since the Ice Age, and the world-wide build-up of ice led to a fall of some 0.6 m (2 ft) in sea level.

The "Little Ice Age" reached a maximum in the period between A.D. 1740 and 1750. Crops failed to ripen, famine set in and the death rate in Norway, for example, far exceeded the birth rate. Historical records from the European Alps chronicle extensive glacier advances during the "Little Ice Age". Villages and farms built in what had been considered safe places were overwhelmed or nearly over-run by advancing ice at this time. Several of these villages are still ice-covered today. There was a vast increase in the amount of sea ice near Iceland and southern Greenland, and on at least one occasion the edge of the sea ice reached the Faroe Islands, only 400 km (250 miles) north of Scotland. It was during this interval that several Eskimos reached Scotland by kayak, presumably after travelling along the edge of the Arctic pack ice.

Analysis of the rate of lime deposition in stalactites in New Zealand caves has made it possible to calculate the major temperature fluctuations in New Zealand since the Ice Age. This work has shown that New Zealand had a history of temperature fluctuations similar to that of Europe. It is thought, for example, that the warm peak of A.D. 800-1000, that facilitated Viking exploration and colonisation in the North Atlantic, also stimulated similar activities at about the same time by Polynesians in the South-west Pacific.

After the initial Maori colonisation of New Zealand, however, climate deteriorated and after A.D. 1200 colder temperatures and an increase in the frequency and severity of ocean storms put an end to the great Polynesian voyages of exploration, just as the Viking voyages of exploration came to an end at about the same time, and probably for the same reasons.

Deterioration in weather, leading to the "Little Ice Age", could have been the crucial factor which prevented the Maoris from pushing on to Australia and Tasmania. The belief that New Zealand got colder after the initial Maori colonisation has also received support from archaeological discoveries that show that the early Maori settlers were able to grow kumara in Otago in places where it would be impossible to grow it today.

All the indications are that, like the Viking

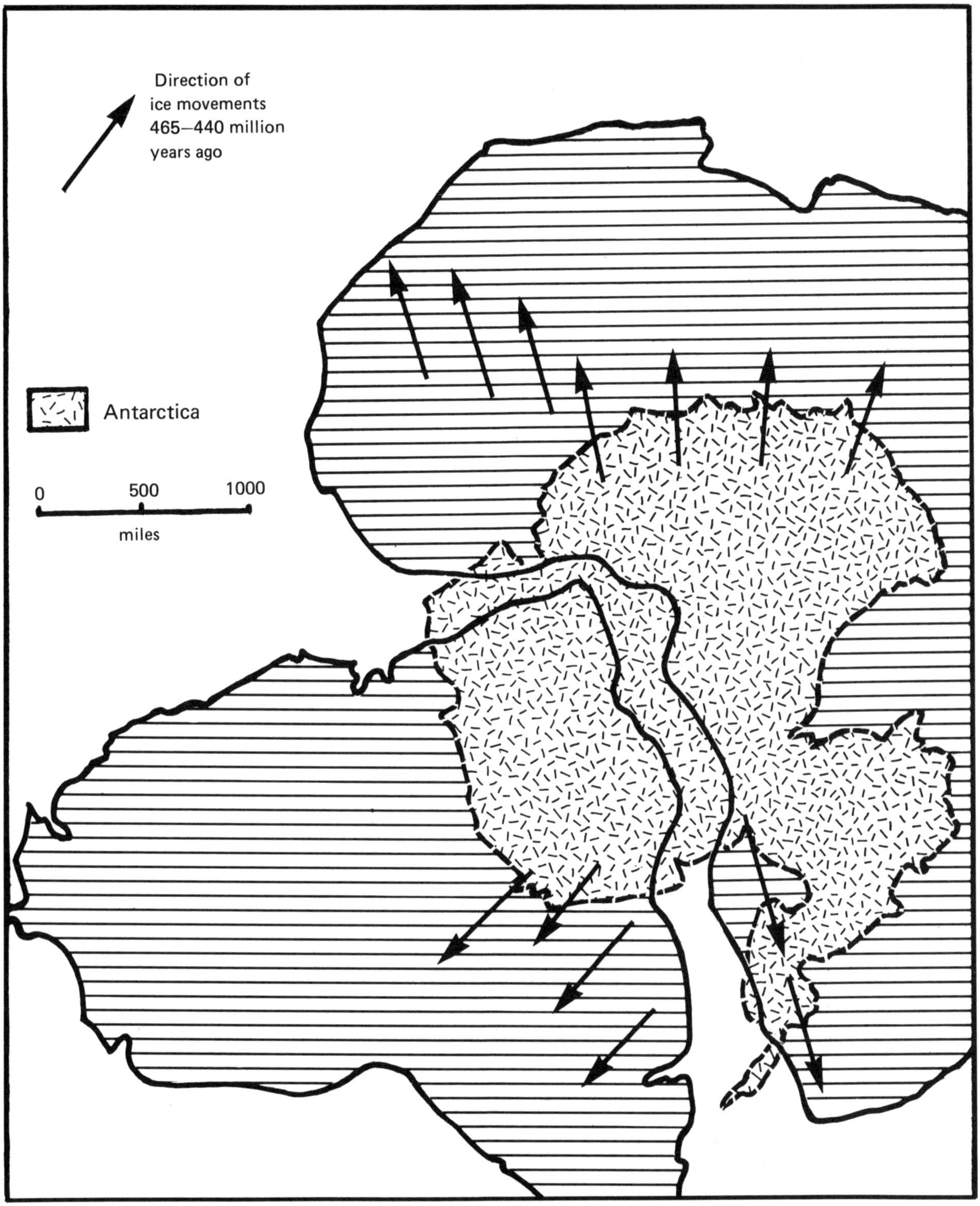


**Fig. 17.10.** This diagram shows the continents of Africa and South America positioned where they are believed to have been during the late Ordovician period (465-440 million years ago), before continental drift carried them to their present locations.

Magnetic and fossil evidence places the South Pole of that time, ice cap and all, in or near north-west Africa. Arrows indicate glacial grooves, ice-rafted boulders and other evidence of Ordovician glaciation. When today's south polar ice cap is superimposed, it is seen to cover an equivalent area.

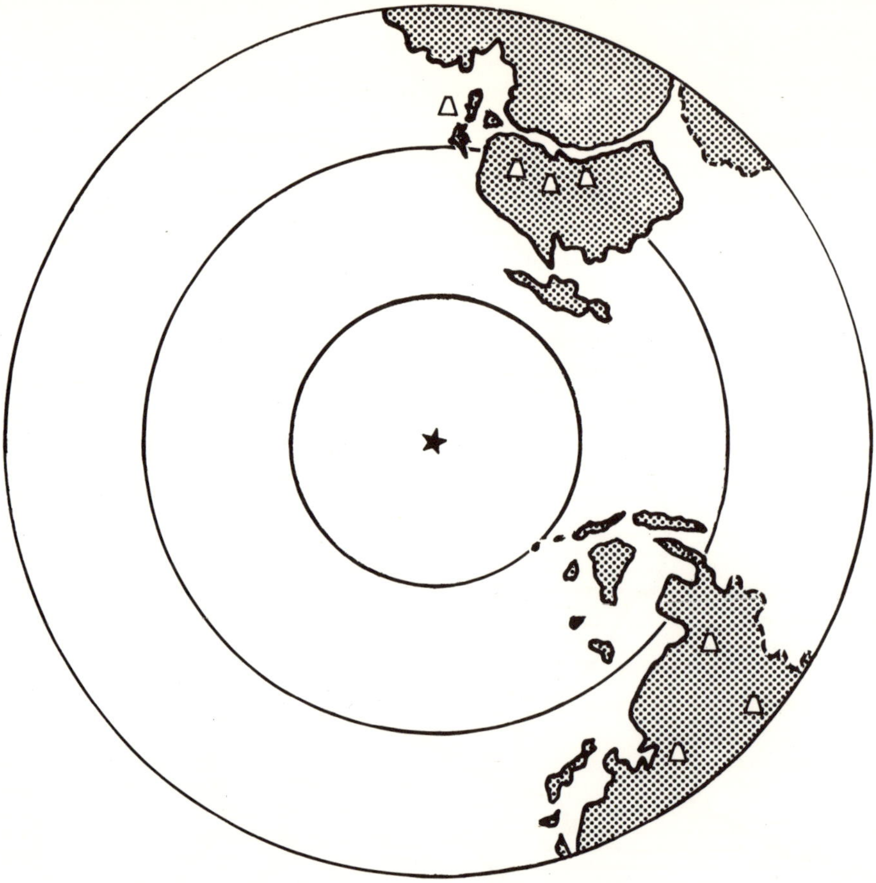

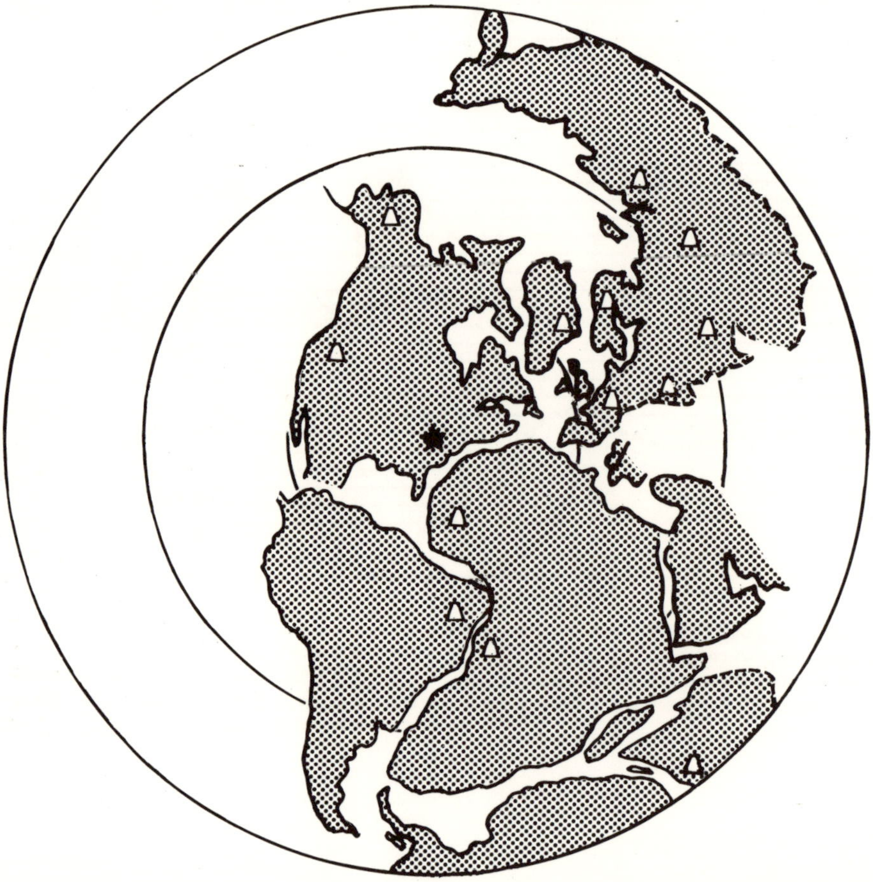

settlers in the North Atlantic region and the people in Europe, the early Maori settlers, particularly those in more southerly areas, must have had a pretty rough time in the "Little Ice Age".

After A.D. 1800, the climate warmed up again, to produce a warm peak in the 1920s, 1930s and early 1940s, but it is again on a downward trend and climatologists have made long-range predictions, based on the study of weather patterns and ice cores, that temperatures may continue to drop for another twenty-five years before a warming trend sets in that may bring us to a warm peak about A.D. 2015.

**Ice ages of the past**

Going back into geological time we can distinguish four major periods during which parts of the earth experienced continental-wide glaciations. They are:

(i) Pleistocene, between 2 million and 10,000 years ago.

(ii) Late Carboniferous and early Permian, between about 305 and 255 million years ago.

(iii) Late Ordovician, between about 465 and 440 million years ago.

(iv) Late Precambrian, between about 680 and 650 million years ago, but with the likelihood of still earlier glaciations, not yet fully distinguished.

The most recent, or Pleistocene, ice age affected both Northern and Southern Hemispheres, and is by no means over. The present day is merely part of an interglacial pause. There is therefore every possibility that some time in the future glaciers will once more advance down to the Canterbury Plains (if the Milankovitch theory is right this could be sometime in the next 1000 years; see p. 342).

**Fig. 17.11.** Ice deposits of late Precambrian age, 680-650 million years ago, are shown on these maps. The North Pole is shown by the star on the top map and similarly the South Pole on the lower map. The reconstructed positions of the continents are based on palaeomagnetism. Note that at this time Australia, New Zealand and South-east Asia were thought to have been in the Northern Hemisphere.

Owing to the difficulty of being absolutely certain about the identification of ice deposits of such antiquity (as they have often been altered by erosion processes and by heat and pressure), some of the ice deposits shown on this map are in the uncertain category. The possible Precambrian ice deposits in North-west Nelson and Westland have been altered by heat and pressure and for this reason are difficult to determine as being definitely of glacial origin.

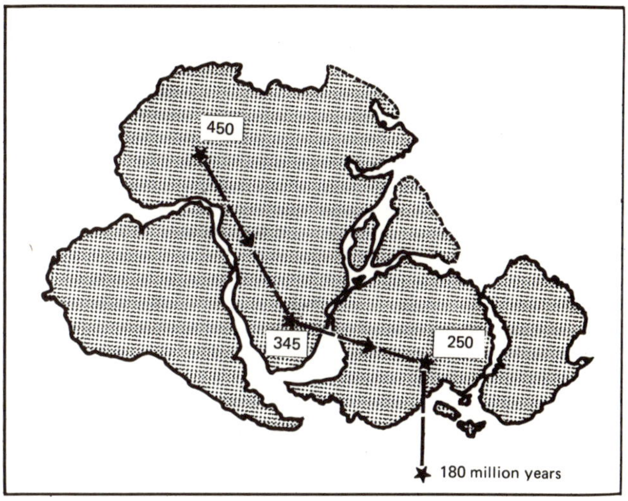


**Fig. 17.12.** Palaeomagnetism has chronicled the slow movement of the Gondwana landmasses across the South Pole, leading to a succession of glacial periods.

In the late Ordovician, about 450 million years ago, the South Pole was in the modern Sahara and Northern and Central Africa experienced glaciation. After this the slow drift of Africa rotated Southern Africa over the South Pole and glacial activity migrated into this region, culminating in the Carboniferous-Permian ice ages. In the early Carboniferous (about 345 million years ago) the centre of glaciation, and presumably the South Pole, was in South Africa, but during the remainder of the Carboniferous and in the succeeding Permian, it migrated to East Antarctica.

Continuing continental drift movements then gradually swung Gondwanaland away from the South Pole and for most of the Triassic and Jurassic the pole was probably sited in the ocean some distance from land (therefore there was no continental glaciation) (e.g. Fig. 12.26).

The diagram shows successive positions for the South Pole, as determined by palaeomagnetism, beginning in the late Ordovician, 450 million years ago, and ending in the early Jurassic, 180 million years ago.

The ice age in the late Carboniferous and early Permian was a Gondwanaland phenomenon (pp. 35–38) and, as we have seen, its traces, now scattered across the Gondwana continents, provide striking testimony of continental drift.

The first signs of the Ordovician ice age were picked up in 1961 by French geologists prospecting for oil in the desolate wastes of the central Sahara. In succeeding years other expeditions found similar glacial formations all across North Africa—from Morocco, Mauritania, Algeria and Chad—a spread of 5000 km (3000 miles). This could scarcely be a local phenomenon. It had to be a continental polar glaciation comparable in size to that of Antarctica today. Then independent palaeomagnetic studies showed that the

South Pole in the Ordovician had been somewhere between the latitudes of Morocco and the Cape Verde Islands. Thus the South Pole, ice cap and all, had been in what is now the Sahara (Fig. 17.10).

Relics of Ordovician glaciation are also known from Argentina, Bolivia, South-west Africa and Cape Province, providing evidence for the other half, as it were, of the polar ice cap.

Evidence for the late Precambrian ice age has only recently been assembled, as this study has been beset by problems of dating such old rocks and arguments over whether deposits now classified as glacial had in fact such an origin.

It has now become apparent, however, that the Precambrian ice age was probably the greatest of all the ice ages and its deposits are found on all the major continents of the world except Antarctica (where critical evidence probably lies under today's ice cap). Even New Zealand has sediments (in north Westland and north-west Nelson) of this ice age (Fig. 17.11).

This does not mean however that the entire earth was covered with ice—this would be most unlikely. During glacial periods ice sheets developed on land in high and middle latitudes, but increasing severity of climate would extend the total area of sea ice. Thus sea ice sheets, and detached icebergs, each loaded with glacial debris, would, on melting, spread such material far and wide in the open oceans of low latitudes, and as many of the deposits of the Precambrian ice age are thought to be waterlain, their geographic distribution is not necessarily the same as that of land-based ice sheets.

The major climatic change that came at the end of the Precambrian ice age may have triggered the burst of evolution that occurred at the very beginning of the Cambrian, 570 million years ago.

Plant fossils—bacteria, algae etc.—can be traced back into Precambrian time, to some 2000 million years ago (p. 257). However it is only in latest Precambrian, 600-700 million years ago, that the first undoubted animal fossils appear—but all that are known from this time are of soft-bodied creatures (e.g. jellyfish, worms etc.). Then quite dramatically the first shelled animals appeared at the beginning of the Cambrian, 570 million years ago, followed, in the remainder of Cambrian time, by a tremendous burgeoning of animal life. During this time appeared representatives of nearly every major animal group that possessed skeletons capable of being fossilised. The only important exception is the chordate group, which includes the vertebrates.

In the past it had been assumed that Precambrian rocks have been too greatly altered by heat and pressure to allow the ancestors of the Cambrian animals to be preserved. Another idea is that the ancestors were all soft-bodied and that a sudden change in the chemistry of sea water allowed them to extract lime, construct shells and thus be preserved as fossils. Some geologists believe however that if the ancestors had been soft-bodied, and therefore rarely preserved, far more abundant traces of their activities (trails, tracks and other markings) should have been found in the Precambrian rocks than has proved to be the case.

Whether soft-bodied ancestors existed or not it is clear that at the end of the Precambrian there was a phase of rapid and radical evolutionary change in animal life—perhaps the greatest in earth's history, and probably surpassing that of the Mesozoic-Tertiary boundary (p. 278).

Whatever other factors may have been involved in this evolutionary event, the major trigger mechanism may have been the dramatic change in climate that came at the end of the Precambrian ice age.

The ice age itself would have created extremely adverse conditions for life. In particular the lowering of sea level, accompanying the accumulation of ice on land, would have sharply reduced the area of shallow seas, which include many of the most favourable habitats for marine life. In contrast, at the end of the ice age the improvement in climate and rise of sea level would have re-created a variety of favourable but biologically empty environments in which opportunity would exist for radical evolutionary changes to take place.

The ice age probably also had a major influence on the chemistry of the oceans and may have made conditions right for marine animals to build shells—perhaps another reason for the sudden burst of shelled animals.

### Causes of ice ages

During each of the four ice ages of the past, world climate varied between that of the hot tropics and that of the icy cold poles, and climatic zones (subtropical, warm and cool temperate etc.), ranging between these two extremes, could be distinguished but with very broad Arctic and Antarctic zones.

In the intervals between the four ice ages the climate of the world was apparently fairly equable, judging from what we know about the distributions of fossil plants and animals. There was a broad tropical-subtropical-warm temperate climatic zone that may have extended north and south from the equator to 50°–60° Latitude, and then there was a relatively narrow cool and cold-temperate zone, extending from about 60° to the poles. However, polar ice caps, with the accompanying Arctic/Sub-Arctic and Antarctic/Sub-Antarctic climatic zones, did not apparently exist.

Therefore over the past 650 million years the world has experienced long periods of equable climate (some 88 percent of time over the last 650 million years), broken by ice ages at intervals of about 250 million years.

The reasons for this have been the subject of much pondering by scientists. However, the problem of ice ages is two-fold. First, we have to explain why the world had ice ages at various times in its history (i.e. how they are started). Second, we must provide an explanation for the maintenance of an ice age and for the see-sawing of climate that occurred within each ice age.

In casting about for likely explanations of how ice ages can be started, some scientists have looked towards events out in the solar system, whilst others have sought causes on the earth itself.

Of the former, extra-terrestrial, explanations we can mention some that in the past have received attention. Perhaps most naturally, some scientists have sought an explanation in possible variations in the heat radiated from the sun. It has been suggested, for example, that there may be periodic changes in the rate at which the sun burns its nuclear fuel. No firm evidence, however, has been found to sustain this idea. Other extra-terrestrial theories involve the blocking of a portion of the sun's rays so that the earth gradually cools. Injections of great quantities of volcanic ash into the atmosphere would be one way of doing this, and another way would be passage of the earth through a cloud of interstellar gas and dust. Although some cooling could be expected from both these mechanisms, no one has produced good evidence of extensive periods of volcanic activity preceding glacial episodes, and passage through clouds of interstellar gas and dust remains in the realm of speculation.

A perhaps more likely theory for the initiation of ice ages relates to events on the globe itself and involves past continental movements. This is where the theory of Continental Drift comes in.

Today there are large land areas in the broad cold regions surrounding both north and south poles, particularly the South Pole, but we know from palaeomagnetic studies, and from fossil plants and animals, that often in the geological past one or other pole was in the middle of the oceans, where currents and general movement of ocean waters allowed sufficient exchange of heat with the atmosphere to make the distribution of temperature relatively uniform from poles to equator. However, once land areas were near or at the poles, the oceanic circulation system was restricted, the difference in warmth between the poles and equator was increased, and the polar regions became very cold. If this theory is correct, the events leading up to the last ice age must have been set in motion some 20 million years ago, for that was when the continents approached their present positions relative to the poles. Indeed, the *Glomar Challenger*'s drill has now provided evidence of glaciations that started well before the main, or Pleistocene, ice age (p. 320).

As understanding of the ice ages has increased, so has the realisation that Antarctica, and further back in geological time, the Gondwana landmasses as a whole, have probably played a key role in triggering ice ages.

Palaeomagnetism has shown us that throughout geological time there have probably never been substantial masses of land straddling the North Pole. But yet at the times Antarctica and various of the Gondwanaland continents have been astride the South Pole, the earth has experienced glaciation—hence the idea that large continents hold the key to the ice ages.

If sea occupies a polar position any snow or ice that falls merely drops into the ocean and melts, or at most contributes to the build-up of pack ice. It is physically not possible to build up thick ice sheets on the sea. What is required is land in the polar region—but it is not enough to have flat land—it must have a fair amount of relief, with highland mountains.

Snow falls in the highland regions, and

gradually builds up in hollows in the landscape. This piling-up process steadily compacts the snow and turns it into hard glacier ice. Once formed, the glacier ice flows out into surrounding regions under the influence of gravity and builds up thick ice sheets. Glacier ice, once it is made, is a very tough rock-like substance and as it moves it flows inexorably over land and sea alike. The presence of highland topography to trap and compact snow into glacier ice is therefore a vital prerequisite to glaciation and it seems that Antarctica has had the right combination of relief, land area and closeness to the pole at the various times in the recent past that the earth has experienced glaciation.

We probably all realise that light-coloured objects reflect the sun's heat, whereas dark-coloured objects absorb heat. Thus solar heaters, for example, use black plastic or black-painted metals to absorb as much of the sun's heat as possible.

Similarly on a world scale, the dark-coloured landmasses and oceans tend to absorb the sun's heat and, like an electric night-storage heater, will tend to distribute it evenly (through ocean currents) and release it slowly during cloudy or cool periods.

If, however, large areas of land and sea are covered by clouds, ice and snow, most of the sun's energy falling on these areas is reflected straight back into space, there to be dissipated. These areas therefore get progressively colder, and as they do so more and more ice and snow accumulate. It has been estimated that the reflectivity of clouds is 90 percent (i.e. 90 percent of solar radiation is reflected back into space), and that of snow and ice 80-60 percent, compared with land and water areas that have values less than 10 percent.

Therefore as cloudiness, snow and ice increase, a chain reaction sets in, as a greater and greater proportion of heat is lost to the earth. This results in progressive cooling, and the earth sinks into an ice age.

Thus a whole set of conditions have to be "just right" before an ice age can be initiated, and it is thought that continental drift is a major force "behind the scenes", manoeuvring landmasses and rucking up mountains, and in this way creating favourable conditions for glaciation.

Just as continental drift can move landmasses towards the pole and initiate an ice age, it can also move them away and end an ice age. Thus while palaeomagnetism has chronicled the southward drift of Antarctica, closer and closer to the South Pole, picking up more and more ice as it moved (p. 67), so it has also chronicled the slow drift of Gondwanaland across the South Pole in the Carboniferous and Permian (Fig. 17.12). In the early Carboniferous southern Africa was astride the South Pole, whereas by early Permian it was eastern Antarctica's turn. However, by Triassic and Jurassic times continental drift had moved the Gondwana countries away from the South Pole, which at that stage was sited in the ancestral Pacific Ocean.

Continental drift cannot, however, explain the alternating glacial and interglacial climates that seem to typify all ice ages.

**Climatic see-saw**

A welter of theories have been proposed to explain the see-sawing of climate during ice ages but up to the last few years a major stumbling block has been the lack of information on the number and timing of glacials and interglacials.

As we have already noted (pp. 324–325), the idea of four glacials and interglacials had become greatly entrenched over the years and as also happened with the theory of continental drift, it took evidence from deep-sea drilling, combined with new techniques in the study of magnetism, palaeontology, radioactive dating, oxygen isotopes etc. to inspire the revolt that has led to the overthrow of this dogma.

Interestingly enough, Alfred Wegener, the pioneer of the continental drift theory, also had a great interest in the ice ages of geological history. While believing that continental drift could go a long way towards explaining the actual initiation of ice ages, he also studied astronomical phenomena to discover mechanisms for the fluctuations within ice ages. For this reasn he was very interested in modern ice sheets, and, as we have seen, perished in 1930 whilst on an expedition to study the Greenland ice cap.

Wegener's death left a Yugoslav, Milutin Milankovitch, to carry the torch for the theory of astronomical fluctuations as a controlling mechanism for alternating glacials and interglacials. However, Milankovitch's ideas, published in 1938, received little recognition in his own lifetime as they tended to run counter to the then accepted number and timetable of glacials. Now,

Milankovitch's theory is the front-runner and it seems very likely that it can account for long-term climatic changes, and furthermore, can be used to predict future trends.

Astronomical circumstances affecting the earth are continually changing—as demonstrated by the often marked mis-alignment today of many of the ancient monuments originally built to incorporate various astronomical effects. The ancient Egyptians, for example, planned the entrance passage of the Great Pyramid of Cheops so that the bright northern star Thubar would shine down it for ever more. Now the star is 20 degrees out of alignment. Similarly Stonehenge and other megalithic stone circles in Europe built with particular stones to mark sunrises, sunsets and so on are now mis-aligned.

Today the earth's axis is not upright with respect to the sun. It inclines at an angle of 23.4 degrees and tips the Northern Hemisphere towards the sun in July and the Southern Hemisphere towards the sun in January. However, because the earth's orbit around the sun is not exactly circular, the earth is three million miles closer to the sun in January than it is in July, which means that sunshine is weaker by 7 percent during the northern summer compared with the southern summer.

All these factors however can change and at least three things can happen.

First the earth's orbit can "stretch" by departing much further from a circle and then gradually changing back to almost true circularity. It is thought that in the case of extreme stretching of the earth's orbit, the intensity of sunshine reaching the earth may vary by up to 30 percent from its strongest to its weakest, in the course of a year.

Second, the earth wobbles like a spinning top, and swivels its axis around and around. Thus there is a change in the season of the closest approach to the sun—the "precession of the equinoxes".

Third, the earth also rolls like a ship, altering the tilt of its axis between 21.8 degrees (more upright) and 24.4 degrees (more tilted). The greater the tilt the more pronounced is the difference between summer and winter.

The three changes—stretch, wobble and roll—occur at different speeds. Thus a complete cycle of stretching of the earth's orbit (i.e. near-circular to stretched and back to near-circular) takes 100,000 years, a wobble cycle 21,000 years and a roll cycle 40,000 years. Combining these cycles and working out calculations of sunshine changes all over the world for say the last two million years calls for considerable mathematical expertise! However when all the sums have been done, the resulting peaks and troughs of the sunshine record correspond very closely to the warm peaks and cold troughs obtained from analyses of deep-sea cores.

Since the 1960s deep-sea cores, obtained both by piston coring and by the *Glomar Challenger* (pp. 89–102), have been studied from the aspects of changing plant and animal life (preserved in the sediments as minute skeletons of plankton life), changing oxygen isotopes and mineral content—all of which reflect changes in sea-water temperatures, as climate oscillates between glacials and interglacials. Dating remained a problem, however, until reversals of magnetism were picked up in the sea-bed cores and could be tied to the magnetic time scales obtained from lavas (pp. 80–85). Meanwhile since the early 1960s Czech geologists had been studying bands of light-coloured dusty soil (loess) deposited when Czechoslovakia was a frigid wasteland hemmed in between the ice sheets of the Alps and northern Scandinavia. It was found that different species of snails and plants obtained from the loess bands kept changing repeatedly as the cold came and went during the Ice Age. Some ten complete cycles of rhythmical cooling and warming were distinguished. The correspondence between the loess cycles and those from the ocean-bed cores was noted in the late 1960s, but then in the early 1970s magnetic studies revealed that the last reversal of magnetism, dated at 700,000 years ago, and detected in the ocean-bed cores eight cold periods down, could be found in the Czech loess layers, also eight cold periods down.

The match between land and sea was thus assured at last, and since that time supporting evidence for the number and dating of glacials has come from studies in two climatic extremes: (i) the climatic record preserved in ice cores, obtained from the Greenland and Antarctic ice caps; (ii) sea-level changes, corresponding to glacials and interglacials, determined from dated coral reefs on rising lands being actively pushed up by plate movements, and as widely separated

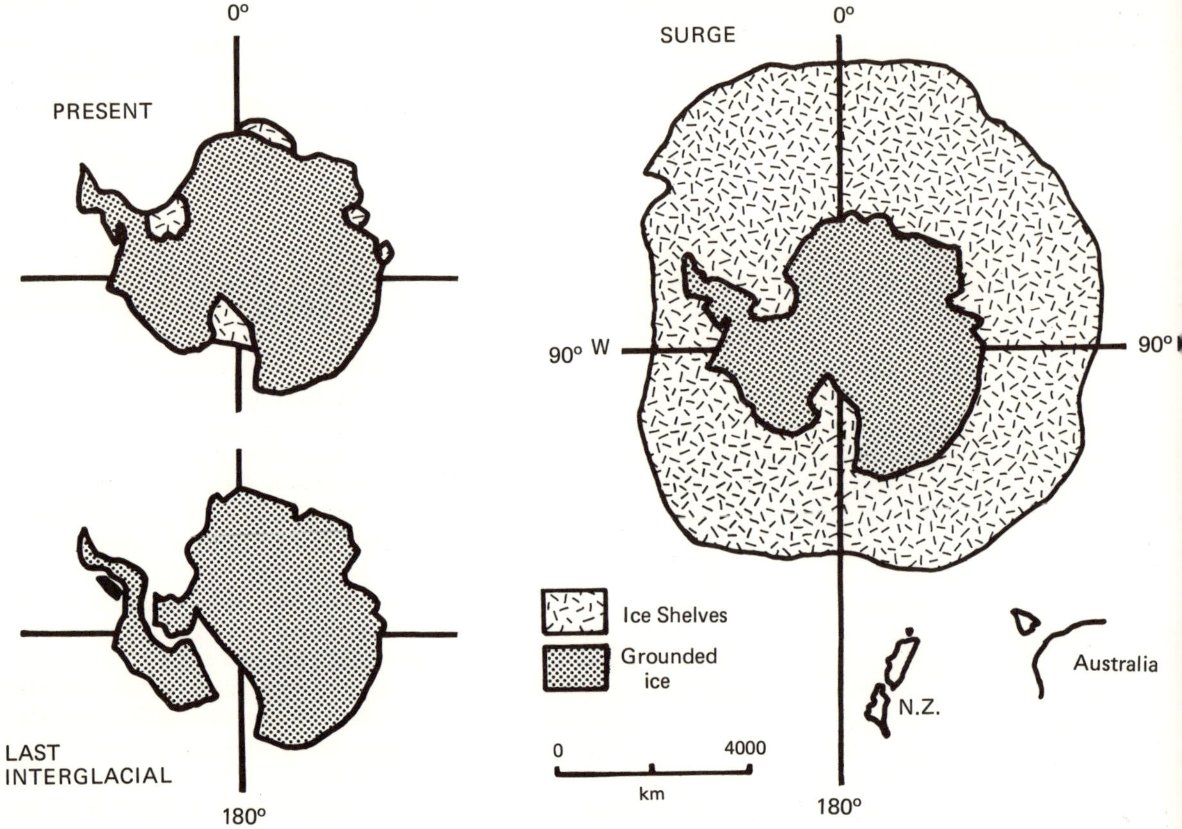


**Fig. 17.13.** One of the theories to account for the alternation of glacial and interglacial periods during Ice Ages visualises the build-up of ice on Antarctica to such an extent that instability of the ice cap occurs, and great amounts of ice slide out and surge across the adjacent oceans, forming vast ice shelves (righthand diagram).

The creation of such a large area of ice is sufficient to reflect back substantial quantities of the sun's heat and, as the earth's climate deteriorates, a glacial phase is ushered in.

After the surge, the Antarctic ice sheet stabilises itself again and the huge ice shelves are gradually broken up by wind and wave action and are not renewed. The ice shelves disappear completely, sea level rises (as water is returned to the oceans), and an interglacial phase begins (diagram in bottom lefthand corner). The top lefthand diagram shows the situation at present, with much of the Antarctic ice sheet stabilised and grounded on continental basement, and only some small areas floating as ice shelves.

as New Guinea, New Hebrides, Barbados and the Ryukyu Islands.

The new studies of cores from both the sea floor and ice caps, with their sophisticated techniques to monitor both time and temperature, have shown that climate can change very quickly.

Thus 90,000 years ago the earth went from a warm period, with a climate similar to today's, straight into the full severity of a glacial period—and this sudden cooling took a mere few hundred years.

Although astronomical factors may set the scene for such a change, some rapid-acting mechanism must be sought. Apart from massive volcanic eruptions, or an extra-terrestrial event, a possible mechanism involves surging of the Antarctic ice sheet, proposed in 1964 by Alex T. Wilson, of Victoria University of Wellington (now of Waikato University, Hamilton) (Fig. 17.13).

According to Alex Wilson the huge ice sheet that covers Antarctica, even in our present fairly warm period, can act like a climatic time-bomb. As the ice sheet gradually builds up from the snow that falls on it, the pressure at its base reaches a point where ice melts. Until that moment the ice has been able to travel over the underlying ground by a slow process of deforming itself ("plastic flow") in which the tiny ice crystals rearrange themselves.

But once its base is lubricated by water the ice sheet can slide bodily, and much more rapidly. The friction of the ground now helps rather than hinders the progress of the ice, by causing further melting at the base (through frictional heating).

At the present time the edges of the Antarctic ice sheet are colder than the pressure melting point of ice and therefore the sheet is frozen to its base, and can only creep by slow internal deformation. However there are areas in the interior of Antarctica that are now at the melting point and hence have water at the base of the ice sheet. At Byrd Station (80°S Latitude, 120°W Longitude) a drill hole reached bedrock after passing through 2164 m (7000 ft) of ice. It was found that the lowest temperatures (–28.8°C, –20.2°F) were not at the bottom of the bore hole—but rather at a depth of 800 m (2600 ft)—and that the temperature at the bottom was –1.6°C (–29°F), with melt water present.

It is thought that when the water layer at the base of the ice sheet spreads and reaches the sheet's edge over wide areas, and is sufficiently thick, very rapid basal sliding of the entire ice cap may occur, producing a surge.

Small-scale ice surges do occur, and are probably produced by the same cause (i.e. basal melting). For example, the ice cap of Spitzbergen surged some 21 km (13 miles) in the 1930s. Glaciers in Alaska have been known to stand almost still for half a century and then move downhill at 3-4 m (10-13 ft) per hour.

Today the Antarctic ice sheet creeps sedately into the sea, making icebergs at a rate roughly in keeping with the accumulation of new snow. However, various ice shelves around its margin are floating today and a massive surge of ice from inland would push them even further northwards.

The Ross Sea ice shelf, due south of New Zealand, is a floating mass of ice the size of France. Today the Ross ice shelf advances at a rate of half a mile per year, but as it does so, its bottom layers melt and icebergs break off and drift away, so that the shelf front itself remains almost stationary.

If an ice surge occurs, floating ice is pictured as moving out from Antarctica and spreading over the Southern Ocean as far north as the Antarctic Convergence (55°S Latitude), where relatively warm (8–10°C, 46–50°F) south-flowing water is encountered. The vast ice sheet would be grounded on islands at many points and this might provide the sheet with sufficient support to prevent its being broken up by oceanic waves (Fig. 17.13).

It has been estimated that if a surge occurred from an Antarctic ice cap averaging 2500 m (8500 ft) in thickness, and it advanced at a rate of 110 m (360 ft) per day, it would take only 40 years for the edge of the ice sheet to reach 55°S Latitude and the floating ice would be some 200 m (675 ft) thick. The huge ice shelf would increase the reflectivity of the Southern Ocean from 8 percent to 80 percent and would decrease the heat received by the earth by 4 percent. This reduction in heat would trigger the formation of ice sheets in the Northern Hemisphere.

Wilson also envisaged ice surging as an explanation of alternating glacials and interglacials. As the ice cap thins through surging, colder ice moves down to form the basal layers, and becomes frozen to the underlying rocks. Also, snow-fall in the centre of the ice sheet would decrease as incoming storms lost their moisture above the huge ice shelves. Supply of ice to the shelves would thus slacken. The shelves would then be rapidly broken up by wave action. The sun's energy that was formerly being reflected by the ice would now warm the Southern Ocean. Some of this heat would be transferred to the oceans of the Northern Hemisphere, and the earth would gradually emerge from an ice age. Once again ice begins to build up on Antarctica and the ice sheet grows until it surges again to push the earth into the next glacial (Fig. 17.13).

Wilson's theory is an attractive one as it accounts for the glacial/interglacial alternation and has its own built-in regulator. The present-day consensus, however, is that astronomical variations *a la* Milankovitch probably provide the major driving forces, with ice-cap surging being relegated to a secondary role.

## Ice ages to come?

Although continental drift will gradually move Antarctica away from the South Pole and draw the present ice age to a close, this will take many tens of millions of years. Therefore alternating glacials and interglacials are likely to be part of the earth's scene for a long time to come.

If, as many scientists believe, the Milankovitch theory can be used for prediction, the picture is rather gloomy. According to the Milankovitch timetable, interglacials are a mere 10,000 years long. Thus today we have already had our 10,000 years, and the ice is overdue.

Judging from cores obtained from both the sea floor and Arctic and Antarctic ice caps, it is the Southern Hemisphere which leads the way into an ice age, with the Northern Hemisphere lagging some 3000 years behind. According to some scientists the Antarctic region has gone more than halfway back from its most recent climatic high point (the Climatic Optimum, p. 331), towards the conditions of a full ice age and although the Northern Hemisphere has not yet caught up, we can expect a global ice age to begin in the near future, perhaps within about 1000 years.

In the meantime many scientists believe that climate will fluctuate and we will have little troughs and peaks—and perhaps another "Little Ice Age" (p. 332)—before plunging into the next glacial, that is calculated to continue through to 119,000 years from now. If we are on our way back into an ice age, just a small deterioration (even a return to a Little Ice Age, let alone the big event), will put an intolerable strain on our capacity to feed our expanding populations. In this way Nature, by a small natural change, could destroy the habitats of millions of humans. Climatic change therefore presents a major challenge—perhaps *the* major challenge—to civilised man.

Needless to say, the entire subject of climatic change is a most contentious issue and the truth of the matter is that we need a lot more data—on meteorology, climatology, oceanography, glaciology, astronomy, etc.—before we can work towards getting a clearer picture.

Hopefully much of this data will be obtained by the numerous international research programmes now under way, involving computers, satellites, floating ice stations, giant balloons and high-flying jet planes.

**Fig. 17.14.** During the last Ice Age, or Pleistocene period, between about 2 million and 10,000 years ago, extinction was the fate of many of the tropical and subtropical plants and animals that had migrated to New Zealand in earlier times.

For example, large warmth-loving cockles (*Maoricardium*) lived in New Zealand in Pliocene and Miocene times and the upper specimen is from the late Pliocene, three million years ago, at Waverley Beach, South Taranaki.

While the cold ice-age climates wiped out these and other organisms in New Zealand waters, they continued to survive in tropical waters and the lower specimen is a living *Maoricardium* from Pemba Island, off the Kenya coast, Indian Ocean.

## Ice age decimation and after

As we have seen (p. 317), the steady deterioration of climate in late Tertiary times thinned out many of the warmth-loving immigrants New Zealand had received earlier. However, the

**Fig. 17.15.** As the New Zealand mountains began to be pushed up in the early Pleistocene and Holocene (i.e. the last 1½-2 million years), certain animals and plants moved up with them and populated the newly-created sub-alpine and alpine environments. In this way the kea (*Nestor notabilis*), a playful and cheeky inhabitant of the South Island high country, developed from the same stock as the kaka, now living in lowland forests.

severity of the repeated glacials of the Pleistocene, coupled with the fact that there was no escape north of 34°S, decimated our flora and fauna and left gaps, only a few of which were later filled by temperate organisms riding the West Wind Drift (Fig. 17.14).

Our podocarp forests were pushed into areas roughly north of Hamilton and elsewhere into a few refuges at sea level. The survival of the podocarps, and the consequent continuation of supply of their fleshy seeds, very attractive to birds and other animals, in turn ensured the survival of many of our primitive land animals dating back to the time we were part of Gondwanaland (p. 248).

Earth movements in the Circum-Pacific mobile belt (p. 309) pushed up high mountains in New Zealand, Australia, New Guinea and Malaysia in times as recent as Pleistocene or late Pliocene—when all indications are that these lands were largely separate (although during glacials organisms were able to migrate between New Guinea and Australia). Thus the fact that New Zealand has well-developed alpine and sub-alpine communities of plants and animals, but obviously has been isolated from other such communities for a long time, has been a puzzle to zoologists and botanists alike.

In general New Zealand has had to develop its own alpine and sub-alpine races from organisms of Malayo-Pacific (i.e. subtropical) origin. In the early Pleistocene therefore, as many of our mountains began to be pushed up, certain plants and animals moved up with them—especially those adapted to life in dry stony river beds, rocky slopes etc. Thus the alpine rock wren, alpine wetas and alpine cicadas were developed from lowland groups and the New Zealand parrots produced the alpine kea (the kaka being its equivalent in lowland forests) (Fig. 17.15).

The changing patterns of land and sea that were a feature of Pleistocene time encouraged development of distinctive varieties of plants and animals, as a result of isolation. Distinct geographic races developed in native land snails, for example, as populations became cut off by earth or sea level movements. The Three Kings Islands, immediately north of New Zealand, remained completely isolated, even during the times of lowest glacial sea level, and as a result now have a range of distinctive plants and animals.

At the close of the last glacial, 10,000 years ago, forest became re-established throughout New Zealand but its recovery from the repeated disruptions during the successive glacials was a long slow process and it is thought that even within the span of human occupation of New Zealand, vegetation changes have been occurring that are related to this long-term recovery process. Coupled with this, however, has been the effect of climatic changes—notably the Climatic Optimum and Little Ice Age (pp. 331, 332).

However, without a doubt the arrival of Polynesians about a thousand years ago initiated a long train of biological events that continued even more rapidly after the visits of Tasman and Cook.

The early Polynesians (the "Moa Hunters") used the moas and other birds as easily hunted

**Fig. 17.16.** Andesite volcanoes were active in Taranaki throughout the Ice Age. The eroded stumps of Kaitake volcano (foreground) (683 m; 2243 ft), active 575,000 years ago, and Pouakai volcano (middle distance) (1399 m; 4591 ft), active 240,000 years ago, are seen in this photo.

The cone of Egmont (2518 m; 8260 ft), seen in the background (cf. Fig. 1.6), has been built up over the last 70,000 years and was very active in the closing phases of the Ice Age. Egmont last erupted in A.D. 1755.

protein sources, and so robbed the New Zealand fauna of many of its older distinctive elements—including moas, native New Zealand geese, swans, eagles, crows etc. The fire brought by human beings, and used by them in hunting and agriculture, destroyed large areas of forest in Northland, central North Island and eastern South Island and reactivated many areas of hitherto stable sandy country—so that sand dunes invaded fertile land in many coastal regions.

The Polynesian rat and dog, also introduced, added to the effects of hunting and use of fire. In particular, New Zealand's abundant bird life suffered local or total extinction. Many birds disappeared altogether, while others became restricted to Fiordland and various islands off the New Zealand coast.

Other animals such as tuataras, native frogs, wetas and native land snails, formerly widely distributed over both North and South Islands, became greatly reduced both in numbers and geographic distribution.

Thus the arrival of the human race, with fire, rats and dogs, coming on top of the effects of the ice age, was the last straw for many primaeval groups, some dating all the way back to Gondwana times.

### Ice-age earth movements

When Alfred Wegener first formulated his continental drift theory, he thought that substantial movements between continents had occurred as recently as the Pleistocene period. As a glance at Fig. 3.2 will show, Wegener envisaged up to two-thirds of the width of the North Atlantic, for example, as having been opened up during this time. Many contemporary geologists, however, did not accept that movements of such magnitude could have occurred in this, the last and most recent of the geological periods. These reservations apply with even greater force today, as we now know that the Pleistocene lasted only between 2 million and 10,000 years ago, whereas Wegener thought it to be two or three times longer. Furthermore, study of magnetic stripes and sea-floor spreading have shown us that continental drift is a process involving very small movements, between 10 and 90 mm (½-4 in) per year, taking place relentlessly over millions of years.

**Fig. 17.17.** This photo looks across the three andesite volcanoes making up the central North Island volcanic centre, various parts of which were active throughout most of the Ice Age.

Tongariro (1968 m; 6517 ft), in the foreground, was built in the early Pleistocene 1½-2 million years ago, but ended its activity with a violent explosion that blew its top off, leaving a truncated mountain.

The scene of activity then shifted to Ruapehu (2797 m; 9175 ft), in the background, about 750,000 years ago, but then in the post-glacial period it too blew its top.

Ngauruhoe (2290 m; 7515 ft), in the centre, then became active and its cone, (900m; 3000ft) high, has been built up over the last 2500 years (cf. Fig. 18.21). Note the fault line traversing the flanks of Tongariro in the right foreground.

This is not to say, however, that very little in the way of earth movements had occurred during Pleistocene time. Very considerable movements had been going on, but mainly as a result of, rather than contributing to, continental drift. In the active volcanic regions of the world, related to plate margins, many large volcanoes were built up during Pleistocene time.

In New Zealand, for example, the large volcanoes of Egmont, Tongariro and Ruapehu are largely of Pleistocene age. Egmont and Ruapehu are believed to have been built up over the last 0.75 million years and Tongariro over the last 1.5 million years (Figs 17.16, 17.17). Ngauruhoe, 900m (3000ft) high, post-dates the Ice Age and is 2500 years old.

These "andesite" volcanoes, and others throughout the world, are related to areas where plate edges over-ride one another, and melting occurs at depth in the crust. Other volcanoes, "basaltic" in type, form along the rifts where plates are being pulled apart. The volcanoes Surtsey and Heimaey, recently active in Iceland, are examples of very rapid volcanic build-up.

The other type of earth movement that has been occurring over the Pleistocene period has been that associated with mountain building. The new mountains of the world—the Rockies, Andes, Himalayas and our New Zealand mountains—have all been created very recently in geological time. Like the volcanoes their creation is related to plate boundaries, and rocks are being actively buckled up in response to enormous compressive forces.

However a mountain range is not born overnight. We know from studies of past earth movements that mountains are created by a series of successive earth movements, along fault lines, that are individually very small, but when taken together over a million years amount to a substantial elevation.

Wellington's Rimutaka and Tararua ranges (941m and 1571m; 3090ft, 5155ft, at their respective highest points) are thought by many geologists to have been largely created during Pleistocene times by means of a series of small sudden upward movements of perhaps a few metres at a time, spaced at intervals measurable in centuries. These movements, occurring along fault lines, and extending over a long period of time, have raised the mountains high above the seas. The latest of these movements in the

**Fig. 17.18.** In 1855 the West Wairarapa Fault, running along the western side of the Wairarapa Valley, moved 2.7m (9ft) vertically and 12.2m (40ft) horizontally. This movement raised the entire Rimutaka-Tararua Range by amounts varying from 2.7m to 1m (9ft to 3ft) and was the latest in a series of uplifts that have pushed up the ranges during the past one million years.

The upper photo shows the line of movement (fault scarplet, indicated by arrow) of the 1855 earthquake, traversing the river gravels at the foot of the Rimutaka Range.

The lower photo shows the 1855 fault scarplet at Pigeon Bush, southern Wairarapa.

Wellington region occurred in 1855, when the West Wairarapa Fault moved 2.7 m (9 ft) vertically and 12.2 m (40 ft) horizontally (Fig. 17.18). The entire Rimutaka-Tararua Range was uplifted by amounts varying from 2.7 m to 1 m (9 ft to 3 ft). Although this amount does not sound very much compared with the hundreds of metres of altitude of the ranges today, if such a movement occurs every 300 or 500 years, then in one million years we arrive at a total uplift of about 5000 m (17,000 ft).

We know from studies of beaches that have been left high and dry by past land uplifts that movements in the Wellington region have occurred with similar frequency (Figs 17.19, 17.20) and this has led us to believe that these mountains, and many others in New Zealand, are very young. Of course, as the mountains go up, so the rivers and streams cut into and lower the bare eroding slopes. Thus to keep pace with simultaneous erosion, a rising mountain has to be pushed up to heights far in excess of its present elevation. The Southern Alps, for example, have been pushed up some 18,500 m (60,000 ft) over about 12 million years, but yet today the highest

**Fig. 17.19.** Raised beaches at Turakirae Head, where the Rimutaka Range heads out into Cook Strait, chronicle the past movements that have steadily uplifted the range during the Ice Age. This view is looking towards the entrance of Port Nicholson.

The raised beaches show up as light-coloured strips roughly parallel to the coastline. These beaches were forming at the dates indicated, but were then suddenly elevated by earth movements (the last in A.D. 1855). The sandy and gravelly beach deposits were left as a ridge, and a strip of rocky sea bed was exposed below them. A new beach began to form at the next lower level.

The Turakirae beaches provide a record extending back to 4900 years ago. Still older beaches are found, however, in Palliser Bay and these date back to 6500 years ago.

The flattened step-like features on the hills between Baring Head and the Orongorongo River (middle distance) are marine terraces cut by the sea during interglacial phases, 100,000 and 250,000 years ago, and now raised to heights of 180-210 m (600-700 ft) and 260-285 m (850-950 ft).

**Fig. 17.20.** Interglacial marine terraces similar to those at Baring Head (Fig. 17.19) have been raised to even greater heights (440 and 790 m; 1450 and 2600 ft) at Cape Palliser.

peak, Mt Cook, is only 3764m (12,349ft)—the difference being the amount eroded off the mountains over the elapsed time.

Our mountainous regions are traversed by numerous large faults (Fig. 18.17). These faults, besides having substantial vertical movements, have also large horizontal movements. The Alpine Fault, with almost 480km (300 miles) of movement, is the largest of these faults and has literally torn the South Island in two (pp. 25–26).

All the Pleistocene earth movements in New Zealand are seen as part of the Kaikoura Orogeny, that began in late Miocene times (p. 314; Fig. 17.21)—and, like the ice ages themselves, they show no signs of coming to an end—so New Zealand is likely to be the "shaky isles" for some time yet.

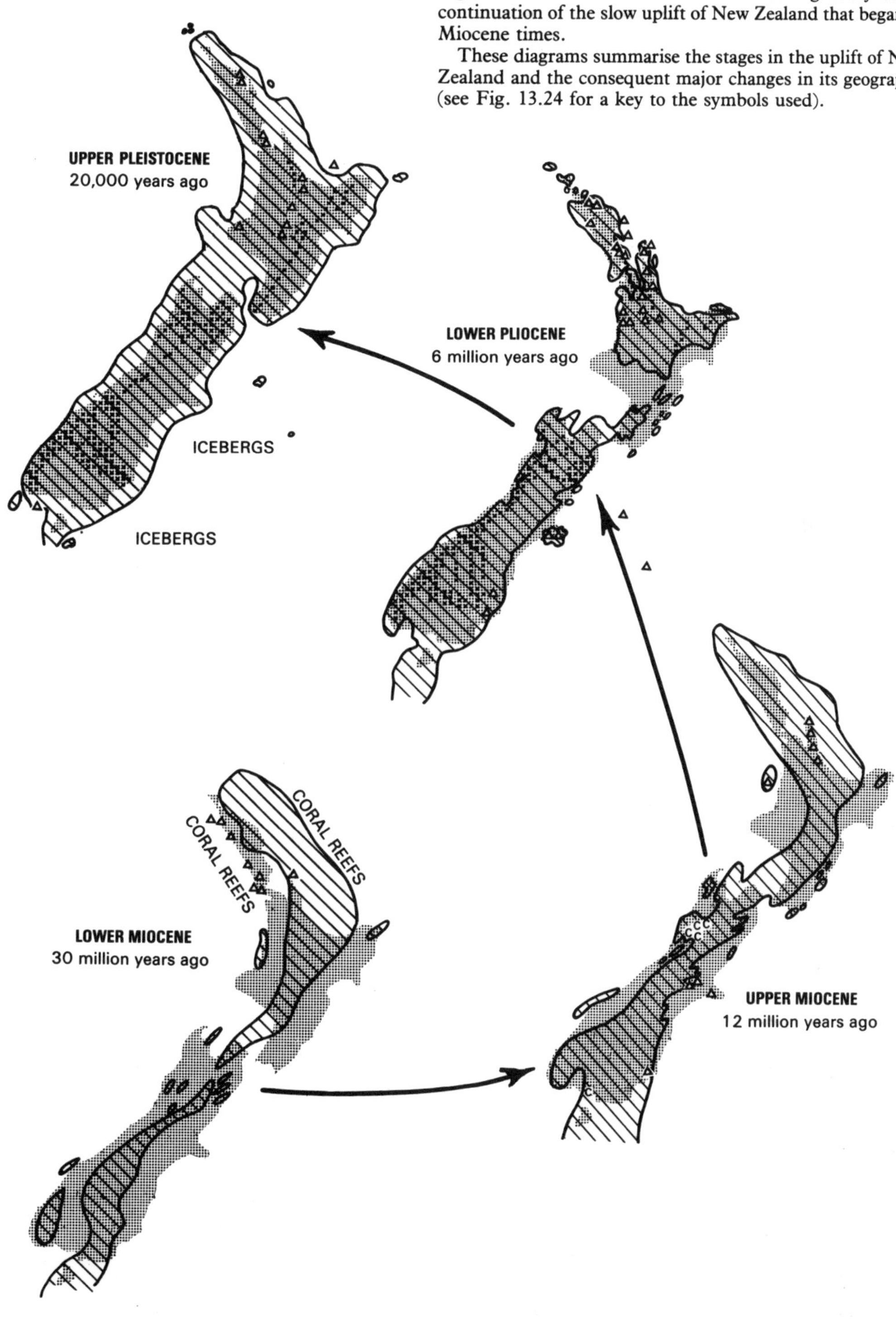


**Fig. 17.21.** The earth movements occurring today are a continuation of the slow uplift of New Zealand that began in Miocene times.

These diagrams summarise the stages in the uplift of New Zealand and the consequent major changes in its geography (see Fig. 13.24 for a key to the symbols used).

CHAPTER EIGHTEEN

# THE MAKING OF MODERN NEW ZEALAND

AFTER our long voyage through the aeons of geological time we at last arrive at modern New Zealand.

New Zealand today is part of a large continental mass about one-half the size of Australia but only one-tenth of which appears above the sea (as New Zealand, and outlying islands such as Auckland, Campbell, Antipodes, Bounty, Chatham, Norfolk and Lord Howe) (Fig. 18.1).

New Zealand itself is surrounded by a generally broad continental shelf, much of which was cut by the sea during times of low sea level in the Ice Age, when great quantities of water were trapped on the land as ice sheets, and thus withheld from the sea (pp. 327–330). The edge of the continental shelf is generally at a depth of about 200 m (660 ft). Beyond this edge are broad submarine platforms, plateaux and ridge-like oceanic rises.

South and south-east of New Zealand extends the Campbell Plateau, with its lower edge at 1000 m (3500 ft), cut into by various large and small troughs (Solander and Bounty troughs). Large mounds (submarine rises) diversify its otherwise plateau-like appearance and some of these are capped with islands, such as Bounty, Campbell and Auckland islands.

East of Christchurch a long submarine ridge-like feature links the Chatham Islands with the Canterbury coast. This is called the Chatham Rise.

Beyond the Campbell Plateau and Chatham Rise the sea floor sinks to abyssal depths, beyond 5000 m (17,000 ft), into the South-western Pacific Basin.

Between the Chatham Islands and East Cape lies a pie-segment-shaped piece of moderately deep (2000-3000 m; 7000-10,000 ft) sea floor with the apex of the segment heading into the Cook Strait-Marlborough region (Fig. 18.1). This segment of ocean floor, the Hikurangi Trench, is thought by many people to represent the tail-end of the giant Tonga-Kermadec Trench, that reaches depths of over 10,000 m (34,000 ft) and extends some 2500 km (1600 miles) north-eastwards, to just short of Samoa.

Parallel to the Tonga-Kermadec Trench is a varied sea-floor topography of ridges and troughs, essentially parallel to the trench and lying between it and the South Fiji Basin. This topography abuts upon New Zealand in the vicinity of the Bay of Plenty.

New Zealand is flanked to the north and north-west by three oceanic rises. The Three Kings Rise is, as its name suggests, a submarine extension of the northern tip of New Zealand northwards into the South Fiji Basin (Fig. 18.1). Then trending in a N.N.W. direction from the West Auckland coast is the narrow Norfolk Ridge, on which is sited Norfolk Island and New Caledonia. The third rise, Lord Howe Rise, takes off from the New Zealand coast between New Plymouth and Hokitika and extends as a broad feature north-westwards towards the Coral Sea, east of Queensland. Lord Howe Island and Ball's Pyramid are situated on the westernmost edge of the Lord Howe Rise.

Between the Lord Howe Rise and the edge of the continental shelf of Australia lie the abyssal depths of the Tasman Basin (maximum depth 5604 m, 18,400 ft), the only area of true oceanic crust in the Tasman Sea, except for a narrow strip of new sea floor between the Lord Howe Rise and Norfolk Ridge. The floor of the Tasman Basin is covered with magnetic stripes, the characteristic "signatures" of which indicate that the basin was created by sea-floor spreading between 80 and 60 million years ago (pp. 260–262) (Fig. 18.2).

Therefore, as may be appreciated, the Tasman Sea, lying between the Australian and New Zealand coastlines, is not all true "ocean" in the geological sense, but rather is composite and contains substantial areas of sunken continental material—the Lord Howe Rise and Norfolk

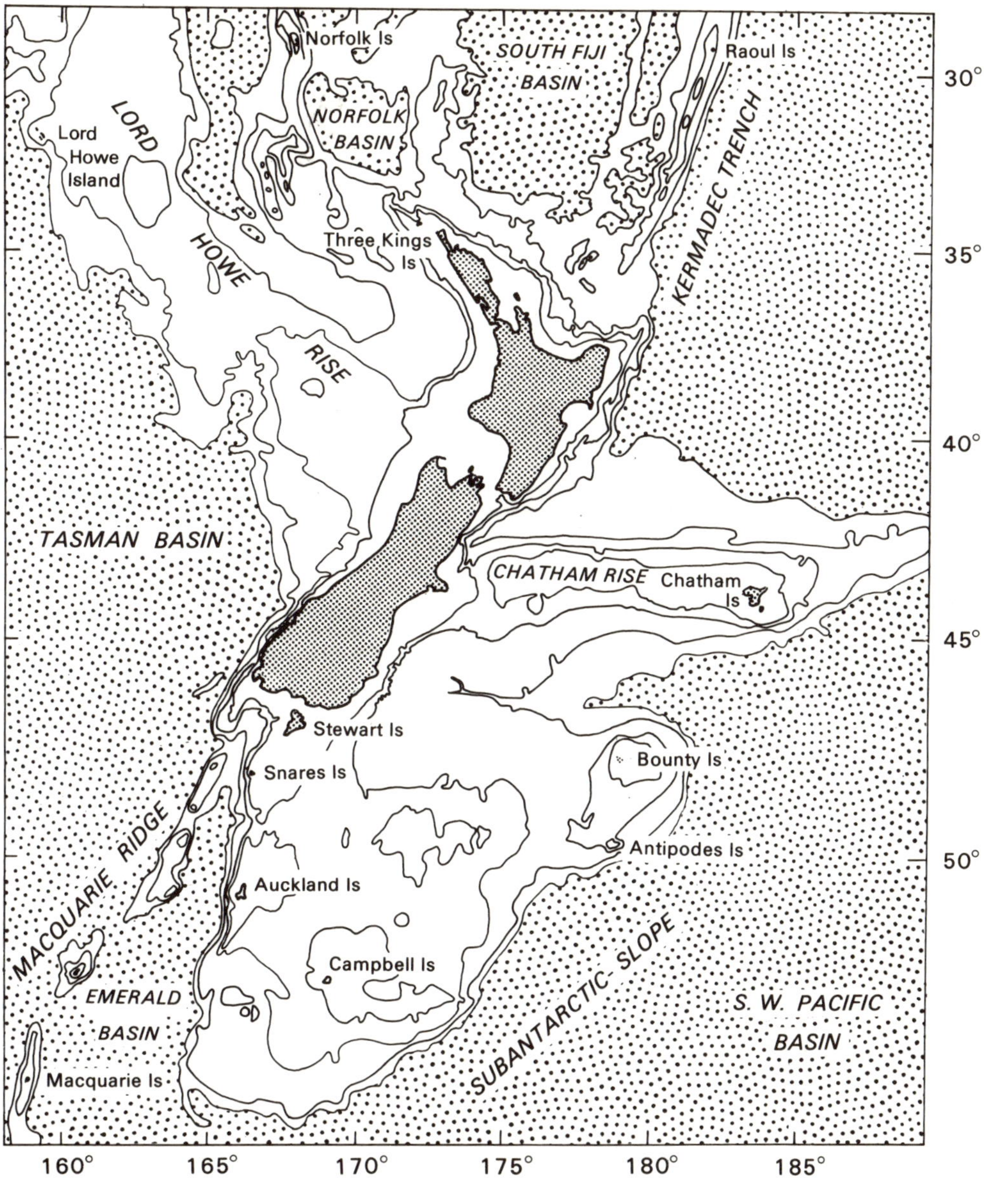


**Fig. 18.1.** Bathymetry of the New Zealand region. Isobaths (submarine contours) are drawn at 500 m, 1000 m, 2000 m and 3000 m (1650 ft, 3500 ft, 7000 ft, 10,000 ft). Oceanic areas deeper than 3000 m (10,000 ft) are stippled. To avoid cluttering the map some major submarine features are not individually named.

The Campbell Plateau includes the region south and south-east of New Zealand on which stand Auckland, Campbell, Antipodes, Bounty and the Snares Islands. The Bounty Trough lies between Bounty Islands and the Chatham Rise. The Solander Trough, a northern continuation of the Emerald Basin, lies to the west (left) of Stewart Island.

The Norfolk Ridge swings in an arc between New Zealand's North Cape and Norfolk Island and continues onto New Caledonia. The Norfolk Ridge is flanked by the New Caledonia Basin to the west and the Norfolk Basin to the east.

The Three Kings Rise extends northwards from the Three Kings Islands to lie between the Norfolk Basin and South Fiji Basin.

The Kermadec Trench continues northwards into the Tonga Trench and together the two are often referred to as the one feature: i.e. "Tonga-Kermadec Trench".

The Kermadec Trench fades southwards towards East Cape, but a trough-like feature, the Hikurangi Trench (or Trough), continues down the East Coast of the North Island, to end off the Marlborough coast, against the northern flank of the Chatham Rise.

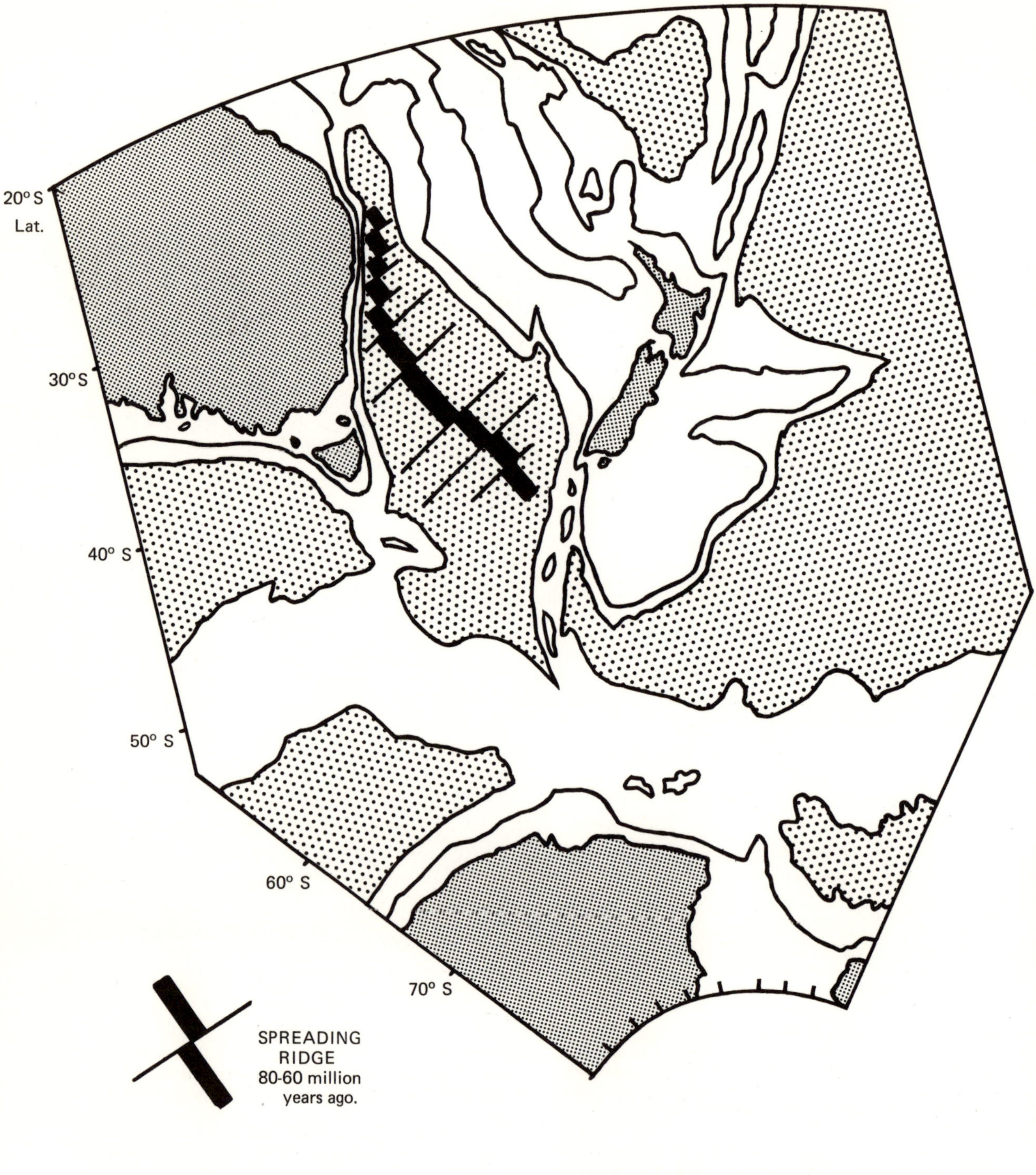


**Fig. 18.2.** The floor of the Tasman Basin, lying between Australia and the Lord Howe Rise, is covered with magnetic stripes, the characteristic "signatures" of which indicate that the basin was created by sea-floor spreading between 80 and 60 million years ago. Furthermore, the magnetic signatures indicate that the opening of the Tasman took the form of a wedging-apart, with an opening first appearing in the south and then gradually extending northwards to the Queensland coast. The generalised submarine contours (isobaths) are at 2000 m (7000 ft) and 4000 m (14,000 ft). Depths below 4000 m are stippled.

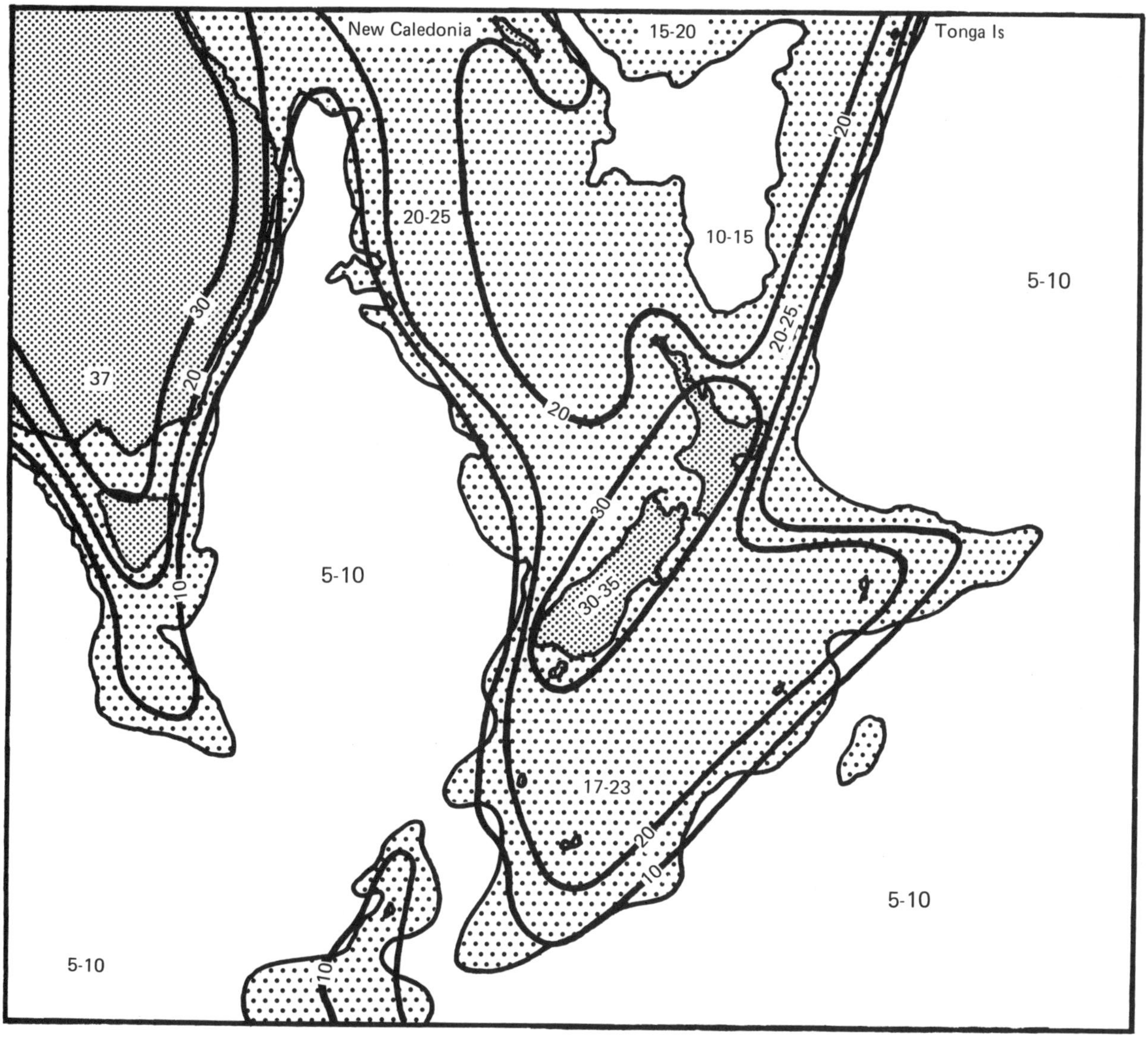


**Fig. 18.3.** The thickness of the earth's crust in the South-west Pacific varies between truly continental (30-37 km; 19-23 miles), underlying Australia and New Zealand, and truly oceanic (5-10 km; 3-6 miles), under the Tasman Basin, in which sea-floor spreading took place 80-60 million years ago. The Campbell Plateau, Chatham Rise, Three Kings Rise, Norfolk Ridge and Lord Howe Rise are the eroded stumps of an ancestral landmass created some 140-110 million years ago, at the close of the New Zealand Geosyncline. These features are underlain by thinned and stretched continental crust 25-17 km (16-11 miles) thick. This map shows the generalised crustal thicknesses (in km, heavy lines) and the generalised 2000 m (7000 ft) submarine contour (depths between 0 and 2000 m are indicated by the large dot pattern).

Ridge—that take up almost three-quarters of its width (Fig. 18.3).

The Tasman Basin expands in width southwards so that in south-western New Zealand its abyssal depths lie very close to the shorelines of south Westland and Fiordland, and here the continental shelf is very narrow. South-west of Fiordland, however, the abyssal oceanic depths are traversed by a narrow-gutted ridge, the Macquarie Ridge, on which is sited Macquarie Island, and three short trenches, the Puysegur, Macquarie and Hjort trenches. The Emerald Basin and Solander Trough separate the Macquarie Ridge from the Campbell Plateau.

The Macquarie Ridge and Puysegur Trench continue the trend of the Alpine Fault, while the

Solander Trough continues into New Zealand as the Waiau Basin, the low-lying area in western Southland flanking Fiordland.

At its southern end the Macquarie Ridge joins the globe-encircling oceanic rise system (here the South-east Indian and East Pacific rises) north of the Balleny Islands (Fig. 8.1).

**Sunken New Zealand**

Deep-sea drilling, magnetic data and studies of off-shore islands all point to the fact that the Campbell Plateau, Chatham Rise, Three Kings Rise, Norfolk Ridge and Lord Howe Rise are submerged extensions of the New Zealand continent, but with very much thinner crust (Fig. 18.3).

As you may recall (pp. 241–245), these submarine features and their continuations towards New Caledonia are all sunken remnants of the once extensive ancestral New Zealand landmass created in late Jurassic and early Cretaceous times, some 140-110 million years ago. This landmass, as we have seen, was created largely out of material that had been eroded off Australia and Antarctica and dumped in a trough in the sea floor—the New Zealand Geosyncline (p. 188). The contents of the geosyncline were eventually compressed, folded and faulted and raised above the sea to form land.

The geosyncline was edged to the west, and partially underlain by pre-existing old continental material, formerly continuous with Australia and Antarctica. As the Tasman Sea opened up, a strip of this continental margin was torn off and left behind in New Zealand, where these rocks are seen today along the West Coast of the South Island and off-shore from the North Island. It is quite likely therefore that the submarine ridges, plateaux and rises extending beyond New Zealand are composed of similar remnants of the continental edges of both Australia and Antarctica, as well as the eroded stumps of the old geosynclinal belt.

As we have seen, New Zealand did not simply drift away from Gondwanaland as a single undistorted unit. Instead it was subjected to twisting movements and extensive fracturing (seen on land in the twisting of the North and South Islands and the great fracture system linked with the Alpine Fault). These movements partially detached, rotated and rifted apart once-continuous pieces of the old landmass. Thus the Chatham Rise, Three Kings Rise, Norfolk Ridge and Lord Howe Rise have been separated in this way, and new sea floor has formed in between them.

The submerged continental remnants—Campbell Plateau, Chatham Rise, Three Kings Rise, Norfolk Ridge and Lord Howe Rise—all have a very much thinner crust than Australia and New Zealand and, as may be seen from Fig. 18.3, their thickness lies between true oceanic and true continental crust. At a stage when these features were above the sea and part of the ancestral New Zealand landmass, they presumably had crust of continental thickness, comparable to that of New Zealand today. But long-continued erosion, culminating in development of the New Zealand peneplain (p. 273), markedly thinned the crust and prepared the way for its later submergence.

Then as continental drift occurred and New Zealand twisted and swung away out into the South-west Pacific, the already thinned crust was stretched out and thinned still further so that its natural buoyancy was reduced to such an extent that it took up a shallow submerged position below sea level. Thus today the submerged continental remnants are stable, geologically "dead" features, as indicated by their low incidence of earthquakes.

On the other hand, the Tonga-Kermadec Ridge and Trench system and the Macquarie Ridge/Puysegur Trench system are very much geologically "alive", as indicated by their seismic activity, and, in the case of the Tonga-Kermadec system, have many active volcanoes.

**Sub-crustal forces at work**

Having set the modern scene in the New Zealand region, we can now look at the processes that are thought to be occurring deep beneath us.

At the moment New Zealand straddles the boundary between the Pacific plate to the east and the Indian-Australian plate to the west.

Reconstruction of plate movements in the South-west Pacific going back into geological time is fraught with uncertainties, as at least three separate major plates have been involved (with the possibility of other minor plates) and the consequent rotations are difficult to work out. A number of solutions are thus possible. The complexities of the South-west Pacific may be

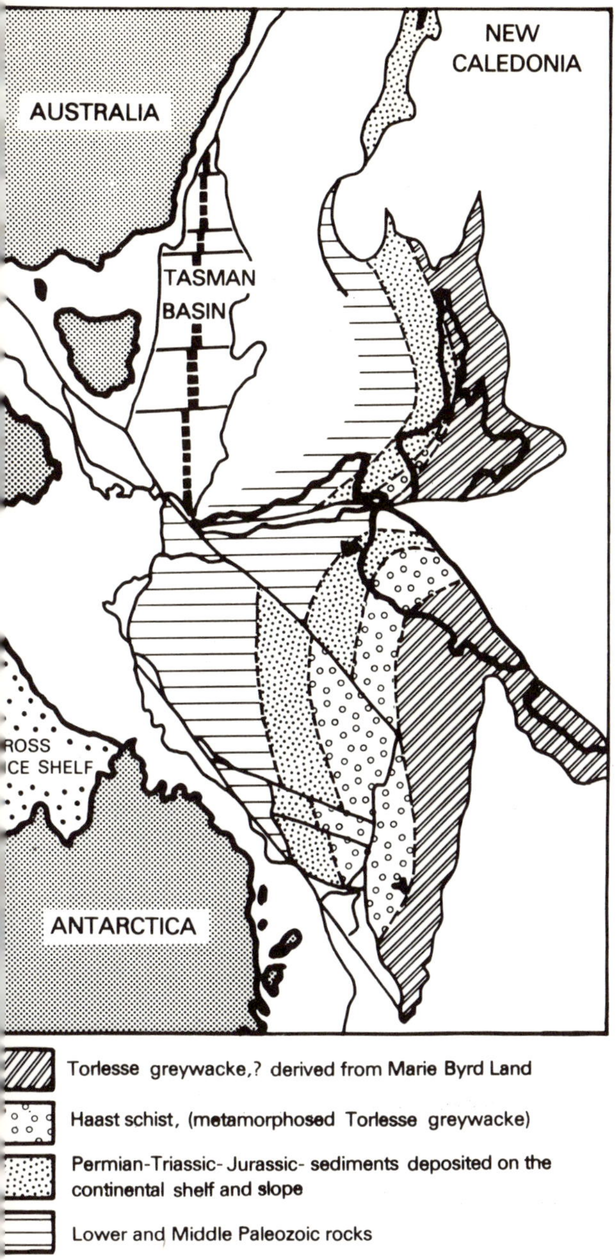


**Fig. 18.4.** Reconstruction of South-east Australia, New Zealand and adjoining regions of Antarctica in the late Cretaceous, as the Tasman Basin was beginning to open.

Opinions differ as to how exactly New Zealand fitted against Australia and Antarctica (cf. Figs 13.14, 15.1-15.5). This reconstruction has reversed movement on the Alpine Fault, closed the New Caledonia Basin and Bounty Trough, reversed sea-floor spreading using oceanic magnetic stripes (Fig. 18.5) and fracture zones, and used computer-assisted bathymetric and magnetic matching techniques.

The continuations of New Zealand rock structure shown in the diagram are greatly generalised and based on meagre off-shore evidence.

contrasted with the simplicity of the situation in the South Atlantic, for example, where there has been an almost straight-line pulling apart of Africa and South America (Figs 8.39-8.46).

As we have seen (pp. 237–243), it is believed that the greywacke (or Torlesse) rocks of New Zealand have been rafted from Marie Byrd Land, Antarctica, by a late Palaeozoic and early Mesozoic episode of sea-floor spreading. The sea floor formed at that time has been steadily consumed by a subduction zone somewhere between Antarctica and New Zealand (Figs 9.21, 16.29A).

The arrival of the Torlesse sediments at the subduction zone and their subsequent collision with and welding to the remainder of New Zealand (during the Rangitata Orogeny) choked the subduction zone and forced it to migrate elsewhere. This phase was no doubt accompanied by a fundamental rearrangement of plates in the South-west Pacific, but at the moment we have no firm ideas on exactly where the plate boundaries were at that time.

Soon afterwards, in the middle Cretaceous, New Zealand started to move independently of Australia for the first time in its long history (except for relatively minor movements, see pp. 184–187) and the Tasman Sea opened up between 80 and 60 million years ago (Fig. 18.4).

At the same time New Zealand started to rotate away from Antarctica and the infant Southern Ocean began to form. These movements are recorded by magnetic striping on the sea floor adjacent to New Zealand (Fig. 18.5). However, there is some uncertainty over how far the present situation of three interacting plates (Antarctica, Pacific, Indian-Australian) can be extended back into geological time. Some geologists maintain that at the time New Zealand was rotating away from Antarctica and Australia a fourth plate came into play—Antarctica being divided into two plates, roughly corresponding to the modern East and West Antarctica. The reconstructions illustrated in Fig. 18.6 make use of this proposition and between 80-40 million years ago some 500 km (310 miles) of movement is postulated along this plate boundary. Other geologists maintain however that the twisting and distortion of New Zealand has taken up all the necessary movement and that a fourth plate is not necessary (Figs 9.20B, 13.14).

Adoption of the concept of an expanding earth,

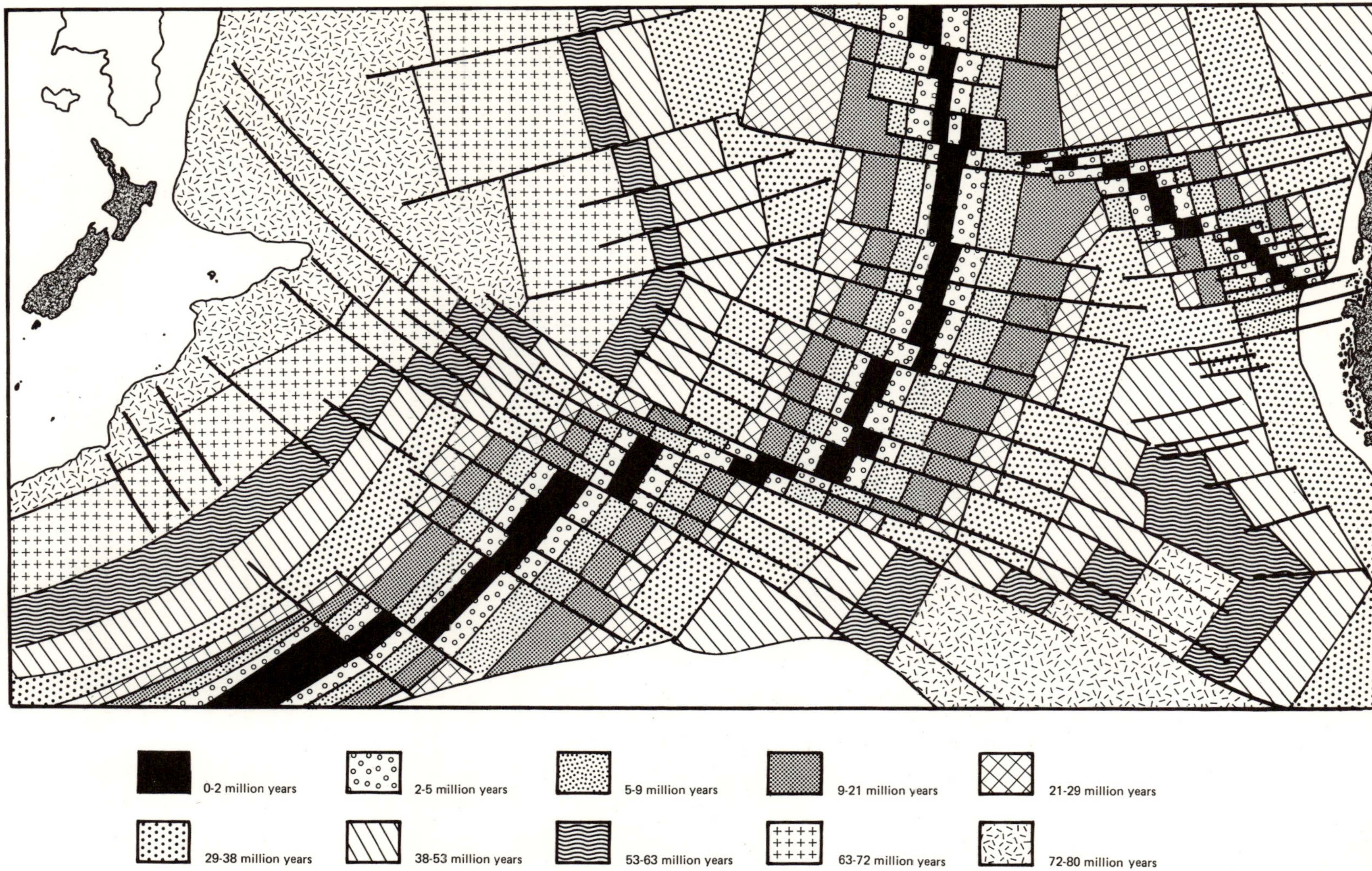


**Fig. 18.5.** The floor of the Southern Pacific Ocean between the continental shelves of New Zealand (left), Antarctica (bottom) and South America (right) is paved with characteristic magnetic signatures, left by episodes of sea floor spreading. This map shows their generalised distribution, in relation to the present-day active spreading ridge (East Pacific Rise), and their assigned ages.

These magnetic patterns may be used as one of the bases for reconstructions (e.g. Fig. 18.4), by reversing the sea-floor spreading and rotating back the landmasses.

as in Figs 15.1-15.5, also avoids the need for a fourth plate.

If we are pressed into giving a definite date for initiation of the present plate system, the best guess is somewhere after 55 million years ago—when sea-floor spreading started between Australia and Antarctica. As Australia drifted steadily northwards at an inexorable 40-50 mm (1½-2 in) per year, the magnetic record preserved on the sea floor tells us that New Zealand and its surrounding oceanic features (Lord Howe Rise, Chatham Rise and Campbell Plateau) also began to move as part of the Indian-Australian plate. Like Australia, we were steadily rafted northwards under the influence of the sea-floor spreading occurring to either side of the Southeast Indian and Pacific-Antarctic oceanic ridges.

During the Tertiary, subduction along the Pacific edge of the Indian-Australian plate has steadily involved New Zealand so that now it is sitting astride the plate boundary (Figs 18.7, 18.8).

The reconstructions depicted in Fig. 18.6 rely on the assumption that the plate boundary was initiated at first along the Macquarie Ridge, between 38-21 million years ago, and then gradually extended northwards into New Zealand, the Alpine Fault becoming the plate boundary about 20-10 million years ago.

Thus, although New Zealand has had more than its share of earth movements in the geological past, it will undoubtedly continue to do so in the future.

North-east of New Zealand the plate boundary is very clear and simple (Fig. 18.9). The Tonga-Kermadec Trench is a classic example of a subduction zone. The trench attains depths in excess of 10,000 m (34,000 ft) and is associated with a large negative gravity anomaly, indicating that light material is present deep under the earth's surface. A zone of earthquakes extends westwards from the trench—at first the individual earthquake shocks come from shallow depths, about 10-12 km (6-7 miles) under the trench, but then, as the zone is traced to the west, the individual shocks get progressively deeper, until those furthest away from the trench originate at some 700 km (450 miles). Thus a westerly-dipping earthquake zone, or Benioff Zone, can be distinguished.

The Tonga-Kermadec Trench is bordered to the west by a well-defined island arc, having the underwater form of a fairly continuous oceanic ridge, but periodically breaking the surface of the sea as islands. The islands dotting the Tonga-Kermadec island arc are all volcanic, composed of andesitic (i.e. "intermediate") rocks.

The Tonga-Kermadec Trench and Arc system is interpreted in terms of modern theory as being a typical zone of subduction. The edge of the Indian-Australian plate is being pushed over the oceanic Pacific plate. The Pacific plate, as it descends into the dense mantle, generates earthquake shocks as it does so—hence the origin of the down-dipping Benioff Zone. The contrast in density between the descending Pacific plate and the dense mantle rocks produces the marked negative gravity anomaly.

As material is swallowed up by subduction it is dragged down into hotter and hotter regions and begins to melt. The resulting molten rock (magma) is a mixture of both oceanic and continental material, derived from the original oceanic slab and the sedimentary material (largely of continental origin) lying on it. Thus the molten rock fed to the active volcanoes is a mixture and is classed as an intermediate rock or "andesite", produced by the mixing and melting occurring under subduction zones.

The Kermadec Ridge, with its active volcanoes, is bordered to the west by the Havre Trough and another ridge, the Colville Ridge.

The Rotorua-Taupo Volcanic Zone, the focus of modern volcanic activity in New Zealand, lies on the extension of the Havre Trough and Kermadec Ridge into New Zealand. On the other hand, the Coromandel Peninsula, the scene of volcanic activity in Pliocene times, 5-2 million years ago, lies on the extension of the Colville Ridge, that presumably was also active at that time, like the Kermadec Ridge today. Thus there has been an eastwards migration of volcanism since the Pliocene.

Structures that are in some respects similar to those of the Tonga-Kermadec system extend in a south-west direction from the tip of Fiordland (Fig. 18.10). An oceanic ridge, the Macquarie Ridge, is present, but unlike the Tonga-Kermadec Ridge active volcanoes are absent, although Solander Island, lying off the Fiordland coast between Puysegur Point and Stewart Island, is the eroded stump of an andesite volcano that was active in the Pliocene and early Pleistocene, some 5-1½ million years ago.

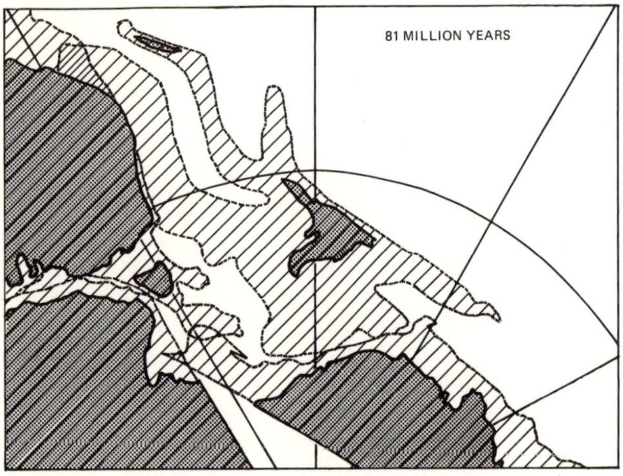


*Fig. 18.6A*

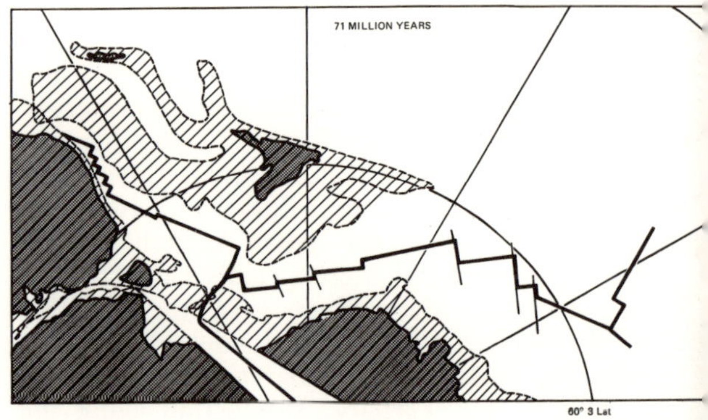


*Fig. 18.6B*

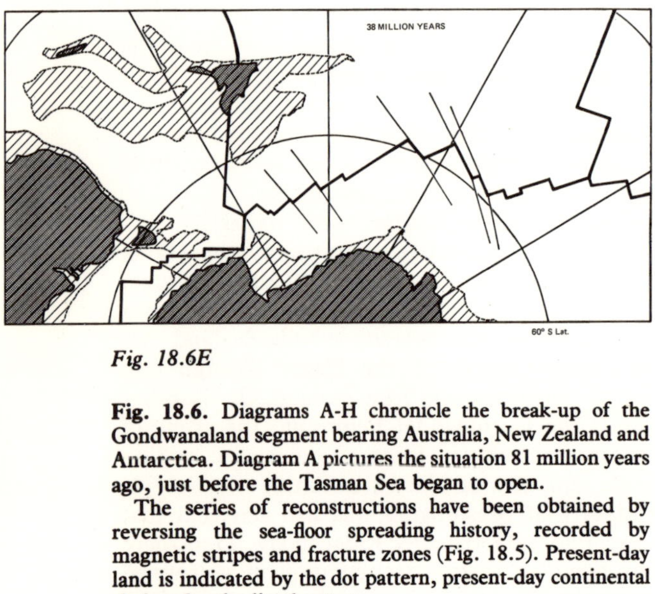


*Fig. 18.6E*

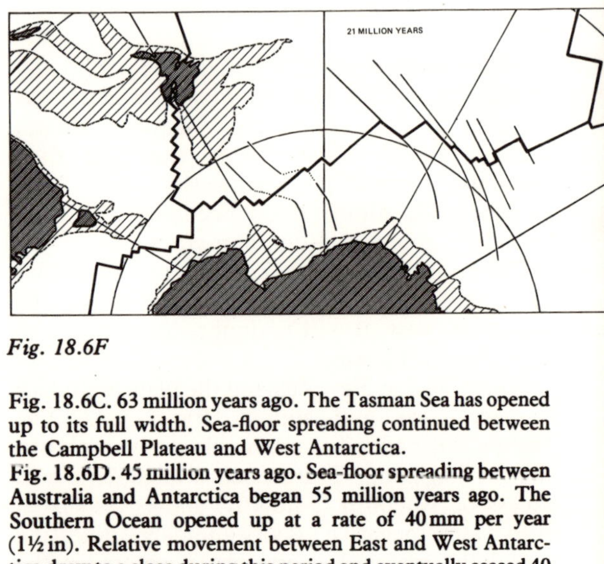


*Fig. 18.6F*

**Fig. 18.6.** Diagrams A-H chronicle the break-up of the Gondwanaland segment bearing Australia, New Zealand and Antarctica. Diagram A pictures the situation 81 million years ago, just before the Tasman Sea began to open.

The series of reconstructions have been obtained by reversing the sea-floor spreading history, recorded by magnetic stripes and fracture zones (Fig. 18.5). Present-day land is indicated by the dot pattern, present-day continental shelves by the lined pattern.

Fig. 18.6A. 81 million years ago, before the opening of the Tasman Sea and Southern Ocean, a closer fit is possible by closing up oceanic basins and by rotating various segments (cf. Figs 9.20B, 13.14). A plate boundary is placed between East and West Antarctica.

Fig. 18.6B. 71 million years ago. Sea-floor spreading between Australia and the Lord Howe Rise began 80 million years ago, the rate of spreading being 60-90 mm per year (2½-3½ in). Spreading between the Campbell Plateau and West Antarctica also began 80 million years ago, at a rate of 60 mm per year (2½ in). Relative motion of about 500 km (310 miles) occurred between East and West Antarctica 80-40 million years ago (although some geologists dispute this).

Fig. 18.6C. 63 million years ago. The Tasman Sea has opened up to its full width. Sea-floor spreading continued between the Campbell Plateau and West Antarctica.

Fig. 18.6D. 45 million years ago. Sea-floor spreading between Australia and Antarctica began 55 million years ago. The Southern Ocean opened up at a rate of 40 mm per year (1½ in). Relative movement between East and West Antarctica drew to a close during this period and eventually ceased 40 million years ago.

Macquarie Island itself is composed of basalt and serpentine and is thought to be an uplifted part of oceanic crust.

Again the Macquarie system, unlike that of Tonga-Kermadec, has no continuous trench. The Puysegur Trench, which reaches depths of some 6000 m (20,000 ft), extends for about 400 km (250 miles) south-westwards from the edge of the continental shelf off the tip of S.W. Fiordland. Close to Macquarie Island is the

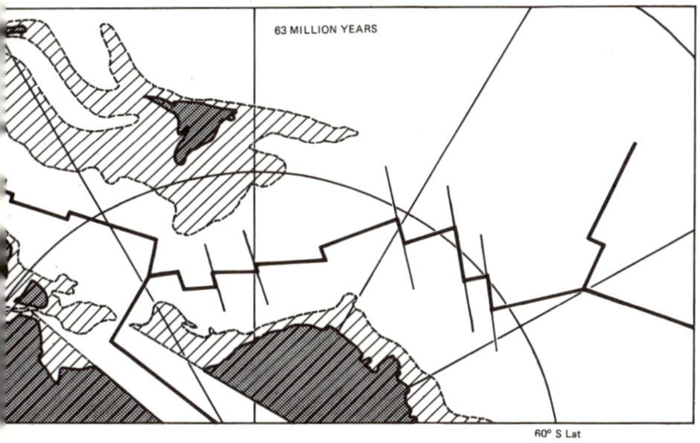


*Fig. 18.6C*

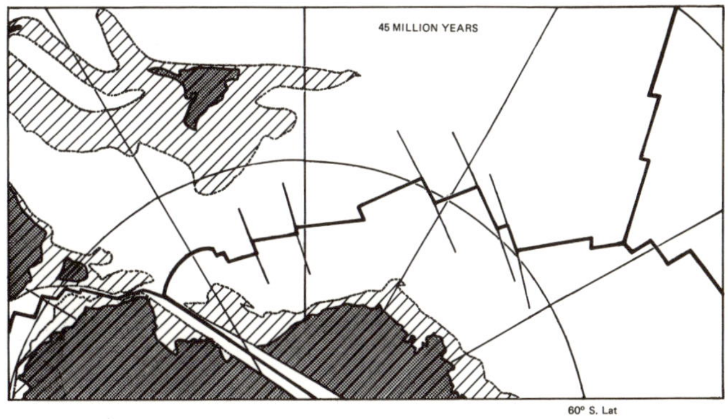


*Fig. 18.6D*

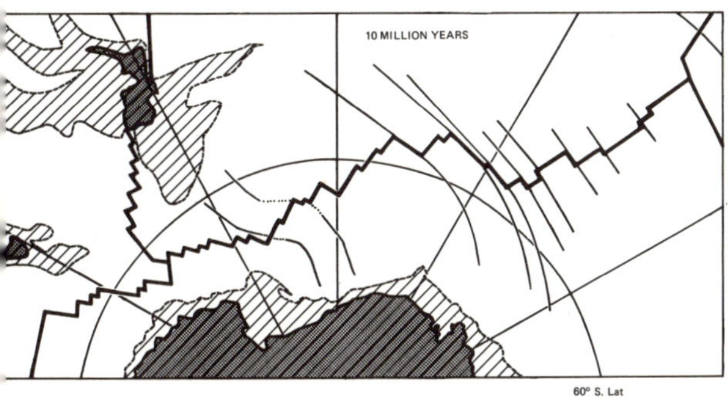


*Fig. 18.6G*

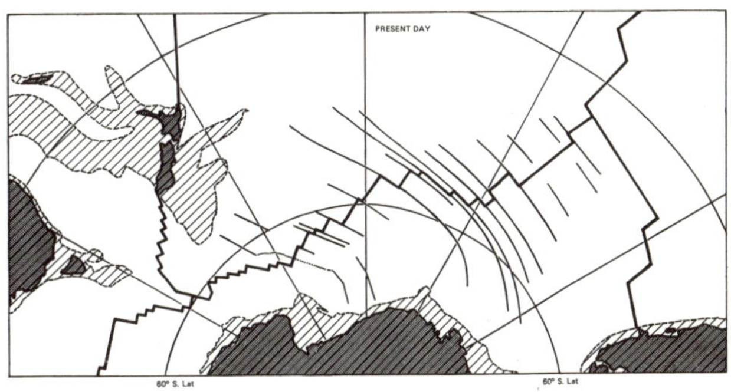


*Fig. 18.6H*

Fig. 18.6E. 38 million years ago. Sea-floor spreading continued at a rate of 40 mm per year (1½ in) between Australia and Antarctica.

Australia was still joined to Antarctica via the South Tasman Rise.

The plate boundary between East and West Antarctica became extinct 40 million years ago and a new plate boundary was at this time being actively propagated through New Zealand—generation of the Macquarie system being the first result.

Fig. 18.6F. 21 million years ago. Australia has now separated from Antarctica. Horizontal movement began on the Alpine Fault, as the plate boundary extended into New Zealand.

Fig. 18.6G. 10 million years ago. Horizontal movement continued along the Alpine Fault, and New Zealand began to approach its present configuration.

Fig. 18.6H. Present day. The plate boundary between the Indian-Australian and Pacific plates traverses New Zealand and continues southwards to form a triple junction with the Antarctic plate. Converging movements across the New Zealand sector of the plate boundary vary between 100 mm per year (4 in) in the north-east and 60 mm per year (2½ in) in the south-west.

450 km-long (280 miles) Macquarie Trench and a trench of similar length (Hjort Trench) is present at the end of the ridge near the Balleny Islands. These, however, are the only trenches developed along the 1750 km (1100 miles) length of the Macquarie Ridge. Furthermore, the trench-ridge arrangement faces the opposite way to that of Tonga-Kermadec.

Earthquakes are concentrated along the Macquarie Ridge, although not to the same extent as on the Tonga-Kermadec Ridge, and they do not seem to indicate the presence of a well-developed Benioff Zone, although one is developed under Fiordland, where the Macquarie system runs into New Zealand.

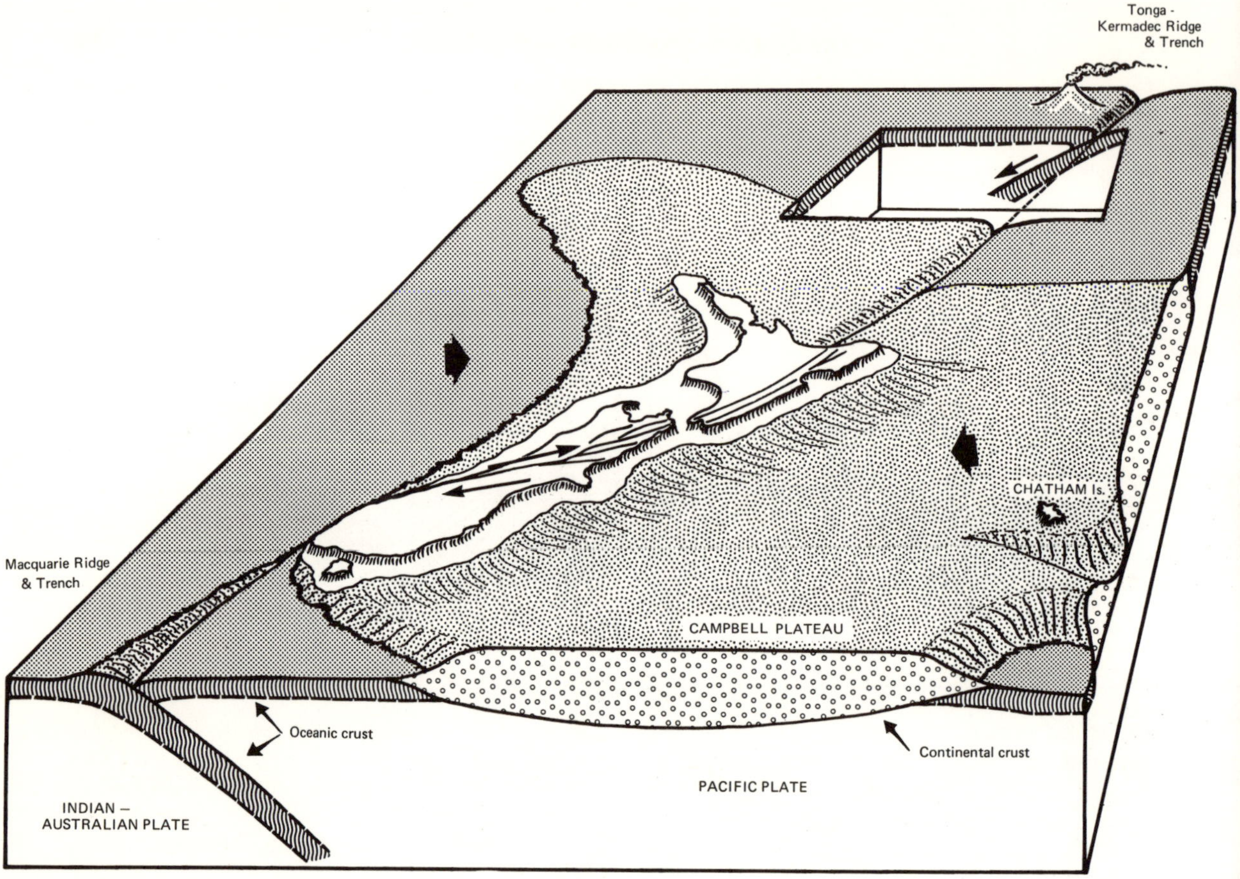


**Fig. 18.7.** The shape and structure of modern New Zealand in three dimensions.

The Tonga-Kermadec Trench and Ridge system is developed along a subduction zone, with the Pacific plate being pushed *under* the Indian-Australian plate. In the Macquarie system, on the other hand, the relationships are reversed and instead the Pacific plate is being pushed *over* the Indian-Australian plate (Fig. 18.7).

Judging, however, from the absence of modern volcanic activity, of a well-defined Benioff Zone, and of a continuous trench, the Macquarie system does not seem to be very active at present. This leads some geologists to conclude that the plate boundary now traversing New Zealand was first initiated along the line of the Macquarie Complex, as a splinter coming off the South-east Indian and East Pacific rises, and then gradually extended onto New Zealand. The reconstructions depicted in Fig. 18.6, for example, adopt this viewpoint and date the initiation of the Macquarie Complex as early Miocene (38-21 million years ago). The Macquarie Complex today is thus seen as being in a waning situation, the focus of activity having shifted northwards into New Zealand. An alternative viewpoint, however, is that we have not been recording data for long enough to enable us to say whether the modern situation is in fact a true representation of the entire story—it might be merely a passing phase.

It is nonetheless clear that two subduction systems—Tonga-Kermadec and Macquarie—run into the New Zealand continental mass from north and south. It is also clear that as New Zealand straddles the boundary between the Pacific and Indian-Australian plates it is being affected by the twisting and shearing movements of both plates. The real situation however, has been complicated by the presence of thickened rigid continental crust.

In most areas of New Zealand the crust has been involved in past mountain-building activities (pp. 241, 314), that have piled up thick masses of compressed rigid sediment. The crust underlying New Zealand is some 30-35 km (19-22 miles) thick (Fig. 18.3). It is therefore thought that the forces acting along a plate boundary, such as the one traversing New Zealand, which in an oceanic situation would produce a clear structure such as the Tonga-Kermadec system, would act differently in the presence of rigid thick continental crust. Instead of producing a volcanic arc, trench etc., the interacting forces cause thrusting and slicing of the land; and this is what we are seeing in the New Zealand landscape.

The Tonga-Kermadec structures can be traced in a general way into the North Island of New Zealand as far south as northern Wellington Province (Fig. 18.11). The Colville Ridge, which was probably the Pliocene version of the presently active Kermadec Ridge, can be traced into Coromandel Peninsula and southwards into the Waikato, both scenes of extensive Pliocene volcanic activity.

The Havre Trough can be traced into New Zealand as the Rotorua-Taupo depression, in which the ancient "basement" rocks have subsided along faults, and the resulting trough filled with volcanic material (Fig. 18.16). The Rotorua-Taupo depression continues off-shore towards the Havre Trough as the White Island Trench.

The volcanic activity along the Kermadec Ridge continues into New Zealand as the andesite volcanoes of Mts Egmont, Ruapehu, Tongariro and Ngauruhoe, as well as other cones in the Rotorua-Taupo region (Figs 18.17-18.21). Until a few years ago it was thought that a large gap existed between White Island, the northernmost New Zealand volcanic activity, and the southernmost Kermadec activity—a gap of some 740 km (460 miles). However, naval research on underwater sounds has revealed the presence of four submarine peaks—named Rumble I, II, III and IV—generating volcano-like noises. Two other nearby peaks were found not to be emitting noises, but yet had the same shape and magnetism and were interpreted as extinct volcanic cones (called Silent I, II). Thus while the gap between White Island and the Kermadec activity has been narrowed, there is still a gap of 350 km (220 miles) without any apparent activity (Fig. 18.11).

A gap also seems to exist in the negative gravity anomaly belt—the New Zealand belt diverging and then ending at East Cape (Fig. 18.17). At first sight the Tonga-Kermadec Trench seems to continue into the New Zealand region as the Hikurangi Trench. This, however, lacks the negative anomalies that characterise the Tonga-Kermadec Trench, and has been shown by oil prospecting work to be full of young sediments (in fact some geologists call it the Hikurangi "Trough", rather than "Trench"). Thus the Hikurangi Trough (or Trench) seems to be a completely different animal from its near neighbour, the Tonga-Kermadec Trench.

Similarly, the Benioff Zone so well developed in the Tonga-Kermadec region can only be traced in a general way into New Zealand. A gap comparable to the volcanic gap occurs between the Benioff Zones of both regions, and the New Zealand one has a different character.

In the North Island a Benioff Zone can be traced extending westwards from the Hikurangi Trench—shallow earthquakes occurring under East Cape, Gisborne and Hawkes Bay—and progressively deeper shocks being encountered in a westwards direction, until shocks originating at depths of 600 km (373 miles) are recorded from beneath northern Taranaki (Figs 18.12-18.16). A continuous Benioff Zone is not developed, however, and there is a break between about 350 km (218 miles) and 600 km (373 miles) below the surface.

The earthquake shocks clustering around a depth of 600 km are thought to be coming from a piece of lithosphere that has become detached from the subducted edge of the Pacific plate, and has moved, by itself, deeper into the mantle (cf. Fig. 18.15). Detachment of this piece of lithosphere has probably been brought about by some major geological disturbance in the New Zealand region.

Studies of magnetic anomalies preserved on the sea floor adjacent to New Zealand have enabled geologists to chronicle the sea-floor-spreading rates affecting the crustal plates in the South-west Pacific (Fig. 18.5). Calculations based on these rates have enabled a date of 3.8 million years ago to be placed on the time of detachment, which coincides with views of vulcanologists on the time of the switch in volcanic activity from the Northland-Coromandel volcanic arc (active in the Miocene and Pliocene), to the line of modern volcanic activity extending through the Taupo-Rotorua region. The switch in volcanic activity is probably related

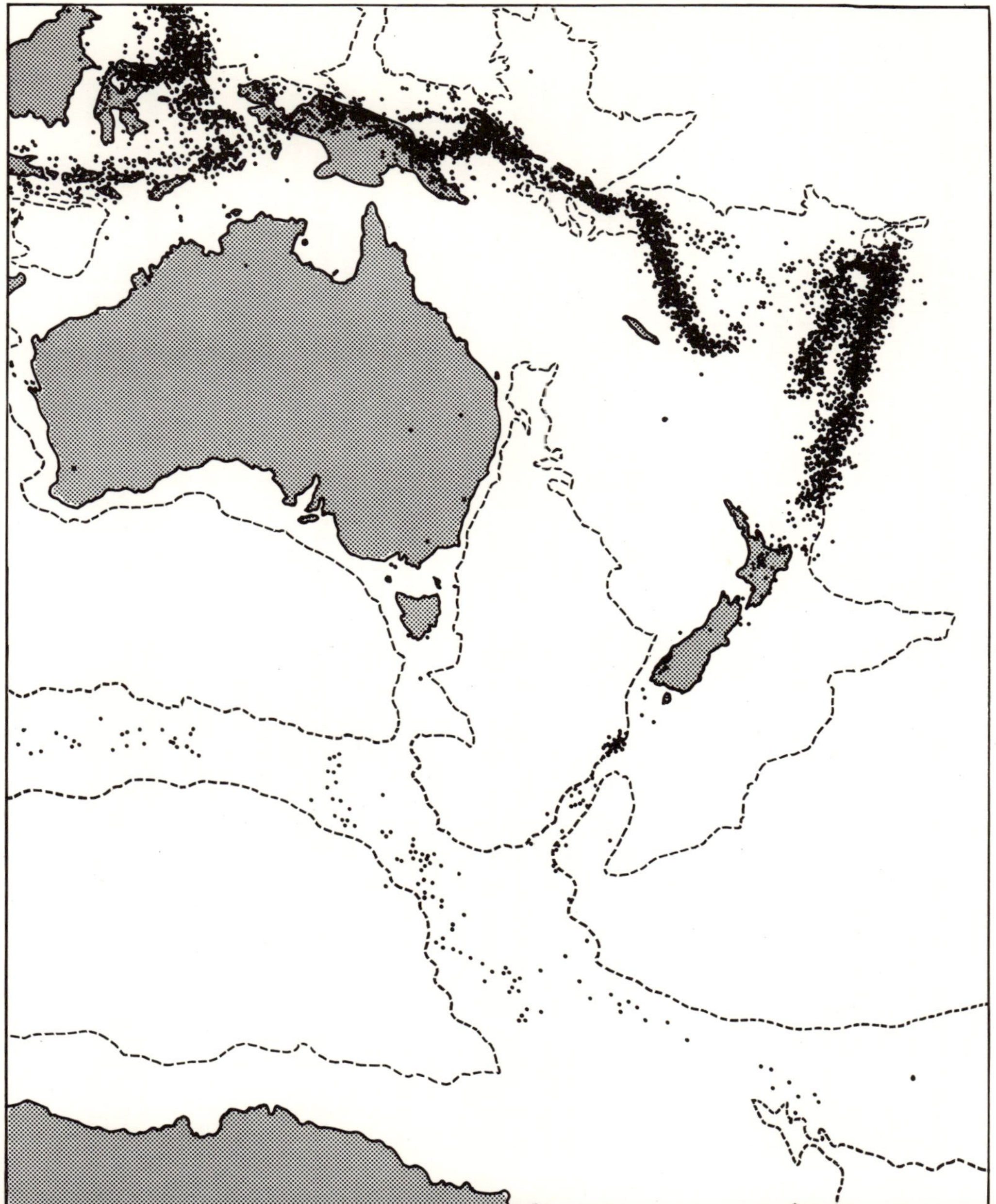

**Fig. 18.8.** The active regions of the earth are clearly indicated by earthquakes, generated as plates draw apart and new crustal material is created, or as plates collide and push over or scrape past one another.

This map shows the earthquakes for 1960-1970 recorded in the South-west Pacific. The generalised 3600 m submarine contour (12,000 ft; 2000 fathoms) is shown as the broken line.

The zone of overthrusting between the Pacific and Indian-Australian plates is shown clearly by a zone of intense earthquake activity extending from Indonesia and New Guinea to the Tonga-Kermadec system.

In New Zealand there is an apparent decrease in earthquake activity, as the underthrusting structure typical of the Tonga-Kermadec system gives way to the Alpine Fault system, acting as a transform fault (Fig. 9.24), enabling the plates to slip sideways.

The Macquarie system, south-west of New Zealand, is regarded by many geologists as a waning island arc system (its acme of development was 38–10 million years ago; see Fig. 18.6E)—hence the lower level of earthquake activity, compared with that of the Tonga-Kermadec system.

Between the Macquarie system and Antarctica (at bottom of map), earthquake shocks are generated by rifting along the mid-oceanic ridge (South-east Indian Rise and Pacific-Antarctic Ridge), marking the divergent boundary between the Antarctic plate and the Pacific and Indian-Australian plates.

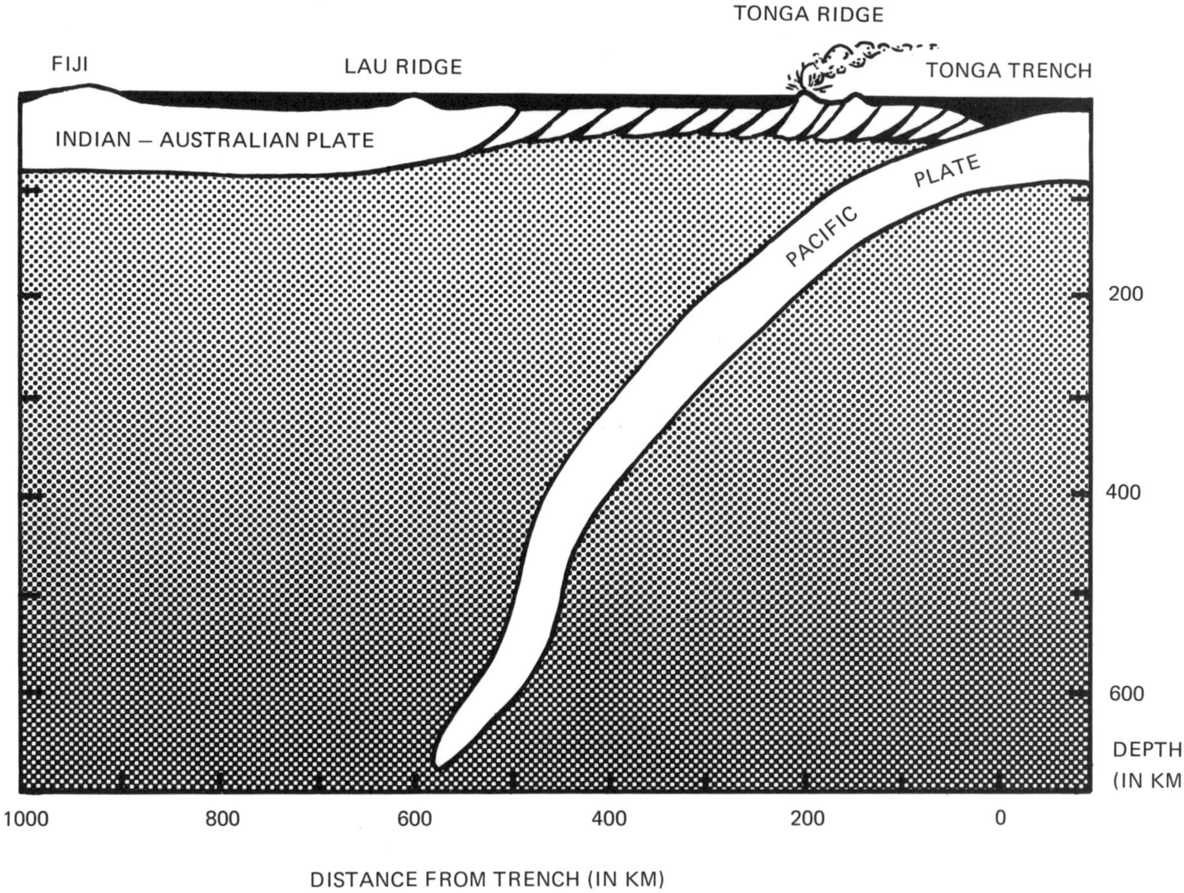


**Fig. 18.9.** The Tonga-Kermadec ridge and trench structure is a "classic" example of an active plate margin, with a clearly developed island arc, submarine trench and Benioff Zone.

This east-west cross-section, extending from Tonga to Fiji, is based on seismic studies and shows the subducted plate plunging at an ever-steepening angle as it encounters progressively denser mantle material. Essentially the same picture applies to the North Island of New Zealand, although complicated by the presence of thick continental crust (cf. Fig. 18.14).

The region between the Lau Ridge and the Tonga Ridge is believed to consist of broken segments of crust, forming the leading edge of the Indian-Australian plate.

to a re-orientation of plate movements in the South-west Pacific, perhaps involving the passing of dominance from the Macquarie system to that of Tonga-Kermadec. Some geologists also place the twisting of New Zealand at this time (see pp. 370–376).

Propagation of the plate boundary through the Taupo-Rotorua region led to massive rifting and vast eruptions of lava and ash extending over the last two million years. The rifting movements have broken a continental fragment (the East Cape block—sometimes called the "Hawke's Bay Microplate") from the nose of the Indian-Australian plate, and thrust it over the position of the original Hikurangi Trench.

In this way, we have arrived at the present situation—with a non-active sediment-filled trench separated by some 300 km (187 miles) from the related volcanic arc and negative anomaly belt that otherwise are closely adjacent, for example, in the Tonga-Kermadec situation (Fig. 18.9).

The Benioff Zone extends westwards from the landward wall of the Hikurangi Trench, running at a shallow angle under the East Cape block and impinging on the underside of the crust for many kilometres, before descending into the mantle under the Taupo-Rotorua rift zone. This situation has all the makings for creation of a broad zone of major earthquake activity—as underthrust and overthrust slabs interact with one another.

Whatever the true situation, it is clear that the simple convergent plate boundary situation, as exemplified by the Tonga-Kermadec system, has been greatly complicated in New Zealand by the presence of thick continental crust.

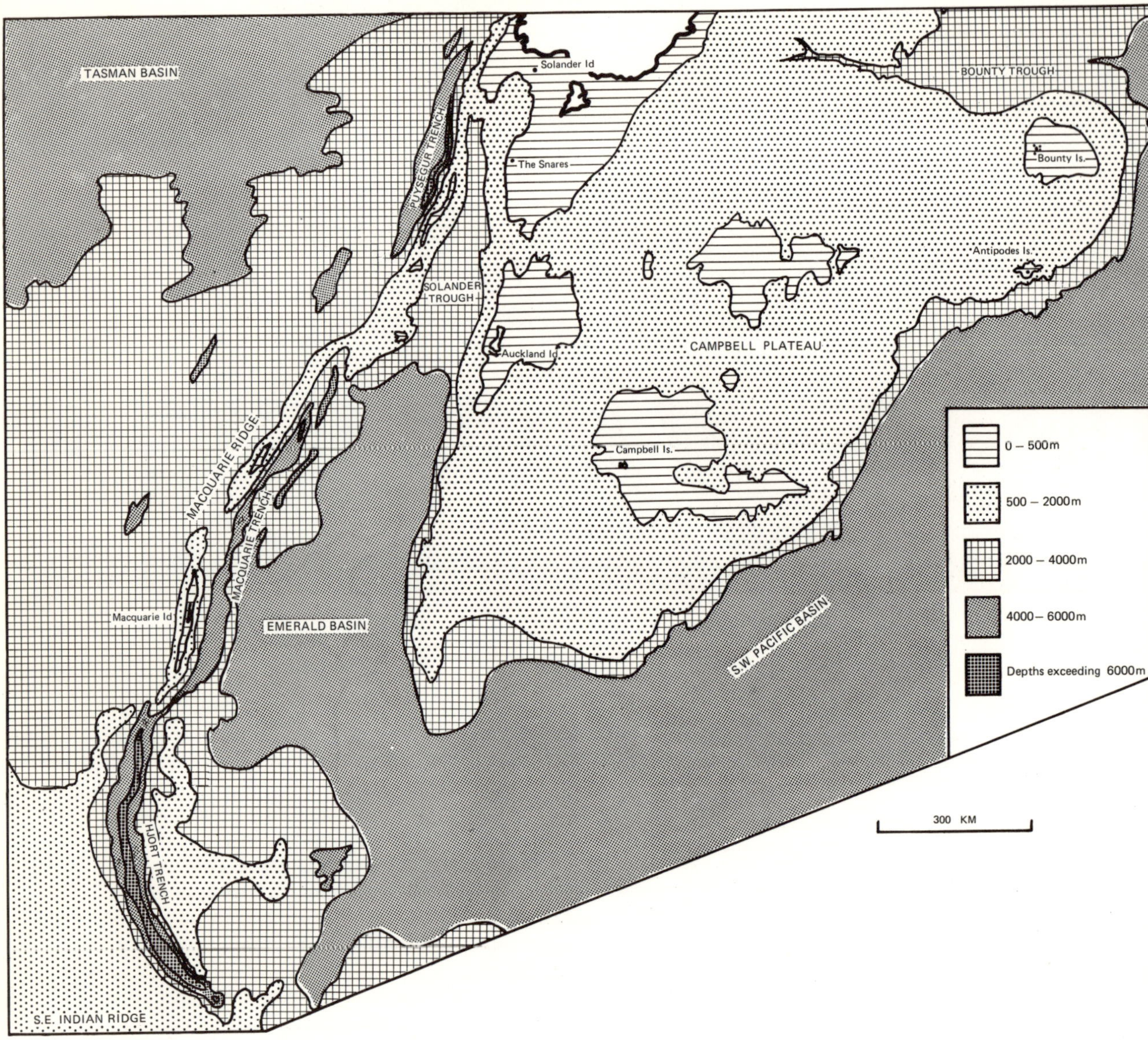


**Fig. 18.10.** South of New Zealand lies a huge submerged landmass—the Campbell Plateau. It is submerged because it is an area of thinner continental crust (17-23 km; 11-14 miles), compared with New Zealand's 30-35 km (19-22 miles), and therefore does not ride as high as New Zealand.

This thinning of the crust is probably the aftermath of long-continued erosion in the Tertiary, and stretching of the crust during continental drifting phases.

Nonetheless the geology of the sub-Antarctic islands and that of the adjacent sea bed (judged from drilling, dredging, and magnetism) show clearly that the geology of the South Island continues across the Campbell Plateau to its very edge (Fig. 18.4).

To the east the Campbell Plateau falls away into the abyssal depths of the South Pacific Ocean. To the west it is separated from the Macquarie Ridge by the Emerald Basin and Solander Trough.

The Macquarie ridge and trench system links Fiordland with the South-east Indian Rise, the active oceanic ridge developed along the boundary between the Indian-Australian and Antarctic plates.

West of Fiordland and the Puysegur Trench oceanic depths decline rapidly into those of the abyssal Tasman Basin.

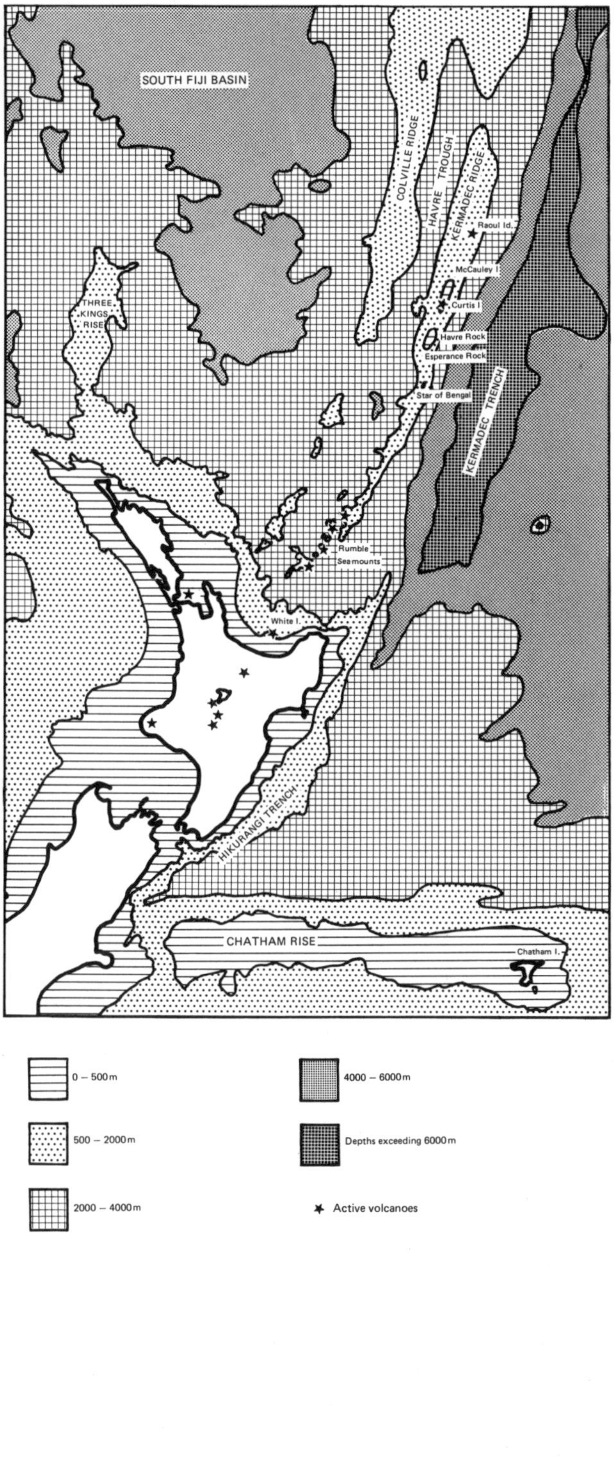


**Fig. 18.11.** Post-glacial volcanic activity in New Zealand and out-lying islands has been confined mainly to centres in the Taupo-Rotorua volcanic zone and along the Tonga-Kermadec ridge, that together constitute the modern active plate margin.

Isolated eruptive episodes, related to the deep crustal processes of the plate margin, have occurred in Taranaki and Auckland provinces.

Raoul Island, the largest and most northerly member of the Kermadec Group, is the main eruptive centre of the Kermadec Ridge. It lies immediately to the west of the Kermadec Trench and is situated some 100 km (62 miles) above the westward-dipping Benioff seismic zone.

In the rest of the Kermadec Group volcanic activity is restricted to thermal areas on Curtis Island.

The volcanic trend is thought, however, to continue south-west towards White Island. Havre Rock, Esperance Rock, and Star of Bengal Shoal are of volcanic origin, although not active, and between them and White Island are a number of seamounts, many of them conical, rising above the sea floor. In one particular region, about 230 km (143 miles) north-east of White Island, six large seamounts have been identified, rising up to some 2400 m (8000 ft) from an otherwise flat featureless sea floor lying at a depth of 2750 m (9000 ft).

Acoustic studies have pin-pointed underwater noises, probably related to volcanic explosions and emissions of gases, coming from four of the seamounts (named Rumble I, II, III, IV). It is therefore believed that, while there is essential continuity between the Taupo-Rotorua and Tonga-Kermadec volcanism, there is nonetheless a narrow gap (350 km; 220 miles) between the two volcanic systems. The same feature is shown by the seismicity (cf. Fig. 18.8).

The New Zealand continent, composed of light sialic rocks, now straddles the plate boundary and, because of its inherent lightness, is interfering with the normal subduction processes, as the Pacific plate is inexorably thrust under the Indian-Australian plate. There has thus been massive splintering and rifting of the part of the New Zealand landmass astride the plate boundary, culminating in the breaking-off of a fragment (the East Cape block) that has been thrust over the Hikurangi Trench, choking it and putting an end to its activity.

What of the future? Judging from theory and from what is known of the history of extinct subduction zones (e.g. Fig. 9.16), the arrival of the New Zealand continental crust at the edge of the subduction zone may eventually see an end to active subduction, in the New Zealand land area at least. Some geologists believe therefore that, like the Macquarie system to the south-west, the New Zealand system may be in a waning phase and that at the present time we are in a

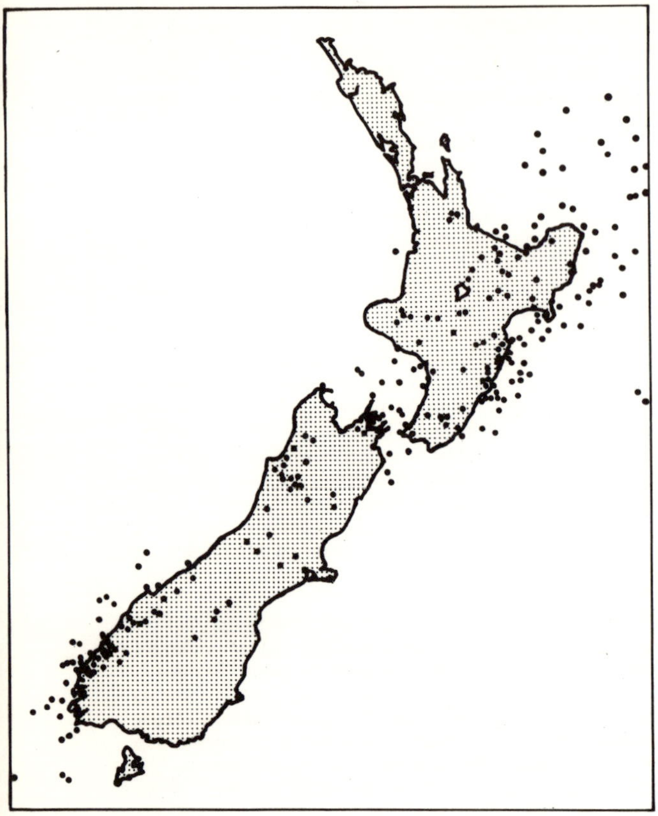

Fig. 18.12. A typical year, 1971-72, showing the distribution of shallow earthquake shocks (i.e. those originating at depths down to 40 km; 25 miles) in the New Zealand region. Note the general north-east/south-west trend, corresponding to the active plate margin running through New Zealand. Most of these shallow shocks are probably related to volcanic activity (sub-surface migration of magma, and related swelling and contracting movements) and surface faulting, related to northern continuations of the Alpine Fault (cf. Fig. 18.16).

Fig. 18.13. A typical year, 1971-72, showing the distribution of earthquake shocks originating from depths below 40 km (25 miles).

The northern group of earthquakes are related to the down-dipping Benioff Zone under the North Island, a continuation of the Tonga-Kermadec structure (cf. Figs 18.9, 18.11).

The depth of origin of the earthquakes gets progressively greater in the north-west direction and shocks deeper than 100 km (63 miles) all lie to the north-west side of a diagonal line extending from NW Nelson through Lake Taupo. The deepest recorded shocks in New Zealand are from depths of 600 km (373 miles) under northern Taranaki (Fig. 18.14).

The deep earthquakes under Fiordland are related to the Benioff Zone of the Macquarie system (Fig. 18.10).

transitional phase—with many of the volcanoes in the volcanic arc dormant or extinct, and a lessening of earthquake activity.

The next stage, in theory, is for the plate boundary to side-step New Zealand (e.g. Fig. 9.16) and for a new subduction zone to be established in the oceanic floor to the east of New Zealand. If this is correct, we are now in a changeover period between the dying out of the old and creation of the new subduction zone. This, however, is very much in the category of crystal-ball gazing, and the true situation will only become apparent after many many more years of research.

In a way the same thing may be happening along the Macquarie Ridge. The present ridge-trench system may be in the process of dying out and in time the plate boundary may develop elsewhere in the Tasman Basin.

In the South Island the fracturing and thrusting of the crust, that in the North Island is distributed across a broad belt, is mainly concentrated along the Alpine Fault, which has

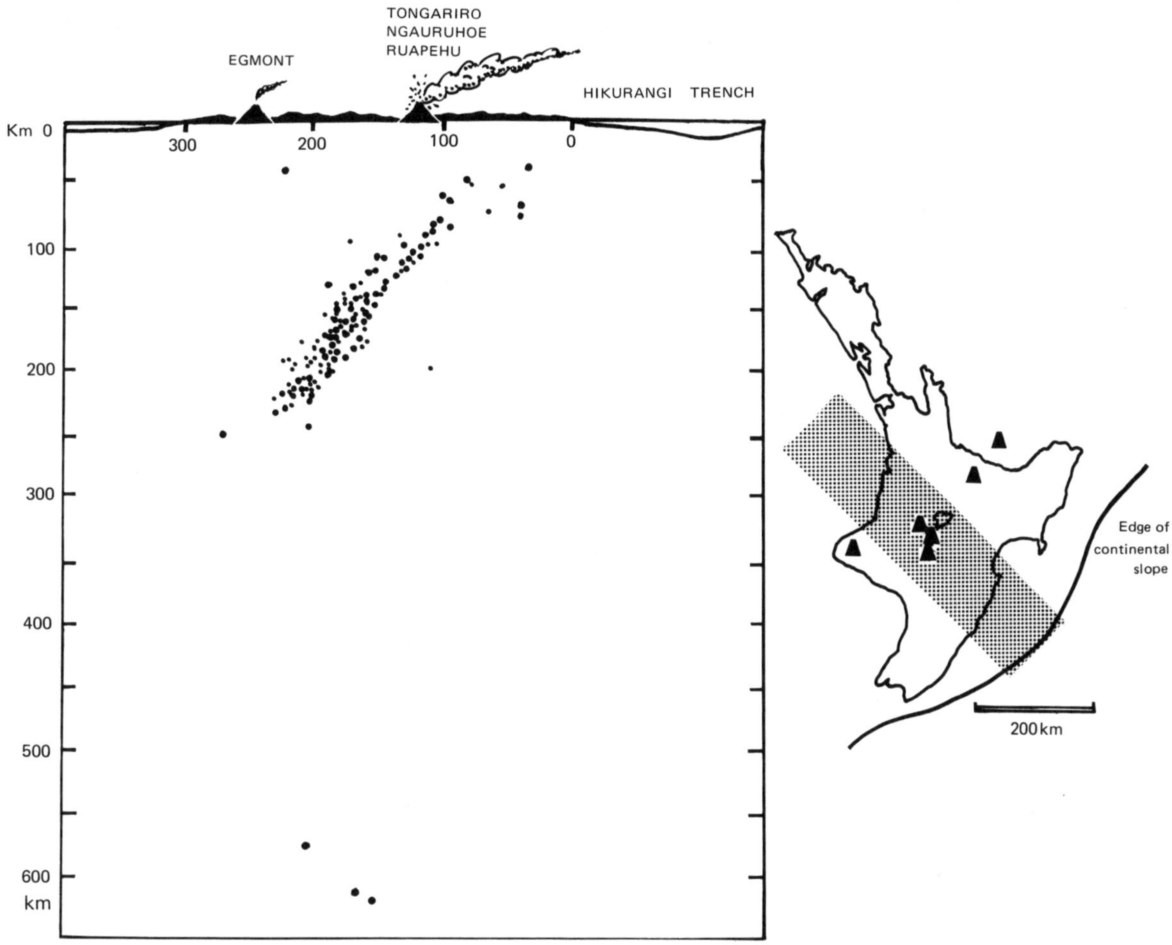


**Fig. 18.14.** A crustal structure, similar to that of the Tonga-Kermadec system (Fig. 18.9), continues into the North Island of New Zealand, but is complicated by the presence of thick continental crust and the northern termination of the Alpine Fault system.

The seismic cross-section depicted in this diagram is based on the earthquake shocks recorded from the shaded region in the key map. Only those shocks that are precisely located have been used.

The horizontal scale for the cross-section (i.e. 0, 100, 200, 300 km on upper line) has been measured from a reference line drawn from the edge of the continental slope. The surface topography has been slightly exaggerated for clarity.

In this and other seismic sections through the North Island a zone of diffuse earthquake activity exists down to a depth of about 100 km (63 miles). This is believed to be related to volcanic activity (movement of magma in the crust and upper mantle and related expansion and contraction adjustments) and faulting within the volcanic belt and the continuation of the Alpine Fault zone ("New Zealand Shear Belt") (Fig. 18.17).

Below 100 km (63 miles), however, the pattern of earthquake sources tends to lie on a well-defined line—producing an essentially planar Benioff Zone, dipping 50° to the north-west. Thus we are viewing at first hand the precise contact between the Indian-Australian plate and the down-dipping Pacific plate (cf. Fig. 18.9, 18.15).

Under the North Island of New Zealand, however, a series of deep shocks have been pin-pointed under Taranaki coming from depths of 600 km (373 miles).

As may be seen from this diagram, these shocks form a distinct group, separated by some 350 km (218 miles) from the other shocks of the Benioff Zone. It has been postulated that these deep shocks originate from a piece of lithosphere that has become detached and has moved by itself deeper into the mantle (cf. Fig. 18.15).

Detachment may have been precipitated by some type of disturbance in the New Zealand region. Calculations from sea-floor-spreading rates place the disturbance at about 3.8 million years ago, coinciding with the switch from the Miocene-Pliocene Northland-Coromandel volcanic arc to that of Taupo-Rotorua. This switch may have been a consequence of a change in the direction of rotation of New Zealand and the waning of the Macquarie plate boundary (Fig. 18.6).

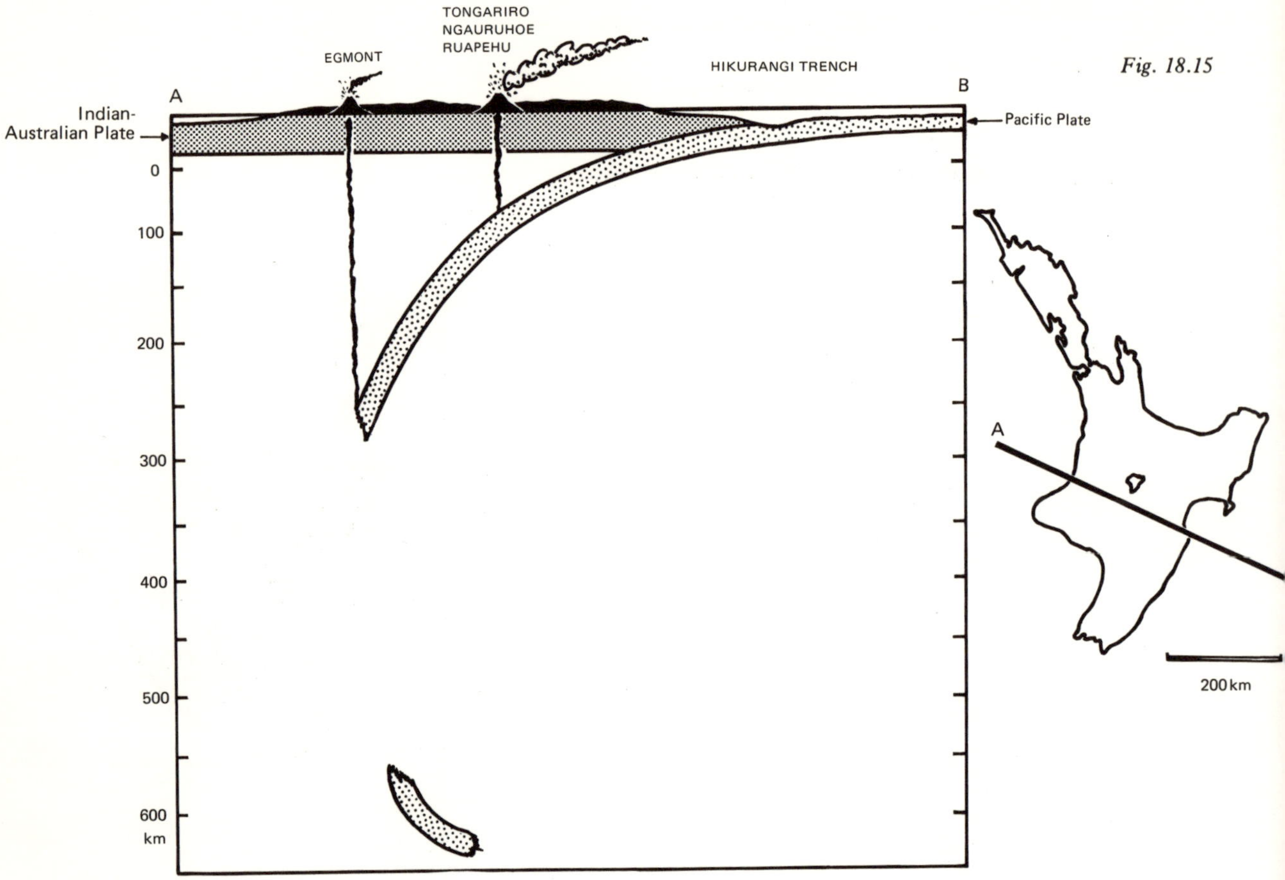


*Fig. 18.15*

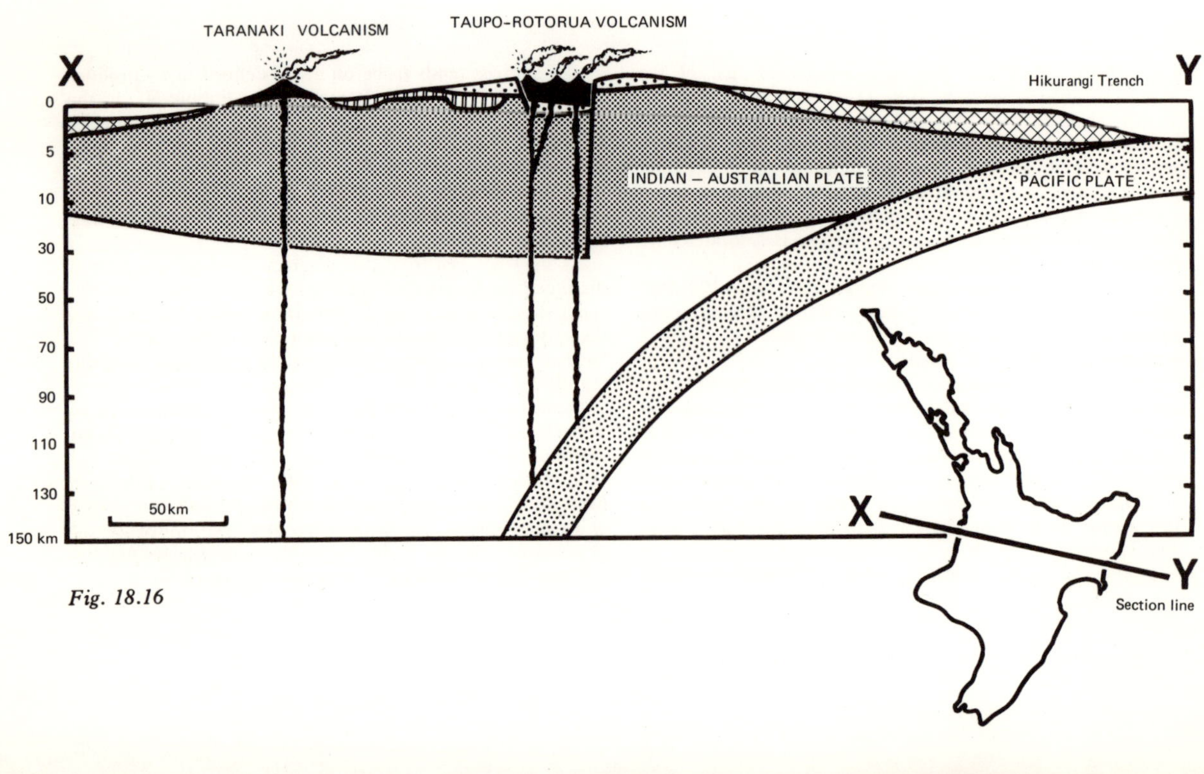


*Fig. 18.16*

◁ **Fig. 18.15.** A sketch of the likely structure beneath the North Island of New Zealand, along the line A-B, based on an interpretation of the seismic section in Fig. 18.14.

The isolated piece of the Pacific plate at depths of 600 km may have become detached some 3.8 million years ago when a fundamental rearrangement occurred in the New Zealand region. At this time the line of volcanic activity in the North Island shifted from Northland-Coromandel to Taupo-Rotorua, and the Macquarie island arc system may have entered a waning phase.

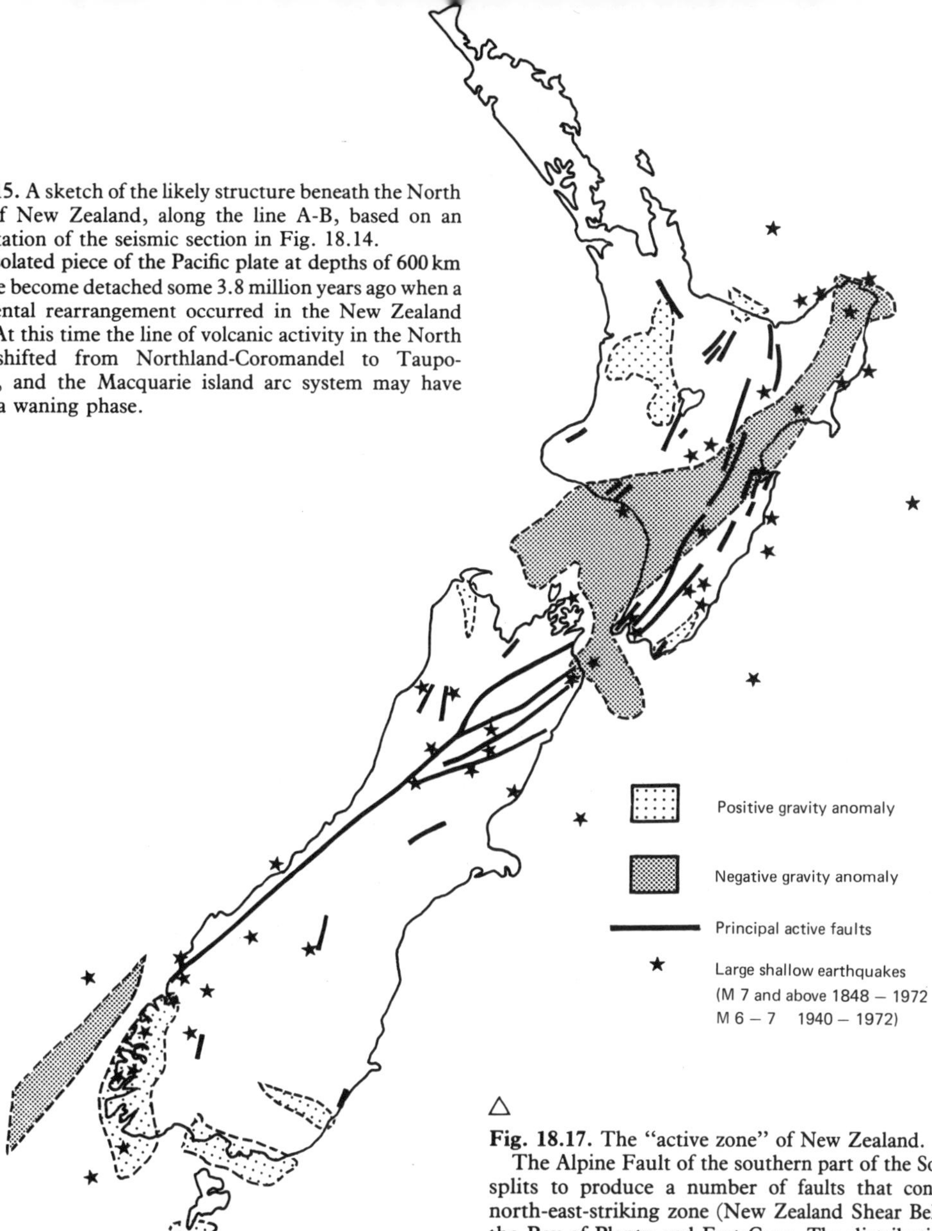


◁ **Fig. 18.16.** An enlargement of the upper part of Fig. 18.15, showing the relationship of the Taupo-Rotorua and Taranaki volcanic activity to the down-dipping Benioff Zone.

As the subducted Pacific plate (Fig. 18.15) is subjected to progressively hotter temperatures the crustal materials are melted to give magmas of various compositions. Thus the large Taranaki andesite cones, of which Mt Egmont (2518 m; 8263 ft) is the latest, are believed to be related to deep melting, whereas the Taupo-Rotorua volcanic activity relates to comparatively shallow melting.

The Taupo-Rotorua region is a fractured and pulled-apart part of the crust and it is believed that stretching of this zone has progressively moved the East Cape crustal block eastwards, so that now some 300 km (187 miles) separates the Benioff Zone and volcanic zone from the submarine trench (Hikurangi Trench), compared with 200 km (125 miles) separation in the Tonga-Kermadec region (Fig. 18.9), where the subduction structure has not been complicated in this way.

△

**Fig. 18.17.** The "active zone" of New Zealand.

The Alpine Fault of the southern part of the South Island splits to produce a number of faults that continue as a north-east-striking zone (New Zealand Shear Belt) towards the Bay of Plenty and East Cape. The distribution of large shallow earthquakes (stars) shows the same south-west/north-east trend and is presumably related to crustal movements within the Shear Belt.

Large gravity anomalies are also related to the "active zone", indicating that parts of the crust have been pushed up or down.

The positive gravity anomalies indicate that the crust has been rising. The negative gravity anomalies, on the other hand, show that the crust has been sinking.

The negative anomalies are closely associated with the subduction zones lying to the south-west and north-east of New Zealand and, although not shown on this map, are essentially continuous with the Macquarie and Tonga-Kermadec island arc and trench systems (Figs 18.10, 18.11). They can therefore be interpreted as representing belts of crust down-buckled along the subduction zones and filled with light sediment. The positive anomalies are in general related to areas of crust upwarped to either side of the subduction zones—producing marked "bulges".

Fig. 18.18. The entire centre of the North Island is cloaked with material thrown out by massive eruptions emanating from the Rotorua-Taupo region over about the past 1-1½ million years. The last major ash shower from this region occurred in A.D. 120.

This photo shows ash and pumice layers near Lake Taupo.

some 480 km (300 miles) of horizontal movement and some 18,288 m (60,000 ft) of vertical movement. This fault may be pictured as running between the Macquarie and Tonga-Kermadec systems and probably provides linkage. If this is so, the Alpine Fault is acting as a transform fault, linking two trench systems (Fig. 9.24). The horizontal movements along the Alpine Fault have dislocated and stretched out New Zealand, but major twisting movements have also occurred.

## Twisted New Zealand

New Zealand has not always had the sinuous twisted shape it has today—it was once a lot straighter.

**Fig. 18.19**. White Island, 52 km (32 miles) off-shore from Whakatane, Bay of Plenty coast, is almost continuously active. It and the submarine Rumble volcanoes (Fig. 18.11) provide a link between the Taupo-Rotorua and Tonga-Kermadec lines of volcanic activity. ▷

The upper view shows the entire island, 2.5 km (1½ miles) wide and 320 m (1053 ft) high. The crater wall has been blown away on the side facing the camera. The present volcano has been built on the collapsed remnants of two older volcanoes, dating back to Pliocene times (2-5 million years ago).

The lower view shows intense steam and hot spring activity on the rear wall of the crater, presumably fed by deep fractures.

In 1914 a fiery avalanche exploded from this region and swept along the crater floor, demolishing a sulphur works and killing 12 people. The hummocky deposits left by this catastrophe are seen in the foreground and middle distance.

In December 1976 and during most of 1977 White Island was extremely active and massive eruptions of new lava occurred—the most massive since the first recorded observations, in 1826.

*Fig. 18.19* △ ▽

**Fig. 18.20.** Mount Ruapehu (2796m; 9175ft) is only occasionally active. Its crater lake poses a considerable danger, however, as hot material can rise quickly and expel hot lake sediments down the mountain side. Such an episode occurred in 1969 when a kiosk was demolished and mud and rocks strewn across a ski field (similar but smaller events also occurred in 1971 and 1975).

Ruapehu's last major eruption was in 1945, when a lava mass rose, formed an island in the crater lake, and then proceeded to grow at the expense of the lake.

The upper photo, taken on 20 April 1945, shows the tip of the lava island and the ash-covered snow.

The lower photo, taken on 22 August 1945, shows the lava mass virtually filling the crater lake and giving off great quantities of ash and gas.

**Fig. 18.21.** The 900 m (3000 ft) andesite cone of Ngauruhoe has been built up over the last 2500 years. Many eruptions have been observed over the last century.

Two major eruptive episodes took place in 1948-49 and 1954-55, when lava flowed down the flanks of the volcano. Since 1955 there have been only small intermittent eruptions, related to local re-heating of the solidified lava plugging the volcanic pipe. The most recent and most violent of these episodes occurred in January and March 1974, as depicted in this photo, when fiery avalanches of re-heated lava cascaded down the mountainside. A sample from this material is illustrated in Fig. 1.5.

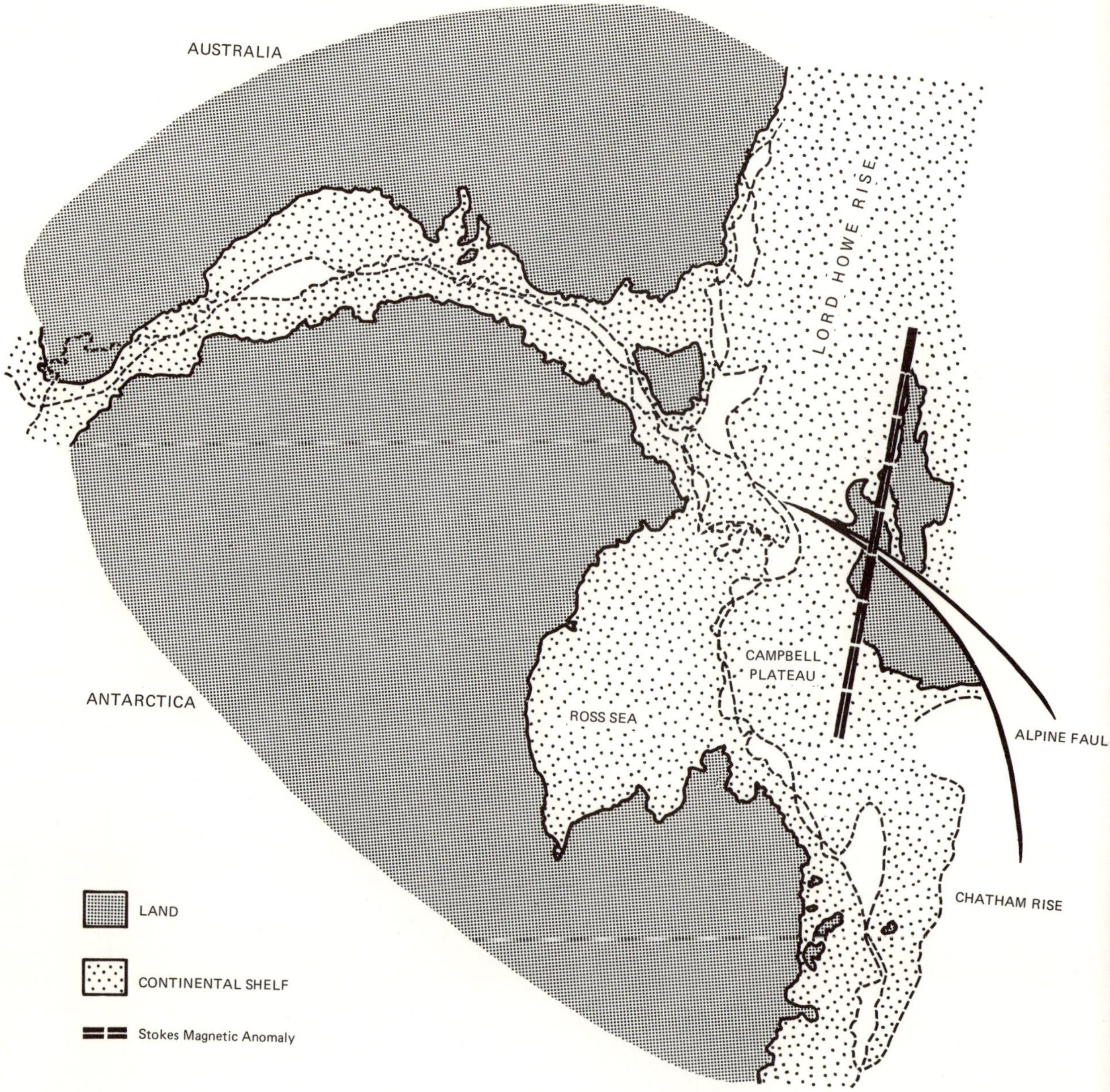


**Fig. 18.22.** The Stokes Magnetic Anomaly is a belt of high earth magnetism that can be traced from Northland to Otago and across the sea floor to the north and south of New Zealand (Fig. 9.20). On land the high magnetic readings are related to outcrops of distinctive rocks (Figs 18.23, 18.24) and it is thought that a belt of comparable rocks is responsible for the anomaly both on-shore and off-shore. The rocks have been termed the Dun Mountain Ophiolite Belt. They consist essentially of sea floor sediments, volcanic rocks, peridotite and serpentine (altered peridotite) and are interpreted as distorted slices of oceanic crust and mantle ("ophiolites"; see Figs 9.18-9.20, 10.14), that were caught up in the earth movements of the Rangitata Orogeny some 140-120 million years ago (Fig. 9.21). Today the Dun Mountain Ophiolite Belt is displaced by faults and twisted into a sinuous shape by the various movements that have distorted New Zealand since Mesozoic times. If, however, it is assumed that it was once essentially straight, New Zealand and its off-shore regions can be rotated back to what may have been their original positions. This has been done in this diagram (cf. Fig. 13.14).

**Figs 18.23, 18.24.** Dun Mountain (Fig. 18.23), in Nelson, and Red Mountain (Fig. 18.24), in NW Otago, are truncated parts of the Dun Mountain Ophiolite Belt, cut and moved apart by movements along the Alpine Fault, so that now they are separated by some 450km (280 miles) (Figs 4.11, 9.20). The peridotite and serpentine of the Ophiolite Belt contain quantities of nickel, chromium, manganese, cobalt, molybdenum and other elements that in high concentrations are lethal to most plant life. Thus the exposed parts of the belt are readily recognised, as they stand out as rugged bare massifs.

△ Fig. 18.23 shows the Dun mountain massif, with the areas of peridotite and serpentine, bare of vegetation, contrasting markedly with the adjoining forested areas of Permian sediments.

Fig. 18.24 is a view of Red Mountain (centre), adjacent to the Alpine Fault (top left). As in Dun Mountain the peridotite areas are free of vegetation, while the surrounding Permian sediments support forest. ▽

It is generally thought that the New Zealand Geosyncline was originally a roughly linear feature. The best indication for this comes from the Stokes Magnetic Anomaly, a distinctive belt of highly magnetic rocks (Figs 9.20, 18.22-18.24). As we have seen (pp. 140–143), this feature can be traced for 3000 km (1900 miles): from 700 km (450 miles) south of New Caledonia to the eastern edge of the Campbell Plateau. The anomaly is thought to be caused by mantle rocks (ultra-basic varieties such as dunite, peridotite, serpentine etc.) which have been dragged up along an ancient plate boundary.

The anomaly belt is thought to have been once continuous and straighter than at present and is thus a valuable reference line, providing a clue to the amount of twisting New Zealand has undergone (Fig. 18.22).

It follows, therefore, that the twisting probably occurred after the end of the New Zealand Geosyncline, during the drifting of ancestral New Zealand. Some of the more recent pre-drift reconstructions show New Zealand in a "straightened-out" position (Figs 13.4, 18.22). Some of the twisting occurred as New Zealand became wedged between active sea-floor spreading to west and south-east and was gradually swivelled out into the South-west Pacific. A certain amount of twisting also accompanied the dislocating movements along the Alpine Fault, as splintering and rotation occurred, particularly in the North Island.

Lastly, some geologists believe that the North Island got its twist some 3-4 million years ago, during a rearrangement of plate motions in the South-west Pacific (pp. 361–363).

However, whatever actually happened, as time went on New Zealand became more and more S-shaped, until today the shape of our country is so distinctive that no one should confuse it with similar-sized islands on the world map.

# CHAPTER NINETEEN

# CONTINENTAL DRIFT—SO WHAT?

YOU MAY be inclined to ask: "What is the use of continental drift? It's just a nice geological theory, when all is said and done! What are the practical applications of a mere theory?"

The very basis of science is the observation and recording of natural phenomena and the placing of them within the framework of a generally accepted (or "working") theory. The theory is then put to the test in interpretation and prediction.

Working theories come and go. Scientists are continually comparing sets of observations and predictions with the theoretical framework within which they are currently working. If the theory is found to be deficient it is discarded or modified and so the process goes on, and with a bit of luck the theoretical basis of any particular branch of science becomes more and more refined. Scientists are human, however, and more often than not have their pet theories, which they tend to hang on to through thick and thin.

Thus as we have seen (pp. 8–17), it took a long time for the theory of continental drift to overthrow the previous "ruling" theories of geology. Now a revolution has occurred and continental drift has the status of a "ruling" theory. The danger of such a situation, however, is that almost everybody jumps on the bandwagon and it is sometimes completely forgotten that continental drift and plate tectonics are still only theories—*not* facts. Some authors have stretched continental drift and plate tectonics to really outrageous limits, and as a consequence a great deal of geology has been rammed willy-nilly into the Wegenerian mould.

As we learn more and more about our earth, and compare the observations with the continental drift and plate tectonic theory, a number of discrepancies have appeared. These may either prove to be the result of over-enthusiastic use of the theory, or alternatively, may indicate the need for amendments of some of the concepts within the overall theory. However, only time and a lot more work will tell.

Nevertheless, it is fair to say that as scientific theories continental drift and plate tectonics have answered more questions than they have raised. For decades geology had been stuck in the groove of blind acceptance of the unyielding permanence of oceans and continents, whereas now most earth scientists are at least prepared to look seriously at the possibility that continents have moved and that oceans have opened and closed in their wake.

The theories of continental drift and plate tectonics have provided earth scientists with an entirely new way of looking at things that occur both on and inside the earth. This has led to a whole series of "new looks" at a great range of practical problems, spanning fields as diverse as exploration for oil, gas, metals, non-metallic industrial minerals (Fig. 19.1) and prediction of long-term climatic changes, earth movements, earthquakes and volcanic eruptions.

## Helping the oil search

As Wegener's ideas gained acceptance the practical implications of continental drift were quickly appreciated by oilmen.

Oilmen are constantly on the look-out for large accumulations of organic-rich young marine sediments. Thus they are very interested in any earth processes involving movements which lead to the formation of deep sedimentary basins. The processes of continental drift and plate tectonics provide a means of creating large depressions, filling them quickly with marine deposits, transforming such deposits into sedimentary rocks by heat and pressure and then raising them to form parts of the continents (Fig. 19.1).

As we have seen (pp. 282, 283), oil and gas are thought to have mainly originated in marine sediments. The oily and waxy substances of

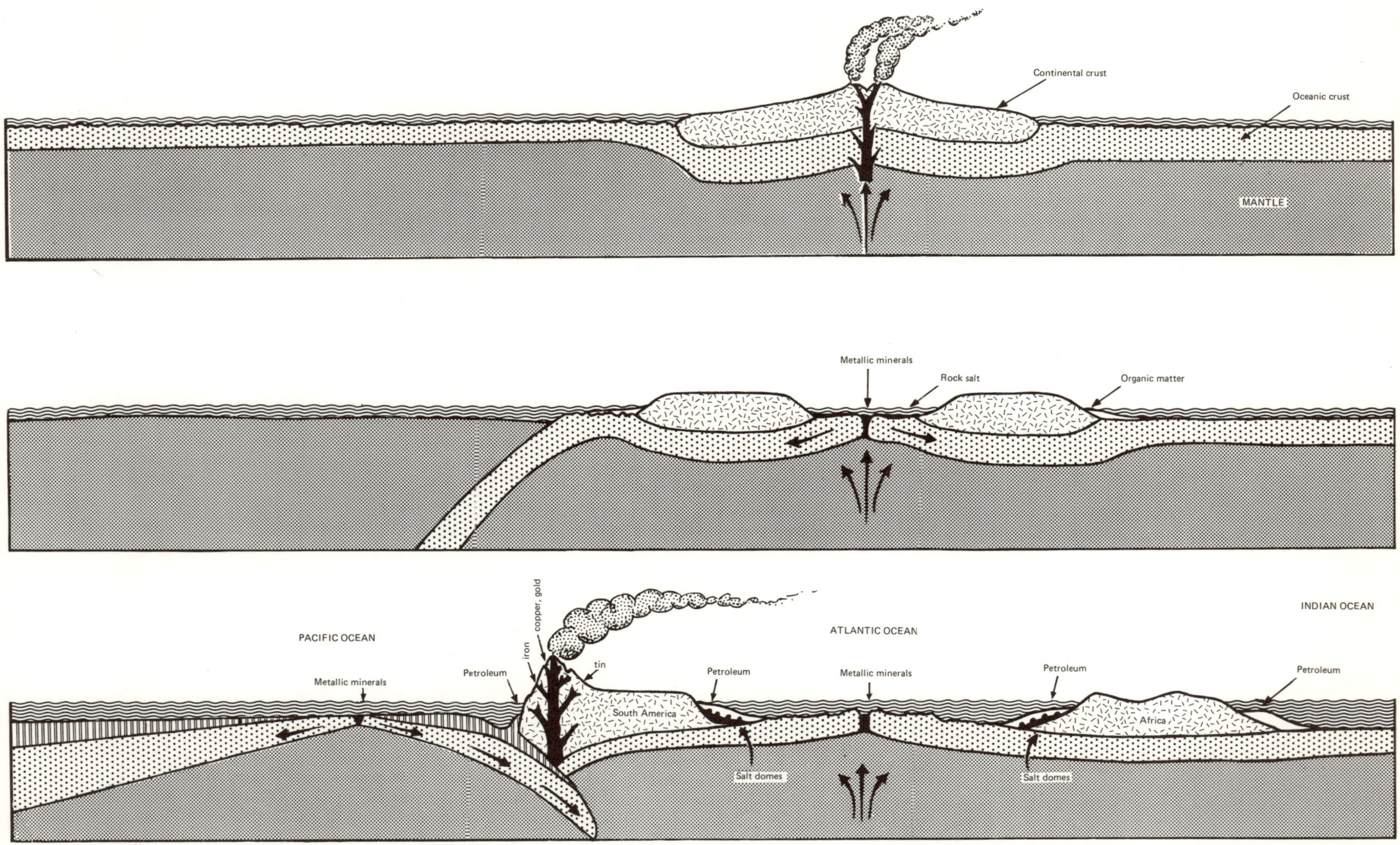
Continental crust
Oceanic crust
MANTLE
Metallic minerals
Rock salt
Organic matter
PACIFIC OCEAN
ATLANTIC OCEAN
INDIAN OCEAN
iron
copper, gold
tin
Petroleum
Metallic minerals
South America
Petroleum
Metallic minerals
Petroleum
Africa
Petroleum
Salt domes
Salt domes

**Fig. 19.1.** The influence of plate boundaries on mineral deposits is exemplified by these diagrams showing development of the South Atlantic Ocean.

The uppermost diagram depicts the super-continent Gondwanaland, in the process of being split in two, to form South America (left, west) and Africa (right, east).

In the middle diagram the embryonic Atlantic Ocean spans a diverging plate boundary. As the two plates pull apart, magma rises along the central rift zone.

Hot fluids and gases rich in a great variety of chemical compounds emanate from the hot magma and pass into the sea-floor sediment and ocean waters. Metallic and other deposits are formed in the sediments and on the sea floor. The highly mineralised water promotes blooms of organic matter in the surrounding sea.

As the landmasses are eroded to either side, quantities of organic-rich material are deposited on the continental shelves.

The lower diagram depicts the modern situation, with the South Atlantic Ocean at its full width and South America overthrust onto the Pacific plate.

Metallic minerals, melted from the Pacific plate as it plunges under South America, rise through the overlying crustal layers and are deposited near the surface to form the ores of the Andes.

In the Atlantic Ocean, metallic and other chemical compounds continue to be deposited around the mid-Atlantic ridge.

Salt beds deposited in earlier layers flow under pressure and rise as salt domes in weakened areas, serving to trap the oil and gas generated from the organic matter present in the early sediments.

**Fig. 19.2.** The giant oil fields of the world (i.e. those capable of yielding at least 500 million barrels of oil).

Many other smaller fields and some large gas fields, unaccompanied by oil, are dotted around the globe, but have not been shown on this map.

marine organisms that had originally lived in the sea, and on death had fallen into the sea-floor deposits, are squeezed out by the pressure exerted by overlying sediments and at the same time altered and "cooked" by the heat flowing out from the deeper parts of the crust.

The oil and gas then migrates through the rock until it is trapped in the porous rocks (sandstones, limestones etc.), where it accumulates under pressure. These "reservoir" rocks then need to be pushed up towards the surface, within reach of an oilman's drill.

These "requirements" for oil and gas generation and accumulation set limits on where it is best to look for commercial quantities of these much-prized commodities.

We know from studies of modern environments that continental margins (i.e. the continental shelves and continental slopes, pp. 3, 49–51) generally favour prolific marine life. It follows therefore that sediments accumulated in this environment offer the most favourable prospects for oil and gas.

Thus sediments that have been deposited on old continental margins in the geological past and have now been raised and incorporated into the continents are prime targets for exploration.

At least 95 percent of the oil and gas discovered to date has been found in these types of sediments (Fig. 19.2), accumulated over the past 250 million years and intimately linked with the basins, island arcs and folded mountain chains formed as a result of continental and plate movements over this period of time. Other smaller quantities of oil and gas have been found, for example, in areas of fractured ancient "basement" rocks, and other at-first-sight unpromising environments, into which they have migrated from nearby younger marine sediments.

As the need for oil and gas has increased, so has recognition of the need to take a "new look" at the land areas that have hitherto provided the bulk of the supplies. The new ideas on the structure of continents, generated by the theories of continental drift and plate tectonics, have produced guidelines for these reassessments.

However, just as the sediments deposited on the continental margins of the past have provided the best hunting grounds during the first 100 years of oil exploration, there is a growing realisation that the continental margins of today offer perhaps the best prospects for the next 100 years.

The modern continental margins of the world—the submerged extensions of the modern continents—going down to water depths of some 4900 m (15,000 ft), far exceed in area the regions on land now being explored or already producing oil and gas. The sheer vastness of this unexplored domain, let alone the magnitude of the technical problems and the cost (an average hole, drilled to depths of 3600 m, 12,000 ft, may cost NZ$10 million off-shore, compared with NZ$1 million on land) is rather daunting, and searches for oil in the future will need all the help they can get. This is where the new ideas involving continental drift and plate tectonics have a major part to play, as they provide for the first time an integrated explanation for the origin and structure of past and present continental margins, giving guidelines for exploration, and allowing oilmen to narrow down the possibilities by indicating likely or unlikely areas with the least expenditure of time, manpower and capital.

Each situation involving moving continents and plates presents the oilman with a different set of challenges—but in each instance "knowing what is going on" (or a fair approximation to it!) allows him to plan exploration accordingly.

**Divergent oil**

Divergent continental margins, where continents or plates are moving apart (p. 121), are encouraging areas for exploration. This is because high earth temperatures, associated with rising magmas and volcanic activity, that favour generation and expulsion of oil and gas, prevail at the same time and place as rifting and subsidence (that assist formation of reservoirs and traps).

Thus thicknesses of potential oil-bearing sediments have accumulated in the gaps created when Africa and South America moved apart—the oil fields of Eastern Brazil and West Africa are in this category. Both these examples are also associated with extensive pull-apart fractures and splinters, extending some distance inland, that have also allowed metal deposits, particularly lead, to form, deposited from hot solutions (Fig. 19.5).

The same processes acting to pull apart continents have also led to the detaching of islands and peninsulas of continental material. This splitting-off of isolated fragments has left

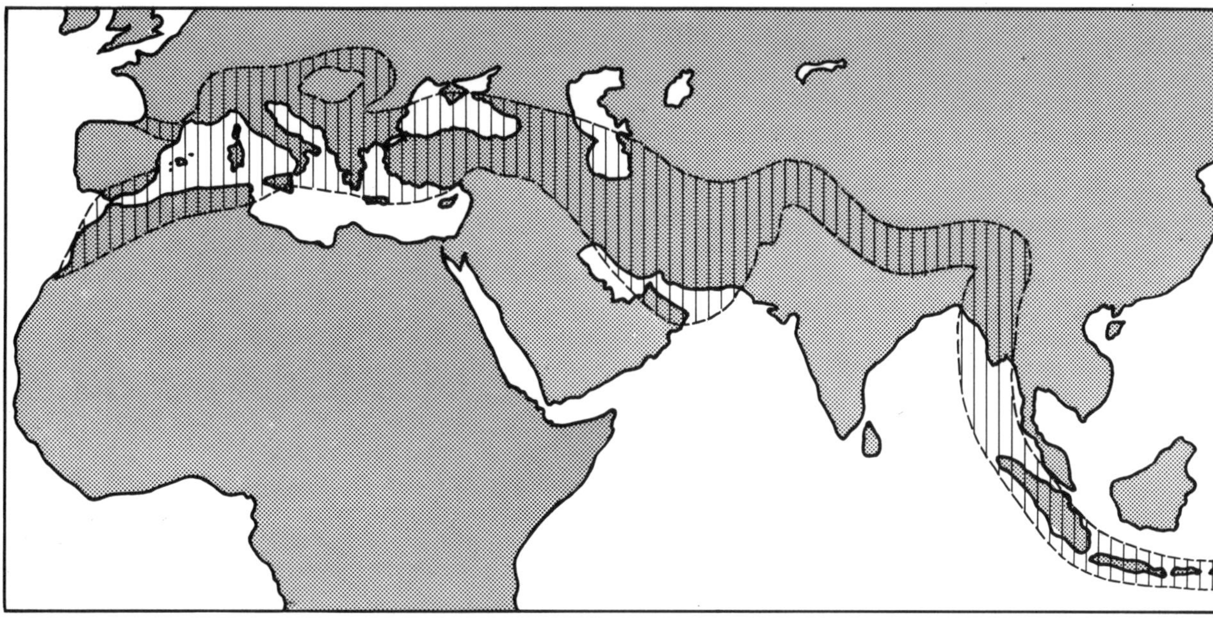

Fig. 19.3. Many of the giant oil fields of the world are related to the Tethys Sea, a tropical-subtropical ocean that existed for almost 200 million years, before caught up in the collisions between Laurasia and Gondwanaland. The old sea-floor sediments of the Tethys, rich in organic material (a legacy of the teeming warm-water life) became pushed up to form the chains of alpine mountains that extend from the Atlantic to South-east Asia (indicated by the lined areas in this diagram).

The folding movements that created the mountains squeezed out and distilled oil from the organic material and at the same time provided ready-made dome-shaped folds ("oil traps") in which the "liquid gold" accumulated.

"holes" in continental margins. These are soon filled with sediments, and as the splitting-off is also often accompanied by injection of magma up through the splintered crust, there is a distillation of oil and gas from the sediments.

Examples of island and peninsula split-offs, leading to the creation of oil and gas regions, are Malagasy, Sri Lanka, Baja California, the North Sea and Tasmania. The oil-prolific Bass Strait area is sited in a "hole" left between Australia and Tasmania and the Maui field off New Zealand is its exact counterpart, on the other side of the Tasman, sited in sediment-filled holes created by the pulling-apart of the North and South Islands (Fig. 13.8).

A similar "hole" exists where New Zealand pulled away from Antarctica 80 million years ago (Fig. 13.14) and it is noteworthy that in this region the *Glomar Challenger* (Fig. 8.18) found signs of oil and gas in three drill holes in the Ross Sea.

The edges of continents that have subsided in the wake of continental pull-aparts also provide good prospects for oil and gas. This is because these regions often contain porous limestones, sandstones, and other rocks, often sealed by salt beds, that provide ideal "traps" for any oil and gas that may migrate in their direction from nearby thicknesses of organic-rich young sediments. Oil fields in the Gulf of Mexico have originated in this way (Fig. 19.1).

**Convergent oil**

Convergent continental margins, where plates or continents are being pushed together (p. 131), in many respects offer the most favourable conditions for generation of oil and gas.

The deep trenches and other areas of subsiding sea floor that occur along the subduction zones developed where plates converge provide natural repositories for sediment, particularly where such zones are close to a continent. Subduction zones far removed from continents, and hence sources of sediment, may not be as favourable, although sediment supply may improve as the subduction zone evolves and the associated island arcs become uplifted and begin to shed sediment as in the West and South-west Pacific (e.g. Tonga-Kermadec, New Hebrides, Solomons,

New Guinea, Indonesia, Moluccas, Philippines etc.).

Subduction zones sited close to a continent may result in the generation of oil and gas in sediments as young as the Pleistocene (2 million-10,000 years ago), as in the Gulf of Alaska, adjacent to the Aleutian Trench. This is because mountainous regions on the adjacent continent ensure an abundant supply of organic-bearing sediments and there is also rapid burial, followed by equally rapid uplift.

The submarine trenches themselves may also generate favourable conditions for marine life on the adjacent shelves, as they provide a source of upwelling cold water, rich in mineral ingredients. The *Glomar Challenger*'s drill has shown, for example, that sediments in or near oceanic trenches commonly have concentrations of 0.5 percent or more of organic carbon, which, when "distilled" by the earth's heat, is more than adequate for the production of oil and gas in commercial quantities.

Rapid burial of dead organic matter before it can be broken up by decay is another vital factor in oil and gas generation. Here again, converging plate boundaries provide optimum conditions for rapid burial, because mountainous regions on the nearby continent provide a ready source of sediment and steeply-dipping continental slopes accelerate the transport of such sediment into the trenches and facilitate their rapid burial.

The down-thrusting movements that occur along a convergent margin as crust is subducted may carry down to depths of several kilometres sediments originally accumulated in shallow waters, and rich in organic material. The sediments would then be compacted into rock and this squeezing, together with the increased temperatures at this depth (5000m, 15,000ft), would encourage the generation and expulsion of oil and gas, derived from the organic matter disseminated throughout the sediment. The natural distillation process would be aided by the abnormally high earth temperatures generated by the volcanic activity associated with the Benioff Zone.

Continued subduction may then cause massive flexing and splintering of the crust so that the compacted sediments (and the oil and gas accumulations) are uplifted to depths within the reach of drill holes.

Many of the convergent margins around the Pacific Ocean provide the requisite conditions for oil and gas generation—examples being the Gulf of Alaska and the Indonesian island arcs. The oil fields of Trinidad, Curacao and Venezuela are similarly linked with convergence, associated with subduction in the Caribbean.

Continental collision, the end-result of continental convergence (pp. 132–149), also pushes up the once deeply buried sediments and exposes them to the oil prospector's drill. The Timor Sea is an example of an oil and gas area produced in the early stage of such a collision, i.e. north-west Australia converging with Indonesia.

Continental collisions result in a jumbled landmass, involving oil and gas-bearing rocks in intricate folding and faulting. It is here that knowledge of processes involving the modern continental margins helps in understanding and unravelling the complexities of the ancient continental margins that now form large areas of the land.

The various continental collisions that occurred as the north and south shores of the Tethys Ocean moved together produced some of the world's greatest known oil and gas fields, as well as the most impressive mountain chains (Fig. 19.3). Thus we have oil and gas fields associated with the French, Swiss and Austrian Alps, the Carpathians, Anatolian Mountains, Caucasus, Elburz and Zagros Mountains, the Himalayan chain and its continuation into Burma, Western China and Thailand.

**Stretching the world's dwindling metal supplies**

If from the new theories of continental drift and plate tectonics we understand better why oil and gas are to be found in some places and not in others, this is even truer for metallic ore deposits (i.e. the primary metal ores, found in veins and fissures; other ore deposits—placer deposits—have been derived by erosion from such primary ores and have accumulated in river gravels, beach shingle etc.). It is now realised that most of the continental veins of metallic ore upon which our civilisations have been built have originated from past and present plate situations.

Exploration work over the years brought geologists to the realisation that in general there was an association of metal ores with past and present volcanic and igneous activity and it was assumed that metals had been carried into the crust and upper mantle during the differentiation processes accompanying the early cooling of the earth (p. 159–163). It was also assumed that the metals were then re-mobilised and concentrated at the surface either by hot waters heated by igneous intrusions, or by separation direct from magmas. Thus although it was known that certain ores were associated with certain geological conditions, the reasons for this were unknown and prediction of potential ore deposits on this basis was therefore rather a hit-or-miss affair.

Recognition, however, of the role of continental drift and plate tectonics in world geology has provided ore searchers with a new conceptual framework and a greater clarity of vision in deciding where to look (and where not to look!) for new resources. Without knowing how ore bodies are formed it is difficult to seek out those that lie buried with only meagre surface clues to their presence.

As we have seen in the instance of oil and gas, conditions favouring deposition of various metallic ores are intimately related to interactions of crustal plates, continents and oceans, either in the past or at present. Thus the type of ore developed depends on whether the geological situation is one of diverging or converging plates, or continental collision. In all situations however, as we saw with oil and gas, the heat of the earth plays a vital part in "boiling out" metallic compounds, that are then transported to within our reach by rising hot fluids.

Modern divergent plate boundaries, formed by the pulling-apart of plates in the central portions of oceanic basins, are the site of mineralisation today. The *Glomar Challenger*'s drill has shown that sediments near the modern active oceanic ridges in the Atlantic, Pacific and Indian oceans are locally enriched in various metals, notably copper and molybdenum. But yet the volcanic activity actually along the ridge crust (e.g. in Iceland) is not accompanied by significant mineralisation. It has been suggested therefore that hot fluids pass out to either side of the newly injected magma and, on making contact with cold sea water in the overlying sediments and older lavas, precipitate ores, particularly those of copper.

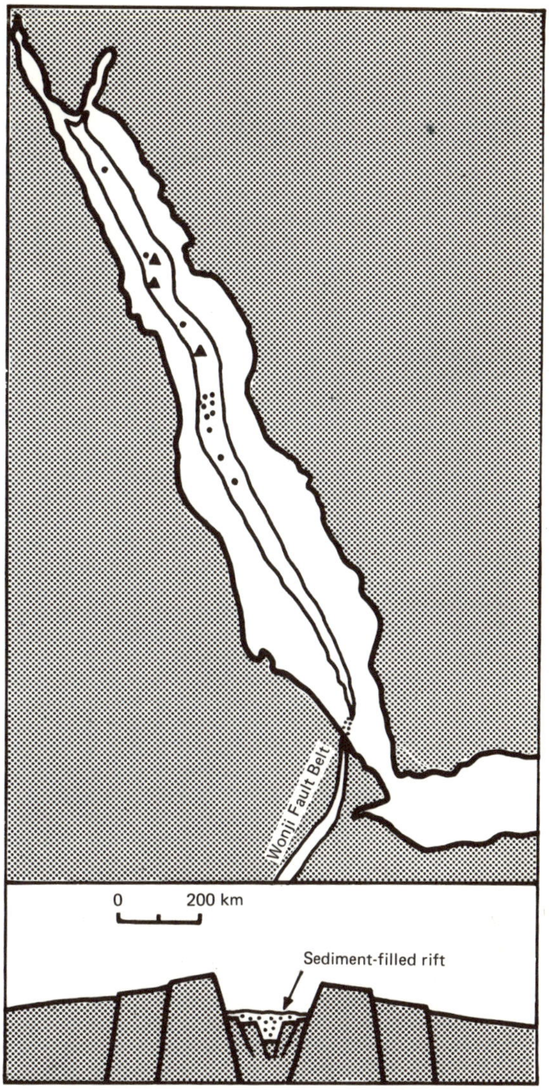


**Fig. 19.4.** The Red Sea is an ocean in the making. It is being formed as the African and Arabian plates pull apart. As they do so, cracking and splitting occur along the central rift valley of the Red Sea and lavas rise towards the surface. Sea water flows down the cracks, encounters the hot lava and flows out again, carrying various metals and salts. The metal-laden hot brines accumulate in various low-lying areas along the floor of the rift valley. Hot brine pools have been found, four of which contain considerable quantities of metal-enriched sediment. The diagram depicts the Red Sea rift zone, showing the axial trench (outlined), with known hot brine pools indicated by dots and metalliferous ones indicated by triangles. The Wonji Fault Belt continues southwards into the faults of the East African Rift Valley.

At the foot of the diagram is an idealised section across the Red Sea and neighbouring land. The Red Sea rift is filled in by a considerable thickness of sediment.

## Hot holes

Some idea of the mineral processes going on at divergent plate boundaries can be gauged from

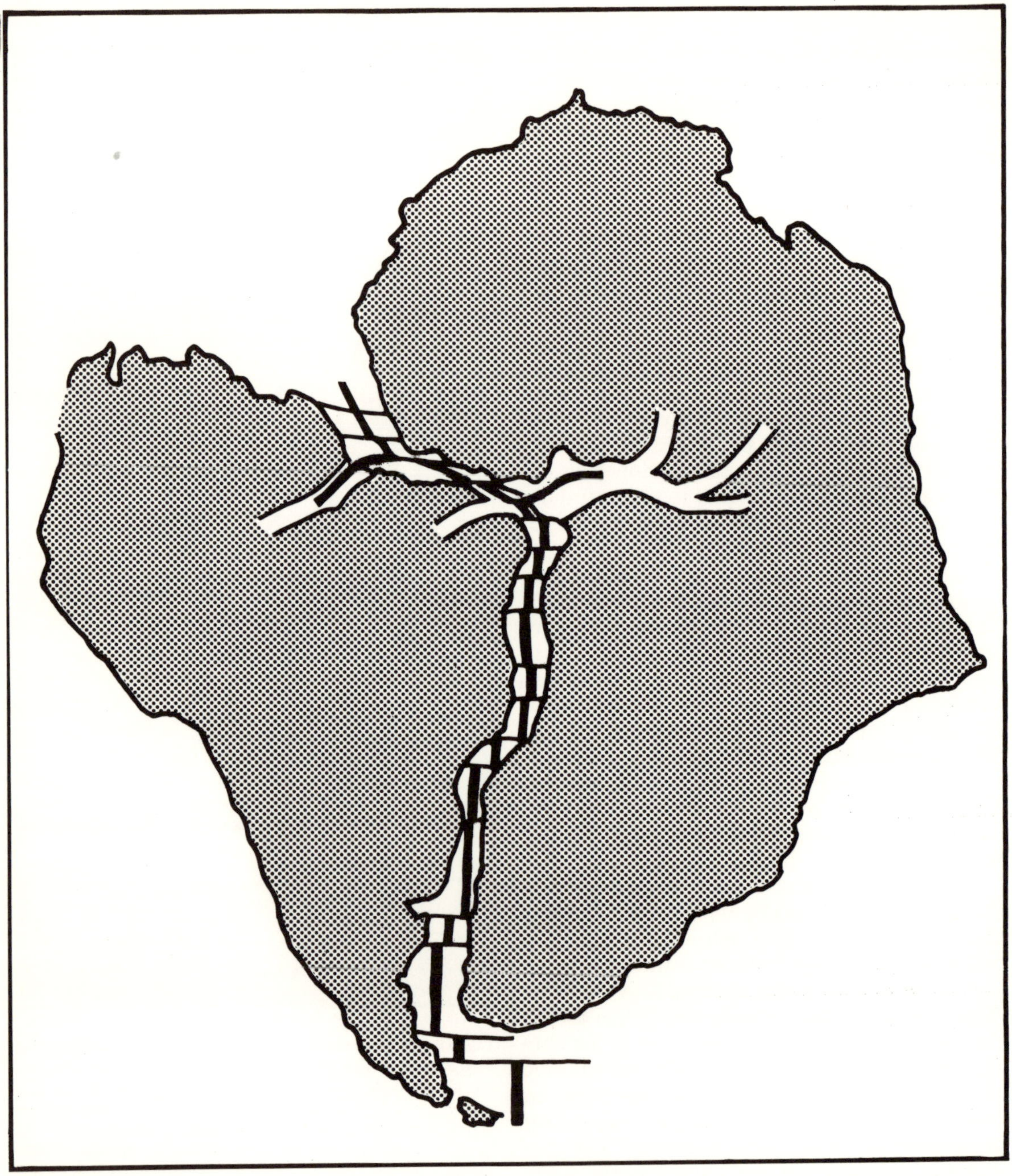

**Fig. 19.5.** As the Atlantic Ocean opened, a great deal of splintering occurred to either side of the line that eventually was to become the main rift. Particularly massive splintering, with formation of rift-valley-like troughs, occurred at the right-angled bend in the coastline, now occupied by the Gulf of Guinea and Brazil's Recife shore.

Initially these troughs were like miniature versions of today's Red Sea (cf. Fig. 19.4), with large fractures penetrating deep into the earth's crust and with magma rising and giving off mineral-bearing gases and solutions. Even today hot sulphur and brine springs are active in some of these troughs, both in the Amazon basin and West Africa.

As the South Atlantic Ocean opened, the troughs became filled with sediment and were buckled by later earth movements. Today, however, they are rich sources of lead, zinc, barium and fluorine and oil-bearing sediments are also present (e.g. Nigeria, Amazon basin).

the modern Red Sea, which in this sense forms an accessible natural laboratory. The Red Sea marks the site of a developing plate boundary, formed as the Arabian and African plates pull slowly away from each other (Fig. 19.4).

Deep-sea drilling has indicated that the Red Sea itself has probably opened up over the past 20 million years—and today resembles a miniature version of the Atlantic Ocean of 200 million years ago. However, about two million years ago a speeding-up in spreading rates opened up an axial rift valley in which water depths plunge to 3000 m (10,000 ft).

A striking feature of the axial valley is that in certain low areas along its length there are concentrations of some of the hottest, most saline water found anywhere in the world. *In situ* temperatures up to 60°C (140°F), inflow temperatures up to 104°C (220°F), and a salt concentration of 256 parts of salt to 744 parts of water have been recorded. By contrast, ocean water salinity is typically 35 parts of salt to 965 parts of water.

Altogether 16 saline basins ("brine pools") have been found along the Red Sea's axial valley (Fig. 19.4). In four of these the hot brines were found to be considerably enriched in heavy metals such as copper, lead, zinc, cadmium, iron, silver and gold, and the underlying sediments when drilled were also found to have considerable quantities of these metals.

The largest of the metalliferous brine pools, named the Atlantis II Deep after the oceanographic survey ship that discovered it in 1963, is about 15 km (9.5 miles) long and 5 km (3 miles) wide and is in 3600 m (12,000 ft) of water.

So far detailed exploration has been confined to the top 8.5 m (28 ft) of sediment in the southern part of the Atlantis Deep. But it has been estimated that this alone contains some 83 million tonnes of dry sediment containing in every tonne 30 percent iron, 6 percent zinc, 1.3 percent copper, 0.1 percent lead and 110 g (4 oz) of silver, and a smaller amount of gold, as well as cadmium and manganese.

If the metals could be extracted they would be worth almost NZ$6000 million. And of course the drill holes have only "scratched the surface" as it were.

The practical problems of recovering the sediments and extracting the metals are enormous, but nonetheless contracts were signed in 1977 to carry out expanded exploration programmes and to develop by 1982 a pilot plant to test the proposed extraction processes.

However, although at the moment it may prove to be uneconomic to raise the material from one mile below sea level and then to refine it, the resource may nevertheless be a valuable metal storehouse for the future—when perhaps we might be desperately short of copper, for example.

As for the origin of the Red Sea ores, it is thought that as the Arabian and African plates move apart and the earth's crust is split along the axis of the Red Sea's central valley, hot magma rises and heats water contained in the surrounding rocks.

These hot waters rise towards the surface and as they do so, they leach out metals and salts from the rocks through which they pass and also may carry metals derived from the magma itself. Then, as the hot salty fluids mix with cold sea water, they deposit a variety of metal minerals in the sedimentary layers lining the sea floor.

The time taken to form the sediments in the hot holes has been measured radioactively and is only 10,000-20,000 years. Thus what we are seeing is just the latest series of hot holes along the Red Sea rift. After some time the lavas in the cracks along the rift zone will cool and the mineralisation process will stop. The ordinary processes of sedimentation will slowly cover the ore deposits with sand, mud and the skeletons of minute marine plants and animals, and soon there will be nothing to show that the ore is there.

Meanwhile, the cracking will start elsewhere along the axis of the sea and new brine pools will form. As the embryonic sea widens still further, the ore bodies will be carried away from the axis by the motions of the African and Arabian plates to either side.

Thus on each side of the Red Sea's central rift it is likely that explorers may find "fossil" hot holes which are no longer hot and are buried under a few centimetres or metres of normal marine sediment. They may have also been covered by later lava flows.

Embryo oceans, such as the Red Sea, are rare. In fact, the only other clear examples are the Gulf of Aden, which is an extension of the Red Sea, and the Gulf of California. However, fossil rifts, that did not become oceans, are common in a number of continental areas—for example, the

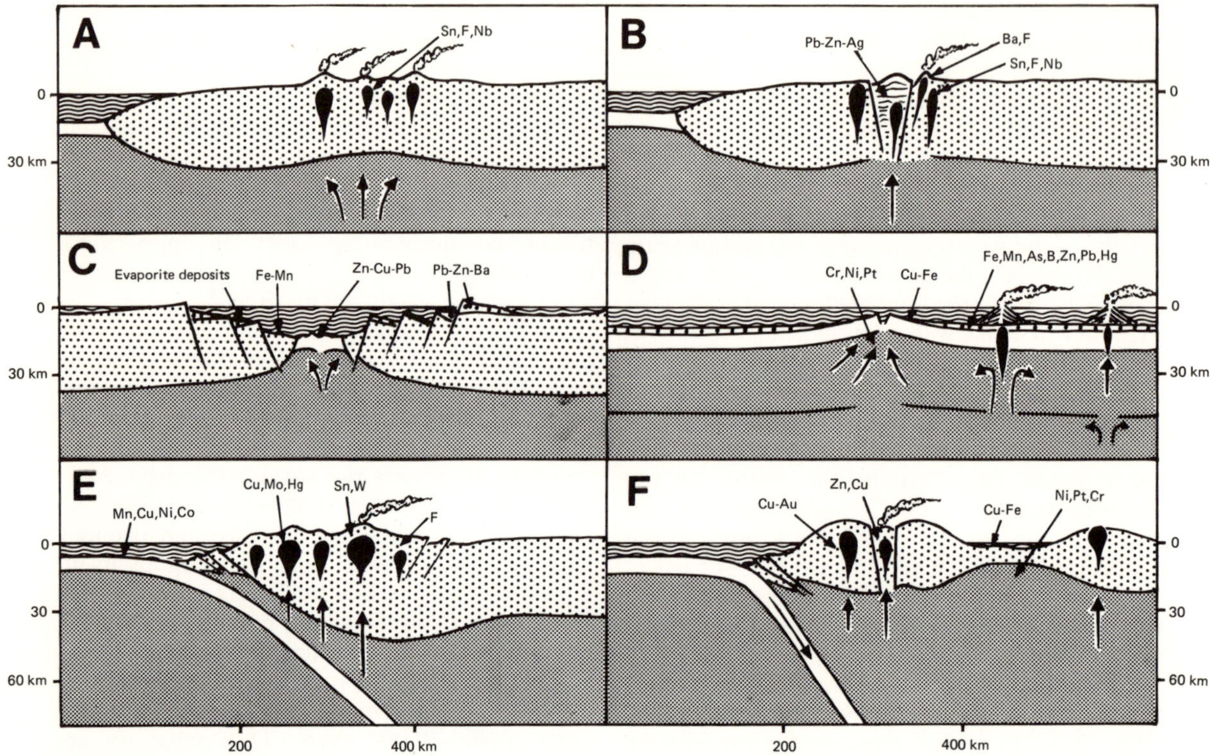


**Fig. 19.6.** Diagrammatic cross-sections through plate boundary situations that are likely to generate ore deposits.
A: A "hot-spot" zone within a continent. Magma rises up a weakened region of the crust. Ores of tin, fluorine and niobium are deposited by hot fluids during the ensuing volcanic activity.
B: An aborted rift zone within a continent—i.e. the continent began to split apart, but although a rift structure formed, an actual break did not occur. Lead, silver and zinc ores are laid down in the sediments and volcanics of the rift valley. Tin, fluorine and niobium ores are deposited by hot fluids emanating from the granite intrusions.
C: A Red Sea-type rift zone (cf. Fig. 19.4). Hot fluids rising up the deep fractures in the crust deposit lead, zinc, silver and copper ores in the rift sediments. Manganese-rich sediments and evaporites form on the sea floor.
D: Oceanic rise and Hawaii-type volcanic chains. Copper and iron deposits, like those of Cyprus, form on the oceanic rise and chrome, nickel and platinum in the rift zone.
E: Convergent plate margin with subduction of oceanic crust. As the oceanic crust and its covering layers of sea-floor sediment are progressively melted, various metallic ores distil off and rise to the surface in magma bodies, to produce a zonation of ores (Fig. 19.8), as seen in the Andes.
Copper, molybdenum and mercury deposits are found close to the subduction zone, followed in order by tin, tungsten and fluorine.
F: Subduction zone with island arc and inter-arc basin. Porphyry copper-gold deposits are related to the volcanic activity of the arc. Copper-zinc deposits form in incipient rifts. Copper-iron deposits form in the inter-arc basin.
A key to the chemical symbols used in the diagrams is provided in the caption to fig. 19.7.

rifts splaying out from the angular separation point of Brazil and Nigeria (Fig. 19.5)—and these are the sites of rich mineral deposits—as are the modern rift valleys of East Africa, extensions of the Red Sea rift.

While no major ore deposits have been found so far in the Gulf of California, bore holes near the Salton Sea (sited in the Imperial Valley, a northward extension of the Gulf of California) yield hot mineralised brines generally similar to those in the Red Sea's hot holes, but containing a higher concentration of metals and more lead and silver relative to copper than do the Red Sea brines.

Closer to home, it has recently been discovered that lead, zinc and copper sulphides are being deposited from hot waters in several bores in the Broadlands geothermal region (24 km, 15 miles, north-east of Wairakei and about 1.6 km, 1 mile, east of the main Taupo-Rotorua highway).

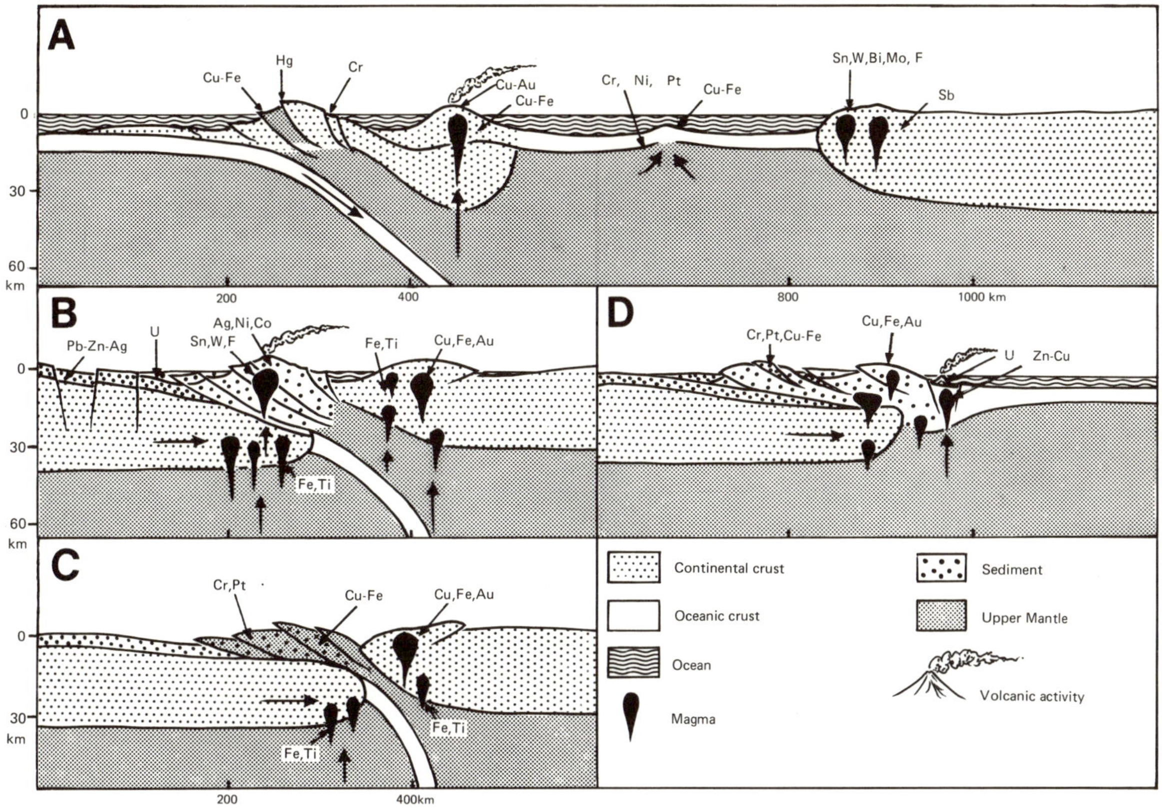


**Fig. 19.7.** Tectonic and volcanic activity following continental collision has a major influence on the type of ore deposits. A: Subduction zone, with island arc, and marginal basin. Porphyry copper-gold deposits are related to the island arc and copper-iron deposits to the adjacent sliced-up sea floor. B: Continental collision belt. Rich mineral deposits occur in the intrusions that are related to volcanic activity and also in the sea floor squeezed up in the collision zone.
C: Continental collision with over-thrust ophiolites (pp. 135–143). The force of the collision between the two continents has pushed oceanic and upper mantle material onto the continental edge.

Cyprus-type copper and iron deposits, as well as chrome and platinum, are related to such geological environments. D: An island arc system has collided with a continent and become welded onto the continental margin. The associated volcanic activity leads to chromium-platinum-copper and iron mineralisation.

Key to chemical symbols used in Figs 19.6, 19.7 (in alphabetical order of symbols).
Ag: silver; Au: gold; Ba: barium; Bi: bismuth; Co: cobalt; Cr: chromium; Cu: copper; F: fluorine; Fe: iron; Hg: mercury; Mn: manganese; Mo: molybdenum; Nb: niobium; Ni: nickel; Pb: lead; Pt: platinum; Sb: antimony; Sn: tin; Ti: titanium; U: uranium; W: tungsten

As we have seen (pp. 361–369; Fig. 18.16), the Taupo-Rotorua region is a trough-like rift valley, filled with volcanic material, representing the southern extension of the Havre Trough into New Zealand. The rift feature has developed as the East Cape block (containing the Ikawhenua, Huiarau, Urewera and Raukumara ranges) was pulled away from the remainder of New Zealand. Thus structurally it has much in common with the East African rift valleys.

At Broadlands the lead, zinc and copper sulphides plus traces of cobalt, nickel and arsenic are being deposited at depths between 300-2300 m (980–7600ft) at temperatures between 140° and 300°C (284–572°F). At the surface, however, the waters deposit a reddish-orange mixture (seen coating the edges of the Ohaki Pool and several bore outlets) containing antimony, mercury, thallium, silver, arsenic and gold. This suggests that a metal separation process, or zoning, is operating, with copper, lead and zinc minerals being deposited from hot fluids at some considerable depth and the other metals from cooler surface waters.

As this type of zoning is typical of many ore deposits that have been formed in past hydrothermal areas, the Broadlands geothermal field provides the chance to measure conditions under which some metal ores can form. In turn this information may be used in interpreting ore deposits in "fossil" geothermal areas, such as Coromandel, that, as we have seen (p. 361), was a Wairakei-like area in Pliocene and Pleistocene times.

**Fig. 19.8.** A situation similar to that of Fig. 19.6E occurred in eastern Australia in Palaeozoic times, when a subduction zone formed, extending through Victoria, New South Wales and Queensland. As oceanic crust and sea-floor sediments were consumed and melted in the subduction zone, various metals "boiled off", as it were, and worked their way up to the surface, via rising magmas and hot fluids.

Thus a zonation of metallic ores is apparent and the diagram depicts the situation in New South Wales during this early subduction phase, as the rich mineral deposits were being formed that are now seen, for example, in Mt Isa and Broken Hill.

A by-product of the release of great quantities of heat at active plate boundaries is the creation of potential areas for development of geothermal energy. Energy from such sources is being tapped all around the Pacific, and in a zone from Spain and Italy through Yugoslavia, Turkey, the Himalayas and Indonesia. Other areas are Iceland and the East African rift valleys. As a glance at Fig. 9.4 will show, all these areas are related to plate boundaries.

## Ores and converging plates

Past and present convergent plate boundaries, where oceanic plates are being carried down into the mantle along subduction zones (pp. 131–133), are the site of many of our major metallic resources of today (Figs 19.6, 19.7).

As with ore deposits formed in divergent plate environments, it is thought that hot water solutions play a vital role in the concentration, transport and deposition of the metals. Again, most of the metals are combined with sulphur in

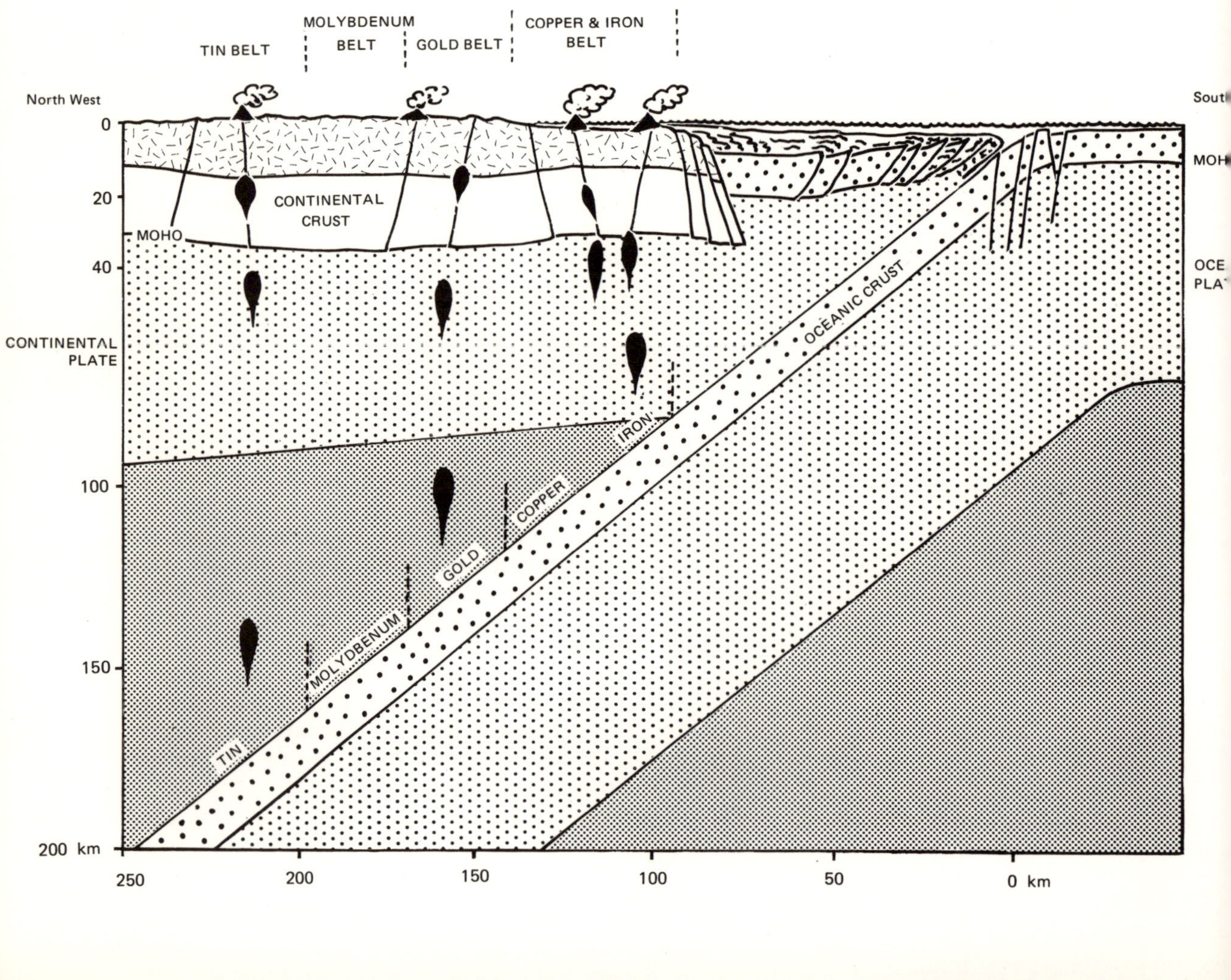

various forms and form the so-called "sulphide" ores.

In the converging plate situation it is thought that as the over-ridden lithospheric plate plunges into the mantle, and is progressively heated up, the water contained in the rocks of the plate and the overlying sediments rises to the surface as hot solutions. As these solutions migrate upwards, the various metals contained in the rocks through which they pass are leached out and transported towards the surface. Eventually, as the rising hot solutions encounter cooler temperatures near the surface, ore minerals are precipitated in cracks and cavities in the surrounding rocks.

In situations where island arcs are developed above descending slabs (Fig. 19.6), the rising metal-bearing hot solutions and magmas come to rest in the island arcs and adjacent sea floor, that thus become the sites of extensive volcanic and hydrothermal activity. The mineral deposits of greatest economic importance in island arcs are so-called "porphyry" copper deposits—very large low-grade deposits of copper sulphides, often associated with gold and sometimes molybdenum. Such deposits supply more than half of the world's copper production.

The porphyry copper deposits are associated with a volcanic rock known as a "porphyry"—an andesite rock with unusually large feldspar crystals. The host porphyry rocks are generally full of holes, cracks and fissures and it has been suggested that this is a by-product of the explosive formation of craters. The mineralisation is widely disseminated throughout the porphyry and adjacent rocks and has probably originated by the percolation of hot fluids carrying metals derived from the magma at depth and leached from surrounding rocks.

Thus the picture that emerges is that the porphyry copper deposits in past and present island arcs represent hot-water ("hydrothermal") mineralisation associated with volcanic activity.

Modern island arcs present in the northern, western and south-western Pacific, eastern Indian Ocean, the Caribbean, Mediterranean and Scotia seas are among the most highly mineralised segments of the earth's crust.

Ancient island arcs, now incorporated within continents, are similarly rich in mineral deposits, although sometimes they may be deeply eroded and the porphyry ores removed. However porphyry deposits dating back to an early Palaeozoic island arc are mined in North Wales, for example.

Gold and mercury deposits are also related to volcanic activity in an island arc situation—examples being Guadalcanal (Solomon Islands), Fiji and the Coromandel Peninsula.

Although, as we have seen, the porphyry ores are thought to be related to terrestrial volcanoes sitting on island arcs, volcanic activity, with associated emission of mineralised fluids, was also occurring at the same time on the sea floor adjacent to island arcs. Such underwater eruptions are accompanied by violent reactions as hot lava encounters cold sea water and quantities of superheated steam and shattered and pulverised rock full of gas holes are produced. Leaching of the lava releases great quantities of iron and manganese, that stain the surrounding sediments and may fan out across the sea floor as dense clouds, that eventually coagulate and settle to form layers of nut-sized nodules (pp. 394–396).

Other metals derived both from leaching of the lava and from deeper crustal rocks are deposited in the holes and crevices of the volcanic rock. Copper, zinc and lead deposits that have been formed in this way in the Tertiary period are known from the modern island arcs of Japan, the Philippines and Fiji. Older deposits of this type, now incorporated into continents, occur in New South Wales, Newfoundland, Egypt and Spain (the famous Rio Tinto mines).

As we have seen (pp. 132, 133), if a continent is involved in a subduction zone, chains of volcanoes develop on the subducting continental margin, as oceanic crust and sedimentary materials are dragged down under the continent and melted.

These volcanoes on continental margins are believed to develop porphyry copper deposits in the same way as volcanoes on island arcs and we find therefore that the entire mountain chain bordering the Americas, from Alaska to Patagonia, is peppered with porphyry copper deposits, related to late Mesozoic and Tertiary subduction (Figs 19.6, 19.7).

A wide variety of other metal ores, apart from copper, are also found in such continental margins and as exploration proceeded, it was found that these occurred in belts parallel to past or present subduction zones.

This zonation is thought to reflect a process of "boiling-off" of metals at various temperatures, as the subducted crustal slab, and its overlying sediments, travels deeper and deeper into the subduction zone and gets hotter and hotter.

Thus it is thought that iron minerals are released first, while the subducted crust is at fairly shallow levels. Then, as it gets deeper and hotter, copper and then gold pass off, followed by silver-lead-zinc and finally, above the deepest sections of subducted crust, tin, tungsten and molybdenum. The metals are carried upwards in magmas and associated fluids and are deposited in the volcanic rocks that are erupted as a result, or in the cracked and broken surrounding rock.

The extensive ore deposits in New South Wales (Fig. 19.8) and South-east Asia, as well as those along the entire alpine chain of North, Central and South America, are examples of such zonation.

**Ores and colliding continents**

Collision of continents, the end-result of converging plate boundaries, crams together a wide range of potential ore-bearing environments.

Old island arcs, with porphyry coppers and other ores thus find themselves wedged between the colliding continents. Ore deposits related to submarine eruptions on the ancient sea floor that once existed between the two continents are also squeezed up and may be thrust onto the continental margins. Then the oceanic crust itself may be sliced up and pieces thrust onto the adjoining continents, so that rocks of the deep crust and upper mantle, often rich in chromium and nickel, may be found on the land. As we have seen on pp. 135–143, the dunite, serpentine and peridotite rocks forming the "mineral belts" of both New Caledonia and New Zealand are thought to have had this origin.

Ancient oceanic ridges may also be caught up between the colliding continents and their materials squeezed up to form land. Ore bodies in Turkey, Cyprus (Fig. 19.9), California and Newfoundland are thought to have originated in oceanic rift zones, like those of the modern Red Sea (pp. 127–131, 383–387).

**Solving jig-saw puzzles to find ores**

As we have seen (pp. 143–147, 161–163), prior to the latest phase of continental drifting that began 200-100 million years ago, there had been a series of continental collisions leading to the building-up of super-continents.

Each continental collision in the past has left in its wake a graveyard of ancient island arcs, continental margins, oceanic rifts and chunks of sea floor, all with their associated ore minerals. Thus as the latest of the super-continents—Laurasia and Gondwanaland—became aggregated, piece after piece of these old mineralised rocks became welded onto the old continental nuclei, forming distinct belts. However, fragmentation of Laurasia and Gondwanaland during the latest drift episode has meant that many of these ore belts are now disjointed and separated by broad oceans.

The most spectacular examples of truncated ore belts are those on either side of the South Atlantic Ocean—notably the separation of the once-continuous copper and tin belts (Fig. 19.10). In some instances while on one continent the ores may be clearly exposed and accessible, on the other continent, on which the truncated ore belt continues, it may be concealed beneath other rocks and the jigsaw-pattern fit of the two continents may provide the only clue to its hidden presence.

Other ores have been eroded out of pre-existing rocks and deposited in materials filling broad sedimentary basins developed in low-lying areas of continents. The rich iron deposits of South Africa and Western Australia are found in sedimentary basins that formed some 2000-2500 million years ago in gentle downwarps on their respective continents (Fig. 19.11). The Transvaal Basin (South Africa) and Hamersley Basin (Western Australia) are identical in age, geological structure, sediments and mineralisation. Apart from iron, both basins are also rich in gold, uranium, manganese and blue asbestos—these minerals having been laid down in sediments or deposited from percolating fluids.

Similar basins, also rich in iron and other minerals, occur in southern India and possibly under Antarctica's ice.

All these basins probably formed at about the same time, under the same conditions and in close proximity to one another, but yet are today separated by a vast stretch of open ocean. A knowledge of their past close association means, however, that exploration tips learnt in the Transvaal Basin, for example, can be applied to the Hamersley and other basins and may also lead to the discovery of hitherto unsuspected ores.

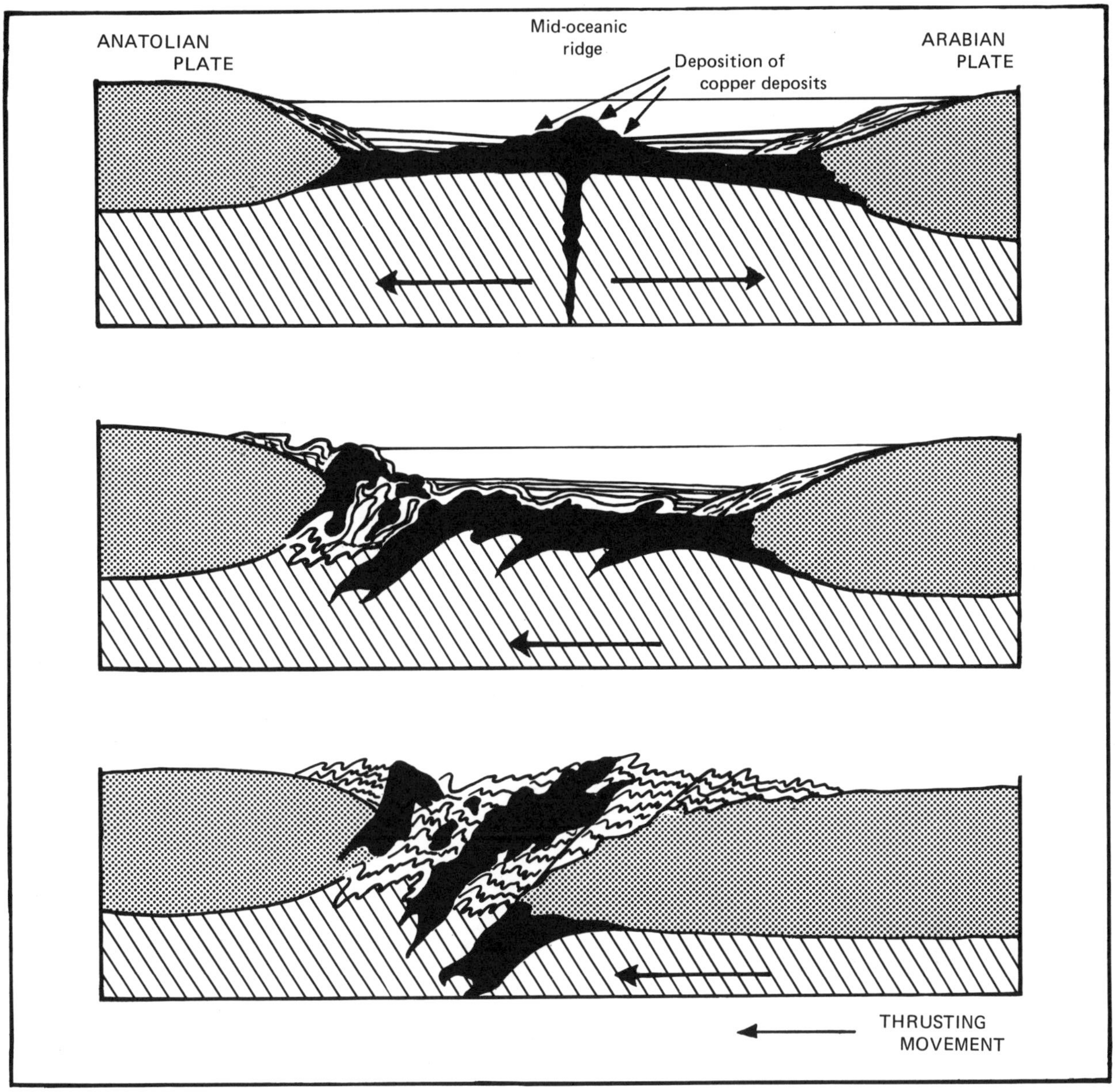


**Fig. 19.9.** The vice-like effect that occurs when continents collide is responsible for thrusting up great slices of normally inaccessible metal-rich material from the lower levels of the oceanic crust and the upper mantle (cf. Fig. 19.7C). This example illustrates the origin of the copper deposits of Turkey and in general is also thought to apply to those of Cyprus.

## The glitter of diamonds

Of less importance to our well-being than ore deposits, but of great interest to gem-hunters, has been the application of continental drift theory to the hunt for precious stones, notably diamonds.

It has long been known that diamonds originate in unique pipe-like structures deeply penetrating the earth and filled with a jumbled mixture of shattered and altered rocks collectively called "kimberlite" (Fig. 19.12) (p. 172). Studies have also shown that great pressures and temperatures are required to convert the rather open structure of carbon atoms found in

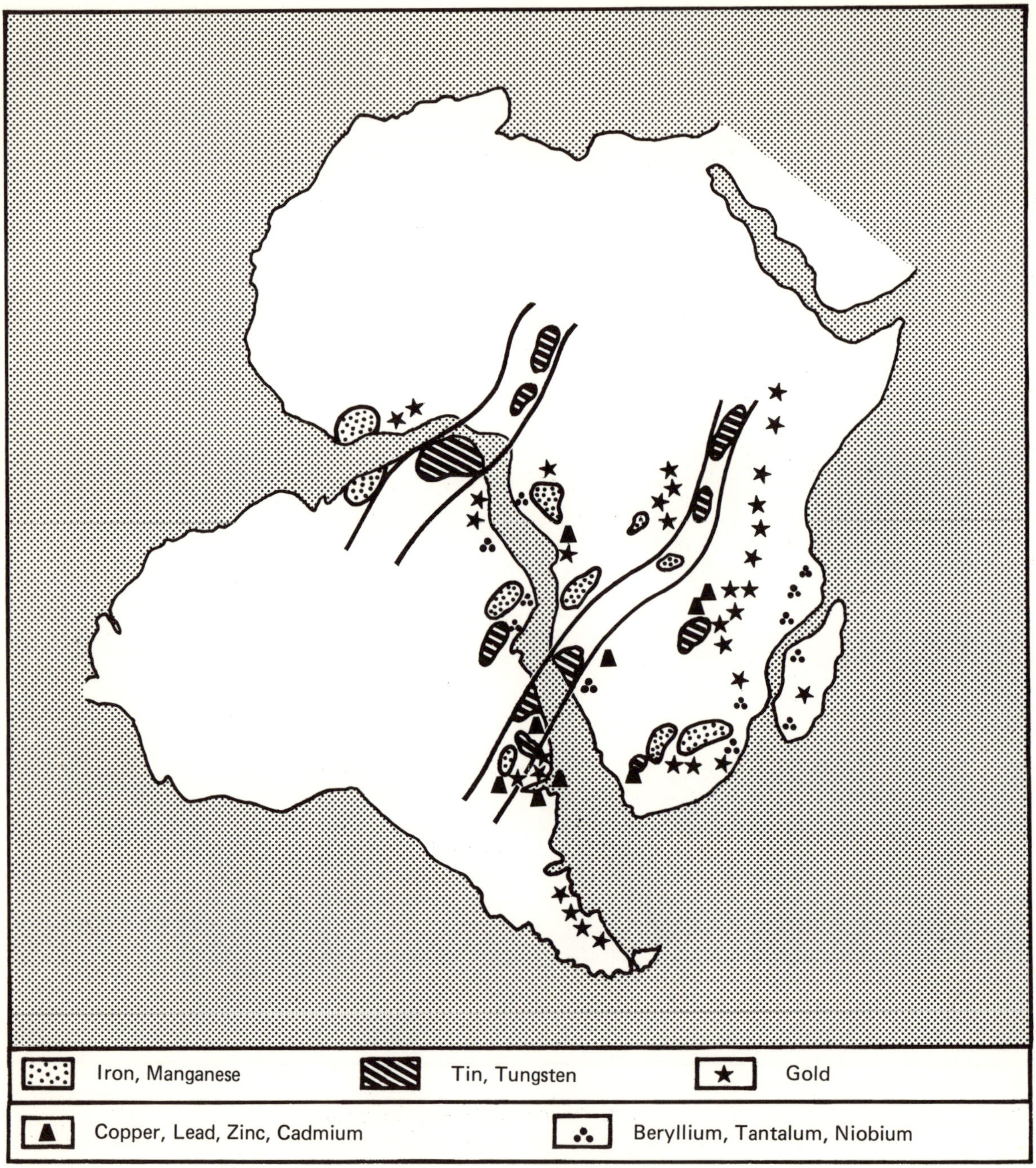


**Fig. 19.10.** The mineral belts of Africa and South America were once continuous (when their respective landmasses were part of Gondwanaland), but they are now separated by the South Atlantic Ocean. Similar relationships exist in the North Atlantic.

By knowing how the pieces of the continental jig-saw puzzle originally fitted together, the ore-seeker can arm himself with "educated guesses" as to where potential ore deposits might lie, as yet undiscovered, below covering layers.

charcoal, coal etc. into the very tightly knit structure of diamond. For example, analyses of garnets found in diamond pipes have shown that they formed at depths of 150-200km (95-125 miles) and at temperatures of 1150-1360°C (2100-2500°F). Analyses have also shown that calcium carbonate (limestone) and water are

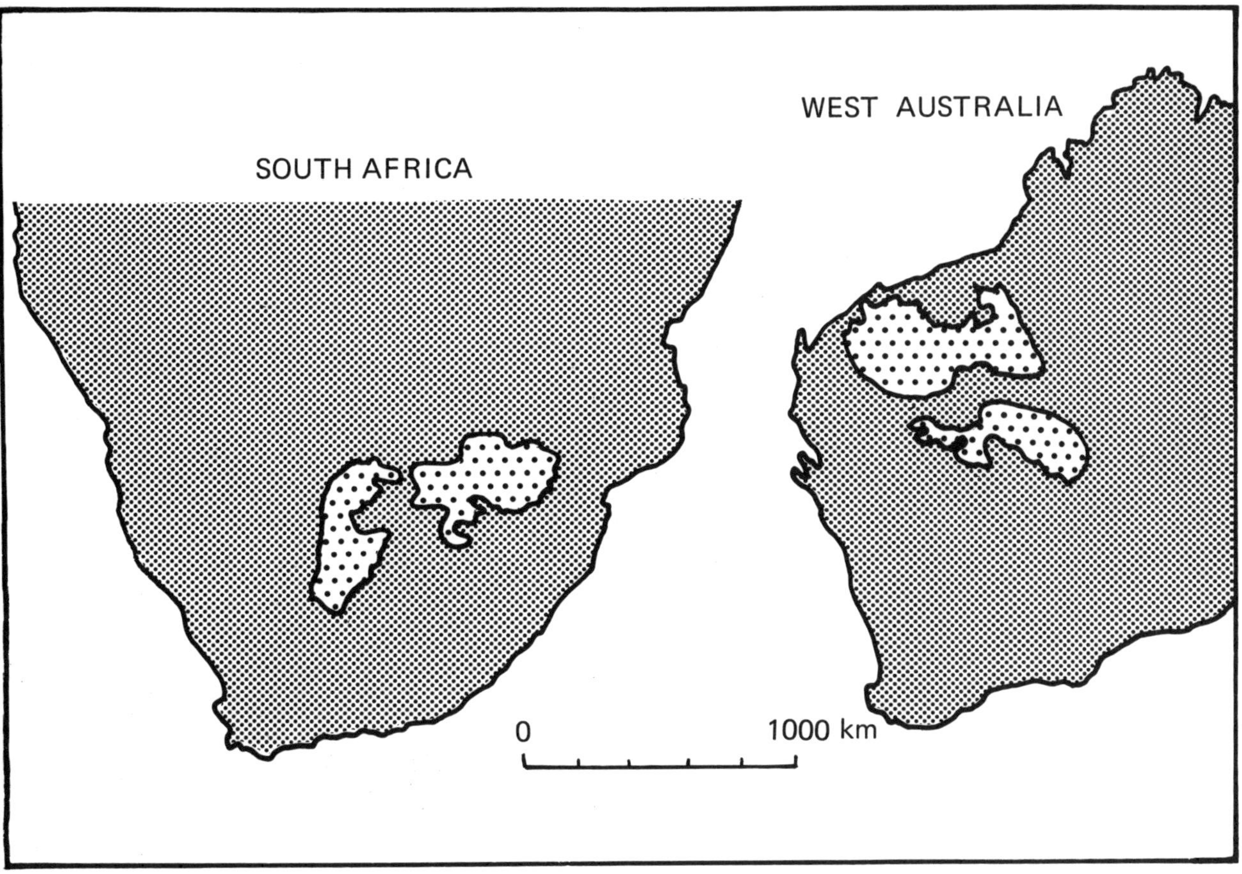


**Fig. 19.11.** The Transvaal Basin (South Africa) and the Hamersley Basin (West Australia) (both indicated on this map by the coarse dot pattern) show many remarkable parallels. They were both formed 2300-2100 million years ago as gentle downwarps of their respective continents and contain very similar sediments. These geological similarities probably date back to a time in the Precambrian when both continents were close together, and the two basins virtually continuous.

The basins have a number of mineral deposits in common including iron, manganese and asbestos. Some mineral deposits however are now known only from the Transvaal Basin, including uranium and the world-renowned Witwatersrand gold reefs. Knowledge, however, of the past close relationship of the two basins will play a key role in planning future exploration programmes.

usually associated with diamonds in the kimberlite pipes.

Thus it has been assumed that diamond pipes represent volcanic eruptions originating deep below the crust, and a mechanism has been proposed using carbon dioxide and steam (given off by the limestone and water when heated at great depths) to propel the kimberlites with great velocity from the depths so that they punch through the surface rocks with a roar like a thousand supersonic jet engines. A deep volcanic pipe is thus created, filled with products of the explosive eruption. Major stumbling blocks, however, have been the presence of carbon (to form the diamonds), water and limestone at such depths, as they are substances normally associated with surface deposits. Mechanisms to account for this, and the rationale for occurrences of diamond pipes, have hitherto been cloaked in obscurity. Plate tectonics offers a solution to both problems.

A possible mechanism, using the principles of plate tectonics, is as follows: sea-floor plates plunging downwards under continental slabs along subduction zones carry both basalt rocks and sea-floor sediments to depths as great as 700 km (435 miles). Here the high temperatures melt the crustal materials and mix them with mantle materials to form a kimberlite magma. Lime and water originally in the sediments are heated up to form carbon dioxide and steam. Carbon, originally present as peat, coal or unaltered plant material forms diamond.

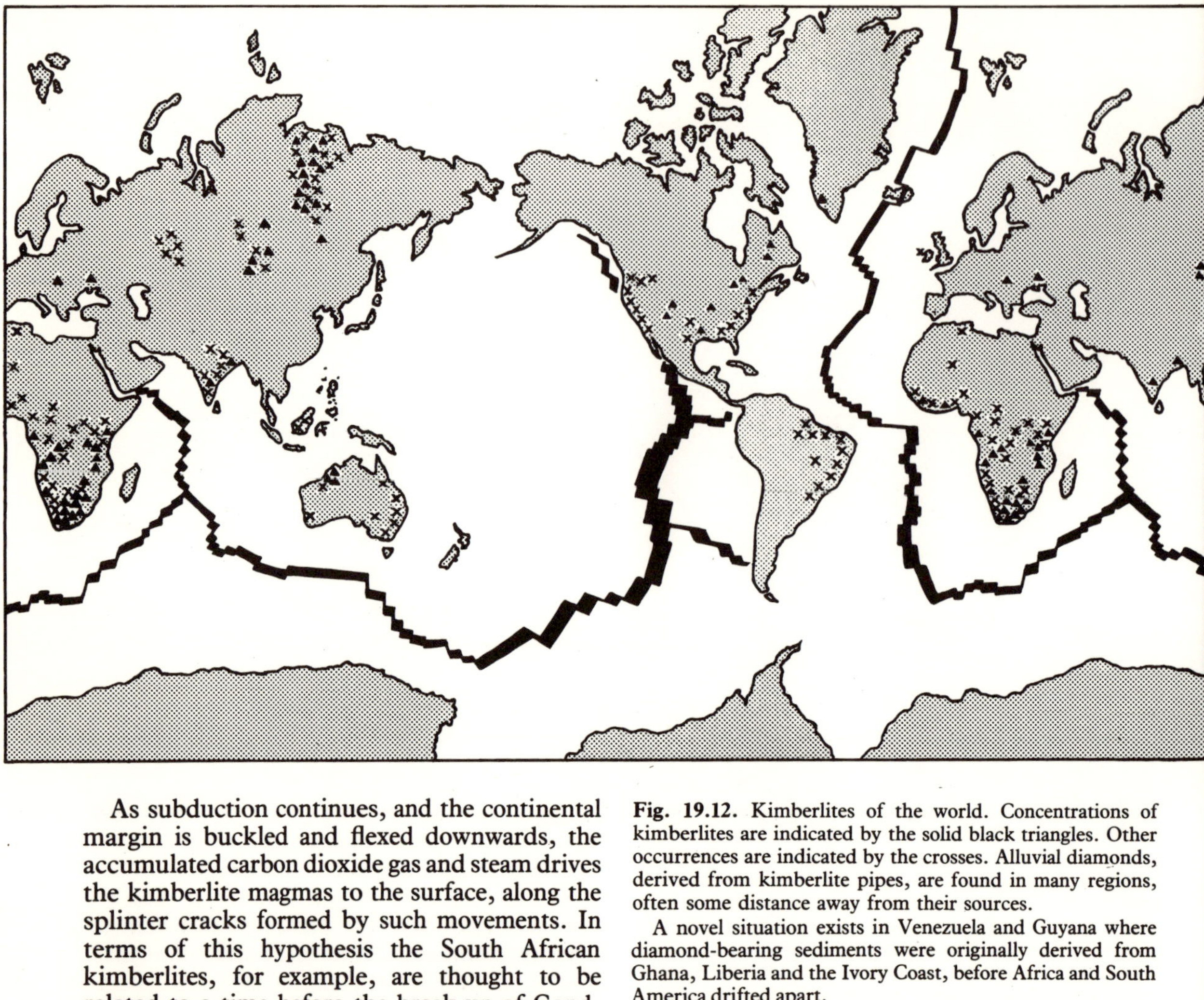

Fig. 19.12. Kimberlites of the world. Concentrations of kimberlites are indicated by the solid black triangles. Other occurrences are indicated by the crosses. Alluvial diamonds, derived from kimberlite pipes, are found in many regions, often some distance away from their sources.

A novel situation exists in Venezuela and Guyana where diamond-bearing sediments were originally derived from Ghana, Liberia and the Ivory Coast, before Africa and South America drifted apart.

As subduction continues, and the continental margin is buckled and flexed downwards, the accumulated carbon dioxide gas and steam drives the kimberlite magmas to the surface, along the splinter cracks formed by such movements. In terms of this hypothesis the South African kimberlites, for example, are thought to be related to a time before the break-up of Gondwanaland, when the Cape Town region was a continental margin, over-riding an oceanic plate. The oceanic plate dived to great depths and went hundreds of kilometres inland under the former continent. Flexing of the continental edge produced a prominent belt of folded rocks behind Cape Town (Cape Ranges, Fig. 19.13) and the associated fractures provided channel ways along which the kimberlite derived from the underlying plate travelled. The pulling-away of Antarctica and South America, as Gondwanaland broke up, may also have opened up fracture zones.

If this story holds water, kimberlites may be expected to occur in belts parallel to fold mountains associated with old subduction zones. This pattern appears to be true for the South African kimberlites and for those associated with the Appalachian and Siberian mountain chains. Other occurrences in the world may relate to still older plate boundaries.

Thus kimberlites may be in the process of being cooked up under the Tonga-Kermadec-New Zealand subduction zone and, given time, may someday pop up in the adjacent ocean.

## Sea-floor wealth

Some of the metals in short supply on land can be found on the sea floor in relative abundance. High hopes have been expressed for the walnut- and tennis-ball-sized manganese nodules that

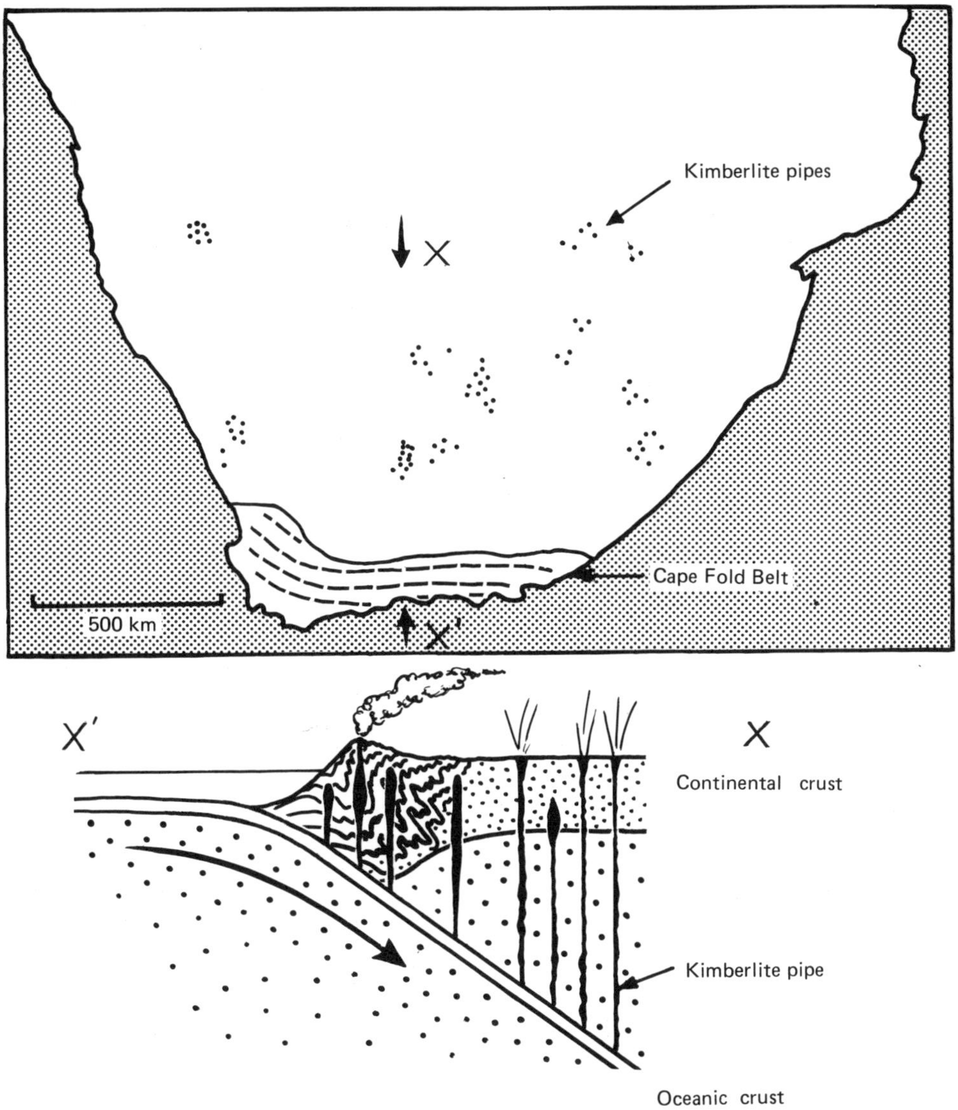


**Fig. 19.13.** A current theory postulates that diamond-bearing kimberlite pipes (Fig. 19.12) are the result of processes occurring at the deep end of a subducted oceanic plate.

As an oceanic plate descends into a subduction zone it is believed that great quantities of sea-floor sediment are also consumed. This sediment is wet and rich in lime, sulphur and carbonaceous material (derived from plant remains). High temperatures below the crust generate steam, carbon dioxide and sulphur dioxide from these ingredients and the temperatures combined with high pressures change the carbon into diamond. If enough gas is generated, a violent gas eruption may occur, and a vent is punched through the crust. The vent then becomes filled with a jumble of diamond-bearing material.

The diagram illustrates the theory as applied to South Africa. The geological history of South Africa suggests that before the break-up of Gondwanaland an oceanic plate was colliding with and under-riding the southern part of Africa. The surface collision between the plates produced a belt of fold mountains of Permian and Triassic age, the remnants of which are now the Cape ranges of South Africa. Meanwhile the subducted oceanic plate dived to great depths and extended many hundreds of kilometres inland underneath the continental plate.

As we have seen with metallic deposits (Fig. 19.8), various components of the oceanic plate and associated sea-floor sediments "boil off" at various depths. It is thought however that diamondiferous kimberlites are generated at great depths, as indicated in this diagram.

Subduction under South Africa is believed to have ceased along with the start of the breakup of Gondwanaland, in the early Jurassic, some 180 million years ago. As the super-continent began to separate, extensive fracturing occurred, reaching a maximum during mid-Cretaceous, 100 million years ago. These fractures provided some of the channel ways for the kimberlites derived from the underthrust plate.

The upper diagram shows the relation of kimberlite pipes in South Africa to the Cape fold belt, and the lower diagram a theoretical cross-section (along the line x-x$^{1}$) of the original subduction zone.

carpet the sea floor in some regions of the world. These nodules, usually 1-5 cm (½-2 in) in diameter, often contain 25 percent of manganese, an essential ingredient in steel-making. Every tonne of steel contains 6 kg (14 lb) of manganese, and no satisfactory substitute has ever been found. Apart from manganese, the nodules also contain significant amounts of iron, nickel, copper, cobalt and smaller amounts of molybdenum, tin, strontium, titanium, vanadium, zinc and barium.

The New Zealand oceanographic research ship *Tangaroa* (Fig. 8.31) and other vessels have mapped large areas of nodules in the South Pacific and similar deposits exist in other oceans. In the Atlantic an American ship *Deepsea Miner* has attempted to "vacuum" the sea floor to test the economics of such deep-sea mining.

These deep-sea nodule deposits are possibly related to mineral-laden hot waters emanating from oceanic rifts which may flow out widely across the sea floor, precipitating metallic compounds as they travel.

**The raw materials of industry**

The new understanding of the earth is also having a profound impact on the search for the many types of non-metallic minerals required by our technological society.

Although strictly speaking non-metallic minerals also include fossil fuels (oil, gas, coal etc.), and water supplies, they also include two other major groups, the importance of which should not be under-estimated:

(i) raw materials for fertiliser and chemical industries (e.g. salts, phosphates, sulphur etc.)

(ii) raw materials for building and construction industries (e.g. sand, gravel, gypsum, cement etc.)

Salts of various types form the basis of many processes in the fertiliser and chemical industries. The major source of these salts are special sedimentary beds known as marine evaporites, resulting from evaporation of sea water. Sodium chloride, or common salt, is the most abundant constituent of sea water (78 percent of total solids), followed by magnesium chloride (9.21 percent), magnesium sulphate (6.53 percent), calcium (lime) sulphate (3.48 percent), potassium chloride (2.11 percent), calcium carbonate (0.33 percent) and magnesium bromide (0.25 percent).

When sea water is evaporated, water is removed and a brine results that becomes increasingly concentrated. As it does so, the various salts precipitate as they reach their saturation points. Thus it is not necessarily the most abundant compound that precipitates first. Saturation of the relatively insoluble compounds is usually reached well before saturation of those that are highly soluble. Calcium carbonate is the first compound to precipitate from evaporating sea water. The next is calcium sulphate (gypsum or anhydrite), which starts to separate when the evaporating body of sea water has been reduced to 19 percent of its original volume. Sodium chloride (halite or rock salt) begins to precipitate when the sea water is reduced to 9.5 percent of its initial volume. Then at about 4 percent of the original volume the first potassium and magnesium salts begin to precipitate.

When we examine evaporite deposits on a world-wide scale we find that the early-formed compounds, lime carbonate and gypsum, are much more abundant and the late-formed potassium and magnesium salts are comparatively rare. It is clear therefore that the complete evaporation of an isolated body of sea water is not all that common, and it has been calculated, for example, that drying-up of the Mediterranean, which averages about 1350 m (4500 ft) in depth, would produce only 24 m (78 ft) of halite and 1.3 m (4.5 ft) of gypsum. However, beds of halite and gypsum hundreds of metres thick are known geologically, and clearly some mechanism other than evaporation of a totally enclosed basin must occur.

Two environments have been envisaged for the deposition of evaporites—the barred basin and the sabkha.

The barred basin consists of a coastal depression edged by a submerged bar, so that water can only flow in, but not out. As these trapped waters evaporate, saline brines form and, being denser, sink to the bottom where the least soluble salts are eventually precipitated. But the salinity only occasionally approaches the concentration at which the more soluble potassium and magnesium salts settle out.

Sabkha regions form when low-lying tidal flats are periodically flooded by the sea. Extensive areas of shallow marine waters are thus formed and as they evaporate, dense brines form from which soluble salts precipitate. Broad sabkhas ring the Persian Gulf area, where they have been intensively studied in order to throw some light on how evaporite deposits form (sabkha is an Arabic term applied to salt flats or salt marshes, occasionally inundated by the sea).

In both environments, barred basin and sabkha, some sinking relative to the barriers and to sea level must have occurred to account for the observed thicknesses of evaporites in the geological record. On the other hand, however, too rapid sinking would lead to flooding and solution of previously deposited beds.

The prolonged slow subsidence that is implied can probably best be matched with movements that occur as the crust and upper mantle adjust themselves after continents have split apart (i.e. isostatic movements, see pp. 2–3).

For continued deposition of evaporites to occur, the barriers of a barred basin or sabkha must be maintained over a long period of time.

Sandbars could possibly be maintained relative to sea level, but it is thought that coral reefs offer the most persistent type of shoreline barrier.

Coral reefs tie in with yet another requirement: thick deposits of evaporites probably need very high rates of evaporation, i.e. places where hot dry conditions are accompanied by strong on-shore and off-shore winds. Coral reefs and the requisite climatic conditions together point to the tropics and, as we have seen, the positions of the continents relative to the equator have varied considerably over geological time. Thus studies of past continental drift movements (involving palaeomagnetism, oxygen isotopes, sea-floor spreading etc.) and the palaeoclimatic implications of such movements, provide the key to pin-pointing likely areas of evaporites and their associated mineral wealth.

Another bonus of evaporite-hunting is linked to the fact that the concentration of nutrients in warm saline waters results in rich blooms of micro-organisms. Thus organic-rich debris, ideal source material for oil and gas, is often formed in close association with evaporites. Yet another bonus is that evaporites tend to flow under pressure and, as they do so, they lubricate earth movements and help create and seal off structures to trap and retain oil and gas. We find, therefore, that many of the major oil and gas fields of the world are associated with evaporites, e.g. the North Sea, continental U.S.A. and Canada, Gulf of Mexico, Iran, Iraq and the Persian Gulf.

During the late Palaeozoic and early Mesozoic exceptionally thick evaporite sequences were formed in North America, Western Europe and the U.S.S.R. and when we look at continental reconstructions for this time (e.g. Figs 5.8, 5.10), we see that continental drift had created favourable conditions for the accumulation of evaporites by grouping these landmasses so that they straddled the tropics. Furthermore, the aggregated landmasses increased the "continentality" of the climate (producing drying winds etc.). At the same time the continents had little topographic relief—so that inland lakes and broad shallow extensions of the sea were common. The warm sea teemed with life—so again the scene was set for a close association of evaporites, oil and gas.

The thick evaporite sequences of North America, Western Europe and the U.S.S.R. provide most of the supplies of potassium (potash), one of the three essential fertilisers (the others being phosphorus and nitrogen). With increasing food demands it is not surprising that consumption of potassium salts has been increasing at a rate of almost 10 percent per year—hence its importance on the world scene.

Another important evaporite mineral is sodium chloride (rock salt or halite). Although salt is an essential ingredient in our diets, human consumption accounts for only a small portion of the total production, as most goes into the manufacture of chlorine and soda ash (sodium carbonate) in the chemical industry and for de-icing of roads.

Sulphur, used in both chemical and fertiliser industries (for the manufacture of insecticides and fungicides; and also superphosphate and ammonium sulphate fertilisers), is derived from evaporite deposits, although it can also be obtained as a by-product of the reduction of various sulphide ores.

Magnesium sulphate (epsom salt) and barium sulphate (heavy spar) are other examples of important evaporite minerals.

Evaporites also contribute to the building and construction industry. Plaster, made by heating gypsum, is one of man's oldest building materials—yet its consumption continues to grow world-wide. Heating gypsum at 177$^{o}$C (350$^{o}$F) drives off 75 percent of the water and changes it to "plaster of paris". When plaster of paris is mixed with water it reverts to a finely interlocked mass of tiny gypsum crystals. Plaster is a versatile material and can be used on its own, mixed with sand, or incorporated in wallboards.

Still other industrially vital minerals such as sodium sulphate, sodium carbonate and boron minerals (e.g. borax) are also found in evaporites, but only in those formed by evaporation of large inland freshwater lakes.

Rift features on the earth's surface can also serve as the focus for non-metallic as well as metallic minerals (pp. 383–386), and the East African rift valleys are rich sources of a wide variety of sodium, potassium and magnesium salts, as well as gypsum and phosphorus compounds.

**Picking the next earthquake**

The modern continental drift theory supplies new enlightenment in the field of earthquake studies. Thus although the horror of earthquakes has not been dispelled, there is no longer very much uncertainty about how most of them originate.

In every part of the world most earthquakes can now be related to interactions of the moving plates of the earth's surface (Figs 9.1, 9.4). In regions where plates converge, earthquakes are generated by the collisions between plates and the descent of plates into subduction zones. In regions where plates diverge, pulling-apart forces associated with rifting generate still other earthquakes. All these earthquakes, related to large-scale straining and fracturing of rocks at active plate boundaries, can be grouped together as plate-edge earthquakes.

In general the earthquakes associated with converging plate boundaries are more hazardous than those associated with divergent plate boundaries.

This is because the earthquakes generated by pulling-apart movements (or tension) of the crust are often related to simple "snapping-apart" breaks or collapses. Also, as upwelling of magma and outbursts of volcanic activity accompany such movements, many of the tensional earthquake shocks are related to deep-seated volcanic explosions, to magma moving upwards towards the surface, or to swelling and contraction of the surface rocks above magma reservoirs. Earthquakes in Iceland and along the rift valleys of East Africa, for example, are related to pulling-apart movements and volcanism.

On the other hand, the sliding, crushing and shearing movements associated with crustal plates being pushed together, pushed under, or dragged sideways, often produce large devastating earthquakes. This is because the forces driving the plates together continue inexorably, so that stresses and strains build up along the plate edges to such an extent that elastic energy is stored up in the rocks—so that they act rather like a compressed coiled spring. Sooner or later the strains reach levels greater than can be endured by the rock. Ruptures then occur, concentrated along fault lines, and the strained rock rebounds until the strain is partly or wholly relieved. The rupturing movements spread in all directions along the fault surface in a series of erratic movements caused by the uneven strength of the rocks along the tear. This leads to bursts of high-frequency waves which travel through the earth to produce the shaking that causes damage to buildings.

The very hazardous earthquakes that often occur in Chile, Peru, the eastern Caribbean, Central America, Mexico, California, southern Alaska, the Aleutians, the Kuriles, Japan, Taiwan, the Philippines, Indonesia, New Guinea, Tonga, New Zealand and the Alpine-Carpathian-Caucasian-Himalayan belt are of the type generated by the edges of converging plates. Volcanoes, often of a very violent type, are also associated with such convergent plate boundaries, and are a source of earthquakes too, but generally of less severity than those related to actual plate-edge movements.

Notable recent examples of very severe plate-edge earthquakes are the Alaskan Earthquake (1964), related to the Aleutian subduction zone, and the Guatemala Earthquake (1976), related to the Central America subduction zone.

The various large earthquakes that have struck different parts of New Zealand in historic times (Fig. 18.17) are all related to the fact that New Zealand straddles the boundary between the Indian-Australian and Pacific plates, although as we have seen (pp. 366–370), the plate movements are spread across a broad zone of individual faults.

As the mechanics of lithospheric plates become better understood, long-term predictions may be possible for plate-edge earthquakes. For

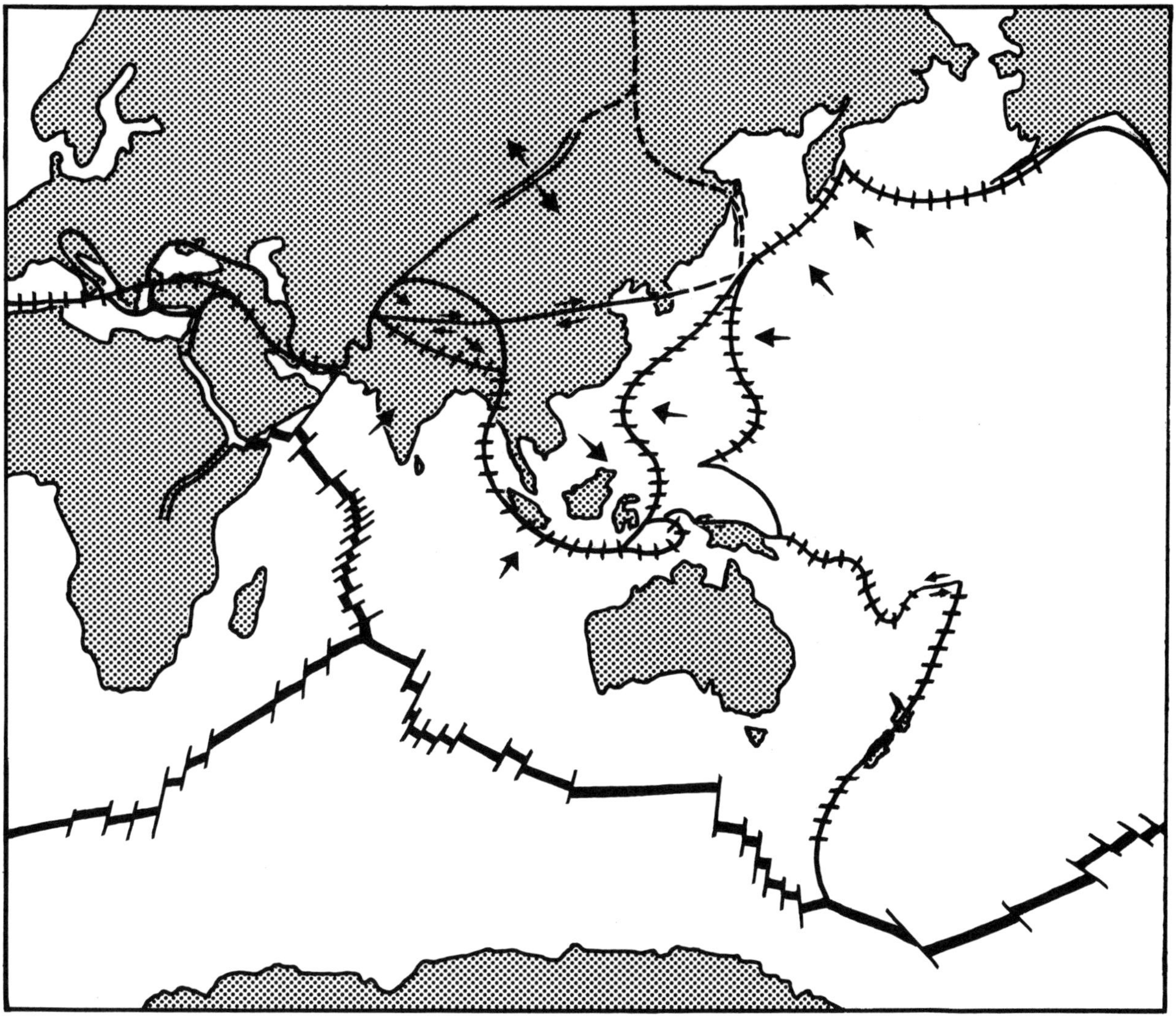

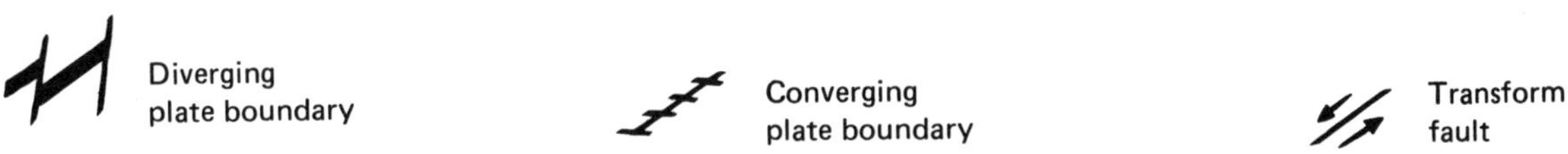


**Fig. 19.14.** Many large earthquakes are related to major plate boundaries, particularly where plates are being pushed together, or slid past one another (Figs 9.1, 9.4).

Other large earthquakes, however, are related to smaller "buffer" plates which in places hold apart the jostling major plates of the world, and in this way help to ease their movements.

This map depicts the oceanic ridge system (divergent plate boundaries) and the consuming (convergent) plate boundaries. Buffer plates are believed to exist in Tibet and China, helping to relieve the crustal stresses accumulating in these regions as the Indian-Australian plate encroaches onto Asia.

The arrows on this diagram indicate directions of crustal movements around China. Large earthquakes are generated by the two Chinese plates and those of Tibet and Sinkiang, as they rotate like ball bearings between the major plates.

example, analysis of sea-floor magnetic stripes indicates that many plates move towards subduction zones at rates of from 20-50 mm (1-2 in) or more per year. Therefore in particularly active regions knowledge of the history of large earthquakes, and the overall plate movements, *may* help in identifying areas that currently lag in earthquake activity (and hence may be in a locked condition and storing energy) and could be prime candidates for future large earthquakes.

Short-term earthquake prediction, however, although set within the general framework of expected plate movements, depends more on the detection of various tell-tale signs that precede a major shock. Although these studies are still very much in their infancy, it looks as though there is a certain amount of leakage of energy before a large earthquake. This manifests itself in a variety of ways—twisting of the land, generation of swarms of micro-earthquakes, disturbance of water levels and other physical properties of the earth's surface etc. With the benefit of hindsight, we can now see that some of these effects preceded the 1968 Inangahua Earthquake.

Current earthquake prediction methods rely heavily on the monitoring of these precursor events and both the Chinese and Japanese have had some notable successes—but also failures—so obviously we have a long way to go yet! In the meantime, however, our new understanding of the earth is providing new avenues for research.

**Intra-plate earthquakes**

While the plate-tectonic theory is an important one for a general understanding of earthquakes, it does not tell the whole story, for sometimes large devastating earthquakes occur well within continents, away from plate edges.

In the United States the most famous of these intra-plate earthquakes are those that occurred in 1811-1812 in the New Madrid area of Missouri, along the Mississippi River. However, the absence of an explanation for intra-plate earthquakes may merely reflect our current ignorance. Thus although a world-encircling system of plates has been worked out (Fig. 9.4), it has only lately been realised that small "buffer plates" exist between many of the larger crustal plates and that often they act as roller bearings, easing the progress of their ponderous neighbours. Movements related to such buffer plates may be responsible for some intra-plate earthquakes.

Small chips of the earth's surface have been identified along the west coast of North America, the rotation and distortion of which has facilitated the movement of the Pacific and North American plates.

The Mediterranean is another buffer region, where smaller plates (such as the Aegean and Turkish plates) have accommodated the encroachment of Africa on Europe (Fig. 19.18). On a larger scale, it is believed that buffer plates in Sinkiang and Tibet help to alleviate the crustal stresses set up between India and Asia (Fig. 19.14).

Thus the disastrous Chinese earthquakes that have occurred recently (one of which was predicted, so that many casualties were averted, but another, in Tangshan, in July 1976 killed, according to the Western press, an estimated 650,000 people and injured 780,000), supposedly in the centre of the Asiatic plate, are now seen as being related to movements of buffer plates, between Eurasia to the north, Sinkiang and Tibet to the west, India to the south, the Philippines to the south-east and the Pacific to the east (Fig. 19.14).

Another crustal feature that has recently been demonstrated, and one with possible significance for intra-plate earthquakes, is the existence of detached plates, dating back to past subduction, but yet still in existence under continents. Detailed earthquake studies linked with detection of Soviet nuclear bomb testing have revealed the presence of one such relic slab lying deep below Novaya Zemlya. Although it is surprising that it has survived for so long, this slab is nonetheless interpreted as being the remains of an old oceanic plate that was swallowed up some 250 million years ago when Asia collided with Europe, the two becoming welded together along the line of the modern Ural Mountains.

Other as yet undetected slabs may lie under other continents, their periodic movements generating infrequent and often unexpected intra-plate earthquakes.

**Spotting the next volcanic eruption**

Earthquakes and volcanoes often go hand-in-hand and, as we have said on p. 398, these twin hazards are intimately related to on-going plate movements.

Like earthquakes, volcanoes are generally a phenomenon of plate edges, although some are related to intra-plate "hot spots" (like Hawaii etc; see p. 127). Again, like earthquakes, the volcanoes associated with converging plate edges are generally more hazardous than those of divergent plates.

The "pulling-apart" that occurs when plates diverge allows melting to occur in the sima and as a consequence magma works its way up to the surface, where it wells out to form basalt volcanoes.

Basalt lava normally comes up very hot and very fluid and therefore usually flows very readily and does not cause obstructions in the crater or volcanic pipe. The fluidity of the lava means that the gases accompanying the lava can expand and be released without violent explosive activity. Also as basalt often originates as direct melting of the sima it is usually uncontaminated by continental rocks containing quantities of water, oxygen, carbon, sulphur, lime etc. that, when activated in a magma, can produce potentially explosive mixtures. Thus although basalt eruptions are often very spectacular—with vivid "fire-fountains" (jets of hot gas being liberated from the lava), and rivers of hot lava—they are usually reasonably safe, as far as loss of human life is concerned.

The Heimaey eruption of 1973, for example, occurred virtually in the suburbs of Vestmannaeyjar, Iceland's most important fishing port. But yet, although lava flowed into part of the harbour (and incidentally created a natural breakwater), and ash covered much of the city, no deaths occurred, the streets and houses were soon cleared, and life now goes on as usual.

A different situation occurs, however, when two plates converge and one plate sinks and is melted up. The sinking plate carries down with it a lot of continental material, in the form of wet sea-floor sediment, and this, when melted and mixed with the basalt of the sinking plate forms a very explosive "brew".

The admixture of continental material contains all the ingredients to produce vast quantities of steam, carbon and sulphur dioxide gases, acid vapours etc. As well, the mixture gives rise to magmas that are often very stiff and viscous, but are at the same time full of gas.

Thus volcanoes of a hybrid type ("intermediate" or "andesitic") are built up along convergent plate boundaries. Depending on the type of "brew" being produced down in the depths, volcanoes of these types can be potentially very dangerous. Sometimes stiff lava flows emerge from the crater as occurred from Ngauruhoe in 1949 and 1954. At other times, the lava is blown to pieces in the volcanic pipe and is spread far and wide as volcanic "ash" (although it is not a product of actual burning). This happened in the 1945 eruption of Ruapehu.

On yet other occasions the lava is so stiff that it completely clogs the volcanic pipe, sealing in all the gases and heat. Enormous volumes of gas build up and temperatures are raised to white-heat. Then suddenly the volcano "blows its top", producing stupendous explosions, blowing the lava plug to smithereens, and sending out fiery clouds of searing hot gases and ash that travel across the landscape at speeds up to 100km (60 miles) per hour. Boiling hot ashes and mud blanket vast areas.

In the Pleistocene (between about 2 million and 10,000 years ago), and up to times as recent as A.D. 130, such eruptions, accompanied by vast outpourings of fiery clouds, have occurred repeatedly in the Taupo-Rotorua region. These eruptions virtually filled the entire centre of the North Island with vast sheets of rock, composed of fragments of volcanic material, welded together to produce an often glassy rock, called ignimbrite (i.e. "fire-rock").

In more recent times, in 1902 on the island of Martinique in the Lesser Antilles, the volcano Mont Pelée loosed off an enormous fiery cloud (called a *nuée ardente*) that in a matter of moments destroyed the capital city of St. Pierre, killing 35,000 people. Closer to home, the volcano Mt Lamington in New Guinea exploded in 1951, and fiery clouds killed some 3000 people.

Sometimes the continental materials carried down subduction zones to great depths under continents generate so much gas in the form of steam, carbon dioxide and sulphur dioxide etc. that violent gas eruptions come roaring up from below the base of the crust, punching their way through the overlying rocks. Such high-velocity gas eruptions sweep up fragments from along their entire route, and after being thrown high in the air, they fall back to form a jumbled mass of material filling the throat of the volcanic pipe. Such gas eruptions (one occurred at Nilahue, Chile, in 1956) are thought to generate the diamond-bearing "kimberlite" volcanic pipes that punctuate the surface of some of the older continents (pp. 391–394).

From the last few pages it may be appreciated that recognition of the plate motion involved in any particular region—whether it is convergent or divergent—provides a guide to the nature of likely volcanic activity. However, while modern plate boundaries are readily defined, fossil boundaries are less obvious, and these may present hidden dangers, as isolated slabs of crust may exist deep under continents and, as they are slowly digested, may generate volcanic outbursts very widely spaced in time. Rome, for example, lies on ignimbrites deposited by prehistoric eruptions of enormous magnitude, related to the collision between Italy and the rest of Europe (Fig. 9.17) that began some 60 million years ago, and it is not at all certain that we have seen the last of such eruptions.

Presently active volcanoes are usually recognised as such, and more often than not are closely watched by all and sundry. Nevertheless the tremendous forces generated by a volcanic eruption, and the apparent inevitability of their consequences, have often led to a feeling of fatalism—the volcano would have its way! However, although the horror of volcanoes, like that of earthquakes, has not been dispelled, there is hope that, armed with our new knowledge of their underlying causes, we can lessen or perhaps even eliminate some volcanic disasters.

As we have seen in the study of earthquakes, the new concepts of earth processes and structure have provided vulcanologists with a fresh theoretical framework and a better understanding of how volcanoes work. These new concepts are contributing towards solving the problem of how humanity can learn to live with volcanoes. Throughout the world millions of people live in the shadow of volcanoes—often attracted by the broad expanses of fertile land around their base—and undoubtedly will continue to live there. It is therefore imperative that vulcanologists learn how to warn of the time, type and place of coming eruptions, how to avoid or alleviate their effects and how to speed the recovery of devastated land. Indeed in our energy-conscious times we should go beyond this and learn to make direct use of the volcano and its prodigious energy.

A better appreciation of fundamental crustal processes that impinge upon a volcano will help modern vulcanology achieve these practical goals. To assist in these studies we need an adequate distribution of instruments (to measure earthquakes, heat flow, magnetism, earth electricity, tilting and swelling of the ground, etc.) and alert, trained personnel to interpret the records. Above all, we need to thoroughly know the surrounding geology, so that the volcano can be viewed in its regional setting.

**Active, dormant, extinct . . .????**

One of the very practical problems of vulcanology is expressed by the simple question: "When is a volcano dead?" If it really is *extinct* it presents no further risk. But volcanoes can remain quite *dormant* for many thousands of years, only to return to activity, and often the activity that marks the return is exceptionally violent. Some of our most destructive eruptions have come from volcanoes that had no previous eruptions during all of historic time, and therefore were not regarded as active. These include, for example, the eruptions of Santorini (Thera), 1400 B.C. (pp. 402–406); Vesuvius, A.D. 79; Mt Lamington, A.D. 1951. It is often said that these returns to activity were unheralded, but in most instances the warnings (earthquakes, unusual hot spring activity etc.) were simply not recognised as being warnings of impending eruptions.

Although study of such short-term warnings is a prerequisite for forecasting such "unexpected" activity, the concepts of continental drift and plate tectonics help provide a long-term rationale so that the vulcanologist can visualise better what is going on deep in the earth. For example, the Cascade Range in the north-western United States has a string of at least 13 major volcanoes, extending through California, Oregon and Washington states. Of these only Lassen Peak, in northern California, has erupted in this century. Another five have been active within the last 150 years. It is quite clear, however, from the geological record of past eruptions that all of the volcanoes have been active over the last 12,000 years and are in the "dormant" class rather than "extinct". This appraisal is confirmed by what we know about the plate-tectonic situation—these volcanoes all being related to the melting-up of the Pacific plate, as it is subducted under the North American plate. As we know from studies of earthquakes, sea-floor magnetic stripes etc. that this subduction is continuing, it follows that the volcanism will continue into the future also—the same type, frequency etc. as in the past. It is reasonable therefore to expect a major eruption to occur sometime in the future. Meanwhile the sleeping volcanoes overlook cities such as Tacoma and Seattle, together with increasingly popular national parks, and regions being utilised for forestry and hydro-electric power generation. As these areas are expanding rapidly in terms both of population and facilities, when the next eruption occurs, more lives and property than ever before will be endangered. Although there is no way now to predict when

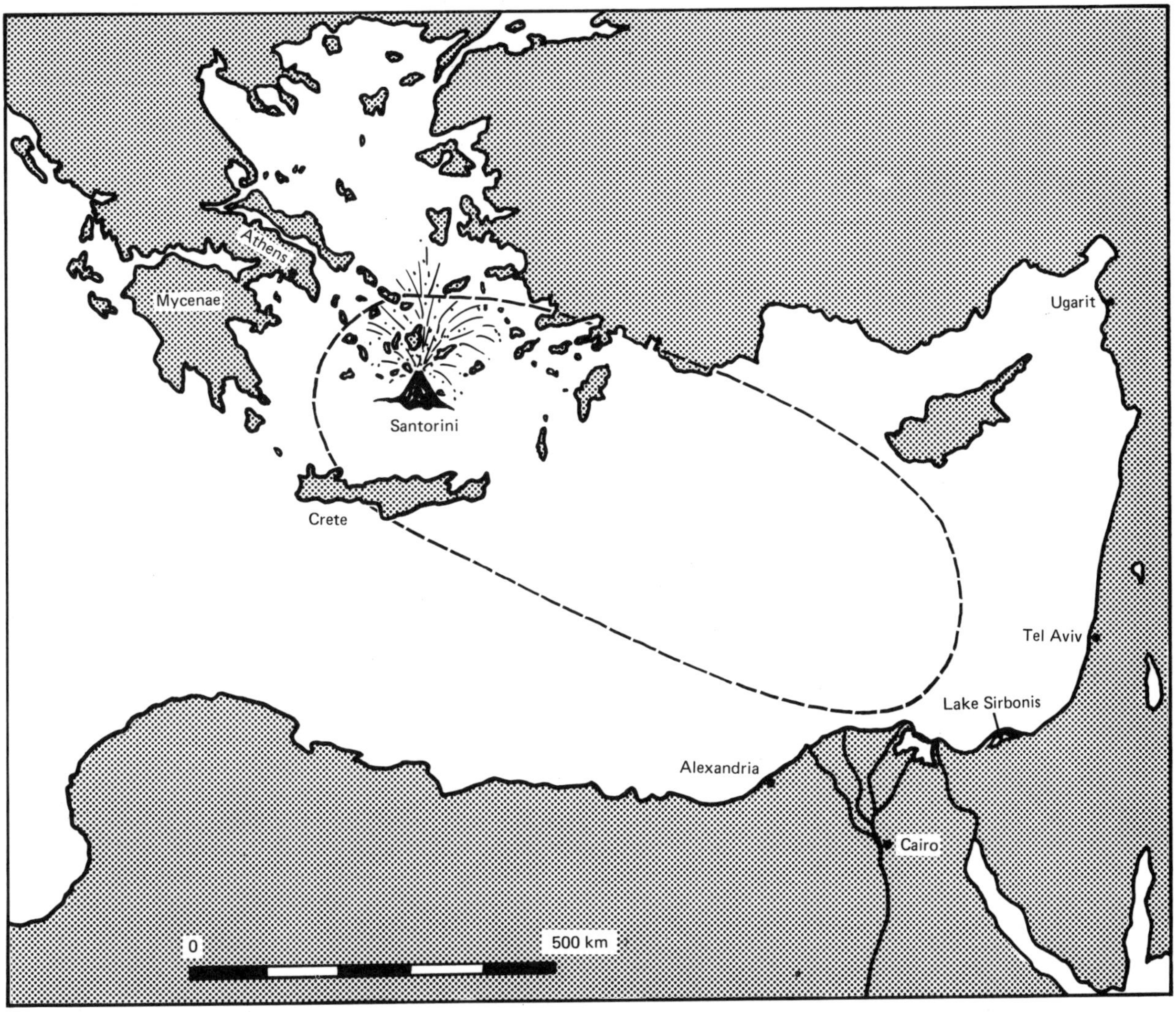


**Fig. 19.15.** The cataclysmic eruption of Santorini, an island in the Cyclades group of Aegean islands, in 1400 B.C. wiped out the Minoan civilisation and in doing so shaped the destiny of the Western world. Santorini itself was blown apart (see Fig. 19.16), giving rise, it is thought, to the legend of Atlantis.

Volcanic ash was widely spread throughout the eastern Mediterranean and has been identified both on land and on the sea floor. Its approximate distribution is indicated by the broken line—although in some parts of the sea floor it has been removed by swift-flowing currents, or deeply buried by great thicknesses of sediments (e.g. off the Nile delta). To judge from historical records, great quantities of ash reached the Nile delta of Egypt, causing deaths and famine. Tidal waves following the eruption swept the entire Mediterranean—being recorded, for example, from Tunisia and the Syrian coast, where the ancient port of Ugarit was drowned.

activity will resume at specific volcanoes, nevertheless the potential volcanic hazards of the future should be borne in mind when patterns of land use are developed, because once such patterns are established they are difficult to change.

### The volcanic paroxysm that changed world history

The volcanoes of the north-western U.S.A. are examples of dormant volcanoes related to a relatively active plate boundary. We know from sea-floor-spreading data that the Pacific is being subducted under the American plate at a rate of 12 cm (5 in) per year—the fastest known—and situations comparable to that in north-western U.S.A. exist all along the American cordillera, from Alaska to Patagonia.

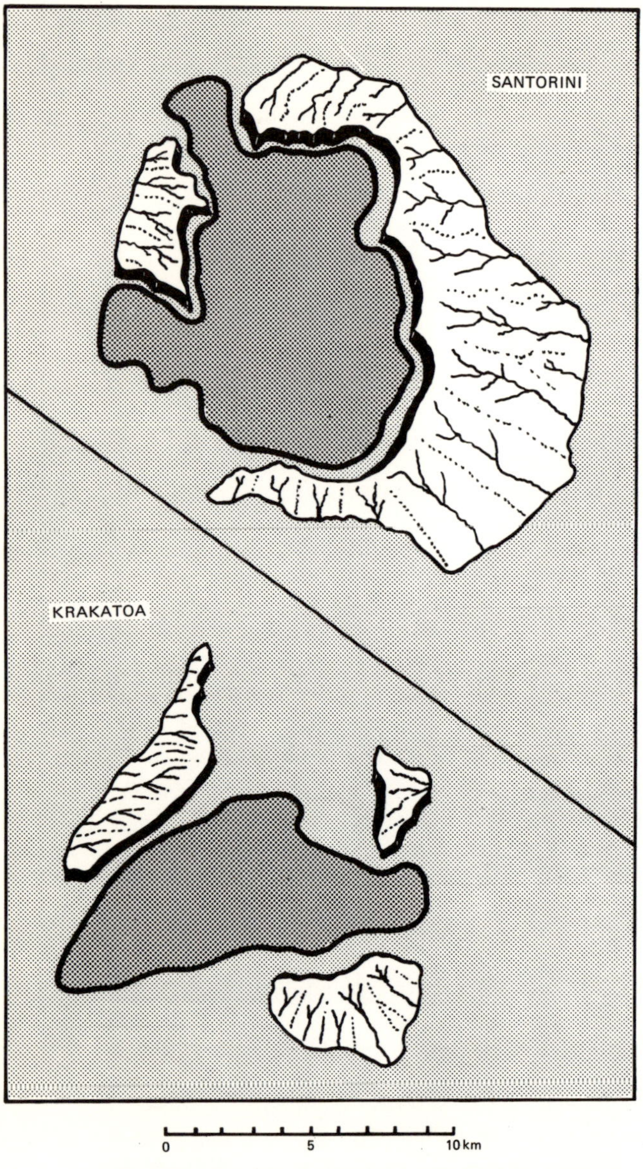


**Fig. 19.16.** The island of Santorini was a well-rounded island about 16 km (10 miles) in diameter and some 1500 m (5000 ft) high before being blown apart in the catastrophic eruption of 1400 B.C. Now ash-covered fragments surround a caldera (a collapsed area above an exhausted magma chamber) 83 sq km (32 sq miles) in area and some 400 m (1300 ft) deep, created when the centre of the volcano collapsed after showering the eastern Mediterranean with vast quantities of ash. The caldera is shown in the top diagram by a dense shading.

By comparison, Krakatoa was an island some 440 m (1460 ft) high before exploding in 1883. Now it consists of three fragments surrounding a caldera 22 sq km (8 sq miles) in area and some 200 m (660 ft) deep (bottom diagram). By comparing the amount of island blasted away in each case, scientists estimate the Santorini eruption to have been much larger than that of Krakatoa—at least four times as large.

Many of the other plates in the world, however, are moving very sluggishly, the rate of subduction is much slower, and the volcanic activity very widely spaced. Some of the volcanoes in these situations may lie dormant for 10,000 or 20,000 years, or longer. This long period of dormancy often sets the scene for disaster.

As the signs and memory of the last eruption fade, people tend to regard the sleeping volcano as a perfectly harmless and "ordinary" mountain and begin to populate and cultivate its fertile well-drained slopes (volcanic rocks often break down to give excellent soils), without giving a moment's thought to impending danger. Then suddenly the volcano springs to life, with catastrophic results.

A prime example of this is the eastern Mediterranean, an area often ignored when talking about volcanic risk. But yet some 3500 years ago it was the site of perhaps the largest volcanic calamity known to humanity, and one moreover, that changed the course of history.

In 1900 an English archaeologist, Sir Arthur Evans, uncovered the ruins of a magnificent city, Knossos, on the north coast of Crete. The excavations revealed sumptuous colonnaded palaces and villas with balconies, staircases, swimming baths, piped water supplies and flush toilets, and air conditioning (obtained by channelling cool breezes). The people who built and inhabited these splendid structures are called Minoans.

The copper and bronze age Minoan civilisation, that first began about 3000 B.C., preceded the flowering of classical Greek culture. Throughout Minoan times ancient Greece was inhabited by primitive tribes. Then about 1400 B.C. the Minoan empire suddenly collapsed and cultural dominance passed to mainland Greece.

The Minoans used a sophisticated form of writing and enjoyed a variety of sports, including boxing, wrestling, archery and bull games in which contestants vaulted over the horns of the charging animals. They created superb vases, exquisitely crafted gold and copper ornaments and beautiful wall paintings—all of which would be classed as high fashion today. The Minoan Empire dominated the eastern Mediterranean from Crete to the Peloponnese, and Minoan fleets ranged far and wide and even traded with far-off Egypt.

With the passage of time the remains of other Minoan settlements were found on islands between Crete and mainland Greece. Then in 1956 Minoan remains were found on Thera, the main island of the Santorini group of islands, a part of the Cyclades, in the Aegean Sea, some 100 km (62 miles) north of Crete. These remains were found under pumice and ash deposits in places up to 66 m (216 ft) thick. Earlier, while the Suez Canal was being built, the pumice of Santorini was highly sought after, as it could be used to make an impermeable cement. In those days great masses of it were removed, eventually laying bare an entire ancient land surface under the pumice: farming lands once tilled by the plough, and remains of olive trees and houses.

The 1956 discoveries were found in the old Suez Canal quarries and further excavations exposed a group of fire-blackened ruins of stone houses. Charred roof timbers lying inside the houses enabled a radio-carbon date to be obtained—disclosing that the ruins dated back to 1400 B.C.—i.e. the fifteenth century B.C. At about the same time and also extending into 1964 a series of borings in the sea floor between Crete and Santorini revealed widespread deposits of volcanic ash, thickening towards Santorini. Dating of these deposits coincided with the age of the Minoan remains on Santorini. It thus became clear that a vast volcanic eruption, centred on Santorini, had occurred in Minoan times.

Later work found pumice and ash derived from Santorini mantling the floor of the entire eastern Mediterranean and much of Crete and the range and depth of these deposits spoke volumes for the terrific power of the eruption.

Prior to its destruction the island of Santorini had been a jewel in the crown of the Minoan Empire. It was a large rounded island, rising to 1500 m (5000 ft). Its harbour was crowded with ships and its fertile slopes were terraced with vineyards and olive groves. People bathed in the warm springs that gushed from the sacred mountain in the centre of the island.

To judge from the ash layers preserved on the sea floor, Santorini had last erupted some 25,000 years ago—so the Minoans had no idea that they were in fact living on a resting volcano.

Then the internal fires that had lain dormant under the island for so long suddenly sprang into life and the mountain heaved, roared and then blew up in a volcanic eruption of immense violence (Fig. 19.15).

The entire central part of the island was blown away, leaving a deep hole in the sea. The pieces that remained—the islands of Santorini today—were buried under thick deposits of pumice and ash.

The volcanic disaster of Santorini must have been very similar to the Krakatoa eruption in Indonesia in 1883 (Fig. 6.9). When Krakatoa erupted, the roar could be heard 4800 km (3000 miles) away (Fig. 6.9) and vast quantities of ash were carried around the world. An hour after the eruption Java and Sumatra, 160 km (1000 miles) away, were plunged into total darkness that lasted two days. Then four days later the sun was blotted out in Japan.

Collapse of the roof of the Krakatoa volcano generated a huge tidal wave. The crest of this wall of water was 36 m (118 ft) high as it crashed along the nearby coastline, razing woods, plantations and villages and dragging everything in its path back into the sea. On shorelines in Sumatra, 88 km (55 miles) away, the tidal wave was still 15 m (49 ft) high and even on the coast of Ceylon it arrived at a height of 2-5 m (6-16 ft). Thirty-six thousand people perished in the tidal wave.

Although Santorini's eruption probably paralleled that of Krakatoa there were major differences in scale. The original Santorini island was higher (1500 m, 5000 ft, compared with Krakatoa's 440 m, 1460 ft), the collapsed crater in the sea bed four times as big and the thickest pumice deposits 66 m (216 ft), compared with 30 cm (less than 1 ft) deposited by Krakatoa. Therefore all the indications are that the Santorini eruption was considerably more violent and extensive than that of Krakatoa—and may have been *at least* four times its size (Fig. 19.16).

Now all that is left of Santorini is a crescent-shaped skeleton of three islands—with settlements and cultivation on their green outward-facing slopes, but on the inside, circular walls of high cliffs exposing layers of black, red and white ash and pumice, plunging into an enormous submerged crater, ranging down to depths of 400 m (1300 ft), the remnant of the collapsed central volcanic pipe.

The volcanic cataclysm swept the entire Minoan civilisation from the face of the earth. The city of Atlantis, on Santorini itself, died first and in so doing gave birth to the celebrated myth.

Meanwhile elsewhere the explosion, ash falls and earthquakes all contributed their share of disaster and a tidal wave following hard on the volcano's collapse devastated the entire coast of the eastern Mediterranean. The catastrophe caused so much confusion over its nature and cause that the ancients never knew what had hit them.

Until the recent geological discoveries, the obliteration of the glittering Minoan civilisation was an intriguing mystery, attributed to revolution or invasions. Now many historians link the destruction with Santorini's eruption and the ensuing tidal waves.

The Minoan settlements on scattered islands and coastlines were overwhelmed by the combination of blast, hot ashes, poisonous gas, and tidal waves. On Crete itself the heavy mantling of volcanic ash made agriculture impossible and survivors soon starved or died of illnesses (conjunctivitis, angina, bronchitis, digestive disturbances) as a result of inhalation and ingestion of volcanic ash.

Some Minoan survivors did, however, manage to reach the Greek coast. Although battered by tidal waves, Greece had not suffered from the volcanic ash falls, thanks to the north-west wind prevailing at the time that had instead carried the ash towards Crete and onwards to Egypt and Israel.

Legend has it that Deucalion and his wife Pyrrha, ancestors of the Hellenic race, came to Greece after floating for nine days—the only survivors of a deluge sent by Zeus to punish humanity.

The results of the Minoan immigration to mainland Greece were quickly apparent in the flowering of Mycenae about 1400 B.C., when the written history of Greece begins. The refugees taught the Greeks their alphabet and games, and helped develop the art and architecture that became the glory of classical Greece.

Many people identify the lost city Atlantis with the Minoan city that was on Santorini before the eruption. The legend of Atlantis came to the Greeks from the Egyptians who told of the overnight disappearance of the island city, leaving a handful of islets. Over the years successive generations have placed Atlantis in the Atlantic, beyond the "Pillars of Hercules" (i.e. the Straits of Gibraltar), but now there is evidence to suggest that the measurements quoted in the legend of Atlantis have been miscalculated during translations from language to language and that in fact Atlantis was closer to Greece, and was in the eastern Mediterranean. Santorini is thus a prime candidate.

Another great historical consequence of the Santorini eruption is the effect it had on northern Egypt, 725 km (450 miles) away, where the Israelites were labouring as slaves at the time.

The north-west wind blowing at the time of the eruption sent clouds of ash in the direction of Egypt and Israel. Tidal waves swept through the whole Mediterranean, unleashing their fury even on quite distant shorelines. Waterlain pumice from Santorini, deposited by the tidal waves, has been found at heights of 240 m (800 ft) on nearby islands in the Cyclades, 30 m (100 ft) on Crete, 5 m (16 ft) on the Israeli coast near Tel Aviv, and history records the overwhelming of the ancient port of Ugarit on the Syrian coast, 1025 km (640 miles) away from Santorini.

Egypt and Israel were therefore affected by noise, vibrations, tidal waves, ash, poisonous gases and darkness.

Contemporary Egyptian documents say:

"The land is utterly perished ... the sun is veiled and shines not."

"O that the earth would cease from noise, and tumult be no more!"

"The towns are destroyed ... no fruit nor herbs are found ... plague is throughout the land."

The plagues of Egypt, as recorded in the Bible, which preceded the exodus of the Israelites from Egypt, may all be considered as direct or indirect consequences of Santorini's prodigious eruption.

According to Exodus 7-11, the plagues were:

1. The waters turned to blood
2. Frogs covered the land
3. Lice afflicted man and beast
4. Swarms of flies
5. Murrain (a cattle disease) attacked livestock
6. Boils and blains
7. Thunder and hail
8. Locusts
9. Darkness for three days
10. Death of the first-born.

The plagues may possibly be interpreted as follows:

The waters of Egypt turned red as blood (with volcanic ash)—killing fish and driving frogs on shore (plagues 1, 2). Ash blotted out the sun for three days (plague 9).

Particles of ash high in the atmosphere brought about the formation of ice crystals which in turn caused cloudbursts, with thunder and hail—very unusual for a normally arid country such as Egypt (plague 7).

Meteorological disturbances created strong winds that brought locusts, which destroyed what crops remained (plague 8).

Insects and flies, bred in the rotting bodies and the swamps, brought disease to beasts and humans (plagues 3-6). Death was so rampant over the land as to amount to the killing of the "first-born" of all beasts and humans (plague 10).

There is also the hint of disaster in other places in the Old Testament:

> That day is a day of wrath, a day of trouble and distress . . . a day of darkness and gloominess, a day of clouds and thick darkness . . . And I will bring distress upon men, that they shall walk like blind men . . . woe unto the inhabitants of the sea coast . . . the land of the Philistines*. I will even destroy thee, that there shall be no inhabitant. (Zephaniah 1:15, 17 and 2:5, written in 7th century B.C.)
>
> Behold, waters rise up out of the north, and shall be an overflowing flood, and shall overflow the land . . . Because of the day that cometh to spoil all the Philistines* . . . for the Lord will spoil the Philistines*, the remnant of the country of Caphtor*.
>
> (Jeremiah 47: 2, 4, written in 6th century B.C.)

The question may be posed: did the enslaved Israelites take advantage of the confusion and begin their epic migration to the Promised Land?

I Kings 6:1 says: "And it came to pass, in the 480th year after the children of Israel were come out of the land of Egypt, in the fourth year of Solomon's reign over Israel." Since Solomon reigned from 970-930 B.C. that puts the Exodus at the time that Santorini erupted.

Exodus 13:21 states that when Moses led the Israelites out of Egypt ". . . the Lord went before them by day in a pillar of cloud . . . and by night in a pillar of fire." This *could* be explained by particles of ash reaching far into the stratosphere forming high clouds over today's Israel. The "pillar of fire" might be luminous night clouds similar to those observed in the high stratosphere after the eruption of Krakatoa.

There remains the Israelites' miraculous crossing of the Red Sea. At first sight crossing the Red Sea seems a rather tall order! However, a mis-identification may have crept in, as the Hebrew words *Yam Suf* can mean either "Red Sea" or "Reed Sea". The "Reed Sea" is a large swampy lake, Sirbonis Lake, between the Nile and Palestine, separated from the Mediterranean by a narrow spit of land (Fig. 19.17). If in fact the Israelites crossed the Reed Sea—far more of a geographical possibility—they may have fled across this spit, in between tidal waves, and the Egyptians were caught in the huge returning wave.

(*The Biblical name of Crete was Caphtor and the Cretans were called Philistines.)

Coming back to geology: at first the existence of a huge volcano in the Aegean Sea didn't make sense, as there was no obvious reason for such violent activity. But then after close study of earthquakes, the Mediterranean ocean floor and the geology of surrounding countries, it was recognised that Greece and Turkey are sitting on two of a number of micro-plates, or buffer plates—small chips of the earth's crust caught up between the giant Eurasian, African and Arabian plates (Fig. 19.18).

The Turkish plate is travelling westwards, pushing the Aegean plate. One of its boundaries is the Anatolian Fault, running through northern Turkey, and the source of many terrible earthquakes. This fault is moving at about 114 mm (4½ in) per year.

The Aegean plate, carrying southern Greece, Crete and the Aegean islands, is being pushed south-westwards by the Turkish plate. This movement is tending to tear Greece apart and is creating a sediment-filled "hole" behind it in the floor of the northern Aegean (the petroleum prospects of which are being hotly contested by Greeks and Turks).

As the Aegean plate is pushed south-westwards it over-rides the leading edge of the African plate, which in this region is the floor of the eastern Mediterranean Sea. A subduction zone has thus been created, with a submarine trench off Crete. The subducted ocean floor has been driven at a steep angle into the earth's mantle, deep under the Aegean plate, and is being slowly melted and consumed under the Cyclades Islands—the great Santorini eruption was, unfortunately, an inevitable result.

**New Zealand's volcanoes**

Although Banks Peninsula, Otago Peninsula,

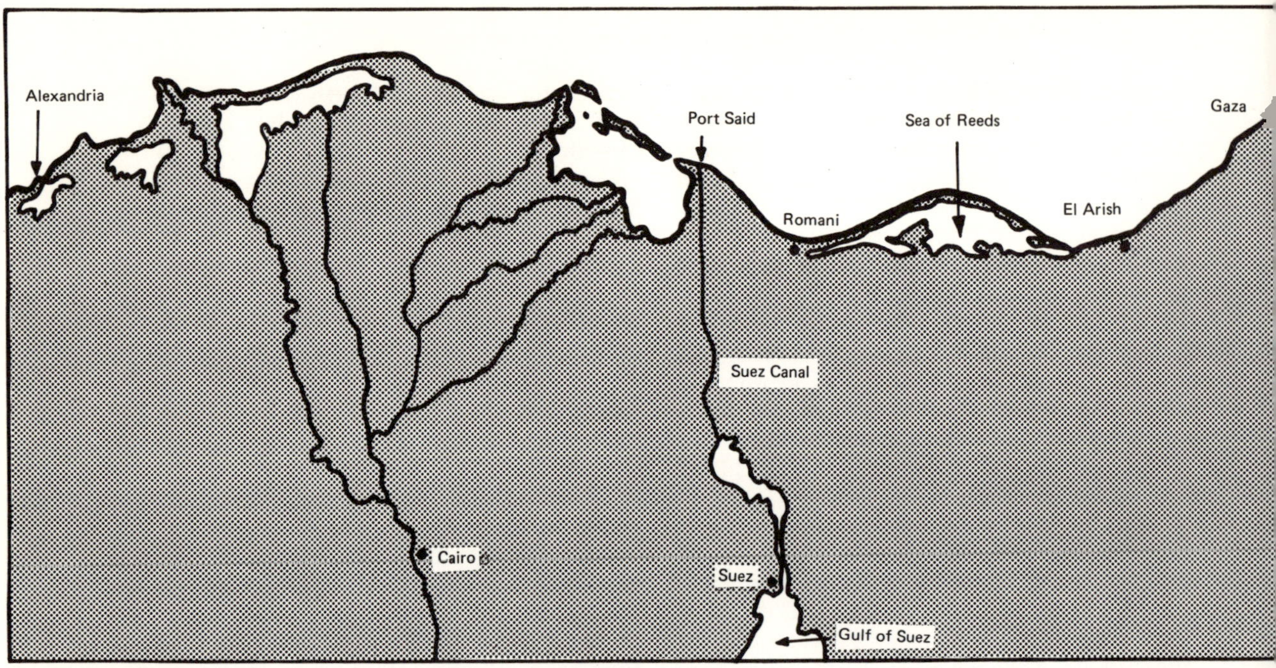


**Fig. 19.17.** The narrow spit separating the Reed Sea (Sirbonis Lake) from the Mediterranean may have been the route taken by the Israelites under Moses in their exodus from Egypt.

Solander Island and the Auckland Islands have been the sites of active volcanoes in geologically "relatively recent" times (i.e. the last 20 million years), these areas can now properly be classed as "extinct" volcanic centres. Active volcanism has now shifted further north, to lie in the Taranaki-Taupo-Rotorua-Auckland-Northland regions, and all available evidence suggests that future activity will occur there also.

New Zealand's previous volcanic history has been related to the inter actions of the Tonga-Kermadec and Macquarie island arcs, and the way they have responded to movements along the boundary between the Indian-Australian and Pacific plates.

As we have seen (pp. 357–360), the influence of the Macquarie island arc seems to have rather diminished in recent times, whilst that of the Tonga-Kermadec arc has increased. The extension of this latter arc into New Zealand, with the accompanying down-dipping Benioff Zone, is clearly reflected in the geology and structure of the North Island. The down-going crustal slab under the North Island has been actively feeding a great number of volcanoes and there is every indication that this will continue—with rhyolite and andesite types in the Rotorua-Taupo region (derived from both continental and oceanic sources) and andesite in Taranaki. Meanwhile, associated melting of the sima layer well behind the line of the volcanic arc will probably continue to produce basalt in Auckland and Northland.

Geological mapping in Northland has shown that at least 26 volcanoes have been active at various times in Northland (mainly around the Bay of Islands and Whangarei) since the late Pliocene, three million years ago. Te Puke, the youngest volcano, erupted between A.D. 200-700. In the Auckland metropolitan region at least 50 volcanoes have erupted over the last 50,000 years, and Rangitoto, the youngest, was last active between 750 and 150 years ago. Thus as far as future volcanic risk is concerned, both Northland and Auckland should certainly be regarded as "dormant" rather than "extinct" volcanic regions, and there is every possibility that new cones may appear in the next 500–1000 years. In both regions, however, volcanic activity has usually formed small basalt cones, with locally distributed ash showers, and it is probably fair to say that future eruptions may follow this pattern. However, while a small basalt cone and a bit of ash may possibly be accommodated on

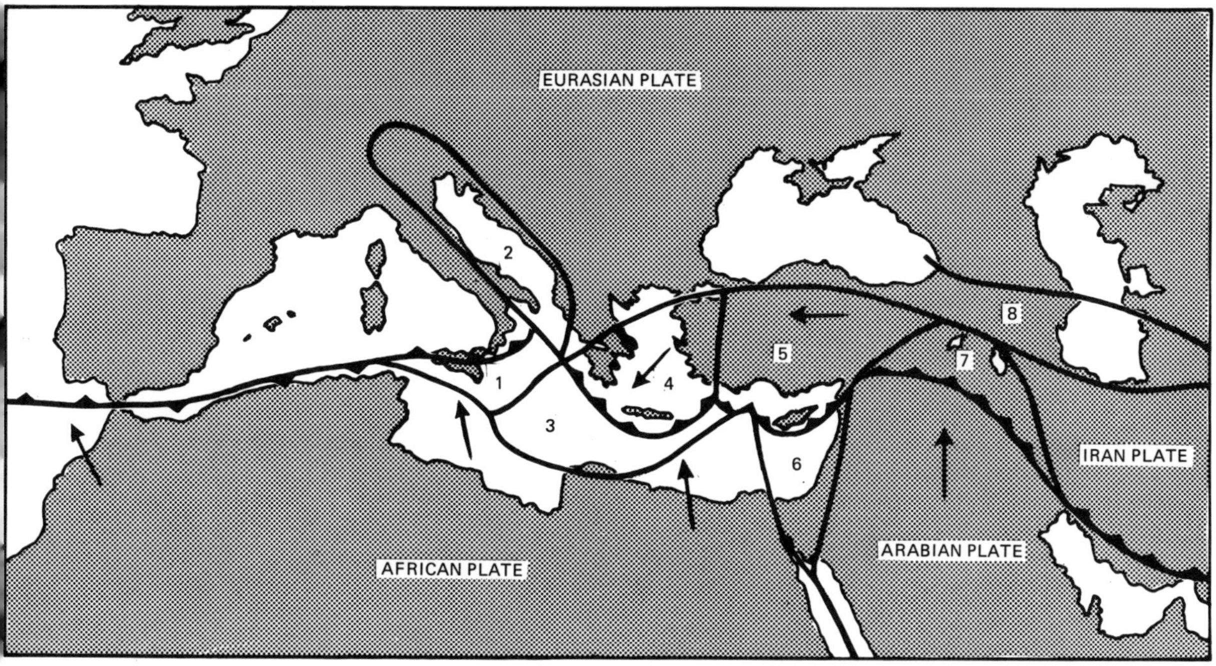


**Fig. 19.18.** The existence of Santorini, the only active volcano in the normally tranquil Aegean, is related to a number of small chips of the earth's crust caught up between the jostling movements of the giant Eurasian, African and Arabian plates. Similar "buffer plates", acting rather like ball bearings, and in this way serving to ease the movements of the world's giant plates, have been recognised in Tibet and China (Fig. 19.14), the western coast of North America, and, as shown in this map, in the Mediterranean. The Mediterranean buffer plates are as follows:

1 — Messina plate; 2 — Adriatic plate;
3 — Ionian plate; 4 — Aegean plate;
5 — Turkish plate; 6 — Levantine plate;
7 — Van plate; 8 — Caspian plate.

Overthrusting plate boundaries are shown by the triangle symbols and arrows show directions of plate movement with respect to the Eurasian plate.

The Aegean plate, carrying southern Greece, Crete and the Aegean islands, is being pushed south-westwards by the Turkish plate. As the Aegean plate is pushed it over-rides the leading edge of the African plate. A subduction zone has thus been created and the floor of the eastern Mediterranean Sea has been driven deeply under the Aegean plate. The subducted material is thus being slowly melted and consumed under the Cyclades Islands and the resulting volcanic material is finding its way up to the surface at Santorini.

farmland, it would be bad news in an urban area! More serious, and perhaps more likely, is the possibility of a rising magma body locally heating ground water or sea water, to produce a steam-generated explosion.

In the Miocene and Pliocene (some 20-2 million years ago) Coromandel Peninsula was the site of a volcanic arc similar to that of the modern Rotorua-Taupo arc. It is thought by some scientists that this line of volcanism through Coromandel marked the New Zealand plate boundary at this time. However, in the late Pliocene and early Pleistocene the Coromandel region, along with Auckland and Northland, was rotated away from the plate boundary, giving the characteristic twist to the North Island's shape. Volcanism thus gradually died out in Coromandel and the volcanic arc was then propagated through the Rotorua-Taupo region.

Taranaki has had a long history of andesite volcanic activity, related to magmas generated from the deepest parts of the down-dipping Benioff Zone underlying the central North Island. Taranaki volcanism began in late Pliocene or early Pleistocene, 1.75 million years ago, when the Sugar Loaves and Paritutu were volcanoes, and progressed through centres at Kaitake (575,000 years ago), Pouakai (240,000 years ago) and lastly Egmont (that began erupting 70,000 years ago)—all in line and 8-10 km (5-6 miles) apart. The latest eruptions of Egmont occurred in A.D. 1500, A.D. 1604, A.D. 1665 and A.D. 1755—so that volcanic activity is certainly still in the "dormant" stage and *not* "extinct". The next eruption may continue the trend-line towards the south, and a new cone *may* form near Kaponga in south-west Taranaki. Or alternatively, Egmont may one day blow its top off, devastating a very wide area. Although this latter possibility is not considered likely, it should not be ignored—hence the need for systems of volcanic surveillance.

In the centre of the North Island, Tongariro and Ruapehu are both volcanoes that have lost their tops in explosions—Tongariro having been reduced virtually to a stump. Activity at Tongariro is probably in a dying phase, but Ruapehu is still going strong. Both Tongariro and Ruapehu date back to early and middle Pleistocene time, whereas Ngauruhoe is a comparative youngster, being only 2500 years old. Ngauruhoe has had two major eruptive episodes lately: 1948-49 and 1954-55. However, although most spectacular, the lava flows and ash emissions accompanying these eruptions did not pose any significant danger. In January and March 1974 and February 1975 Ngauruhoe erupted incandescent material but no lava.

On the other hand, Ruapehu's crater lake poses a very substantial risk, as even very slight activity can send vast quantities of debris-laden water roaring down the volcanic slopes. Immense flood avalanches and mud flows (called "lahars") have been a feature of previous activity and if one occurs in the future, extensive damage may extend down valleys from the volcano all the way to the sea. In 1953 a lahar carried away the Tangiwai railway bridge and 151 people died when a crowded express train plunged through the gap soon after. In April 1975 a small lahar covered part of the ski field and demolished two small buildings, fortunately without loss of life. (Lahars have also been a feature of Egmont's past activity, but they appear to be related to times in the Ice Age when eruptions occurred under ice caps, melting the ice and releasing vast quantities of water.)

Elsewhere in the Taupo-Rotorua region 163 basalt and andesite volcanic vents have been identified, and many more are probably concealed beneath the vast sheets of ignimbrite (welded volcanic particles, deposited by fiery cloud eruptions, see p. 401). Some 16,700 cubic kilometres (4000 cubic miles) of material has been erupted from the Taupo-Rotorua zone during the past 1–1½ million years. The ejection of such large quantities of material from beneath has caused repeated subsidence of the surface—leading to formation of the present volcanic trough (or graben). The last major outburst from the Taupo-Rotorua region was in A.D. 130, although smaller eruptions occurred in A.D. 500, A.D. 700 and A.D. 1020. Mount Tarawera erupted in 1886 and White Island is continuously active.

Without exaggeration it may be said that the greatest degree of volcanic risk in New Zealand is probably located in the Taupo-Rotorua volcanic zone—closely related to the down-dipping Benioff Zone—and it is in this area that the volcanic research and surveillance efforts of the D.S.I.R. and the universities are concentrated.

**Climate for tomorrow**

As we have seen (pp. 150–153), continental movements are incredibly minute when compared with a human's life span. Thus the theory of continental drift can only provide long-*long* range forecasts for the world, as global weather systems respond to new continental positions and oceanic current patterns. The theory of ice ages is perhaps of more immediate relevance to mankind—although even then, forecasts are in terms of hundreds and thousands of years (pp. 341–342).

However, for what it is worth, the motions of continental drift as causes of radical climatic changes can be projected 50 million years into the future (Fig. 19.19). Frigid winters are forecast for the Mediterranean, balmy weather in Greenland and fewer cyclones in the South Pacific.

Because by this time the Atlantic will have widened considerably, more of the warm Gulf Current will flow northward, perhaps melting parts of the north polar ice cap and the icebergs that drift south, and ameliorating the bitter northern climate.

The on-going northward movement of Africa will close up the Mediterranean Sea, removing the weather buffer that gives the surrounding countries their present enviable conditions.

A new cyclone corridor will develop in Europe and Asia, but the opposite is likely to occur in the Pacific Ocean. The reason is that Australia, also moving northward, will not only change from its present climate to a moist tropical one, but will reduce the number of cyclones and typhoons in the region by occupying the part of the ocean where they now originate.

As they did in the past, North America and the

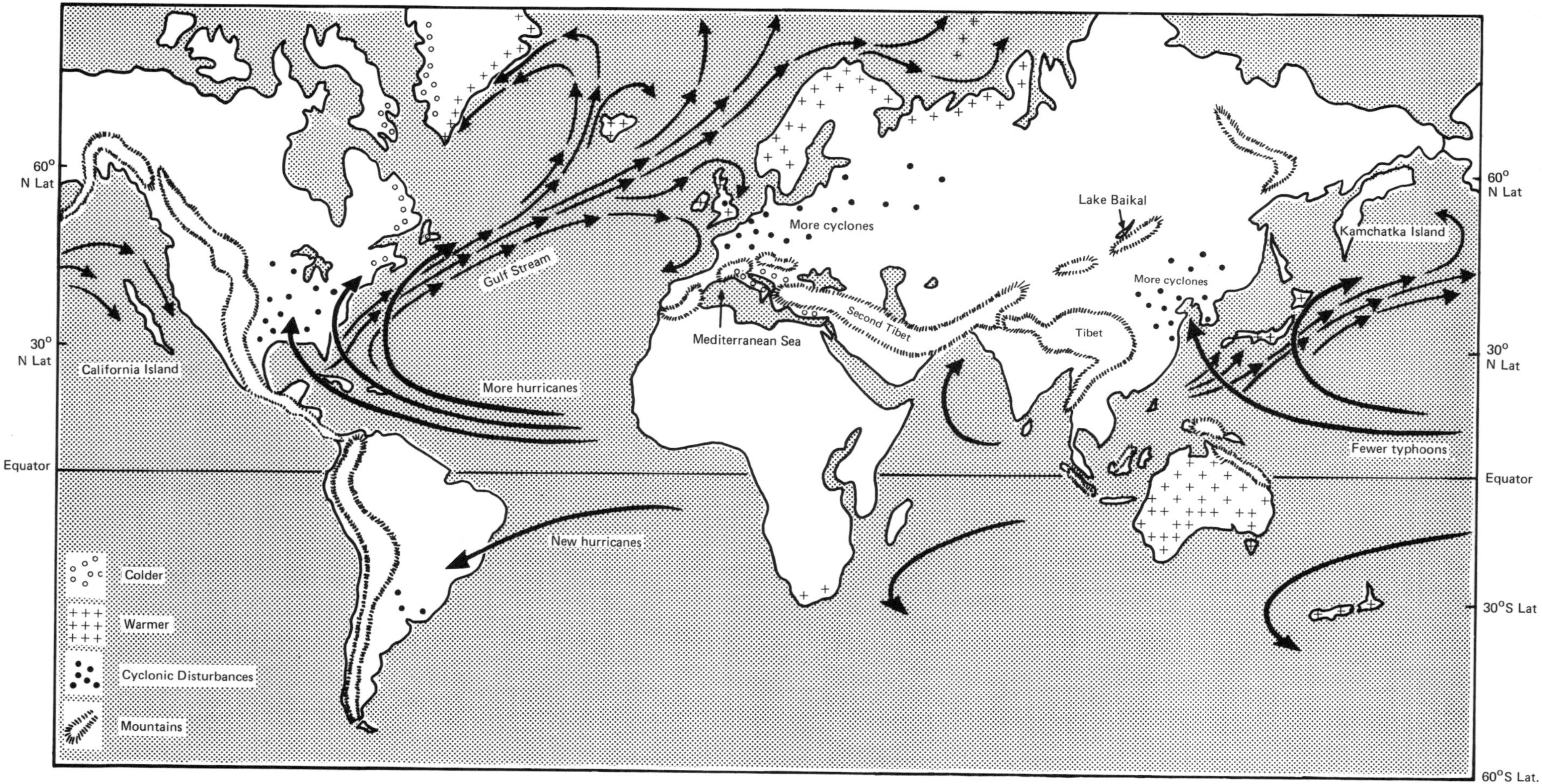


**Fig. 19.19.** The theory of continental drift provides us with the opportunity of looking ahead to A.D. 50,001,979. The slow and steady movements of the world's continents will lead to radical climatic changes.

This world weather map for 50 million years into the future forecasts balmy weather in Greenland, colder weather in the Mediterranean and fewer cyclonic storms in the South Pacific.

Soviet Union will again join hands, forming a land bridge across what are now the Aleutian Islands, and blocking the entry of cold Arctic water into the North Pacific.

**Facing the future**

The message coming through loud and clear from our new understanding of the earth is that humanity will have to face up to many changes in the future.

Although we now know that natural processes are continually extracting coal, oil, gas and other mineral resources from sediments, oceanic crust and mantle, and moving them into more accessible regions, we also now know that they do so at an incredibly slow rate. We thus have to face the fact that humanity is living on a storehouse of resources that have been accumulated over millions of years.

Admittedly we may dig deeper and deeper, or use new energy sources (e.g. atomic fusion) to process low-grade ores. Ultimately industrial raw materials may be extracted from ordinary rocks. Most of the elements industry needs are present in granite, for example, together with the nuclear fuels, uranium and thorium, which could supply the energy for extracting the other materials. However, the required scale of operations is staggering. To sustain our present usage of copper, for example, using ordinary granite, would require the quarrying and processing of 50,000 million tonnes of rock per year!

Perhaps a far better way is for our civilisation, as it gets older and (it is to be hoped) wiser, to reduce its demands for fresh resources in order to match more closely the slow rate at which they are replenished.

Thus although the new geology is bound to have a profound impact on the search for new materials—opening up fresh opportunities for discovery on both land and sea—it does nonetheless underscore the very limited state of our resources. Global industry must therefore be prepared to change—using low-grade ores and recycling the existing stocks of materials much more carefully.

Continental drift and ice age theories clearly indicate that major climatic changes are ahead and these may present humanity with one of its most forbidding challenges. It is clear that human affairs will need to be geared to long-term changes in environments as well as in industries—just as we are now adjusted to the short-term changes of seasons, earth's daily rotations etc. In this we will be aided by our new perceptions of the forces that interact in our environment.

The twin hazards of volcanic eruptions and earthquakes are intimately related to on-going plate movements and in the past have had profound effects on human affairs. Thus the explosion of the volcanic island of Santorini altered the balance of power in the Aegean region some 34 centuries ago. And the repeated disruption of Middle East cities and irrigation systems by earthquakes influenced the rise and fall of ancient dynasties. No doubt other similar natural disasters may be around the corner in the future.

However, armed with our new understanding of volcanoes and earthquakes we may with a bit of luck be able to devise methods of prediction, and even perhaps full or partial prevention, and to pin-point likely trouble spots. It is clear that in this and many other aspects of our globe we are entering into a new and significant phase of geology in the service of humanity.

Lastly, to end on a local note, New Zealand's fate is to accompany Australia on its northerly drift. Therefore in the long-*long* term its climate may slowly improve—although in the short term, if the ice ages theory is correct, New Zealand may be heading towards a colder and stormier phase.

As New Zealand drifts northwards it may be torn apart along the line of the Alpine Fault and its branches, as the Indian-Australian and Pacific plates interact. Or if, as some people believe, the plate boundary presently running through New Zealand is getting ready to side-step into the Pacific, we may find ourselves arching up on the margin of an actively forming new subduction zone and being affected by new variations of the present faulting and volcanism.

Whatever happens, it is clear that we have certainly not seen the last of geological activity in our "shaky isles".

CHAPTER TWENTY

# SUGGESTIONS FOR FURTHER READING

THERE IS an extensive, and rather daunting, literature on continental drift and plate tectonics, but Tina Kasbeer, librarian, University of California, has provided a most useful bibliography, with the entries classified, annotated and indexed:

KASBEER, T., 1973:
Bibliography of Continental Drift and Plate Tectonics. Part 1. Geological Society of America Special Paper 142.

KASBEER, T., 1975:
Bibliography of Continental Drift and Plate Tectonics. Part 2. Geological Society of America Special Paper 164.

In the references provided below I have attempted to present a selection of the more significant works, as well as the more readable.

Particularly noteworthy in this latter respect are the numerous articles published in "Scientific American", many of which have also been reprinted in special-topic works (e.g. Moore, 1972; Revelle *et al.* 1969; Wilson 1972, 1976.)

The reference lists are not exhaustive, and readers requiring additional details may like to consult Tina Kasbeer's bibliography. In the New Zealand sections, however (Chapters 11-18), I have tried to give a reasonably comprehensive coverage.

The foreign language literature is also large, but is not included in this compilation. Nonetheless, some excellent works are available in languages other than English, and for example a beautifully illustrated booklet was published in 1973 by the Swedish series "Forskning och Framsteg".

The following publications provide a general background to the range of topics covered in "New Zealand Adrift":

BICKFORD, M. E. *et al.* 1974:
"Geology Today".
CRM Books, Del Mar, California.

BIRD, J. M., ISACKS, B. (Editors) 1972:
"Plate Tectonics".
American Geophysical Union, Washington D.C.

BLACKETT, P. M. S. *et al.* 1965:
"A Symposium on Continental Drift".
Royal Society, London.

CALDER, N. 1972
"Restless Earth".
British Broadcasting Corporation, London.

FRASER, R. 1967:
"Understanding the Earth".
Pelican, London.

GASS, I. G.; SMITH, P. J.; WILSON, R. C. L. (Editors) 1971:
"Understanding the Earth".
Artemis Press.

GOLDEN, F. 1972:
"The Moving Continents".
Charles Scribner, New York (also published by Methuen, London).

HALLAM, A. 1973:
"A Revolution in the Earth Sciences: From Continental Drift to Plate Tectonics".
Oxford University Press.

——— *et al.* 1977:
"Planet Earth. An Encyclopaedia of Geology".
Elsevier-Phaidon, Oxford.

HOLMES, A. 1965:
"Principles of Physical Geology" (Second Edition).
Thomas Nelson, London.

INSTITUTE OF GEOLOGICAL SCIENCES, LONDON.
"The Story of the Earth".
Her Majesty's Stationery Office, London.

MARVIN, U. B. 1973:
"Continental Drift: The Evolution of a Concept".
Smithsonian Institution Press, Washington, D.C.

MATTHEWS, S. W. 1973:
This Changing Earth.
National Geographic Magazine 143 (1) : 1-37.

MAYNARD, C. *et al.* 1974:
"Planet Earth".
New Horizon Library, Sampson Low, Maidenhead.

MOORE, J. R. (Editor) 1972:
"Oceanography".
Scientific American Books (W. H. Freeman, San Francisco).

PHINNEY, R. A. (Editor) 1968:
"The History of the Earth's Crust: A Symposium".
Princeton University Press.

PRESS, F.; SIEVER, R. 1974.
"Earth".
W. H. Freeman, San Francisco.

REVELLE, R. *et al.* 1969:
"The Ocean".
Scientific American Books (W. H. Freeman, San Francisco).

RUNCORN, S. K. (Editor) 1962:
"Continental Drift".
Academic Press, London.

SEYFERT, C. K.; SIRKIN, L. A. 1973:
"Earth History and Plate Tectonics".
Harper & Row, New York.

SMITH, P. J. 1973:
"Topics in Geophysics".
The Open University Press.

SULLIVAN, W. 1974:
"Continents in Motion: The New Earth Debate".
McGraw-Hill, New York, & Macmillan, London.

TARLING, D. H. & M. P. 1971:
"Continental Drift".
Bell, London.

TARLING, D. H. & M. P. 1973:
"Continental drift: a study of the Earth's changing surface".
Pelican, London.

TARLING, D. H.; RUNCORN, S. K. (Editors) 1973:
"Continental Drift, Sea Floor Spreading and Plate Tectonics: Implications to the Earth Sciences" (2 Volumes).
Academic Press, London.

WILSON, J. T. *et al.* 1972:
"Continents Adrift".
Scientific American Books
(W. H. Freeman, San Francisco).

WILSON, J. T. (Editor) 1976:
"Continents Adrift and Continents Aground".
Scientific American Books (W. H. Freeman, San Francisco).

WYLLIE, P. J. 1971:
"The Dynamic Earth".
Wiley, New York.

——— 1974:
Plate tectonics, sea-floor spreading and continental drift: an introduction.
American Association of Petroleum Geologists
Memoir 23: 5-15.

——— 1976:
"The Way the Earth Works: An Introduction to the New Global Geology and its Revolutionary Development".
Wiley, New York.

YORK, D. 1975:
"Planet Earth".
McGraw-Hill, New York.

---

**THE REMAINING references comprise those that supplement the material presented in individual chapters. Some of them, however, are applicable to more than one chapter. To conserve space these are not repeated but rather are placed under the chapter judged to be the most relevant.**

## CHAPTERS 1-3

BULLARD, E. 1975:
The Emergence of Plate Tectonics: a personal view.
Annual Review of Earth and Planetary Sciences 3: 1-30.

CAREY, S. W. (Editor) 1958:
"Continental Drift: A Symposium".
Geology Department, University of Tasmania.

DU TOIT, A. L. 1937:
"Our Wandering Continents".
Oliver & Boyd, Edinburgh.

HALLAM, A. 1975:
Alfred Wegener and the hypothesis of continental drift.
Scientific American 233 (2): 88-97.

KING, L. C. 1972:
"The Morphology of the Earth".
Oliver & Boyd, Edinburgh.

VINE, F. J. 1977:
The continental drift debate.
Nature 266: 19-22.

WEGENER, A. 1915:
"Die Entstehung der Kontinente und Ozeane".
Vieweg, Braunschweig (translated into English, Methuen, 1924, and also published as "Origin of Continents and Oceans", Dover Publications, 1966).

## CHAPTER 4

ANDERSON, D. L. 1971:
The San Andreas Fault.
Scientific American 225 (11): 52-68.

CANBY, T. Y. 1973:
California's San Andreas Fault.
National Geographic Magazine 143 (1): 38-54.

EIBY, G. A. 1967:
"Earthquakes".
Muller, London.

KENNEDY, W. Q. 1947:
The Great Glen Fault.
Quarterly Journal Geological Society
London 102: 41-76.

SUGGATE, R. P. 1963:
The Alpine Fault.
Transactions Royal Society of New Zealand Geology 2: 105-129.

——— 1978:
The Kaikoura Orogeny
pp. 671–701 IN: Suggate, R. P.; Stevens, G. R.; Te Punga, M. T. (Editors)
"The Geology of New Zealand" N.Z. Geological Survey & N.Z. Government Printer.

**CHAPTER 5**

BULLARD, E. C.; EVERETT, J. E.; SMITH, A. G. 1965:
The fit of the continents around the Atlantic.
pp. 41-76 IN:
Blackett, P. M. S. *et al.*,
"A Symposium on Continental Drift"
Royal Society, London.
COLBERT, E. H. 1972:
Antarctic fossils and the reconstruction of Gondwanaland.
Natural History 81: 66-73.
——— 1973:
"Wandering Lands and Animals".
W. P. Dutton, New York.
CRADDOCK, C. 1970:
Antarctic geology and Gondwanaland.
Antarctic Journal U.S. 5 (3): 53-57.
HALLAM, A. 1972:
Continental drift and the fossil record.
Scientific American 227 (5): 56-66.
HURLEY, P. M. 1968:
The confirmation of continental drift.
Scientific American 218 (4): 52-64.
——— 1970:
Distribution of age provinces in Laurasia.
Earth & Planetary Science Letters 8: 189-196.
MCPHERSON, J. G. 1975:
"Footprints on a Frozen Continent".
Hicks Smith, Wellington.
MILDENHALL, D. C. 1970:
Discovery of a New Zealand member of the Permian Glossopteris flora.
Australian Journal of Science 32: 474-475.
SCHOPF, J. M. 1970:
Gondwana palaeobotany.
Antarctic Journal U.S. 5 (3): 62-66.
SMITH, A. G.; HALLAM, A. 1970:
The fit of the southern continents.
Nature 225: 139-144.
STEVENS, G. R. 1974:
A Land is Made.
New Zealand's Nature Heritage: 14-23.

**CHAPTER 6**

DARLINGTON, P. J. 1957:
"Zoogeography: The Geographical Distribution of Animals".
Wiley, New York.
DYER, A. J. 1975:
An international initiative in observing the atmosphere.
Search 6: 29-33.
FELL, H. B. 1962:
West-Wind-Drift Dispersal of Echinoderms in the Southern Hemisphere.
Nature 193: 759-761.
GILL, E. D. 1952:
Palaeogeography of the New Zealand region in Lower Devonian time.
Transactions Royal Society of New Zealand 80: 171-185.
GODLEY, E. J. 1967:
Widely distributed species, land bridges and continental drift.
Nature 214: 74-75.
HALLAM, A 1967:
The bearing of certain palaeozoogeographic data on continental drift.
Palaeogeography, Palaeoclimatology, Palaeoecology 3: 201-241.
HUGHES, N. F. (Editor) 1973:
Organisms and continents through time.
Palaeontological Association Special Paper 12.
JARDINE, N.; MCKENZIE, D. P. 1972:
Continental drift and the dispersal and origin of organisms.
Nature 235: 20-24.
KURTEN, B. 1969:
Continental drift and evolution.
Scientific American 220 (3): 54-65.
MASON, B. J. 1970:
Global atmospheric research programme.
Nature 233: 382-388.
SIMPSON, G. G.; PITTENDRIGH, C. S.; TIFFANY, L. H. 1957:
"Life: An Introduction to Biology".
Harcourt, Brace and World, Philadelphia.
VALENTINE, J. W.; MOORES, E. M. 1974:
Plate tectonics and the history of life in the oceans.
Scientific American 230 (4): 80-89.

**CHAPTER 7**

BRIDEN, J. C. 1967:
Recurrent Continental Drift of Gondwanaland.
Nature 215: 1334-1339.
CREER, K. M. 1973:
A discussion of the arrangement of palaeomagnetic poles on the map of Pangaea for epochs in the Phanerozoic.
pp. 47-76 IN: Tarling, D. H.; Runcorn, S. K. (Editors)
"Implications of Continental Drift to the Earth Sciences".
Academic Press, London.
EMBLETON, B. J. J. 1972:
On the unity of Gondwanaland during the lower Palaeozoic.
Search 3: 338-339.
———; VALENCIO, D. A. 1977:
Palaeomagnetism and the reconstruction of Gondwanaland.
Tectonophysics 40: 1-12.
IRVING, E. 1964:
"Palaeomagnetism and its Application to Geological and Geophysical Problems".
Wiley, New York.
MCELHINNY, M. W. 1973:
"Palaeomagnetism and Plate Tectonics".
Cambridge University Press.
RUNCORN, S. K. 1966:
Rock magnetism.
Science Progress 54: 467-482.
TARLING, D. H. 1971:
"Principles and Applications of Palaeomagnetism".
Chapman and Hall, London.
THOMPSON, R. 1974:
Palaeomagnetism.
Science Progress 61: 349-373.
VILAS, J. F.; VALENCIO, D. A. 1970:
Palaeogeographic Reconstructions of the Gondwanic Continents Based on Palaeomagnetic and Sea-Floor Spreading Data.
Earth & Planetary Science Letters 7: 397-405.

## CHAPTER 8


BLACK, D. I. 1967:
Cosmic ray effects and faunal extinctions at geomagnetic field reversals.
Earth and Planetary Science Letters 3: 225-236.

BULLARD, E. C. 1969:
The Origin of the Oceans.
Scientific American 221 (3): 66-75.

COLEMAN, P. J. (Editor) 1973:
"The Western Pacific: Island Arcs, Marginal Seas, Geochemistry".
University of Western Australia Press, Perth.

COLEMAN, P. J. 1975:
On island arcs.
Earth Science Reviews 11: 47-80.

COX, A. 1972:
Geomagnetic reversals — their frequency, their origin and some problems of correlation.
pp. 93-105 IN: Bishop, W. W.; Miller, J. A. (Editors)
"Calibration of Hominoid Evolution: Recent Advances in Isotopic and other Dating Methods".
Scottish Academic Press for Wenner-Gren Foundation.

——— (Editor) 1973:
"Plate Tectonics and Geomagnetic Reversals".
W. H. Freeman, San Francisco.

———; DALRYMPLE, G. B.; DOELL, R. R. 1967:
Reversals of the earth's magnetic field.
Scientific American 216 (2): 44-54.

CRAIN, I. K. 1971:
Possible direct causal relation between geomagnetic reversals and biological extinctions.
Bulletin Geological Society of America 82: 2603-2606.

FISHER, R. L. 1974:
Pacific-type continental margins
pp. 25-41 IN: Burk, C. A.; Drake, C. L. (Editors)
"The Geology of Continental Margins".
Springer-Verlag, Berlin & Heidelberg.

GLASS, B. P.; HEEZEN, B. C. 1967:
Tektites and geomagnetic reversals.
Scientific American 217 (1): 32-38.

HALLAM, A. 1974:
Mass extinctions in the fossil record.
Nature 251: 568-569.

HAY, W. W. 1973:
Significance of palaeontologic results of Deep Sea Drilling Project Legs 1-9.
American Assocation of Petroleum Geologists Bulletin 57: 55-62.

HAYS, J. D. 1971:
Faunal extinctions and reversals of the Earth's magnetic field.
Bulletin Geological Society of America 82: 2443-2447.

HEEZEN, B. C.; MACGREGOR, I. D. 1972:
Riddles chalked on the ocean floor.
Saturday Review 55 (8): 54-59.

HEIRTZLER, J. R. 1968:
Sea-floor spreading.
Scientific American 219 (12): 60-70.

———; DICKSON, G. O.; HERRON, E. M.; PITMAN, W. C. III; LE PICHON, X. 1968:
Marine magnetic anomalies, geomagnetic field reversals and motions of the ocean floor and continents.
Journal of Geophysical Research 73: 2119-2136.

HEIRTZLER, J. R.; LE PICHON, X. 1965:
Crustal structure of the mid-ocean ridges.
Journal of Geophysical Research 70: 4013-4033.

HORSFIELD, B.; STONE, P. B. 1972:
"The Great Ocean Business".
Hodder and Stoughton, London.

KNOPOFF, L.; HEEZEN, B. C.; MACDONALD, G. F. J. (Editors) 1968:
The world rift system.
Tectonophysics 8: 265-577.

LAUGHTON, A. S. 1977:
Deep drill holes in the ocean floor.
Geographical Journal 143: 235-248.

LE PICHON, X. 1968:
Sea floor spreading and continental drift.
Journal of Geophysical Research 73: 3661-3697.

MCDOUGALL, I.; TARLING, D. H. 1963:
Dating of polarity zones in the Hawaiian islands.
Nature 200: 54-56.

MACDOUGALL, J. D. 1976:
Fission-track dating.
Scientific American 235 (6): 114-122.

MENARD, H. W. 1964:
"Marine Geology of the Pacific".
McGraw-Hill, New York.

OROWAN, E. 1969:
The origin of the oceanic ridges.
Scientific American 221 (5): 103-119.

PITMAN, W. C. III; HERRON, E. M.; HEIRTZLER, J. R. 1968:
Magnetic anomalies in the Pacific and sea-floor spreading.
Journal of Geophysical Research 73: 2069-2085.

———; LARSON, R. L.; HERRON, E. M. 1974:
Age of the ocean basins determined from magnetic anomaly lineations.
Charts published by Geological Society of America.

RAFF, A. D. 1961:
The magnetism of the sea floor.
Scientific American 205 (4): 146-156.

WADDINGTON, C. J. 1967:
Palaeomagnetic field reversals and cosmic radiation.
Science 158: 913-915.

WILSON, J. T. (Editor) 1971:
Global tectonics and sea-floor spreading.
Tectonophysics 12: 171-270.

UFFEN, R. J. 1963:
Influence of the earth's core on the origin and evolution of life.
Nature 198: 143-144.

VACQUIER, V. 1972:
"Geomagnetism in Marine Geology".
Elsevier, Amsterdam.

———; RAFF, A. D.; WARREN, R. E. 1961:
Horizontal displacements in the floor of the northeastern Pacific Ocean.
Bulletin Geological Society of America 72: 1251-1258.

VINE, F. J.; MATTHEWS, D. H. 1963:
Magnetic anomalies over oceanic ridges.
Nature 199: 947-949.

WERTENBAKER, W. 1976:
"The Floor of the Sea—Maurice Ewing and The Search to Understand the Earth".
Little, Brown, Boston.

## CHAPTER 9

BALLARD, R. D. 1975:
Dive into the great rift.
National Geographic 147 (5): 604-615.
——— 1976:
Window on Earth's interior.
National Geographic 150 (2): 228-249.
BENDER, P. L.; SILVERBERG, E. C. 1975:
Present tectonic-plate motions from lunar ranging.
Tectonophysics 29: 1-7.
BONNIN, J.; DIETZ, R. S. (Editors) 1977:
The present state of plate tectonics.
Tectonophysics 38 (1-2).
BRANDER, J. 1976:
The measure of plate tectonics.
New Scientist 69: 110-113.
BRIDEN, J. C.; DREWRY, G. E.; SMITH, A. G. 1974:
Phanerozoic equal-area world maps.
Journal of Geology 82: 555-574.
BROCKIE, W. J. 1974:
Landform Systems.
pp. 15-46 IN: Johnston, R. J. (Editor).
"Society and Environment in New Zealand".
Whitcombe and Tombs Ltd.
BURKE, K. C.; WILSON, J. T. 1976:
Hot spots on the Earth's surface.
Scientific American 235 (2): 46-59.
CARR, A.; COLEMAN, P. J. 1974:
Sea-floor spreading theory and the odyssey of the green turtle.
Nature 249: 128-130.
COOMBS, D. S.; LANDIS, C. A.; NORRIS, R. J.; SINTON, J. M.; BORNS, D. J.; CRAW, D. 1976:
The Dun Mountain Ophiolite Belt, New Zealand, its tectonic setting, constitution and origin, with special reference to the southern portion.
American Journal of Science 276: 561-603.
DALRYMPLE, G. B.; SILVER, E. A.; JACKSON, E. D. 1973:
Origin of the Hawaiian Islands.
American Scientist 61: 294-308.
DEWEY, J. F. 1972:
Plate tectonics.
Scientific American 226 (5): 56-66.
———; BIRD, J. M. 1970:
Mountain belts and the new global tectonics.
Journal of Geophysical Research 75: 2625-2647.
——— ; 1970:
Plate tectonics and geosynclines.
Tectonophysics 10: 625-638.
———; PITMAN, W. C. III; RYAN, W. B. F.; BONNIN, J. 1973:
Plate tectonics and the evolution of the alpine system.
Bulletin Geological Society of America 84: 3137-3180.
DICKINSON, W. R. (Editor) 1972:
Plate tectonics.
American Journal of Science 272: 549-676.
DIETZ, R. S. 1972:
Geosynclines, mountains and continent-building.
Scientific American 226 (3): 30-38.
———; HOLDEN, J. C. 1970:
The breakup of Pangaea.
Scientific American 223 (4): 30-41.
DIETZ, R. S.; HOLDEN, J. C. 1970:
Reconstruction of Pangaea: breakup and dispersion of continents, Permian to present.
Journal of Geophysical Research 75: 4939-4956.
ENGEL, A. E. J. 1963:
Geologic evolution of North America.
Science 140: 143-152.
FILSON, J. 1975:
Array Seismology.
Annual Review of Earth and Planetary Sciences 3: 157-166.
FLEMING, C. A. 1972:
Continents in dispersion.
New Zealand's Heritage: 12-21.
GASS, I. G.; SMITH, A. G.; VINE, F. J. 1975:
Origin and emplacement of ophiolites.
pp. 54-65 IN: Sutton, J. *et al.*,
"Geodynamics Today — A Review of the Earth's Dynamic Processes".
Royal Society, London.
GIRDLER, R. W. 1958:
Relationship of the Red Sea to the East African Rift System.
Quarterly Journal Geological Society of London 114: 79-102.
GIRDLER, R. W. 1972 (Editor):
East African rifts.
Tectonophysics 15 (1-2).
HEEZEN, B.C.; MACGREGOR, I. D. 1973:
Evolution of the Pacific.
Scientific American 229 (5): 102-112.
HEIRTZLER, J. R. 1975:
Where the Earth turns inside out.
National Geographic 147 (5): 586-603.
——— *et al.* 1977:
Project FAMOUS.
Bulletin Geological Society of America 88 (4–5): 481–678
———; GRASSLE, J. F. 1976:
Deep-sea research by manned submersibles.
Science 194: 294-299.
HORSFIELD, B.; DEWEY, J. 1971:
How continents are made and moved.
Science Journal 7: 43-48.
HSU, K. J. 1972:
When the Mediterranean dried up.
Scientific American 227 (6): 27-36.
ISACKS, B. J.; OLIVER, J.: SYKES, L. R. 1968:
Seismology and the new global tectonics.
Journal of Geophysical Research 73: 5855-5899.
JONES, D. E. 1973:
The evolution of the Andes.
Scientific American 229 (2): 61-69.
MCDOUGALL, I. 1973:
Volcano.
McGraw-Hill Yearbook of Science and Technology.
MCKENZIE, D. P. 1970:
Plate tectonics and continental drift.
Endeavour 29: 39-44.
——— 1972:
Plate tectonics and sea-floor spreading.
American Scientist 60: 425-435.
———; SCLATER, J. G. 1973:
The evolution of the Indian Ocean.
Scientific American 228 (5): 62-72.

MOLNAR, P.; TAPPONNIER, P. 1977:
The collision between India and Eurasia.
Scientific American 236 (4): 30-41.
MORGAN, W. J. 1972:
Deep mantle convection plumes and plate motions
American Association of Petroleum Geologists
Bulletin 56: 203-213.
OLIVER, J. 1972:
Contributions of seismology to plate tectonics.
American Association of Petroleum Geologists
Bulletin 56: 214-225.
———; SYKES, L.; ISACKS, B. 1969:
Seismology and the new global tectonics.
Tectonophysics 7: 527-541.
OXBURGH, E. R. 1974:
The plain man's guide to plate tectonics.
Proceedings Geologists Association 85: 299-357.
PENTZ, M. J. 1971:
"Major features of the Earth's surface: continental movement, sea-floor spreading and plate tectonics".
The Open University Press.
POLLACK, H. N.; CHAPMAN, D. S. 1977:
The flow of heat from the Earth's interior.
Scientific American 237 (2): 60-76.
WILSON, J. T. 1963:
Continental Drift.
Scientific American 208 (4): 86-100.
——— 1963:
Evidence from islands on the spreading of ocean floors.
Nature 197: 536-538.
——— 1965:
New class of faults and their bearing on continental drift.
Nature 207: 343-347.
TAZIEFF, H. 1970:
The Afar Triangle.
Scientific American 222 (2): 32-51.

**CHAPTER 10**

CAREY, S. W. 1972:
The face of the Earth.
Australian Natural History 17: 254-257.
——— 1975:
The expanding Earth—an essay review.
Earth Science Reviews 11: 105-143.
——— 1977:
"The Expanding Earth."
Elsevier, Amsterdam.
CREER, K. M. 1965:
An expanding Earth?
Nature 205: 539-544.
DOOLEY, J. C. 1973:
Is the Earth expanding?
Search 4: 9-15.
EGYED, L. 1961:
Expanding Earth.
Transactions New York Academy of Sciences 23: 424-432.
GASS, I. G. *et al.* 1972:
"Geology: Internal Processes".
Open University Press.
GIRDLER, R. W. 1962:
Initiation of continental drift.
Nature 194: 521-524.
IRVING, E. (Editor) 1973:
Mechanisms of plate tectonics.
Tectonophysics 19: 87-194.
JACOBS, J. A. 1973:
The Earth's Core.
Transactions Leicester Literary &
Philosophical Society 67: 25-38.
KAHLE, C. E. (Editor) 1974:
Plate Tectonics: Assessments and Reassessments.
American Association of Petroleum Geologists Memoir 23.
MACDONALD, J. F. 1959:
Calculations on the thermal history of the Earth.
Journal of Geophysical Research 64: 1967-2000.
MCKENZIE, D. P.; RICHTER, F. 1976:
Convection currents in the Earth's mantle.
Scientific American 235 (5): 72-89.
MEYERHOFF, A. A. 1970:
Continental Drift: implications of palaeomagnetic studies, meteorology, physical oceanography and climatology.
Journal of Geology 78: 1-51; 78: 406-444.
——— & H. A. 1972:
The new global tectonics: major inconsistencies.
American Association of Petroleum Geologists
Bulletin 56: 269-336.
———; HARDING, J. L. 1971:
Some problems in current concepts of continental drift.
Tectonophysics 12: 235-260.
———; TEICHERT, C. 1971:
Continental drift: late Palaeozoic glacial centres and Devonian-Eocene coal distribution.
Journal of Geology 79: 285-321.
MOORBATH, S. 1977:
The oldest rocks and the growth of continents.
Scientific American 236 (3): 92-104.
MORGAN, W. J. 1971:
Convection plumes in the lower mantle.
Nature 230: 42-43.
OWEN, H. G. 1976:
Continental displacement and expansion of the Earth during the Mesozoic and Cenozoic.
Philosophical Transactions, Royal Society, London.
A 281: 223-291.
PENTZ, M. J. *et al.* 1971:
"The Earth, its shape, internal structure and composition".
Open University Press.
POLLACK, H. N.; CHAPMAN, D. S. 1977:
The flow of heat from the Earth's surface.
Scientific American 237 (2): 60-76.
ROBERTSON, E. C. (Editor) 1970:
"Nature of the Solid Earth".
McGraw-Hill, New York.
RUNCORN, S. K. 1962:
Towards a theory of continental drift.
Nature 193: 311-314.
WESSON, P. S. 1972:
Objections to continental drift and plate tectonics.
Journal of Geology 80: 185-197.
WHITE, P. T. 1974:
The Eternal Treasure—Gold.
National Geographic Magazine 145 (1): 1-51.

WYLLIE, P. J. 1975:
The Earth's mantle.
Scientific American 233 (3): 50-63.

TOKSOZ, M. N. 1975:
The subduction of the lithosphere.
Scientific American 233 (5): 89-98.

THE FOLLOWING references relate to the faunal and floral development and plate tectonic setting of New Zealand and the South-west Pacific region and are of general application to chapters 11-18:

ANDREWS, J. E.; PACKHAM, G. *et al.* 1975:
"Initial Reports of the Deep Sea Drilling Project, Volume 30".
U.S. Government Printing Office, Washington.

AUSTIN, P. M.; SPRIGG, R. C.; BRAITHWAITE, J. C. 1973a:
Structural Development of the Eastern Chatham Rise and of the New Zealand region pp. 201-215 IN:
Fraser, R. (Compiler) "Oceanography of the South Pacific 1972".
New Zealand National Commission for UNESCO, Wellington.

———; SPRIGG, R. C.; BRAITHWAITE, J. C. 1973b:
Structure and petroleum potential of Eastern Chatham Rise, New Zealand.
American Association of Petroleum Geologists, Bulletin, 57: 477-97.

——— 1975:
Palaeogeographic and Palaeotectonic Models for the New Zealand Geosyncline in Eastern Gondwanaland.
Geological Society of America Bulletin 86: 1230-1234.

AXELROD, D. I. 1975:
Plate tectonics and problems of Angiosperm history.
Memoires du Museum National d'Histoire Naturelle Serie A (Zoologie) 88: 72-86.

——— 1976:
Evolution of the India-Pacific plate boundary in North Island, New Zealand.
Bulletin Australian Society Exploration Geophysicists, 6: 58-59.

BALLANCE, P. F. 1976:
Evolution of the Upper Cenozoic magmatic arc and plate boundary in northern New Zealand.
Earth & Planetary Science Letters 28: 356-370.

BURNS, R. E.; ANDREWS, J. E. *et al.* 1973:
"Initial Reports of the Deep Sea Drilling Project, Volume 21".
U.S. Government Printing Office, Washington.

CARTER, R. M.; NORRIS, R. J. 1976:
Cainozoic history of southern New Zealand: an accord between geological observations and plate-tectonic predictions.
Earth and Planetary Science Letters 31: 85-94.

CHRISTOFFEL, D.; FALCONER, R. K. H. 1972:
Marine magnetic measurements in the South-west Pacific Ocean and the identification of new tectonic features. pp. 197-209 IN: Hayes, D. E. (Editor) Antarctic Oceanology II. The Australian-New Zealand Sector.
American Geophysical Union Antarctic Research Series Volume 19.

CHRISTOFFEL, D.; FALCONER, R. K. H. 1973:
Changes in the direction of sea-floor spreading in the South-west Pacific. pp. 241-247 IN: Fraser, R. (Compiler) "Oceanography of the South Pacific 1972".
New Zealand National Commission for UNESCO, Wellington.

COOPER, R. A.; LANDIS, C. A.; LE MASURIER, W. E.; SPEDEN, I. G. 1978:
Geological history and regional patterns in New Zealand and West Antarctica; their palaeotectonic and palaeogeographic significance. Proceedings Third Symposium on Antarctic Geology and Geophysics (Madison, Wisconsin).

CRACRAFT, J. 1973:
Continental drift, palaeoclimatology and the evolution and biogeography of birds.
Journal of Zoology 169: 455-545.

——— 1975:
Mesozoic dispersal of terrestrial faunas around the southern end of the world.
Memoires du Museum National d'Histoire Naturelle Serie A (Zoologie) 88: 29-54.

FLEMING, C. A. 1967:
Biogeographic change related to Mesozoic orogenic history in the South-west Pacific.
Tectonophysics 4: 419-427.

——— 1971:
Continents in Dispersion.
New Zealand's Heritage: 12-21.

——— 1975:
The Geological History of New Zealand and its Biota. pp. 1-86 IN: Kuschel G. (Editor)
"Biogeography and Ecology in New Zealand".
W. Junk, The Hague.

——— 1977:
The history of life in New Zealand forests.
New Zealand Journal of Forestry 22: 249-262.

——— 1978:
"The Geological History of New Zealand and its Life".
Auckland University Press, Auckland.

GASKIN, D. E. 1970:
The Origins of the New Zealand Fauna and Flora: A Review.
Geographical Review 60: 414-434.

——— 1975:
Reappraisal of the New Zealand Mesozoic with respect to sea-floor spreading and modern tectonic plate theory.
Memoires du Museum National d'Histoire Naturelle Serie A (Zoologie) 88: 87–97.

GRIFFITHS, J. R. 1971:
Reconstruction of the South-west Pacific margin of Gondwanaland.
Nature 234: 203-7.

——— 1973:
Plate Tectonics and the Evolution of New Zealand. pp. 263-269 IN: Fraser, R. (Compiler)
"Oceanography of the South Pacific 1972".
New Zealand National Commission for UNESCO, Wellington.

GRIFFITHS, J. R. 1974:
Revised continental fit of Australia and Antarctica.
Nature 249: 336-337.

——— 1975:
New Zealand and the South-west Pacific Margin of Gondwanaland.
pp. 619-637 IN: Campbell, K.S.W. (Editor) "Gondwana Geology" Australian National University Press, Canberra.

——— & VARNE, R. 1972:
Evolution of the Tasman Sea, Macquarie Ridge and Alpine Fault.
Nature 235: 83-86.

GRINDLEY, G. W.; DAVEY, F. J. 1978:
Some aspects of the reconstruction of New Zealand, Australia and Antarctica.
Proceedings Third Symposium on Antarctic Geology and Geophysics (Madison, Wisconsin).

HAYES, D. E.; FRAKES, L. A. *et al.* 1975:
"Initial Reports of the Deep Sea Drilling Project. Volume 28".
U.S. Government Printing Office, Washington.

———; RINGIS, J. 1973:
Sea-floor spreading in the Tasman Sea.
Nature 243: 454-458.

KATZ, H. R. 1974c:
Margins of the South-west Pacific.
pp. 549-565 IN: Burk, C. A.; Drake, C. L. (Editors)
"The Geology of Continental Margins".
Springer-Verlag, Berlin.

KEAST, A. 1971:
Continental drift and the evolution of the biota on southern continents.
Quarterly Review of Biology 46: 335-378.

——— 1973:
Contemporary biotas and the separation sequence of the southern continents.
pp. 309-343 IN: Tarling, D. H.; Runcorn, S. K. (Editors)
"Implications of Continental Drift to the Earth Sciences" Vol. 1.
Academic Press, London.

KENNETT, J. P. 1974:
Circum-Antarctic current studies on Leg 29, Deep Sea Drilling Project.
Antarctic Journal of the United States 9: 222-224.

——— *et al.* 1972:
Australian-Antarctic continental drift, palaeocirculation changes and Oligocene deep-sea erosion.
Nature Physical Science 239 (91): 51-55.

———; HOUTZ, R. E. *et al.* 1974:
"Initial Reports of the Deep Sea Drilling Project, Volume 29".
U.S. Government Printing Office, Washington.

——— *et al.* 1974:
Development of the Circum-Antarctic current.
Science 186: 144-147.

LEITCH, E. C. 1975:
Mesozoic-Middle Tertiary tectonic development of Northern New Zealand.
Bulletin Australian Society Exploration Geophysicists 6: 56-58.

MOLNAR, P.; ATWATER, T.; MAMMERICKX, J.; SMITH, S. M. 1975:
Magnetic Anomalies, Bathymetry and the Tectonic Evolution of the South Pacific since the Late Cretaceous.
Geophysical Journal Royal Astronomical Society 40: 383-420.

PACKHAM, G. H. 1973:
A speculative Phanerozoic history of the South-west Pacific. pp. 369–388 IN: Coleman, P. J. (Editor)
"The Western Pacific: Island Arcs, Marginal Seas, Geochemistry". University of Western Australia Press, Perth.

RAVEN, P. H.; AXELROD, D. I. 1972:
Plate tectonics and Australasian palaeobiogeography.
Science 176: 1379-1386.

SKIPWORTH, J. P. 1974:
Continental Drift and the New Zealand Biota.
N.Z. Journal Geography 57: 1-13.

STEVENS, G. R. 1967:
Upper Jurassic Fossils from Ellsworth Land, West Antarctica, and Notes on Upper Jurassic Biogeography of the South Pacific Region.
N.Z. Journal of Geology and Geophysics 10: 345-93.

——— 1974a:
"Rugged Landscape: The Geology of Central New Zealand". A. H. & A. W. Reed, Wellington.

——— 1974b:
The Palaeogeographic history of New Zealand.
New Zealand Entomologist 5: 230-239.

——— 1974c:
A Land is Made.
New Zealand's Nature Heritage: 14-23.

——— 1975:
Drifting Continents.
New Zealand Cartographic Journal 5 (3): 2-20.

——— 1977:
Mesozoic biogeography of the South-west Pacific and its relationship to plate tectonics.
pp. 309-326 IN: "International Symposium on Geodynamics in the South-west Pacific" Editions Technip, Paris.

SUGGATE, R. P. 1972:
Mesozoic-Cenozoic development of the New Zealand region.
Pacific Geology 4: 113-120.

———; STEVENS, G. R.; TE PUNGA, M. T. (Editors) 1978:
"The Geology of New Zealand".
N.Z. Geological Survey and N.Z. Government Printer.

WEISSEL, J. K.; HAYES, D. E. 1971:
Asymetric seafloor spreading south of Australia.
Nature 231: 518-522.

——— 1972:
Magnetic Anomalies in the South-east Indian Ocean.
pp. 165-196 IN: Hayes, D. E. (Editor) Antarctic Oceanology II: The Australian-New Zealand Sector. American Geophysical Union, Antarctic Research Series 19: 165–196.

———; HAYES, D. E. 1977:
Evolution of the Tasman Sea reappraised.
Earth and Planetary Science Letters 36: 77–84.

WEISSEL, J. K.; HAYES, D. E.; HERRON, E. M. 1977:
Plate tectonics synthesis: the displacements between Australia, New Zealand and Antarctica since the late Cretaceous.
Marine Geology 25: 231-277.

WELLMAN, H. W. 1975:
New Zealand 60 M.Y. Ago.
Bulletin Australian Society Exploration Geophysicists 6: 55-56.

---

REFERENCES RELATING more specifically to individual chapters are as follows:

**CHAPTER 11**

COLBERT, E. H. 1970:
The fossil tetrapods of Coalsack Bluff.
Antarctic Journal U.S. 5: 57-61.

COOPER, R. A. 1975:
New Zealand and South-east Australia in the early Palaeozoic.
N.Z. Journal of Geology and Geophysics 18: 1-20.

LAIRD, M. G.; COOPER, R. A.; JAGO, J. B. 1977:
New data on the lower Palaeozoic sequence of northern Victoria Land, Antarctica, and its significance for Australian-Antarctic relations in the Palaeozoic.
Nature 265: 107-110.

**CHAPTER 12**

ARKELL, W. J. *et al.* 1957:
Ammonoidea "Treatise on Invertebrate Palaeontology", Part L.
University of Kansas Press.

GREGG, D. R. 1974:
Sea dragons of the Mesozoic.
New Zealand's Nature Heritage: 84-88.

STEVENS, G. R. 1965:
The Jurassic and Cretaceous belemnites of New Zealand and a review of the Jurassic and Cretaceous belemnites of the Indo-Pacific region.
New Zealand Geological Survey Palaeontological Bulletin 36.

——— 1971:
Relationship of isotopic temperatures and faunal realms to Jurassic and Cretaceous palaeogeography, particularly of the South-west Pacific.
Journal of the Royal Society of N.Z. 1: 145-58.

——— 1973a:
Jurassic Belemnites.
pp. 259-274 IN: Hallam, A. (Editor)
"Atlas of Palaeobiogeography". Elsevier, Amsterdam.

——— 1973b:
Cretaceous Belemnites.
pp. 385-401 IN: Hallam, A. (Editor)
"Atlas of Palaeobiogeography" Elsevier, Amsterdam.

——— 1974a:
Biogeographic changes in the Upper Jurassic of the South Pacific.
Memoires Bureau Recherches Geologiques et Minieres 75: 163-177.

——— 1974b:
Squids with outside shells.
New Zealand's Nature Heritage: 75-79.

STEVENS, G. R. ; CLAYTON, R. N. 1971:
Oxygen isotope studies in Jurassic and Cretaceous belemnites from New Zealand and their biogeographic significance.
N.Z. Journal of Geology and Geophysics 14: 829-897.

**CHAPTER 13**

BLAKE, M. C.; JONES, D. L.; LANDIS, C. A. 1974:
Active continental margins: contrasts between California and New Zealand pp. 853-872 IN:
Burk, C. A.; Drake, C. L. (Editors):
"The Geology of Continental Margins".
Springer-Verlag, Berlin.

BRADSHAW, J. D.; ANDREWS, P. F. 1973:
Geotectonics and the New Zealand Geosyncline.
Nature, Physical Science 241: 14-16.

COOMBS, D. S.; LANDIS, C. A.; NORRIS, R. J.; SINTON, J. M.; BORNS, D. 1976:
The Dun Mountain ophiolite belt, New Zealand, its tectonic setting, constitution and origin, with special reference to the southern portion.
American Journal of Science 276: 561-603.

GRIFFITHS, J. R.; AUSTIN, P. M. 1977:
Palaeogeographic and palaeotectonic models for the New Zealand Geosyncline: discussion and reply.
Geological Society of America Bulletin 88: 1203-1210.

GRINDLEY, G. W. 1974:
New Zealand.
pp. 387-416 IN: Spencer, A. M. (Editor):
Mesozoic-Cenozoic Orogenic Belts: Data for Orogenic Studies.
Geological Society of London Special Publication 4.

WODZICKI, A. 1974:
Geology of the pre-Cenozoic basement of the Taranaki-Cook Strait-Westland area, New Zealand, based on recent drillhole data.
N.Z. Journal of Geology and Geophysics 17: 747-757.

**CHAPTER 14**

CHARIG, A.; HORSFIELD, B. 1975:
"Before the Ark".
British Broadcasting Corporation, London.

CRACRAFT, J. 1974:
Phylogeny and evolution of the ratite birds.
Ibis 116: 494-521.

DAWSON, J. W. 1974:
How plants came to New Zealand.
New Zealand's Nature Heritage: 48-55.

FLEMING, C. A. 1962:
History of the New Zealand land bird fauna.
Notornis 9: 270-274.

FLEMING, C. A. 1974:
The coming of the birds.
New Zealand's Nature Heritage: 61-68.
——— *et al.* 1976:
Special issue: New Zealand.
Australian Natural History 18: 345-380.
HARDIE, R. 1975:
The riddle of Peripatus.
Australian Natural History 18: 180-185.
KUSCHEL, G. (Editor) 1975:
"Biogeography & Ecology in New Zealand".
W. Junk, The Hague.
SMITH, J. L. B. 1956:
"Old Four Legs".
Longmans, London.

**CHAPTER 15**

COX, C. B. 1970:
Migrating marsupials and drifting continents.
Nature 226: 767-770.
CRACRAFT, J. 1972:
Continental drift and Australian avian biogeography.
Emu 72: 171-174.
CRANWELL, L. M. 1964:
Antarctica: cradle or grave for its *Nothofagus*?
pp. 87-93 IN: Cranwell, L. M. (Editor):
"Ancient Pacific Floras: The Pollen Story".
University of Hawaii Press, Honolulu.
FOODEN, J. 1972:
Breakup of Pangaea and isolation of relict mammals in Australia, South America and Madagascar.
Science 175: 894-898.
CRESSITT, J. L. (Editor) 1963:
"Pacific Basin Biogeography — A Symposium".
Bishop Museum Press, Honolulu.
HUGHES, N. F. 1976:
"Palaeobiology of Angiosperm Origins".
Cambridge University Press.
KEAST, A.; ERK, F. C.; GLASS, B. (Editors) 1972:
"Evolution, Mammals and Southern Continents".
State University of New York Press, Albany, New York.
MAYR, E. 1972:
Continental drift and the history of the Australian bird fauna.
Emu 72: 26-28.
MCGOWRAN, B. 1973:
Rifting and drift of Australia and the migration of mammals.
Science 180: 759-761.
RAVEN, P. H.; AXELROD, D. I. 1974:
Angiosperm biogeography and past continental movements.
Annals Missouri Botanical Garden 61: 539-673.
RICH, P. V. 1975:
Antarctic dispersal routes, wandering continents, and the origin of Australia's non-passeriform avifauna.
Memoirs of the National Museum of Victoria 36: 63-125.
TEDFORD, R. H. 1974:
Marsupials and the new palaeogeography.
pp. 109-126 IN: Ross, C. A. (Editor):
Palaeogeographic Provinces and Provinciality.
Society of Economic Palaeontologists and Mineralogists Special Publication 21.
WARDLE, J. 1974:
Beech forests.
New Zealand's Nature Heritage: 173-181.

**CHAPTER 16**

ADIE, R. J. (Editor) 1964:
"Antarctic Geology".
North-Holland, Amsterdam.
——— (Editor) 1972:
"Antarctic Geology and Geophysics".
Universitetsforlaget, Oslo.
CAMPBELL, K. S. W. (Editor) 1975:
"Gondwana Geology".
Australian National University Press, Canberra.
DAWSON, J. W. 1963:
New Caledonia and New Zealand — a botanical comparison.
Tuatara 11: 178-193.
——— 1966:
Observations on *Nothofagus* in New Caledonia.
Tuatara 14: 1-7.
PARIS, J.-P.; BRADSHAW, J. D. 1976:
Palaeogeography and geotectonics of New Caledonia and New Zealand in the Triassic and Jurassic.
pp. 209-216 IN:
"International Symposium on Geodynamics in the South-west Pacific".
Editions Technip, Paris.
LILLE, R. 1976:
New Caledonia: evolution from Permian to Miocene.
pp. 195-208 IN:
"International Symposium on Geodynamics in the Southwest Pacific".
Editions Technip, Paris.
SIMPSON, G. G. 1976:
"Penguins: Past and Present, Here and There".
Yale University Press, New Haven.
SPROLL, W. P.; DIETZ, R. S. 1969:
Morphological continental drift fit of Australia and Antarctica.
Nature 222: 345-348.
STONEHOUSE, B. (Editor) 1975:
"The Biology of Penguins".
University Park Press, Baltimore.
THORN, R. F. 1965:
Floristic relationships of New Caledonia.
Iowa State University, Studies in Natural History 20: 1-14.

**CHAPTER 17**

CALDER, N. 1974:
"The Weather Machine and the Threat of Ice".
British Broadcasting Corporation, London.
COX, A.; DOELL, R. R.; DALRYMPLE, G. B. 1963:
Geomagnetic polarity epochs and Pleistocene geochronometry.
Nature 198: 1049-1051.
CROWELL, J. C.; FRAKES, L. A. 1970:
Phanerozoic glaciation and the causes of ice ages.
American Journal of Science 268: 193-224.
DANSGAARD, W. *et al.* 1975:
Climatic changes, Norsemen and modern man.
Nature 255: 24-28.

DENTON, G.; PORTER, S. C. 1970:
Neoglaciation.
Scientific American 222 (6): 100-110.
FLINT, R. F. 1971:
"Glacial and Quaternary Geology".
Wiley, New York.
HARLAND, W. B.; RUDWICK, M. J. S. 1964:
The great infra-Cambrian ice age.
Scientific American 211 (2): 28-36.
KENNETT, J. P.; HOUTZ, R. E. *et al.* 1975:
Cenozoic palaeo-oceanography in the South-west Pacific Ocean, Antarctic glaciation, and the development of the Circum-Antarctic Current.
Initial Reports of the Deep Sea Drilling Project 29: 1155-1169.
U.S. Government Printing Office, Washington.
MARGOLIS, S. V.; KROOPNICK, P. M.; GOODNEY, D. E. 1977:
Cenozoic and late Mesozoic palaeo-oceanographic and palaeo-glacial history recorded in circum-Antarctic deep-sea sediments.
Marine Geology 25: 131-147.
MATTHEWS, S. W. 1976:
What's happening to our climate?
National Geographic Magazine 150 (5): 576-615.
McELHINNY, M. W.; GIDDINGS, J. W.; EMBLETON, B. J. J. 1974:
Palaeomagnetic results and late Precambrian glaciations.
Nature 284: 557-561.
SELBY, M. J. 1973:
Antarctica: The key to the Ice Age.
New Zealand Geographer 29: 134-150.
TRENBERTH, K. E. 1977:
Climate and Climatic Change: A New Zealand Perspective.
N.Z. Meteorological Service Miscellaneous Publication 161.
TUREKIAN, K. K. (Editor) 1971:
"Late Cenozoic Glacial Ages".
Yale University Press, New Haven.
WRIGHT, A. E.; MOSELEY, F. (Editors) 1975:
"Ice Ages: Ancient and Modern".
Seel House Press, Liverpool.

## CHAPTER 18

ADAMS, R. D.; WARE, D. E. 1977:
Subcrustal earthquakes beneath New Zealand; locations determined with a laterally inhomogeneous velocity model.
N.Z. Journal of Geology and Geophysics 20: 59-83.
BARAZANGI, M.; ISACKS, B. 1971:
Lateral variations of seismic wave attenuation in the upper mantle above the inclined earthquake zone of the Tonga Island arc.
Journal of Geophysical Research 76: 8493-8516.
CHRISTOFFEL, D. A.; VAN DER LINDEN, W. J. M. 1972:
Macquarie Ridge-New Zealand Alpine Fault transition.
pp. 235-242 IN: Hayes, D. E. (Editor):
Antarctic Oceanology: The Australian-New Zealand sector.
American Geophysical Union Antarctic Research Series Volume 19.
DAVEY, F. J.; HOUTZ, R. E. 1977:
The Campbell Plateau and its relationship with the Ross Sea, Antarctica.
Marine Geology: 25: 61-72.
EIBY, G. A. 1975:
Seismology in New Zealand.
Geophysical Surveys 2: 55-72.
——— 1976:
The junction of the main New Zealand and Kermadec seismic regions.
pp. 167-178 IN: "International Symposium on Geodynamics in the South-west Pacific".
Editions Technip, Paris.
HATHERTON, T. 1970:
Gravity, seismicity and tectonics of the North Island, New Zealand.
N.Z. Journal of Geology and Geophysics 13: 126-144.
——— 1974:
Active continental margins and island arcs.
pp. 93-103 IN: Burk, C. A.; Drake, C. L. (Editors):
"The Geology of Continental Margins".
Springer-Verlag, Berlin.
——— ; DICKINSON, W. R. 1968:
Andesite volcanism and seismicity in New Zealand.
Journal Geophysical Research 73: 4615-4619.
KIBBLEWHITE, A. C. 1966a:
Detection and location of a new underwater volcano.
Nature 210: 938-939.
——— 1966b:
The acoustic detection and location of an underwater volcano.
N.Z. Journal of Science 9: 178-199.
——— 1967:
Note on another active seamount in the South Kermadec Ridge Group.
N.Z. Journal of Science 10: 68-69.
LAWRENCE, P. 1967:
New Zealand Region: Bathymetry, 1: 6,000,000.
N.Z. Oceanographic Institute Chart, Miscellaneous Series 15.

## CHAPTER 19

BARBERI, F.; GASPARINI, P. 1976:
Volcanic hazards.
Bulletin International Association of Engineering Geology 14: 217-232.
BONATTI, E. 1975:
Metallogenesis at oceanic spreading centres.
Annual Review of Earth and Planetary Science 3: 401-431.
BRANDER, J. 1976:
When do earthquakes occur?
New Scientist 69: 342-344.
BROWNE, P. R. L. 1969:
Sulphide mineralisation in a Broadlands geothermal drillhole, Taupo Volcanic Zone, New Zealand.
Economic Geology 64: 156-159.
BURK, C. A. 1972:
Global tectonics and world resources.
American Association of Petroleum Geologists Bulletin 56: 196-202.
BULLARD, E. C. 1974:
Minerals from the deep sea.
Endeavour 33: 80-85.

BUTTON, A. 1976:
Transvaal and Hamersley basins—review of basin development and mineral deposits.
Minerals Science and Engineering 8: 262-293.
BUZZALINI, A. D. *et al.* 1969:
Evaporites and petroleum.
American Association of Petroleum Geologists Bulletin 53 (4).
COLLINS, B. W. (Editor) 1964:
Volcanism and volcanic risk.
New Zealand Journal of Geology and Geophysics 7 (1).
DAWSON, J. B. 1971:
Advances in kimberlite geology.
Earth Science Reviews 7: 187-214.
DEGENS, E. T.; ROSS, D. A. 1969:
"Hot Brines and Recent Heavy Mineral Deposits in the Red Sea".
Springer-Verlag, New York.
FISCHER, A. G.; JUDSON, S. (Editors) 1975:
"Petroleum and Global Tectonics".
Princeton University Press.
FRANCIS, P. 1976:
Fire and water.
New Scientist 69: 329-331.
GALANOPOULOS, A. G.; BACON, E. 1969:
"Atlantis: The Truth behind the Legend".
Thomas Nelson, London.
KATZ, H. R. 1976:
Sedimentary basins and petroleum prospects, onshore and offshore New Zealand.
American Association of Petroleum Geologists Memoir 25: 217-228.
LUCE, J. V. 1969:
"The End of Atlantis: New Light on an Old Legend".
Thames and Hudson, London.
MARINATOS, S. *et al.* 1971-1973:
Acta of the 1st International Scientific Congress on the Volcano of Thera.
(2 Volumes).
MITCHELL, A. H. G.; BELL, J. D. 1973:
Island-arc evolution and related mineral deposits.
Journal of Geology 81: 381-405.
———; GARSON, M. S. 1976:
Mineralisation at plate boundaries.
Minerals Science and Engineering 8: 129-169.
NEALE, V. E. 1974:
"The Volcanic History of Taranaki".
Egmont National Park Board.
——— 1976:
Lahars as major geological hazards.
Bulletin International Association of Engineering Geology 14: 233-240.
NINKOVICH, D.; HEEZEN, B. C. 1965:
Santorini Tephra.
pp. 413-453 IN: "Submarine Geology and Geophysics".
Vol. 17 Colston Papers, University of Bristol.
Butterworths, London.
OAKESHOTT, G. B. 1976:
"Volcanoes and Earthquakes; Geologic Violence".
McGraw-Hill, New York.
PETRASCHECK, W. E. 1973:
Some aspects of the relations between continental drift and metallogenic provinces.
pp. 567-572. IN: Tarling, D. H.; Runcorn, S. K. (Editors):
"Implications of Continental Drift to the Earth Sciences".
Academic Press, London.
PRESS, F. 1975:
Earthquake prediction.
Scientific American 232 (5): 14-23.
RONA, P. A. 1973:
Plate tectonics and mineral resources.
Scientific American 229 (1): 86-95.
VAN ROSE, S.; MERCER, I. 1974:
"Volcanoes".
Institute of Geological Sciences, London.
SCHOLTZ, C. 1974:
Toward infallible earthquake prediction.
Natural History 83 (5): 54-59.
SHARP, W. E. 1974:
A plate-tectonic origin for diamond-bearing kimberlites.
Earth & Planetary Science Letters 21: 351-354.
SILLITOE, R. H. 1973:
Formation of certain massive sulphide deposits at sites of sea-floor spreading.
Transactions Institution Mining and Metallurgy, Section B, 82 pp: 40-45.
SKINNER, B. J. 1969:
"Earth Resources".
Prentice-Hall, New Jersey.
STRONG, D. F. (Editor) 1976:
Metallogeny and plate tectonics.
Geological Association of Canada Special Paper 14.
TAMRAZYAN, G. P. 1970:
Continental drift and thermal fields.
Geothermics Special Issue 2, Part 2: 1212-1225.
TARLING, D. H. 1973a:
Metallic ore deposits and continental drift.
Nature 243: 193-196.
——— 1973b:
Continental drift and reserves of oil and natural gas.
Nature 243: 277-279.
TAZIEFF, H. 1974:
"The Making of the Earth: Volcanoes and Continental Drift".
Saxon House, Farnborough.
THOMPSON, T. L. 1976a:
Plate tectonics in oil and gas exploration of continental margins.
American Association of Petroleum Geologists Bulletin 60: 1463-1501.
——— 1976b:
Plate tectonics in petroleum exploration of convergent continental margins.
American Association of Petroleum Geologists Memoir 25: 177-188.
WEBER, C. R.; SCHEIBNER, E. 1977:
The origin of some Permo-Triassic metal zones in the New England region, New South Wales.
Quarterly Notes N.S.W. Geological Survey 26: 1-14.

# GLOSSARY

THROUGHOUT this book I have tried to keep the technical terms to a minimum, as it has been written for the reader's enjoyment and for dipping into. Some have been used, however, as they provide convenient shorthand "labels" for geological ideas. Although all the terms used are explained at the time of their introduction into the text, the following glossary is provided for the readers who would like to dip into the text, and who might find themselves confronted with a completely unintelligible term.

The definitions provided do not pretend to be all-embracing, and have been tailored to fit the needs of the book. Detailed definitions of geological terms are provided in various dictionaries of geology, two of which are as follows: *Glossary of Geology* edited by M. Gary, R. McAfee, C. L. Wolf (American Geological Institute, Washington, D.C., 1972); *A Dictionary of Geology* (4th edition) by J. Challinor (University of Wales Press, Cardiff, 1973).

**Ammonite.** Extinct "sea snails" with coiled, chambered shells. Zoologically they are cephalopod molluscs, distantly related to the modern Pearly Nautilus, squid and octopus. They lived from early Devonian to late Cretaceous (390-65 million years ago). Their coiled shells were subdivided internally into chambers that assisted the animal's buoyancy.

**Amphibolite.** A type of metamorphic rock (*q.v.*), mostly containing feldspar and the dark mineral amphibole (silicates of magnesium, calcium and iron).

**Andesite.** A type of volcanic rock intermediate in composition between basalt (*q.v.*) and rhyolite (*q.v.*). Andesite lavas, when erupted onto the earth's surface, usually flow only very slowly. Such flows are often accompanied by violent explosions and the emission of great quantities of volcanic ash (*q.v.*). The ash and lava flows pile up to form a volcanic cone.

**Angiosperm.** The flowering plants, comprising one of the main classes of the plant kingdom. They range from the early Cretaceous (110 million years ago) and, from the end of the Cretaceous, dominated the world's floras. In the angiosperms the seeds are enclosed in an ovary, comprising the fruit.

**Argillite.** A rock essentially composed of mud. The argillite was formerly mud lying on the sea bed, but over geological time the mud has been covered by other material and compacted. During this process water has been squeezed out and the individual grains of clay have been pushed together, to form a reasonably hard rock.

**Asthenosphere.** An internal layer of the earth, lying beneath the lithosphere (*q.v.*). Earthquake waves slow down as they pass through the asthenosphere, suggesting that it is a soft layer, probably partially molten, or a mush of crystals. The weak asthenosphere is thought to provide a lubricating zone on which move the rigid lithosphere plates (bearing the continents and oceans). The asthenosphere lies at an average depth of 50-250 km (32-160 miles) below the earth's surface.

**Atoll.** A continuous or broken circle of coral reef and low coral islands surrounding a central lagoon. It is believed that atolls have originally been formed around oceanic volcanic cones that have gradually subsided, allowing the living coral reefs to grow upwards.

**Austral.** A name applied to the distinctive animals and plants living in the cool- and cold-temperate regions in the Southern Hemisphere. Their counterpart in the Northern Hemisphere are the "Boreal" animals and plants (*q.v.*).

**Awaruite.** A mineral consisting of a natural alloy of nickel and iron. First described from Awarua, Southland.

**Basalt.** A type of volcanic rock very rich in iron, magnesium and manganese but comparatively poor in silica (cf. rhyolite, *q.v.*). Basalt lava erupted on the earth's surface usually flows very readily—often with almost stream-like rapidity. Because of the high content of iron, manganese etc. fresh basalt is normally dark red or black, but weathers to light red or pink.

**Basement.** The oldest rocks recognised in a given area (i.e. the "foundation" rocks).

**Bed.** A term usually applied to sedimentary rocks. A bed (or stratum) is an original layer that was laid down on the sea floor—a layer of mud, sand or gravel. A sedimentary rock is spoken of as being "bedded" or "stratified". Rocks usually break along the bedding surfaces and along joints (*q.v.*).

**Basic rock.** An igneous rock (*q.v.*) containing little or no free quartz (*q.v.*).

**Bathymetric contours.** Contours drawn on the sea floor (i.e. lines drawn to join points of the same depth).

**Belemnite.** Extinct cousin of the modern squids. Like the ammonites (*q.v.*), belemnites are, zoologically speaking, cephalopod molluscs. They lived in Jurassic and Cretaceous times (195-65 million years ago), although they were preceded and succeeded by related forms. Unlike the ammonites, the belemnites had straight bodies, but with comparable buoyancy chambers. To keep the animal on an even keel in the water, a solid shelly rod (or "guard") was secreted internally to act as a counter-weight to the head and arms, and this is usually the part that is preserved as a fossil in the rocks.

**Benioff Zone.** A downward dipping zone of earthquakes, often extending to depths of 700 km (450 miles). These earthquake shocks are believed to define the zone of movement between a plate that is being overthrust and a plate being underthrust and pushed at an angle into the mantle, forming a subduction zone (*q.v.*).

**Benthic.** A term applied to organisms that live on, or very near to, the ocean floor.

**Bivalve.** Marine molluscs having a shell in two halves (valves) but, unlike the Brachiopods (*q.v.*), having the valves positioned to left and right (rather than top and bottom as in Brachiopods). Bivalves appeared in the middle Cambrian, 550 million years ago, and form the majority of shellfish living in modern seas.

**Boreal.** A name applied to the distinctive animals and plants living in the cool- and cold-temperate regions of the Northern Hemisphere. Their counterpart in the Southern Hemisphere are the "Austral" animals and plants (*q.v.*).

**Brachiopod.** A phylum of marine animals having a shell in two pieces (valves), upper and lower, which open along a hinge line. Brachiopods appeared in the early Cambrian (570 million years ago) and were very common in the Palaeozoic and Mesozoic seas. They are less common in modern seas, but small colonies exist off Banks Peninsula and Fiordland, for example.

**Cainozoic** (see Cenozoic).

**Cambrian.** The lowest of the geological systems of the Palaeozoic era. Its age has been determined as 570-500 million years ago. The first shelly fossils appear in the early Cambrian—although soft-bodied organisms are known from earlier (Precambrian) strata.

**Carboniferous.** A geological system embracing those rocks laid down 345-280 million years ago. The name is derived from "carbon-bearing", as the system includes in its upper part nearly all the true coal beds of the world (excluding lignite)—although the coals of New Zealand, for example, are appreciably younger.

**Cenozoic** (Cainozoic, Kainozoic). This era of geological time spans the age of recent life and is applied to the last 65 million years of the geological record (i.e. post-Cretaceous). This era saw the rise of the modern land plants (angiosperms, *q.v.*) and the mammals (*q.v.*) and although such organisms occur only rarely as fossils (because of their dominant land habitat), the era is often called the "age of mammals" or the "age of angiosperms".

**Cephalopoda.** One of the three main classes of the Phylum Mollusca. Includes the octopus, squid, ammonites, belemnites and nautilus.

**Chordata.** The phylum of animals having throughout life or, at least, from an early stage of development, an internal supporting structure in the form of either a continuous flexible rod (notochord), or a series of jointed pieces (vertebrae) running dorsally along the mid-line of the body.

**Chromite.** A brownish to iron-black mineral of the spinel group, containing chromium, iron, magnesium and aluminium. It occurs in basic and ultrabasic rocks (*q.v.*), often as isolated crystals, but sometimes as massive deposits.

**Chronology.** The science of time and putting things in time-sequence. Geological chronology is "geochronology" and in this a distinction is drawn between a succession of events in their time order (relative chronology), obtained through the study of rock sequences and fossil successions, and the placing of these events on an absolute time scale (absolute chronology), obtained via radioactive dating.

**Climatic Optimum.** See Post-Glacial Climatic Optimum.

**Conrad Discontinuity.** A sudden change ("discontinuity") in the speed of earthquakes passing through the earth's crust (*q.v.*). The discontinuity divides the crust into an upper part, believed to correspond to the sial (*q.v.*) and a lower part, believed to correspond to the sima (*q.v.*).

**Continental drift.** The horizontal displacement or rotation of continents relative to one another.

**Continental shelf.** The submarine shelf that surrounds most landmasses. Its width varies greatly, but its edge (generally at 100 fathoms, 182 m, 600 ft) is the true edge of the continents and its outline is thus important in questions of continental drift.

**Continental slope.** The submarine slope bordering the continental shelf (*q.v.*) on its seaward edge. The slope extends between the shelf and the abyssal oceanic basins.

**Convection.** The transfer of heat through a liquid in which hot material from the bottom rises, because of its lesser density, while cool surface material sinks. The resulting currents are called "convection currents".

**Convection current.** In geology this term is applied to hypothetical currents within the earth—either within the asthenosphere, or within the mantle (*q.v.*)—that are thought to provide a driving mechanism for continental drift.

**Coriolis force.** The centrifugal force generated by the spinning motion of the earth. The velocity exerted by the earth's rotation is greatest in those regions near the equator.

**Cretaceous.** The uppermost of the three Mesozoic geological systems. It lasted from 136-65 million years ago. It includes in its upper part the Chalk, a characteristic lime formation in many parts of the world (e.g. the "White Cliffs of Dover"). Many organisms, including the dinosaurs, belemnites and ammonites, became extinct at the close of Cretaceous time.

**Crust** (of the earth). The base of the earth's crust is generally placed at the Moho (Mohorovicic Discontinuity; *q.v.*). Under the ocean basins the crust is thin, averaging 5 km (3 miles) in most places, whereas under the continents it varies between 20 and 65 km (13-40 miles). Within the crust another seismic break, the Conrad Discontinuity (*q.v.*), divides the crust into an upper part, the sial (*q.v.*), comprising the bulk of continental material, and a lower part, the sima (*q.v.*), comprising the oceanic material.

**Curie point.** The temperature above which a mineral cannot retain any permanent magnetism.

**Devonian.** A geological system preceding the Carboniferous (*q.v.*). It lasted from 390-345 million years ago. The Devonian is particularly noteworthy as it is believed that some time during its period the first fish-like animals ventured onto the land—giving rise to ancestral amphibia.

**Dinosaur.** A collective name for some types of land reptiles, many of them large, that lived in Mesozoic times. Dinosaurs dominated the land between 200 and 65 million years ago and gave rise to a great variety of types, including some of immense size and others with bizarre armour of spikes and plates. Their sudden disappearance at the end of Cretaceous time is the subject of much geological speculation.

**Dunite.** A variety of peridotite (*q.v.*) composed essentially of the mineral olivine. Named after Dun Mountain, Nelson.

**Echinoderm.** A phylum of marine animals that includes the starfish, sea eggs (sea urchins), brittle stars, sea lilies and sea cucumbers. They first appeared in early Cambrian, 570 million years ago, although soft-bodied varieties may have existed in Precambrian times.

**Eclogite.** A rock composed of red garnet and a bright green variety of pyroxene. It is believed to form under conditions of high pressure and temperature deep in the earth's mantle. Eclogite has the same composition as a basalt or gabbro, but the atoms of its minerals are more densely packed.

**Eocene.** The second to lowest series of the Tertiary succession. It lasted from 53-37 million years ago.

**Epicentre.** The point or small area on the earth's surface directly above the origin (focus) of an earthquake.

**Epiplanktonic.** Organisms that have "hitched a ride" on a planktonic (i.e. free floating) organism. Thus barnacles, for example, may attach themselves to floating seaweed and by this means drift for considerable distances.

**Fault.** A surface of fracture, often nearly a plane, in a rock mass along which there has been movement, resulting in permanent displacement.

**Feldspar.** The most important group of rock-forming minerals. They consist of silicates of aluminium, potassium, sodium, calcium and (rarely) barium.

**Fission-track dating.** A method of dating rocks based on measurement of the amount of damage suffered by crystals of volcanic glass, contained in the rocks, when exposed to radiation being given off by minute grains of radioactive material incorporated in the rocks at the time they were deposited.

**Fossil.** The remains, traces or impressions, in the rocks, of living organisms that inhabited the earth during past ages.

**Gabbro.** A coarse-grained basic igneous rock composed essentially of calcium feldspar and pyroxene or olivine or both.

**Geomagnetic Polarity Epoch.** A period of the earth's history characterised by normal polarity (i.e. like that of the present day) or reversed polarity of the earth's magnetic field. These polarity changes, or "flips" of polarity, can be used to date sequences of rocks.

**Geophysics.** The study of the physics of the earth.

**Geosyncline.** An elongated downwarp of the earth's crust, measured in scores of kilometres, which is subsiding as sedimentary and volcanic rocks accumulate to thicknesses of thousands of metres.

**Glacial.** Characterised by the presence of ice.

**Glacial period.** A period in earth history when glacial conditions were so much more extensive than they are today that they covered considerable areas, and extended down to levels now having a temperate climate.

**Glaciation.** Glacial action.

**Glauconite.** A silicate of iron and potassium, combined with water. Light or dark-green in colour.

**Glossopteris.** A distinctive seed fern that lived in Gondwanaland (*q.v.*) during the Permian period (280-230 million years ago).

**Gneiss.** A coarse-grained foliated (banded) crystalline rock produced by heat and pressure (metamorphism, *q.v.*).

**Gondwanaland.** The primaeval super-continent that existed in the Southern Hemisphere up to Triassic-Jurassic times. In the Palaeozoic, Triassic and Jurassic modern South America, Africa, Madagascar, India, Australia, New Zealand and Antarctica were all part of Gondwanaland, and have been moved to their present positions by the processes of continental drift. A corresponding super-continent (Laurasia, *q.v.*) existed in the Northern Hemisphere.

**Graben.** A trough-like feature (rift-valley).

**Granite.** In the broad sense a coarse-grained acid igneous (i.e. once-molten) rock, consisting essentially of quartz and feldspar (both light coloured) and dark minerals such as mica and hornblende.

**Graptolite.** Extinct colonial organisms ranging from Cambrian to Carboniferous (but chiefly Ordovician and Silurian). The skeleton is in the form of rows of cups or small tubes arranged along one or more branches. The branches commonly look like a minute saw. They are thought to be primitive Chordates (*q.v.*).

**Gravity anomaly.** The difference between the observed value for the earth's gravity in a locality from that calculated from a hypothetical globe. The presence of light rocks below the land surface lessens the pull of gravity (negative gravity anomaly) and heavy rocks produce an increase (positive anomaly).

**Greensand.** A sandstone containing considerable amounts of the mineral glauconite (*q.v.*).

**Greenstone.** An informal name for a greenish gemstone. In New Zealand it is applied to varieties of nephrite ("New Zealand jade" or "pounamu" of the Maori), an exceptionally tough, compact, fine-grained greenish amphibole, found in association with serpentines (*q.v.*) in the Southern Alps. A related "greenstone", bowenite ("tangiwai"), is found as flakes or thin sheets in dunite (*q.v.*) in Anita Bay, at the entrance to Milford Sound.

**Greywacke sandstone.** A special type of sandstone (*q.v.*). The word greywacke is derived from the German "grauwacke" (= "grey rock"). Greywacke sandstones contain a lot of sand (rounded quartz grains), but also have many rock fragments of volcanic and igneous rocks. Greywacke sandstones are characteristic of geosynclines (*q.v.*) into which turbidity currents (*q.v.*) transport much material.

**Gutenberg low-velocity layer.** See low-velocity layer .

**Guyot.** See seamount.

**Gypsum.** A white rather soft mineral composed of hydrated sulphate of calcium.

**Hot spot.** A zone of melting within the mantle, perhaps related to upwelling convection currents, that remains as a fixed point, relative to the overlying moving lithosphere plate. Thus in time there is a progressive migration of volcanic activity as the plate moves slowly across the hot spot (synonym: "mantle plume").

**Hydrogrossular.** A hydrated lime-garnet (containing calcium-aluminium silicate). It is found as the main mineral constituent of the distinctive rock called "rodingite" (named after the Roding River, Nelson) (*q.v.*), which occurs as masses in the serpentine rocks of the South Island. Hydrogrossular is hard and pale green in colour and has potential as a gemstone (in South Africa similar material is called "South African jade").

**Ichthyosaur.** One of the two orders of Mesozoic swimming reptiles (see also plesiosaur). The body was shaped like that of a modern dolphin or porpoise, with paddle-like limbs.

**Ignimbrite.** A rock formed by the deposition and consolidation of ash flows and nuées ardentes (*q.v.*). The individual particles are often welded together by the contained heat. The name was coined in New Zealand and means "fire-rock".

**Igneous.** A rock or mineral that has solidified from molten or partly molten material, i.e. from a magma (*q.v.*). Igneous rocks can either crystallise deep in the earth's crust to form "plutonic" rocks or erupt on the surface as "volcanic" rocks. Igneous rocks constitute one of the three main classes into which all rocks are divided: i.e. igneous, metamorphic, sedimentary.

**Interglacial.** The time interval between two successive glacial periods.

**Isostasy.** The theoretical tendency of the earth's crust (*q.v.*) to maintain a state of balance (equilibrium) as it floats on the mantle (*q.v.*). It is assumed to result from the flowage of the dense plastic sima (*q.v.*) in the lower crust in response to changes in the position and thickness of the lighter sial (*q.v.*) above.

**Isotope.** One or more types of the same chemical element. Although the isotopes of an element have the same chemical properties, they differ in their atomic structure (e.g. by having more neutrons).

**Isthmian link.** A land bridge between two continents (e.g. the isthmus of Panama, linking North and South America).

**Joint.** A fracture in a rock distinct from a fault (*q.v.*) in that there has been no lateral displacement between the sides.

**Jurassic.** The second period of the Mesozoic era, i.e. after the Triassic and before the Cretaceous. It lasted from 195-136 million years ago.

**Kainozoic.** See Cenozoic.

**Kimberlite.** A type of peridotite (*q.v.*) containing garnet and olivine, found in deeply penetrating volcanic pipes, by means of which it probably came from the upper mantle. Frequently diamond-bearing.

**Lahar.** A mudflow down the flanks of a volcano, often generated by the overflow of a crater lake, as lava rises within the volcano and domes-up the lake's floor.

**Laurasia.** The protocontinent of the Northern Hemisphere, corresponding to Gondwanaland (*q.v.*) in the Southern Hemisphere. The protocontinent included North America, Greenland, Europe and most of Asia, excluding India. The main zone of separation was along the North Atlantic, across which geological features can be matched.

**Laurentia.** The Canadian shield rocks (mainly Precambrian).

**Lava.** A molten rock that has been extruded onto the earth's surface; also the resulting solidified rock.

**Limestone.** A sedimentary rock composed mainly of calcium carbonate, but sometimes with magnesium carbonate.

**Lithosphere.** The solid outer shell of the earth, as compared with the atmosphere and hydrosphere. The lithosphere (literally "rock sphere") is regarded as including the crust (sima and sial; *q.v.*) and the upper mantle and its lower boundary is the Low-Velocity Layer (or asthenosphere; *q.v.*), starting at a depth of about 50 km (32 miles). Plates of lithosphere are believed to move differentially on the weaker asthenosphere—leading to continental drift and plate tectonics.

**Little Ice Age.** In A.D. 350-1200 the world experienced a warm phase similar to that of the Post-Glacial Climatic Optimum (*q.v.*), but very much shorter. This was followed by a gradual deterioration in climate until in AD 1550-1800 the world passed through the "Little Ice Age". This period produced the lowest temperatures on land and sea since the Pleistocene period (*q.v.*), and the world-wide build-up of ice led to a fall of some 0.6 m (2 ft) in sea level.

**Lodestone.** A magnetic variety of natural iron ore.

**Low-velocity layer.** A zone in the upper mantle, ranging between depths of 50-250 km (32-160 miles), in which the speed of earthquake waves is 6 per cent lower than in the surrounding rocks. This is interpreted as being related to the reduced strength of the material in the layer, which is thought to be partially molten. (Synonym: Gutenberg Low-Velocity Layer.)

**Magma.** Molten rock material generated within the earth. It may contain suspended solids (crystals and rock fragments), and substantial quantities of gas. If magma crystallises below the earth's surface the resulting rocks are termed "plutonic" (e.g. granite). If it erupts on the surface the resulting rocks are "volcanic".

**Magnetic anomaly.** The value of the local magnetic field remaining after the regional magnetism of the earth has been subtracted. The magnetism of a group of rocks may thus be stronger or weaker compared with the regional magnetic field.

**Magnetic flip.** A change in the direction (or polarity) of the earth's magnetic field, e.g. a change from a north-polarity field to one of south-polarity.

**Magnetic polarity scale.** See geomagnetic polarity scale.

**Magnetic stripes.** Systematic variations of the earth's magnetism, revealed on the sea floor. Bands of normal and reversed magnetism have been mapped, interpreted as being related to "flips" of the earth's magnetic field.

**Magnetism** (of the earth—geomagnetism). It is believed that the spinning of the earth imparts different spinning rates to the earth's solid inner core, as compared with the fluid outer core and solid inner mantle. This creates a dynamo effect, generating electric currents that form a magnetic field enveloping the earth.

**Magnetite.** Magnetic iron oxide.

**Mammal.** An animal that feeds its young on milk, secreted by mammary glands.

**Mantle.** The part of the earth's interior extending between the outer core, at a depth of 2900 km (1800 miles), and the base of the crust, at depths of between 5 km (3 miles) (under the oceans) and 20-65 km (13-40 miles) (under the continents). The Mohorovicic Discontinuity (*q.v.*) marks the upper boundary of the mantle. The upper part of the mantle is divided into two portions by the Low-Velocity Layer (*q.v.*) (ranging between depths of 50-250 km, 32-160 miles).

**Mantle plume.** See hot spot.

**Marsupial.** A type of mammal (*q.v.*) whose young are born while at a very immature stage. The young crawl from the birth canal into a pouch ("marsupium") where they attach themselves to a nipple. The rest of the young animal's development is carried out in the pouch (e.g. kangaroo, koala, opossum).

**Mesosphere.** The part of the earth's crust between the outer core (*q.v.*) and the asthenosphere (*q.v.*), i.e. the lower mantle. Many geologists believe that the mesosphere is not involved in the earth's tectonic activity. Others believe, however, that convection currents (*q.v.*) stir the entire mantle.

**Mesozoic.** The middle age of the succession of life in the geological past. It lies between the ages of ancient (Palaeozoic) and modern (Cenozoic) life and spans the period of time between 230 and 65 million years ago. The main fossil groups are ammonites, belemnites, bivalves, brachiopods, corals and reptiles. The various groups of reptiles, on land (dinosaurs), and also in the sea and air, are so unusual and bizarre that often the era is called the "Age of Reptiles".

**Metamorphism.** The changing of the character of solid rock by the effects of heat and pressure.

**Mica.** A group of rock-forming minerals, composed of silicates of aluminium and potassium (white mica, muscovite) or with the addition of iron and magnesium (black mica, biotite). Mica crystals have a pearly lustre and break into very thin laminae.

**Miocene.** An epoch of the Tertiary period, after the Oligocene and before the Pliocene. It lasted from 25-5 million years ago.

**Mohorovicic Discontinuity.** Often abbreviated to "Moho". The boundary that separates the earth's crust from the mantle. It is marked by a sharp discontinuity, or break, in the speed of earthquake waves. Its depth varies from about 5 km (3 miles) beneath the ocean floor to about 30-35 km (19-22 miles) below the continents, although it may reach 65 km (40 miles) under some mountain ranges. It is thought to represent a chemical change between the basalt materials above (sima, *q.v.*) to peridotite or dunite (*q.v.*) below.

**Monotreme.** A primitive type of mammal that lays eggs but feeds its young with milk provided by mammary glands (e.g. echidna, platypus).

**Nautilus.** A living cephalopod with a coiled chambered external shell. Today it lives in the vicinity of New Caledonia, New Hebrides, Solomon Islands, New Guinea and the Philippines. It is the only living cousin of the ammonites (*q.v.*), so common in Mesozoic seas. Nautiloids as a group reached their acme in the Ordovician and Silurian, 500-390 million years ago.

**Nektoplanktonic.** A general name for organisms that swim (as distinct from float, see planktonic, *q.v.*) in the upper levels of the sea.

**Normal magnetism.** The earth's magnetism as it is today, i.e. with a north polarity (i.e. compass needles are north-seeking).

**Nothofagus.** The southern beech.

**Nuée ardente.** A swiftly flowing, turbulent fiery cloud of hot volcanic gas and ash particles.

**Obduction.** The pushing up of crustal material.

**Oceanic crust.** The type of earth's crust that underlies the oceans. It is equivalent to the sima (*q.v.*) and is basaltic in composition. It is about 5 km (3 miles) thick and its lower boundary, where it adjoins the mantle, is marked by the Mohorovicic Discontinuity (*q.v.*).

**Oceanic ridge** (mid-oceanic ridge or rise). A continuous median submarine mountain range extending through the major oceans of the world. It is a broad region of fractured, rugged terrain. It has a central rift valley along its crest, and is the source of numerous earthquakes, high heat flow and volcanic eruptions. The ridges stand about 1-3 km (½-2 miles) above the surrounding ocean basins. According to the theory of sea-floor spreading the mid-oceanic ridges are the sources of new crustal material.

**Oligocene.** An epoch of the Tertiary period, after the Eocene and before the Miocene. Spans the period of time between 37-25 million years ago.

**Olivine.** An olive-green, greyish-green or grey-brown mineral containing silicates of iron and magnesium. Olivine is a common component of basic and ultra-basic rocks (e.g. basalt, gabbro, peridotite, dunite). It is metamorphosed to produce serpentine.

**Ophiolite.** A group of basic and ultra-basic rocks (e.g. basalt, gabbro, peridotite, dunite etc.), including masses rich in serpentine derived from them by metamorphism, that have often been thrust, sometimes as over-riding sheets, or as slices, onto the land, during continental collisions. Together they are regarded as displaced fragments of the sea floor and upper mantle.

**Ordovician.** The second-earliest period of the Palaeozoic era. It lasted from 500-440 million years ago.

**Orogeny.** The process of formation of mountains. Includes the formation of structures within mountain chains—thrusting, folding and faulting in the outer (higher) layers—and plastic flowage and folding, metamorphism and igneous activity in the deeper (lower) layers.

**Palaeo-.** A prefix meaning "ancient".

**Palaeomagnetism.** The study of the "fossil" magnetism of the earth. A record of the magnetic field of the earth in past ages is preserved in certain rocks and study of this record has been of prime importance in helping validate the theory of continental drift.

**Palaeontology.** The science of the life of past geological ages, as revealed by fossils.

**Palaeozoic.** The age of ancient life in the geological record. Spans the periods from Cambrian to Permian, 570-230 million years ago. The main fossil groups of the early Palaeozoic are trilobites, graptolites, brachiopods, corals and nautiloids; of the late Palaeozoic: brachiopods, corals, fishes, bivalves.

**Paleocene.** The earliest epoch of the Tertiary period, following the Cretaceous. Includes rocks deposited 65-53 million years ago.

**Pangaea.** The primaeval super-continent which split up to give Laurasia (*q.v.*) to the north and Gondwanaland (*q.v.*) to the south. Pangaea formed in the mid and late Palaeozoic and gradually disintegrated during the succeeding Mesozoic periods.

**Peneplain.** The end-stage of a cycle of erosion on the land. The original terrain has been worn down by erosion, resulting in the formation of a gently undulating plain, lying almost at sea level.

**Peridotite.** A coarse-grained plutonic rock composed chiefly of olivine and containing little or no feldspar. Peridotite is metamorphosed to serpentine. It is believed that peridotite is a major constituent of the earth's mantle.

**Permian.** The last period of the Palaeozoic era. It lasted from 280-230 million years ago.

**Permo-Carboniferous.** The entire Permian and Carboniferous considered as a single unit.

**Petrified.** Literally "turned to stone". A process of fossilisation in which organic matter is converted into rock by the infiltration of water containing dissolved inorganic matter (silica, lime, etc.) which replaces the organic material, often atom by atom, retaining details of the original structure.

**Photosynthesis.** The process by which plants take water, carbon dioxide and carbon monoxide and use the sun's energy (and chlorophyll as a catalyst) to combine them into starches, sugars and other carbohydrates.

**Phragmocone.** The chambered part of a belemnite skeleton.

**Pillow lava.** When hot lava is erupted onto the sea floor, or flows from the land into the sea, it reacts violently with the cold ocean water. The lava explodes violently, forming tuff (volcanic fragments), gives off great quantities of iron, silica, etc. and also swells up and breaks into pillow-like blobs that pile up on one another. Individual pillows range in size from a few centimetres to a metre or more in diameter (generally 30-60 cm, 12-24 in).

**Placental.** A mammal that retains its developing young inside its body (in a uterus) and provides nutrient via a placenta.

**Planktonic.** Organisms that float (usually at the surface) in marine or other waters, and generally have only weak swimming capabilities.

**Plate tectonics.** A fundamental theory of geology, to explain continental drift, earthquake and volcano distributions, sea-floor spreading and mountain building.

The earth's crust is viewed as consisting of a number of semi-rigid plates in motion relative to each other. The plates are composed of both continental and oceanic crust and mantle, and "float" on a plastic underlayer in the mantle (the asthenosphere, *q.v.*).

Where plates meet, one edge is over-ridden (subduction, *q.v.*) and results in oceanic trenches, deep earthquake activity and arcs of volcanic islands. At continental margins similar over-riding results in orogenies (*q.v.*). Complex orogeny results when lighter continental blocks collide. The plates pull apart along mid-oceanic ridges, where new crustal material is generated by the upwelling of basalt from the underlying sima.

**Platform.** Part of a continent underlain by ancient basement rocks, consolidated during early mountain-building phases.

**Pleistocene.** The "Ice Age". The period of time that lasted from about 2 million years ago to 10,000 years ago. During this time the earth experienced a series of cold glacial periods, separated by warm interglacials.

**Plesiosaur.** A long-necked extinct marine reptile, with paddle-like flattened limbs. Characteristic of the Jurassic and Cretaceous periods.

**Plume.** See hot spot.

**Pliocene.** The uppermost epoch of the Tertiary period. Spanned the period of time 5-2 million years ago.

**Plutonic.** The action of heat deep within the earth's crust, particularly as applied to the "plutonic" rocks so formed.

**Podocarp.** The group of southern conifers; members of an ancient Southern Hemisphere lineage of cone-bearing plants that evolved early in Mesozoic times. The name means "fleshy fruit", and refers to the fact that the fruit are borne in cones, with each fruit having a fleshy stalk.

**Polarity** (magnetic). The sense of orientation of the earth's magnetic field—i.e. whether it is north-orientated or south-orientated.

**Porphyry.** An igneous rock characterised by large feldspar crystals, set in a fine-grained matrix.

**Post-glacial climatic optimum.** About 7000 years ago the earth's climate began to become progressively warmer than that of the present day, and from about 7000 to about 4000 years ago the world experienced a warm phase—the "Post-Glacial Climatic Optimum". The warming probably reached a peak about 5500 years ago and world temperatures were then 2-3°C higher than at present.

**Potassium-argon technique.** See radiometric dating.

**Precambrian.** Geological time before the Cambrian (i.e. before 570 million years ago). Modern usage is to recognise a number of subdivisions in Precambrian time.

**Pyroxene.** A group of rock-forming minerals, mainly silicates of magnesium, calcium and iron.

**Pumice.** See rhyolite.

**Quartz.** Crystalline silica (silicon dioxide). A hard glassy-looking mineral. A constituent of all acid igneous rocks and some intermediate and basic types. Common in metamorphic rocks and in veins. Dominant constituent of sandstones (the sand grains being quartz), and a wide range of other sedimentary rocks.

**Quartzite.** A rock almost completely quartz, i.e. a nearly pure quartz sandstone cemented by silica into a hard mass, or recrystallised by metamorphism into a mass of interlocking quartz crystals.

**Quaternary.** The post-Pliocene (i.e. post-Tertiary) geological time. Includes the Pleistocene (2 million-10,000 years ago) and the Recent (or Holocene), spanning the last 10,000 years.

**Radiometric dating.** A technique in which the age of a rock or artifact may be determined by comparison of the amount of a particular radioactive substance present there, and the amount of its decay products. Best known is the radiocarbon, or carbon-14 technique, although it can only be applied to rocks up to 50,000 years old. Techniques utilising Potassium/Argon and Rubium/Strontium are applied to older rocks.

**Ratite.** An order of birds lacking a keel on their breast bone (to which the massive flight muscles are attached in other birds). They are flightless, or virtually so, and include the moa, kiwi, emu and ostrich.

**Reversed magnetism.** The reverse of the modern earth's magnetism (i.e. with a south polarity so that compass needles point towards the south).

**Rhyolite.** A type of volcanic rock rich in silica, but poor in iron, magnesium, manganese etc. (cf. basalt, *q.v.*). Rhyolite lava is extremely viscous (stiff) and often gets stuck in the volcanic orifice, in which case hot gases whip it into a frothy mass which eventually is blown clear by violent explosions. The frothy material, full of gas bubbles, is called pumice.

**Rift valley.** A trough-like valley, bounded by fault scarps. Rift valleys characterise regions of the earth's crust that are being pulled apart by plate movements (e.g. East Africa, the Red Sea and the crests of mid-oceanic ridges).

**Rodingite.** A calcium-enriched gabbro rock, often associated with serpentines (*q.v.*). Is sometimes called the "spotted dog rock" as it is often white with black spots (rather like a dalmatian dog). Named after the Roding River, Nelson.

**Sandstone.** During weathering and erosion processes most of the constituents of rocks decompose to form mud, but minerals like quartz are resistant. These resistant mineral grains are deposited as beds of sand. Geological processes may bury the sand deeply under other deposits and gradually the water is squeezed out and the individual sand grains are compressed together. The resulting rock is a sandstone.

**Schist.** A type of metamorphic rock (*q.v.*)—a rock altered by heat and pressure. New minerals, stable at high temperature and pressure, have grown in the original rock. Some of the minerals are various forms of mica (*q.v.*) and these form platey silvery crystals in the schist.

**Sea-floor spreading.** A geological hypothesis that the oceanic crust is steadily increasing by means of the upwelling of magma along the mid-oceanic ridges (*q.v.*) of the world. This continuous injection of new material produces a conveyor-belt-like movement, thought to be a major driving force for plate tectonics (*q.v.*).

**Seamount.** An isolated tall mountain rising from the seafloor, but not breaking the surface of the water. Often they are flat-topped; these are differentiated as guyots.

**Sedimentary rocks.** Rocks consisting of weathered particles of igneous, metamorphic or even former sedimentary rock transported, usually by water, and deposited in distinct layers (beds or strata). They may also be of organic origin (chalk), or composed of volcanic debris (pyroclastic rocks). Sedimentary rocks contain traces of past life, preserved as fossils. They comprise one of the three major types of rock of the earth's crust (i.e. igneous, sedimentary, metamorphic).

**Seismograph.** A machine used to record the passage of an earthquake. The machine consists of two parts—the seismometer, sensitive to ground movements, and the recorder (generally a revolving drum), which provides a permanent record (or seismogram).

**Seismology.** The science of the study of earthquakes.

**Serpentine.** A group of minerals composed of silicates of magnesium and iron. Serpentines have a greasy or silky lustre, a slightly soapy feel (because of the presence of talc) and are commonly green, greenish yellow or greenish-grey. They have been derived by metamorphism (*q.v.*) of magnesium-iron rich rocks, especially peridotites, dunites etc.

**Shield.** A large area of exposed ancient basement rocks (*q.v.*), commonly with a gently convex surface, surrounded by sediment-covered platforms (*q.v.*).

**Sial** (*si*lica-*al*uminium). A collective term for those rocks lighter and more rigid than those of the sima (*q.v.*) and composed to a great extent of silica and aluminium that form the bulk of the continents.

**Silicate.** Mineral containing one atom of silicon and four of oxygen. The silicates are the most important class of minerals and comprise about 90 per cent of the rocks of the earth's crust.

**Silicon dioxide.** See quartz.

**Silurian.** The second-oldest period of the Palaeozoic. It lasted from 440-390 million years ago.

**Sima** (*si*lica-*ma*gnesium). A collective term for those rocks denser and more plastic than the sial (*q.v.*), and corresponding to a basalt composition, that form the lower portion of the earth's crust and underlie the oceans.

**Strata** (singular: stratum). See bed.

**Subduction.** The forcing of one crustal plate under another (see plate tectonics).

**Tectonic.** An adjective applying to deformation of the earth's crust and the processes accompanying such deformation.

**Tertiary.** A term used to include the Paleocene, Eocene, Oligocene, Miocene and Pliocene epochs of geological time, ranging in age from 65-2 million years ago.

**Tethys.** The sea that lay between the super-continents of Laurasia and Gondwanaland. As the continents evolved toward their present form, the Tethys was progressively closed off, the modern Mediterranean being an isolated remnant. The sediments that accumulated in the Tethys were folded to produce mountain chains such as the Caucasus, Carpathians, Himalayas, etc.

**Transcurrent fault.** A fault in which the dominant movement is horizontal.

**Transform fault.** A type of transcurrent fault (*q.v.*) characteristic of mid-oceanic ridges (*q.v.*) and along which the ridges and associated magnetic patterns have been offset.

**Triassic.** The lowest of the Mesozoic periods. Spans the time interval of 230-195 million years ago.

**Tsunami.** A large sea wave produced by a disturbance of the sea floor (submarine landsliding, volcanic eruption, or earthquake). Although a tsunami can pass almost unnoticed in the open sea, it can pile up to great heights (30m, 100 ft, or more) and cause considerable damage along shallow coastlines, often thousands of kilometres from the source.

**Ultrabasic rock.** An igneous rock, usually plutonic, but sometimes volcanic, containing little or no quartz or feldspar, but rich in iron and magnesium silicates.

**Volcanic ash.** Finely fragmented volcanic rock.

**Weathering.** The breaking-down of rocks through contact with the atmosphere (e.g. wind and rain). There are two types of weathering: chemical (action of acids and alkalis) and physical (action of heat and cold, wetting and drying, etc).

**Zeolite.** A soft white silicate mineral, composed of calcium, aluminium, sodium, barium, strontium and silica, formed at low pressures and temperatures. It often fills cracks and crevices in the rocks. Zeolite can be confused with quartz (*q.v.*), but quartz is harder and cannot be scratched with a knife. (Zeolite on the other hand can be readily scratched with a knife.)

# INDEX

250 million years

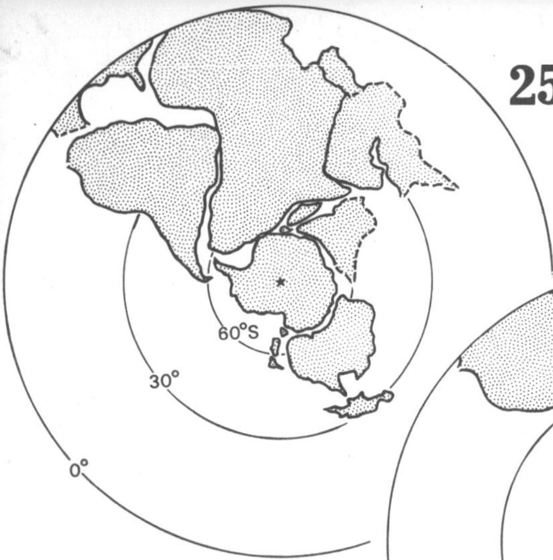


200 million years

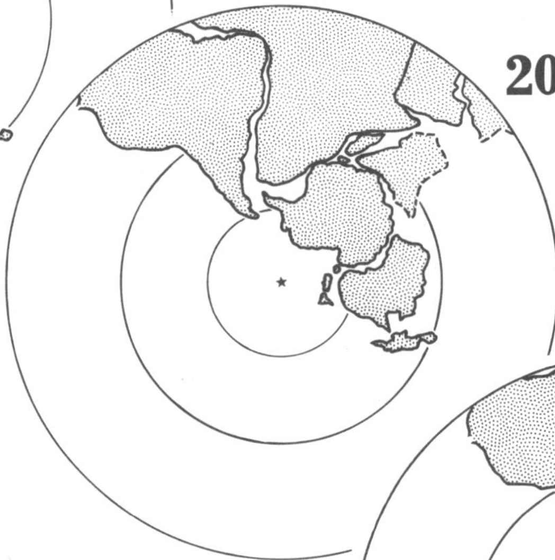

150 million years

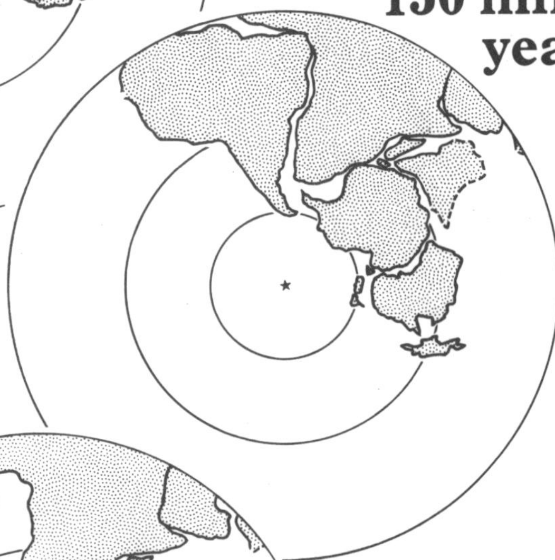

100 million years

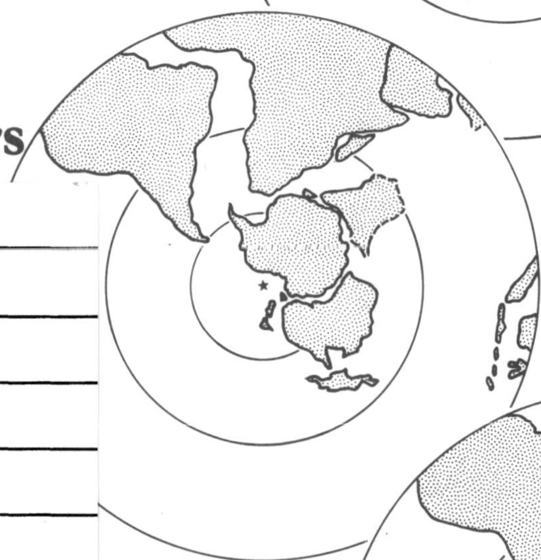



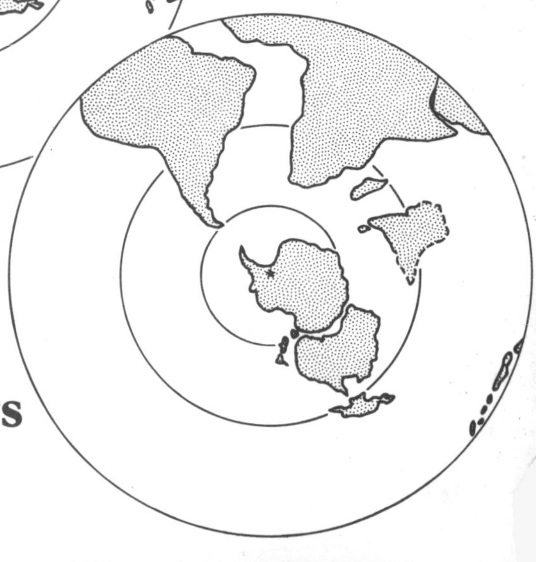

llion years